GEOTECHNICAL ENGINEERING OF DAMS

GEOTECHNICAL ENGINEERING OF DAMS

Robin Fell
School of Civil and Environmental Engineering, The University of New South Wales, Australia

Patrick MacGregor
Consulting Engineering Geologist

David Stapledon
Consulting Engineering Geologist

Graeme Bell
Consulting Dams Engineer

A.A. BALKEMA PUBLISHERS   LEIDEN / LONDON / NEW YORK / PHILADELPHIA / SINGAPORE

Cover illustration: Cethana Dam, Tasmania, reproduced with the permission of Hydro Tasmania

Published by: A.A. Balkema Publishers Leiden, The Netherlands, a member of Taylor & Francis Group plc
www.balkema.nl and www.tandf.co.uk

Library of Congress Cataloging-in-Publication Data

British Library Cataloguing in Publication Data

ISBN 04 1536 440 x

Printed and bound in England by Antony Rowe Ltd, Chippenham, Wiltshire

Table of Contents

About the Authors

Robin Fell is Professor of Civil Engineering at the University of New South Wales and also works as a consultant. He has 38 years of experience in geotechnical engineering of dams, landslides and civil and mining projects in Australia and Asia. He has worked on over 100 dams worldwide and has been involved in all aspects of planning, site investigation, design and construction of embankment dams.

Patrick MacGregor is a Consulting Engineering Geologist with more than 40 years experience in the assessment of geological constraints for major civil engineering projects in a number of countries. He has been involved in dam investigation, design and construction, and particularly worked on hydroelectric developments at all stages from inception to operation.

David Stapledon spent many years on the investigation of large dam construction sites in various countries. He was a Professor of Applied Geology at the University of South Australia (1964–1993) and worked as a Consultant in Engineering Geology, contributing to major dam projects in Australia, New Zealand and South East Asia. He has 51 years of experience and was awarded the John Jaeger Memorial Medal for Contributions to Geomechanics in 1995.

Graeme Bell has been a Consulting Dam Engineer since 1962. His role has varied from providing the full technical input, design management and construction advice for new dams to the preparation of complex structural analyses of existing dams. From 1979, he has acted as an independent reviewer on many dam projects, mainly in Australia, but also on several overseas locations.

1

Introduction

1.1 OUTLINE OF THE BOOK

The book sets out to present a state of practice of the geotechnical engineering of embankment dams and their foundations and the geotechnical engineering of the foundations of concrete dams.

It assumes that the reader is trained in Civil Engineering or Engineering Geology, with knowledge of soil and rock mechanics.

There is an emphasis on the assessment of existing dams, as well as investigation, design and construction of new dams. We have set out to give the background to design methods, as well as the methods themselves, so the reader can develop a proper understanding of them. The book is largely written about large dams, which ICOLD (1974) define as:

Large Dam: A dam which is more than 15 metres in height (measured from the lowest point in the general foundations to the crest of the dam), or any dam between 10 metres and 15 metres in height which meets one of the following conditions:

- the crest length is not less than 500 metres;
- the capacity of the reservoir formed by the dam is not less than one million cubic metres;
- the maximum flood discharge dealt with by the dam is not less than 2000 cubic metres per second;
- the dam is of unusual design.

Three of the authors wrote "Geotechnical Engineering of Embankment Dams" (Fell, MacGregor and Stapledon, 1992). This book has some common elements with that book but has been extensively revised to reflect the current state of practice and the knowledge the authors have gained since that book was written, particularly about how existing dams behave.

1.2 TYPES OF EMBANKMENT DAMS AND THEIR MAIN FEATURES

There are several types of embankment dam. The designs have varying degrees of in-built conservatism, usually relating to the degree to which seepage within the dam is controlled by provision of filters and drains, the use of free draining rockfill in the embankment and the control of foundation seepage by grouting, drainage and cutoff construction.

Table 1.1 lists the common embankment dam zones and their functions. The zone numbering system shown is used throughout this book. There is no universally adopted numbering system for zones in embankment dams. Some dams will have several rockfill zones to accommodate the materials available from the quarry and excavations required for the spillway.

Figures 1.1 and 1.2 show schematic cross sections of the most common types of embankment dams now being constructed. Figure 1.3 shows some earlier types of dams which are

Table 1.1.   Embankment dam zones description and function.

Zone	Description	Function
1	Earthfill ("core")	Controls seepage through the dam
2A	Fine filter (or filter drain)	(a) Controls erosion of Zone 1 by seepage water, (b) Controls erosion of the dam foundation (where used as horizontal drain), (c) Controls buildup of pore pressure in downstream face when used as vertical drain
2B	Coarse filter (or filter drain)	(a) Controls erosion of Zone 2A into rockfill, (b) Discharge seepage water collected in vertical or horizontal drain
2C	(i) Upstream filter	Controls erosion of Zone 1 into rockfill upstream of dam core
	(ii) Filter under rip rap	Controls erosion of Zone 1 through rip rap
2D	Fine cushion layer	Provides uniform support for concrete face; limit leakage in the event of the concrete face cracking or joints opening
2E	Coarse cushion layer	Provides uniform layer support for concrete face. Prevents erosion of Zone 2D into rockfill in the event of leakage in the face
1–3	Earth-rockfill	Provides stability and has some ability to control erosion
3A	Rockfill	Provides stability, commonly free draining to allow discharge of seepage through and under the dam. Prevents erosion of Zone 2B into coarse rockfill
3B	Coarse rockfill	Provides stability, commonly free draining to allow discharge of seepage through and under the dam
4	Rip rap	Controls erosion of the upstream face by wave action, and may be used to control erosion of the downstream toe from backwater flows from spillways

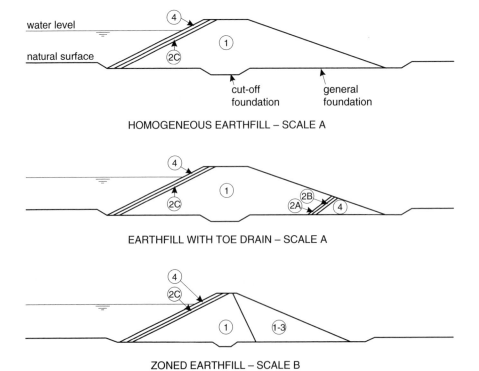

Figure 1.1.   Schematic cross sections of typical earthfill dams.

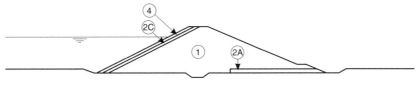

EARTHFILL WITH HORIZONTAL DRAIN – SCALE B

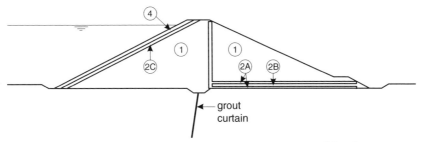

EARTHFILL WITH VERTICAL AND HORIZONTAL DRAIN – SCALE B

NOTES:
1. Crest detailing and downstream slope protection not shown.
2. Scales relate to overall size, details are not drawn to scale.

Scale A 0   10   20m

Scale B 0   20   40m

Figure 1.1.   (Continued).

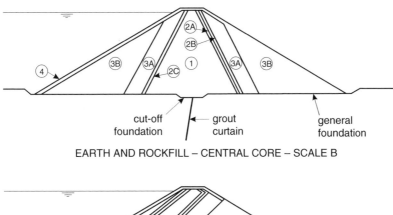

EARTH AND ROCKFILL – CENTRAL CORE – SCALE B

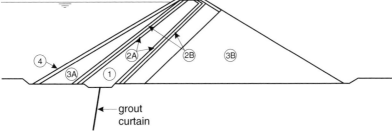

EARTH AND ROCKFILL – SLOPING UPSTREAM CORE – SCALE B

Figure 1.2.   Schematic cross sections of typical earth and rockfill and concrete face rockfill dams.

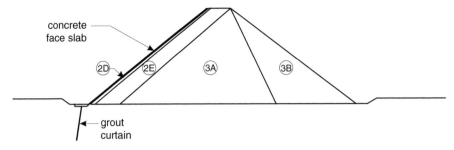

CONCRETE FACE ROCKFILL – SCALE B

NOTES:
1. Crest detailing and downstream slope protection not shown.
2. Scales relate to overall size, details are not drawn to scale.

Scale B

Figure 1.2.   (Continued).

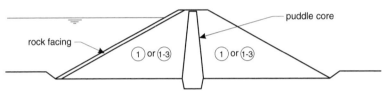

PUDDLE CORE EARTHFILL – SCALE B

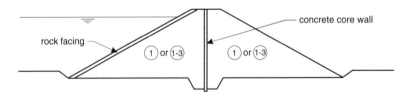

EARTHFILL WITH CONCRETE CORE WALL – SCALE B

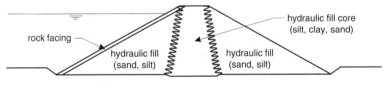

HYDRAULIC FILL – SCALE B

NOTES:
1. Crest detailing and downstream slope protection not shown.
2. Scales relate to overall size, details are not drawn to scale.

Scale B

Figure 1.3.   Schematic cross sections of some earlier dam types.

Table 1.2. Embankment dam foundation treatment.

Item	Description
General foundation excavation	Excavation of compressible and low strength soil and weathered rock as is necessary to form a surface sufficiently strong to support the dam and to limit settlement to acceptable values
Cutoff foundation excavation	Excavation below general foundation level to remove highly permeable and/or erodible soil and rock
Curtain grouting	Drilling of holes into the foundation and injecting grout (usually cement slurry) under pressure to reduce the permeability of the rock
Consolidation grouting (also called "blanket grouting")	Grouting carried out in the upper part of the cut-off foundation to reduce permeability (of the rock)

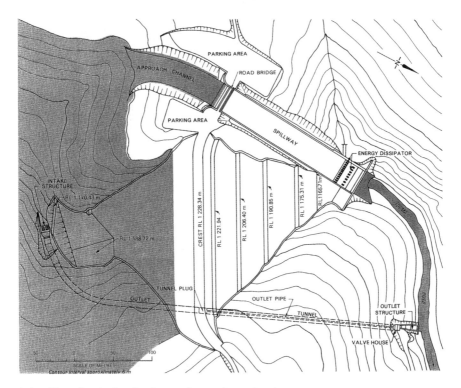

Figure 1.4.   Plan of typical embankment dam and associated structures.

no longer being built. There are other types of embankment dams, but these are not specifically covered in this book including:

- – Concrete face earthfill;
- – Asphaltic core rockfill;
- – Bituminous concrete face earth and rockfill;
- – Steel face rockfill;
- – Geomembrane face earth and rockfill.

Table 1.2 describes the foundation treatment commonly used in embankment dams.

Figure 1.4 shows a fairly typical plan of an embankment dam showing the spillway, outlet tunnel (which is commonly used for river diversion during construction of the dam) and outlet structures.

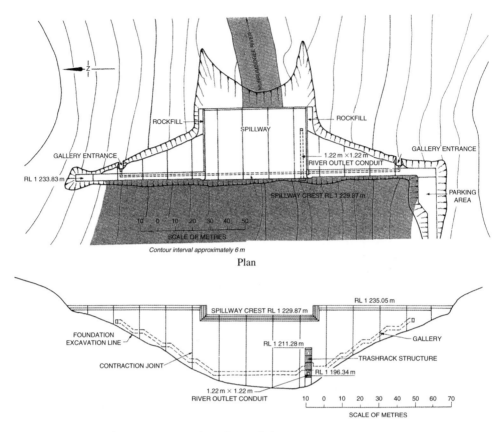

Figure 1.5(a). Typical concrete gravity dam plan and elevation.

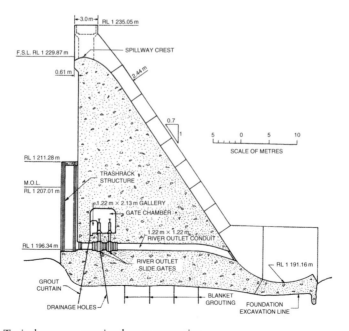

Figure 1.5(b). Typical concrete gravity dam cross section.

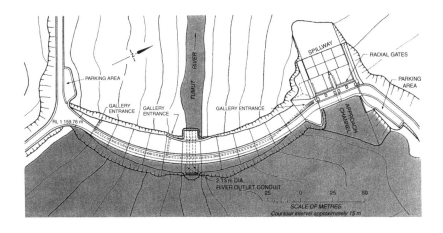

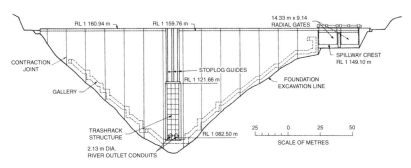

Figure 1.6(a). Typical concrete arch dam plan and elevation.

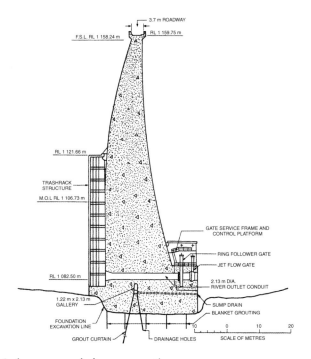

Figure 1.6(b). Typical concrete arch dam cross section.

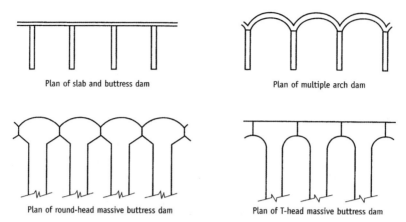

Figure 1.7.   Different types of buttress dam (ANCOLD 2000).

1.3   TYPES OF CONCRETE DAMS AND THEIR MAIN FEATURES

Figures 1.5 and 1.6 show examples of concrete gravity and arch dams. Concrete gravity dams rely on the weight of the concrete to withstand the forces imposed on the dam. Concrete arch dams transpose these forces into the abutment foundation by the arching action and generally impose higher loads on the foundations.

Modern concrete gravity and arch dams have a gallery in the dam from which grout holes are drilled to reduce the rock mass permeability of the foundation. Grouting, combined with drainage holes reduces the uplift pressures within the dam foundation and at the dam foundation contact.

Figure 1.7 shows different types of concrete buttress dams.

Dependent on the configuration, these may impose large stresses on the dam foundation.

2

Key geological issues

Most dams and reservoirs are located at or near the bases of river valleys. The topography and geological situation at every site will have developed by the interaction of many geological and related processes during vast periods of time. The resulting geological structure at dam and reservoir sites can be complex, and no two sites will be the same. Some of the processes which formed the site may still be active, or may be reactivated by the project and capable of influencing the feasibility of its construction or operation. It is vital therefore, that sites are investigated using all appropriate knowledge and methods of classical geology. These are essential for good engineering geology (Baynes, 1999; Fookes, 1997; Fookes et al., 2000; Hutchinson, 2001).

The engineering team responsible for construction of a dam at each site must be able to make reliable predictions about suitable construction methods and how the dam and its foundation will interact and perform, under every envisaged operating condition. The predictions will usually involve both judgements and quantitative analyses, based on data provided by the site investigation team. This data is provided as a geotechnical model, which consists of the following:

- a sufficiently detailed three-dimensional picture or model of the site geological situation, using geotechnical descriptive terms, and an assessment of the history of the site and the effects of any processes which are still active, and
- adopted values of parameters for critical parts of the model, as required for the analyses.

This chapter introduces geotechnical descriptive terms, and uses them while discussing processes found to be important in dam engineering. Defects or discontinuities and their modes of formation are defined and discussed first, because understanding of them is needed throughout the book. Some aspects of the valley-forming processes of weathering (breakdown) and erosion (removal) of rocks and soils are then covered in detail because they are the youngest and most active processes at most sites. For further study of these processes the reader is referred to Selby (1982), Bell (1983b), Hunt (1984), Fookes and Vaughan (1986) and Fookes (1997).

2.1   BASIC DEFINITIONS

The following definitions and descriptive terms are recommended, and are used throughout this book.

Material – a non-specific general term to describe any sample or body of soil or rock.

Substance – material which is effectively homogeneous in its engineering properties, but may be isotropic or anisotropic. The term "effectively" here means within the tolerances applicable to the project in hand.

Rock substance – a homogeneous cohesive aggregate of mineral grains which has such strength that it cannot be broken up or remoulded by hand when dry, or when immersed in water.

Soil substance – a homogeneous aggregate of mineral grains, either non-cohesive, or cohesive but which disintegrates or can be remoulded by hand, when immersed in water.

Defect – a discontinuity, or break in continuity of a body of rock or soil substance.

Mass – any body or rock or soil which is not effectively homogeneous; it can comprise two or more substances without defects, or one or more substances with one or more defects.

In most situations the engineering behaviour of rock masses is dominated by the number, type and configuration of defects within them. The behaviour of soil masses is usually influenced, and sometimes dominated, by defects present in them.

2.2   TYPES OF ANISOTROPIC FABRICS

Any of the following types of fabric, developed uniformly within a body of rock or soil substance, will cause that substance to be anisotropic with respect to its engineering properties – strength, stiffness and permeability.

Bedding   –   layered or parallel arrangement of grains, developed during deposition as sediment

Foliation   –   layered or parallel arrangement of grains (often tabular or flakey in shape) developed either by viscous flow (in igneous rocks) or by pressure and heat (in metamorphic rocks)

Cleavage   –   foliation in which many surfaces have developed along which the substance splits readily

Lineation   –   linear arrangement of (often elongated) grains, developed by viscous flow (in igneous rocks) or pressure with or without heat (in metamorphic rocks); the lines of grains may or may not lie within surfaces or layers of foliation.

If a body has any of these fabrics developed in a non-uniform way, that body is mass, by definition.

2.3   DEFECTS IN ROCK MASSES

Figure 2.1 illustrates and defines the most common and important types of defect met in rock foundations for dams. It also relates characteristics of the defect types to ways in which they were formed. The joints and two types of faults have developed directly from the stress histories of the rocks. Soil infill and extremely weathered seams usually occur along or within the above defect types, and hence most of them also result (indirectly, at least) from the stress histories. As a consequence of this, the defects invariably occur in more or less regular patterns, which can usually be distinguished at each site by the use of structural geology.

Further discussion of these defects and their significance in dam engineering is given in Sections 2.3.1 to 2.3.5. The characteristics of defects which occur in particular rock types or geological environments are discussed in Chapter 3.

2.3.1   *Joints*

Joints generally cause a rock mass to be less strong, less stiff and more permeable than an equivalent body of the rock substance. Where joints have been acting as conduits for ground waters, the rock next to them is often weakened by weathering (see Section 2.6). As joints are typically of limited extent, they do not usually present serious problems in embankment dam foundations but may be more significant in concrete gravity and arch dam foundations. Treatments for jointed rock usually include excavation, grouting and drainage (see Chapters 17 and 18).

DEFECT NAME	DESCRIPTION	TYPICAL EXTENT/OCCURRENCE	FORMED BY
JOINT	Almost planar surface or crack, across which the rock usually has little tensile strength. May be open, air or water-filled, or filled by soil substance, or rock substance which acts as cement. Joint surfaces may be rough, smooth or slickensided.	From less than 1 metre, up to 10 metres.	Shrinkage on cooling (igneous rocks). Extension or shear (any rocks) due to tectonic stresses, or unloading.
SHEARED ZONE (FAULT) (1)	Zone with roughly parallel almost planar boundaries, of rock substance cut by closely spaced (often <50mm) joints and or cleavage surfaces. The surfaces are usually smooth or slickensided and curved, intersecting to divide the mass into lenticular or wedge-shaped blocks.	More than 10 metres, up to kilometres.	Faulting, during which small displacements occurred along slickensided surfaces distributed across the width of the zone.
CRUSHED SEAM (FAULT) (1)	Seam with roughly parallel almost planar boundaries, composed of usually angular fragments of the host rock substance. The fragments may be of gravel, sand, silt or clay sizes, often mixtures of these. The finest fragments are often concentrated in thin layers or 'slivers' next to and parallel to the adjacent rock. Slickensides are common in the sliver zones and along the rock boundaries.	More than 10 metres, up to kilometres.	Faulting, during which relatively large displacements occurred within the zone.
SOIL INFILL SEAM	Seam of soil substance, usually with very distinct roughly parallel boundaries. Thin seams usually CH; thick near-surface seams sometimes GC or GC-CL (2).	Less than 10 metres, within near-surface mechanically weathered zone, or in rock disturbed by creep or landsliding.	Migration of soil into an open joint or cavity.
EXTREMELY WEATHERED (OR ALTERED) SEAMS	Seam of soil substance, often with gradational boundaries. The type of soil depends on the composition of the original rock.	Varies widely, but extremely weathered seams are usually within or near the near-surface weathered zone.	Weathering (or alteration) in place, of rock substance, usually next to or within a pre-existing defect.

NOTES
(1) In the geological sense,
(2) Group symbols as defined in the United States Bureau of Reclamation Earth Manual, 1963, Figure 3.1,
(3) A 'marker' feature, showing the nature of the displacements. Such features are observed only rarely in sheared zones.

Figure 2.1.   Common defects in rock masses.

2.3.2 *Sheared and crushed zones (faults)*

Regardless of whether its displacements have been normal, reverse or transcurrent, a fault may be a sheared zone, a crushed zone, or some combination of these two. In crushed zones the rock in the displacement zone has suffered brittle failure. In sheared zones the rock failure may have ranged from brittle to pseudo-ductile, where the shear displacements have been microscopic in size. Rarely, a single slickensided joint may prove to be a continuous and significant fault. This same fault could be a thin crushed seam (gravelly clay, CL-GC) where it passes through a shale bed, and a wider crushed zone (fine gravel GP) where it passes through a sandstone bed. Faults are important in dam foundations because they contain material which is usually

– of low strength and stiffness in shear;
– compressible and erodible, and
– of large extent.

Some faults are more permeable than the average rock, or have open-jointed zones of high permeability next to them. Some clay or gouge filled fault zones are less permeable than the surrounding rock, and may act as low permeability barriers to groundwater movement. Often, the rock in and next to faults is weakened further by weathering or alteration.

Because of these characteristics, if present in foundations but undetected or inadequately treated or allowed for in designs, faults have the potential to seriously disrupt dam construction or operation, or even to cause or contribute to dam failure. A thin crushed seam contributed to the failure of Malpasset Dam in 1958. The role of this seam is discussed in Gosselin et al. (1960), Terzaghi (1962), Jaeger (1963), Londe (1967), Stapledon (1976) and James and Wood (1984).

The nature of treatment applied to a fault in a dam foundation will depend upon the designer's assessment of the capacity of the fault to adversely affect the behaviour of the foundation and its interaction with the dam, during operation. Examples of treatments of faults are described in Chapter 17.

2.3.3 *Soil infill seams (or just infill seams)*

Soil infill seams are formed by soil which has gravitated, washed or squeezed into slots. Most of the slots would have been gaping joints in rock masses which have undergone either appreciable dilation due to stress relief (Figures 2.1 and 2.6) or disruption due to creep or landsliding (see Sections 2.10 and 3.10). In soluble rocks (carbonates or evaporates) infill materials occur mainly in irregular cavities, but also as seams in slots resulting from widening of joints or thin beds, by solution (see Sections 3.7.1 and 3.8).

The infill materials are commonly clay (CH), and often contain roots or tubes remaining after the rotting of roots. Although individual infill seams are rarely of large lateral extent, they are likely to have low shear strength and to be compressible and erodible.

2.3.4 *Extremely weathered (or altered) seams*

Extremely weathered (or altered) rock is material which was once rock but has been converted by weathering (or alteration), in place, to soil material (see Sections 2.6.3, 2.7, 2.8.1 and Tables 2.3 and 2.6). The extremely weathered seam (a) in Figure 2.1 has formed by the weathering of a thin bed of shale next to its boundary with a sandstone bed. Some such seams in this type of situation show slickensided surfaces, which can indicate either:

– The seam was, or has formed next to, a bedding-surface fault (Figure 3.19), or
– It is simply a weathered bed within which some displacement has occurred since it became weathered.

Extremely weathered seam (b) has developed along or next to a joint.

Extremely weathered (or altered) seams can be of large or small extent. Their shear strengths, shear stiffnesses, compressibilities and erodibilities, will depend largely on the compositions and fabrics of the parent rocks. Weathered seams normally occur within or not far below the near-surface weathered zone (see Section 2.6.3), but altered seams can occur at any depth.

2.3.5   The importance of using the above terms to describe defects in rock

Use of the terms defined in Figure 2.1 and discussed above can enable site investigators to compile geotechnical models in an efficient and economical manner, due to the following:

– Individually, they are based on geological understanding, and used together they allow the history of the site to be understood;
– They form more reliable bases for correlation between data points (exposures in boreholes or pits) than purely engineering descriptions of the materials which form them;
– Some of their engineering properties, and their probable shapes and extents, can be predicted from the defect names;
– They can provide evidence that a rock mass has been mechanically loosened, e.g. by mechanical weathering, landsliding, or volcanic explosion.

ISRM (1985) use the general terms "infilling" or "filling" to describe any materials (usu-ally with soil properties) which form, or occur within the boundaries of, defects (or dis-continuities) in rock masses. This generalization may have been a useful shorthand for some engineers, or those involved in mathematical modeling, but its use in engineering for dams is not recommended because

– It provides none of the above-listed benefits of the terms on Figure 2.1, and
– It can be confused with the "infill" or "soil infill" term of Figure 2.1.

The authors recommend that if a seam or zone cannot be confidently identified in terms of Figure 2.1, then the material in it should be described as "seam material" rather than "infilling".

2.4   DEFECTS IN SOIL MASSES

Soils formed by the extreme weathering of rocks can inherit any of the fabrics described in Section 2.2. Also they may contain remnant joints, crushed zones or sheared zones. The con-trast between the strengths of these remnant defects and that of the soil (extremely weathered rock substance) is usually lower than in a comparable mass of less weathered or fresh rock.

Soils of sedimentary origin can contain individual thin beds which behave effectively as defects. For example, a thin silty or sandy bed in a clay deposit can represent a leakage path with permeability several orders higher than that of the adjacent soil. Also, a thin clay bed in a sandy soil can form a local barrier to water flow, and a surface of low shear strength.

Soils of any origin type can contain defects of the following types:

– Fissures (or joints) – near-planar surfaces, usually polished or slickensided, formed in clayey soils by (usually) small shear displacements associated with swelling;
– Tension cracks (or joints) – near-planar surfaces formed by extension due to shrinkage, subsidence, landsliding or tectonic displacements;
– Remoulded zones – zones of remoulded and softened soil, often containing fissures, formed by appreciable shear displacements due to subsidence, landsliding or tectonics;
– Tubular or irregular holes – formed by burrowing animals, rotted roots, or erosion by seeping or flowing water;
– Infilled features – previously open joints or holes, wholly or partly filled with younger soil.

All of these defect types generally lower the strength and stiffness, and increase the permeability of soil masses (Stapledon, 1971; Walker et al., 1987; MacGregor et al., 1990; Moon, 1992).

2.5   STRESSES IN ROCK MASSES

Measurements of *in situ* rock stresses at shallow depths in many geological environments throughout the world have shown horizontal stresses which are generally higher than can be explained theoretically from the weight of the present overburden. Hast (1967) published the results of a large number of stress measurements in relatively intact granitic and metamorphic rocks in Scandinavia. Hast showed that at all Scandinavian sites, the major and intermediate principal stresses were horizontal, and that the sum of these two stresses (Line A on Figure 2.2) increased linearly with depth. Also plotted on Figure 2.2 are the mean horizontal stress (Line B) and the theoretical vertical stress due to overburden weight (Line C). It can be seen that within about 50 m of the ground surface the mean horizontal stress appears likely to be more than ten times the vertical stress.

Brown & Hoek (1978) compiled Figure 2.3 and Figure 2.4 from stress measurements at mining and civil engineering sites throughout the world. Figure 2.3 shows that the vertical components of the measured stresses are in fair agreement with the calculated vertical stress due to overburden. Figure 2.4 shows the variation with depth, of the ratio of the

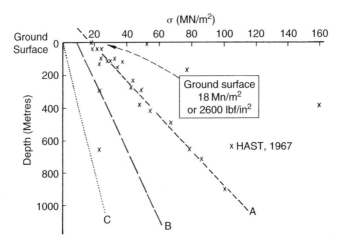

Figure 2.2.   Plot of horizontal stresses against depth in Scandinavian mines (based on Hast, 1967).

mean horizontal stress to the vertical stress. It can be seen that horizontal stresses are generally significantly greater than vertical stresses, at depths of less than 500 m, and that at greater depths the stresses tend to equalize.

2.5.1   *Probable source of high horizontal stresses*

The prime source of high horizontal stresses is believed to be tectonic forces, i.e. the forces which drive and resist the motion of the earth's crustal plates. Evidence for this comes mainly from depths of more than a kilometre below the surface, and is provided by analyses of spalling (breakouts) and tensile fractures in the rock around deep wells drilled for petroleum.

The horizontal stress fields inferred at these great depths have been related generally to plate motion, and a world stress map has been produced (Zoback, 1992). There is also a stress map covering Australia and Papua New Guinea (Hillis, 1998).

At the much shallower depths relevant to dam projects, the local stress fields often cannot be related directly to the regional tectonic stress field at depth. The differences are probably caused by local effects such as variations in topography or structure of the bedrock.

It is possible also in some situations that the near-surface stresses result from strain energy which has been locked into rocks during their formation, often, but not necessarily, at great depth. This may have occurred in igneous rocks during their solidification, and in sedimentary and metamorphic rocks, during compaction, cementation or recrystallization; Emery (1964), Savage (1978), Nicholls (1980), Brown and Windsor (1990) and Brady and Brown (1985). At a microscopic level the strain energy is considered to be

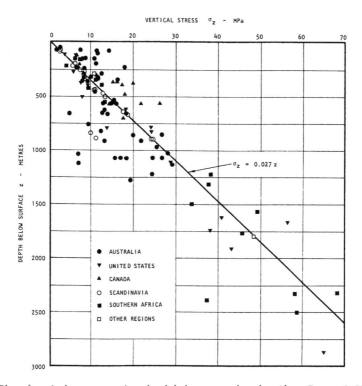

Figure 2.3.   Plot of vertical stresses against depth below ground surface (from Brown & Hoek, 1978, by permission of Pergamon Press).

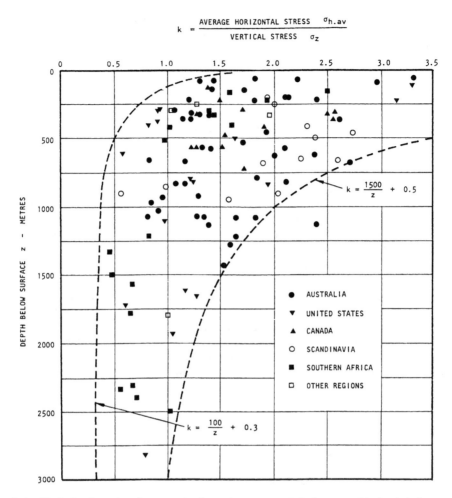

Figure 2.4.   Variation in ratio of average horizontal stress to vertical stress with depth below surface (from Brown & Hoek, 1978, by permission of Pergamon Press).

locked into mineral crystals or grains by cementation and interlocking. An analogous model would be compressed springs embedded in plastic. As the vertical load on high stressed rock is slowly lowered by erosion, vertical stresses are relieved progressively by upward expansion. However, because the rock remains confined laterally, the horizontal stresses decrease in accord with Poisson's Ratio, i.e., at about one-third of the rate of the vertical stresses. This results in the recorded near-surface imbalance.

2.5.2   Stress relief effects in natural rock exposures

Field evidence of the existence of the high horizontal stresses at shallow depths is seen most clearly in areas of massive igneous rocks which contain very few tectonically induced fractures. Most of Hast's measurements plotted on Figure 2.2 were made in such rocks. These rocks usually contain "sheet joints" near-parallel to the ground surface, as shown in Figure 2.5. The sheet joints show rough, irregular, plumose surfaces indicating that the rock failed in tension, by buckling or spalling, the tension being induced by the high horizontal compressive stress. Some sheet joints show slickensides near their extremities indicating local shear failure which should be expected here (Figure 2.5). The spacing of sheet

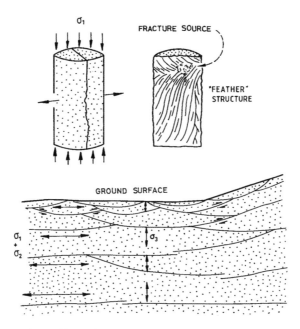

FRACTURE SOURCE

"FEATHER" STRUCTURE

GROUND SURFACE

Figure 2.5.   Sheet-joints, formed by induced tensile failure (as in an unconfined compression test with zero friction at the plattens).

joints is often 0.3 to 1 m near the ground surface, and becomes progressively wider with depth.

From all of these characteristics, sheet joints are considered to be stress-relief features. The linear decrease in horizontal stresses as the ground surface is approached (Lines A and B, Figure 2.2) is due to the progressive relief of horizontal stresses partly by buckling and spalling, as the overlying rock load is removed by erosion.

Holzhausen (1989) provides a comprehensive account of the characteristics and origin of sheet joints.

In fractured rock masses, i.e. those which are already weakened by defects of tectonic origin, the effects of horizontal stress relief are not so obvious but are always present, as opening up of the existing defects, as shown in Figure 2.6b. The destressing effects usually extend to greater depths in jointed rock than in massive rock. However, in most geological situations it can be assumed that beyond the near-surface effects, the tectonically induced joints will become progressively tighter (less open) with greater depth (Snow, 1970).

2.5.3   *Effects in claystones and shales*

Claystones and shales (i.e. "mudrocks") are formed mainly by consolidation of clay-rich sediments, but may be strengthened further by partial recrystallization and cementation. Such rocks have much higher porosities than igneous and metamorphic rocks. They usually show expansion, spalling and fretting on unloading and exposure. It is believed that these effects are due partly to the release of stored strain energy (Bjerrum, 1967) but also to the absorption of water and subsequent swelling of the clays. (See Section 2.9.1 and Section 3.5.1).

2.5.4   *Special effects in valleys*

Gentle anticlines, in some cases with associated thrust faults as shown in Figure 2.7, have been recorded across many river valleys cutting through near-horizontal sedimentary

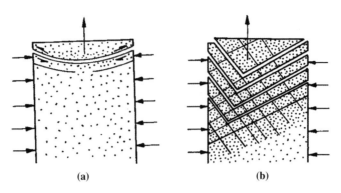

Figure 2.6. Effects of destressing in (a) intact rock, and (b) jointed rock.

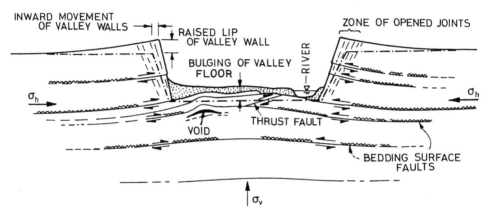

Figure 2.7. Complex valley structures related to stress release in weak, flat-lying rocks (based on Patton & Hendron, 1972).

rocks of moderate to low strength. The phenomenon is referred to as "valley bulging" or "valley rebound", and has been described by Zaruba (1956), Simmons (1966), Ferguson (1967), Patton and Hendron (1972), Matheson and Thompson (1973), Horswill and Horton (1976), McNally (1981) and Hutchinson (1988).

Most of the features shown on Figure 2.7 have clearly developed as a result of buckling and shear failure under high horizontal compressive stresses. The stresses were concentrated beneath the valley floor as a result of load transfer as the excavation of the valley removed lateral support from the rock layers above the floor, and vertical load from the rock beneath the floor. The steeply-dipping joints next to the cliff faces probably opened up due to expansion of the rock layers under the influence of horizontal stresses both across and parallel to the valley.

All of the effects shown on Figure 2.7 were present at the site for Mangrove Creek Dam near Gosford, New South Wales. This 80 m high concrete faced rockfill dam is located in a valley 200 m to 300 m deep, cut through an interbedded sequence of sandstones, silt-stones and claystones (Figure 2.8). Away from the river the rock layers in the valley sides generally show joints at wide to very wide spacings.

Near and beneath the river bed there is a broad, gentle, "valley bulge" as shown on Figure 2.8. Although the shape of this feature is not very pronounced, the rock within it was intensely disrupted down to 15 m below the river bed. The sandstone unit E (Figure 2.9) showed extension joints which were open or clay-filled, to a maximum of 100 mm,

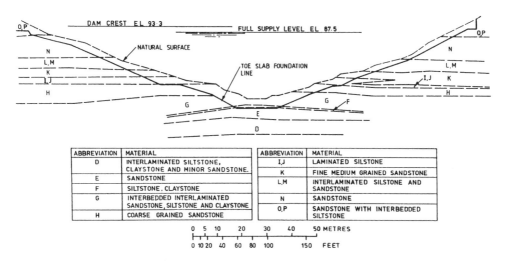

ABBREVIATION	MATERIAL
D	INTERLAMINATED SILTSTONE, CLAYSTONE AND MINOR SANDSTONE.
E	SANDSTONE
F	SILTSTONE, CLAYSTONE
G	INTERBEDDED INTERLAMINATED SANDSTONE, SILTSTONE AND CLAYSTONE
H	COARSE GRAINED SANDSTONE

ABBREVIATION	MATERIAL
I,J	LAMINATED SILTSTONE
K	FINE MEDIUM GRAINED SANDSTONE
L,M	INTERLAMINATED SILTSTONE AND SANDSTONE
N	SANDSTONE
O,P	SANDSTONE WITH INTERBEDDED SILTSTONE

```
0  5  10      20      30    40      50 METRES
0 10 20  40      60  80  100        150   FEET
```

Figure 2.8.   Cross-section along the grout cap foundation at Mangrove Creek Dam. (MacKenzie & McDonald, 1985).

Figure 2.9.   Upper surface of Unit E sandstone at Mangrove Creek Dam valley bulge, showing joints open as much as 100 mm, clay-filled in part.

and the underlying unit D (interbedded) contained zones of crushed rock and clay up to 600 mm thick, apparently produced by overthrust faulting (Figure 2.10). This unit also contained gaping joints parallel to bedding.

To provide a stable foundation for the grout cap and to prevent possible erosion of the clay filling from joints, cable anchors were installed and a concrete diaphragm wall 30 m long and 15 m deep was constructed beneath the valley floor (MacKenzie and McDonald 1985).

Although usually not as pronounced as in Figure 2.8 to Figure 2.10, destressing effects are invariably found at and near the floors of valleys in strong to extremely strong rocks, in areas where high horizontal stresses are known to exist. The effects include sheet joints and gaping or soil-infilled joints of tectonic origin. These joints usually occur both in the valley walls and beneath the floor. It is clear from Figure 2.11a that large stress concentrations are

Figure 2.10.   Crushed zone in Unit D (Interbedded siltstone and claystone) formed by overthrusting. Photograph courtesy G. McNally.

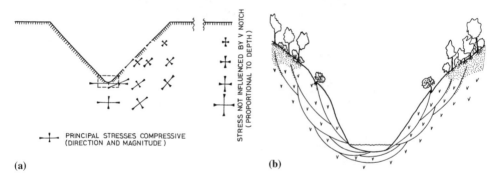

Figure 2.11.   High stresses developed in valley floor and lower sides, and resulting hypothetical pattern of sheet joints in strong rock without "tectonic" joints (model results from Alexander, 1960).

likely to occur in rock beneath the bottom of deep, youthful gorges, and so such destressing effects, even if not visible, are likely to be present, as shown in Figure 2.11b.

2.5.5   Rock movements in excavations

Spalling, buckling upwards, and the formation of new sheet joints are relatively common occurrences in near-surface excavations in very high strength rocks in North America, Scandinavia and Australia. Emery (1964) reports that an anticlinal fold about 100 m long with crest elevation of almost 5 m was formed overnight in a limestone quarry in Kingston, Ontario. Lee et al. (1979) describe rock bursts in shallow excavations in granite and gneiss in Maine, U.S.A. Ward (1972) describes a 30 m deep excavation in gneiss at New York City, where inward movement of the walls caused rock bolt failures.

Bowling and Woodward (1979) describe "rockbursts" and the formation of new sheet joints at Copeton Dam in New South Wales. Relatively minor spillway discharges of up

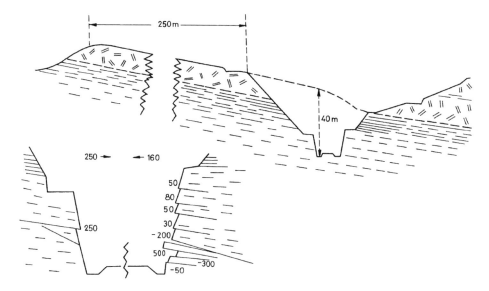

Figure 2.12.   Convergence indicated by traces of presplit holes, near base of railway cutting in fresh, strong siltstone.

to 460 m³/s caused erosion of a channel about 20 m deep in fresh, massive granite. The granite apparently failed progressively by buckling upwards. As each slab of rock was carried away a new sheet joint formed, by buckling. Isolated rockbursts continued to occur in the channel floor for more than 6 months after the floods. The channel was eventually stabilized by rock anchors and dental concrete.

Inward movement or "convergence" of the walls of relatively shallow excavations in rock is another common destressing effect. Figure 2.12 shows part of a railway cutting 40 m deep and about 500 m long near Paraburdoo, Western Australia. After excavation of the lowest 15 m of the cutting, by presplitting and trench-blasting methods, a maximum possible convergence of 410 mm was indicated by displacement of traces of presplit holes. Although part of this convergence was undoubtedly caused by the development of blast-initiated cracks behind the rock faces, it is believed that much of the 410 mm was caused by destressing of the gently dipping siltstone. This belief is based on many other similar situations where the authors have observed inward movements to continue for several days after completion of excavations. Similar observations have been reported by Wilson (1970).

2.6   WEATHERING OF ROCKS

Broadly speaking, weathering of a rock is its response to the change from the pressure, temperature, moisture and chemical environments in which it was formed, to its new environment at and near the ground surface. Weathering processes are of two fairly distinct types, namely mechanical, and chemical.

2.6.1   *Mechanical weathering*

Mechanical weathering includes all of the near-surface physical processes which break rock masses down to progressively smaller rigid blocks or fragments, and cause those blocks to separate. Mechanical weathering generally precedes chemical weathering. It renders the

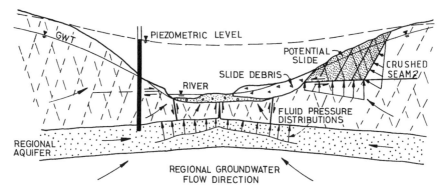

Figure 2.13.   Possible effects of high fluid pressures on valleys in groundwater discharge areas (from Patton & Hendron, 1972).

rock mass more permeable and facilitates access for groundwater to large surface areas of rock substance.

Destressing, in particular the formation of sheet joints and the opening up of existing "tectonic" joints near the ground surface, is the primary and generally the most significant mechanical weathering process (See Figures 2.6, 2.7, 2.9 and 2.11b). The other processes, in order of their (generally) decreasing significance, are as follows:

– gravitational creep (e.g. of slabs and wedges, and toppling);
– joint water thrusting and uplifting during extreme rainfall events;
– earthquake induced displacements;
– growth of tree roots in joints;
– expansion of clays in joints;
– freezing of water in joints;
– extreme temperature changes causing differential expansion and contraction of exposed rock faces.

Patton and Hendron (1972) suggested possible mechanical weathering effects beneath the floor and lower sides of valleys in groundwater discharge areas. This type of situation (Figure 2.13) may be relatively rare, but if it occurs the hydraulic uplift and thrust effects shown are clearly possible.

2.6.2   Chemical decomposition

The term chemical decomposition as used here includes all of the chemical (and to a minor extent physical) processes which cause mineral changes resulting generally in weakening of rock substances, so that eventually they assume soil properties.

Throughout this book, the products of chemical decomposition are described using the terms defined in Section 2.8.1, Tables 2.3 to 2.6.

Chemical decomposition of rocks can be caused by either near-surface (weathering) processes or deep-seated (alteration) processes. Recognition of this distinction is important in civil and mining engineering because the nature and distribution of weathered materials are generally different from those of altered materials (Figure 2.14).

2.6.3   Chemical weathering

Chemical weathering is caused mainly by circulating groundwater which gains access to low-porosity rock substances via cleavage micro-cracks, open joints and fractures associated

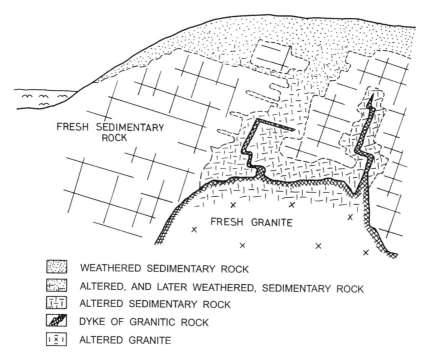

FRESH SEDIMENTARY
ROCK

FRESH GRANITE

▦ WEATHERED SEDIMENTARY ROCK

▦ ALTERED, AND LATER WEATHERED, SEDIMENTARY ROCK

▦ ALTERED SEDIMENTARY ROCK

▦ DYKE OF GRANITIC ROCK

▦ ALTERED GRANITE

Figure 2.14.   Diagrammatic cross section showing hydrothermally altered zone near a granite intrusion, with the uppermost part of the altered zone being weathered subsequently.

with faults. In the case of more porous rocks, e.g. some sandstones and limestones, groundwater can also enter through intergranular pores. Most chemical weathering occurs at extremely slow rates, such that the changes to the strength of high strength, non-porous rocks (e.g. granite) are likely to be insignificant during the operating life of most civil engineering projects. There are however some minerals and rocks which decompose, weaken or disintegrate within a few months or years of exposure. These effects will be discussed separately under Section 2.9, Rapid weathering.

Chemical weathering involves the more or less continuous operation of all or most of the following:

1. Chemical reactions between the minerals in the rock, and water, oxygen, carbon dioxide, and organic acids. These reactions cause decomposition of the minerals to form new products, some of which are soluble;
2. Removal of the soluble decomposition products by leaching;
3. Development of microcracks in some rocks, probably due to some decomposition products having larger volumes than the original minerals, or to destressing, or capillary or osmotic suction effects;
4. Deposition of some decomposition products, in pores or microcracks.

Processes 3 and 4 were illustrated by Dixon (1969) who made microscopic studies of granite and schist in a range of weathered conditions. He found that in slightly weathered samples, i.e. rocks which showed only slight discolouration in hand specimen, the first weathering effects visible microscopically were slight discolouration of the felspars and mica minerals, and the presence of many microcracks, some open and others filled with an opaque mineral assumed to be limonite or clay (Figure 2.15).

Baynes and Dearman (1978a) describe microfabric changes which occur in granites during various stages of weathering, using a scanning electron microscope. In Baynes and

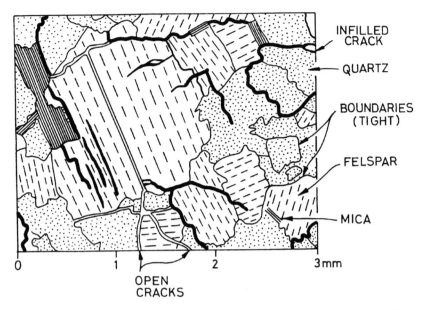

Figure 2.15.   Microscopic view of the structure of slightly weathered granite (Dixon, 1969).

Dearman (1978b) they relate changes in engineering properties to the microfabric changes.

The chemical reactions involved in chemical weathering include carbonation, hydrolysis, solution, oxidation and reduction. As most cannot be actually observed, details of the reactions, as published by various workers, are in part speculative. Useful accounts are given by Selby (1993), Price (1995) and the Geological Society (1995).

2.6.3.1   *Susceptibility of common minerals to chemical weathering*

As would be expected, the susceptibility to weathering (or the "weatherability") of minerals in igneous rocks varies in accordance with the temperatures at which they were formed. This is illustrated in Table 2.1. With the exception of quartz, the most stable mineral, all of the others on Table 2.1 weather eventually to clay minerals. Quartz is slightly soluble in water. It is almost unaffected by weathering except under tropical conditions when it is readily dissolved (in a geological time-frame).

The susceptibility of other common minerals to weathering is indicated in Table 2.2. The minerals in the carbonate and evaporite groups (shown in this table) may occur in any of three different ways, namely:

– as rocks, e.g. calcite as limestone, or dolomite as dolomite (rock) or dolostone, or
– as cements in sedimentary rocks, e.g. calcite as cement in sandstone composed mainly of quartz grains, or
– as veins or joint fillings or coatings in any rock mass.

2.6.3.2   *Susceptibility of rock substances to chemical weathering*

The susceptibility of a rock substance to weathering depends upon the following:

– the susceptibility to weathering of its component minerals, and
– the nature of its fabric, i.e. the degree of interlock and/or cementation of the mineral grains, and
– its porosity and permeability.

Table 2.1. Susceptibility of igneous rock-forming minerals to weathering.

Temperature of formation	Susceptibility to weathering	Mineral	Common igneous rock types
Highest	Highest	Olivine	
			Basalt, dolerite, gabbro
		Calcic felspar	
		Augite	Andesite, diorite
		Hornblende	
		Sodic felspar	
		Biotite	
			Rhyolite, granite
		Muscovite	
Lowest	Lowest	Quartz	

Table 2.2. Susceptibility of other common minerals to weathering.

Group	Mineral	Effects of weathering
Carbonates	Calcite	Readily soluble in acidic waters
	Dolomite	Soluble in acidic waters
Evaporites	Gypsum	Highly soluble
	Anhydrite	Highly soluble
	Halite (common salt)	Highly soluble
Sulphides	Pyrite and various other pyritic minerals	Weather readily to form sulphates, sulphuric acid and limonite
Clay minerals	Chlorite	Weathers readily to other clay minerals and limonite
	Vermiculite	Weathers to kaolinite or montmorillonite*
	Illite	Weathers to kaolinite or montmorillonite*
	Montmorillonite	Weathers to kaolinite
	Kaolinite	Stable**
Oxides	Haematite	Weathers to limonite
	Ilmenite	Stable
	Limonite	Stable

* These minerals expand and contract with wetting and drying and this can cause large disruptive forces and disintegration of some rocks.
** Softens on wetting.

For example, both rhyolite and granite contain sodic felspar, micas and quartz, but in rhyolite the crystals are much more fine grained and more tightly interlocked than in granite. Hence granite is much more susceptible to weathering than rhyolite (Figure 2.16).

Also, a dense, non-porous limestone comprising almost 100% calcite is likely to be less susceptible to weathering (solution) than a porous sandstone comprising 80% quartz grains which are durable but cemented by calcite.

2.6.4 Weathered rock profiles and their development

The following are the main factors which contribute to the development of weathered profiles:

– climate and vegetation;
– rock substance types;
– defect types and patterns;

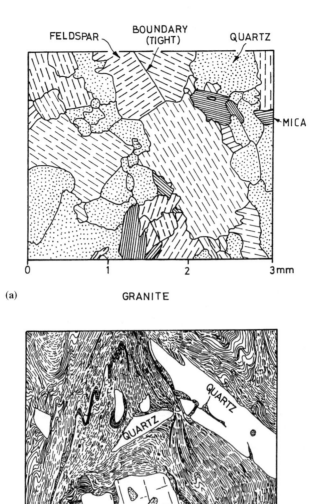

Figure 2.16.   Microscopic views of the structure of (a) granite and (b) rhyolite.

– erosion;
– time;
– topography;
– groundwater.

The influence of each of these factors is discussed in the sections which follow. Although the factors are discussed separately or in pairs, they usually all interact and influence the development of weathered profiles. The terms used to describe weathered rocks are defined in Section 2.8.

2.6.4.1 *Climate and vegetation*

Climate is the dominant factor. At one extreme, in a desert situation, chemical weathering effects are usually almost negligible. Mechanical weathering effects may include opening up of joints due to destressing and some fragmentation of exposed rock surfaces due to extreme temperature changes. At the other extreme, under hot, humid (e.g. tropical or sub-tropical) conditions chemical weathering proceeds relatively rapidly, due to the ready availability of water containing oxygen, carbon dioxide and organic acids derived from the vegetation. Regardless of the composition and structure of the parent rock mass, the near-surface weathered profile is usually of the lateritic type, the upper part of which may consist almost entirely of oxides of iron and aluminium. Figure 3.40 shows a typical near-surface profile this type, and the processes believed to be involved in its development. It must be appreciated that the weathered profile at a particular site may not have been developed under the present climatic conditions. Throughout some arid areas of Australia and South Africa there are deep weathered profiles which developed under tropical or semi tropical conditions, largely during Tertiary time.

2.6.4.2   *Rock substance types, and defect types and pattern*

Many fresh rock masses are relatively complex in their composition and structure. They may contain several different rock types, with widely differing substance strengths and sus-ceptibilities to chemical weathering, and the rocks may be folded and intersected by defects such as joints and faults. The material in the fault zones may be crushed rock which has essentially "soil" properties in the fresh (unweathered) state. Chemical weathering generally proceeds from the joints and faults, which act as groundwater conduits. Because of this, the distribution of intensely weathered rock is usually governed as much or more by the pattern of occurrence of these defects, than by the depth below the ground surface.

Figures 2.17 to 2.24 show a range of weathered rock profiles, illustrating these effects.

Figure 2.17 is a profile developed under tropical conditions in a gneiss rock mass with very simple structure – sheet joints parallel to the original (sloping) ground surface. The uppermost 13 m comprises a laterite soil profile overlying extremely weathered gneiss. The extremely weathered gneiss is a gravelly clay (CL-GC). Its upper boundary with the pallid horizon of the laterite profile is gradational, but its lower boundary with slightly weathered gneiss is sharp, coinciding with Sheet Joint (1). It is likely that several other sheet joints were present initially at shallower depth, but all trace of them has been obscured by the extreme weathering or by disturbance of the exposed surface of the cut. Between Sheet Joints (1) and (2) the gneiss is slightly weathered except for a narrow extremely weathered zone which sur-rounds Sheet Joint (2). Below this zone the gneiss is fresh. It is clear from this simple profile that chemical weathering proceeded both from the ground surface and from the sheet joints.

Figure 2.18 shows a more complex profile developed under tropical conditions in the same region as the profile in Figure 2.17. The profile steps downwards (i.e. deepens) to the east because of the presence of three different types of granite which are progressively less siliceous and hence are more susceptible to chemical weathering. The dolerite dyke has a high resistance to weathering and so forms a prominent ridge in the upper surface of the dominantly fresh rock zone. The contact between the dyke and quartz feldspar granite is sheared, and weathering of the sheared rock has resulted in a deep, narrow slot of highly to extremely weathered rock. A local depression in the fresh rock zone occurs also along a sheared zone near the eastern side.

Figure 2.19 shows an idealised weathered profile through granitic rocks in Hong Kong described by Ruxton and Berry (1957). Broadly similar profiles are found in many other areas underlain by granitic and other igneous rocks, and this profile has been widely accepted as typical for such rocks. It is clear that the main controls on the distribution of weathered materials have been the depth below the ground surface and the pattern of the joints in the rock mass.

The corestones shown on Figure 2.19 usually display spheroidal weathering effects, that is, they comprise fresh or slightly weathered rock surrounded by concentric shells of rock which becomes progressively more weathered away from the core, as shown on Figure 2.20.

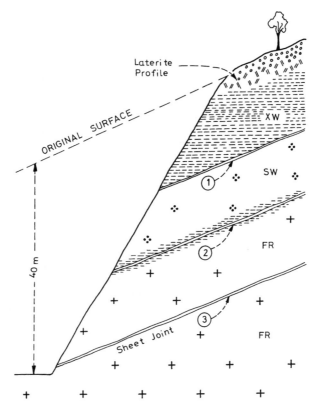

Figure 2.17.   Weathered profile developed previously on sheet-jointed gneiss, Darling Range, Western Australia (Gordon, 1966).

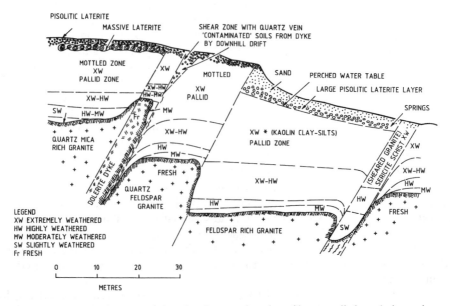

Figure 2.18.   Diagrammatic cross section showing weathered profile controlled partly by rock type and partly by sheared zones, Darling Ranges, Western Australia (Gordon, 1984).

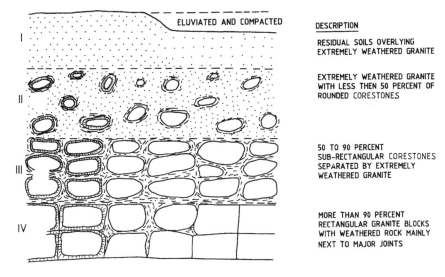

Figure 2.19.   Idealised weathered profile in granitic rocks. (Ruxton & Berry, 1957).

The cracks which isolate the shells are like small scale sheet joints and are probably caused by relief of residual stresses and other stresses set up by capillary, osmotic and other chemical weathering processes.

Figure 2.21 shows what initially appears to be an excellent example of the idealised weathered granitic profile of Ruxton and Berry (1957), exposed in elevation view, by erosion. The upper slopes show mainly extremely weathered granite (soil properties). In the lower, rock slope this material occurs between corestones and becomes progressively less abundant, and is eventually absent in the outcrops close to water level.

The actual situation on this hillside is not so simple. Near the centre of the photograph, on the skyline and elsewhere at intermediate levels on the slope, hidden by trees, there are granite boulders and areas of what appear to be outcrops of essentially fresh granite.

Excavations made into slopes showing similar surface evidence have shown subsurface profiles as on Figure 2.22. At its right hand edge this figure shows a weathered profile similar to that of Ruxton and Berry (1957) except that the extremely weathered material extends locally much deeper along and next to the fault. Near the centre the profile at depth is also similar to that of Ruxton and Berry (1957). However, outcrop area A and the boulders B occur at the ground surface and are underlain by 7–10 m of extremely weathered material. On the left side, outcrop area C looks similar to area A but is continuous downwards into mainly fresh bedrock.

The differences between the profiles A, B and C raise the following questions:

– Are outcrop A and boulders B really *in situ*?
– If they are *in situ*, then why don't they continue downwards into progressively less weathered rock mass as is the case at outcrop area C?

It can usually be shown from the continuity of the joint and fault pattern within the extremely weathered rock, whether or not masses of rock such as A and B are *in situ*. In Figure 2.22 the continuity of coated joints is clear evidence that both are *in situ*. It is often not possible to provide an unequivocal answer to the second question. One possibility is that the extremely weathered rock under A and B was partly decomposed by chemical alteration (Section 2.7) before becoming weathered. Another is that the rock substance

Figure 2.20.   Granitic boulders or corestones showing spheroidal weathering effects. Photo courtesy of
Dr. R. Twidale.

forming A and B may be for some reason more resistant to chemical weathering than that
which occurred initially adjacent to and below them. Yet another is simply that the joints
which surrounded these rock masses were less permeable than those elsewhere, allowing
much less groundwater percolation.

Although it is not often possible to understand the reason for such anomalies in weathered
profiles in apparently "uniform" igneous rocks, it is very important to appreciate that they
occur quite commonly, and hence to expect them and allow for them in site investigations.

Figure 2.23 shows the nature of part of the weathered profile developed at the spillway of
Split Rock Dam, New South Wales. The spillway excavation below the depth of ripping
refusal was the designated quarry to supply rockfill for the dam. Alternating beds of
greywacke, greywacke breccia and siltstone dip gently into the hillslope. The greywacke is
extremely strong when fresh and relatively resistant to chemical weathering. The siltstone is
strong to very strong when fresh, but is more susceptible to chemical weathering than the
other rocks. The resulting profile, with beds of fresh greywacke (extremely strong) overlying variably weathered siltstone (weak rock grading to soil properties) made it difficult to
meet some rockfill requirements, with normal quarrying methods.

Figure 2.21.   Granite corestones and outcrops below granitic soils on Granite Island, Victor Harbor, South Australia.

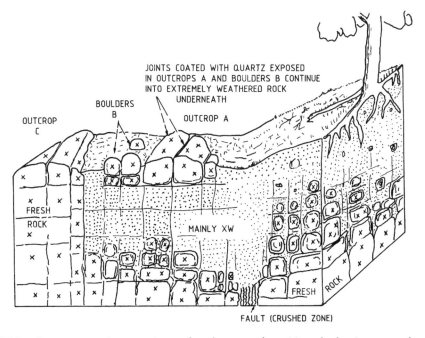

Figure 2.22.   Features sometimes seen in weathered masses of granitic and other igneous rocks.

2.6.4.3   Time and erosion

The depth of a weathered profile depends upon the amount of time during which the rocks have been exposed. However, this factor (time) must be considered together with erosion, because erosion will have been continually lowering the ground surface, during

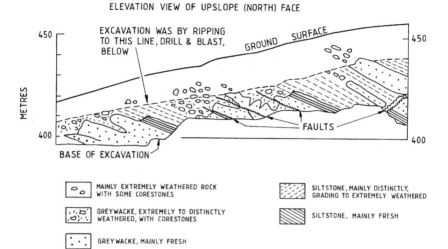

Figure 2.23.   Distribution of fresh and variably weathered rock, controlled largely by variation in rock substance type, at Split Rock Dam, New South Wales.

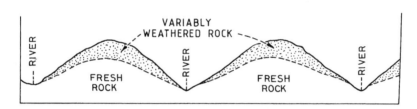

Figure 2.24.   Shape of upper surface of fresh rock in a situation where the rate of valley erosion has been exceeding the rate of lowering of the chemically weathered zone.

the development of the profile. In tectonically stable areas where the effects of erosion have been relatively small, great depths of weathered rock have developed. For example, large areas of Western Australia which have been relatively stable and mainly exposed since Precambrian time, are now almost flat "peneplains", underlain by extremely weathered rock to depths of 30 m to 50 m. In tectonically active areas uplift and subsequent erosion occur, and result in much shallower weathered profiles in comparable rocks. In the areas of North America and Scandinavia eroded by ice-sheets during the Pleistocene glaciation, there is generally no chemically weathered profile.

Figure 2.24 shows a situation which is frequently seen: weathered profiles which are deepest beneath hilltops, ridges and plateaus, and shallow or absent at or close to the floors of valleys. This situation may have arisen simply because the rate of valley erosion (i.e. removal of weathered materials) has exceeded the rate of deepening of the weathered profile. However, in some places situations similar to that in Figure 2.24 can have a different origin as indicated in Figure 2.25. This shows the situation in parts of southern and eastern Australia, where mountain ranges were formed during late Tertiary time, by the faulting and uplifting of deeply weathered rock. Relatively rapid erosion of river valleys since late Tertiary time has resulted in composite weathered profiles as shown on Figure 2.25. The rock near the valley floors is mainly fresh but is usually somewhat mechanically loosened due to destressing.

Ancient weathered profiles can be buried and preserved beneath younger sediments or even sedimentary rocks. Figure 2.26 is a diagrammatic cross section showing the geological

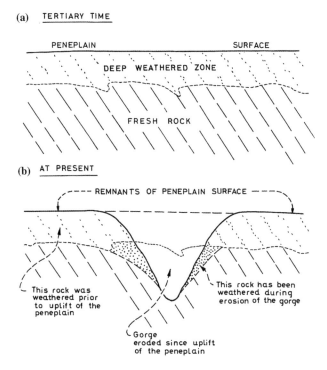

Figure 2.25. Type of valley weathered profile where a river has cut down through an uplifted block which has been deeply weathered previously.

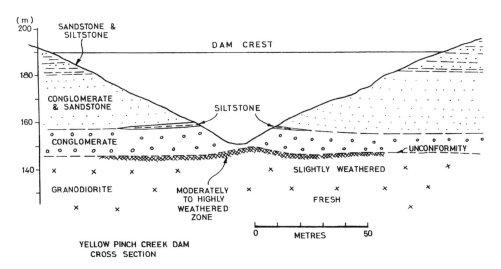

Figure 2.26. Cross section at Yellow Pinch Dam, New South Wales, showing buried land surface with weathered zone.

situation at Yellow Pinch dam in New South Wales. The valley walls at this site are formed by a near-horizontal sequence of mainly conglomerates and sandstones, of Permian age. These rocks are slightly weathered. However, at and just below river level, they unconformably overlie granodiorite of Devonian age. The granodiorite is moderately to highly weathered in its uppermost 2–5 m. The upper surface of the granodiorite is an eroded and weathered land surface of Permian Age.

2.6.4.4   Groundwater and topography

Optimum conditions for chemical weathering occur where vertical movement of aerated groundwater is at a maximum, i.e. beneath elevated flat ground or gentle slopes, above the main water table. Beneath steep slopes there is less infiltration of surface waters due to rapid runoff.

Field evidence in many places indicates that very little weathering has occurred beneath the main or fundamental water table. This is probably due to less oxygen and slower ground-water movement, below the water surface. In some places it is due to the fact that the main water table is in fact perched above fresh rock which is effectively impervious. However, where suitably permeable rocks occur beneath the main water table, rapid groundwater flow can occur well below the water table, and chemical weathering (or solution, in soluble rocks) occurs.

It must be appreciated that due to changes in climate, the present water table may be higher or lower than the water table which dominated when a particular weathered pro-file was formed.

2.6.4.5   Features of weathered profiles near valley floors

Figure 2.27 is a hypothetical cross section through the lower part of a relatively steep val-ley in relatively strong, jointed rocks. It shows diagrammatically a number of features found commonly when excavating foundations for dams.

The river is flowing in an entrenched meander channel and at this site it has been migrat-ing towards the right bank. As a result of this the left bank slopes are flatter, and show less rock outcrop, than the right bank slopes. The river channel occupies only half of the valley floor, the left side being formed by a well established (overgrown) alluvial terrace.

There are three bedrock types: granite (1) and microgranite (2) which are fresh and very strong at depth, and basalt (3) in the form of an intrusive dyke which is distinctly altered, weak. The granite contains two faults (4) in which the rock is sheared and partly crushed to soil properties. Joints at depth (5) are mainly tightly closed, but as the valley floor is

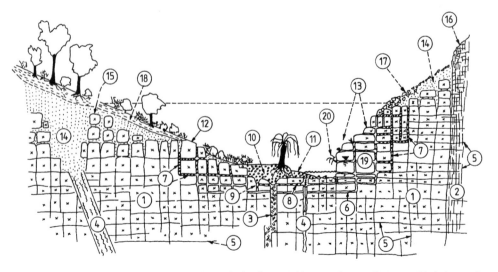

Figure 2.27.   Diagrammatic cross section through the floor and lowest slopes of a steep-sided river valley, showing the types of feature developed by mechanical and chemical weathering processes.

approached they are either slightly open (6) due to destressing, or else (7) have been open but are now partly or wholly infilled with clay which has migrated down from the chemically weathered zone. In the bedrock just below the river, some joints (8) have become infilled with alluvial silt which has migrated down from the river.

It can be seen that although the valley in profile is V-shaped, the profile of the base of the main mechanically weathered zone is closer to U-shaped. This is often found to be the case, and is due to pronounced destressing at the highly stressed base of the V-notch (see Figure 2.11a). In this very strong rock there is no obvious bulge or antiform structure across the valley floor, but the opened joints beneath and next to the river bed (including that part beneath the alluvial terrace) indicate that significant rebound movements have occurred.

The actual river bed below the alluvium (10) and (11) contains a slot (9), eroded differentially by the river, along an infilled joint. Similar eroded slots occur over the altered dyke (3) and fault (4).

Upslope from the river there are small granite outcrops (12) on the left bank and larger and steeper granite outcrops (13) on the right bank.

Above the levels of these outcrops the granite is variably weathered. Extremely ranging to distinctly weathered rock (14) locally contains corestones (15) of fresh and slightly weathered rock, and extends downwards next to some joints in the dominantly fresh rock. Extremely weathered rock extends down to about river level, along the fault zone (4).

The microgranite is very resistant to chemical weathering, and forms fresh and slightly weathered steep outcrops (16) at the ground surface. This rock has closely spaced joints, however, and is mechanically weathered near the surface. Toppled and fallen blocks of microgranite have formed the scree (17) which covers the slope below.

On the left bank, slopewash soil (18) derived from extremely weathered granite occurs to shallow depth beneath the ground surface in most places, becoming deeper locally over the fault (4) and beneath gullies.

The water table (19) on the right bank, daylights at a seeping, partly infilled joint. Ferns and swampy type grasses (20) are established along the outcrop of this joint.

The alluvial gravels (10) beneath the terrace are very old (of Pleistocene age, probably). Gravels of this age are commonly partly weathered. Sands and gravels of the present day point bar (11) are not weathered – they are deposited during the dying stages of present-day floods.

2.6.5   Complications due to cementation

Chemical weathering does not always result in weakening of rock substances. Detailed observations at many sites have indicated that cements such as limonite or other iron oxide minerals have been leached out from one zone, to produce weaker, more porous substance, and redeposited in another to produce stronger, denser substance. The latter substance in some cases has been noticeably stronger than the (assumed) fresh substance. Joints and faults have also been found to have been strengthened greatly by such deposition during weathering.

These effects are seen commonly in all rock types in the deeply weathered mantle of western and southern Australia, and in many siltstones and sandstones in Victoria and New South Wales. They are believed to result from weathering during warm, humid, ranging to tropical conditions.

Similar effects have been observed in siltstones and sandstones containing small amounts of carbonate minerals in their matrices.

2.7   CHEMICAL ALTERATION

Some igneous rocks have been partly or wholly decomposed by hot waters and gases during late stages of their solidification. The hot waters may be derived from the igneous magma or else they may be groundwaters which have penetrated to great depth and become heated by the magma. Sedimentary and metamorphic rocks in the vicinity may also become partly or wholly decomposed (Figure 2.14).

This type of deep-seated decomposition is called hydrothermal alteration. It involves the chemical breakdown of some of the primary (original) minerals to form secondary minerals, which are generally weaker and less stable in water, than the primary minerals. The secondary minerals include serpentine and montmorillonite (from olivine), chlorite (from biotite) and kaolin (from felspars). The altered rocks are therefore generally weaker, and more susceptible to chemical weathering, than the same rocks when unaltered.

Altered rocks can often be distinguished from weathered rocks by their colour – usually green or white, whereas weathered rocks are usually yellow-brown or brown due to oxidation which has produced limonite in most of them. It is important in dam engineering to be able to distinguish between altered and weathered rock. This is because in an altered rock situation the altered materials may be overlain by fresh rock, and become progressively more altered and weaker, with increasing depth. It is possible to find both altered and weathered rock at the same site, with some of the weathered rock being previously altered (See Section 2.6.4.2 and Figure 2.22).

2.8   CLASSIFICATION OF WEATHERED ROCK

From the discussion in Sections 2.5 to 2.7 it is not surprising that the products of rock weathering are often highly complex and variable over short distances. To assist in mapping and understanding them, some simplification and classification into groups with like characteristics has been required. Two main types of classification have been developed:

– Rock substance types, which define fresh substance and 4 or 5 "grades" of weathered substance, each grade showing greater effects of weathering than the previous one: examples include Moye (1955) and McMahon et al. (1975);
– Rock mass types, in which rock masses are classified into zones showing progressively greater weathering effects, as indicated by the percentages of various grades of weathered rock substances within them: examples include Ruxton and Berry (1957), Deere and Patton (1971), International Society for Rock Mechanics (1978), and British Standards Institution (1981).

The substance type systems recommended by the authors are set out below in Section 2.8.1.

The authors consider that mass type classifications are best designed specifically for individual sites, because they need to be tailored to the project needs as well as the site conditions. In practice, sites for dams are usually classified into zones taking into account variations in the overall structure (rock types, folds, faults etc) of the rock mass, and the nature and distribution of at least the following:

– mechanically loosened rock mass;
– infilled seams, and
– rock substances at various weathered grades.

The Geological Society (1995) provides useful reviews of various aspects of rock weathering and of classification systems for weathered rocks. It highlights the difficulties which arise in classification, and concludes that any attempt to devise a single standard system, to cover all rock types would be "futile". It includes a flow chart containing a suggested classification system or "Approach", for each of four different weathered rock situations. The explanatory paragraphs following the chart include this statement:

"The recommended classifications are intended as general guidelines which can form the basis for site specific or problem specific approaches"

The authors consider that the chart can be useful if used in this way. As pointed out above, they consider that in most cases systems for weathered rock mass should be site-specific. Unfortunately the Geological Society chart is included as Figure 19 in British Standards Institution (1999), without the above note. However, in Section 44.4.2.2 the Standard notes that *"Formal classification may often not be appropriate, and so is not mandatory"*.

2.8.1 *Recommended system for classification of weathered rock substance*

Any system should be used in conjunction with appropriate geological and engineering descriptions, and should fulfill the following requirements:

1. It should provide a "shorthand" of descriptive terms and abbreviations, to facilitate recording (during logging and mapping) of the distribution of rock substance at various recognisable stages of change due to weathering;
2. This distribution can then be used to provide an understanding of the local (site) relationships between the intensity of weathering effects and the common controlling factors discussed in Sections 2.3 to 2.7, in particular
   – the distribution of rock substance types, and
   – the pattern of defects in the mass.
   Such understanding should assist the site investigator to make soundly based correlations, and predictions about the distribution of rock in various weathered conditions in other parts of the site, not directly explored;
3. It must allow unambiguous communication of the descriptions and predictions, verbally, and in logs, drawings and reports;
4. It should enable determination of any site relationships which may exist between the weathered conditions of the rock substances present, and their strengths and other engineering properties. If at a particular site such relationships can be adequately proven, it may be possible for them to be used as part of a system of acceptance criteria e.g. for rockfill or for foundation levels.

The classification system for weathered rock substance (Table 2.3), when used in conjunction with the ISRM substance strength classification (Table 2.4) has been shown to meet these requirements, for many common rock types, and is recommended by the authors. Using this approach, rock substance is classified as in the examples in Table 2.5.

Table 2.3.   Recommended descriptive terms for weathered condition of rock substance. Modified from McMahon, Douglas & Burgess (1975).

Term	Symbol	Definition
Residual soil	RS	Soil developed on extremely weathered rock; the mass structure and substance fabric are no longer evident; there is a large change in volume but the soil has not been significantly transported.
Extremely weathered	XW	Rock is weathered to such an extent that it has soil properties, i.e. it either disintegrates or can be remoulded, in water.
Distinctly weathered	DW	Rock strength usually changed by weathering. The rock may be highly discoloured, usually by iron-staining. Porosity may be increased by leaching, or may be decreased due to deposition of weathering products in pores.
Slightly weathered	SW	Rock is slightly discoloured but shows little or no change of strength from fresh rock.
Fresh rock	FR	Rock shows no sign of decomposition or staining.

Note that altered rock is classified in the same way as weathered rock except that "weathered" is replaced by "altered" and in the abbreviation, "W" is replaced by "A".

In some granitic and similar rocks which are very strong when fresh it is commonly possible and useful to subdivide the distinctly weathered category further, on the basis of dry strength of 50 mm diamond drill cores. The terms used are based on those of Moye (1955) and Hosking (1990), defined as in Table 2.6. The rock fabric will be apparent in all weathering grades.

The distribution of rock substance at various weathering grades can be shown on borehole logs, and on plans, sections or elevation views of rock exposures, either by the abbreviations, or graphically by suitable colour or black and white overprinting of the rock type symbols.

Table 2.4.   Recommended descriptive terms for strength of rock substance.

Rock strength[1]	Symbol	Point load strength index[2][3] $I_{s(50)}$	Approximate unconfined compressive strength Qu(MPa)[2][3]
Very weak	VW	<0.2	<5
Weak	W	0.2–1	5–25
Medium strong	MS	1–2	25–50
Strong	S	2–4	50–100
Very strong	VS	4–10	100–250
Extremely strong	ES	>10	>250

[1] Based on ISRM (1985): Suggested methods for the quantitative description of discontinuities in rock masses. [2] In rocks with planar anisotropy, the strength terms and numbers refer to the strengths when tested normal to the anisotropy. [3] The numbers assume that Qu is 25 times $I_{s(50)}$. In practice the ratio ranges typically from 10 to 25, with the weaker rocks towards the lower end.

Table 2.5.   Description of weathered or altered rock substance: Examples.

Rock type term	Rock condition term	Rock strength term	Abbreviated form
Granite	Fresh	Very strong	Granite FR (VS)
Granite	Extremely weathered	Soil properties (GW-GC)	Granite XW (GW-GC)
Granite	Distinctly weathered	Medium strong	Granite DW (MS)
Sandstone	Fresh	Weak	Sandstone FR (W)
Basalt	Distinctly altered	Medium strong	Basalt DA (MS)

Table 2.6.   Weathered rock substance classification for granite and similar rocks. Modified from Moye (1955) & Hosking (1990).

Term	Abbreviation	Description
Fresh	FR	Rock shows no evidence of chemical weathering
Slightly weathered	SW	Rock is slightly discoloured but rings when struck by a hammer; not noticeably weaker than fresh rock
Moderately weathered	MW	Rock is discoloured, produces only a dull thud when struck by a hammer; noticeably weaker than fresh or slightly weathered rock but dry samples about 50 mm across cannot be broken across the fabric by unaided hands
Highly weathered	HW	Rock is discoloured, can be broken and crumbled by hand, but does not readily disintegrate in water
Extremely weathered	XW	Material disintegrates when gently shaken in water, i.e. has soil properties

2.8.2 *Limitations on classification systems for weathered rock*

Quite often, particularly in sedimentary rocks, it is difficult or even impossible to determine which, if any of the rock substance at a site is fresh (unweathered). This is because such rocks can vary greatly in colour, porosity and strength due to past processes involved in their formation, rather than weathering processes.

Some rocks which are weak or very weak when fresh (e.g. some shales and poorly cemented sandstones) assume soil properties (i.e. they classify as "extremely weathered") when only slightly affected by chemical weathering. Intermediate weathered conditions cannot be defined in these cases.

Weathering products of carbonate rocks and rocks cemented by gypsum or anhydrite range from cavities or soils (due to solution) to very strong rock substances (due to redeposition of material dissolved elsewhere in the mass). Such weathering products usually cannot be described adequately using the system on Table 2.3. Site-specific classifications are preferable in such cases.

2.9 RAPID WEATHERING

Rapid weathering processes are defined here as those processes which cause exposed rocks to be significantly weakened during periods of only days, months or years. In general the rocks most affected by rapid weathering are relatively porous and not of high strength. However even some very high strength rocks can be affected and so careful observational studies and testing are needed whenever rocks are to be used as construction materials.

2.9.1 *Slaking of mudrocks*

The weakest claystones and shales, which can barely be described as rock, have usually been strengthened only by consolidation. When such "compaction shales" are exposed in excavations they usually develop fine cracks at the exposed surface due to destressing and drying out. Swelling, further cracking and sometimes complete disintegration back to clay occurs, when the destressed shale or claystone is allowed to absorb water.

Stronger claystones, siltstones and shales (i.e. in the weak to strong rock range) have usually received some strength by cementation and/or recrystallisation of clay minerals to form micas, as well as by consolidation. These rocks also generally develop fine drying out cracks when exposed by excavation but the cracking is much less severe and occurs more slowly than in the "compaction shales". Once the fine cracks appear on the rock surface, addition of water causes relatively rapid deterioration of the rock. The reasons for this are believed to be as follows. Water is absorbed rapidly into the cracks and adjacent rock substances, by capillary suction. The water compresses the air trapped in the cracks and allows clay minerals in the adjacent rock to swell slightly. As a result the cracks widen and propagate.

The mechanisms of slaking are discussed in more detail by Taylor and Smith (1986), Taylor (1988) and Olivier (1990).

Many strong siltstones develop cracks eventually with repeated wetting and drying. However in constant humidity environments such as in rockfills or earthfill, such rocks have been found to have suffered little or no breakdown over periods of up to 90 years.

The weathering of mudrocks and its significance in dam construction is discussed in Chapter 3, Sections 3.5.1 and 3.5.4.

Some poorly cemented sandstones which contain a high proportion of clay minerals in their matrices, show similar slaking properties to mudrocks.

Laboratory tests aimed at distinguishing between durable and slaking rocks include the ISRM Slake Durability Test (ISRM, 1979) and the Modified Jar Slake Index Test (Czerewko & Cripps, 2001).

2.9.2   Crystal growth in pores

Where rock is exposed in saline environments, e.g. as rockfill or riprap on marine break-waters and some tailings dams, some salt is absorbed by the rock either by periodic inundation or from spray. During warm, dry conditions the rock dries out and salt crystals grow in pores near its surface, disrupting mineral grains (as in the Sodium Sulphate Soundness Test). Periodic wetting up with rainwater causes further disruption. It appears that fresh water is drawn rapidly into the pores by osmotic and capillary suction, compressing entrained air. The amount and rate of degradation which occurs by the above process depends largely upon the porosity, texture and strength of the rock.

2.9.3   Expansion of secondary minerals

Some rocks appear fresh and strong but in fact have been chemically altered (see Section 2.7) and contain highly expansive secondary minerals. If these are in sufficient quantities they can cause the rocks to be significantly weakened or even to disintegrate, on inundation, or on exposure to the weather. The rocks most commonly affected are basic igneous rocks (basalt, dolerite and gabbro), and the most common expansive secondary mineral is montmorillonite (see Chapter 3, Section 3.2.3).

2.9.4   Oxidation of sulphide minerals

Metallic sulphide minerals (e.g. pyrite) occur in small amounts in many rock types of all ages. They can occur in many ways, the most common being:

– grains from microscopic to a few millimetres across, scattered through the rock;
– grains as above, concentrated within particular beds;
– veins of any size, and
– coatings on joints or other defects.

In outcrops and in weathered rock, sulphide minerals are invariably absent, having been oxidised to form limonite or similar minerals. Due to their lack of surface exposure, small grain sizes, and often complex distribution, the sulphide mineral content in rock is usually difficult to quantify, and sometimes not recognised.

When exposed to oxygen and moisture, e.g. in excavations, or in rock processed for rockfill, filters or concrete aggregates, sulphides can oxidise rapidly and release sulphuric acid and metallic sulphates. Bacteria assist in the oxidation. The acid can attack other minerals in the rocks, weakening or dissolving them and producing other sulphates which can cause swelling, heave and sulphate attack on rock or concrete (see also Section 3.5.1, Engineering properties of mudrocks).

The effects listed above can render sulphide-bearing rocks unsuitable for rockfill, filters or concrete aggregates. Also, acid and metallic salts can be released into the environment. Taylor (1998) describes serious problems of this kind, referred to in the mining industry as "acid drainage". This has occurred in many places due to oxidation in waste rock dumps and tailings from mining operations. Abandoned mine excavations which exposed sulphide minerals have also caused acid drainage problems.

The severe effects described by Taylor (1998) have come mainly from materials with high contents of sulphide minerals. However the following examples indicate that oxidation of much smaller contents of sulphides in rocks used commonly in dam construction may produce undesirable effects.

2.9.4.1   *Sulphide oxidation effects in rockfill dams – some examples*

Corin Dam (Australian Capital Territory). Sandstone and quartzite which formed most of the rockfill and filters in this 75 m high earth and rockfill dam contained about 1% of pyrite (Haldane et al., 1971). The reservoir started to fill in 1968, with water of pH about 6.5. During and after filling, seepage from the downstream rockfill was found to be highly acidic (pH around 4). Both acidity and conductivity were highest after heavy rainfall. It was concluded that oxidation of pyrite in the fill was occurring due to the percolation of rainfall as well as underseepage from the dam. Haldane et al. (1971) reported that the acid waters (base flow rate about 25 l/s) were diluted sufficiently by other waters to prevent any problems downstream.

Testing of the reservoir water in the 1980's and 1990's showed it to be slightly alkaline and to have an average sulphate content of 0.6 mg/l. Sampling of the seepage showed that there had been a gradual decrease in its acidity, pH being around 5 by March 1999 (Tabatabaei, 1999, and Tabatabaei, Pers.comm.).

Kangaroo Creek Dam (South Australia). Under Section 2.4.4 Decomposition of sulphide minerals, Fell et al. (1992) reported on observed oxidation of metallic sulphides and associated fretting in schist exposed in excavations at the site of this concrete-faced rockfill dam. During the early site studies the schist was judged to contain less than 1 percent of sulphides. Based on the absence of deterioration in 40-year old schist rockfill near the site, the schist was adopted as the main source of rockfill, with some dolomite in the uppermost quarter of the bank (Trudinger, 1973). It was assumed that pyrite oxidation would not occur in the constant moisture environment at depth in the dam.

As at Corin Dam, the underseepage includes rainwater which enters the rockfill. The seepage rate fluctuates widely with rainfall and storage level, but is usually between 0.5 l/s and 2 l/s. Chemical analyses of the seepage have been made only since 1999 (29 years after first storing of water). The seepage has a pH of around 7 (slightly lower than that of the reservoir) but much higher concentrations of calcium and magnesium sulphates than the reservoir water. These concentrations have been decreasing since 1999 and it can be inferred that they were higher during the early years of storage. It is considered that oxidation of the pyrite, and acid attack on the dolomite, have been occurring since completion of the dam in 1970. However, the embankment is performing satisfactorily, and the effects of the seepage on the quality of the water downstream have not been significant.

Carsington Dam (Derbyshire, UK). This dam failed by upstream sliding during construction, in 1984. It was a 38 m high zoned rockfill structure (Cripps et al., 1993, and Banyard et al., 1992). Its clay core (extremely weathered mudstone) contained 0.8% pyrite and 0.25% calcite. The shoulders, of moderately weathered mudstone, had 5% pyrite and 7% calcite. Drainage blankets were of limestone. The dam was reconstructed between 1989 and 1991. During the reconstruction it was revealed that sulphuric acid generated from the pyritic fills had attacked the limestone and resulted in blockage of the drains by precipitates of gypsum and iron hydroxide. Also, carbon dioxide had accumulated in drains, manholes and excavations. Seepage from the dam had high concentrations of sulphates, iron and other metals. While these features had not been responsible for the failure, they were undesirable, and the new design included the following features aimed at reducing or counteracting the effects of pyrite oxidation:

– Allowance for degradation, in the embankment shear strength parameters;
– Non-calcareous fill used for the drainage blankets and filters;
– Concrete structures protected by coatings of bitumen;

– Stringent safety precautions against gas in confined spaces;
– Downstream lagoons for long-term treatment of seepage waters.

Roadford Dam (Devon, UK). This is an asphaltic concrete faced rockfill embankment 45 m high. It was built in 1988–89 (Cripps et al., 1993). The rockfill materials are sandstone and mudstone with average pyrite content of 1.12%. A basal drainage blanket was formed by altered dolerite, which contained 11% calcite. Some degradation of the fill was allowed for in the selection of shear strength parameters and of freeboard.

On first filling in 1989, drainage water was found to have unacceptably high levels of sulphate, calcium, iron and other minerals. Rapid pyrite oxidation and sulphuric acid attack on the calcite was indicated. From the results of tests on the rockfill and blanket materials about 30 months later, Cripps et al. (1993) conclude that the amounts of calcite removed were so small that the effect on the geotechnical properties of the materials would be negligible. However they note that the drainage water requires treatment, probably "for some years".

2.9.4.2   Possible effects of sulphide oxidation in rockfill dams

The four cases discussed above indicate that if materials used in rockfill dams contain sulphide materials, the following can occur:

1. Seepage waters containing sulphuric acid and metallic sulphates, which may require ongoing remedial treatment;
2. Attack on concrete by sulphuric acid and/or sulphates;
3. Lowering of strength of individual particles in rockfill, giving rise to settlement or lowered shear strength of the embankment.

If an embankment also contains carbonate rocks or minerals, in its rockfill, drains, filters, or in its foundation, these could be attacked by sulphuric acid to cause any of the following:

4. Clogging or cementation of drains or filters, by sulphate minerals;
5. Loss of material and changes to gradings, in carbonate rockfill and/or filter zones, giving rise to settlement and/or impaired filter performance.

Whether or not any of the above effects occur to an extent affecting the successful operation of a project would depend on the interaction of many site-specific factors. It is possible also that some of the above effects could occur when a dam of any type is built on a foundation (or with a storage area) containing sufficiently large amounts of sulphide minerals. See further discussion and recommendations in Section 3.7.7.

2.9.4.3   Sulphide oxidation – implications for site studies

In the light of all of the above it is important that site investigations for dam projects include checking for the presence of sulphide minerals and assessment of the risk of possible problems associated with their breakdown. Such studies should include:

(a) checking for any known sulphide mineralization or acid drainage problems in the region;
(b) checking for sulphides and for past and current oxidation effects, during geological mapping and during logging of test excavations and drill samples;
(c) testing streams and groundwaters for pH, electrical conductivity and chemical composition;
(d) examining samples petrographically to assess the types, concentrations and modes of occurrence of any sulphide minerals found during (a) or (b);
(e) accelerated weathering tests or other tests to simulate the proposed in-service environment.

2.9.5   *Rapid solution*

Rocks which contain appreciable amounts of evaporite minerals (gypsum, anhydrite or halite) as cement or matrix, may be weakened rapidly if water is allowed to pass through them. James and Lupton (1978), James and Kirkpatrick (1980) and James (1992) provide predictive models for solution rates, and guidelines for investigation of dam sites where these minerals are present. These are discussed briefly in Chapter 3, Section 3.8.

2.9.6   *Surface fretting due to electro-static moisture absorption*

Within 3 years of the completion of Kangaroo Creek Dam, it was observed that fretting was occurring on the undersides of some large blocks of quartz-sericite schist exposed on the downstream face. Many of these blocks did not appear to contain sulphide minerals, and the fretted flakes did not have the characteristic taste of sulphate salts.

West (1978) confirmed that much of the deterioration of the exposed schist blocks was not caused by the sulphide/sulphate effects described by Trudinger (1973). He demonstrated that the likely cause was cyclic adsorption-desorption by the near-surface few millimetres of the schist, of moisture from the air, in a semi-confined or sheltered environment. The moisture changes resulted in expansion and contraction, probably of the sericite, in this near-surface layer.

2.10   LANDSLIDING AT DAM SITES

At the site of any dam located across a river it is quite common to find evidence of past landsliding on at least one side of the valley. This is not surprising because most river valleys will have developed by some combination of the following processes:

(a)   erosive downcutting by the stream, which unloads the materials underlying the valley floor and slides and causes the valley sides to be steepened;
(b)   mechanical weathering processes (resulting from a) which weaken and usually increase the permeability of the materials under the valley floor and sides;
(c)   chemical weathering processes which further weaken the materials forming the valley sides, converting near-surface rocks partly or wholly to soils and generally lowering their mass permeability, and
(d)   soil creep, erosion, deposition, and rockfalls on the valley sides, giving rise to deposits of slopewash and scree, and to overall flattenning of the valley slopes.

The overall effect of (a), (b) and (c) is to reduce the stability of the valley sides.

The valley-bottom profile on Figure 2.27 was developed by these processes, in granitic rocks and in a mediterranean or semi-arid climate. Because of the relatively dry conditions and a pattern of joints which favours stability, no landsliding has occurred or appears likely to occur.

However, if the joints were in less favourable orientations and the climate wetter, resulting in a higher water table, landsliding would be more likely to occur. Figure 2.28, taken from Patton and Hendron (1973), shows such a situation.

Wedge A B D, bounded by relict joints in the low permeability zone of residual soil and extremely weathered rock, may be subject to excessive water pressures from the underlying more permeable zone, causing it to slide.

A much larger, deeper slide would be possible if a continuous thin seam of extremely weathered granite parallel to surface AB was present within the more rocky Zone IIA, daylighting at the lower line of springs(S).

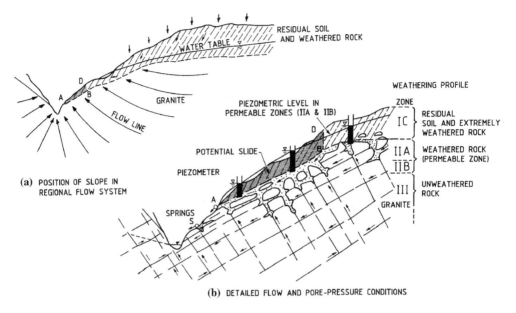

Figure 2.28.   Typical conditions likely to result in landsliding of slope in weathered granite (based on Patton & Hendron, 1973).

Deere and Patton (1971) describe weathering profiles developed in most common rock types and show how landsliding often occurs within them.

Many valleys (particularly in Europe and North America) have been subjected to glaciation, in relatively recent geological times (see Section 3.12). The resulting deepening and steepening has in some cases contributed to rockfills and landslides.

There are many different forms and mechanisms of landsliding and it is beyond the scope of this book to describe them. For useful accounts readers are referred to Turner and Schuster (1996), Varnes (1958, 1978), Selby (1982), Hunt (1984) and Hutchinson (1988). Stapledon (1995) provides some guidelines for geological modelling of landslides, and examples of their application.

2.10.1   First-time and "reactivated" slides

A distinction can be made between "first-time" landslide activity which develops at "intact" sites like that shown on Figure 2.28, and that which is simply reactivation or increased activity of an old landslipped mass which has become stabilised or is subject to only small occasionally active movements.

This distinction is important because there is a very large difference between the predictability and characteristics of the two kinds of sliding.

2.10.1.1   Reactivated slides

ICOLD (2002) has shown that in at least 75 percent of cases, disturbance of an old inactive or occasionally active landslide e.g. by cutting into or inundating its toe, causes reactivation or increased rates of movement. Hence if there is clear evidence that all or part of a slope is an old slide mass, a high risk of reactivation by construction activities or reservoir operation must be assumed, at least until geotechnical conditions in the slope are sufficiently well known to prove otherwise.

2.10.1.2 *First-time slides*

First time slides are much more difficult to predict. Reliable prediction is not possible without considerable knowledge of the geotechnical conditions in the slope, and of the nature of the possible disturbing activities. For example, there is not enough subsurface data on Figure 2.28 for one to say definitely that Wedge A B D will slide, with or without some excavation above Point A. The prediction of first time slides is discussed further in Section 2.11.

2.10.2 *Importance of early recognition of evidence of past slope instability at dam sites*

It is very important to recognize any evidence of past instability at a dam site, during the Prefeasibility or Feasibility and Site selection Stage of the project planning (see Chapter 4). The reasons for this are as follows:

1. Landslide debris, comprising broken rock, soil or mixtures of soil and rock in an uncompacted state is usually not acceptable anywhere in the foundation of a dam. This is because the compressibility and permeability of such materials are potentially high, variable and often unpredictable (see Sections 3.10.1.3 and 3.10.2.3). If these materials occur in only in a relatively shallow deposit, it might be possible to remove them and expose a satisfactory foundation. However, the presence of a small deposit of landslide debris may indicate the presence of a much larger volume of unstable or potentially unstable material nearby or upslope.
2. Also, as mentioned above in Section 2.10.1.1, experience shows that where landslipped materials on slopes are affected by excavation or inundation, renewed landsliding often occurs either in the materials themselves or from the source areas of the deposits. Some potential consequences of landsliding within dam storage areas are listed and discussed in Section 2.11.

In the light of the above, if all or a large part of one bank at a potential dam site is found to be underlain by landslipped materials, it is usually best to adopt, or look for, an alternative site. If it is decided to persevere with the original site, a very thorough site investigation will be needed, the results of which may well cause the site to be abandoned.

Comprehensive site studies would be necessary to determine the slide models and mechanisms and to assess the risks and hazards involved.

It should be clear that the presence of old landslides in any of situations above raises questions which ideally should be answered during the feasibility stage and certainly by the end of the design stage of the project planning (see Chapter 4, Section 4.3).

Many old landslides are easily recognised during stereoscopic examination of aerial photographs, by their characteristic surface form and by anomalies in the local vegetation, soils or rocks (see Chapter 5).

2.10.3 *Dams and landslides: Some experiences*

This section describes some experiences with dams at sites showing evidence of past landsliding.

2.10.3.1 *Talbingo Dam*

At the original site for Talbingo Dam, in the Snowy Mountains of New South Wales (Figure 2.29), the right bank was formed almost entirely by the remnants of a mass of variably weathered basalt which had slipped more than a kilometre downslope and buried the valley of a tributary creek.

The geological picture in Figure 2.29 was deduced almost entirely from air photos and geological outcrop mapping on 1:2400 scale. It was confirmed by refraction seismic traverses

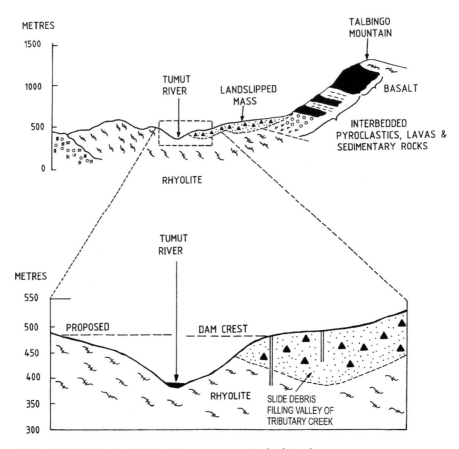

Figure 2.29.   Original site for Talbingo Dam, cross section looking downstream.

and diamond drilling, which showed the slide mass to consist of extremely weathered basalt containing a small percentage of irregularly distributed blocks of less weathered basalt from gravel size up to several metres across.

An alternative site free of landslide evidence was adopted. At this site, more than 100 m upstream, the 161 m high Talbingo Dam (rockfill) was built, using 2.3 million m³ of the basaltic landslide deposit for its impervious core. During operation of the borrow pit, some renewed sliding movements occurred within the deposit.

2.10.3.2   Tooma Dam

Tooma Dam is a 68 m high earth and rockfill structure, constructed during 1958–1961 in the Snowy Mountains of New South Wales.

The dam is located in a steep-sided valley which has been entrenched about 80 m below an older, broad valley. Prior to construction, outcrops of granitic rocks were present along both banks of the river and extended locally to 10 m to 40 m above river level. The largest outcrop area formed a cliff about 30 m high on the right bank beneath the upstream shoulder of the proposed dam (Figure 2.30 and Figure 2.31).

This outcrop and a lower outcrop area directly opposite on the left bank, were terminated on their upstream sides by soil-filled gullies. The rock outcrops above river level showed near-vertical joints roughly parallel and normal to river direction.

During the planning stages subsurface exploration of the site included trenches up to 1.5 m deep cut by hand methods, and 17 diamond boreholes with water pressure testing.

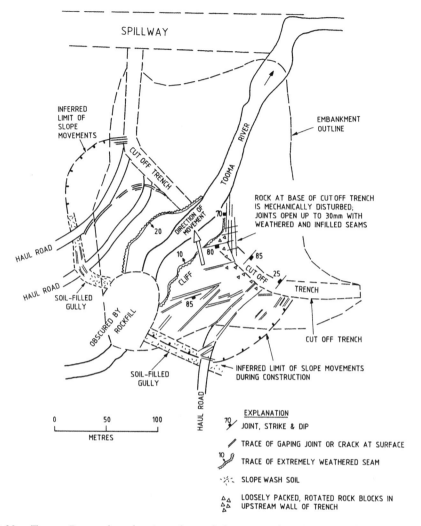

SPILLWAY

INFERRED
LIMIT OF
SLOPE
MOVEMENTS

EMBANKMENT
OUTLINE

CUT OFF TRENCH

TOOMA RIVER

ROCK AT BASE OF CUTOFF TRENCH
IS MECHANICALLY DISTURBED;
JOINTS OPEN UP TO 30mm WITH
WEATHERED AND INFILLED SEAMS

DIRECTION OF MOVEMENT

70

20

10   80

85

CLIFF   CUT OFF   25

HAUL ROAD   85   TRENCH

HAUL ROAD
SOIL-FILLED
GULLY

OBSCURED BY ROCKFILL

CUT OFF TRENCH

SOIL-FILLED
GULLY

INFERRED LIMIT OF SLOPE MOVEMENTS
DURING CONSTRUCTION

HAUL ROAD

0   50   100

METRES

EXPLANATION

70 ⟋   JOINT, STRIKE & DIP

⟋   TRACE OF GAPING JOINT OR CRACK AT SURFACE

10 ⟋   TRACE OF EXTREMELY WEATHERED SEAM

⋰⋱   SLOPE WASH SOIL

△△
△△   LOOSELY PACKED, ROTATED ROCK BLOCKS IN
△   UPSTREAM WALL OF TRENCH

Figure 2.30. Tooma Dam, plan showing observed features and movements which occurred during
              construction.

It was concluded from the results that the site would show wide variations in weathered
profile, but no special problems were envisaged.

During stripping of the foundation, the variations in weathered profile were confirmed
(Figure 2.31). However the exposed rock was much more mechanically loosened than
predicted. In the valley sides, the floor of the cutoff trench showed near-vertical joints parallel to the river which were open as much as 30 mm, apparently due to downslope block
movements (in the past) along prominent joints dipping towards the river at 10–30°.
Clay-infilled joints and weathered seams next to joints were abundant.

The last specified stripping was removal of 2 m to 3 m depth of boulders and gravel
from the river channel. During this operation, many cracks open from 10 mm up to
500 mm appeared in and above the cliffs on the right bank (Figure 2.30 and Figure 2.31).
The cracks were mapped to the upstream side of the cutoff trench where they became ill-
defined in a steeply dipping zone of open jointed, clearly disturbed and weathered rock
striking obliquely to the river (Figure 2.30). Lesser cracks and scarps appeared in
extremely weathered granite on the left bank.

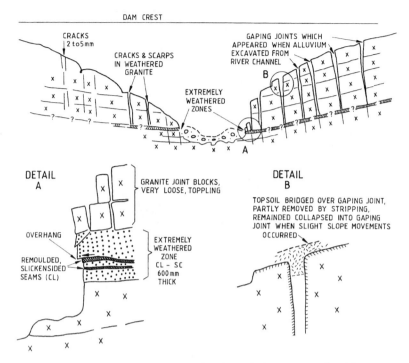

Figure 2.31.   Tooma Dam, cross section and sketches through downstream shoulder.

At the same time, 400–700 mm wide zones of extremely to highly weathered granite which were wet and showed local seepages became exposed close to river bed level on both banks. Both zones were undulating but dipping generally between 10 degrees and 20 degrees towards the river. Each zone contained one or more remoulded seams which included surfaces with slickensides (Figure 2.31A). In the right bank seams these were trending downslope and obliquely downstream. Displacements of 50 mm were observed across one slickenside seam during a period of two days. Monitoring across this seam ceased when several cubic metres of rock toppled from immediately above it.

When all of these features appeared the contractor withdrew from the upstream part of the site until it could be demonstrated that large-scale landsliding was not imminent.

Close examination of the largest cracks on the right bank showed that these had formed by the collapse of surface soils into pre-existing gaping joints which appeared to extend almost down to the extremely weathered zone (Figure 2.31B). Thus, although the cumulative past downslope extension appeared to be between 1 m and 2 m, the amount of current movement appeared to be much smaller, probably just enough to cause the bridging soils to crack and fall into the already gaping joints. This was confirmed by monitoring which was started after the appearance of the cracks, and showed only about 150 mm of further movement over a period of about 2 weeks.

It was concluded that the gently dipping joints were probably sheet joints and that the basal weathered zones had been formed by weathering along the lowest of these joints. The pre-construction block movements towards the river were probably initiated by stress relief, and continued due to gravity.

On each bank the largest of the old movements appeared to have occurred between the upstream soil-filled gullies and the cutoff trench (Figure 2.30).

Movements of the right bank cliff area were stopped by early placement of rockfill in the river bed. Some of this is visible at the right hand edge of Figure 2.32.

Figure 2.32.   Right bank of Tooma Dam after completion of foundation stripping and placement of dental concrete.

Changes to the dam design were made during construction, to allow for the potential for water to penetrate along open joints under the upstream part of the earth core, to minimise the possibility of adverse deformation of the core, and to allow the unstable slope to be quickly restrained. The adopted measures included:

− Steepening of the upstream face of the earth core from 1:1 to 0.25 H:1 V (Figure 2.33);
− addition of a secondary downstream grout curtain, and blanket grouting between the two curtains (Figure 2.33);
− extending the lower half of the cutoff trench about 25 m downstream, and immediate commencement of placing the upstream rockfill, in advance of the earth core zone placement.

In addition to the above, extensive dental treatment of open joints (with mortar) was carried out in the cutoff trench, and backfill concrete was placed over the badly disturbed and weathered rock at the upstream edge of the cutoff trench on the right bank, shown on Figures 2.31 and 2.33.

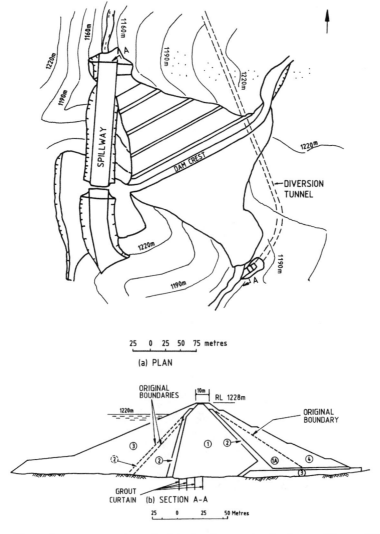

Figure 2.33.   Tooma Dam, cross sections of original and rezoned embankment (Hunter, 1982).

Further details of the above are provided in Hunter (1980, 1982) and Hunter and Hartwig (1962).

Pinkerton and McConnell (1964) describe the performance of the embankment during construction and its first 2 years of operation and show that it behaved well.

Although it was possible to modify the design during construction and a good result was achieved, there must have been some cost to the project because the evidence of past slope movements was not recognised and allowed for during the early planning. It is considered that adequate evidence was there, but was not found or recognised, mainly because the question of possible valley-side movements was not addressed. In later projects deep trenches excavated by bulldozer and ripper have usually been successful in locating the gaping or infilled joints indicative of this kind of slope instability.

2.10.3.3   *Wungong Dam*

Wungong Dam is a 66 m high earth and rockfill structure built during 1976–1979 near Perth, Western Australia. The site is underlain by granite containing several intrusive dykes of dolerite. The rocks are variably weathered. Lilly (1986) describes landslides which occurred during construction.

During early site investigations, a rounded scarp was recognised on the right bank above the upstream shoulder of the proposed dam, and noted as a "suspected old land-slide" (Figure 2.34). Trenches below this feature confirmed that the ground was disturbed, but the areal extent and depth of the disturbed mass was not determined.

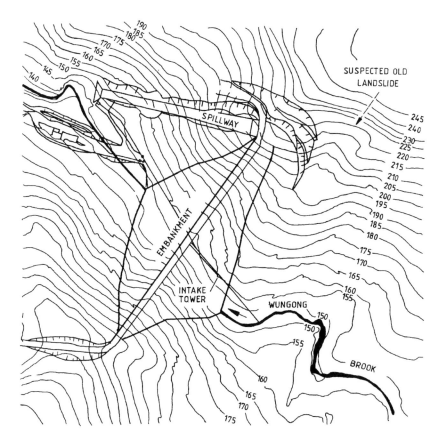

Figure 2.34.   Wungong Dam, plan showing adopted layout and inferred approximate position of old landslide (Lilly, 1986).

To allow for this suspected landslide the dam axis was rotated to bring the upper right shoulder downstream of it (Figure 2.34). It was planned to investigate the area further during construction, and it was expected that much of the landslipped material near the embankment would be removed during excavation of the spillway approach channel.

During stripping of the foundation and placement of fill, four landslides occurred from the area below the suspected old slide. The slides cut through a construction access road, and the fourth and largest slide (about 100,000 m^3) cut through the spillway approach channel and moved into the upstream rockfill and filter zones (Lilly, 1986).

At this stage it was appreciated that the stability of the dam embankment was also in question, and a major investigation involving 17 diamond drillholes was carried out.

Excavation of the slide material in the foundation area showed that the basal failure surface here was a seam less than 50 mm thick of micaceous clayey silt, probably a weathered and sheared, basic dyke. The seam was slightly undulating and dipping almost directly downstream (i.e. beneath the dam) at about 15°. The slide mass here was a wedge bounded by this seam and a joint striking roughly at right angles to the river, dipping 80° upstream (Figure 2.35). Movement had been directly towards the river at the wedge intersection plunge angle of about 5° (Figure 2.36, Section BB). Although water pressures

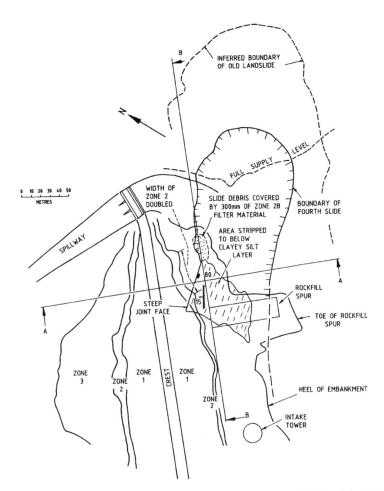

Figure 2.35.   Wungong Dam, right abutment, plan showing the extent of the fourth landslide and remedial works. (based on Lilly, 1986).

obviously contributed to the sliding, an extremely low shear strength was indicated for the seam.

After removal of some rockfill which had already been placed, a 40 m by 70 m area of the rockfill foundation was excavated down to below the seam level (Figure 2.35 and Figure 2.36). The slide was then stabilised by a spur of rockfill which later formed a projecting part of the dam shoulder. Other modifications to the dam included a local widening of the downstream filter zones and placement of a filter blanket over an area of slide material left in place beneath the rockfill (Figure 2.35 and Figure 2.36).

Monitoring during filling showed local downslope movements of up to 90 mm in and upstream from the spillway approach area. Lilly (1986) explains that these were to be expected, as cracks up to 150 mm previously noted in these areas would be closing up. No significant unusual movements of the dam embankment were recorded.

Although a satisfactory dam was the end result, the landsliding and discovery of disturbed materials extending into the rockfill and filter foundations caused serious questions to be raised, and delays to construction. With hindsight, more thorough investigation and delineation of the slide during the planning stages might have resulted in the dam being located well away from the affected area.

2.10.3.4 Sugarloaf Dam

Sugarloaf Dam is an 85 m high, 1000 m long concrete faced rockfill structure, constructed near Melbourne, Australia during 1976–1979. It impounds an offstream storage of 95,000 megalitres.

The site geology and its influence on the design and construction of the dam are discussed in Stapledon and Casinader (1977), Casinader and Stapledon (1979) and Casinader (1982).

The site is in a region of broadly folded siltstones and sandstones of Devonian age which are distinctly weathered to depths of more than 30 m below ridges and plateaus.

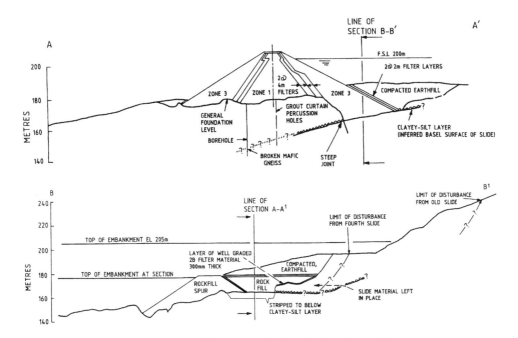

Figure 2.36.   Wungong Dam, sections showing remedial works (based on Lilly, 1986).

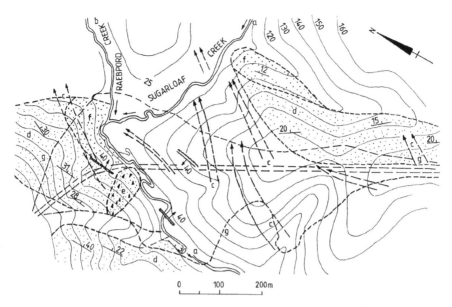

Figure 2.37.   Sugarloaf Dam, geological plan showing dipslopes. (c) axes of minor plunging folds; (d) dipslope, slightly disrupted; (e) scar of old landslide; (f) dipslope, very disrupted; (g) outline of embankment.

Figure 2.37 shows the layout of the embankment which extends across the valley of Sugarloaf Creek and then for 500 m along the crest of a ridge. No other site was acceptable topographically.

During the feasibility stage, detailed geological mapping of surface features and trenches dug by bulldozer and ripper were the main investigation activities.

After initial excavation the floors of the trenches were cleaned carefully by small backhoe and hand tools to ensure that all minor defects such as open and clay-filled joints and thin crushed seams were clearly exposed.

This work showed that much of the foundation area on both banks was formed by dipslopes; thinly interbedded and weathered siltstone and sandstone were dipping parallel to the ground surface (Figure 2.37). Dip and slope angles ranged from 12° to 30°. On the right bank a suspected old landslide was confirmed by the trenching to be underlain by deep, disturbed soil. With the exception of this feature the valley slopes in the foundation area showed no topographic evidence of past landsliding or of significant undercutting of the dipslopes.

Trenches cut straight down the dipslopes showed the bedding surfaces to be very smooth and that slickensided seams of low plasticity clay 2 mm to 20 mm thick were present along them at spacings of 1 m to 2 m. The seams were highly dispersive. Diamond core drilling showed that the clay seams were present to depths of more than 1 m and that they were the weathered equivalents of bedding surface faults (crushed seams, see Chapter 3, Figure 3.19 and Section 3.5.2), which were present at about the same spacings, at greater depths.

Present in some parts of the trenches were near-vertical features of the kind shown on Figure 2.38. These were mainly joints which were open or infilled with gravel or up to 20 mm of high plasticity clay. Also present near the base of a nearby dipslope was a large but ancient "dropfold" structure where upper beds had clearly collapsed into a gaping slot in a lower bed. Figure 5.10 in Chapter 5 shows this feature.

It is judged that these near-vertical and "dropfold" features had been formed by extension of the near-surface beds during downslope sliding movements along weathered bedding surface seams. Cross-cutting defects near the base of the slope and in some places

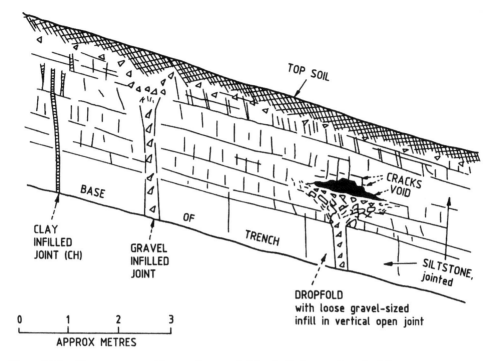

Figure 2.38.   Features exposed in trenches on a left bank dipslope at Sugarloaf Dam.

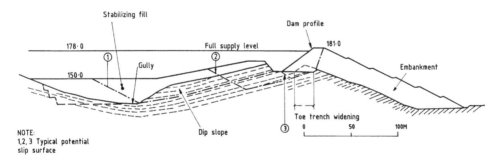

Figure 2.39.   Sugarloaf Dam, cross section through left abutment ridge section.

past undercutting by erosion had apparently provided freedom for the slope movements to occur.

Laboratory direct shear tests showed the weathered bedding surface seams to have effective residual strengths of about 10°. The residual value was adopted in stability analyses because the seams were initially near-planar faults and the small displacements during slope movements would have been enough to reduce their strength to this value. The combination of very low strength dispersive seams and dipslopes provided interesting design questions which are discussed in Casinader and Stapledon (1979).

Figure 2.39 shows the arrangement adopted for the section of the dam on the left bank ridge. The plinth was located close to the crest of the ridge, to minimise loading of the dip-slope. This resulted in much of the embankment being located on the downstream side of the ridge. The dipslope was stabilised by covering it, and filling the gully upstream from it, with waste material from a nearby excavation.

During construction the conditions found were essentially as predicted. Minor landsliding occurred on the right bank dipslope when the surface soil and vegetation was stripped, and again when a 4 m deep trench was cut directly down the slope (Regan 1980, Stapledon 1995). These occurrences under very low normal stress conditions confirmed the need to use residual strength in the design.

Stability problems on the right bank proved more difficult than predicted, but early recognition of this during construction enabled the design here to be modified with minor disruption of the works.

The evidence of thin seams and past slope movements at Sugarloaf was quite subtle and may not have been found if it had not been consciously looked for and exposed in well-prepared, very long, deep trenches.

2.10.3.5   *Thomson Dam*

The 166 m high Thomson Dam is an earth and rockfill structure located 120 km east of Melbourne, Australia. It was constructed between 1977 and 1984.

The site is underlain by a folded, interbedded sequence of siltstone and sandstone, similar to that at Sugarloaf. However the folds are tighter and more complicated than at Sugarloaf, there are many normal and thrust faults, and bedding-surface sheared and crushed seams are spaced generally at less than 1 m intervals.

Figure 2.40 shows the major folds and faults at the site. Fold axes trend generally north-south and plunge northwards. A major feature at the site is the Thomson Syncline which plunges to the north at about 12°.

Although this fold pattern was recognised during the feasibility and pre-construction design stages, its potential to contribute to instability of slopes and foundations was not appreciated at that time.

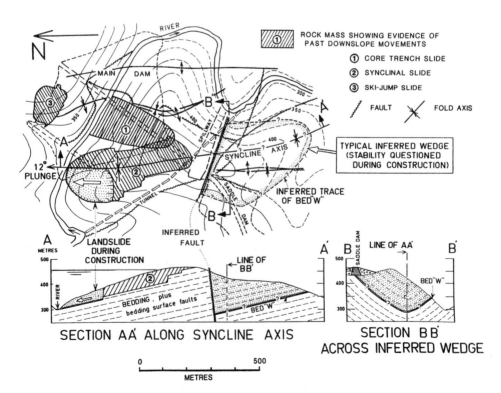

Figure 2.40.   Thomson Dam site, geological plan and sections showing the main structural features.

As at Sugarloaf, the site was trenched extensively during the feasibility investigations but the trenches were not so deep, and the same standard of cleanup was not achieved. However the Ski-jump slide (Figure 2.40) on the upstream left abutment was identified from highly disturbed rock in a trench cut into an area showing typical landslide topography.

Diamond drilling in the upper right bank core foundation area indicated closely jointed, seamy and partly weathered rock down to 70 m. High losses occurred during water testing of this zone and it was suspected that it may have been disrupted by past landsliding. This was allowed for in the designed depth (up to 30 m) of the cutoff trench on the upper right bank. Also, the owner, Melbourne and Metropolitan Board of Works, decided to excavate the whole of the cutoff trench on this bank, prior to awarding a contract for construction of the dam. During this excavation, and excavation of associated haul roads, it was found that past landsliding in the site area had been more widespread than expected and had caused disruption of the rock mass in the three areas shown on Figure 2.40. In each area the sliding had occurred where the combined effects of valley erosion and the fold shapes had caused downslope dipping beds to daylight. Details of each slide and the nature of the disturbed materials are given in Marshall (1985).

The Ski-jump Slide proved to be a relatively small feature. The folded rock here daylighted into a gully immediately upstream (Figure 2.40). The slide mass was buttressed effectively by placing rockfill in the gully and the rockfill shoulder of the dam was built over it.

The Core Trench Slide proved deeper than expected, downslope movements having occurred along folded beds daylighting at river level. The trench excavation ranged from 10 m to more than 40 m deep and involved more than a million cubic metres of excavation, all by Caterpillar D9 bulldozer without ripping. In the upper half of the trench deeper excavation appeared likely to give rise to instability of the sides and so it was decided to leave the deeper, moderately disturbed rock in place and treat it with a 5-row grout curtain. This rock contained many joints which were open or clay-filled, from 1 mm to 20 mm. It was appreciated that most of the clay would remain after the grouting but could be washed out in the long term, when the reservoir was filled. To allow for this possibility a reinforced concrete gallery was built in a slot cut into the upslope half of the cutoff trench. The gallery provides access for monitoring and for drilling and grouting equipment if regrouting should ever be needed.

The Synclinal Slide (Figure 2.40, Section AA) was discovered during construction when cracks and displacements appeared in haul roads near the toe of the ridge at the right bank upstream shoulder of the dam in mid-1977. The estimated slide mass was about 350,000 m³, but, once the slide had been defined geologically, the potential slide mass was estimated at well over 3 million m³, with a basal surface that would have sliced through the right abutment of the main dam and extended downstream to affect the top of the spillway. Movements in the toe area ranged from 1 mm/day in dry weather to 3 mm/day after rain. Movements stopped after 9 months (a total of 300 mm cumulative movement) on completion of a 50 m high stabilising rockfill built against the ridge toe. This required diversion of the river into a channel cut into the left bank, as the diversion tunnel was still under construction.

Detailed investigations showed that past movements in the ridge had extended right up to its crest. The sliding had occurred directly down the 12° plunging axis of the Thomson Syncline. Upslope from the toe area the base of the disturbed rock zone was stepped as shown diagrammatically on Figure 2.40, Section AA. In the ridge crest area rock disturbed by this slide occurred in the top few metres of the gate shaft (alongside and just to the left of the spillway) and also at the top of the cutoff trench. The inferred basal surface of the potential slide was estimated to be nearly 75 m below the top of the gateshaft and generally 60–70 m deep below the crest of the dam on the right abutment.

Changes to the project resulting from this slide included construction of a permanent stabilising fill of more than 2.6 million m^3, that involved filling in the valley upstream of the dam for a depth of 50 m with rockfill, constructing a larger supporting fill on the right abutment upstream from the dam and elimination of an access bridge from the gate shaft area to the upper outlet tower.

During detailed studies of the Core Trench Slide and Synclinal Slide the succession of rock types and the shape of the Thomson Syncline became well known in those areas. Projection of this known geological picture southwards raised serious questions about the possibility of first-time landsliding within the ridge which formed the reservoir rim, imme-diately south of the right abutment of the dam. The spillway and part of the saddle dam were located on this ridge (Figure 2.40, Section BB). Assuming a full storage and using residual strengths obtained from back analysis of the Synclinal Slide, this ridge appeared to have safety factors of less than unity against sliding downstream along any of the numerous bedding surface seams which daylighted above river level on the downstream side. A major new site investigation was carried out over a period of 16 months, including 1.6 km of trenches, adits 250 m and 460 m long, diamond drilling and field and laboratory testing. This work is described in detail by Marshall (1985) and summarised by Stapledon (1995). It showed that irregularities in the fold shape and rotational displacements on faults effec-tively precluded failure of the ridge. The following additional works and changes were undertaken, to further guarantee its stability:

- Weathered, potentially unstable rock was removed from the top of the ridge;
- The axis and grout curtain of the saddle dam were moved 10 m upstream, to reduce the embankment load on the active wedge;
- An additional drainage gallery 70m below dam crest level was driven into the ridge and connected by a drainage curtain of vertical boreholes to the previously constructed exploratory gallery that was some 115m below dam crest level.

The site for Thomson Dam is geologically very complex and was difficult to explore due to its steepness, dense vegetation and often deep soil cover. Some evidence of the slope sta-bility problems at the site was found during the early planning and used in the design of the dam. The owner's decision to excavate the right bank cutoff trench prior to awarding the main construction contract (Hunter, 1982) was a good one. As Hunter (1982) points out, this decision, although driven as much by concern over limiting potential industrial prob-lems as to geotechnical aspects was more or less in accord with the approach of Terzaghi and Leps (1958) at Vermillion Dam. Terzaghi and Leps concluded that the foundations at Vermilion were so complex that further more detailed exploration would still leave a wide margin of interpretation. They therefore advocated proceeding with construction of the dam as designed on the best interpretations and assumptions, with careful monitoring of the con-struction, and modification of the design to suit the conditions as found.

However, at Thomson, after construction was well advanced the expenditure of money, technical manpower and time on geotechnical site investigations far exceeded that spent in the feasibility and design stages. Most of this effort went into investigation of the down-stream ridge, the stability of which was a question that should have been answered in the feasibility stage.

If the construction stage studies had found this ridge to be unsafe, the options for reme-dial works were very limited, extremely expensive and would themselves have required further detailed site investigations to prove their feasibility. As discussed in Section 2.11.1 and Section 4.6, the authors consider that for all dams, large and small it is vital that suffi-cient funds are made available at the feasibility stage to ensure that all questions affecting feasibility are asked and answered satisfactorily. For dams of the size of Thomson the exploration will often include adits as well as extensive and carefully cleaned trenches.

Dam owners, dam designers and all others involved in the implementation of a large dam project ignore the geological/geotechnical matters of their damsite and dam at their peril. All, including the geological/geotechnical team members, must ask the awkward question about potential deep-seated instability early in the process. One can only imagine the disaster at Thomson if the lingering doubts of the dam designer in 1977 about the "disturbed rock and potential instability at the damsite" had been lost in the desire to get the "job moving" and, for whatever reason or luck, the 350,000 m³ of the right abutment that did move did not do so. That the project was set back by 5 years or so was a major concern to the owner and the designer, but that concern could be nothing compared to a major failure during initial reservoir filling, which is very likely what would have happened had the major stabilising fill works not been constructed.

2.11 STABILITY OF SLOPES AROUND STORAGES

The slopes around a proposed storage area may include some which are intact and stable, others intact but less stable, others formed by dormant landslides, and still others formed by active landslides. The local raising of the water table which occurs when the storage is filled, and lowering when the storage level is drawn down, may reduce the stability of any of these slopes. Sometimes the reduced stability causes landsliding in a previously intact slope. This landslide is called a "first-time" slide (see 2.10.1). For a dormant landslide, or an active or occasionally active slide, the reduced stability can cause reactivation of the former, or increased movement rates in the latter. Affected slopes are normally within the storage area (Figures 2.41 and 2.42). However, slope failure is also possible (but rare) on the outer side of a storage rim (Figure 2.43).

The consequences of landsliding within the storage area can include

– Damage to existing property or infrastructure;
– Constraints on operation of the storage;
– Damage to roads;
– Damage to the dam, or its foundations;
– Damage to or blocking of intake or outlet works;

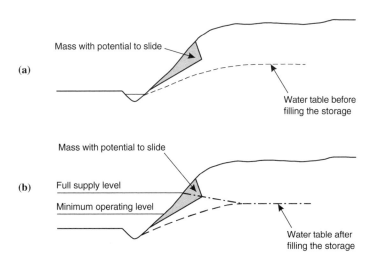

Figure 2.41.   Effect of filling and operating a storage, on the stability of a jointed rock slope partly inundated by it. (a) Before filling the storage (b) During operation.

- Damage to a spillway;
- Damage to a saddle dam or its foundations;
- Partial or complete blockage of the storage, where this is narrow, to form a landslide dam (or the perceived potential for this to occur). Flooding upstream and overtopping and eventual failure of this landslide dam are further possible consequences. Failures of such dams can be disastrous (Schuster & Costa 1986, and King et al., 1987; 1989), or
- Generation of a large, destructive wave by rapid sliding into the reservoir, such as that which overtopped Vaiont Dam in Italy in 1963, and claimed 2600 lives (Hendron & Patton, 1985; Muller 1964, 1968; Kiersch 1965).

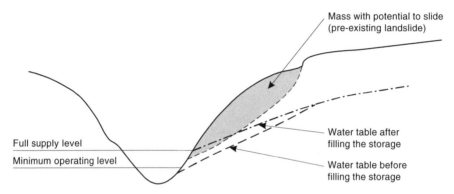

Figure 2.42.   Effect of filling and operating a storage at the narrow base of a deep, steep-sided valley.

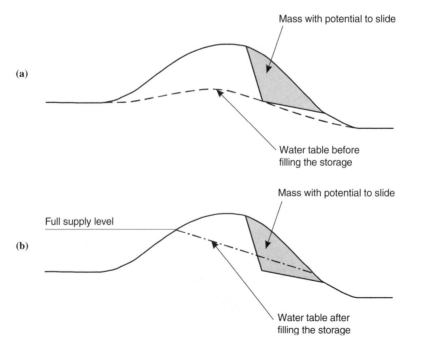

Figure 2.43.   Effect of filling a storage, on the stability of a low, narrow part of its rim, formed by jointed rock (a) Before filling the storage (b) After filling the storage.

The consequences of landsliding on the outer parts of the storage rim may include:

- Constraints on operation of the storage;
- Damage to existing property or infrastructure;
- Damage to a saddle dam or its foundations (Cabrera, 1992), or
- In an extreme case, breaching of the rim, causing loss of storage and flooding of land.

It is clear that the possibility of landsliding is an important feasibility issue for storage projects. However, from examination of 145 reservoir landslides case histories, ICOLD(2002) has reported that:

- Only about 36% of the known reservoir landslides were recognised during the planning stage investigations;
- At least 75% of the known reservoir landslides were pre-existing dormant or occasionally active or active features, and
- Such pre-existing slides cause the most problems for reservoir owners.

These figures indicate that the planning stage studies for some reservoir projects have been seriously ineffective. Recognising this, ICOLD (2002) has compiled guidelines on the identification and treatment of reservoir landslides, and management of the risks they creat. The guidelines are aimed at both owners of existing projects, and planners of future projects. They include 89 pages of text, figures and tables, 9 pages of references and 4 appendices. They cover all of the relevant issues, as follow:

1. International experience – case histories in Appendix A;
2. Management;
3. Geological and geotechnical investigations;
4. Slope assessment and analysis;
5. Impulse waves and valley blockage;
6. Risk management and risk mitigation;
7. Monitoring;
8. Operational requirements.

The authors commend the guidelines to readers. Assuming their availability and use by those involved with storage projects, the following text is limited to views on the timing of storage area stability studies, and some strategies for managing instability issues should they become evident.

2.11.1   *Vital slope stability questions for the Feasibility and Site Selection Stages*

The authors consider it vital that the following questions be answered with confidence, by the end of Stage 2, Feasibility and Site Selection (see Chapter 4).

1. Which features (existing or proposed) and/or parts of the storage would be most vulnerable to landsliding, should this occur when the storage is filled or operated?
2. Which, if any, parts of the storage area rim show evidence of a) currently active landsliding, or b) old, dormant landslides which may become reactivated?
3. Which parts of the storage area rim may be susceptible to first time landsliding?
4. For each feature or area defined by the answers to 1 to 3 above, what is the assessed probability of landsliding, when the storage is filled and operated?
5. For each potential landslide, what is the likely volume, velocity and travel distance?
6. For each such feature or area, what are the assessed consequences of landsliding, under those conditions?

7. In the light of these assessments, what needs to be done, e.g. in the location and design of structures, in treatment of the ground, or in operation of the storage?

The sections which follow discuss when and how Questions 1 to 5 should be answered, and include suggested strategies for action, in respect to Questions 6 and 7.

2.11.1.1   *Most vulnerable existing or proposed project features, and parts of storage area? – Question 1*

This question can usually be largely answered in Stage 1 (Pre-feasibility), by examining contour plans showing existing infrastructure and buildings, the proposed storage in outline, and a preliminary layout of the proposed dam and associated works. From the lists in 2.11 above it is clear that generally, the project features most vulnerable to the effects of landsliding will be the dam, saddle dams, inlet and outlet works, spillways and roads. Also, topographically, the most vulnerable parts of the storage area will be the highest, steepest slopes within narrow sections (Figure 2.42), and steep, narrow ridges forming its rim (Figure 2.43). The exception may be existing landslides which are often not the steepest slopes.

2.11.1.2   *Currently active or old dormant landslides? – Questions 2 and 4 to 7*

Tentative answers to Question 2 should come also during Stage 1, from existing historical records, examination of air photos, and from inspections from the air and on the ground. Currently active landslides, and the scars or remnants of old, dormant landslides can usually be recognized and their plan boundaries delineated, during air-photo interpretation (Section 5.2.2.2), and/or geomorphological mapping (Section 5.3) and geotechnical mapping (Section 5.4.2.2). Methods for their recognition and subsequent investigation are available in Rib and Liang (1978), Sowers and Royster (1978), Hunt (1984), Stapledon (1995), and Soeters and van Westen (1996).

In view of the ICOLD (2002) statistics quoted in Section 2.11, any pre-existing landslide feature found must carry a substantial risk of reactivation. The significance of a major pre-existing slide of either type, in the rim of a storage area, will depend on its size and location. Two possible situations are considered here.

Situation A – a slide which is judged to be sufficiently large and located so that its future movements could endanger a vulnerable existing feature, or proposed project feature. For this situation the following approach might be followed, in Stage 2, Feasibility and Site Selection.

1. Do enough surface (and if necessary, subsurface) investigations to more accurately assess the size of the mass and its likely movements during storage operation;
2. If this confirms the size of the slide and that its predicted future movements are unacceptable, the following options should be considered:
   - Relocation of the existing feature, or proposed project feature, to a proven stable area, or
   - Design of systems aimed at stabilizing the slide or limiting its displacements to acceptable velocities/amounts. Such systems are listed and discussed Section 9.3 of ICOLD (2002).

   From experience, the authors believe that relocation is likely to be the better option, in most cases.

Situation B – a slide which because of its apparent size and location (e.g. Figure 2.42) is judged to present a serious risk to the feasibility of the whole storage project. For this situation the following approach might be followed.

1. Do enough surface and subsurface investigations to prove whether or not the landslide is likely to become reactivated, and is large enough to present an unacceptable risk to the project feasibility, when reactivated;
2. If this shows that the slide is large enough and the risk unacceptable, then the following options would have to be considered:
   - Design of systems to stabilise it, or
   - Abandonment of the proposed storage area.
3. If from the investigations in 1, it is concluded that the slide is not likely to be reactivated, or that the reactivation will be slow moving and/or is not large enough for its activation to present an unacceptable risk to the project, it would still be prudent to confirm that conclusion by continuing to monitoring its behaviour through commissioning and early operation.

Stabilising systems for either situation A or B would have to be based on a geotechnical model of the landslide, capable of providing predictions of its future behaviour. As landslides are characterized by extreme variability in composition, structure and permeability (see Sections 2.10.2, 3.10.1.3 and 3.10.2.3), to produce such a model would require a comprehensive and time-consuming investigation. Questions to be answered, and investigation methods which can be used to answer them, are discussed in ICOLD (2002) and Stapledon (1995). However, it must be appreciated that at the start of such an investigation there could be no guarantee that its completion would answer all of the questions and allow the design of an adequate and economically feasible stabilizing system. This is made clear by ICOLD (2002) under Key Management Issues, as follows:

"*Advanced techniques of field investigation, laboratory testing and stability analysis are now available. However, because of the complex nature of landslides and reservoir interaction, it is often difficult to define specific failure mechanisms and complete a meaningful stability analysis. This may leave substantial uncertainty in the determination of hazard and risk*".

2.11.1.3   *Areas where first-time landsliding may be induced (Questions 3 to 7)*

As pointed out in Section 2.10.1.2 first-time landslides are much more difficult to predict than reactivated landslides. They are often more important than existing landslides because they are more likely to travel rapidly.

a) *Rock slopes*. For rock slopes, a common type of first-time slide occurs where joints, bedding, or weak seams dip out of the slope and form all or part of a translational slab or wedge with kinematic freedom to move. The shaded wedges on Figures 2.41 and 2.43 are examples. Other possible first-time failure types include rockfalls, buckling and toppling. See also Section 2.10.3.5, and Section B on Figure 2.40 which shows the suspected potential first-time slide across the storage rim near Thomson Dam.

To assess the probability of a first-time slide from any rock slope requires (at least) knowledge of the pattern of defects within it. For a rock slope showing abundant outcrops, or whose structure is otherwise well exposed (e.g. in exploratory trenches, shafts or tunnels) the plotted and analysed results of geotechnical mapping can indicate the probability that a kinematically unstable wedge or slab exists, and its potential size. To assess the probability that such a kinematically unstable mass would actually fail, due to construction activities, operation of the storage, or natural processes, would require substantial further site investigations. Some suggested questions to be answered in such investigations are set out in Appendix B of ICOLD (2002) and in Stapledon (1995). The assessment of the likelihood of such sliding is a technically complex matter and should be addressed by persons with expertise in rock mechanics and engineering geology as applied to large landslides.

Where such a suspect mass is located at or near a vulnerable project feature, the site investigation could form part of the site studies for that feature. However it would be preferable to look for an alternative, less suspect, site for the project feature, and spend the time and money confirming its stability.

If the investigations show that the probability of landsliding is high, and its consequences could affect the project feasibility, then the options for action would be:

- Design of stabilizing systems, or
- Abandonment of the proposed storage area.

b) *Soil slopes*. An apparently intact slope formed by soils would be suspect for first-time landsliding during filling or operation of a storage, if the slope was known to have marginal stability and:

- It was close to areas also underlain by soils, but showing evidence of currently active or past landsliding;
- The underlying soils were currently well drained, at low moisture contents, and the water table in them was low (well below the proposed storage levels);
- The soils were known to contain swelling clays or soluble materials such as halite or gypsum.

A large slope meeting any of those criteria would be best treated in the same way as a suspect rock slope, that is, avoided if possible during the location of any project feature. Should its potential size and location be such that its failure would endanger existing infrastructure or the project feasibility, then the options would be the same as those for a rock slope.

2.11.1.4.   *What is the likely post failure velocity and travel distance?*

For landslides into the reservoir, it is large, rapid slides which are of most concern because they have the potential to cause waves which can overtop the dam, even breaching the dam.

Glastonbury (2002), Glastonbury and Fell (2002a, b, c) have developed a decision analysis framework for assessment of the post failure velocity of large natural rock slope failures based on the study of a large number of rapid and slow landslides. This shows that for a landslide to travel rapidly after failure there has to be a significant loss of shear strength on the surface of rupture and/or internally or on the lateral margins of the landslide; or the factor of safety has to be maintained below 1.0 after failure by high groundwater pressures.

A slide having these characteristics is often considered as being "brittle", with the strength of a surface being brittle if there is a large loss of strength. Table 2.7 summarises the dominant sources of brittleness for the landslides studied by Glastonbury (2002). All the "rapid" slide cases examined in the database involved relatively high strength rock masses. Many rupture surfaces were pre-sheared, yet brittle collapse still occurred due to either brittleness on lateral margins or brittle internal deformation.

Table 2.8 and Figure 2.44 summarise the typical characteristics of the rapid landslides. The study also showed that many of the more obvious landslides associated with reservoirs

Table 2.7.   Landslide mechanism classes and dominant source of brittleness (Glastonbury, 2002; Glastonbury & Fell, 2002c).

Brittle Buttress Deformation	Brittle Internal Deformation	Brittle Basal Rupture Surface	Brittle Lateral Rupture Surface
Toe Buttress Compound	Bi-Planar Compound Curved Compound Toe-Buckling Translation	Rough Translational ⟵ Rock Collapse ⟶ Planar Translational	Large Rock Glide
⟵		Irregular Compound	⟶

Table 2.8. Typical characteristics of large rapid natural rock landslides (Glastonbury 2002, Glastonbury & Fell 2002c).

Characteristics	Large rock glide	Translational slide class			Compound slide classes				
		Rough translational	Planar translational	Toe buckling	Bi-planar compound	Curved compound	Toe buttress compound	Irregular compound	Rock collapse
General geology	Sedimentary (esp. carbonate)	Anisotropic rock mass	Sedimentary (esp. carbonate)	Sedimentary	Variable – anisotropic	Folded sedimentary terrains	Anisotropic rock mass	Irregular – disturbed rock mass	Brittle cap rock over argillaceous strata
Slide Mass characteristics	Very thick, (several hundreds of metres) generally intact	Some disaggregation	Intact	Intact	Often highly fractured	Structure to slope	Often highly fractured	Often partly fractured	Intact
Rupture surface characteristics	Pre-sheared, bedding defined	Stress relief joints	Pre-sheared, bedding defined	Pre-sheared, bedding defined	Bi-planar, pre-sheared	Curved, pre-sheared	Upper surface pre-sheared	Irregular geometry	Rotational geometry in low strength rock
Rupture surface inclination	Very low $(10-20°)$	$40°+$	Typ $20-30°$	Typ $30-40°$	$\alpha_L = 20-30°$	$\alpha_L <30°$	$\alpha_L = 10-25°$	$\alpha_{av} = 5-50°+$	Variable – rotational
Rupture surface inclination v ϕ_b	$\alpha \approx \phi_b$	α typ $10-30°$ greater than ϕ_b	α typ within $5°$ of ϕ_b	α typ within $5°$ of ϕ_b	$\alpha_L \approx \phi_b$	α_L typ $2-3°$ less than ϕ_b	α_L typ $5°$ less than ϕ_b	α typ within $5°$ of ϕ_b	Variable
Rupture surface asperity influence	Minimal	High	Minimal	Minimal	Some	Minimal	Minimal	High	Likely some
Lateral margins	High influence	Moderate influence	Moderate influence	Low influence	Variable	Unknown	Unknown	Unknown	Moderate
Water influence	Unknown	Some	Common	Some	Some	Some	Limited	Unknown	Some
First time/reactivated	First-time	Predominantly first-time	Both	First-time	Both	Both	Both	Both	First-time
Slope history (Dominant cause)	Glacial debuttressing	Fluvial and glacial	Fluvial erosion	Fluvial erosion	Glacial	Fluvial and glacial	Fluvial and glacial	Fluvial and glacial	Fluvial
Pre-collapse signs	Very limited	Limited	Limited	Limited	Unknown	Some	Common	Limited	Common

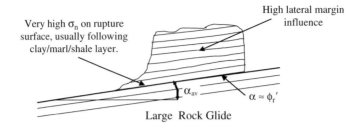

Very high σ_n on rupture surface, usually following clay/marl/shale layer.

High lateral margin influence

α_{av}

$\alpha \approx \phi_r'$

Large Rock Glide

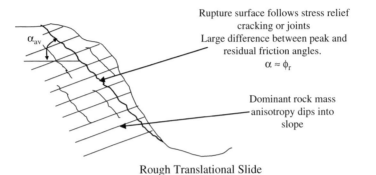

α_{av}

Rupture surface follows stress relief cracking or joints
Large difference between peak and residual friction angles.
$\alpha \approx \phi_r$

Dominant rock mass anisotropy dips into slope

Rough Translational Slide

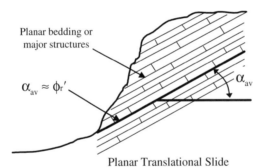

Planar bedding or major structures

$\alpha_{av} \approx \phi_r'$

α_{av}

Planar Translational Slide

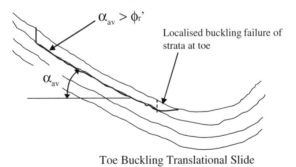

$\alpha_{av} > \phi_r'$

Localised buckling failure of strata at toe

α_{av}

Toe Buckling Translational Slide

Figure 2.44.   Summary of features of large rapid rock landslides – Sheet 1 of 2 (Glastonbury 2002, Glastonbury & Fell 2002c).

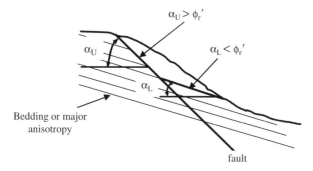

$\alpha_U > \phi_r'$

α_U

$\alpha_L < \phi_r'$

α_L

Bedding or major
anisotropy

fault

Bi-Planar Compound Slide

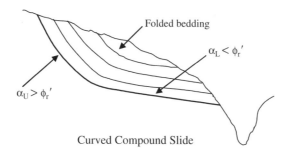

Folded bedding

$\alpha_L < \phi_r'$

$\alpha_U > \phi_r'$

Curved Compound Slide

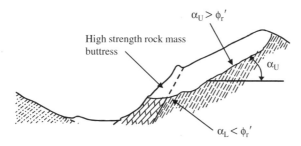

$\alpha_U > \phi_r'$

High strength rock mass
buttress

α_U

$\alpha_L < \phi_r'$

Toe Buttress Compound Slide

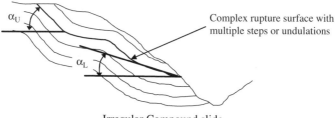

α_U

Complex rupture surface with
multiple steps or undulations

α_L

Irregular Compound slide

Figure 2.44.   (Continued) Summary of features of large rapid rock landslides – Sheet 2 of 2 (Glastonbury 2002, Glastonbury & Fell 2002c).

which have been studied and monitored are likely to be slow moving because they are active or reactivated slides with little likelihood of brittle failure.

2.12   WATERTIGHTNESS OF STORAGES

Absolute watertightness is unlikely to be achieved in most natural storages. However it is usually desirable that the rates of any leakage are minimized, for any of the following reasons:

- The cost of water lost can be unacceptably large e.g. in arid areas, or in offstream storages fed by pumping;
- In some situations leakage may cause raising of water tables, development of swamps, or flooding, in areas adjacent or downstream.

In a few cases leakage has caused instability of slopes on the outer edge of the storage rim (see Section 2.11).The watertightness of a storage basin will depend upon

- The permeability of the underlying rock or soil mass;
- The permeability of the rock or soil mass surrounding it;
- The groundwater situations in those masses, and
- The lengths and gradients of potential leakage paths.

Masses of soluble rocks (carbonates and evaporites, see Sections 3.7 and 3.8) are often cavernous and have permeabilities many orders higher than those of non-soluble rocks. This discussion on storages will therefore consider non-soluble rocks separately from those on soluble rocks.

2.12.1   *Models for watertightness of storages in many areas of non-soluble rocks*

Most non-soluble rock substances are effectively impervious, and rock masses formed by them are only permeable if sufficient numbers of interconnected open joints are present. As pointed out in Section 2.5.2, joints generally become more tightly closed with increasing depth, so most masses of non-soluble rock will become less permeable with increasing depth. Also, their effective porosities (or storage capacities) decrease with increasing depth and are usually very low (i.e. 10^{-3} to 10^{-6}). The diagrams on Figure 2.45 show the effects of creating storages in such areas of non-soluble rock, in five relatively common topographic and groundwater situations. In the discussion of each it has been assumed that:

(i)   The rock becomes less permeable with increasing depth;
(ii)   Across the floor of the storage area, the water table lies close to the ground surface;
(iii)   The water tables shown represent end of dry season conditions;
(iv)   Groundwater pressures measured below the water table as shown will all confirm it as the piezometric surface, and
(v)   Flow beneath the high ground or ridges forming the rims will be essentially laterally to or from the storage, unlike the isotropic material case, proposed by Hubbert (1940).

The perimeter of any particular storage area may include sections topographically similar to any or all of Models a/b, c/d or e.

Model (a). In this case the rainfall is high and the storage is in an entrenched valley underlain by rock and surrounded by wide areas of high ground underlain by rock. The rock in the valley floor may be partly or wholly concealed by soils of alluvial or glacial origin. Beneath the high ground, the water table rises above the proposed FSL. There is no

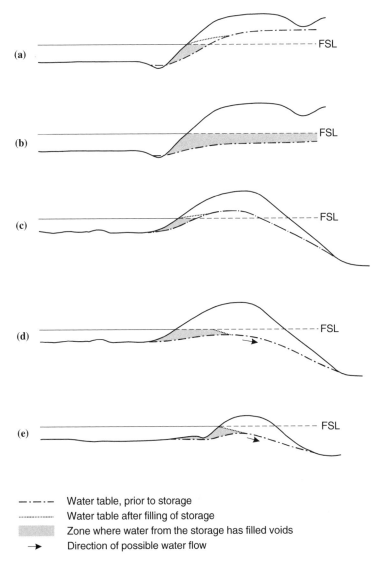

Figure 2.45.   Storages formed by non-soluble rocks – watertightness models.

potential for leakage. The storage should be essentially watertight, and will be fed partly by groundwater. As the storage fills, a very small amount of water will fill the voids in the adjacent rock mass (shaded). When the storage is full, the water table beneath the valley side will have risen slightly and will meet it at the FSL (see dotted line).

There are many storages fitting this model in non-soluble rock areas with high rainfall.

Model (b). The topographic and geological situations are similar to those for Model (a), but the water table lies below the proposed FSL. There may be some bodies of perched water, above this true water table. The reasons for the water table being so much lower than for Model (a), may include any or all of the following:

1. The climate is drier;
2. The surface runoff is greater;
3. The rock is more permeable and drains more rapidly towards the valleys.

In this situation there is potential for leakage, but leakage paths would be so long and at such low gradients, that loss from the storage would probably be negligible. The shaded area indicates where small amounts of the storage water will enter, fill voids and cause the water table to rise locally.

It is likely that this model represents many storages in non-soluble rock areas with low to moderate rainfall.

Model (c). This storage is elevated well above an adjacent valley, and separated from it by a high, broad, ridge. The water table within the ridge rises above the proposed FSL. There is no potential for leakage from the storage to the valley, so the storage should be essentially watertight. A small amount of storage water will fill voids above the water table in the adjacent valley side (shaded) and the water table will rise slightly and meet this valley side at FSL.

Model (d). The topography and geology are the same as Model (c), but the water table does not rise above the proposed FSL. The reasons for this may include any or all of the following:

– The climate is drier;
– The surface runoff is greater;
– The rock is more permeable and drains more rapidly towards the valleys.

There is potential for leakage. However, because of the high rock cover and long potential leakage path, significant leakage would be unlikely. A small amount of storage water will fill voids in the rock above the water table in the adjacent valley side (shaded) and should cause some rise in the water table.

Model (e). The storage is elevated, as for Models (c) and (d), but the ridge forming its rim is low and narrow. As a result of this, the rock beneath the ridge and below FSL is likely to be affected by de-stressing, and hence more permeable than that at Models (c) and (d). If the rainfall is high, as in the case shown, the water table may rise slightly beneath the ridge, but would be unlikely to rise above the proposed FSL, as at Model (c). When the storage is filled, seepage from it will cause the water table to rise, to at least the position of the dotted line. There is potential for leakage from the storage across to the lower valley (see arrow).

2.12.2   *Watertightness of storage areas formed by soluble rocks*

As discussed in Chapter 3, Section 3.7 Carbonates and Section 3.8 Evaporites, areas underlain by these soluble rocks often exhibit karst topography and may be cavernous to great depths. In carbonate rocks, depths of more than 300 m has been proven at Attaturk Dam (Riemer et al., 1997) and about 600 m at Lar Dam (Salambier et al., 1998). Ruquing (1981) records cavities developed to about 1200 m below the Yilihe River in China.

The evaporites are more soluble than the carbonates and pose much more difficult problems in dam engineering. The factors affecting watertightness of storages are essentially the same for both, but there have been many more dams built in carbonate areas. Therefore, the following discussion will deal with principles and experiences from carbonate rock areas.

The carbonate rocks can occur in beds ranging in thickness from a metre or so up to several hundred metres, e.g. see Figure 3.29, a cross section at Lar Dam. They commonly occur together with beds of non-soluble rocks, in uplifted sequences which have been folded, jointed and displaced and locally crushed by faulting. Such complex rock masses will have been affected by tectonic movements and dissolution at various times during their long geological history. In the more recent episodes they have been selectively dissolved and eroded, both at the surface, producing deep river valleys and underground producing interconnected caverns and complex flow paths.

There can be many types of direct connections between surface and underground streams. For example, during the wet season a surface river may flow into a tunnel or shaft, flow underground for a kilometer or more and emerge as a spring, either high on the side of another valley or lower down, close to another surface river. This flow pattern might be quite different during the dry season.

A further complication at some sites is blanketing of the valley floor by either residual soils, or sediments deposited by rivers or lakes. Obvious sinkholes or subtle depressions on these valley floors indicates that material is or has been eroding or collapsing into cavities in the underlying karst. The river and storage area at Lar are underlain by up to 600 m of lake sediments, and before impounding the water table was about 200 m below the river bed. Such rivers and storages are referred to as "supported" or "hanging".

As a result of this degree of complexity, it is not possible to present a set of models for storages in karst areas, equivalent to those on Figure 2.45. Also, it can rarely be safely assumed that a karstic rock mass will become less permeable at depth. This means that assumption i) on which the Figure 2.45 models are based, does not usually apply. It might appear that the right abutment ridge at Khao Laem Dam fits Model (b) (see Section 3.7.2). The ridge was formed by cavernous limestone with a dry season water table above the valley floor but below the proposed FSL. However, if the dry season water level in Khao Laem ridge had been above the proposed FSL, it would not have been safe to assume a Model (a) (i.e. watertight) situation, unless it could be proved that the limestone:

– Became free of interconnecting cavities, and effectively impermeable, at depth, or
– Was not within a bed which emerged at a lower level, either downstream, or in another valley.

A classification of karst river valleys has been compiled by Ruquing (1981) using experience gained during engineering works in the Yunnan and Guizhow Provinces of Southwest China. The classification includes notes on the potential of each valley type for the storage of water. Ruquing (1981) recognizes 5 categories of 10 types, but notes that the classification is far from complete.

2.12.3 *Features which may form local zones of high leakage, from any storage area*

Some proposed storage areas underlain mainly by non-soluble rocks may contain localised individual features of very high permeability, which could potentially act as drains. Such features may include:

– Buried channels, alluvial or glacial (see Sections 3.9.1 and 3.12.1);
– Lava flows with clinker (Section 3.2.2.1);
– Lava tubes (see Section 3.2.2.1);
– Some open-jointed fault zones (see Section 2.3.2);
– Abandoned underground mine workings.

It is probable that such features would rarely cause significant leakage from storages fitting Models (a) or (b), because leakage paths would be too long and gradients too low. However they could form effective drains, if located within or through the outer ridges in Models (c), (d) and (e).

Any of these features could occur also within a storage area underlain by soluble rocks and might influence its watertightness. It is also possible that an isolated bed of cavernous, soluble rock might occur within a storage area formed almost entirely by non-soluble rock and influence its watertightness.

2.12.4 *Watertightness of storages underlain by soils*

Many dams and weirs are built in the valleys of large rivers and the reservoir watertightness controlled by the alluvial or glacial soils. Often the water tables are relatively high and graded towards the river, so the issue is essentially one of the limiting seepage under and around the ends of the embankment as discussed in Chapter 10. Some storages have to rely on liner systems for watertightness if the underlying soils are very permeable, deep and/or the water table is low.

2.12.5   Assessment of watertightness

During the planning stages of the project (Stages 1 to 3 on Table 4.3) the following questions should be addressed and answered:

1. Will the proposed storage area be essentially watertight?
2. If not, where are the areas of potential leakage?
3. What is the estimated total leakage rate?
4. What are the consequences of this leakage?
5. What, if any, treatment is needed?

The amount and types of investigation needed to answer these questions will depend largely on whether the storage area is formed in essentially non-soluble rocks or largely on soluble rocks, or soils. This "overall geological situation" will usually be evident from examination of existing regional geological maps. If these are not available, new mapping will be needed. It is important also that the existing maps or new mapping provide an understanding of the geological (and modern) history of the area.

2.12.5.1   Storages in non-soluble rock areas – assessment of watertightness

It is usually possible to get preliminary answers to Questions 1 to 3 using existing contour plans, regional geological maps, air photos and climatic and groundwater records. The perimeter of the proposed storage area is divided into a number of watertightness regimes, based on differences in topography and assumed geology and groundwater situations. A judgement on the probable watertightness of each regime is made by comparison with the models on Figure 2.41. Regimes which appear to fit Models (a), (b) and (c) are assumed to be watertight. For regimes which appear to fit Models (d) or (e), ranges of possible leakage rates through each can be obtained by simple calculations using wide ranges of possible rock mass permeabilities and assumed flow paths, cross sectional areas and gradients. Construction of realistic flow nets is unlikely to be possible, or necessary.

If the geological and modern history suggests that high permeability features as listed in Section 2.12.3 may occur, but none are shown on the geological maps, they can be looked for on the air photos. They are often indicated by areas of anomalous vegetation. If such a feature is recognised and confirmed by ground inspection, ranges of possible leakage rates through it can be found be similar simple calculations.

The results of the above preliminary assessment are considered in relation to the consequences of leakage, which will usually be known. If the largest calculated leakage rates are shown to be of no consequence and the geological and groundwater data known to be reliable, then the watertightness studies could be considered complete, with no further action needed.

If some of the calculated leakages are seen to be significant, and there are doubts about the reliability or sufficiency of the geological or groundwater data, further field and office studies would be required. The field studies may include any or all of the following:

- Further regional geological mapping;
- Plotting the positions and levels of springs or seepages, and measuring flow rates;
- Plotting the positions and levels of existing water wells or bores and recording of water levels in them;

– Detailed studies of any recognized high permeability features, and of parts of the storage rim which are suspect for leakage because of, for example, their lack of rock exposures, low and narrow shapes, suspected low water tables, and steep potential leakage
gradients. Such studies have typically included detailed surface mapping, excavation
and mapping of trenches, and drilling and permeability testing of cored boreholes. The
positions of water tables may be judged during the progress of drilling, using the DVD
method (see Chapter 5), but would have to be confirmed by piezometers.

The results of the above, presented in plans, sections and diagrams, would be used for
new calculations of potential leakage rates. It is unlikely that the structure-related permeability of any jointed rock zones would be sufficiently well known to justify more than the
simplest flow nets.

In cases where the calculated rates are deemed to be unacceptable, consideration might
be given to alternative storage sites, or to measures such as blanketing or construction of
a grout curtain or slurry trench (see Sections 2.12.6 and 10.4). Further investigation
would be needed before adoption of any of these.

2.12.5.2   *Storages in soluble rock areas – assessment of watertightness*

From the discussion in Section 2.12.2 it is clear that the flow paths developed in a soluble
rock area may have developed as results of:

– Past, and in some cases presently active, tectonic processes;
– Dissolution and erosion by surface and underground waters, at times when the geological model and topography were quite different to those of today, and
– Dissolution and erosion by surface and underground waters, in the present day geology
and topography.

The keys to watertightness assessment therefore lie in:

– Knowing the regional topography and geological picture;
– Understanding the geological history, and
– Understanding the flow pattern of the present day surface and underground waters,
and their relationship to the geological structure and history.

Accordingly, the initial activities would usually include:

– Using existing contour and regional geological plans (supplemented where necessary
by new mapping) to compile cross sections and/or diagrams providing a 3-dimensional
picture of the geological structure beneath and surrounding the storage area;
– Compiling a table and diagrams explaining the assessed geological history, and
– Hydrological studies, including recording wet and dry season flow rates of surface
streams and springs, water levels in wells and boreholes.

The likely overall watertightness, and likely areas and models for leakage, would be
assessed from the results of the above. Consideration of the range of models described by
Ruquing (1981) and of experiences at the dams listed in Tables 3.3 and 3.4 should be
helpful. When likely areas of leakage are recognised, they can be investigated further
using any of the following methods:

– Detailed geological surface mapping, plus trenching or test pitting;
– Mapping of caves;
– Geophysical methods, which might include microgravity, resistivity, or ground probing
radar (from the surface or between boreholes);

- Core- and non-core drilling, preferably with continuous recording of penetration rates and pressures (to assist in recognition of cavities);
- Permeability testing, by the Lugeon method or (preferably) by pump out tests;
- Installation and monitoring of piezometers;
- Further hydrological studies, including water analyses, tritium dating, tracer studies (Gasper and Onescu, 1972, Habic, 1976, 1992 and Eriksson, 1985), temperature measurements (Fuxing and Changhua, 1981), spring blocking studies (Fuxing and Changhua, 1981) and pulse analysis (Williams, 1977). Breznik (1998, Page 55) describes the use of "geobombs" or large explosive devices, to locate the positions of underground channels unreachable by speleologists.

2.12.5.3   Storages formed in soils – assessment of watertightness

The watertightness of a storage formed by a soil mass will depend on the permeabilities and resistance to piping erosion of its component soils and their boundaries and configurations in relation to potential leakage paths. Understanding the site history and the soil origin type or types can provide clues to the likely nature and orientations of soil boundaries, and to the presence of defects and other features likely to affect permeability (see Sections 3.9 to 3.13). Assessment should therefore begin with examination of existing geological maps and reports and making detailed observations and mapping of soils exposed in erosion gullies, river banks or excavations. Other initial activities would include simple tests for dispersive soils, collection of published data on groundwater, plotting the position of bores, measuring water levels and plotting water level contours.

A preliminary geotechnical model for the proposed storage floor and rim, or a model for each of several regimes, can be developed from the above activities. Simple calculations using assumed parameters and possible flow paths should indicate whether a storage is likely to be feasible, with or without lining or extensive treatment, or both.

More detailed studies, planned to answer questions raised from the results of the above, might include the following:

- Excavation of trenches or test pits, and detailed logging of exposed faces after cleanup by hand tools and compressed air;
- Drilling and undisturbed sampling of boreholes, geological and engineering logging of the samples, testing them for moisture contents and grading and if appropriate, Atterberg Limits and erodibility;
- Installation and monitoring of piezometers, located at positions indicated by the logs and moisture profiles;
- Permeability tests in boreholes;
- Excavation and conducting infiltrometer tests in uniformly shaped pits, (1) after cleanup, logging and photographing the floors and (2) after remoulding the floor soil to a standard degree and depth;
- Construction of trial ponds or tanks and using these for large-scale permeability tests, and to test the applicability of various soil treatments and linings.

2.12.6   Methods used to prevent or limit leakages from storages

Localised features shown to cause unacceptable leakage can be treated by cement grout curtains, concrete diaphragm walls or mining and backfilling with concrete. In karst carbonates it is common to carry out extensive grouting and excavation/backfilling from adits into the abutments. These may extend considerable distances.

A clay blanket together with a soil-cement grout curtain was successful in reducing leakage through weathered gneiss forming a critical part of the rim of Lajes Reservoir, in Brazil (Cabrera, 1991).

Breznik (1998, Page 131) reports that sealing of rocky slopes in karst areas, by shotcreting and plugging of individual caves with concrete, has generally been successful in preventing leakage. These methods may be applicable also for lava tubes, lava flows or old mine workings.

Breznik (1998, Pages 126–130) describes how sediments covering the floor of the Popovo "hanging" storage in Herzegovina were treated locally by blanketing with low permeability soils, covered in parts by plastic sheeting. "Chimney" pipes were installed to prevent disruption of the blanket by air expelled from cavities when the water table rose. The treatment was successful at this site but similar treatments at four other sites were not.

As noted in Section 3.7.2.1, shotcreting, concreting and a geomembrane blanket have been recommended for the Lar storage area adjacent to the dam. Riemer (2003) has advised that these works have not been carried out.

Breznik (1998) notes that localized areas of sinkholes near the margins of some storage areas have been isolated from the storages by construction of low embankment dams.

Chengjie and Shuyong (1981) describe the effectiveness of an 8 m high concrete masonry dam which was used to isolate an area of sinkholes from the Baihua hydroelectric storage in Southwest China.

3

Geotechnical questions associated with various geological environments

As explained in Chapter 2 and later in Chapter 4, site investigations for a dam need to be undertaken with a good understanding of the local and regional geological environment and the investigations should be aimed at answering all questions known to be of relevance to dam construction and operation in that environment. This chapter discusses twelve common geological environments in which dams have been built and derives check lists of geotechnical questions of specific relevance to each.

It is important that readers appreciate the limitations of these generalisations and check lists. The lists refer simply to features that might be present because they have been found during construction at many other sites in similar environments and because geological reasoning suggests that they could be present. At any particular site the actual geological conditions found will have been developed as a result of many geological processes acting at different times over vast periods of geological time. If some of these processes have been very different from those assumed in the "general" case, then some or even all of the generalisations may not be valid at that particular site.

For further accounts of regional and local geological environments, with derived predictions of site conditions, see Fookes (1997) and Fookes et al. (2000).

3.1   GRANITIC ROCKS

Included under this heading are granite and other medium or coarse grained igneous rocks. Most rocks of these types have been formed by the cooling and solidification of large masses of viscous magma, generally at depths of greater than 5 km below the ground surface.

3.1.1   *Fresh granitic rocks, properties and uses*

In unweathered (fresh) exposures, granitic rocks are usually highly durable, strong to extremely strong (substances) and contain very widely spaced (greater than 2 m) tectonic joints in a roughly rectangular pattern (Figure 3.1). Many of these joints are wholly or partly healed by thin veins of quartz, or quartz/felspar mixtures. Sheet joints are common but, as they are almost parallel to the ground surface, they may be difficult to detect during surface mapping.

Fresh granitic rocks are commonly quarried for rip-rap, rockfill and concrete aggregates, but mica-rich granites may be unsuitable for use as fine aggregates in concrete due to excessive amounts of fine, platy particles in the crushed products.

3.1.2   *Weathered granitic rocks, properties, uses and profiles*

Chemical weathering of granitic substances usually causes cracking at the grain boundaries and decomposition of the felspars and ferromagnesian (dark) minerals, leaving quartz grains essentially unaffected.

Figure 3.1.   Granitic cliff showing 3 sets of joints approximately at right angles to one another.

Table 2.6 is a practical descriptive classification scheme for weathered granitic rocks. When extremely weathered (i.e. soil properties) most granitic materials are silty or clayey fine gravels or sands (GM, GC, SM or SC). *In situ* these materials are usually dense to very dense, but in some tropically weathered areas, where quartz has been partly or wholly removed, they are more clay rich and of low density. Somerford (1991) and Bradbury (1990) describe low density, extremely weathered granitic materials at Harris Dam, Western Australia.

The extremely weathered materials often make good core or earth fill materials and where the parent rock is very coarse grained the resulting gravels can make good quality road sub-base for sealed roads or base course for haul roads.

The silty nature of some extremely weathered granitic rocks often causes them to be highly erodible, when exposed in excavation and when used in fills. At Cardinia earth and rockfill dam near Melbourne extremely weathered granite is dispersive and where exposed in the storage area shoreline has required blanketing with rockfill to prevent erosion and subsequent water turbidity problems.

Lumb (1982) describes engineering properties of granitic rocks in various weathered conditions.

Typical weathered profiles in granitic rock masses are shown on Figures 2.17 to 2.22. The chemical weathering is initiated at and proceeds from the ground surface and from sheet-joints, tectonic joints and faults, causing the roughly rectangular joint-blocks to become smaller, rounded and separated by weathered materials. Thus the profile grades usually from residual granitic soil near the surface to fresh rock at depth, with varying amounts of residual "boulders" of fresh or partly weathered rock occurring at any level. Fresh outcrops or large fresh boulders at the ground surface may or may not be underlain by fresh rock. It is not uncommon to find that weathering has occurred beneath such outcrops, along sheet joints, gently dipping tectonic joints, or within previously altered granitic rock (Figure 2.22). Understanding of this potential for variability in weathered granite profiles is important not only for dam foundations, but also when planning and operating either a quarry for rockfill, rip-rap, filters or aggregate, or a borrow pit for earth fill or core materials.

3.1.3   Stability of slopes in granitic rocks

Active landsliding, or evidence of past landsliding, is relatively common in steep country underlain by weathered granitic rocks, particularly in areas with high rainfall. Brand (1984) describes the widespread occurrence of natural and man-induced landslides in weathered granitic rocks in Hong Kong.

Although some landslides in weathered granitic rocks no doubt occur by failure through the fabric of extremely weathered material, it is probable that in many landslides failure occurs wholly or partly along relict joints or other defects (see Chapter 2, Section 2.10 and Figure 2.28).

The landsliding at Tooma and Wungong dams (Sections 2.10.3.2 and 2.10.3.3) occurred along localised weathered zones along pre-existing defects and past movements had occurred well into rock masses which were dominantly slightly weathered.

3.1.4   Granitic rocks: check list

- Concealed sheet joints?
- Fresh rock outcrop; does it extend down into fresh rock?
- Chemically altered zone(s)?
- Fresh granite "boulders" within extremely weathered materials?
- Extremely weathered materials, suitable for impervious core? road pavements? highly erodible? low density in situ?
- Past landsliding? stability of extremely weathered materials in cuts?

3.2   VOLCANIC ROCKS (INTRUSIVE AND FLOW)

The common rocks in this group range from basalt (basic) through andesite, dacite, trachyte, to rhyolite (acidic). Basalt is the most common. All are formed from molten magma and are very fine grained, usually very strong to extremely strong when fresh. In this fresh condition the rocks generally are also very durable and are used commonly as sources of materials for filters, concrete aggregates, rockfill and road base courses. However volcanic rocks, particularly basalts and andesites, often show subtle alteration effects, which in some cases render them unsuitable for some or all of these purposes. This matter will be discussed further in Section 3.2.3. Also most volcanic rocks have initially contained some glass. In rocks of Mesozoic age and older the glass has usually "devitrified" or crystallized. However, in rocks of Tertiary and younger age the glass is usually still present today and, if the rock is used as concrete aggregate, it may react with alkalis in the cement and cause the concrete to deteriorate (see also Section 3.2.6).

The shape and other field characteristics of a body of volcanic rock depends upon the circumstances in which it solidified, i.e. as a plug, dyke, sill or flow.

3.2.1   Intrusive plugs, dykes and sills

In these types of bodies the magma has been confined within other rocks (or soils), and has flowed against them and eventually solidified against them. As a result of this mode of formation, any of the following characteristics shown in Figure 3.2 are commonly seen:

(a) The host-rock (or soil) close to the contacts may be stronger and more durable than elsewhere due to being subjected to very high temperatures;
(b) The intrusive rock has "chilled" margins, i.e. it is extremely fine-grained or even glassy, close to its contacts, due to a faster rate of cooling than in the interior of the mass;

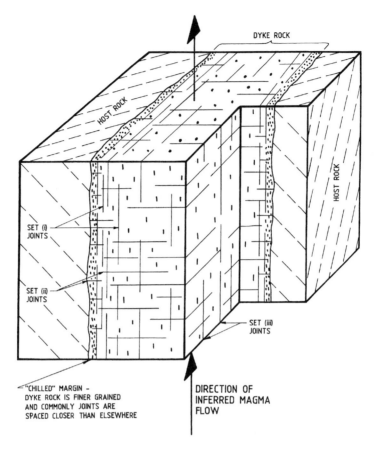

Figure 3.2.   Some features commonly seen in dyke of intrusive rock.

(c) The intrusive rock has developed a "planar" foliation parallel to its contacts and, within this, a lineation, or linear arrangement of mineral grains, parallel to the direction of flow during intrusion;

(d) Joints in the intrusive body occur in at least 3 sets, as shown on Figure 3.2. Set (i) joints are parallel to the contacts (and to the foliation). Set (ii) joints are normal to the lineation i.e. to the direction of magma flow and also to the contacts. Set (iii) joints are normal to the contacts and parallel to the lineation direction. The joints in all sets commonly show extension characteristics (i.e. rough or plumose surfaces) and are either slightly open or infilled with secondary minerals including calcite and zeolite minerals. It can be inferred that shrinkage was an important factor in their formation, although extension during viscous flow seems a likely initiating factor for Set (ii).

It is sometimes found that both the host and intrusive rock are sheared or crushed, along and near the contact zone. In some cases this appears to be as a result of viscous drag, but more generally tectonically induced movements after solidification appear to be the likely cause.

In some cases, open joints in the body and contact zones may render the intrusive mass highly permeable. If continuous, such permeable masses represent potential leakage zones beneath dams or from storages.

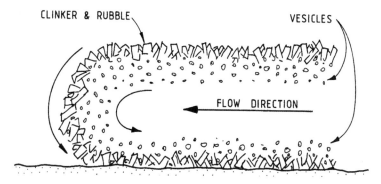

Figure 3.3.   Diagrammatic longitudinal section through flowing lava showing the "caterpillar track" mechanism which results in upper and lower layers of vesicular lava and in some cases clinker or breccia.

3.2.2   Flows

Lava "flows" are bodies which have been formed by molten lava which has been extruded at the ground surface or the sea floor and has flowed over a pre-existing surface of rock, soil or sediment.

3.2.2.1   Flows on land

Structures developed in lava flows on land are described in some detail in Hess and Poldervaart (1967), Francis (1976) and Bell (1983b). The flows move forward in a manner similar to that of a caterpillar tractor, as shown in Figure 3.3. The lava near the exposed, upper surface of the flow usually develops many small holes or vesicles, formed by bubbles of expanding gases becoming trapped as the magma solidifies. The upper surfaces of some flows develop an extremely rough, fragmented structure, comprising sharp, irregular fragments up to 150 mm across, termed clinker. Bell reports that clinker layers several million years old and buried at 500 m to 1000 m depths in Hawaii show little or no sign of compaction and are highly permeable. As the flow (Figure 3.3) moves forward the clinker-covered surface and vesicular layer are carried forward, deposited over the front and eventually buried beneath the flow. Thus the solidified flow comprises an inner layer of massive rock sandwiched between two layers of clinker and vesicular rock which are highly permeable.

Other lava flows develop a hummocky and sometimes twisted ropy structure at their upper surfaces, due to viscous drag on the surface crust while this is still plastic.

Where successive lava flows have occurred to form a continuous layered sequence, the individual flows and their boundaries may be distinguishable by the following:

– The development of a weathered or soil profile on the upper surface of a flow which was exposed to weathering for some time before the next flow occurred;
– The presence of chilled and vesicular zones, or clinkered, brecciated or ropy zones near flow boundaries.

At Foz do Areia dam in Brazil, breccia (clinker) zones at the boundaries of basalt flows were locally weathered and generally highly permeable (Figure 3.4). Treatment of these zones by excavation, dental concrete and grouting is described by Pinto et al. (1985).

It is common for lava flows to be interbedded with pyroclastic materials (ash and lava fragments) or tuff and agglomerate in rock sequences derived from volcanic explosions. Also flows may occur interbedded with alluvial or other sediments.

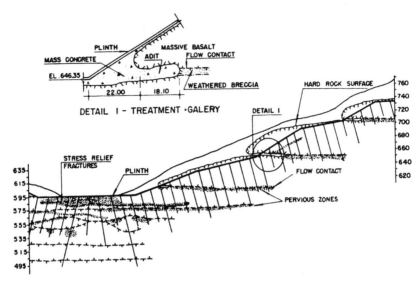

Figure 3.4.   Geological section along Plinth (right bank) at Foz Do Areia Dam (Pinto et al., 1985).

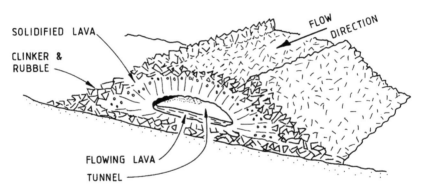

Figure 3.5.   Perspective view of, and cross section through a lava flow, showing a lava tunnel developed when the lava flows forward faster than the supply.

Thick lava flows sometimes contain "lava tunnels", which are circular, ovoid or lenticular in cross section, up to 20 m across and some kilometres in length. These are formed when the supply of lava to the flow is exhausted, and the internal lava "stream" (Figure 3.5) drains away. Figure 3.6 shows a small lava tunnel exposed in basalt forming the wall of a 30 m diameter shaft at Hoppers Crossing, Victoria, Australia.

Most lava flows show a hexagonal columnar joint pattern, with the columns being interrupted by near-planar or saucer-shaped cross joints. These joints have developed as a result of shrinkage on cooling and are usually either slightly open or filled with secondary minerals or alteration products. Figure 3.7 shows columnar-jointed rhyolite exposed during construction at High Island dam, in Hong Kong. The rhyolite has been intruded by a younger, darker rock forming a dyke up to 1.5 m thick.

In some flows there are closely spaced joints parallel and near to the margins, apparently caused by shearing associated with viscous flow. Where columnar joints in lava flows are open, e.g. due to mechanical weathering close to steep valley sides, the rock mass permeability can be high.

Figure 3.6.   Lava tunnel in basalt, Hoppers Crossing, Victoria, Australia.

Figure 3.7.   Columnar joint pattern in rhyolite, and younger intrusive dyke. High Island Dam, Hong Kong.

Figure 3.8.   Pillow structure in basalt, Pembrokeshire, U.K.

As stated at the start of this chapter, fresh volcanic rocks are used widely as construction materials. However, columnar jointed rock can be difficult to quarry, especially when the columns are almost vertical. Blastholes drilled vertically tend to jam in the open joints and explosive gases vent into them. The columns tend to topple over rather than fragment, resulting in a poorly graded (one-sized) quarry-run product. If the column diameter is large (i.e. more than 1 metre) then the product may be too large for the crusher. Regardless of the column size the poorly graded quarry product does not make good rockfill because it is difficult to compact and large voids remain between the rock blocks after compaction. This rockfill often has relatively low modulus and because of its large voids, requires special attention to filter design.

3.2.2.2   *Undersea flows*

Lava which has been extruded from the sea-bed is often in the form of distorted globular masses up to 3 m long known as pillows (Figure 3.8).

The pillows may be welded together, separated by fine cracks, or separated by sedimentary detritus derived from the lava. The long axes of the pillows are generally roughly parallel to the boundaries of the "flow". Bell (1983b) notes that joints, vesicles, and phenocrysts in the pillows are in some cases arranged radially. The joint pattern and the irregular shapes of the cracks at the pillow boundaries produce an overall fracture pattern that might at first appear to be "random".

3.2.3   *Alteration of volcanic rocks*

Fresh volcanic rocks are composed of minute, strong, tough mineral crystals, generally arranged in an extremely dense, interlocking manner. This structure results in negligible porosity and great strength and durability within the lifetime of engineering structures. However, as stated earlier, these rocks are often found to have been altered, probably during the late stages of solidification. In the altered rocks some of the crystals are wholly or partly changed to secondary minerals, including serpentine, calcite, chlorite, zeolites and clay minerals. These minerals are weaker and in some cases larger in volume than the original minerals, resulting in microcracking, increase in porosity and weakening of the rock. The altered rock is usually greenish in colour.

Where these effects are pronounced, and particularly if the clay mineral montmorillonite is produced, the altered rock is obviously much weaker than the original rock and is likely to deteriorate or even disintegrate on exposure to air or immersion in water. However if the alteration effects are relatively minor, e.g. only the margins of the original crystals are changed to secondary minerals, then the visual appearance of the rock substance may be little changed from that of fresh, unaltered rock. The altered rock will usually appear a little dull and have a slightly higher porosity and absorption than the fresh rock. Rocks such as this, showing only very minor and subtle alteration effects, often deteriorate rapidly in pavements, are likely to be unsuitable for crushing to produce filters and may prove to be unsuitable for concrete aggregate and rockfill. Because of this, before adopting volcanic rocks for use as construction materials, they should always be subjected to very thorough checking, by local past performance and field observations, petrographic analysis and laboratory tests. For further details readers are referred to Shayan and Van Atta (1986), Van Atta and Ludowise (1976), Cole and Beresford (1976), Cole and Sandy (1980) and Hosking and Tubey (1969).

As the alteration is caused by hot water and gases which move through permeable features such as joints and highly vesicular zones, the more altered rock is usually located along and adjacent to such features. Quite commonly, secondary minerals occur as veins, seams or irregular masses filling previously gaping joints or voids. Such features, particularly where of large extent and composed of clay, serpentine or chlorite, represent significant rock mass defects with low shear strength.

3.2.4 Weathering of volcanic rocks

Although all unaltered volcanic rocks are highly durable within the life-span of normal engineering structures, the more basic varieties, particularly basalt, are quite susceptible to chemical weathering in a geological time frame. The distribution of weathered materials in volcanic rocks is governed by the distribution of any previously altered material, as well as by the pattern of joints and vesicular zones.

When extremely weathered, all volcanic rocks are clayey soils in the engineering sense. The acidic types tend to produce low or medium plasticity clays and the basic types high plasticity clays. Basalts commonly display spheroidal weathering profiles, with spheroids of fresh to distinctly weathered basalt surrounded by extremely weathered basalt, which is clay of high plasticity. Extremely weathered basalts and the surface residual soils developed on them are usually highly expansive and fissured.

3.2.5 Landsliding on slopes underlain by weathered basalt

The presence of a well developed pattern of slickensided fissures causes the shear strength of the mass to be significantly lower than that of the intact material. Slopes steeper than about 10 degrees which are underlain by such materials often show geomorphological evidence of past or current landsliding.

Such landsliding occurs commonly at the steep margins of plateaus or hills capped by basalt flows which overlie old weathered land surfaces, as shown in Figure 3.9. The sliding occurs in some cases simply as a result of over-steepening of the hillside by erosion, and in others due to pressure from groundwater exiting from a permeable zone beneath the extremely weathered basalt. The permeable zone may be jointed, less weathered basalt as in Figure 3.9, or alluvial sands or gravels on the old buried land surface. Examples are given in Fell (1992).

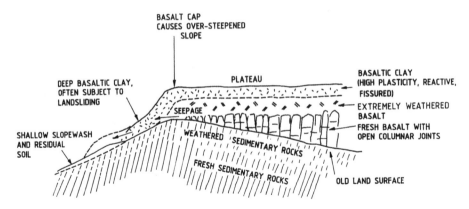

Figure 3.9.   Typical profile through margin of basalt plateau, showing conditions which lead to slope instability.

Landsliding in fissured, extremely weathered basalt in Victoria, Australia is described
by MacGregor et al. (1990).

3.2.6   Alkali-aggregate reaction

As discussed in Section 3.2 above, most volcanic rocks of Tertiary and younger ages contain appreciable amounts of glass, which may react with alkalis in Portland cement if the
materials are used as aggregates in concrete. In older rocks, much or all of the glassy materials have usually developed a very fine crystalline structure and are less likely to be reactive. Zeolite minerals which occur in many volcanic rocks may also react with alkalis in
cement.
    Guillott (1975, 1986) and McConnell et al. (1950) describe observed effects of concrete
expansion due to alkali aggregate reaction and discuss the mechanisms involved. Stark
and De Puy (1987) describe observations and tests on affected concrete at five dams in
USA. Cole and Horswill (1988) describe the deleterious effects and remedial work carried
out at Val de la Mare dam, in Jersey, Channel Islands.
    Shayan (1987) and Carse and Dux (1988) discuss some limitations of chemical and
mortar bar tests for the prediction of the actual performance of aggregates in concrete.
    The deleterious effects can be largely avoided by the use of low-alkali cement or poz-
zolanic additives in the concrete.
    If volcanic rock is to be used as aggregate careful checking for possible alkali-aggregate
activity is advisable, particularly if low-alkali cement is not available. Studies usually
include checking of past performance, petrographic examination (ASTM, 1974a), and
laboratory testing including the Quick Chemical Test (ASTM, 1974b or Standards
Association of Australia, 1974a), the Gel-Pat Test (Building Research Station, 1958) and
the Mortar Bar Test (ASTM, 1974c or Standards Association of Australia, 1974b).

3.2.7   Volcanic rocks (intrusive and flow) check list of questions

– Vesicular zones?
– "Clinker" or "breccia" zones?
– Lava tunnels?
– Old weathered/soil profiles?
– High mass permeability?
– Interbedded pyroclastic or sedimentary materials?

- Columnar joint pattern?
- Toppling failure?
- Difficulties in blasthole drilling?
- Poor fragmentation during blasting?
- Irregular joint pattern and "pillow" structure?
- Alteration effects – secondary minerals?
- Fresh, extremely strong boulders in extremely weathered materials (high plasticity clay)?
- Very high plasticity soils, expansive, fissured?
- Unstable slopes?
- Alkali-aggregate reaction?

3.3   PYROCLASTICS

Pyroclastic or "fire-broken" deposits are those which have been formed by the accummulation of solid fragments of volcanic rock, shot into the air during volcanic eruptions. The rock fragments include dense, solidified lava, highly vesicular lava (termed scoria) and extremely vesicular lava (termed pumice). Pumice is formed only from acidic lavas (e.g. rhyolite) and is so porous that it will float on water. Francis (1976) provides a detailed account of the ways in which pyroclastic materials are formed, based mainly on historical accounts of modern eruptions. Prebble (1983) describes the pyroclastic deposits of the Taupo Volcanic Zone in New Zealand and the difficulties they present in dam and canal engineering.

3.3.1   *Variability of pyroclastic materials and masses*

Pyroclastic deposits are characterized by extreme variability in engineering properties over short distances laterally and vertically. They range from extremely low density "collapsing" type soils to extremely strong rocks. This wide range in properties results from differences between the ways in which they were initially deposited and also from the ways they have been modified since deposition.

There are four main types of deposit, based on initial mode of deposition:

(a) Air fall deposits in which the fragments have simply been shot up into the air and fallen down again. Where they have "soil" properties such deposits are termed ash (sand sizes and smaller), or lapilli and bombs (gravel sizes and larger). Where welded, compacted or cemented to form rocks, they are termed tuff (sand sizes and smaller) or agglomerate (gravel sizes and larger in a matrix of ash or tuff).

(b) *Water-sorted deposits* in which the fragments fall into the sea or a lake and become intermixed and interbedded with marine or lake deposits. These also may be "soils" or rocks depending upon their subsequent history.

(c) *Air-flow, or "nuées ardentes", deposits* in which the fragments are white-hot and mixed with large volumes of hot gases, to form fluidized mixtures which can travel large distances across the countryside at speeds of probably several hundreds of kilometres per hour. The resulting materials, known as ignimbrites, range from extremely low density soils with void ratios as high as 5 (Prebble, 1983) to extremely strong rocks. The latter are formed when the white-hot fragments become welded together to form rocks almost indistinguishable from solidified lavas. These rocks are called welded ignimbrites or welded tuffs.

(d) Hot avalanche deposits which are formed by the gravitational breakup and collapse of molten lavas on steep slopes. The deposits comprise loosely packed but partly welded boulders, often showing prismatic fracture patterns which indicates that they were chilled rapidly after deposition (Francis, 1976).

Following or during their initial deposition, any of these types of deposit may be modified greatly by any or all of the following, which prevail in active volcanic environments:

– Faulting;
– Intrusion of further igneous material in plugs or dykes;
– Lava flows;
– Intense thunderstorm activity which causes erosion and redeposition, landsliding and mudflows;
– Hydrothermal alteration and chemical weathering.

The alteration and weathering produce clay minerals which may include montmorillonite, noted for its low shear strength, nontronite and allophane and halloysite which, in some cases, are highly sensitive.

The air-flow and air-fall deposits have commonly been deposited on old land surfaces with variable relief and covered by residual soil and weathered rock profiles. Prebble (1983) notes that such deposition in the Taupo Volcanic Zone of New Zealand resulted in permeable sand and gravel sized materials burying weak and relatively impermeable residual clays in old valleys. Subsequent cycles of deposition and near-surface weathering have resulted in a "valley-upon-valley" sequence of aquifers, interlayered with clayey aquicludes which tend to be sensitive and collapsible (see Figure 3.10).

Pyroclastic materials are also found in near-vertical pipes or necks, called diatremes. The Kimberley diamond pipe in South Africa and the Prospect Diatreme in Sydney, New South Wales (Herbert, 1983) are examples. The materials in the diatremes comprise angular fragments of volcanic rock plus fragments of the underlying and surrounding rocks. It seems likely that diatremes have been formed by explosive volcanic eruptions.

All of the above processes result in the extreme variability found in modern volcanic deposits. In geologically old deposits, which have been deeply buried by later sedimentation and folded, faulted and uplifted, new defects and variabilities are introduced, but the effects of compaction and consolidation can cause the strength contrasts between the various pyroclastic substances to be greatly reduced.

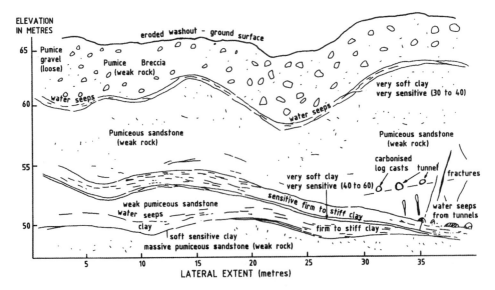

Figure 3.10.   Exposure of pyroclastic materials in the collapsed area of Ruahihi Canal, New Zealand (Prebble, 1983).

3.3.2   Particular construction issues in pyroclastics

Prebble (1983), Jones (1988) and Oborn (1988) describe problems which were encountered at the Ruahihi and Wheao hydroelectric projects, constructed in the Taupo Volcanic
Zone. Headrace canals for each project were constructed by cut and fill methods, on and
through highly variable ash and ignimbrite deposits. Both canals failed by piping and subsequent collapse during early operation, apparently due to the high erodibility of some of
the soils, both *in situ* and when compacted, and to their brittle, non-healing nature which
enabled the development of erosion tunnels. Oborn (1988) suggests that some of the soils
at Ruahihi were probably dispersive and that accelerated rates of settlement after canal filling may have been due to the collapse of very low density soils on saturation after loading.
At the Wheao failure area, erodible ash soils were located above very high strength welded
ignimbrite with a columnar joint pattern. Near the upper surface of the ignimbrite, the
joints were "bridged" by infill soils from above, but below this they were open as much as
50 mm. This feature was apparently missed during the construction stage cleanup. During
operation, erodible ash soils were washed into these gaping joints, close to the penstock
intake structure as shown on Figure 3.11, taken from Jones (1988). Prebble (1983) and
Oborn (1988). Note also that the extreme sensitivity (up to about 60) of some of the alteration products (allophane and halloysite clays) caused problems during construction of the
canals. Prebble predicts that these soils could collapse and liquefy when disturbed by earthquake loading or changed groundwater levels.

Jacquet (1990) describes the results of comprehensive laboratory tests on andesitic ash
soils from seven sites in New Zealand. The soils contained high proportions of allophane
or halloysite and all classified as MH in the Unified System. Sensitivities ranged from 5 to
55. Jacquet concluded that the sensitivity was associated with irreversible rupture of the
structural fabric of the soils and was not directly related to the clay mineralogy or classification characteristics of the soils.

Not all weathered pyroclastic materials are sensitive. In weathered agglomerates at the
site of Sirinumu Dam in Papua New Guinea the matrix soils, which are clays of medium
to high plasticity with 40 to 50 percent moisture content, are very resistant to erosion.
The clay mineral types include halloysite, kaolin and allophane.

Rouse (1990) describes tropically weathered andesitic and dacitic ash soils from
Dominica, West Indies, which occur on generally stable slopes of 30° to more than 50°.
The soils are mainly allophane and halloysite clays, with very high residual friction angles
(most between 25° and 35°).

Some unweathered non-welded ash and weakly-welded ignimbrite materials can be
used as sand and gravel sized embankment filling, the weakly welded materials breaking
down readily during compaction. However based on the experience at the 60 m high
Matahina rockfill dam in New Zealand (Sherard, 1973) it is suggested that such materials should not be used for filter zones unless it can be shown that they will remain cohesionless in the long term. At Matahina weakly welded, partly weathered ignimbrite was
compacted to form "transition" zones between the impervious core and rockfill zones.
Subsequent excavation through the compacted ignimbrite showed it to have developed
appreciable cohesion and it appeared to behave like a very low strength rock. This
strength developed due to the interlocking of needle-shaped particles of glass. This cohesive, brittle behaviour was an important contributing factor to the piping incident which
occurred during the first filling of the reservoir in January 1967. The main cause of this
incident, which resulted in the loss of more than 100 m^3 of core and transition materials
into the rockfill, was the failure to remove a 1.8 m projecting bench from the steeply sloping foundation. This projection caused cracks to develop in the core and transition zones
as the embankment settled. Other factors, as described by Sherard (1972, 1973) and
Gillon and Newton (1988), included the possible reinforcement of the transition zones by

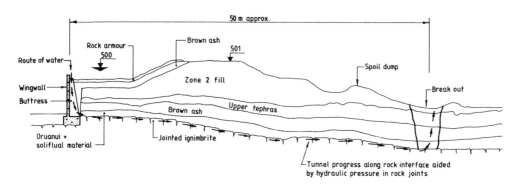

GENERAL ARRANGEMENT LOOKING UPSTREAM

Scale:
1 0 2 4 6 8 10 12 14 16 18m

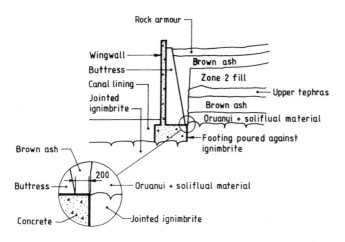

DETAIL OF WINGWALL & BUTTRESS

Scale:
1 0 1 2 3 4 5 6 7 8 9m

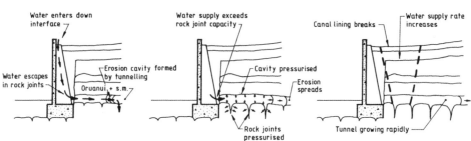

PHASE 1. (early Dec. 1982)

Water exploits interface; penetrates remainder of lining and starts to erode Oruanui ash into rock joints; tunnels form.

PHASE 2. (late Dec. 1982)

Water supply exceeds rock joint capacity; pressure in erosion cavity rises and assists spread of erosion.

PHASE 3. (shortly before 7.55 am 30·12·82)

Lining above tunnel breaks; water supply increases greatly; tunnels and rock exposed to full canal head; tunnel advances rapidly along favourable joints, assisted by water pressure.

STAGES   OF   DEVELOPMENT

Figure 3.11.   Wheao Canal failure. Diagrams showing the general arrangement and how the failures developed (Jones, 1988).

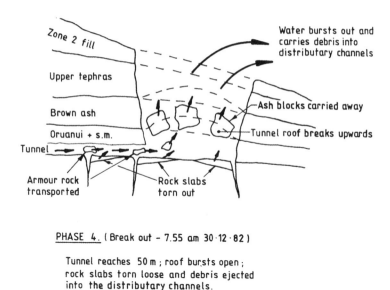

PHASE 4. (Break out - 7.55 am 30·12·82)

Tunnel reaches 50 m ; roof bursts open ;
rock slabs torn loose and debris ejected
into the distributary channels.

STAGES   OF   DEVELOPMENT

Figure 3.11.   (Continued).

means of cement grout, the dispersive nature of the core material and large voids in the
rockfill shoulders which were formed by poorly graded very strong welded ignimbrite,
quarried from a columnar-jointed mass. Most blocks in the rockfill are of blocky shape
and in the range 300 mm to 600 mm (Section 3.2.2.1).

As with lavas, pyroclastic materials, especially those of Tertiary or younger ages, contain glassy materials which may react with alkalis in Portland cement. They should therefore be tested thoroughly before use as aggregates in concrete (Section 3.2.6).

3.3.3   *Pyroclastic materials – check list of questions*

– Extreme variability?
– Very low *in situ* densities – collapse type behaviour?
– High *in situ* permeability?
– Brittle *in situ* and when compacted?
– Highly erodible *in situ* and when compacted?
– Highly to extremely sensitive zones?
– Complex groundwater distribution?
– Welded rocks: gaping joints?
– Columnar jointed welded rocks: poorly graded rockfill, quarrying problems?
– Interbedded lavas?
– Intrusive dykes, sills or plugs?
– Alkali-aggregate reaction?

3.4   SCHISTOSE ROCKS

Included in this group are those metamorphic rocks, e.g. slate, phyllite and schist which
have developed a pronounced cleavage or planar foliation. The cleavage or foliation results

from the parallel arrangement of platy minerals, commonly clays, muscovite, biotite, chlorite and sericite. Also often present and in parallel arrangement, are tabular or elongate clusters of other minerals, usually quartz and felspars and occasionally amphiboles.

Although the foliation is referred to as "planar", the foliae or layers are commonly folded. The folds can range in amplitude and wave-length from microscopic up to hundreds of metres. Small-scale folds, which cause the surfaces of hand-specimens of schist to appear rough or corrugated, are called crenulations.

In some schists the foliation has been so tightly and irregularly folded as to give a contorted appearance. Such rock is called "knotted schist".

Slate, phyllite and most mica schists have been formed by the regional metamorphism of fine grained sedimentary rocks (mudstones or siltstones). This has involved relatively high temperatures and directed pressure over long periods of geological time. Under these conditions the clay minerals present in the original rocks have changed partly or wholly to mica minerals, usually muscovite, chlorite and biotite. These minerals have become aligned normal to the direction of the maximum compressive stress. The proportion of these new minerals is least in slate and greatest in schist.

Greenschists, which are well foliated rocks containing a large proportion of chlorite and other green minerals, have been formed by the regional metamorphism of basic igneous rocks (e.g. basalt and gabbro).

Schists can be formed also as a result of shear stresses applied over long periods of geological time to igneous rocks or to "high-grade" metamorphic rocks, e.g. gneisses. This process is known as retrograde metamorphism and usually produces schist containing abundant sericite and/or chlorite. These are weak minerals and so the rocks produced by this process tend to be weaker than other schists.

3.4.1   *Properties of fresh schistose rock substances*

Most schistose rocks have dry strengths in the very weak to medium strong range (see Table 2.4). Schists containing abundant quartz are generally medium strong or strong, while greenschists which are rich in chlorite are generally very weak.

The most significant engineering characteristic of schistose rocks is their pronounced anisotropy, caused by the cleavage or foliation. Figure 3.12 taken from Trudinger (1973) shows the results of unconfined compressive tests on fresh schist samples from Kangaroo Creek Dam, South Australia. This schist is stronger than most; it contains generally about 40% quartz and felspar and about 60% sericite and chlorite. It is foliated but not exceptionally fissile. It can be seen that the strengths recorded for samples loaded at about 45° to the foliation were about one third of those for samples loaded at right angles, i.e. the anisotropy index of this schist in unconfined compression is about 3.

When tested by the point load method, i.e. in induced tension, the schist at Kangaroo Creek shows anisotropy indices ranging typically from 5 to 10. Failure along the foliation surfaces in this test is by tensile splitting, rather than in shear.

Most other schists, slates and phyllites show similar anisotropic properties (Donath, 1961). It is clear that foliation angles in relation to loading directions should always be carefully recorded during tests and reported with the results.

In weak or very weak schists (often those rich in chlorite) the effective angle of friction along foliation surfaces can be low. Landsliding is prevalent in areas underlain by these rocks.

In knotted schists the foliation surfaces are often so contorted that shearing or splitting along near-planar surfaces is not possible. As a result, knotted schists are usually appreciably stronger than those in which the foliation surfaces are near-planar.

Figure 3.12 also indicates that the schist at Kangaroo Creek Dam showed a 25–65% reduction in strength, after soaking in water for 1–2 weeks. The greatest strength reduction

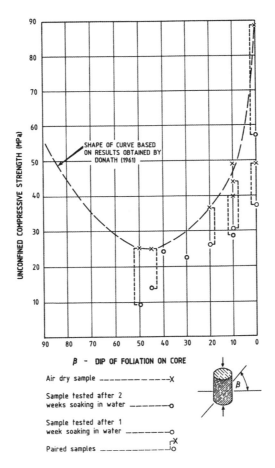

Figure 3.12.   Variation of unconfined compressive strength of schist with angle between the foliation
planes and applied loads (Trudinger, 1973).

occurred in the samples loaded at about 45° to the foliation. This result is typical of schistose rocks.

3.4.2   Weathered products and profiles developed in schistose rock

Schistose rocks vary widely in their susceptibility to chemical weathering. Varieties rich in
quartz are very resistant and, at the other extreme, rocks rich in clay minerals or chlorite
are very susceptible. Because of this, no typical weathered profile exists for all schistose
rocks. However many schistose rock masses consist of layers (parallel to the foliation) of
varying susceptibilities to weathering. Thus weathering often tends to exaggerate the
strength anisotrophy of such masses and the upper surface of fresh rock often has a deeply
slotted or serrated shape, as shown in Figure 3.13.

Schists which are rich in micaceous minerals (biotite, muscovite or chlorite) tend to
form micaceous silty or clayey soils when extremely weathered. The silty varieties are
often of low *in situ* density and are highly erodible by water or wind. Also they tend to be
hydrophobic, making dust control difficult on construction sites.

Even when fresh or only partly weathered, the more micaceous, fissile schists usually
produce much dust due to abrasion during handling and trafficking.

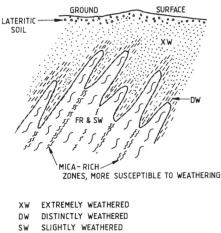

XW   EXTREMELY WEATHERED
DW   DISTINCTLY WEATHERED
SW   SLIGHTLY WEATHERED
FR   FRESH

Figure 3.13.   Weathered profile developed on schistose rock with steeply dipping foliation.

3.4.3   *Suitability of schistose rocks for use as filter materials, concrete aggregates and pavement materials*

Schistose rocks are generally unsuitable for any of these purposes due to the very flaky shapes of the crushed materials and inadequate strengths of the particles. Kammer and Carlson (1941) describe the unsuccessful use of phyllite as aggregate for concrete in a hydro-electric plant. Silicates in the phyllite reacted with alkalis in the cement to cause expansion and disruption of the concrete. The strongest, most siliceous schists have been used successfully in base courses of pavements, where more suitable materials have not been available.

3.4.4   *Suitability of schistose rocks for use as rockfill*

Despite their tendency to produce very platy block shapes, schistose rocks have been used successfully as rockfill on several dams up to 80 m high. At Kanmantoo Mine in South Australia a 28 m high rockfill dam with a thin sloping earth core was built in 1971 for the storage of tailings and water for use at the mine (Stapledon et al., 1978). The rockfill was mainly very weak quartz-biotite schist, the waste rock from the mine. It was placed in 0.6–0.9 m thick layers and compacted dry with a Caterpillar D8 tractor and by trafficking by the 50-tonne dump trucks. Post-construction crest settlements 5.5 years after completion ranged from 0.3% to 0.7% of the embankment height.

The successful use of schist as rockfill at Kangaroo Creek Dam has been described by Good (1976) and Trudinger (1973). Trudinger describes in some detail the construction procedures, the behaviour of the rock and the fill densities obtained. The concrete faced dam as built initially was 60 m high. Zone 3, which contained about 300,000 m^3 of the 443,000 m^3 total rock in the embankment, was built mainly of weak to medium strong, slightly weathered schist compacted in 1 m layers. The maximum specified block size was 1 m, but many blocks were longer than this due to the very platy shapes obtained during quarrying (Figure 3.14). Because the schist suffered a large strength loss on saturation (Figure 3.12) the rock was heavily watered during placement. During compaction by 4 passes of a 10-tonne vibrating roller, the uppermost 50–300 mm of most layers were crushed into a gravelly sandy silt (Figure 3.15). These layers formed thin, lower permeability barriers but were readily penetrated by the basal rocks of the next layer and hence they did not form weak or compressible zones in the dam.

Figure 3.14.   Schist being quarried in the spillway excavation at Kangaroo Creek Dam. Note the platy
shapes of the rock fragments.

Figure 3.15.   Slurry (Gravelly sandy silt) formed on the top of schist rockfill layer as a result of weathering and compaction by 4 passes of a 10-tonne vibrating roller (Trudinger, 1973).

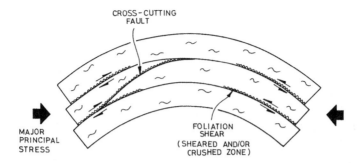

Figure 3.16.   Probable way in which foliation shears and cross-cutting faults are formed in some schistose rocks.

After 20 years of operation the dam height was increased to 64 m by placing a rock-filled reinforced concrete trough across its crest.

In 2003, thirty four years after its initial completion, the maximum creep settlement at the original crest level was 184 mm, which is 0.3% of the initial height and 0.29% of the revised height.

3.4.5   Structural defects of particular significance in schistose rocks

Three defect types are discussed below.

3.4.5.1   Minor faults developed parallel and at acute angles to the foliation

Schists commonly contain minor faults (narrow, sheared zones or crushed seams, or both) parallel to the foliation. Deere (1973) refers to these features as foliation shears. In folded schists, the foliation shears have probably been formed by inter-layer slip, as shown on Figure 3.16.

Also present in many folded schists are similar "shears" cutting across the foliation at acute angles, generally less than 20° (Figure 3.16). In some cases these are thrust faults. Residual shear strengths of both foliation shears and cross-cutting faults have been found in laboratory tests (i.e. excluding the effects of large scale roughness), to lie in the range 7° to 15°. Such defects commonly form the initiating failure surfaces of landslides in schistose rocks (Figure 3.17) and may provide potential sliding surfaces into spillway or foundation excavations, or within the foundations of an embankment dam.

As these features are often only 50 mm or so in thickness, they can escape detection during site investigations unless the investigator sets out to look for them, using appropriate techniques, e.g. well cleaned up trenches and high quality core drilling. The sheared zones can be particularly difficult to detect. In these zones the rock is more intensely foliated than elsewhere and is usually rich in chlorite and/or sericite. The sheared material is therefore appreciably weaker than the normal schist and is readily recognizable when the rock is fresh. However, in distinctly or extremely weathered exposures in which both sheared and unsheared materials are greatly weakened by weathering, it can be quite difficult to recognize the sheared zones, because the strength contrast is much reduced, and the shear-induced cleavage or foliation is similar in appearance to, and may be parallel to, the foliation in the normal schist. Stapledon (1967) describes how initial, poor quality core drilling at Kangaroo Creek (South Australia) failed to indicate the presence of foliation shears and associated infill clay seams. Discovery of these in a later exploration programme led to abandonment of a thin concrete arch design in favour of a decked rockfill dam. Paterson et al. (1983) describe how surface mapping and diamond drilling were not adequate to define the full extent and frequency of foliation shears at the site for Clyde

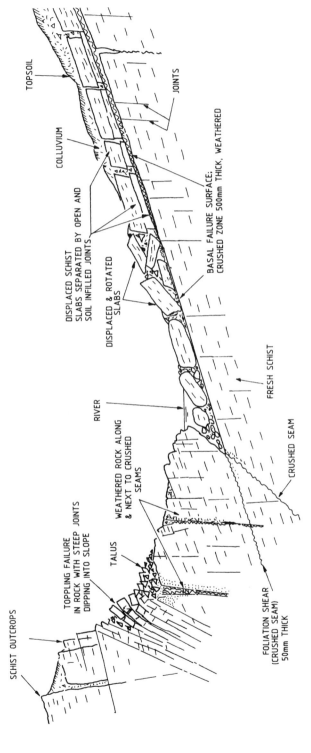

Figure 3.17.   Typical valley profile in gently dipping schist affected by past landsliding along foliation shear.

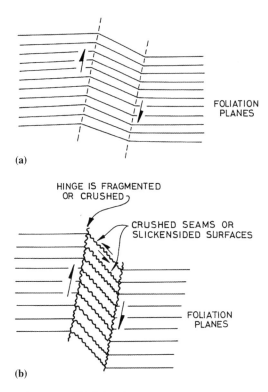

FOLIATION
PLANES

(a)

HINGE IS FRAGMENTED
OR CRUSHED

CRUSHED SEAMS OR
SLICKENSIDED SURFACES

FOLIATION
PLANES

(b)

Figure 3.18.   Formation and nature of kink bands.

Dam in New Zealand. The location and treatment of some of these shears during construction of the dam are discussed in Section 17.8.3.

3.4.5.2   *Kink bands*

Schistose rocks often also contain "kink bands" within which the foliation layers have been displaced to form features similar to monoclinal folds but with sharp, angular hinges (Figure 3.18a).

   In some cases continuous near-planar joints have developed along the hinges and the foliation layers within the band have parted to form open or infilled joints. In other cases crushed seams have developed along the hinges and the foliation layers within the band are either slickensided or partly crushed (Figure 3.18b). Such kink bands can be considered a special class of fault.

3.4.5.3   *Mica-rich layers*

Some schists contain layers or zones which consist almost entirely of micaceous minerals, e.g. biotite, muscovite, sericite or chlorite. Such layers or zones are usually much weaker than the normal schist. It is good practice to consider them as individual defects of low shear strength.

3.4.6   *Stability of slopes formed by schistose rocks*

Landsliding is relatively common on slopes underlain by schistose rocks. Well-known examples are the Madison Canyon Rockslide in Montana, U.S.A. (Hadley, 1978), the

Downie Slide in British Columbia (Piteau et al., 1978) and the Tablachaca Slide in Peru (Arnao et al., 1984; Deere and Perez, 1985).

The landsliding usually occurs by slope failure along weathered foliation surfaces or foliation shears as shown on Figure 3.17. Bell (1976, 1982) describes how the development of a river valley in New Zealand has involved landsliding along dipping foliation surfaces.

In steeply dipping schistose rocks failure by toppling is also relatively common, the toppled slabs or columns being separated by joints or shears along the foliation direction. Riemer et al. (1988) describe a complex toppling failure at San Pablo, Peru.

Where the schistose rocks contain abundant tectonically-formed defects in other orientations, many other failure models have been recorded. However in most of these it is likely that the low shear strength of the schistose rocks along their foliation surfaces has contributed to the development of slope failure.

In long, high slopes in mountainous areas some failures of schistose rocks appear to have occurred by buckling of, and eventually shearing through, the foliation. The buckling is facilitated by the low shear strength of the foliation surfaces which allows multiple shear displacements to occur along them. Examples of this type of slope failure are discussed in Beetham et al. (1991), Riemer et al. (1988), Radbruch-Hall et al. (1976), Nemcok (1972) and Zischinski (1966, 1969).

Examples of landsliding in schists forming slopes around the reservoir of Clyde Dam in New Zealand are discussed in Gillon and Hancox (1992), Riddolls et al. (1992) and Stapledon (1995).

3.4.7   *Schistose rocks – check list of questions*

- Degree of anisotropy, and its effect on the project?
- Low durability in exposed faces?
- Particle shapes and strengths inadequate for filter, concrete or pavement materials?
- Suitability for use as rockfill?
- Foliation shears?
- Kink bands?
- Mica-rich layers?
- Unstable slopes?

3.5   MUDROCKS

Included under this heading are all sedimentary rocks formed by the consolidation and cementation of sediments which are predominantly clays or silts or clay-silt admixtures. The common rock types are:

- Claystone (predominantly clay sizes);
- Siltstone (predominantly silt sizes);
- Mudstone (clay-silt admixtures);
- Shale (any of the above, but fissile due to well-developed cleavage parallel to the bedding).

The sediments may have been deposited in either marine or fresh water conditions and usually have been derived from erosion of older rocks. In some cases they contain particles of volcanic origin. Possible cementing agents include calcite, silica, iron oxides and evaporite minerals such as gypsum, anhydrite and halite (common salt). Mudrocks occurring in or associated with coal-bearing sequences often contain abundant carbonaceous material and sulphide minerals, e.g. iron pyrite.

3.5.1   *Engineering properties of mudrocks*

Most mudrocks when fresh lie in the weak to very weak range as defined on Table 2.4. The very weak claystones grade into hard, overconsolidated clays. The strongest mudrocks lie in the medium strong and strong ranges and in most cases these owe their greater strength to cementation by calcite or silica.

Because of their relatively high clay contents the porosity and water absorption properties of mudrocks are much higher than those of most other rocks. As a result of this and the expansive nature of clays, all mudrocks swell and develop fine cracks on prolonged exposure to wetting and drying. The strongest siltstones, which contain appreciable amounts of calcite or silica as cement, can be exposed for up to a year before cracks are evident. At the other end of the scale, the weakest claystones and shales develop fine cracks as soon as they dry out (often only hours or days) and disintegrate with further cycles of wetting and drying. Some of these materials also swell noticeably on removal of overburden.

The mechanisms involved in deterioration of mudrocks on exposure have been described already in Sections 2.5.3 and 2.9.1.

Because of their instability when exposed, special care needs to be taken during preparation of foundations on mudrocks. Treatments range from shotcreting or slush concreting immediately after exposure and cleanup, to an initial cleanup followed by a final cleanup immediately before placement of concrete or fill.

Taylor and Spears (1981), Cripps and Taylor (1981) and Spink and Norbury (1993) provide useful information about the engineering properties of mudrocks occurring in the United Kingdom.

Mudrocks containing iron pyrite or other sulphide minerals can cause severe problems for dam projects due to rapid weathering of the sulphides to form sulphuric acid and metallic hydroxides and sulphates. The rapid weathering processes and some of their effects on dam projects are described in Section 2.9.4. Such effects may include:

– Damage to the fabric of concrete due to sulphate attack;
– Heave of foundations or of excavation sides causing damage to concrete slabs or walls (Penner et al., 1973; Hawkins and Pinches, 1987);
– Blockage and/or cementation of filter zones or drains;
– Seepage waters which are acidic and rich in iron or other heavy metals;
– Possible lowering of shear strength, of embankments formed by pyritic mudrocks.

3.5.2   *Bedding-surface faults in mudrocks*

Valley bulging (see Chapter 2, Section 2.5.4 and Figure 2.7) is a common feature in mudrock sequences, where the beds are near-horizontal. In such situations and in any other situation in which mudrock sequences have been disturbed by folding, tilting, or stress relief movements, thin seams of crushed rock develop within the mudrocks, usually at their boundaries with interbedded stiffer rocks (e.g. sandstones or limestones). These seams are developed due to interbed slip during the movements, in the same way as foliation shears develop in schistose rocks (Figures 2.8, 3.16 and 3.19). In mudrocks they are known as bedding-surface faults, bedding-surface shears or bedding-plane shears. They usually consist mainly of clay, are almost planar and have slickensided surfaces both within them and at their boundaries. Residual effective shear strengths of laboratory sized samples (i.e. excluding larger scale roughness effects) are commonly in the range 7° to 12°, with zero cohesion.

Although usually extending over wide areas, bedding surface faults may be only a few millimetres thick. Such defects are difficult to recover and recognize in diamond drill cores. This is particularly so when the defect is normal to the axis of the borehole, as even minor

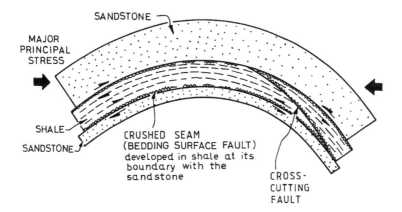

Figure 3.19.   Usual way in which bedding surface faults are formed in mudrocks.

core rotation causes remoulding of the defect, which is then difficult to distinguish from remoulded mudrock at a "drilling break" in previously intact core (see Section 3.5.6).

The influence of bedding surface faults on the design of several embankment dams is described by Casinader (1982). Their effects on the design and construction of Sugarloaf and Thomson Dams are discussed in Chapter 2, Section 2.10.

Maddox et al. (1967) describe their effects at Meadowbanks Dam, a 43 m high massive buttress dam in Tasmania, founded on interbedded sandstones and mudstones. The dam was planned originally to be a concrete gravity structure but, during excavation of the foundation, continuous bedding-surface seams of clayey silt were found within the mudstones. Change to the buttress design and installation of prestressed cables were needed to achieve the required factor of safety against sliding.

3.5.3   *Slickensided joints or fissures*

Many mudrocks, particularly claystones, contain zones in which the rock contains an irregular network of curved, intersecting, slickensided joints. These joints are believed to have originated when the material was still a clay soil and to have developed by any of the following types of process:

– syneresis (Skempton and Northey, 1952; White, 1961);
– shrink and swell movements (Corte and Higashi, 1964);
– differential shear movements during consolidation;
– large lateral stresses (Aitchison, 1953; Terzaghi, 1961).

Due to their lack of continuity and their curved, irregular nature, these joints usually do not form continuous zones of very low shear strength. However, the shear strength of the jointed mass is appreciably lower than that of the intact mudrock. In adopting strength parameters for use in design, the strength of both the intact substance and the joints, and the spacing, orientation and continuity of the joints, need to be taken into account.

3.5.4   *Weathered products and profiles in mudrocks*

Weathering of mudrocks usually involves mechanical disintegration as described in Section 3.5.1 and the removal of cements such as calcite and silica. In the extremely weathered condition all mudrocks are clays or silts. Intermediate weathered conditions (e.g. slightly and distinctly) are often difficult to define in the weaker mudrocks, which when fresh are only a little stronger and more durable than hard clays. The Geological Society (1995) provides a suggested approach for site-specific mass-type classification of weathered profiles in these weaker mudrocks.

Weathered profiles in mudrocks are more uniform, gradational and generally not as deep as those in other rocks. Deere and Patton (1971) describe these types of weathered profiles in shale and point out that the lack of distinct boundaries within such profiles has led to contractual disputes over the depth to acceptable foundations.

3.5.5   Stability of slopes underlain by mudrocks

Slopes underlain by mudrocks commonly show evidence of past instability, even when the slope angles are small (e.g. 10° to 15°). This is not really surprising when we consider that all of the characteristics described above in Sections 3.5.1 to 3.5.4 tend to lower the strength of rock masses containing mudrocks.

Relatively shallow landsliding is common in the residual clay soils developed on the weaker mudrocks. Taylor and Cripps (1987) provide a comprehensive review of slope development and stability in weathered mudrocks and overconsolidated clays, mainly relating to examples in the United Kingdom.

Deere and Patton (1971) give a useful review of experience with unstable slopes on shales and on shales with interbedded sandstones. They point out that, in common with most other rocks, weathering of shales usually produces a low-permeability zone near the surface, underlain by jointed, less weathered shale, which is more permeable. Instability can arise when groundwater transmitted through this lower zone or along sandstone beds causes excessive pore pressures in the near-surface, more weathered shale. (See Section 3.6.4 and Figures 3.23 and 3.24).

The most common situations in which larger scale landsliding occurs, or is likely to occur, are where the bedding "daylights" on a valley slope. This occurred at both Sugarloaf and Thomson dams in Victoria, Australia, as described in Chapter 2, Section 2.10.

3.5.6   Development of unusually high pore pressures

Stroman et al. (1984) and Beene (1967) describe a major slide which occurred during the construction of a 30 m high section of the 5.5 km long Waco Dam in Texas. The dam was located on near-horizontally bedded shales cut through by three steeply dipping normal faults, which displaced the shale units by up to 30 m and caused them to be locally folded (Figure 3.20a).

The originally designed embankment is shown in outline on Figure 3.20(b). The design was based on the assumption that the weakest foundation material was a localized 12 m thick layer with unconsolidated undrained strength of $\phi = 5°$, c = 144 kPa. No potential for pore pressure development was expected (because of the high degree of overconsolidation of the shales) and consequently no piezometers were installed in the original construction.

The failure occurred in a 290 m long section of embankment constructed between two of the normal faults. As can be seen on Figure 3.20c, the failure surface was mainly horizontal within the Pepper Shale about 15 m below the main foundation level. This failure surface broke out to the ground surface at an average distance of 235 m downstream from the dam axis.

Figure 3.21 shows the pore pressure distributions before the sliding and at the end of the reconstruction. Stroman et al. (1984) state that the unusually high pore pressures at the Pepper Shale/del Rio Shale contact were the cause of the slide. However they also report that direct shear tests on precut samples of Pepper Shale gave effective friction angles between 7° and 9°, with zero cohesion. They state that "This is the laboratory test

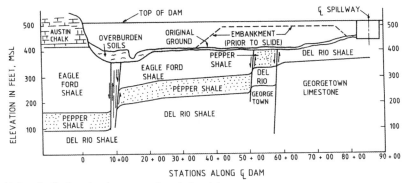

(a) Longitudinal section (exaggerated vertical scale)

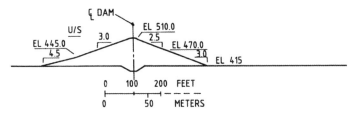

(b) Cross section original design

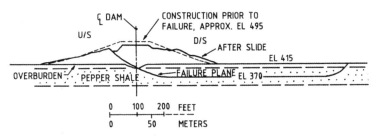

(c) Cross section showing failure surface

Figure 3.20.   Longitudinal and cross sections, Waco Dam (Stroman et al., 1984).

condition that can be related to a material that has been broken prior to construction and to the condition of the Pepper Shale after the slide".

Londe (1982) considers that the sliding at Waco Dam occurred by progressive shear failure (e.g. as described by Bjerrum, 1967, Terzaghi and Peck, 1967, Skempton and Hutchinson, 1969 and subsequently by Skempton and Coates, 1985, to explain the failure of Carsington Dam).

It seems equally possible to the present authors that very thin but continuous bedding surface faults with 7° to 9° residual strength may have existed in the Pepper Shale before the slide and contributed to the movement. Such bedding surface features require very small movements for their development and these could easily have occurred during the formation of the normal faults which bound the slide area. As described in Section 3.5.2, such minor bedding faults can be difficult to recover and recognize in drill cores.

3.5.7   *Suitability of mudrocks for use as construction materials*

Most mudrocks are not suitable for the production of materials for use in concrete, filters or pavements due to their generally low strengths and their slaking properties. However,

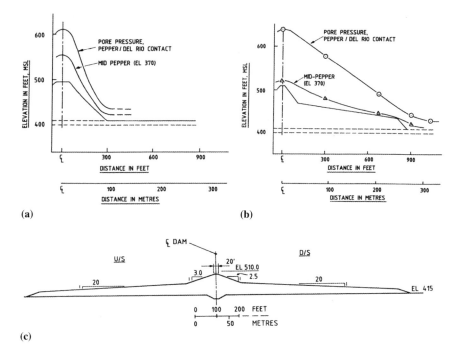

Figure 3.21.   Waco Dam, pore pressure plots and cross section through embankment as rebuilt. (a) pore pressures prior to slide; (b) pore pressures after reconstruction; (c) embankment as reconstructed (Stroman et al., 1984).

near Sydney, a trial road pavement built using ripped and trafficked shale has performed satisfactorily for 20 years (Won, 1985). The shale is weak to medium strong. Re-sampling and testing of the material showed that there had been little breakdown in service.

Siltstones in the medium strong to very strong range have been used successfully as rockfill for some years. Near the site of Sugarloaf Dam excavation of test pits into 100 year old tunnel spoil dumps comprising such materials, showed that slaking had not occurred in the rock below the near-surface two metres, i.e. in a near constant humidity environment.

To the knowledge of the authors no mudrocks have been used successfully as rip-rap. We would not expect it.

Random fills, earthfills and cores for embankment dams have been built successfully using mudrocks in various conditions ranging from fresh to extremely weathered (Vaughan 1994).

3.5.8   Mudrocks – check list of questions

– Slaking or disintegration on exposure?
– Swelling on exposure?
– Valley bulging?
– Soluble minerals in beds or veins?
– Presence of sulphide minerals?
– Slickensided fissures?
– Progressive shear failure?
– Bedding surface faults or shears?

Table 3.1.   Common characteristics of sandstones, arkoses and greywackes.

| Rock name | Particle shapes, grading | Minerals | |
		Most grains	Common matrix/cements
Sandstone	Usually rounded, one-size grains and less than 15% matrix or cement	Quartz, fragments of older rocks	Silica, clay, iron oxides, calcite, gypsum
Arkose	Sub-angular, often well graded, little matrix	Quartz plus at least 25% feldspar; some mica	Clay, iron oxides, silica
Greywacke	Angular, well graded down to clay matrix which is usually >15% of volume	Felspar, quartz hornblende, micas, rock fragments, iron oxides	Clay, and same as grains

- Unstable slopes (shallow, in weathered materials)?
- Unstable slopes (deep-seated, if bedding in folded rocks daylights)?
- Possibility of high pore pressures, in layered sequences?
- Suitability for rockfill, random fill, earthfill and haul roads?

3.6   SANDSTONES AND RELATED SEDIMENTARY ROCKS

The following are the main rock types considered under this heading. They will be referred to as the sandstone group.

- Sandstones;
- Arkoses;
- Greywackes;
- Siltstones;
- Conglomerates.

Table 3.1 sets out some common characteristics of sandstones, arkoses and greywackes. Siltstones and conglomerates have a similar range in composition to those of the other rocks. All except the siltstones occur usually in thick beds. The siltstones may be thickly or thinly bedded.

Also included under this heading are the lightly metamorphosed equivalents of the above, e.g. quartzites, metasiltstones and metaconglomerates. If the original depositional environment of the particular rocks at a site is known, then many generalisations can be made about their mineral contents, fabrics, bed-thicknesses and sedimentary structures and about rock types likely to be associated with them. Detailed discussion of these sedimento-logical aspects is beyond the scope of this book and only a few features of particular impor-tance in dam engineering will be described. For more details readers are referred to Pettijohn (1957), Pettijohn et al. (1972), Selley (1982) and Walker (1984).

3.6.1   Properties of the rock substances

When fresh the sedimentary rocks range from extremely weak, non-durable to very strong and durable. The strengths and durabilities depend upon the strengths and durabilities of the grains and of the cements or matrices and these can vary widely depending upon the environment of deposition and subsequent histories of the rocks.

Quartz-sandstones (and conglomerates in which most grains are quartz) are often stronger and more durable than arkoses and their conglomerate equivalents, because of the superior strength and durability of quartz. However sandstones often have significant porosity (5% to 20%) and may also be slightly permeable.

Greywackes tend to be stronger than sandstones due to the angularity and grading of their particles.

Silica cement usually occurs in strong, durable rocks and at the other extreme rocks cemented by clay or gypsum are usually weak and non-durable.

If gypsum or anhydrite is proven or suspected as a cement in a sandstone forming all or part of the foundation of a dam, its significance needs to be assessed carefully and special testing may be required, as discussed in Section 3.8.

The metamorphic rocks when fresh are usually stronger and more dense and durable than the equivalent sedimentary rocks.

3.6.2   *Suitability for use as construction materials*

Rocks in the sandstone group which lie in the strong to extremely strong range have been used successfully as rockfill and rip-rap in many dams. They are widely used also as aggregates in concrete.

However, in a few cases concrete containing quartzite or strong sandstone as aggregate has suffered expansion and cracking due to alkali-silica reaction. The authors recommend that any silica rich rocks intended for use as concrete aggregate be tested for reactivity (see Section 3.2.6).

The quartzites are often extremely strong and this together with the high content of quartz (Moh hardness = 7) makes them highly abrasive. This can result in high quarrying and handling costs. Also if the quarry-run rock is not well graded it can be difficult to compact, as little breakdown occurs under the roller.

The weaker rocks tend to be more porous and usually lose significant strength on saturation. Mackenzie and McDonald (1981, 1985) describe the use of sandstones and siltstones, mainly medium strong when dry, as rockfill in the 80 m high Mangrove Creek Dam in New South Wales. Both rock types lost about 50% of their strength on saturation.

The rock was compacted with up to 5% water by volume. Higher water quantities caused the material to become unworkable. The fills produced were of high density and moduli but of generally low permeability. The latter was allowed for by the inclusion of drainage zones of basalt and high quality siltstone.

At Sugarloaf (Winneke) Dam thinly interbedded siltstone and sandstone were used for rockfill and random fill in the 85 m high concrete faced embankment (Melbourne and Metropolitan Board of Works, 1981; Regan, 1980).

The rockfill zone material was fresh or slightly weathered and strong to medium strong. Compacted with up to 15% water it produced a dense, free-draining fill.

The rock used in the random fill zone was slightly to highly (or distinctly) weathered and medium strong to weak. Compacted at about 10% moisture content it produced a dense fill which was not free-draining and contained up to 20% of silt and clay fines which were dispersive. It was therefore underlain everywhere by a blanket of rockfill and its outer surfaces were protected by a thin layer of the rockfill.

3.6.3   *Weathering products*

The main effect of chemical weathering on rocks of this group is weakening due to removal or decomposition of the cement or matrix. As a result the rocks with silica or iron oxide cements are the most resistant to weathering. The sandstones and conglomerates become progressively weaker until at the extremely weathered stage they are usually sands or gravels. Quartzites and quartz rich sandstones produce relatively clean quartz sands, while arkoses and greywackes usually produce clayey or silty sands. Siltstones usually produce clayey silts or clays.

Some sandstones (usually weak, porous types) are locally strengthened in the weathered zone, by the deposition of limonite in their pores.

It can be difficult in some sandstones (e.g. those with calcite or limonite cements) to distinguish effects of weathering from effects of processes involved in the formation of the unweathered rocks. This can make it impossible to classify the rocks in the usual weathered condition terms, e.g. those in Table 2.3.

3.6.4   Weathered profiles, and stability of slopes

Weathered profiles in the weaker, more porous rocks commonly show gradational boundaries between rock in various weathered conditions. This is also the case in the stronger, more durable rocks when they are closely jointed.

Where large contrasts occur between the resistance to weathering of interbedded rocks, sharp but irregular sawtooth shaped boundaries can occur as shown in Figure 2.23.

Figure 3.22 shows the type of mechanically weathered profile often developed close to cliffs formed by near-horizontal sequences of thick sandstone beds underlain by or interbedded with siltstones or shales. Crushed seams (bedding surface faults) occur along the bed boundaries and steeply dipping joints have opened up in the sandstones. If these effects only occur close to the side of the valley it is likely that they result mainly from interbed movements due to stress relief – the shales/siltstones expanding further out of the

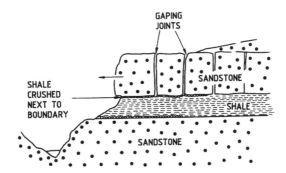

Figure 3.22.   Commonly observed features close to cliffs formed by horizontally bedded sandstones with shale or siltstone interbeds. Based partly on Deere and Patton (1971).

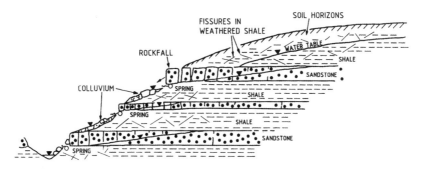

Figure 3.23.   Collapse of outcropping sandstone beds due to removal of support of underlying shales (Based on Figure 13b of Deere and Patton, 1971).

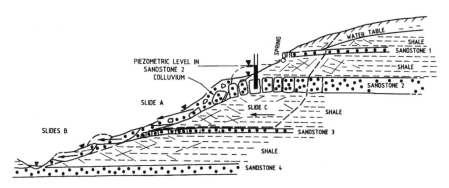

Figure 3.24.   Landslides which can occur in a slope where sandstone beds are covered with colluvium (Based on Figure 13c of Deere and Patton, 1971).

slope than the sandstones. Joint-water pressures during extreme rainfall periods and earthquake forces may also contribute to the slight movements of the sandstone blocks.

Opening up of the joints in the sandstones causes the permeability of layers near the surface to be greatly increased.

Where whole hillsides are underlain entirely by rocks of the sandstone group, these rocks commonly form steep slopes or cliffs. Slope failures are rare and usually occur by rockfalls or toppling from the cliff portions.

More commonly, sandstones occur together with shales or siltstones as shown on Figures 3.23 and 3.24. In these situations weathering extends deeper into the mass and landsliding is more prevalent.

Figure 3.23, taken from Figure 13b of Deere and Patton (1971), shows a slope underlain by horizontally interbedded sandstones and shales. In high rainfall areas or during the wet season in other areas, the water table on Figure 3.23 will be high and of the form shown, due to the relatively free-draining nature of the sandstone beds. Springs occur at the bases of sandstone outcrops and the shale below and between the sandstone beds is either continually wet or alternately wet and dry. Under these conditions chemical weathering of both rock types occurs but is usually more pronounced in the shale. This rock is at least partly weathered to clay and contains clay-coated joints or fissures, often slickensided. The weathered shale either slumps or its bearing capacity is exceeded, allowing large movements and eventual collapse of the outermost sandstone block. Continuation of these processes leads to development of layers of scree and colluvium on the slope, and to "cambering" of the near-surface part of the sandstone bed as shown on Figure 3.24.

Deere and Patton also describe landsliding observed commonly on interbedded sandstone/shale slopes where sandstone beds have become covered by colluvium, as shown on Figure 3.24. The colluvium restricts drainage from the sandstone (Bed 2) which may become a semi-confined aquifer with piezometric surface as indicated. Pore pressures so developed cause sliding to occur, usually along the colluvium-weathered shale contact (Slides A and B on Figure 3.24).

Slide B is a double slide in which the toe of the first slide (No.1) has overloaded the metastable top of the slope below, causing Slide No.2 (Deere and Patton 1971).

Deere and Patton (1971) point out that deepseated slides, such as Slide C on Figure 3.24, can occur when a combination of unfavourable geological conditions exists. These could include bedding surface crushed seams along the shale-sandstone boundaries, high water pressures in sandstones Nos 2 and 3 and high water levels in the affected mass. Deere and Patton point out that after the first small movements, the permeability of the deeper part of the mass will be increased and softening and weathering of the near-surface shale will be accelerated.

In folded sequences of sandstones with interbeds of siltstones or shales, landslides are relatively common where dipping beds daylight on steep slopes. Slides are also common on dipslopes. Examples of these types of sliding involving sandstones are given in Chapter 2, Section 2.10.3.4 and 2.10.3.5.

3.6.5 *Sandstones and similar rocks – list of questions*

- Relatively high porosity, permeable?
- Gypsum or anhydrite present as cement?
- Quartzites: High quarrying and handling costs, difficult to compact?
- Rocks of medium or lower strength may not produce free-draining rockfill?
- Interbeds of shale or claystone?
- Bedding-surface faults at bed boundaries?
- Horizontal beds: Open joints and bedding surface crushed seams near surface due to stress relief?
- Horizontal beds with shale interbeds: Cambering and collapse due to removal of support by weathering shale?
- Landsliding in colluvium developed on weathering sandstone/shale slopes?

3.7   CARBONATE ROCKS

Carbonate rocks are defined here as those which contain significant amounts of the soluble minerals calcite, aragonite or dolomite in their substance fabrics. The most common are sedimentary carbonate rocks, marble (metamorphosed carbonate rock), and calc-silicate rocks formed by the metamorphism of impure carbonate rocks.

It is necessary to consider carbonate rocks in two categories, based on the dominant carbonate minerals present in each category.

Geologically young carbonate rocks (Category Y). Category Y carbonate rocks are usually of Tertiary, or younger, age. Most comprise loosely packed, weakly cemented shell fragments and are porous and weak to very weak. Rarely they can be well cemented and dense. The carbonate minerals are mostly aragonite and high-magnesian calcite (Friedman 1964, 1975; Molenaar and Venmans, 1993; Prothero and Schwab, 1996). These two minerals formed all original marine carbonate sediments and are forming in present day deposits. There is field and laboratory evidence (see Sections 3.7.1.3 and 3.7.7.2) that high-magnesian calcite is more susceptible to dissolution and cementation than aragonite and calcite. With time, exposure to fresh water, compaction and recrystallisation both aragonite and high-magnesian calcite eventually revert to calcite, and the rock becomes *Category O.*

Geologically old carbonate rocks (Category O). Category O carbonate rocks are generally of Mesozoic or older age and are usually dense, non porous and range from strong to extremely strong. They are formed by the minerals calcite or dolomite. They include marble, which comprises coarsely crystalline calcite and is usually dense, non porous and strong to very strong. Also included are calc-silicate rocks, which contain carbonate minerals together with silicate minerals often including olivine, diopside and garnet.

The carbonate rocks at the sites of most large dams are *Category O.* The exceptions mentioned in this book are Kopili Dam, Perdikas, Montejagne and May Dams, each of which failed to store water (Table 3.4).

Table 3.2 proposed by Dearman (1981) is a practical engineering classification for sedimentary carbonate rocks. Dearman (1981), Cruden and Hu (1988) and Bell (1981, 1983b and 1992) present data on physical properties of some carbonate rocks. Fookes and Higginbottom (1975) provide a table which classifies and names various types of

Table 3.2.   Engineering classification of sedimentary carbonate rocks (Dearman, 1981).

Percentage Carbonate	0		10	50	90	100
					LIMESTONE	
		CONGLOMERATE	Calcareous Conglomerate	Gravelly Limestone	Calcirudite	
(grain size 2)		SANDSTONE	Calcareous Sandstone	Sandy Limestone	Calcarenite	LIMESTONE
Predominant grain size (mm) 0.06	MUDSTONE	SILTSTONE	Calcareous Siltstone	Silty Limestone	Calcisiltite	
0.002		CLAYSTONE	Calcareous Claystone	Clayey Limestone	Calcilutite	
			MARLSTONE			

Note : non-carbonate constituents are rock fragments or
quartz, micas, clay minerals.
: predominant grain size implies over 50%.

PERCENTAGE CALCITE	0	10	50	90	100
	Dolomite	Calcitic Dolomite	Dolomitic Limestone	Limestone	
PERCENTAGE DOLOMITE	100	90	50	10	0

original pure carbonate sediments (detrital, chemical and biochemical) and their progressively strengthened counterparts. This table has useful descriptive terms and its accompanying discussion indicates some of the complexities involved in the strengthening of carbonates.

Calcrete (caliche) is a *Category Y* carbonate rock of highly variable strength found in many arid areas. The most common variety, pedogenic calcrete, occurs within a metre or so of the ground surface (Netterberg, 1969, 1971; James and Coquette, 1984; Meyer, 1997).

3.7.1   *Effects of solution*

Solution is one of the processes of chemical weathering which affects all rock types to some extent. It is more severe and causes cavities in carbonate rocks because calcite and dolomite are relatively soluble in acidic waters e.g. carbonic (dissolved carbon dioxide), organic (from vegetation) or sulphuric (from volcanic activity or oxidation of sulphides). They are more soluble in saline water than in pure water and their solubility increases with desceasing water temperatures. Also, their rates of solution are high (James and Kirkpatrick, 1980; James, 1981).

The term karst is used to describe terrain underlain by cavernous rocks (usually carbonates or evaporites) and also to describe the cavernous rocks themselves. The factors involved in the formation of karst are the same as those which contribute to the development of weathered profiles described in Chapter 2, Section 2.6.4. Detailed discussion of karst development, structure and hydrogeology can be found in Sweeting (1972, 1981), Milanovic (1981), Bonacci (1987), Ford and Williams (1989), Beck (1993, 1995) and De Bruyn and Bell (2001).

The effects of solution as seen in the most common types of carbonate rock masses are discussed below.

Figure 3.25.   Karst landscape in New Zealand. The outcrops of limestone are up to 1.5 m high and separated by soil-covered holes and slots up to 1 m across.

3.7.1.1   *Rock masses composed of dense, fine grained rock substances comprising more than 90% of carbonate (usually Category O)*

When fresh and intact these rocks usually have very low porosities and their substance permeabilities are effectively zero. Groundwater flow is confined to joints or other defects and these are widened by solution to form slots, shafts, tunnels and cavities of all shapes and sizes. Important features are:

– The upper surface of rock, both in outcrops and below ground, usually consists of pinnacles of fresh rock separated by deep slots or cavities, which may be empty or soil-filled (Figures 3.25 and 3.26);
– No weathered rock substance is present, either at the ground surface or at depth. This is because dissolution of the relatively pure limestone substance leaves less than 10% of insolubles, which form residual soils at or near the ground surface and infill some cavities below (Figure 3.27). The residual soils are commonly fissured and rich in clay and iron oxides;
– Dam construction experience has shown that cavities capable of accepting or transmitting large inflows of water can be met at any depth down to 300 m below ground surface (see Section 3.7.2.1);
– Sinkholes, usually shaft-like cavities, are often exposed at the ground surface and the formation of new sinkholes can be initiated by natural or man-made activities.

Figure 3.27 represents the uppermost 10 m or so of a karst profile beneath a small area, showing relatively small cavities and solution-widened defects, which would be connected to larger cavities at greater depths. It also illustrates some of the mechanisms by which sinkholes have been formed.

Feature (a) is a sinkhole formed directly by the collapse of the roof of a cavity.

Feature (b) can also be referred to as a sinkhole. It is a gentle, soil-covered depression, roughly circular in plan view. Exploration shows it to be underlain by residual

Figure 3.26.   Close up view of limestone pinnacles.

soil with a concave upwards layered structure. It can be inferred that this sinkhole has developed because the residual soil has migrated slowly into the cavities in the limestone below.

Feature (c) is a large flat area showing residual soil at the surface. It is underlain at depth by cavities developed along joints and a fault zone. Such flat areas may show no obvious evidence of settlement or disturbance, but there is a significant risk of a deep, steep-sided sinkhole forming (with little or no warning) at the ground surface. This could be initiated by migration of soil into the cavities below, causing a large void to develop near the base of the residual soil. The sinkhole could then form by collapse of the overlying soil into the void. A similar dangerous sinkhole could also develop within the gently sloping sinkhole (b).

The mechanisms of sinkhole formation and their significance in dam engineering are discussed further in Section 3.7.3.

3.7.1.2   Rock masses composed of dense fine grained rock substance containing 10% to 90% of carbonate (usually Category O)

In these rocks the pattern of solution cavities is similar to that described in Section 3.7.1.1 and seen in Figure 3.27, but usually the rock substance next to some of the cavities is weathered. Experience and logic show that in general the lower the percentage of carbonate in the fresh rock the higher the proportion of weathered rock formed compared to cavities and the higher the ratio of infilled cavities to open cavities. The weathered rock is usually much weaker and less dense than the fresh rock. However, the proportion and properties of weathered rock formed depend also on the percentage of insoluble particles present in the fresh rock and the degree to which these particles are bound by non-soluble

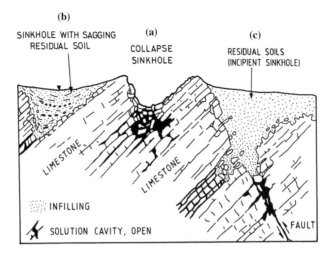

Figure 3.27.   A typical weathering profile in dense, relatively pure carbonate rock (Based on Deere and Patton, 1971).

cements. These effects are illustrated by strong dolomitic siltstones which occur widely in South Australia. Many weathered exposures of these rocks appear to have lost all of their dolomitic cement, but are still weak or medium strong rock, due to some non-carbonate cement. A few beds within such a sequence at and near Little Para Dam appear to lack this non-soluble cement and these beds form very low density, very weak rock, grading to clayey silt and small cavities (see Figure 3.33 in Fell et al., 1992).

3.7.1.3   *Rock masses composed of porous, low density carbonate rock substance (usually Category Y)*

Most of these rocks are weak to very weak calcarenites. In many cases the carbonate grains are recognizable shell fragments. The rocks are usually in horizontal or gently dipping beds and are relatively free from joints or other defects of tectonic origin.

These rocks have sufficient substance permeability to allow surface waters to enter and migrate downwards, causing both solution and redeposition effects. The 8 m high quarry face in Figure 3.28 shows a typical profile. The very weak calcarenite here is a windblown deposit of Pleistocene Age. It is off-white in colour and overlain by yellow-brown quartz which appears dark grey in the photograph. The profile in the calcarenite is characterized by the following:

- Vertical pipes or tubes; each is surrounded by an annular zone of calcarenite which is much more dense and strong than elsewhere in the mass. The tubes have been infilled with the quartz sand, so show up well where exposed on the face;
- Vertical pinnacles of calcarenite which are much more dense and strong than elsewhere in the mass.

The tubes have been formed by solution of the calcarenite. Strengthening of their rocky surrounds and of the pinnacles appears to have been caused by redeposition of calcite derived from that solution process and possibly also from solution of calcarenite which previously existed above the present ground surface.

The porous, low-density carbonates rarely contain large cavities like those in dense, jointed carbonates. However Twidale and Bourne (2000) describe many shallow sink-holes up to 10 m across, and several much larger, on wind-deposited calcarenite in South

Australia. They present evidence which suggests that these have developed due to solution along localized calcarenite zones rendered more permeable by small movements along faults in the bedrock at a depth of 100 m or more.

Chalk is very low strength, low-density calcilutite (Table 3.2, and Table 13.8 in Bell (1983b). It has extremely low substance permeability, but can be permeable where joints are present. Bell (1983b) and Goodman (1993) point out that large cavities are rare in chalk, but solution pipes and sinkholes can occur, near its contact with colluvium or other rocks.

3.7.2   *Watertightness of dam foundations*

Many dams have been built successfully on sites underlain by *Category O* carbonate rocks, despite the fact that solution cavities have been present at most sites and beneath at least part of their storage areas.

Selection of dam foundation levels and treatments can be difficult, due to the highly irregular nature of the uppermost surface of the fresh rock, as shown by Figures 3.25 to 3.28. This applies particularly to embankment dams, because of their large foundation areas and the need to provide stable, non-erodible surfaces for placement of embankment materials.

Treatments to fill cavities and prevent excessive leakage have included cement grouting, concrete curtain walls and selective mining and backfilling with concrete. The presence of clay infilling in cavities presents a problem when cement grouting is proposed. Clay usually prevents grout penetration and is not readily removed by flushing between boreholes. However, if left in place, it may be flushed out later when the dam is filled. High pressure cement grouting designed to cause hydraulic fracturing of clayey infills has been shown by Zhang and Huo (1982) to give significant improvement in their resistance to leakage and piping.

At the 35 m high Bjelke-Peterson earth and rockfill dam in Queensland about 250 m of the left bank was formed by a landslide deposit underlain by limestone. In this section the cutoff trench was excavated down to the water table, 15–20 m below the original surface and 5–15 m below the irregular upper surface of the limestone (See Chapter 8, Figure 8.8). At the trench base the limestone was mainly fresh, dense and very strong, but contained open and infilled solution cavities within and next to sheared and crushed zones (Eadie,

Figure 3.28.   Weathering profile on very weak, porous calcarenite near Perth, Western Australia. Note the extremely irregular nature of the rock surface.

1986; McMahon, 1986). One of these zones was locally deepened by up to 8.5 m and backfilled with concrete. Throughout the limestone sections the base of the trench was covered by a reinforced concrete slab anchored into the rock. Cavities exposed on the downstream face of the trench were sealed with concrete. A single row grout curtain below this grout cap used 32 m deep vertical holes at 0.5 m spacing. Each 8 m stage, before testing and grouting, was flushed by high pressure air and water until the wash water was substantially clear. McMahon (1986) believes that this washing was unsuccessful. Relatively high grout pressures were used (3 times the reservoir head), aimed at achieving hydraulic fracturing of the clayey infill materials, like that reported by Zhang and Hao (1982).

The general foundation levels for the rockfill shoulders were reached by scrapers. In the limestone areas the exposed rock was essentially strong mounds and pinnacles separated by clay-filled depressions. These were cleaned out to the depth at which the clay filling was less than 1 m wide and backfilled with high quality rockfill.

At sites where cavities are numerous and/or large and partly or wholly filled with clayey soils, cement grouting alone is not usually relied on to form the cutoff. Other methods have included

- Cleaning out of some individual caves and backfilling with concrete;
- Walls formed by mining out slots of cavernous rock and backfilling with concrete;
- Diaphragm walls comprising overlapping boreholes backfilled with concrete;
- Closely spaced drilled holes, washed out with compressed air and water and backfilled with high-slump mortar, poured in and needle vibrated.

The last three methods were used, together with cement grouting, at the 85 m high, 1050 m long concrete face rockfill Khao Laem dam in Western Thailand. At the dam they were used to construct the upper 20–60 m of a curtain which extended up to 200 m (but generally less than 100 m) below the plinth. In one section, high pressure flushing with air and water was used to flush out sandy silt (weathered calcareous sandstone) from cavities and open joints before grouting.

Another curtain, 3.5 km long and up to 200 m deep, was constructed beneath the right abutment ridge, which was formed entirely by cavernous limestone, with its water table below the proposed storage level. All four methods were used in parts of this curtain, together with cement grouting. Lek et al. (1982) and Somkuan and Coles (1985) describe details of the methods.

3.7.2.1   *Dams which have experienced significant leakage problems*

Table 3.3 lists dams which have recorded very high leakages, and Table 3.4 lists others which have never stored water because the leakage rates exceeded the inflow. Two recently completed dams which have recorded very high leakages are discussed below.

Attaturk Dam. Riemer et al. (1997) describe cavities met during the planning and construction of this 179 m high, 1800 m long rockfill dam in Turkey, completed in 1990. The site is formed by a folded sequence of *Category O* limestones including marly, cherty and bituminous units. The rock substances are moderately strong to weak, dense and impermeable. The rock mass is cut through by many steeply dipping faults and joints. Solution along these has resulted in a network of chimneys and near-horizontal tunnels. Despite the use of many adits and drill holes, the design stage studies failed to indicate the magnitude of this cavity network and of the grouting program required.

The main 3-row grout curtain is 5.5 km long and extends generally 175 m below the foundation with a local extension down to 300 m. Even at these depths the curtain is described by Riemer et al. (1997) as "suspended". An elaborate monitoring program involving 3.6 km of adits and about 300 piezometers has indicated seepage discharge rates

Table 3.3.   Dams at which very high leakage rates were recorded (Based on Erguvanli, 1979; Riemer et al., 1997; Salembier et al., 1998).

Dam	Country	Rock	Maximum recorded leakage (m³/s)
Hales Bar	USA	Carboniferous limestone and shale	54
Keban	Turkey	Palaeozoic limestone	26
Lar	Iran	Mesozoic limestone, volcanics, lake deposits	16
Attaturk	Turkey	Palaeozoic limestones, folded and faulted	14
Great Falls	USA	Limestone	13
Camarasa	Spain	Jurassic dolomitic limestone	12
Dokan	Iraq	Cretaceous dolomitic limestone	4 to 5
Fodda	Morroco	Jurassic limestone	3 to 5

Table 3.4.   Dams which failed to store water (Erguvanli 1979).

Dam	Country	Rock
Civitella Liciana	Italy	Cretaceous limestone
Cuber	Spain	–
Kopili	India	Eocene limestone*
May	Turkey	Mesozoic and Tertiary limestone *
Montejagne	Spain	Mesozoic and Tertiary limestone*
Perdikas	Greece	Miocene limestone*
Villette Berra	Italy	–

* Geologically young carbonate rock (*Category Y*) present.

are governed mainly by reservoir level. The (1997) total discharge is estimated as 11–14 m³/sec, with reservoir levels ranging from 6–8 m below FSL. These losses are tolerable, as the average flow of the impounded river is 850 m³/sec. The monitoring also shows some high piezometric levels downstream from the curtain and some decrease in head upstream of it. Riemer et al. (1997) point out that these results "could hint at" ongoing erosion, either of grout or clayey infilling and they foreshadow the possible need for further grouting and/or drainage works.

Lar Dam. This is an earthfill embankment dam, 110 m high and 1100 m long, across the valley of the Lar River in Iran. Djalaly (1988) and Salembier et al. (1998) summarise the site geology and describe difficulties met since completion of the dam construction in 1980. Uromeihy (2000) provides more detail on the geology. The site is at the foot of an active volcano. Bedrock in the area is complexly folded, faulted and karstic limestones and mudrocks of Mesozoic Age. The dam and 15 km long reservoir area are located partly on these rocks and partly on near-horizontal materials of Quaternary Age. These include lava, ash, alluvial and lake deposits (mainly sandy and silty). They fill an old valley, which is around 200 m deep at the dam site and up to 600 m deep in the reservoir area. Before construction there were several large sinkholes (up to 25 m across and 30 m deep) in the alluvial/lake deposits in the reservoir area. The water table in the dam area was more than 200 m below the river bed and there was a karstic spring 10 km downstream from the dam, flowing at about 0.5 m³/sec.

Uromeihy (2000) notes that outgassing CO_2 and H_2S from hot springs flanking the volcano would increase the acidity of the streams draining into the reservoir but provides no data on the chemistry of the reservoir water.

After impoundment started in 1981, large new sinkholes appeared in alluvium. These were exposed when the lake was drawn down and many depressions in the lake floor were located by sonar survey. Uromeihy (2000) showed that 8 large sinkholes within 6 km upstream from the dam lie close to the trace of a known fault. The water table near the dam rose by 80–100 m. New springs appeared downstream and flow from the original spring increased to 9.9 m³/sec. The total flow rate from all springs indicated that leakage from the reservoir was about 16 m³/sec. This was more than the average inflow at the site and the highest reservoir level reached was 23 m below Full Supply Level. After studies and analyses of the leakage, site exploration, grouting and other remedial works were commenced in 1983. These involved 55,000 m of drilling and 100,000 tonnes of injected materials. The largest cavern located and treated was 200 m below the river bed, and was 27 m high and 68 m wide. Smaller cavities were located down to 430 m below the river bed (Djalaly, 1988). The treatment works did not lower the leakage rate and were stopped in 1990.

In 1991 site studies started again, but with a new emphasis, as indicated in the following words by Salambier et al. (1998) *"The key to understanding the leakage conditions and consequently proposing the most appropriate measures was geological modelling based on extensive investigation and comprehension of the geological history of the site"*.

The scope of the new studies is summarized by Salambier et al. (1998). They included geological surface mapping, interpretation of air photographs and satellite imagery, an extensive geophysical program and drilling 10 holes totalling 3,750 m. The deepest drill hole was 650 m.

This work resulted in improved geological and hydraulic models from which Salambier et al. (1998) showed that the extent of permeable ground was much greater than previously assumed and that there was a serious problem with potential sinkhole formation (see Section 3.7.3).

They also predicted the extent of ground requiring treatment to solve the leakage and associated problems and were able to delineate this in plan view and on sections. In the bedrock beneath the dam and its right abutment the treatment area continued to 600 m below river bed level as shown on Figure 3.29. Under the storage area the treatment area was shallower, but extended more than 5 km upstream from the dam.

Salambier et al. (1998) concluded that "successful plugging and grouting treatment cannot be guaranteed" and that the cost could not be estimated "within a reasonable range of certainty". They recommended that, as an alternative, blanketing of the ground surface be considered – by shotcrete and poured concrete on the limestone and by geomembrane, over the alluvial/lake deposits. Clay was considered feasible, but too scarce near the site. Reimer (personal communication) advises that as at mid-2003 no work has been done.

3.7.3   *Potential for sinkholes to develop beneath a dam, reservoir or associated works*

Sinkholes which develop naturally (Figure 3.27) are present in most areas underlain by carbonate rocks. In many such areas collapse of the ground surface to form a new sinkhole is a relatively common occurrence. Some of these collapses may happen naturally but more often they are induced or, accelerated by man's activities, as seen at Lar Dam. Possible mechanisms for sinkhole formation include the following:

1. *Dewatering.* This can cause loss of buoyant support of the rock or soil forming the roof of a cavity or, by steepening the hydraulic gradient, cause erosion and collapse of overlying soils into the cavity. It may also cause collapse due to drying out, causing shrinkage and ravelling failure of overlying soils.
2. *Inundation.* This can cause previously dry soil to lose apparent cohesion and collapse or erode into a cavity or can steepen the gradient in previously wet soils, causing them to erode into a cavity or into joints widened by solution.
3. *Vibrations* from machinery or blasting.

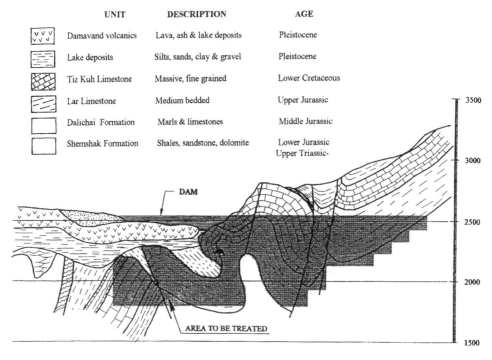

UNIT	DESCRIPTION	AGE
Damavand volcanics	Lava, ash & lake deposits	Pleistocene
Lake deposits	Silts, sands, clay & gravel	Pleistocene
Tiz Kuh Limestone	Massive, fine grained	Lower Cretaceous
Lar Limestone	Medium bedded	Upper Jurassic
Dalichai Formation	Marls & limestones	Middle Jurassic
Shemshak Formation	Shales, sandstone, dolomite	Lower Jurassic Upper Triassic-

Figure 3.29.   Lar Dam, cross section showing the predicted extent of ground treatment needed at the dam site (Salambier et al., 1998).

Examples of most of the above have been described by Sowers (1975), Brink (1979), Newton and Tanner (1987) and Wilson and Beck (1988).

Chen et al. (1995) have suggested that a rapidly rising water table in cavernous rock could cause a "water hammer effect" – pressure waves which might accelerate or induce collapse of overlying soils into cavities. First and subsequent fillings of a reservoir could provide these rapid rises in a water table.

It is clear that for all sites in karst areas there will be some risk of ground surface collapse affecting any part of the project works, including borrow areas and haul roads as well as the dam embankment and associated works. The worst imaginable event of this type would be a major sinkhole forming beneath a dam embankment during reservoir operation.

The possibility of this situation arising at Lar Dam has been accepted by Salambier et al. (1998). The dam has been extensively monitored, with instruments including standpipe, hydraulic and pneumatic piezometers. From a flow net for the dam and its foundation (Figure 3.30) they point out that:

– Vertical gradients exist in the upstream part of the core, in the upstream transition zone and in the alluvial foundation, and
– Piping of any of these materials is possible if the reservoir is operated at higher levels.

They conclude that "...development of a large sinkhole cannot be excluded. If this were to happen, complete failure of the embankment could occur".

Uromeihy (2000) warns that there is a risk of sinkhole development also beneath the downstream shoulder.

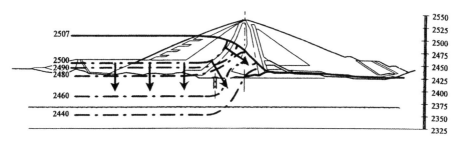

Figure 3.30.   Lar Dam, flow net in the embankment and foundation (Salambier et al., 1998).

3.7.4   *Potential for continuing dissolution of jointed carbonate rock in dam foundations*

The following sections relating to dissolution processes and effects rely heavily on theoretical and experimental material described by James and Lupton (1978, 1985), James (1981) and James and Kirkpatrick (1980).

Most of the material in those papers is included and expanded upon in James (1992), a comprehensive book covering a wide range of aspects ranging from the fundamentals of solubility to the engineering behaviour of all soluble rocks, i.e. carbonates and evaporites. Also included are useful case histories of dams and other structures in which solution effects were investigated.

James and Lupton (1978) discuss the principles governing the dissolution of minerals, develop mathematical models which predict how the minerals gypsum and anhydrite dissolve in the ground and describe laboratory tests which confirm the validity of the predictions. Water flow through both jointed rock and porous granular material are considered. They show that the rate at which the surface of a mineral or rock retreats depends primarily upon two properties of the mineral, namely:

- the solubility (c) of the mineral, which is the amount which can be dissolved in a given quantity of solvent, at equilibrium, and
- the rate of solution of the mineral, which is the speed at which it reaches the equilibrium concentration.

The solution rate constant (K) is further dependent on the flow velocity and temperature of the solvent and concentrations of other dissolved salts.

James and Lupton (1978) point out that for water flowing through a joint in an effectively impervious but soluble rock substance, widening of the joint by solution of the rock walls will occur only until the water becomes saturated with the soluble material. Thus a solution zone (Figure 3.31a) is initiated near the fresh water interface and migrates progressively downstream. For water flow through soluble rock substance with intergranular (fabric) permeability a comparable solution zone is formed and migrates downstream (Figure 3.31b).

James and Kirkpatrick (1980) use the methods and models of James and Lupton (1978) to derive conclusions on foundation treatments needed to prevent dangerous progressive dissolution when dams are built on rocks containing gypsum, anhydrite, halite and limestone. They use the following solution rate equations to predict the ways in which the materials dissolve:

For gypsum, halite and limestone

$$\frac{dM}{dt} = KA(c_s - c)$$
(3.1)

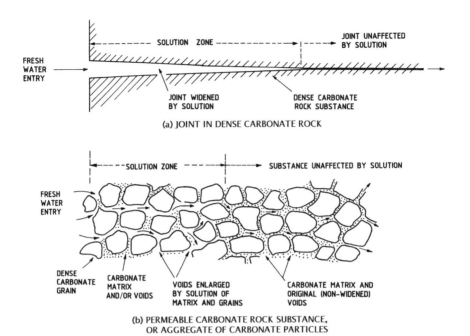

Figure 3.31.   Development of solution zones in (a) a joint and (b) soluble rock substance or aggregate of carbonate particles (Based on James and Lupton 1978).

For anhydrite

$$\frac{dM}{dt} = KA(c_s - c)^2 \tag{3.2}$$

where M is the mass dissolved in time t, A is the area exposed to solution, c_s is the solubility of the material, c is the concentration of material in solution at time t and K is the solution rate constant.

The limestone used in their laboratory tests was Portland stone, a dense, old rock (Jurassic Age), in which the carbonate mineral is calcite. For this rock they showed that K values rose sharply at flow rates close to the onset of turbulence. They note that these "transitional" velocities (around 0.75–0.8 m/s) would be reached by water flow in joints with apertures of about 2.5 mm, with an hydraulic gradient of 0.2.

James and Kirkpatrick (1980) use mathematical modelling to provide predictions of the way joints enlarge as water flows through them. Their Figure 3, reproduced here as Figure 3.32, shows their predictions for the enlargement by pure water of limestone joints with initial apertures ranging from 0.5 mm to 2 mm.

It can be seen that the solution occurs essentially by downstream migration of the solution zone, with no enlargement further downstream. After a period of 100 years, the solution zone of the 0.5 mm open joint has migrated only 13 m from the inlet face. The 0.7 mm joint shows a similar type of behaviour, but the enlarged portion migrates 30 m in only 40 years. During very short periods the larger aperture joints show long tapered enlargements indicating that in real dam situations (i.e. seepage paths often much less than 100 m) seepage flows would accelerate.

James and Kirkpatrick (1980) conclude that the smallest aperture joint which will cause dangerous progressive solution in limestone is 0.5 mm for pure water or 0.4 mm for water

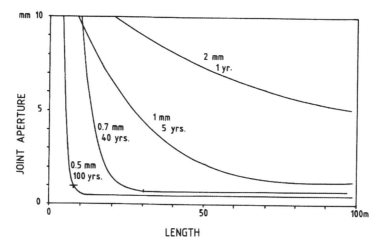

Figure 3.32.   Enlargement of joints in dense *Category O* limestone (Portland stone) by pure flowing water (James and Kirkpatrick, 1980, reproduced by permission of the Geological Society).

Table 3.5.   Solution of jointed rock (from Table 3 of James and Kirkpatrick, 1980).

	Largest joint aperture (mm)		
	For stable inlet face retreat	For rate of retreat of 0.1 m/year	Suggested preventive measures
Gypsum	0.2	0.3	Grouting
Anhydrite	0.1	0.2	Cut-off – e.g. cement-bentonite
Halite	0.05	0.05	Cut-off – e.g. cement-bentonite
Limestone (Category O)	0.5	1.5	Grouting

Note: Values are for pure water; at joint spacing of one per metre and an hydraulic gradient of 0.2.

containing 300 mg/l of dissolved carbon dioxide. They further conclude that if all large cavities are backfilled with grout or concrete, then cement grouting (which can fill joints down to about 0.2 mm aperture) should be adequate to prevent progressive solution of limestone foundations. They point out the need for care in the conducting of Lugeon permeability tests and in estimating the apertures of joints from the results of the tests.

Table 3.5 summarises the conclusions of James and Kirkpatrick for preventing progressive solution along joints in all four soluble rock types.

3.7.5   *Potential for continuing dissolution of aggregates of carbonate rock particles, and of permeable carbonate substances (Category O carbonate, in each case)*

James and Kirkpatrick (1980) also discuss solution rates in "particulate deposits". It is assumed that such deposits might comprise either aggregates of rock fragments (as in some zones of sheared or crushed rock, or in filter zones) or substances rendered permeable by intergranular voids (Figure 3.31b). Using the theoretical models of James and Lupton (1978), James and Kirkpatrick show that the length of the solution zone depends mainly on the solution rate constant and the solubility and that other factors such as the

Table 3.6.   Solution of particulate mineral deposits (James and Kirkpatrick, 1980).

Mineral	Limiting seepage velocity (m/s)	Length of solution zone (m)
Gypsum	1.4×10^{-6}	0.04
Anhydrite	1.6×10^{-6}	0.09
Halite	6.0×10^{-9}	0.002
Limestone (Category O)	3.0×10^{-4}	2.8

Note: Rate of movement of solution zone is 0.1 m/year. Mineral particles diameter 50 mm; pure water.

shape and sizes of the particles and the proportion of soluble minerals present, are of secondary importance. They show further that the downstream migration rate of the solution zone is governed largely by the seepage velocity and by the solubility, the proportion of soluble mineral present being less critical.

Table 6 of James and Kirkpatrick (1980), reproduced here as Table 3.6, shows the calculated seepage velocities which would be required to cause 0.1 m/year downstream migration of solution zones in a range of such "particulate mineral deposits". Based on their opinion that the required seepage rate for "particulate" limestone (3.0×10^{-4} m/s) is so high that it would be "rarely tolerated in dam foundations" they conclude that control of seepage by a good grouting program or other means should prevent dangerous progressive solution of "particulate" forms of dense carbonate rock.

James (1981) determined the solution parameters for 10 carbonate rocks of different compositions – 8 limestones, 1 calcitic dolomite and 1 probable chalk. They ranged in age from Carboniferous to Cretaceous. Using finely crushed samples and pure water, he found their solubilities to be virtually the same as that of "pure calcium carbonate" (presumably calcite) and that differences in their solution rates were too small to be significant in engineering design.

James (1981) showed for one sample of limestone, that small amounts of carbon dioxide dissolved in the water lowered the solution rate by about a factor of 10, but caused large increases in the solubility. Under "Engineering Considerations" he confirmed the conclusions and recommendations of James and Kirkpatrick (1980), but recommended that the following were also needed when assessing the potential for progressive solution of carbonate rock in a dam foundation:

– Chemical analyses of the appropriate reservoir, river or groundwaters, and
– Laboratory tests to determine the solubility of the carbonate rock in these waters.

3.7.6   Discussion – potential for continuing dissolution of carbonate rocks in foundations

3.7.6.1   Category O carbonate rocks

A conclusion which might be drawn from the work of James and Kirkpatrick (1980) and James (1981) as discussed in Sections 3.7.4 and 3.7.5, is that dangerous ongoing solution of old, dense carbonate rocks in a dam foundation can always be avoided by a thoroughly planned and executed grouting program using Portland cement. However, early in their paper, under "Modelling of ground conditions", James and Kirkpatrick include the following warnings against such a general conclusion.

"The conditions analysed below are idealized and do not take account of complex variations which occur in the ground" and

"Eventually it must be a matter for the engineer to judge what reliance to place on site investigation data and other engineering geology considerations, and thence to decide what factors of safety to employ".

The authors note that the individual conclusions of James and Kirkpatrick (1980) and James (1981) assume:

(a) the relatively simple joint and "particulate" models show on Figure 3.31;
(b) the largest joint apertures for limestone, on Table 3.5, and
(c) the limiting seepage velocity of 3×10^{-4} m/s for pure water through particulate or fragmented limestone, on Table 3.6.

With regard to (a) and (b) we note that at sites in faulted and weathered carbonate rocks the geological situation as exposed in detail at the excavated foundation surface and in tunnels, shafts or boreholes below it, can be very complex. In addition to features of the types on Figure 3.31, any or all of the following have been observed, often at great depths below ground surface:

– Open cavities or tunnels of all sizes, regular or irregular shapes, connected or disconnected;
– Cavities as above, partly or wholly filled with soils of variable erodibility;
– Beds or masses of variably weathered and erodible carbonate rock (see Section 3.7.1.2), and
– Beds or masses of weathered and erodible non-carbonate rock.

Achieving a satisfactory grout curtain can be extremely difficult or impossible in a foundation containing such features. During dam operation removal of soil or weathered rock by erosion could provide a flow path wide enough to cause ongoing solution of adjacent limestone. This type of process may have occurred at Hales Bar, Lar and Attaturk Dams (see Section 3.7.2.1 and Table 3.3).

With regard to (c), the maximum leakage rates shown on Table 3.3 suggest that rates much higher than 3×10^{-4} m/s are/were being tolerated, at least locally, through the foundations of some of these dams. Such local flows may have high velocities and could be turbulent. It is noted also that Bozovic et al. (1981) and Riemer et al. (1997) foreshadow the possibility of future increases in leakages at Keban and Attaturk Dams.

However, as discussed previously in Section 3.7.2, many dams have been built in *Category O* limestone sites, with cement grouting for seepage control and have performed well for many years without excessive or increasing seepage. The authors accept that a good grouting program should be adequate to control seepage and prevent ongoing solution at many dams built on such carbonate rocks. However, it is recommended that those responsible for building any dam on these rocks should:

– Pay special attention to the design of monitoring systems (see Chapter 20) and
– Make adequate allowance (e.g. access and ability to lower storage levels) for additional grouting programs which might be required after first filling or later during the lives of the dams.

3.7.6.2 *Category Y carbonate rocks*

The resistance of *Category Y* carbonate rocks to dissolution in fresh water has not been researched to the extent of the work done on *Category O* rocks by James and Kirkpatrick (1980) and James (1981). However, the following reasons suggest strongly that their potential for continuing dissolution, if present in a dam foundation, would be higher than that for *Category O* rocks:

– Their lower strengths and higher porosities;
– The field and laboratory evidence of their rapid solution and redeposition, presented in Section 3.7.7.2, and
– The experiences at Kopili, Montejagne, Perdikas and May dams (Table 3.4).

3.7.7   *Potential problems with filters' composed of carbonate rocks*

Most filters are well graded sands or gravels with few or no fines. They are compacted to 60% to 70% density ratio. Their moisture environments (in chimney zones, in particular) may vary widely. For example, in a filter, months or years of inundation and large or small flow rates may be followed by months or years of unsaturated, ranging to almost dry, conditions. Heavy rainfall at any time can cause water containing carbon dioxide to enter the upper part of a chimney zone. Carbon dioxide developed by the oxidation of methane from the groundwater or reservoir, organic acids derived from rotting vegetation or sulphuric acid derived from oxidation of sulphide minerals may also enter a filter zone. Possible effects of these events on a carbonate rock filter include:

(a) change of grading due to dissolution, or
(b) partial dissolution and recementation, or
(c) interlocking of grains due to pressure solution (see Figure 3.33).

Each of these effects and the likelihood of occurrence are discussed in Sections 3.7.7.1 and 3.7.7.2.

3.7.7.1   *Category O carbonate rocks*

Dense, old carbonate rocks in the strong to very strong range have been crushed and processed to provide coarse and fine filter materials for many existing dams. The authors have found no report of a dam incident or failure due to malfunction of any of these filters.

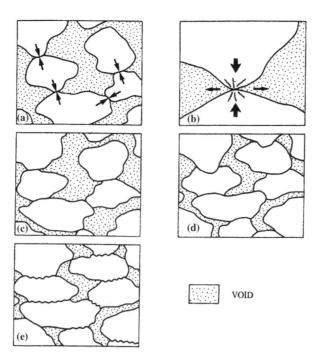

Figure 3.33.   Progressive effects of pressure solution in granular materials (Harwood, 1988). (a) Point grain to grain contacts (arrowed), (b) Stressed grain to grain contact (large arrows) leading to formation of dislocations in crystal lattice and subsequent dissolution, with lateral fluid transport of solutes (small arrows), (c) Planar grain to grain contacts, (d) Interpenetrating grain to grain contacts, (e) Sutural grain to grain contacts.

However, not all dams have instruments sufficient to monitor the performance of all of their filters. Also the critical filters (Figure 9.2) may be seriously tested only in an emergency, e.g. when a crack develops in a dam core or the foundation beneath either of the dam shoulders is disrupted.

(a) *Change of grading due to dissolution.* The critical filter zones in most large dams are often less than 2.8m wide. According to Table 3.6, if pure water was to flow at a rate of 3 3 1024m/s through such a zone formed by 50 mm diameter calcite particles, it would lie entirely within the "solution zone". That is, the water would never become saturated with carbonate, and the particles would be progressively reduced in size, by dissolution. With much smaller particles, as in a fine filter zone, with acidic water and an aggressive reaction the size reduction might become significant during the life of a dam.

(b) *Partial dissolution and recementation.* There have been many accounts of recementation of crushed carbonate materials, due to acids formed by the oxidation of sulphide minerals (See Section 2.9.4). Several examples have been in South Australia where cemented zones have occurred in pavements, embankments and stockpiles formed by crushed rocks of Cambrian and Precambrian Ages. Tests have shown the cementing materials to be calcium sulphate (gypsum) in crushed marble and magnesium sulphate in crushed dolomites. Both the marble and dolomite have performed well as base courses in pavements.

The minus 30 mm crushed marble is well graded and contains 5% to 10% fines (all calcite) with plasticity index of 2. Despite Los Angeles Abrasion Losses of 60% or more, its performance in pavements is remarkably good. Falling Weight Deflectometer tests on a sealed pavement built in 1998 showed its initial modulus to be 700 MPa. The same test after 4.6 years of service showed 1700 MPa. There was no rutting, permanent deflection or change in density (Andrews 2003). The authors consider that the formation and deposition of the gypsum cement, and possibly some associated interlocking of the marble particles due to pressure solution (Figure 3.33) have been the main causes of this increase in stiffness with time.

(c) *Interlocking of grains due to pressure solution.* Both gypsum and magnesium sulphate are highly soluble in water. This raises the following question. Assume that a chimney zone filter has become cemented by either of these during a period of very low and zero seepage. If now increased flow occurs through a crack in the core and cemented filter, how would this filter behave as its cement becomes redissolved? The results of testing by Hazell (2003) suggest that the filter might retain the crack, or "hang up". The test used was developed to assess the effects of chemical stabilising agents in base materials for unsealed roads. Samples are prepared by removal of the plus 2.36 mm fractions and compacting the remainder at OMC to 98% MMDD in 100 mm long by 100 mm diameter moulds. After removal from the mould samples are bench dried for 7 days. Each is then loaded by an annular weight of 1560 g, through which water drips on to it at a rate of about 400 ml/h. Hazell (2003) advises that during this test all other local non-cementing road base materials (most are crushed quartzites) collapsed within 24 hours, but the crushed marble remained intact, although completely saturated and soft to touch, for at least 460 hours. Assuming that by this stage the gypsum cement would have been completely dissolved, then the sample must have gained strength from some other process. The authors suspect that particle interlock may have developed due to pressure-solution at the grain contacts.

The process and effects of pressure solution in quartz sands in geological time frames are described by Dusseault and Morgenstern (1979), Palmer and Barton (1987) and Schmertmann (1991). Harwood (1983) discusses the role of this process in the reduction of intergranular porosity of sands. Pressure solution effects should proceed much more rapidly in carbonate sands than in quartz sands, because the solution rates of the carbonate minerals are much higher than that of quartz. Relatively inexpensive research (e.g. examination of thin sections cut through epoxy-impregnated old pavements) may show whether or not the effects are rapid enough to render filters composed of *Category O* carbonates ineffective.

3.7.7.2   *Category Y carbonate materials*

(a) *Change of grading due to dissolution.* Sedimentary geologists consider that in fresh water high magnesian calcite is more soluble than calcite and aragonite (Prothero and Schwab 1996). Quantitative data on the solubility and solution rate of high magnesian calcite are not presently available. However, from the field and laboratory evidence discussed below, the authors believe that *Category Y* carbonates used as filter materials would suffer more rapid dissolution than equivalent *Category O* carbonates.

(b) *Partial dissolution and recementation.* Netterberg (1975) discussed very high CBRs yielded by pedogenic calcretes in Southern Africa and stated "*It is difficult to ascribe soaked CBRs much above those yielded by crusher-runs to anything other than cementation*". He also included test results showing that the CBRs of some calcretes had more than doubled after a few cycles of wetting and drying. In Netterberg (1969) he described studies aimed at understanding how these calcretes were formed in soil profiles. He concluded that fine carbonate material in the soil is dissolved and redeposited as cement during changes in soil water suction and partial pressure of carbon dioxide. He also provided estimated rates of calcrete formation in nature, ranging from 0.02 mm/year to 1 mm/year (Netterberg 1969, 1978). In Netterberg (1971) he suggested that the apparent "self stabilisation" of calcretes used in pavements may be caused by similar (solution and redeposition) processes. The authors have seen many examples of recemented calcrete pavements in Australia.

Sterns (1944) and Tomlinson (1957) describe evidence of recementation of coral sands and gravels, where used as base materials in airfield pavements.

Coquina limestone of Tertiary age is used widely (when crushed) as base course for highways in Florida. This rock comprises corals, shell fragments and quartz grains. Graves et al. (1988) report that the strength increase which occurs with time in these pavements is due to dissolution and reprecipitation of fine carbonate particles. They support this claim using results of the LBR (limerock bearing ratio) test, a modified version of the CBR. Quartz sands were mixed in different proportions with carbonate sands containing less than 2% of fines. The samples were compacted into LBR moulds and tested after soaking continuously in water for periods of 2, 7, 14, 30 and 60 days. All samples with 40% or more of carbonate sand showed strength increases ranging from 23% to 65%, after the first 14 days of soaking. No further increase was recorded after this time. Graves et al. suggested that the very fine carbonate particles may have been "used up" by the end of 14 days and that the coarser grains were not providing cementing material as efficiently. Tests with crushed carbonate base course materials containing "slightly more carbonate fines" were tested similarly for 30 days and showed increases in strength up to that time.

McClellan et al. (2001) carried out further laboratory research into these effects but were unable to reproduce comparable strengthening of samples within curing times of up to 60 days.

(b) *Interlocking of grains due to pressure solution.* Shoucair (2003) has advised that the Florida Department of Transport has been adding coral sands to their road bases with significant strength gains. However he notes that the gains "practically disappear when the materials are soaked again" and considers that mechanisms described by

Schmertmann (1991) are more likely causes of the initial strength gains than solution and recementation of the carbonate material.

One of the mechanisms described by Schmertmann was interlocking of grains due to pressure solution and the authors consider it possible that this may be the cause of the small remaining strength after resoaking.

Based on their experience with the gypsum-cemented marble (see Section 3.7.7.1) the authors suggested that the initial strength increase might have been caused by the presence of minute amounts of gypsum and that this could have been present in the coquina or in the waters used. McClellan (2003) has advised that the presence of gypsum is possible, even likely, during road construction in Florida. Traces of pyrites occur in some of the limestones and sea sprays (containing sulphates) occur widely. However he believes that gypsum is not a factor in their laboratory tests. SEM/EDX examination of the cemented areas of their specimens has never shown any significant sulphur.

Some very weak, porous calcarenites are sawn into blocks, which are then used for the walls of small buildings. Newly-sawn blocks are soft to touch and easily scratched, gouged or broken. After exposure to the weather (wetting and drying) for several years, the surfaces of the blocks become noticeably stronger and more durable. It appears that this "case-hardening" happens because carbonate derived from dissolution within the blocks becomes deposited near their surfaces. The Gambier Limestone (Tertiary Age) in South Australia is an example. Millard (1993) notes that "soft local limestone" on Malta, shows similar behaviour. Similar "case-hardening" effects can be seen in the faces of many cliffs formed by *Category* Y limestones, and carbonate-cemented sands.

The following conclusions can be drawn about *Category* Y carbonates:

1. When subjected to compaction, followed by soaking or wetting and partial drying, some of them have become strengthened by solution and recementation of carbonate grains in a few days to a few years. In most cases the above has involved some very fine carbonate material. In some cases most of the strength gains have been lost on re-soaking;
2. Exposed surfaces of some very weak, porous carbonate rocks become strengthened in a few years by solution and redeposition of carbonate;
3. In each of the above cases it is possible that the strengthening has been partly or wholly caused by gypsum.

The moisture environments in filter materials (in chimney zones, in particular) may at times be similar to those in which cementation was postulated, in 1 above. A carbonate filter zone would not normally contain significant amounts of fines. However the authors consider that the possibility exists that during the lifetime of a dam, solution and rececementation and/or grain interlock from pressure solution, could render it cohesive and ineffective.

The authors consider that carbonate materials should not be adopted for filters until the possible long-term effects of dissolution have been assessed by geotechnical specialists.

3.7.8   *Suitability of carbonate rocks for embankment materials*

Category O carbonate rocks have been used widely and with apparent success as rockfill, random fill and riprap. However, there is some potential for deleterious effects, if sulphide minerals are present:

– in the carbonate rocks;
– in other materials in the embankment, or
– in the foundation or storage area.

Sulphuric acid formed by oxidation of the sulphides may attack the carbonate rock and form calcium or magnesium sulphates, which could become deposited in adjacent filter zones.

Carbonate rocks commonly occur in association with mudrocks containing sulphide minerals and carbonate rocks themselves often contain sulphide minerals. In Australia sulphide minerals can occur in all rock types and have been found at the sites of many dams. Quite small concentrations (much less than 1%) of these minerals can have deleterious effects.

The authors believe that it is best to avoid using carbonate rocks (of any age) as embankment materials, if sulphide minerals are present in any of the above ways, as the solution products could clog filters. In any rock or at any site, it can be difficult to determine the amount of sulphides present and their distribution.

3.7.9 *Suitability of carbonate rocks for concrete and pavement materials*

Dense carbonate rocks in the strong to extremely strong range have been used extensively for the production of aggregates for concrete, bituminous concrete and pavements. However, a few dolomitic rocks have been found to react with alkalis in Portland cement causing expansion and cracking effects like those from alkali-silica reaction. Experiences at sites where this has occurred are described by Luke (1963), Highway Research Board (1964) and Huganberg (1987). Guillott (1975, 1986) describes petrographic work on reactive carbonate rocks and concludes that only very fine grained dolomitic rocks containing some clay are likely to cause expansion.

Some carbonate rocks contain nodules or beds of chert (extremely fine grained or glassy silica). Alkali-silica reaction is likely to occur if these rocks are used as aggregate with high alkali cement.

The authors recommend that all carbonate rocks intended for use in concrete be assessed for reactivity.

3.7.10 *Stability of slopes underlain by carbonate rocks*

Natural landsliding is not common in areas underlain by pure carbonate rocks. In weathered, solution affected carbonate rocks it is common to find that joints, faults and bedding partings have been partly or wholly "healed" by redeposited calcite. This along with their inherently high frictional strength of joints seems likely to be the reason for the low frequency of landslides.

In the experience of the authors most slides in carbonate rocks have occurred along interbeds of mudstone or shale. Figures 3.34 to 3.36 show a landslide of approximately 2500 million m^3 in folded limestone of Tertiary Age in Papua New Guinea. The slide is believed to have occurred into an abandoned valley of the Mubi River, due to daylighting of a thin mudstone bed within the limestone (Figure 3.36). The slide was probably triggered by earthquakes associated with fault movements, which displaced the Mubi River about 400 m laterally, and by continued uplift and tilting of the limestone to the south of the fault.

The Bairaman landslide, also in Papua New Guinea, occurred in a 200 m thick horizontal bed of limestone (King et al., 1987) and its horizontal basal failure surface was semicircular in plan, covering about 1 km^2. Its basal surface must have been a mudstone bed.

In detailed studies of carbonate rocks in the Vaiont landslide Hendron and Patton (1985) have shown that clay-rich units and clay are present along and near most of the failure surface.

James (1983) draws attention to situations in Sri Lanka where solution of near-horizontal limestone beds near valley floors has caused collapse, undercutting and landsliding in

Figure 3.34.   Air photo showing the scar of an old landslide in folded limestone, near the Mubi River, Papua New Guinea.

overlying beds. Retreat of these beds has produced broad valleys bounded by steep escarpments, with hanging tributary valleys. James considers these almost "glacial-like" valleys to be useful indicators of the presence of limestone in areas devoid of outcrop.

3.7.11   *Dewatering of excavations in carbonate rocks*

Excavations below the water table in carbonate rocks containing cavities are likely to require more continuous pumping at higher rates than equivalent excavations in non-soluble rocks. Also because the sizes and distribution of cavities are usually so variable and irregular, a number of borehole pumping tests at carefully selected locations may be needed to get a reliable estimate of the inflow rates.

3.7.12   *Carbonate rocks – check list of questions*

– Category *O or Y?*
– Cavities, air-filled or water-filled?
– Cavities, soil-filled?

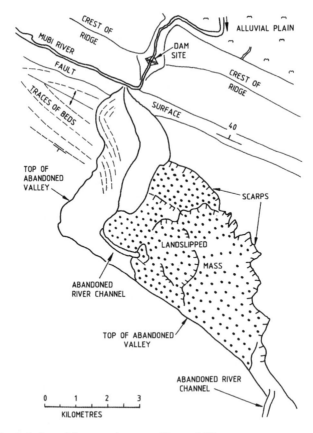

Figure 3.35.   Geological plan of the area shown on Figure 3.34.

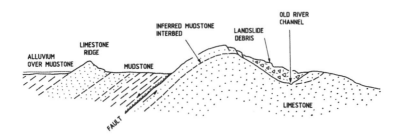

Figure 3.36.   Sketch section through the area shown on Figure 3.35.

– Collapse of cavities?
– Extremely irregular, often pinnacled surface of fresh rock?
– Sharp boundary between residual soils and fresh rock?
– Strong rock around solution tubes and cavities in weak, porous rocks?
– Solution cavities in altered carbonate rocks or metamorphosed impure carbonate rocks?
– Very weak, low density, erodible weathered materials?
– Extremely high permeabilities?
– Extreme variations in permeability?
– Possible deep, major leakage paths out of reservoir?

- Presence of sinkholes, exposed or concealed?
- Composition and pH of the groundwater and reservoir water?
- Presence, amounts and distribution of any sulphide minerals?
- Potential for dangerous ongoing solution in the dam foundation?
- Suitability for use for embankment materials?
- Suitability for use in concrete and pavements?
- Alkali-carbonate reaction?
- Chert present: Alkali-silica reaction?
- Shaley (argillaceous) rocks: Durability?
- Unstable slopes, where interbeds of mudrocks are present?

3.8   EVAPORITES

The common evaporites, gypsum ($CaSO_4.2H_2O$) and halite (NaCl) are formed in arid areas by evaporation from inland seas, from inland and coastal salt lakes and tidal flats (Stewart, 1963; Murray, 1964). They can occur as individual beds in sedimentary rock sequences, often in association with or interbedded with carbonates. They occur also as matrix, cement, nodules, veins or joint fillings, in mudrocks or sandstones. Anhydrite ($CaSO_4$) is a less common evaporite. Most anhydrite is believed to be formed from gypsum, when it is buried at depths greater than 150 m and its chemically attached water is removed due to overburden pressure and heat. Anhydrite also occurs as nodules or infilling cavities in limestone and dolomite (Murray 1964; Brune, 1965).

The evaporites are much more soluble than the carbonates and only outcrop in arid regions. Sequences containing evaporites show a wide range of solution effects related to present and/or past groundwater levels. Brune (1965), Bell (1983b, 1993), Cooper (1988), Thompson et al. (1998), Hawkins and Pinches (1987), James (1992, 1997), Hawkins (1998) and Yilmaz (2001) describe such effects and related ground engineering problems. These include:

- karst topography with extensive systems of caves, tunnels, depressions, chimneys and sinkholes;
- ongoing subsidence, on small and large scales; very slow or sudden collapse;
- leakage through or into solution cavities;
- ground weakening and/or increasing permeability due to ongoing solution;
- heaving ground, due to growth of gypsum crystals;
- explosive heaving/uplifting of ground due to the hydration of anhydrite to become gypsum. Brune (1965) describes how creek channels in Texas were cracked open and uplifted several metres along distances of up to 300 m during such explosions.

3.8.1   *Performance of dams built on rocks containing evaporites*

The authors know of no dam built at a site containing thick beds of halite. Many dams and reservoirs have been built successfully on sites containing gypsum. A few have later suffered some degree of distress, caused by ongoing dissolution of this material from their foundations. James and Kirkpatrick (1980) refer to Poechos Dam in Peru, founded on clay shales with gypsum, noting that seepage from its right "dyke" was saturated in calcium sulphate and also contained 3.5% of sodium chloride. Notable embankment dam examples in the USA include McMillan and Cavalry Creek Dams (Brune, 1965), Tiber Dam (Jabara and Wagner, 1969). San Fernando Dam (concrete gravity) was affected by rapid solution of gypsum cement from very weak conglomerate (Ransome, 1928; Jansen, 1980).

A more recent example with a gypsum-related issue is Caspe Dam in Spain. Cordova and Franco (1997) cite dissolution of gypsum in the foundation as the cause of a large leakage and damage to this 55 m high embankment dam in 1989 during its second filling. Another is Tarbela Dam in Pakistan, where the right abutment rocks are intensely fractured and include gypsum and "sugary" limestone which is friable and erodible. Grouting has been only partially effective in controlling flows which have high contents of dissolved solids. Additional drainage adits have been required (Amjad Agha, 1980).

Reports on the failure of Quail Creek Dike (James et al., 1989; O'Neill and Gourley, 1991) provide useful details on the weathered profile developed on the interbedded dolomite, siltstone and "silty gypsum" which formed the foundation of that 24 m high embankment dam. Its failure was not related to any post-construction solution effects, but was due largely to the failure during design to recognize and take into account the gaping joints in the near-surface rock produced by the past solution of gypsum.

3.8.2   Guidelines for dam construction at sites which contain evaporites

Predictive models derived by James and Lupton (1978) for the rates of dissolution of gypsum and anhydrite have been discussed briefly above in Section 3.7.4. James and Lupton also provide guidelines for the investigation of dam or reservoir sites where these minerals are present and indicate the kinds and levels of risks associated with construction and operation at such sites. They point out that it is important to know the chemical compositions of the waters involved, because the solution rates of both gypsum and anhydrite are increased by the presence of sodium chloride, carbonate and carbon dioxide. They note that conglomerates which are cemented by gypsum or anhydrite "can produce a material which is potentially very dangerous" if small proportions of the cement are removed by solution. They conclude that the risks of accelerating solution effects are higher with anhydrite than with gypsum and that an "efficient cutoff" is the only practical method to reduce seepage velocities to values low enough to provide "complete safety" against solution effects in "massive anhydrite". They also warn about the possibility of anhydrite converting to gypsum and expanding, with the potential for heave.

James and Kirkpatrick (1980) confirm the conclusions of James and Lupton and provide further predictive data for gypsum and anhydrite and comparable data for halite (see Tables 3.5 and 3.6). They stress the need for chemical tests on drilling water and cores, when halite is suspected, and special sampling procedures if halite is found. They conclude "It would be most unwise to build a dam on massive halite. Unconsolidated strata containing sodium chloride may be unavoidable and control measures are feasible, if costly".

The authors acknowledge the valuable work carried out by James and Lupton (1978), James and Kirkpatrick (1980) and James (1992) and endorse their views, subject to the following:

- We would recommend extreme caution to those considering dam construction at sites where thick beds of anhydrite occur at depths shallower than 150 m;
- As for sites on carbonate rocks (see Section 3.7.6) we would have some reservations about the effectiveness of cement grouting at sites containing thick beds of evaporites severely affected by past weathering and solution. Special attention would be needed to seepage monitoring and facilities for re-grouting during operation of the dam;
- The possibility that filter zones could become cemented with gypsum needs to be considered;
- It is our view that engineers intending to build at sites containing evaporites would be wise to get advice from persons with special knowledge of the evaporite minerals, and from others with past experience of construction at such sites.

3.8.3   *Evaporites – checklist of questions*

– Cavities, air-filled or water-filled?
– Cavities, soil-filled?
– Collapse of cavities – subsidence?
– Ground weakening due to ongoing solution?
– Increasing permeability due to ongoing solution?
– Heave due to growth of gypsum crystals?
– Large scale heave due to hydration of anhydrite?
– Chemical composition of groundwaters/reservoir waters?
– Presence of halite – chemical tests?
– Possibility of cementation of filter materials, by gypsum?

3.9   ALLUVIAL SOILS

For the purposes of this discussion "alluvial soils" includes soils which have been deposited in the channels and flood-plains of rivers and in lakes, estuaries and deltas. These soils are characterised by great variability, both vertically and laterally and can range from clays of high plasticity through to coarse sands, gravels and boulders.

Detailed sedimentological studies in recent years have provided one or more "facies models" for the sediments (soils) deposited in each of the above environments. A facies model includes block diagrams which summarise the compositions and configuration of the bodies of soil deposited in a particular environment, together with an indication of the processes by which they were formed. Figure 3.37 is a block diagram showing key features of the "meandering river" model, which will be the only one discussed in detail here. Further information about meandering rivers and details of the deposits formed in other environments can be found in Leopold et al. (1964), Leeder (1982), Selley (1982), Lewis (1984) and Walker (1984).

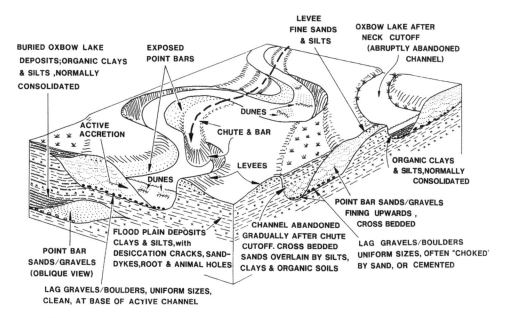

Figure 3.37.   Schematic view of soils deposited by a meandering river (based partly on Figure 1 of Walker, 1984).

The main types of deposits shown on Figure 3.37 are as follows:

- Lag deposits, usually gravel or boulders, occur along the base of the river channel and are moved only during peak flood times. They are usually uniform sized and in the active river channel may have very high voids ratio and permeability. Where preserved at the bases of abandoned or buried channels their voids may be "choked" by sand or by fines.
- Point bar deposits, usually sands and gravels, are deposited on the insides of bends in the stream. During normal flows these materials occur above the lag deposits along the whole of the channel with their upper surfaces in the form of migrating dunes. The cross-bedding seen in the bar deposits results from preservation of some of these dunes. The point bar deposits are usually coarser at depth, becoming finer towards the top.
- Levees of fine sands and silts are formed along the top of the river banks where these coarser materials are deposited more quickly than the fine silts and clays when flood-waters overtop the banks.
- Flood-plain deposits are usually fine silts and clays, deposited in thin horizontal layers during floods. In situations where the soils dry out between floods, desiccation cracks are formed and these may be preserved as sand infilled joints if wind-blown or water-borne sand is deposited in them. Small tubular holes left by burrowing animals or decomposing vegetation are also common in floodplain deposits. These may be preserved open or else infilled by sand or fines.
- Oxbow lake deposits occur in lakes formed during floods when parts of the meandering channel are cut off from the stream. The nature of the deposits depends on whether the channel is abandoned slowly (chute cutoff) or abruptly (neck cutoff) as shown. In each case the fine-grained soils are usually near normally consolidated and are at least partly organic due to the presence of rotted vegetation.

Not shown on Figure 3.37 are bar deposits which can be seen at low flows along straight parts of meandering streams and also in the channels of "braided" streams. Such bars may migrate quite rapidly with aerial photographs taken 10 years apart showing significant changes in the stream bed geometry.

Cary (1950) reports that gravel bars in several fast-flowing rivers in U.S.A. contain elongated lenticular deposits of essentially uniform-sized "open-work" gravel. He reports many open-work gravel lenses in glaciofluvial deposits in the north-western U.S.A., and comments on the large voids and extremely high permeabilities of these materials. The authors have seen open-work gravels and cobbles in which the large voids have become "choked" with sand, which clearly must have migrated into the voids after the original deposition of the gravels.

Cary considers that open work gravels are probably formed in fast-flowing rivers at the downstream ends of rapidly aggrading bars. He suggests that in these situations eddies sometimes occur, which remove finer gravels and sand, leaving only the coarse materials which form the open-work deposits.

In arid or semi-arid climates where stream flows are intermittent, it is common to find cemented layers in the lag and lowest bar deposits of meandering steams and of other e.g. braided streams. This cementation occurs as the waters dry up and the most common cements include gypsum, calcite and limonite. Cementation is relatively common also in the floodplain deposits.

It is not uncommon to find timber, in some cases the remains of large trees, buried in channel or floodplain deposits. The timber is often well preserved, particularly where groundwaters are highly saline. In some situations the timber is partly or wholly rotten and may have left gaping voids in the alluvial deposit.

It is difficult to generalise about the properties of alluvial soils, because of the extremely wide range of soil types. The following are some observations which may be taken as a general guide.

3.9.1   River channel deposits

The sands, gravels, cobbles and boulders are often highly permeable, particularly in the horizontal direction. Layers, often thin, of finer or coarser materials cause marked differences between vertical and horizontal permeability. This is shown diagrammatically in Figures 3.37 and 10.2. Clean gravels can be interlayered with sands or sandy gravels, giving overall horizontal permeabilities 10 times to 1000 times the overall vertical permeability. The relative density of such deposits is variable, but the upper few metres which are most affected by scour and redeposition during flooding, are likely to be loose to medium dense and, hence, will be relatively compressible and have effective friction angles in the range of 28° to 35°. Deeper deposits are more likely to be dense, less compressible and have a high effective friction angle.

3.9.2   Open-work gravels

At the 143 m high, 2740 m long Tarbela Dam in Pakistan extensive deposits of open-work gravels occur in the 190 m deep alluvium which forms the foundation for the embankment. The alluvium comprises sands, open-work gravels and boulders and boulder gravels in which the voids are sand-filled (i.e. extremely gap-graded materials). The design allowed for underseepage to be controlled by an impervious blanket which extended 1500 m upstream from the impervious core. The blanket ranged in thickness from 13 m near the upstream toe of the embankment to 1.5 m at its upstream extremity. For several years after first filling (1974) many "sink-holes" or graben-like craters and depressions appeared in the blanket, apparently due to local zones of cavitation within the underlying alluvium. Some of the sinkholes were repaired in the dry and others by dumping new blanket material over them through water using bottom-dump barges. The local collapse zones which caused the sinkholes are believed to have formed when excessively high flow rates through open-work gravels caused adjacent sandy layers to migrate into their large voids.

3.9.3   Oxbow lake deposits

Where clays, silts and organic soils deposited in oxbow lakes have not dried out they are near normally consolidated and may be highly compressible. McAlexander and Engemoen (1985) describe the occurrences of extensive oxbow lake deposits up to 5 m thick in the foundation of the 29 m high Calamus Dam in Nebraska, USA. These deposits comprised fibrous peat, organic silty sands and clays and were highly variable in thickness and lateral extent. Testing showed that the peat was highly compressible. Because of concern about differential settlements and cracking in the embankment, the organic materials were removed from beneath the impervious core and from beneath extensive parts of the shoulders.

3.9.4   Flood plain, lacustrine and estuarine deposits

The clays and silts in these deposits are likely to show pronounced horizontal stratification, with each flood or period of deposition resulting in an initially relatively coarse layer fining upwards as the flood recedes. This may result in marked anisotropy in permeability, with the horizontal permeability being 10 times or even 100 or 1000 times the vertical permeability.

The permeability of these deposits is often increased by the desiccation cracks, sand filled cracks, fissures and holes left by burrowing animals and rotted vegetation. Where such defects have been backfilled by clay soils, the permeability of the mass can be decreased.

Where desiccated the clay soils are overconsolidated and their shear strengths are affected by the presence of fissures which are often slickensided.

In environments where the water table has remained near the surface, e.g. coastal estuaries, the flood-plain soils may be very soft clays, which are near normally consolidated except for an overconsolidated upper 0.5–2 m. In these cases the soils at depth may have a very low undrained shear strength and very high compressibility.

3.9.5   *Use of alluvial soils for construction*

The filters and concrete aggregate for many dams are obtained from alluvial sand and gravel deposits. In many cases the strict grading requirements and need for low silt and clay content (usually less than 5% or 2% passing 0.075 mm) necessitate washing, screening and regrading. The source of the sediments will have a marked effect on their durability. For example, sand and gravel in a stream fed from areas partly underlain by siltstone is likely to have gravel size particles which will break up readily, rendering the gravel unsuitable for filters or concrete aggregates, whereas those originating from areas underlain by granite, quartzite or other durable rocks are more likely to be suitable.

Alluvial clays, sandy clays and clayey sands (including a proportion of gravel in some cases) can be suitable for earthfill zones in dam construction. Because of their likely variability, the deposits need careful investigation to delineate suitable areas. Borrowing with a shovel and truck operation is sometimes necessary to ensure adequate mixing. Some Australian dams in which alluvial materials have been used for earthfill core zones include Blue Rock, Cairn Curran, Buffalo, Eildon (Victoria), Blowering (N.S.W.), Bjelke-Peterson and Proserpine (Queensland), Hume (NSW – Victoria).

3.9.6   *Alluvial soils, list of questions*

- Vertical and lateral variability related to deposition conditions?
- Lenticular deposits of open-work gravels with extremely high permeability?
- Anisotropy due to layering?
- High kH : kV ratio?
- Oxbow lake deposits, compressible organic soils?
- Cracks, fissures, holes after rotting vegetation or burrowing animals, all either open or backfilled?
- Cemented layers?
- Buried timber, rotten or preserved, large voids?

3.10   COLLUVIAL SOILS

3.10.1   *Occurrence and description*

Included under this heading are all soils which have been eroded and deposited under gravity forces, often with the aid of water flow. They include slopewash, scree (talus), and landslide debris. The soils range from high plasticity clays through to boulder talus deposits, but are characterised by being mixtures of particles of contrasting sizes e.g. clays with embedded gravel and boulders in landslide colluvium and clayey gravelly sand slopewash deposits. They are also commonly variable within each deposit. Figures 3.38 and 3.39 show typical environments in which scree and slopewash deposits are formed.

3.10.1.1   *Scree and talus*
These are deposits of rock fragments which detach from cliffs or areas of steep outcrops and fall by gravity and roll/slide downslope. The upper scree slopes are composed of smaller rock fragments and usually are at slopes of 35° to 38°; the toe of the slope usually

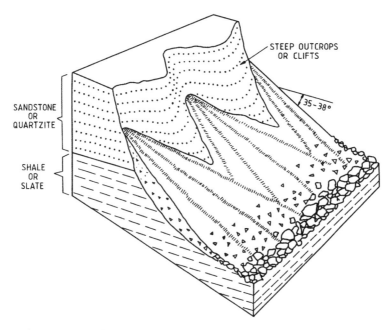

Figure 3.38.   Schematic view of scree deposits.

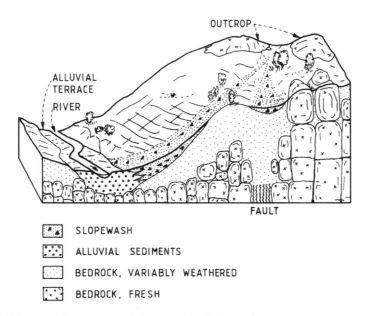

Figure 3.39.   Schematic view of residual, slopewash and alluvial deposits.

comprises large blocks at flatter angles. The deposits are not water-sorted. They are usually very loose (low bulk density) and just stable at the natural angles. When the deposits contain 30% or more of fine-grained soil, they are called talus.

Selby (1982) provides a more detailed discussion on the variety of processes of formation of scree and talus and on the range of fabrics resulting from this. Some talus deposits show poorly developed soil profiles near the surface or at intervals at depth. These indicate periods during which weathering of the deposit has been proceeding at a faster rate than accumulation of new rock fragments.

It is not uncommon to find timber embedded in scree, either rotted or in a preserved condition.

3.10.1.2   *Slopewash soils*

Slopewash soils are admixtures of clay, sand and gravel which have been moved downslope by the combined actions of soil creep (due to gravity forces) and erosion by water. The thickest deposits are developed in depressions or gullies as shown on Figure 3.39. Near the base of steep slopes slopewash soils often overlie or are intertongued with alluvial deposits (also shown on Figure 3.39).

In cold climates (see Section 3.12) freezing and thawing of the ground can be a major contributing factor in soil creep and deep slopewash deposits are common.

Slopewash soils sometimes show indistinct bedding parallel or non parallel to the ground surface. Slopewash usually has low density and often exhibits tubular voids left by rotted vegetation or roots or burrowing animals or caused by erosion of fines from within the deposit.

3.10.1.3   *Landslide debris*

Landslide debris can range from high plasticity clay through to silty sand from ash flows or sand/gravel/boulder soils resulting from avalanches. In most cases the soils are very variable, vertically and laterally, and it is not uncommon to find large boulders embedded in a clay matrix.

Timber (the remains of trees) is often present in modern landslide debris. It may be well preserved or rotting. Voids left by rotted timber are sometimes found.

Open cracks and irregular voids are often found in landslide debris, particularly where the debris has resulted from or has been affected by modern slope movements.

At sites where landslides have dammed and diverted pre-existing rivers it is common to find landslide debris overlying river alluvium.

Deposits of landslide debris are often underlain by a sheared or slickensided zone (the slide surface) and there may be several sliding and shear surfaces at other levels within the debris. In many cases the main slide surface may be in a zone of material which appears to be residual soil or extremely weathered rock and is characterised by a higher clay content than that of most of the debris.

High groundwater tables are common in landslide debris, but this is not always the case.

3.10.2   *Properties of colluvial soils*

As for alluvial soils, it is difficult to generalise because of the extremely wide range of soil types. Some general characteristics which may be present are:

3.10.2.1   *Scree and talus*

These materials are likely to be highly permeable, and compressible. As they are sorted they are likely to be poorly graded.

As these materials occur close to their natural angle of repose, excavation into scree or talus slopes usually causes ravelling failures extending upslope. Entry of excessive water (e.g. by discharge from roads) into talus materials can cause them to develop into debris-flows.

3.10.2.2   *Slopewash*

These soils may be more permeable than expected from their soil classification, reflecting the presence of voids and loose structure. They are also likely to be relatively compressible. Many slopewash soils are highly erodible.

Where they occur on steep slopes (e.g. as in Figure 3.39) slopewash deposits are often only just stable. Construction activities which cause such deposits to be over-steepened or to take up excessive amounts of water can result in landsliding.

3.10.2.3   *Landslide debris*

Many landslide debris soils have relatively low permeabilities, but their mass permeabilities may be high, due to the presence of cracks resulting from sliding movements. The shear strength of the colluvium is often reasonably high, but slide surfaces at the base and within the colluvium will be at or near residual. Almost invariably, the soil at the base of the slide is not the same as the slide debris, so shear strength tests on the slide debris can be misleading and usually overestimate the strength. Where the colluvium is derived from fine grained rocks such as shale, siltstone or claystone, the weathered rock underlying the colluvium often has higher permeability than the colluvium, an important point when considering drainage to reduce pore pressures and improve stability.

Landslide debris deposits are often only marginally stable and slope instability may be initiated by minor changes to the surface topography or to groundwater conditions (see Section 2.10.1).

3.10.3   *Use as construction materials*

Landslide debris and slopewash were used for the impervious core of the 161 m high Talbingo Dam, see Hunter (1982) and Section 2.10.3. The landslide debris, composed mainly of extremely weathered andesite, was used in most of the core. The slopewash, higher plasticity material derived from the landslip deposit, was used as core-abutment contact material.

Slopewash derived from extremely weathered granitic rocks was used for core-abutment contact material in the following dams, for which the parent extremely weathered granite formed the remainder of the core.

- Eucumbene (N.S.W., Australia);
- Dartmouth (Victoria, Australia);
- Thomson (Victoria, Australia);
- Trengganu (Malaysia).

Slopewash derived from extremely weathered rhyodacite was used for the core at Tuggeranong Dam (A.C.T., Australia).

When considering the possible use of landslide debris as earthfill, the critical issues are the potential variability of the soil and the possible need to remove large boulders and cobbles. Also the possibility of renewed slope movements must be considered, where the deposits occur on sloping ground. High groundwater levels and resulting wet conditions may create further difficulties.

3.10.4   *Colluvial soil – list of questions*

(a) Scree and talus:
- High permeability and compressibility?
- Timber debris, rotted or preserved?
- Potential for instability or debris-flow?
(b) Slopewash:
- Tubular voids causing high mass permeability?
- Compressible?
- Erodible?

    – Potential for slope instability?
(c) Landslide debris:
    – Variability in composition and properties – laterally and vertically?
    – Boulders?
    – Large voids?
    – Gaping or infilled cracks?
    – High compressibility?
    – Timber, rotting or preserved?
    – High permeability?
    – High water tables – wet conditions?
    – Old slide surfaces of low strength at the base, or at other levels in the deposit?
    – Potential for renewed sliding movements?

3.11   LATERITES AND LATERITIC WEATHERING PROFILES

Townsend et al. (1982) provide the following definition of laterite:

    "Laterite refers to varied reddish highly weathered soils that have concentrated oxides of iron and aluminium and may contain quartz and kaolinite. Laterite may have hardened either partially or extensively into pisolitic, gravel like, or rock-like masses; it may have cemented other materials into rock-like aggregates or it may be relatively soft but with the property of self-hardening after exposure".

3.11.1   *Composition, thicknesses and origin of lateritic weathering profiles*

The following features of lateritic weathering profiles, some of which are shown also on Figure 3.40, are based mainly on Selby (1982).

Thickness (m)	Description
0 to 2	Soil zone, often sandy and sometimes containing nodules or concretions; this may be eroded away.
2 to 10	Ferricrete or alcrete crust of reddish or brown hardened or slightly hardened material, with vermiform (or vermicular) structures (i.e. having tube-like cavities 20–30 mm in diameter) which may be filled with kaolin; less cemented horizons may be pisolithic (i.e. formed by pea-sized grains of red brown oxides).
1 to 10	Mottled zone; white clayey "kaolinitic" material with patches of yellowish iron and aluminium sesquioxides.
Up to 60 but often <25	Pallid zone; bleached kaolinitic material; the distinction between the mottled and pallid zone is not always apparent and they can be reversed; silicified zone which may be hardened (i.e. silcrete on Figure 3.40).
1 to 60	Weathered rock showing original rock structures.

The "crust" materials are termed ferricrete when they consist mainly of iron oxides. The crust can be either gravel or rock ranging from very weak to very strong, often requiring blasting for its excavation. The very weak rock materials often become stronger when exposed to the weather.

    Laterites are believed to have been formed under tropical or sub-tropical climatic conditions and most laterites of Tertiary Age and younger occur in tropical or sub-tropical areas on both sides of the equator. Lateritic profiles are found also within rocks of Mesozoic and Palaeozoic Ages, mostly at higher latitudes, where their presence indicates ancient tropical conditions which can be explained by continental drift (Bardossy and

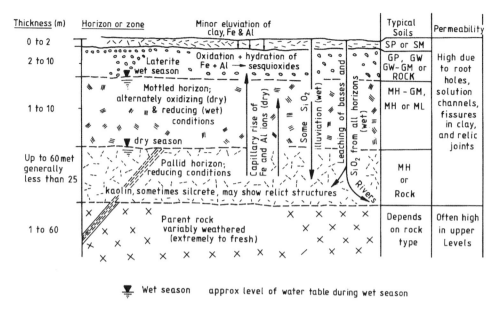

Figure 3.40.   Diagram showing the lateritic weathering profile of Selby (1982) and some of the processes involved in its development (by permission of Oxford University Press).

Aleva, 1990). Laterites have been developed on the full range of common igneous, meta-morphic and sedimentary rocks (Gidigasu, 1976).

There are several theories about the formation of laterites (Bardossy and Aleva, 1990; McFarlane, 1976) but most include the influence of a fluctuating water table to allow solution and transfer of soluble silica, iron and aluminium ions, resulting in iron and alu-minium oxides accumulating in the upper part of the profile (Figure 3.40).

Lateritic profiles may be much shallower than described above e.g. in the Ranger Mine area in northern Australia and in many other exposures in Australia and south-east Asia they are less than 5 m thick. Some of these shallow laterite profiles are of detrital origin i.e. they comprise ferricrete and/or alcrete gravels which have clearly been eroded from an earlier laterite weathering profile and redeposited. These "reworked laterites" show varying degrees of re-cementation and may or may not be underlain by mottled and pallid zones.

Figure 10.3 shows some common features of lateritic profiles in valley situations in northern and western Australia.

3.11.2   Properties of lateritic soils

The most abundant soils in the mottled zone are usually clays, sandy clays or gravelly sandy clays, which behave as soils of medium to high plasticity, but which usually plot below the "A" line in the Casagrande classification chart. Strictly speaking they are there-fore classified as silts according to the Unified Soil Classification. This behaviour is a result of the presence of allophane, kaolin, gibbsite, bauxite and often halloysite.

Other clays have been found in some laterites, for example Gordon (1984) records montmorillonite and illite in profiles developed over dolerite bedrock at Worsley, Western Australia. However this would not be common. Gordon and Smith (1984a,b) describe the results of field and laboratory tests on these and other laterite soils in that area.

As indicated on Figure 3.40 *in situ* laterite profiles are often highly permeable. Many of the structural features which cause the high permeability are near-vertical. In the upper zones vertical tubes, called "channels" or "drains" may have been formed by the decomposition of roots. These may be open or infilled with sand. The mass permeability in these zones may be 10^{-2} m/sec to 10^{-4} m/sec. These near-vertical features are not located readily by conventional drilling and water pressure testing and so in these cases the permeability is best determined by vertical infiltration tests involving relatively large test areas.

3.11.3   Use of lateritic soils for construction

Lateritic soils usually make excellent earthfill construction material. The most notable case of the use of lateritic soil was in Sasumua Dam, where Terzaghi (1958) showed that despite its apparently peculiar classification properties (i.e. plotting below the "A" line), the lateritic soil was an excellent dam building material. When used as fill, laterite soils are characterised by high effective friction angle and medium to low density and permeability. In most cases they are readily compacted despite often having high and poorly defined water content. For example at Sirinumu Dam, lateritic clays were readily compacted at water contents between 40% and 50%. However, some particularly silty laterites with high halloysite contents can be difficult to compact.

The ferricrete gravels and weak rocks in the near-surface crust zone are used in lateritic areas throughout the world as base or sub-base material in pavements for roads and airstrips. Strongly cemented rock from this zone has also been used successfully as rockfill and rip-rap.

In the laterite and mottled zone soils are usually non dispersive and in the authors' experience, very resistant to erosion *in situ* and recompacted.

3.11.4   Karstic features developed in laterite terrain

Sinkholes similar to those seen in karstic limestone are known to occur also in some laterite profiles developed on non-carbonate rocks.

Twidale (1987) describes sinkholes up to 50 m diameter and 15 m deep on the Western Stuart Plateau in Northern Territory, Australia. The plateau is underlain by a sequence of siltstones and quartzites, ranging from 40 m to 230 m in thickness, which is underlain by limestone. The laterite profile is developed at the top of the non-carbonate sequence and may be 20–30 m deep. Twidale (1987) suggests that where the sequence is thin some of the sinkholes may have formed by the lateritic materials collapsing directly into voids in the underlying limestone. However he believes that they most probably developed as follows:

– Voids were formed within the lateritic profile by the removal of silica and silicate minerals by solution, these being carried down joints, deeper into the sequence, and
– The ferricrete cap collapsed into the voids.

Twidale believes that the subsurface voids developed in late Tertiary time and that collapse of the ferricrete cap into the voids has continued intermittently since that time. He provides details of the youngest sinkholes, formed in 1982, and describes how many older ones are much modified by sidewall collapse and infilling with slopewash and alluvial soils.

Twidale (1987) refers also to karstic features in laterite developed on peridotite (ultrabasic rock) in New Caledonia. The karstic features in that area occur in broad plateaus surrounded by steep ridges which are remnants of relatively fresh peridotite. They include:

– internal drainage into sinkholes;
– recently active sinkholes up to 15 m wide, and
– swamps, believed to represent collapsed areas now filled with sediment and water.

Many of the sinkholes occur in lines above steeply dipping defect zones in the peridotite. The collapses appear to have occurred progressively into voids formed where silica and silicates have been removed by both dissolution and erosion and carried away in near-horizontal and near-vertical flow paths (widened defects) within the defect zones.

The authors suggest that, if such features are evident, investigations should be carried out to establish the mechanism and the potential effect on the dam or its storage.

3.11.5   *Recognition and interpretation of silcrete layer*

As shown in Figure 3.40, the silcrete formed by deposition of silica in the pallid horizon can contain "relic" structural features, e.g. bedding or foliation which were present in the original rock before it was weathered. It is important that this is understood, to avoid errors in interpretation of drill cores and exposures in excavations. For example, a typical silcrete layer is very strong rock, similar to quartzite and is near-horizontal. It is usually underlain by weathered rock with soil or very weak rock properties. If the silcrete was formed within weathered material which was previously vertically bedded rock, it will appear to show vertical bedding. It might then be misinterpreted as part of a vertical layer of quartzite bedrock, continuing to depth.

This type of situation was found during the raising of Hinze rockfill dam in Queensland (see Fell et al., 1992, Page 122). Steeply dipping beds of open-jointed and cavernous "quartzite" were found to grade downwards into dolomite and limestone. It is believed that silica released during lateritic weathering of adjacent metamorphic rocks had progressively replaced the uppermost 40 m or so of the dolomite and limestone, to form a cavernous silcrete horizon.

3.11.6   *Lateritic soils and profiles – list of questions*

– Variable laterally and vertically?
– Deeply weathered?
– High *in situ* permeability, vertical and horizontal?
– Low *in situ* density at depth?
– If sinkholes present, their mechanism and effect?
– Fine soils suitable for earth core?
– Gravelly ferricrete or alcrete suitable for pavements?
– Cemented material in crust suitable for rockfill or rip-rap?
– Silcrete horizon or quartzite bed?

3.12   GLACIAL DEPOSITS AND LANDFORMS

During the Pleistocene period, large parts of the earth's surface were covered by sheets of ice, similar to those which occur today in Greenland, Antarctica, parts of Europe and North and South America. The ice moved across the landscape, eroding and reshaping it, and when it melted it deposited the eroded materials. Similar "ice ages" occurred earlier in the earth's history, but with rare exceptions glacial deposits formed at those times have been so modified and strengthened by diagenesis or metamorphism that they are now rocks not greatly different in engineering properties from the surfaces on which they were deposited. The discussion here is therefore limited to effects of the Pleistocene glaciation and of valley glaciers such as still occur commonly in alpine regions (Figure 3.41).

Gaciated landscapes usually have complex histories of erosion and deposition, including for example, the following sequence shown on Figure 3.42.

Figure 3.41.   Aerial view of the Tasman Glacier in New Zealand. Photo courtesy of Mr. Lloyd Homer, DSIR, New Zealand.

– Erosion and shaping by rivers (Figure 3.42a);
– Erosion and reshaping by ice, and deposition of glacially derived materials (Figure 3.42b);
– Erosion and reshaping of the new landscape and the glacial deposits, by subsequent rivers (Figure 3.42c).

Such histories have resulted in a wide variety of landscapes and deposits. The deposits vary widely in their engineering properties. Unfortunately, from examination of the mineral content and texture of a soil it is frequently not even possible to determine whether or not it is of glacial origin. If these characteristics are considered in relation to the landforms on or in which the soil occurs it is often possible to confirm a glacial origin and to make more detailed predictions about its history and thus about its likely distribution and engineering properties.

Boulton and Paul (1976) introduced the "land system" form of terrain evaluation as a means of classifying and mapping of sediment sequences and landforms. This land system approach is used by Eyles (1985) who recognizes three main land systems resulting from glaciation.

– Subglacial and supraglacial – characteristic of lowlands where sediments and landforms were formed by large sheets;
– Glaciated valley – characteristic of areas of high relief in which the ice was restricted to valleys.

Other glacially-related environments are:

– Periglacial – in which intense frost action modified the glacially developed landforms and materials;

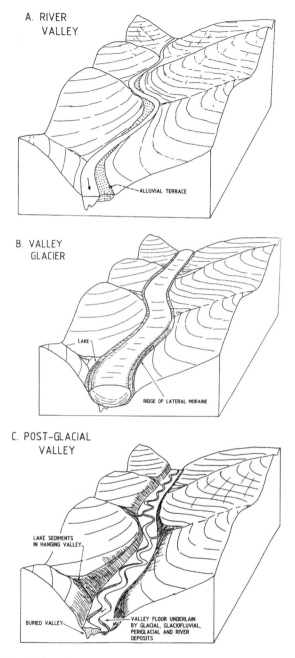

A. RIVER
   VALLEY

~ALLUVIAL TERRACE

B. VALLEY
   GLACIER

LAKE

RIDGE OF LATERAL MORAINE

C. POST-GLACIAL
   VALLEY

LAKE SEDIMENTS
IN HANGING VALLEY

BURIED VALLEY

VALLEY FLOOR UNDERLAIN
BY GLACIAL, GLACIOFLUVIAL,
PERIGLACIAL AND RIVER
DEPOSITS

Figure 3.42.   River valley, before during and after glaciation.

– Glaciofluvial – in which glacially derived sediments were transported and deposited by
  water.

The processes and products of each land system and environment are discussed in some
detail in Eyles (1985). Discussion in this present chapter will be limited to the glaciated valley land system (the most significant in dam engineering) plus brief notes on glaciofluvial
and periglacial aspects.

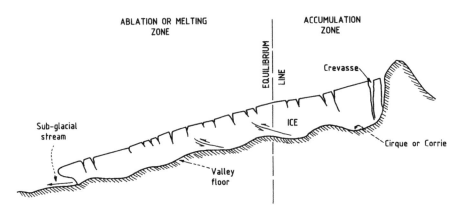

Figure 3.43.   Diagrammatic section along a valley glacier (based on Blyth and de Freitas, 1989).

3.12.1   *Glaciated valleys*

Figure 3.43 is a diagrammatic section along a valley glacier during a period of its "advance" down the valley. It can be seen that the ice tends to move over irregularities in its path, behaving in a viscous manner near its base and failing in shear and tension at higher levels. The resulting valley floor is uneven and often contains deep hollows eroded by the ice. Many such hollows are now occupied by lakes, the water is dammed partly by ridges of rock and partly by glacial debris. Streams of meltwater flow beneath the ice and exit at the snout.

During periods in which winter snows do not survive through summer the glacier diminishes in size and its snout "retreats" up the valley.

Although glacial valleys are generally U-shaped when viewed broadly, in detail the valley floors are often highly irregular as at Parangana Dam in Tasmania (Figures 3.44 and 3.45). The shape of the buried valleys at this site and nature of the infill materials, indicates that they are remnants of pre-glacial river valleys. Figure 3.44 illustrates the way in which such a pre-glacial river channel can be preserved under glacially derived deposits.

In other cases locally deeper valley sections under glaciers may have been differentially eroded by ice or meltwater streams.

3.12.2   *Materials deposited by glaciers*

The general generic term for material deposited by glaciers is "till". Deposits of till are often referred to as "moraine". A further term "drift" has been used extensively and loosely to describe surficial deposits of glacial, alluvial or colluvial origins.

Figure 3.46 shows mechanisms of accumulation, ingestion and transport of debris by a glacier. Upslope from the equilibrium line (Figure 3.46a) windblown dust and rock debris is buried within the snow which feeds the glacier and is transported downglacier forming the basal debris zone and thin basal traction layer. The rock particles in the traction layer abrade, polish and groove the rock floor, generating fine rock particles known as rock flour. The ice and rock blocks within it also pluck or "quarry" rock from the floor and sides. The resulting material developed in contact with the rock floor is known as lodgement till (Figure 3.46b) and usually contains a wide range of particle sizes. The lodgement till is formed under relatively high effective normal pressures and consequently is usually compacted to at least stiff consistency. Its upper surface is grooved or fluted parallel to the direction of flow of the ice.

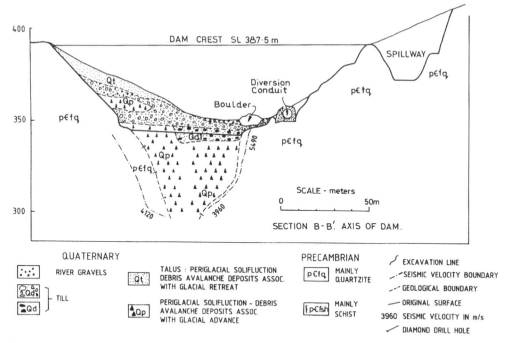

Figure 3.44. Parangana Dam, Tasmania, cross section along dam axis (Paterson, 1971).

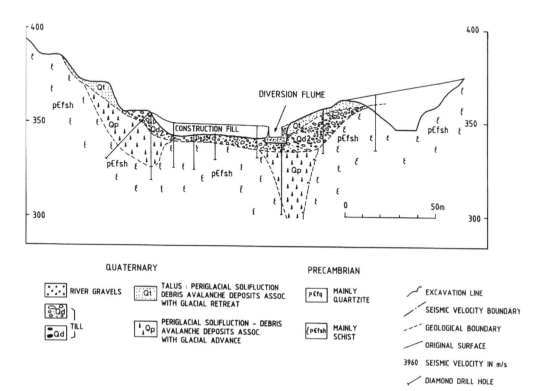

Figure 3.45. Parangana Dam, Tasmania, cross-section 100 m downstream from dam axis (Paterson, 1971).

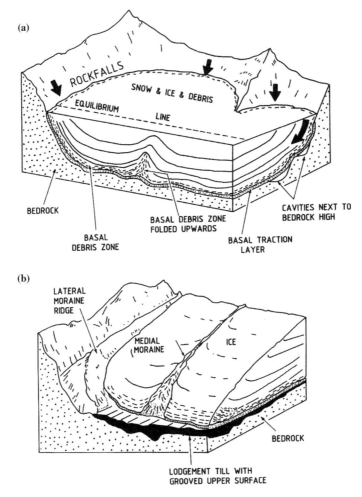

Figure 3.46.   Debris transport by a valley glacier (Eyles 1985, by permission of Pergamon Press).

Where two glaciers converge to form a composite valley glacier, the debris zone may be folded upwards by the compression generated along the glacier contacts and in the ablation zone (Figure 3.46b) the debris may become exposed as a supraglacial medial moraine.

Below the equilibrium line (Figure 3.46b) debris falling onto the glacier surface is not ingested by the glacier because the winter snows do not survive the next summer. The debris is transported along the glacier sides as ridges of supraglacial lateral moraine or as supraglacial medial moraine after the convergence of two glaciers.

Figure 3.47 shows a typical situation at the snout where debris from the glacier surface and within it are deposited as the ice melts.

Figures 3.48 and 3.49 show the development of a ridge of supraglacial lateral moraine and soils associated with it during a single advance/retreat stage of a glacier.

During the advance (Figures 3.48a to c), debris which slides off the glacier surface forms steeply dipping deposits (lateral moraine) analogous to talus or end-dumped fill. The material, more precisely known as supraglacial morainic till, includes a wide range of particle sizes. Outwash materials with initial near-horizontal bedding are deposited by streams in the trough between the lateral moraine and the valley side. At depth the accumulated moraine

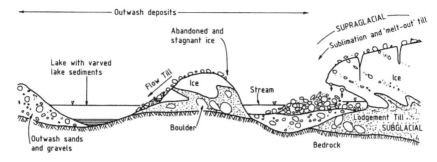

Figure 3.47.   Diagrammatic longitudinal section at a glacier snout, and downstream (Blyth and de Freitas, 1989).

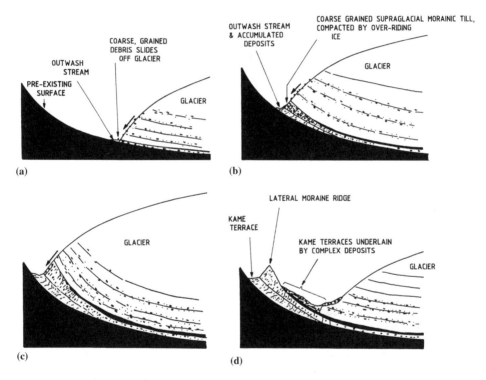

Figure 3.48.   Development of a lateral moraine ridge and associated deposits (Boulton and Eyles, 1979).

and outwash materials are compressed and compacted by the glacier and their dips become steeper.

During the retreat (Figure 3.48d) the glacier shrinks away leaving the lateral moraine as a ridge. Such ridges often remain steep (up to 70°) due to the high degree of compaction and some cementation of the moraine. Between the ridge and the ice a series of terraces develops. These are known as kame terraces and are underlain (Figures 3.48d and 3.49) by complex sequences including glaciofluvial sands and gravels (stream-deposits), laminated clays (lake-deposits), poorly- or non-compacted till (supraglacial morainic till) and lodgement till.

Terraces formed by the outwash next to the valley side are also known as kame terraces (Figure 3.49).

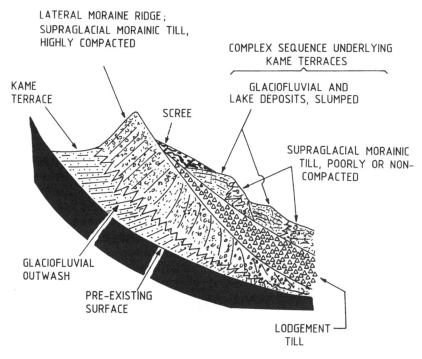

Figure 3.49.   Diagrammatic section through a lateral moraine ridge (Boulton and Eyles 1979).

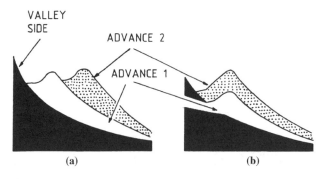

Figure 3.50.   Lateral moraine ridges deposited by successive glacial advances. (Boulton and Eyles, 1979).

The lateral moraine and associated deposits are usually more complex than shown on Figure 3.49 because of repeated glacial advances and retreats. Figure 3.50 shows the relationships between lateral moraine ridges deposited by successive glacial advances. In Diagram (a) Advance 1 was more extensive than Advance 2, and in Diagram (b), Advance 1 was the less extensive.

Figure 3.51 is the complete glaciated valley land system, showing common features of the ablation zone and the deposited materials and landforms. Numbered features on this diagram which need further explanation are discussed below.

The lodgement till (1) at the base of the glacier is poorly sorted, usually containing more than 50% of sand, silt and clay sizes, forming a matrix which supports gravel and larger sized particles. When unweathered, the fines fraction usually comprises finely ground quartz, carbonate and inactive clay minerals. The lodgement till is well compacted

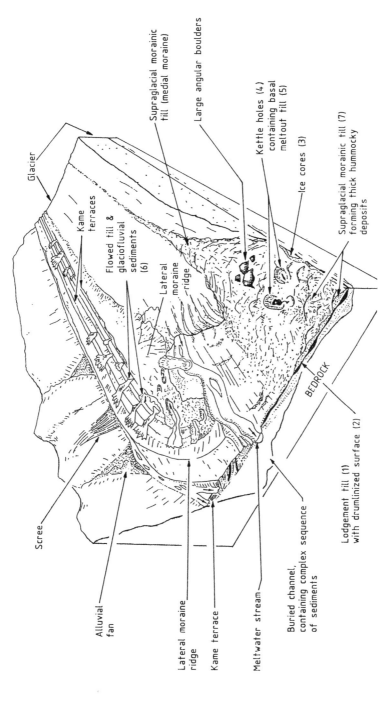

Figure 3.51.   Landforms and sediments associated with the retreat of a valley glacier (based on Eyles 1985, by permission of Pergramon Press).

by the overlying ice, has a well developed fissured fabric parallel to the ice flow direction and is commonly highly deformed (Figure 3.46). It usually has very low permeability, but may be rendered locally permeable in the mass by thin glaciofluvial sand layers deposited in channels of meltwater streams.

The till near the downstream toe at Parangana Dam (Figure 3.45) ranged from "toughly compacted' to weak rock and is considered by Paterson (1971) to be lodgement till.

Drumlins (2) are elongated dune-like mounds on the surface of the lodgement till, apparently moulded to this shape by the moving ice.

Ice-cores (3) are masses of ice left behind by a retreating glacier and buried in till. Melting of ice-cores causes the development of sinkhole-like features known as kettles (4) and deposition of basal melt-out till (5) which is usually crudely stratified.

Flowed till (6) is till which has been reworked and deposited by mudflows, the scars of which can be seen extending from the lateral moraine ridge through its adjacent kame terraces. The flowed till may be stratified due to redistribution of fines during flow. The kame terraces here are underlain by complexly interbedded sediments as shown on Figure 3.49.

Supraglacial morainic till (7) deposited at the snout, as shown on Figure 3.47, contains a large range of gravel and larger fragment sizes and is usually deficient in fines. It is known also as supraglacial melt-out till or ablation till. The younger glacial deposit at the dam axis area at Parangana Dam (Figure 3.44) is believed to be of this type. It comprises up to 70% of gravel to boulder sizes in a matrix of sand, clay and silt. The largest boulders are up to 3 m diameter.

3.12.2.1 *Properties of till materials*

It is clear that basal melt-out till, flowed till and supraglacial morainic till will be poorly consolidated unless they become desiccated or covered by new sediment or ice at a later time in their history. The latter appears to be the case at Parangana Dam where the supraglacial morainic till is well compacted, apparently due to being overridden by ice during subsequent glacier advances (Paterson, 1971). Experience has shown that most tills of all types have low permeabilities (less than 10^{-10} m/s). However, till deposits may contain or be next to bodies of permeable sands and/or gravels of glaciofluvial origin, filling old channels. For example the various outwash deposits shown on Figure 3.48 would normally have relatively high permeabilities. At Parangana Dam the younger till beneath the valley floor contained a layer of glaciofluvial sands several metres thick with measured permeability of 10^{-6} m/s. To allow for this layer the core cutoff trench was excavated down into underlying materials of low permeability (Figure 3.44). The younger till was left in place beneath the downstream shoulder, but was covered with a filter blanket before placement of the rockfill.

Walberg et al. (1985) describe remedial works required at the 23 m high Smithfield Dam in Missouri, USA, after seepages and high piezometric pressures were recorded during first filling. Exploratory drilling using 152 mm diameter cable tool tube sampling showed that glacial outwash sands and gravels beneath the left abutment were more continuous and permeable than assumed from the pre-construction drilling and sampling.

At Cow Green Dam in Britain most of the material filling a buried channel beneath the left side of the valley was found to be lodgement till, described as "stiff, dark brown, poorly-sorted, unstratified, silty, sandy clay of medium plasticity containing subangular to rounded gravel, cobbles and boulders.....". The boulders ranged up to 2 m in mean diameter (Money, 1985).

Sladen and Wrigley (1985) describe further generalisations which can be made about the geotechnical properties of lodgement tills.

3.12.2.2 *Disrupted bedrock surface beneath glaciers*

Knill (1968) describes the open-fractured nature of bedrock beneath glacial materials at several sites and concludes that gaping or infilled joints near-parallel to the rock surface were initiated as shear fractures by the moving ice (i.e. by "glacitectonic thrusting") and then opened up by ice wedging. The authors have found similarly fractured rock with open and infilled joints at the bases of many valleys which have not been subjected to glaciation (see Chapter 2, Section 2.5.4) and suggest that the effects seen by Knill may have been formed largely by stress relief.

Regardless of their origin, the presence of such features at many sites means that the rock next to the base of glacial deposits is likely to be of poor quality e.g. in terms of compressibility, permeability and erodibility. Where the existence of such poor quality rock would be of significance to the stability and/or watertightness of the dam it is important that this "rockhead" zone be investigated thoroughly.

Money (1985) draws attention to the difficulties often encountered while doing this by core drilling and mentions cases where large boulders have been mistaken for bedrock. He refers to UK practice at that time which was to recommend that the *in situ* rock be proven by a minimum of 3 m of cored rock. He states that this figure is likely to be inadequate and that it is certainly not enough to allow for adequate permeability testing of the upper part of the bedrock. The authors agree and suggest that the actual depth of coring needed will depend upon:

– The inherent fabric of the bedrock (i.e. is it massive or too well-cleaved or closely jointed for it to have formed large boulders?);
– The quality of the core samples and the extent to which core orientation can assist in assessing whether it is *in situ* or not. Core orientation may be determined either by impression packer or orientation device or by the presence of bedding or foliation with known, consistent orientation;
– The actual depth of the disturbed zone.

Difficulties in delineation of the top of *in situ* rock can occur also where the rock in this zone has been chemically weathered. This situation exists at the site for Kosciusko Dam in New South Wales (Figure 3.52).

It appears that intense weathering has occurred in both the bedrock and the till. Geological surface mapping, track exposures and boreholes on the right bank show that most of the upper 5–20 m of the bedrock comprises residual "boulders" of fresh to slightly weathered granite set in a matrix of highly to extremely weathered granite which is mainly a very compact silty, clayey sand.

A shaft close to the creek on the left bank (Figure 3.53) showed similarly weathered granite boulders set partly in a matrix of gravelly clay (till) and partly in glaciofluvial sands and clays. It was clear that without very good recovery of little-disturbed core, these materials could not be readily distinguished from the *in situ* weathered "bouldery" sequence. At 8.6 m these materials rested on extremely weathered rock whose mineral content and foliation attitude matched the known bedrock on the right bank. This weathered rock was therefore inferred to be *in situ*.

Holes drilled elsewhere on the left bank recovered about 20% of fresh or slightly weathered granite in "boulder" lengths, but little or no matrix. Hence the upper surface of the *in situ* rock (assuming that it was more than slightly weathered) could not be determined here from the drilling results. The top of mainly fresh granite was inferred from the drill cores together with the results of refraction seismic traverses (Figure 3.52).

Glastonbury (2002), Glastonbury and Fell (2002c) note that many of the large rapid landslides which have occurred in historic times have occurred in valleys which have been glaciated, and attribute this to the stress relief effects on the valley sides as the glaciers retreat.

3.12.3 *Glaciofluvial deposits*

As well as depositing some glaciofluvial materials in their immediate vicinity (as discussed in Section 3.12.2) glaciers release very large meltwater flows giving rise to deposition of

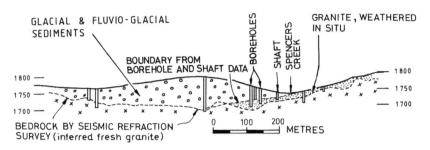

Figure 3.52.   Cross section through the site for Kosciusko Dam.

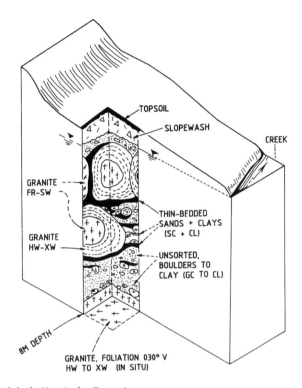

Figure 3.53.   Log of shaft, Kosciusko Dam site.

vast amounts of gravels and sands in braided (multiple-channel) rivers usually extending tens of kilometres downstream, often across broad outwash plains (Figure 3.54). Grain sizes range from gravel-dominated near the glaciers to sands and silts further downstream. In some places lakes are formed, in which laminated silts and clays are deposited.

Deposition, eroding and reworking of the braided river deposits occurs cyclically in phase with glacial advances and retreats, resulting in terraces at various levels across the river valleys and outwash plains. The streams eventually flow into lakes or the sea.

Exposed areas of silt and fine sand are eroded by wind and redeposited as loess on the surrounding country.

Miall (1985) describes sedimentological aspects of the braided stream deposits.

The sands and gravels are usually clean with well-rounded particles and often provide excellent sources of materials for embankments and for concrete aggregate.

Figure 3.54.   Broad valleys downstream from glaciers in New Zealand, underlain by great depths of glaciofluvial gravels and sands. Photo courtesy of Mr. Lloyd Homer, DSIR, New Zealand.

The braided streams and lakes produce very complex, lenticular deposits of sands, gravels, silts and clays. Terzaghi and Leps (1958) describe the design, construction and performance of Vermilion Dam, a 39 m high zoned earthfill structure built on such complex glaciofluvial deposits, with maximum depth of 82 m.

In some deposits there are lenticular beds of "open-work" gravels or boulders which are uniformly sized materials with large voids and extremely high permeability. Cary (1950) describes the widespread occurrence of these materials in glaciofluvial deposits in northwestern USA.

3.12.4   Periglacial features

Periglacial conditions are defined here as those under which frost is the predominant weathering agent. They are often, but not always, associated with glaciers. Permafrost conditions are commonly present but are not essential. Permafrost occurs where winter temperatures are rarely above freezing point and summer temperatures are only high enough to thaw the upper metre or so of the ground.

Figure 3.55, modified slightly from a diagram of Eyles and Paul (1985), shows features which may be developed under periglacial conditions.

1. Deep-seated creep of weak sedimentary rocks into the valley. Competent beds develop widely gaping or infilled extension joints (gulls) and move downslope on the weak materials as complex slides or rafts;
2. Weak rocks contorted and bulged upwards, overlain by terraced gravels;
3. Outcrops showing evidence of toppling and cambering, with scree and rockfall deposits downslope;
4. Outcrops of very strong crystalline bedrock surrounded by blockfield of frost-heaved bedrock. Terraces cut by nivation – free-thaw and slopewash at the margins of snow patches;

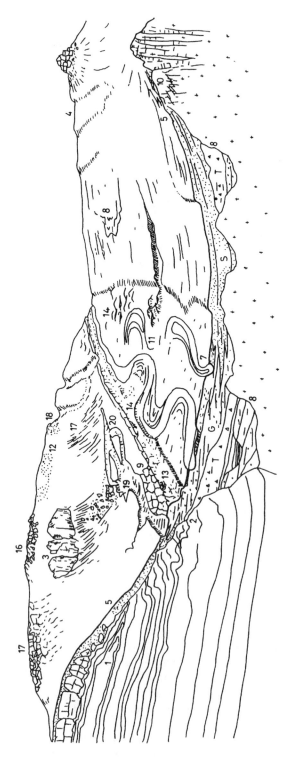

Figure 3.55.   Diagram showing features which may be developed under periglacial conditions (Based on Eyles and Paul, 1985).

5. Solifluction fans or lobes – crudely bedded gravelly or bouldery deposits, thickening downslope, grading into slopewash beneath the lower slope;
6. Mudflow scar underlain by low-angle shear surface;
7. Mantle of solifluction debris (S) intertongued with gravels (G) deposited by periglacial braided stream, and overlain by peats and silts (M) forming the floodplain of the modern meandering stream;
8. Till (T) in the buried valley and in pockets in the rockhead;
9. Polygonal patterned ground and fossil casts of ice-wedged cracks;
10. Frost-shattered bedrock;
11. Doughnut-shaped degraded ramparts left by former pingos;
12. Soil mantle resulting from mechanical churning (frost heaving) of pedological soils and bedrock. The underlying rockhead is often highly irregular;
13. Involutions and contortions in the soil caused by high pore pressures developed during re-freezing of thaw-soaked soil;
14. Faulted soil zones caused by ground contraction during freezing and/or collapse into voids following melting of buried ice;
15. Patterned ground – polygonal nets and stripes;
16. Blockfield;
17. Alluvial fans;
18. Dry Valleys;
19. Terraced gravels (G) deposited by braided periglacial streams. The bedrock surface below may contain scoured channels and hollows;
20. Loess.

Eyles and Paul (1985) point out that most of the features indicated on this diagram may also develop (more slowly) in warmer climates. The authors agree and in particular consider it likely that the "Deep-seated disturbances" Features 1 and 2 would have been developed to a large extent as a result of destressing (as described in Chapter 2) before any cold-climate effects. The following explanation of the main cold-climate processes and features indicated on Figure 3.55 is based also on Eyles and Paul (1985).

Solifluction (7) is the slow flowage or creep of a water-soaked mass of soil and rock debris, either as a true flow or as a slide where most movement occurs over a basal shear surface. Typical flow rates range from 10 mm to 60 mm per year. Solifluction is caused by the generation of excess pore pressures during the thaw. Eyles and Paul (1985) provide a summary of theoretical studies of the process by Nixon and Morgenstern (1973) and McRoberts and Morgenstern (1974).

Another solifluction mechanism is the downslope displacement of soil particles by needle-ice. The ice needles grow normally to the ground surface and so during each freeze-thaw cycle the supported particles are displaced slightly downslope.

Frost-shatter (10) is the mechanical disruption of rock masses by the expansion on freezing of groundwaters.

Cyroturbation (12), (13) and (15) is the churning or mixing of soils which occurs in permafrost conditions at the end of the melt season due mainly to high pore pressures set up towards the base of the thawed layer as the soil re-freezes from the surface down. Expansion due to freezing of porewaters also contributes, as does the upward heave of cobbles and boulders resulting from their greater thermal conductivity. The "involutions" (13) are pseudo-intrusive structures similar to flame structures in sediments and to gilgai in clays. Patterned ground (15) is ground showing a regular hummocky pattern (polygons or circles). It is a surface expression of cyroturbation.

Ice-wedging (9) refers to cracks which occur in soils during intense cold widened by wedging action when water freezes in them. The cracks are preserved as casts by soil which migrates into them.

Pingos (11) are conical mounds of buried ice up to 40 m high and 600 m in diameter. Pingos, which have developed by the freezing of upward-moving groundwater, have caused updoming of the surrounding sediments. Melting of the ice results in the dough-nut-shaped surface features shown on Figure 3.55.

Paterson (1971) provides a useful account of periglacial and other effects of glaciation features at Parangana Dam. Core-drilling into the steep sided buried pre-glacial river channels at the site (Figures 3.44 and 3.45) showed the lower part of each to be filled mainly by angular rock fragments up to 500 mm across set in a sandy clay to clayey sand matrix. The ratio of rock to matrix was roughly 60 : 40. The rock fragments were of quartzite and schist, clearly derived from the bedrock immediately upslope in each case. These deposits were judged to have been formed by "solifluction debris avalanches" from the steep valley sides, under periglacial conditions during the glacial advance period.

In the upstream channel the avalanche material contained a discontinuous bed of clay believed to have been deposited in a lake formed upstream from one of the slides. Also present in the slide deposits were lenticular beds of sand and gravel, inferred by Paterson to be stream deposits in channels eroded through the slide dams. The distribution of sand and gravel beds suggested that at least five major landslides had occurred.

Locally derived talus covers most of the ground surface at the site and extends to about 10 m depth in the foundation area (Figures 3.44c and 3.45). This material is believed to be a periglacial solifluction product formed during the final retreat stage of the glacier.

3.12.5   Glacial environment – list of questions

Only the most significant glacier-related questions are listed here; these and other features which are usually of less significance in dam engineering are shown on Figures 3.51 and 3.55.

- Buried valleys?
- Bedrock surface or boulder?
- Bedrock disrupted near upper surface?
- Wide variety of till types?
- Materials unsorted – clay to boulder sizes?
- Slickensides in clay-rich till?
- Variable compaction and cementation?
- High permeability sands and gravels?
- Loess?
- Landslipped deposits?
- Creeping landslides?

4

Planning, conducting and reporting of geotechnical investigations

4.1 THE NEED TO ASK THE RIGHT QUESTIONS

The French detective, Bertillon, considered by many to the father of modern crime detection, is reputed to have made the following statement.

"We only see what we observe, but we can only observe that which is already in the mind". Experience from analysis of many case histories (Stapledon, 1976, 1979 and 1983) shows that this principle (that we will only find that which we recognise) is equally true in engineering site investigations. In almost every foundation failure and contractual dispute over "changed geological conditions", it is found that a major contributing factor has been the failure of project planners and site investigators to fully understand and define all of the geotechnical questions which needed to be answered by the site investigations. There are two types of questions, namely:

– *engineering questions*, which relate essentially to the design, construction and operation of any structure of the type proposed, and
– *geological questions*, which arise from understanding of the site geological environment and its likely influence on the design, construction and operation of the project (see Chapters 2 and 3).

4.1.1 *Geotechnical engineering questions*

For dams which are intended to store water, or water plus solids, it is obvious that important questions must relate to the permeability of the foundations. However, there are many other equally important questions, because construction and operation of a dam causes much greater changes to a site environment than any other type of engineering activity. Four main processes involved in dam engineering, their principal effects on the site environment, and some resulting questions for the designer and site investigator, are as follows;

1. Excavation: To reach suitable levels for founding the dam, and also for the spillway, outlet works and for construction materials. Excavation causes removal of support from and increases in shear stress in the surrounding material and hence raises questions of stability of the excavations themselves, during construction and/or operation, and of the stability of the adjacent valley slopes.
2. Foundation loading: Imposed by the dam structure, raises questions of compressibility of the foundation and its shear strength against sliding upstream or downstream, before and after filling of the storage and under flood and earthquake loading.
3. Inundation – Filling the storage: Causes changes to the groundwater regime, lowering of strengths of cohesive soils, weak rocks and joint cements, and decreases in effective stress. These effects all add to the questions of stability of the dam and its foundation and also they raise the question of stability of the reservoir sides. These stability questions are more serious when water storage levels are required to fluctuate widely and rapidly.

4. Flood discharges: Have high potential to erode, raising questions about the location and erodibility of discharge areas for the spillway and outlet works.

Table 4.1 is a suggested checklist of geotechnical engineering questions to be answered during site investigations for dam projects.

4.1.2   Geological questions

Typical geological questions for twelve common geological environments have been discussed and listed in Chapter 3. The following notes highlight the importance of asking

Table 4.1.   Geotechnical engineering questions.

1. Sources of materials, for the following purposes:
   - Earthfill, for the core or other zones;
   - Filters;
   - Rockfill;
   - Rip-rap;
   - Concrete aggregates;
   - Road Pavements;
      For each material: Location of alternative sources, qualities/suitabilities, quantities, methods for winning and processing. Overburden and waste materials and quantities. Possible use of materials from required excavations, e.g. spillway, outlet works and dam foundations.
2. Reservoir
   - Watertightness;
   - Effect on regional groundwaters – Levels or quality;
   - Stability of slopes inside and outside of reservoir rim;
   - Erodibility of soils – Possibility of turbidity problems;
   - Siltation rates and likely location of deposits.
3. Dam
   - Location – To suit topographic and geological situations;
   - Alternative* sites, for comparison of costs and of geotechnical and other issues;
   - Type(s) of dam suited to site(s);
   - Depths to suitable foundations for: concrete dam; earthfill; core; filters; rockfill; plinth or grout cap;
   - Nature of materials to be excavated, excavation methods, and possible uses of materials;
   - Stability of excavations, support and dewatering requirements;
   - Permeability, compressibility and erodibility of foundations;
   - Foundations treatment(s) required: grouting; drainage; slurry concrete; dental treatment; filter blanket; other;
   - Embankment zones, methods of placement, and of control of quality, moisture and compaction;
   - Stability of dam, and dam plus foundation in all situations;
   - Monitoring systems: types, siting.
4. Spillway, river diversion works and permanent outlet works
   - Location and type;
   - Excavation method(s), possible use for excavated materials;
   - Stability of excavations, need for temporary/permanent support;
   - Channel, need for lining/drainage;
   - Need for protection of the discharge area, or for excavation of a stilling basin.
5. Tunnels (for river diversion, spillway, or permanent outlet works)
   - Location of sites for portals;
   - Tunnel alignment(s);
   - Portals, excavation and support methods and final treatment;
   - Tunnel excavation, support and final treatment methods.
6. Seismicity of region
   - Design earthquake, annual exceedance probability versus ground motion;
   - Maximum credible earthquake.

* Alternative sites for the dam and associated structures are usually investigated during Stage 2 (Feasibility and site selection).

and finding the answers to such questions, and of relating them to other site factors such as climate and topography and the proposed development.

4.1.2.1   *Questions relating to rock and soil types, climate and topography*

The relative importance of any one of the engineering questions on Table 4.1 and the amount and kind of site investigations needed to get the answer to it will depend on the topographic, geological and climatic environments in which the project is to be located. For example, consider the effects of first filling of a water supply dam in an arid region, in a steep sided valley underlain by a very weak sandstone. The water table is likely to be very low or absent, and the sandstone may owe a large part of its strength and stiffness to cementation by a water-soluble minerals such as gypsum. Filling of the reservoir has the potential to cause dramatic changes to this site – significant raising of the water table, solution of water soluble mineral with resulting weakening and possible increase in permeability of the foundations, and possible instability in the storage area sides. Hence specific geological questions to be answered during investigations of this site would include the following:

– What are the cementing agents in the sandstone?
– How much reduction in strength and stiffness will occur in the sandstone when saturated for long periods?
– Could solution effects during dam operations result in increase in permeability of the foundation?
– Could solution/strength reduction or water table rise result in instability in (a) the foundation or (b) the reservoir sides?

At the other extreme, a site in a high rainfall area with gentle slopes underlain by very strong quartzite, and with a high water table, is likely to be almost unaffected by inundation.

It can be seen from these two very simple examples that certain generalizations can be made about geotechnical conditions likely to be met at a site, when its broad geological setting is known, and this is considered together with the site climate and topography.

4.1.2.2   *Questions relating to geological processes, i.e. to the history of development of the site*

It is not enough, during the design and construction of a major dam, to know simply what rock or soil types are present, their engineering properties and their approximate distribution. Understanding the site environment implies also understanding the geological processes which developed the region and the site. The most important processes are usually the youngest, commonly those relating to the near-surface. This is partly because they will have had a major influence on the strength and stability of the valley slopes, and also because they may still be active, or may be reactivated by the construction or operation of the dam. Table 4.2 is a list of such processes, some of which have been discussed in more detail in Chapters 2 and 3.

Table 4.2.   Processes which may be active enough to affect a dam project.

– Destressing	– Freezing
– Chemical weathering of rocks and/or soils	– Burrowing by animals
– Solution	– Growth of vegetation
– Deposition of cement	– Growth of vegetation
– Erosion by wind or water	– Rotting of roots of vegetation, or buried timber
– Deposition of sediment	– Seismicity, i.e. shaking, or displacement on a fault
– Creep, landsliding	– Vulcanism
– Subsidence	– Glaciation
– Pressure by groundwater	

4.2   GEOTECHNICAL INPUT AT VARIOUS STAGES OF PROJECT DEVELOPMENT

Most dam projects develop in five more or less distinct stages as set out on Table 4.3. Geotechnical input is required in all stages. Table 4.3 summarizes the broad objectives of the geotechnical work and the usual activities and reporting needed, during each stage.

Table 4.3.   Geotechnical input during stages of development of dam project.

Stage No.	Name	Objectives and activities (Geotechnical)
1	Pre-feasibility	Assist in the selection of possible sites and obtain enough understanding of the geological situation to plan the feasibility and site selection studies. Usually includes review of existing data plus a short ground and sometimes air inspection.
2	Feasibility and site selection	Assess the project feasibility and design from the geotechnical viewpoint, considering both the regional and local geological situations. Explore alternative sites for dam and other key structures, and adopt the most promising sites. Explore these further if necessary to confirm feasibility and provide sufficient data for preliminary design and feasibility stage cost estimate. Provide regular progress reports and prepare formal report at the end of Stage 2 with a definite statement confirming (or otherwise) the project feasibility from the geotechnical point of view.
3	Design and specification	Answer any questions outstanding or arising from the feasibility studies, and additional geotechnical questions raised during the design. Further site exploration and testing are usually necessary. Provide regular progress reports, and report for tenderers or construction agency. Provide assistance in the design of the embankment and other structures and preparation of the specification.
4	Construction	Ensure that the geological picture exposed during construction is as assumed in the design and if not, that modifications are made to the design, if necessary. Provide day to day advice on geotechnical matters to the resident engineer. Provide record of geological exposures during construction and of any rock movements, water inflows, etc. in regular progress reports. This data is vital to the Surveillance Group (see Stage 5); should any malfunction develop it may form the main basis for (sometimes rapid) correct action. Activities include detailed mapping, colour photography, review of inspector's records, installation and reading of instruments, monitoring simple tell-tales, etc. Input to design and specification modifications to suit site conditions.
5	Operation	Ensure that the structure is performing as designed, from the geotechnical point of view, and assist in the design of remedial measures, if it is not. Inspect the completed structure, the site area and records from instruments and operator's observations, at regular intervals, preferably as a member of a surveillance committee, including representatives of design, construction and operation branches. Should any malfunction be evident, assist in design of remedial measure. This could involve conducting further site investigation and/or analysis of the geotechnical records from Stages 1 to 5.

It is important during the planning stages (Stages 1 to 3) that the site investigation studies should proceed in phase with other engineering work being carried out at those times, e.g. hydrological and topographic surveys, design and preparation of specifications. This is because most decisions affecting the project design or feasibility from the geotechnical viewpoint will affect the work requirements in the other fields. Conversely, decisions made from the results of these other studies will affect the scope of the geotechnical work.

It is desirable that the dam designer takes a keen interest in the geotechnical work and that field and office consultations with the geotechnical team are frequent. Design drawings (e.g. cross-sections through structures) should include the main features of the foundation (i.e. the portion which Nature has made) as well as details of what Man proposes to build.

It is also vital that the geotechnical investigator be given adequate time to do his or her work. The whole process of geotechnical investigation, and indeed of design, requires 'thinking', not just the mechanical production of drill hole logs, pit logs and the like. Too often, time is reduced by the project planners to a minimum, without realising the risk that action could put the whole project in.

Timing and co-operation are just as critical during the construction stage. During this stage the foundation rock or soil is exposed better than ever before or (hopefully) after, and if the site investigator arrives too late, the exposed area showing a critical clue can be covered by the first layer of fill or concrete. It goes without saying that unless the site investigator has the confidence and co-operation of the construction team, this phase of the work will not be as fruitful as it otherwise could be. In extreme cases, lack of a competent geotechnical observer and advisor during the construction stage can prove disastrous – resulting in either expensive contractual disputes or later failure of the structure.

Similarly, during the operation stage inspections must be made at regular intervals, to ensure that any malfunctions are discovered while there is still time to remedy them. Such inspections are carried out as part of a surveillance programme as discussed in Chapter 20.

4.3   AN ITERATIVE APPROACH TO THE INVESTIGATIONS

During each of the project planning stages (Table 4.3, Stages 1 to 3) the geotechnical studies should follow the iterative approach shown on Figure 4.1, which is modified from ISRM (1975) and Stapledon (1983). The following notes relate to the 4 phases on Figure 4.1.

Phase 1. First, the objectives of the work, or questions to be answered, are defined. As discussed in Section 4.1, these will include both geotechnical engineering and geological questions. The number of geological questions which can be defined clearly at this time might be quite small, if little is known of the geology of the region or site.

Phase 2. Existing geological, geotechnical and other data relevant to the site are collected and compiled to give a tentative geotechnical model (or models). Tentative answers to the questions asked in Phase 1 are obtained where possible, from local knowledge or from rapid analyses, or both. New questions are usually added during this phase, arising from the understanding acquired while assessing the existing data.

Phase 3. The investigations are planned to confirm the tentative answers and tentative geotechnical model and to answer the outstanding questions. The plan is based on the 'broad to particular' approach and types of activities set out in Section 4.4.

It is usual (and desirable) for the work to proceed in stages. For each stage the proposed activities are related to time and money respectively, in an activity chart and cost estimate. The proposals are set out in a report to the client, seeking approval to proceed.

Phase 4. The investigations proceed in stages as planned. A cooperating team of engineering geologist(s) and geotechnical engineer(s) is able to complete most of the studies in

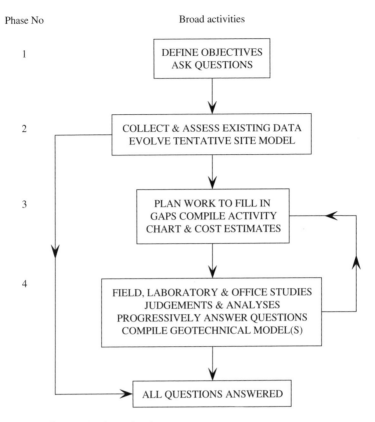

Figure 4.1.   Activity flow in site investigations.

either logical sequences or concurrently and to integrate their results. Answers to the purely geological and geotechnical engineering questions are produced progressively from the results, by combinations of analyses and judgements. Sometimes, new questions arise and extra work has to be planned and approved (see arrow back to Phase 3). The site geotechnical model, and in some cases more detailed models for particular project features e.g. dam, spillway, outlet works and quarry, are compiled.

When all questions have been answered with sufficient confidence, the investigation is complete. Until then, further cycles of investigation are carried out until the required levels of confidence are reached. The principle of 'diminishing returns' is applied.

4.4   PROGRESSION FROM REGIONAL TO LOCAL STUDIES

The geological studies in Phases 1 to 4 start with consideration of the site location in relation to the global tectonic situation, and should include study of the geology of a broad region surrounding the site. This is necessary to assess the effects on the project of large scale processes, some of which (Table 4.2) may have potential to damage it.

The regional geological studies are followed by geological and geotechnical engineering studies at and near the site, on intermediate and detailed scales. The objectives and usual activities of the regional and detailed studies are set out below, generally following ISRM (1975). More detailed descriptions and discussion of the various activities are presented in Chapter 5.

4.4.1   Broad regional studies

4.4.1.1   Objectives

1. To provide an understanding of the geological history of the project area, that is, of the processes which have developed the present geological situation at and in a broad region around the site. In particular it is important to determine the following:
   - Major geological processes which have the potential to cause impact on the project;
   - Any processes that are active or potentially active, e.g. as listed on Table 4.2;
   - The possible effects of any active processes on the proposed works both during construction and in service;
   - Whether any construction activities (e.g. excavation) or operation of the storage are likely to cause such changes to the existing regime (e.g. stress or hydrologic) as to require remedial works or to affect the project feasibility.
2. To determine the regional stratigraphy and geological structure;
3. To explain the geomorphology of the project area in terms of the regional stratigraphy, structure and geological history;
4. To draw attention to important features, e.g. major faults or landslides, occurring at or close to the site, but not exposed or recognisable at the site;
5. To get an appreciation of the regional groundwater conditions;
6. To form a logical basis for the location and proving of sources of construction materials.

4.4.1.2   Activities

The amount and money spent on regional studies will depend upon the size and complexity of the project, its consequence of failure, and the amount and relevance of regional information already available. The activities usually will include some or all of the following:

1. Examination of existing regional geological maps, cross sections and reports;
2. Interpretation of satellite images and aerial photographs;
3. Interpretation of ground photographs;
4. Ground reconnaissance over previously mapped areas, and remapping of important areas with the project objectives in mind;
5. Compilation of regional plans and cross sections showing the proposed project works.

4.4.2   Studies at intermediate and detailed scales

4.4.2.1   Objectives

1. To explain the development of the site topography in terms of the regional and local geology and geological history;
2. To delineate any major features, e.g. folds, faults or landslides, which may or may not be indicated already by the regional studies;
3. To provide geotechnical models for the site and for particular project features, and to provide answers to the geotechnical engineering questions.

4.4.2.2   Activities

1. Geotechnical mapping on intermediate and detailed scales;
2. Studies of the pattern of defects using structural geology and stereographic methods;
3. Application of geophysical methods (commonly seismic refraction) and interpretation of the results in the light of the evolving geological picture;
4. Planning and technical supervision of direct exploration, which can include core (or other) drilling, sluicing, and excavation of pits, trenches, shafts or tunnels;

5. Geotechnical logging and photographing of the cores of drill holes and of exposures in the exploratory excavations;
6. Logging of drill holes using video imaging, cameras, impression devices, etc;
7. Permeability testing in drillholes;
8. Field and/or laboratory testing of soils;
9. Field and/or laboratory testing of rocks;
10. Plotting of all geological and test data, as soon as it is obtained, onto plans, sections and 3-dimensional models (physical or computer generated). In this way it is possible to visualise and analyse the surface and subsurface data, progressively filling in gaps in the subsurface picture until the required degree of detail is obtained. All plans, sections and models show the proposed works, at least in outline. The end results of this activity are the geotechnical model and geotechnical design parameters;
11. Assist in the design and analysis to assess the behaviour of proposed structures and their foundations, under construction and operating conditions.

4.5   REPORTING

It is important during all stages, that the geotechnical facts, interpretations, conclusions and decisions made from them are recorded regularly by a system of formal progress reports. A comprehensive report is essential at the end of each stage, setting out the answers to the questions of that stage and with recommendations for the next stage.

In general terms, a geotechnical report should consist mainly of a carefully planned and ordered sequence of drawings (and preferably some photographs) with relatively short explanatory text, and appended tables and calculations. The drawings and photographs should convey a clear threedimensional picture of the surface and subsurface conditions at the site, and the geotechnical properties of the foundation and construction materials. A clear distinction should be made between factual data and inferences made from them. The report (or a separate one) should include the results of analysis and design e.g. stability analysis, design of filters.

A formal system for checking and certification is needed for all drawings and reports.

4.6   FUNDING OF GEOTECHNICAL STUDIES

In the opinion of the authors, it is most important that dam projects receive adequate funding at the feasibility stage. This is because the question 'Is it economically feasible to build and maintain a safe structure at this site?' is the really vital and difficult one. Usually there are alternative sites but finally one must be adopted and the above question must be answered in relation to it.

Unfortunately, some organisations may not have (or seek?) large amounts of money until they believe they have shown a project to be feasible. Also, they may not wish to, or cannot, acquire the site until feasibility is assured. Because of these matters, and the inevitable pressure from landowners and conservation groups, the investigations at this vital stage often tend to be less logical and less thorough than necessary – too much reliance is placed on indirect methods (geophysical) and drilling of small diameter boreholes. The advantage of these methods is that they cause minimal disturbance of the land. However, without an adequate understanding of the geological situation, which often can be obtained only by large, continuous exposures in deep bulldozer trenches or access track cuttings, the results of geophysical studies and drilling can be difficult to interpret. For major, high hazard dams at geologically complex sites, exploratory adits are often the most appropriate method for answering the feasibility questions.

The consequences of a poorly planned or inadequately funded feasibility stage investigation can be:

– Adoption of the less satisfactory of two alternative sites;
– Adoption of a site which later proves to be non-feasible or for which large cost overruns occur to address unforseen conditions (for examples of such sites see Table 3.4);
– Abandonment of a site (or possibly a whole project) which would have been proven feasible, by well conceived (but expensive) studies.

The authors consider that the 'observational method' as adopted by Terzaghi & Leps (1958) at Vermilion Dam, should not be used as a means of overcoming funding difficulties during the feasibility and design stages of dam projects. They agree with the practice of excavating part or all of the dam foundations during the design stage, or prior to awarding the main construction contract (Hunter, 1980), but believe that this should be done only when all questions relating to feasibility of the project have been adequately answered.

4.7   THE SITE INVESTIGATION TEAM

The following seven attributes for success in site investigations for dams, put forward by Stapledon (1983), suggest that a team approach is necessary, especially for large dam projects:

1. Knowledge of precedents;
2. Knowledge of geology;
3. Knowledge of soil and rock mechanics;
4. Knowledge of geotechnical, dam, and civil engineering design;
5. Knowledge of civil engineering and dams construction;
6. Knowledge of direct and indirect exploratory methods;
7. Above average application.

It should be clear from Section 4.1 to 4.4, that Attributes 1 to 5 are necessary for understanding, defining and answering the geotechnical questions associated with dams.

Attribute 6 refers to direct subsurface exploratory methods (e.g. pits, boreholes) and indirect methods (e.g. geophysical transversing). Knowledge of the application and limitations of methods such as these is necessary if the questions posed are to be answered effectively and economically.

The authors consider that these six attributes are necessary for successful site investigations for dams. They also consider that these attributes can be provided most effectively by a team including engineering geologists and engineers. It is important that all team members have enough 'general' knowledge in all areas to be able to communicate effectively with one another and with engineers involved with the design, construction and operation of dams. As well as having this broad knowledge of the industry, it is desirable that they have Attribute 7, 'above average application' as defined by Stapledon (1983) *"…ability to get things done; this involves effective cooperation with others, enthusiasm and drive and the ability to make decisions in the field."*

5

Site investigation techniques

Several techniques or "tools" may be used in a dam site investigation designed to follow the broad framework explained in Chapter 4. This chapter discusses some of the most common techniques, their applicability and limitations. It is emphasised that use of several techniques is always required, as restriction to a single investigation method would be unlikely to yield correct answers to the site questions in an economical manner.

5.1 TOPOGRAPHIC MAPPING AND SURVEY

A fundamental requirement for the investigation and design of any project is accurate location and level of all relevant data. Topographic maps at suitable scales are essential with establishment on site of clearly identified bench marks. All features recorded during the investigation should be located and levelled, preferably in relation to a regional coordinate system and datum. A local system may be established provided that at some stage during the investigation the relation between the regional and local survey systems is determined.

Survey control for regional studies can utilise Global Positioning System (GPS) techniques but detailed project studies should be based on standard ground survey methods.

These statements may appear obvious but, in many instances, interpretation, construction and related contractual problems have been shown to have originated from poor survey control.

Topographic maps at several scales are required. The actual scale will depend on the size and complexity of the area but the following common scales are given as a guide:

- Regional maps, 1:250 000 with 20–50 m contours to 1:25 000 with 10 m contours;
- Catchment area, 1:25 000 with 10 m contours to 1:2000 with 2 m contours;
- Project area, 1:1000 with 2 m contours to 1:200 with 1 m contours;
- Individual engineering structures, 1:500 with 1 m contours to 1:200 with 0.5 m contours.

Regional topographic maps issued by government agencies in most countries are published at several standard scales which range from 1:250 000 with contours at 20–50 m to 1:25 000 with contours at 10 m interval. The user of these maps should consider their original purpose and the accuracy of the information plotted on them in relation to the project requirements. The notes on the map which indicate the method of compilation – whether aerial photogrammetry or ground survey – and the date of preparation of the map may be relevant.

Photogrammetry can be used in the preparation of project specific plans provided that the aerial photographs have been flown at low level and that accurate ground control points can be identified on the photographs and that vegetation is not dense.

Photogrammetry is often not adequate in steep, tree-covered areas. The authors are aware of two projects where errors of up to 20 m in elevation have occurred due to inadequate allowance for tree cover. This resulted in the requirement for substantially larger saddle dams than estimated in the feasibility studies and significantly affected the viability of the projects.

The field survey team should liaise with the site investigation group, and provide a series of clearly labelled ground control marks throughout the project area to assist in the location of features identified during the geotechnical mapping. All boreholes, pits and trenches should be located, levelled and clearly labelled.

5.2   INTERPRETATION OF SATELLITE IMAGES AERIAL PHOTOGRAPHS AND PHOTOGRAPHS TAKEN DURING CONSTRUCTION

Satellite image maps and aerial photographs are available from government agencies in most countries. For investigations of existing dams, photographs taken during construction are an invaluable aid to assessing the geology and construction of the dam.

5.2.1   Interpretation of satellite images

Standard LANDSAT images are at 1:100 000 scale but enlargements at 1:500 000 and 1:250 000 are also available.

These small scale images or image maps provide a broad view of the region in which the project is to be located. This broad view can indicate correlations between geological features or the position of geological boundaries or faults, when these features are of such great extent that they are not recognisable on larger scale photographs covering smaller areas.

The broad view also provides an indication of relationships between the regional geology and landforms, drainage, soils, vegetation and land-use, which may be useful in:

– planning of access routes to and within the project area;
– location of potential sources of construction materials, and
– assessing reservoir siltation rates.

5.2.2   Interpretation of aerial photographs

5.2.2.1   Coverage

If the existing aerial photographs in the project area are inadequate in quality, coverage or scale, it is best to take new photographs with the following advantages:

– the photographs will show present conditions;
– the required scales and coverage can be specified, and
– the photographs can be used to prepare topographic maps.

Aerial photographs taken at different dates can indicate changes in site conditions.

5.2.2.2   Interpretation

Photogeological interpretation using a stereoscope forms a major part of the initial appraisal of regional and local site conditions during the pre-feasibility and feasibility stages (see Section 4.4.1.2 and Table 4.3). It usually gives a good indication of likely geotechnical constraints on the project and the extent of investigations required.

Much has been written about the techniques of photogeological interpretation and their application in engineering (Rib and Liang, 1978; Rengers and Soeters, 1980; Bell, 1983b; Lillesand, 1987; and Hunt, 1984). Comments will therefore be limited to aspects of special significance in dam engineering.

The stereoscope provides a three-dimensional image with a vertical scale which is exaggerated by a factor which depends on the distance between the photo-centres. This

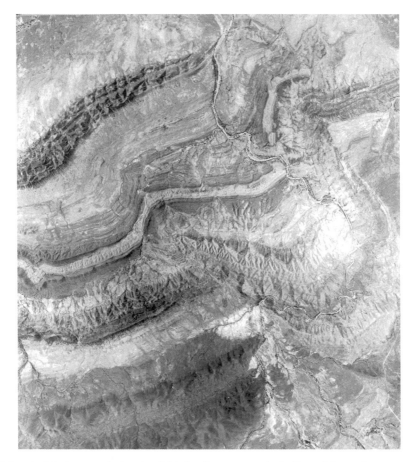

Figure 5.1.   Aerial photograph showing folded sedimentary rocks in Central Australia.

exaggeration can be of benefit as it highlights surface features which are actually more subtle, but interpreters should realise that the ground slopes viewed under the stereoscope appear steeper than they really are.

A major advantage of aerial photographs is that distance of observation is not impeded by relief. It is possible to observe features on both sides of a hill at the same time and thus establish their continuity, if it is present. This is particularly important in the interpretation of geological structure.

Landforms which reflect the structure of folded rocks show up well on aerial photographs. In horizontal strata, mesa forms are common and, in dipping strata, dip slopes and scarps indicate the direction and dip of the bedding (Figure 5.1).

An important part of photo-interpretation for dam engineering is the recognition and plotting of lineaments. Lineaments are simply linear features or linear arrangements of features that are visible on the photographs.

Faults, particularly those which are steeply dipping, usually show up well as lineaments. A lineament indicative of a fault may be a linear arrangement of features which are topographically low, or otherwise indicative of the presence of deep soils or low strength materials. Such features include straight section of rivers or creeks, gullies, saddles, springs, swamps and usually dense vegetation.

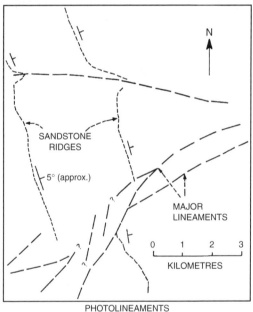

Figure 5.2.   Aerial photograph and sketch showing lineaments in an area underlain by gently
dipping sandstones.

Figure 5.2 shows several lineaments passing through an area in Central Australia underlain by sandstones dipping to the east at about 5 degrees. The major lineaments occur
along sheared zones with little or no displacement of the beds. The minor lineaments visible on the aerial photograph follow joints.

Linear arrangements of other features, e.g. vegetation boundaries, are usually found to
correlate with geological features.

As discussed in Chapter 2, Section 2.10, it is very important that evidence of any currently active or past landsliding in the project area is recognised early. Fortunately the presence of active or ancient landslides is usually indicated by:

- unvegetated scarps, cracks and areas of exposed soil or rock;
- characteristic topographic forms, e.g. scarps, spoon-shaped troughs, hummocky or steep ground and areas of internal drainage;
- areas of anomalous vegetation, e.g. where trees are dead, or younger than elsewhere, or where the vegetation is more dense due to an area of deeper or wet soil;
- evidence of past or current restriction or damming of streams.

Rib and Liang (1978) provide guidance on the recognition of landslides on aerial photographs in a range of geological environments.

In densely vegetated areas, evidence of active or very recent landsliding may not be readily visible on photographs. For example, in some tropical areas shallow landsliding produces subdued topographic forms and revegetation is rapid. As shown in Figure 3.34 large landslides show up well even in tropical rainforest. Fookes et al. (1991) describe how examination of several sets of aerial photographs assisted the analysis of slope movements at Ok Ma damsite, Papua New Guinea.

Another use of photo-interpretation is in terrain analysis where the features observed, including topography, geology, soils, drainage and vegetation are used to divide the area into land units with similar characteristics (Grant 1973, 1974). This type of study is more suited to regional assessment of catchment or reservoir areas or to the location of construction materials, than to localised assessments, e.g. dam sites. Figure 5.3 is an example of a terrain map prepared to assist in the location of construction materials. Such maps form a logical basis for the planning of ground surveys followed by exploration of possible borrow areas. These maps should not be relied on for selection of sites without a field check of site conditions.

5.2.3 *Photographs taken during construction*

Any review of an existing dam should include a thorough search for photographs taken during construction, and systematic review of the photographs. They are particularly valuable as a record of foundation preparation and clean-up, and for assessing matters such as segregation of filter and transition zones.

5.3 GEOMORPHOLOGICAL MAPPING

Before embarking on a subsurface investigation program, which may be (a) expensive and (b) provide limited information of dubious relevance, it is recommended that the surface evidence should be systematically recorded.

The ground surface reflects both the underlying geology and the geomorphological development of the area. Geomorphological mapping of surface features can provide an indication of the distribution of subsurface materials, their structure and areas of possible mass movement, e.g. landslides. Geomorphological mapping should be carried out at two levels – an overall appraisal of the study area and a detailed recording of site data using a topographic map as the base. Hutchinson (2001), Varnes (1978) and Brunsden et al. (1975) describe mapping methods.

The overall appraisal involves an examination of the topographic features of the area, surface shape, drainage, local depressions and areas of distinctive vegetation.

An hour spent on a convenient hilltop reviewing the geomorphology of the whole area before commencing detailed mapping may provide an understanding of the problem

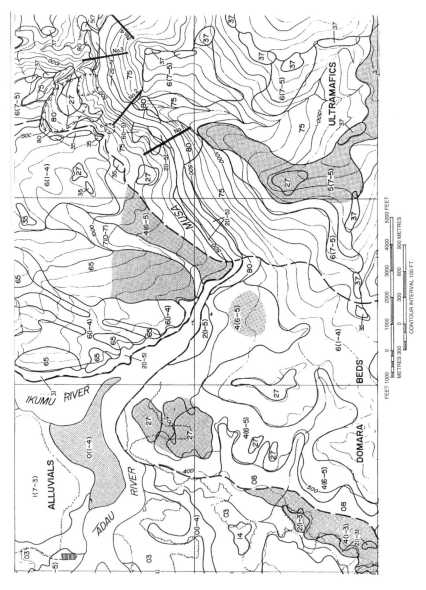

Figure 5.3.   Terrain map for construction materials investigation, Musa Dam, Papua New Guinea (courtesy of Coffey Geosciences).

TERRAIN CLASSIFICATION FRAMEWORK & EXPLANATORY NOTES

GEOLOGY		TERRAIN MAPPING UNITS				
GEOLOGICAL GROUP	LITHOLOGY	No.	PHYSIOGRAPHIC FEATURE	NATURAL SLOPE RANGE	No.	SOILS	U.S.C. SYMBOL

GEOLOGICAL GROUP	LITHOLOGY	No.	PHYSIOGRAPHIC FEATURE	NATURAL SLOPE RANGE	No.	SOILS	U.S.C. SYMBOL
ULTRAMAFICS: [(Kgd) & (Kpd) undifferentiated]	Massive noncumulus peridotite, dunite – commonly layered and in places brecciated; minor gabbro.	0	Level or depressional, poorly drained areas.	Level or very low gradient.	0	Rock outcrop or skeletal soils.	–
		1	Level, very gently undulating or very gently sloping surfaces—well drained areas.	Level or up, to 2°.	1	Gravels – (alluvial or colluvial).	GC-GM-GW
DOMARA BEDS: [Includes Domara River Beds (Qpd) and minor Musa Volcanics (Qpm)]	Conglomerate, siltstone, shale with minor intermediate to basic agglomerate, tuff and basic lava.	2	Undulating or irregular surfaces.	Variable—predominantly long, gentle slopes (to 8°).	2	Sand – Uniform profile.	SW-SM-SP
		3	Strongly undulating, irregular or hummocky surfaces.	Variable—predominantly short moderately steep slopes (8°–18°).	3	Sandy or silty medium textured soils (gradational, uniform or duplex profile).	SM-ML or SC-CL
ALLUVIALS: (Qa)	Unconsolidated gravel, sand, silt and clay, local areas of boulder alluvium.	4	Slopes gentle (hill slopes; foot slopes etc.)	to 8°	4	Silty soils or stratified soils – predominantly silty.	SM-ML-MH
		5	Slopes moderate (hill slopes, escarpment slopes etc.)	8° – 18°	5	Gravelly or stony clay, silty clay, sandy clay etc.	GC-CH & MH
		6	Slope steep (hill slopes, escarpment slopes etc.)	18° – 30°	6	Duplex soils– medium to coarse textured soil over fine textured soil.	SP-SC/CL-CH
		7	Slopes very steep	30° – 45°	7	Clay soils – uniform gradational or duplex (including gley clays).	CL-CH or CH
		8	Precipitous slopes, cliffs, benched slopes, escarpment slopes etc.	>45°	8	Clay soils or soils with very high liquid limit. (Clays, silty clays, volcanic ash soils etc.)	CH or MH
		9	Water ways.		9	Organic soils–peat, peaty clay etc.	OL-OH-Pt

Left margin grouping: Nos. 0–3 = Horizontal or Sub-horizontal Surfaces; Nos. 4–8 = Sloping Surfaces.

*NOTE: A dual symbol (1–5) indicates that both material types 1 and 5 are likely to occur within the unit as mapped.

EXAMPLE:

GEOLOGY —— TERRAIN MAPPING UNIT

ULTRAMAFICS —— 75

GEOLOGICAL GROUP | PHYSIOGRAPHIC FEATURE / SOIL*

LEGEND

- Village
- Track
- River, stream
- Treeline
- /1500⁓ Contour in feet (amsl)
- △RIVER Control point TES Z189/96
- No. 3 Dam site
- Geological boundary
- Terrain boundary
- Landslide area
- Potential borrow area (subject to investigation)

NOTE. The base map is compiled from CDW Drawing Nos. PH68/70 & PH67/225. The reliability of the map position on the grid, north of 335,000N, is fair.

Figure 5.3. (Continued).

TOPOGRAPHY

Symbol   *Ground Profile*

convex ⎫ well defined or angular
concave ⎭ break of slope

convex ⎫ poorly defined or
concave ⎭ smooth change of slope

⊤⊤⊤⊤⊤ breaks of slope ⎫ convex and concave too close together
⊤ ⊤ ⊤ ⊤ changes of slope ⎭ to allow the use of separate symbols

–◇–◇– sharp ⎫
–◇–◇ rounded ⎭ ridge crest

Cliff or escarpment or sharp break
3   40° or more (estimated height in metres)

15→ Uniform Slope ⎫
10(→ Concave Slope ⎬ Slope direction and angle (Degrees)
8)→ Convex Slope ⎭

▼▼▼ Top ⎫
▼▼▼ Bottom ⎭ Cut or fill slope, arrows pointing down slope

~~~ Hummocky or irregular ground

OTHER FEATURES

Boulder

Seepage/spring

Swallow hole for runoff

Natural water course

Open drain, unlined

L→··L→ Open drain, lined

Fenceline

—·—·— Property boundary

O⊃O⊃ Dry Stone Wall

J ——200—— J Major joint in rock face (opening in millimetres)

–T–T– Tension crack
10 (opening in millimetres)

Masonry or concrete wall

Ponding water

Boggy or swampy area

EXAMPLE OF USE OF TOPOGRAPHIC SYMBOLS:

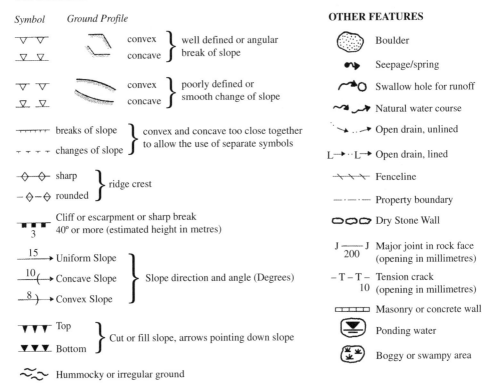

BLOCK DIAGRAM

GEOTECHNICAL PLAN

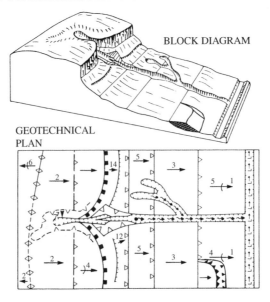

Figure 5.4.   Geomorphological mapping symbols (Courtesy Jeffery and Katauskas).

which is obscured in a closer examination. Binoculars enable closer examination of more remote localities and land-based stereo-photographs have proved a useful method of recording information for subsequent re-evaluation in the office.

The compilation of a geomorphological map as a basis for the succeeding stages in the study involves the identification and location of all relevant features including changes in slope, scarps, cracks, areas of seepage, displaced or rotated vegetation.

These features are recorded on the topographic map using one of the many systems of symbols which have been developed. Most of the systems have a similar basis but differ in detail depending on the site conditions experienced in the original study. The symbols shown in Figure 5.4 are often sufficient and commonly used. Any geomorphological map should be accompanied by a comprehensive legend explaining the symbols.

Of particular relevance is the demarcation of boundaries of areas judged to be disturbed by slope movements. Figure 5.5 is a large scale geomorphological map of a landslipped area above the abutment of O'Shannassy Dam.

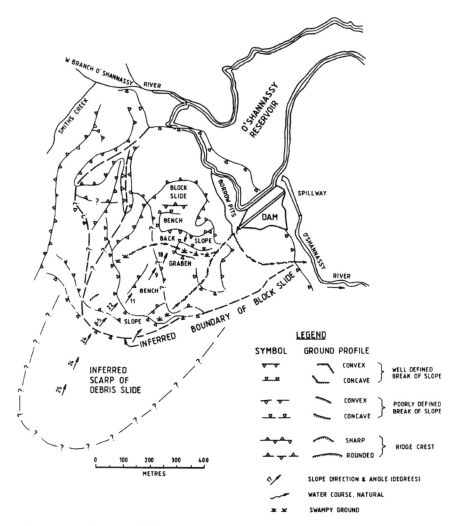

Figure 5.5.   Large scale geomorphological plan of landslide area at O'Shannassy Dam, near Melbourne, Australia (courtesy Jeffery and Katauskas, and MMBW).

5.4   GEOTECHNICAL MAPPING

Geotechnical mapping is essentially geological mapping aimed at answering engineering questions. At any particular site it involves the location and plotting on suitable scales of all data which assists in understanding the geotechnical conditions at that site.

5.4.1   Use of existing maps and reports

Some useful data can often be obtained from existing maps and reports prepared for other purposes.

Maps showing the regional geology on scales ranging from 1:100,000 to 1:1,000,000 are usually available from government agencies. Some mining areas and areas of existing or proposed urban development may have been mapped at larger scales. The regional maps are often accompanied by explanatory notes which are useful.

The regional maps show the inferred distribution of the main rock types and the inferred geological structure. Areas where rock is overlain by unconsolidated sediments, surface soils or talus are not differentiated unless these deposits are known to be widespread and of significant depth. The data plotted on the maps is obtained from examination of surface outcrops, information from stratigraphic and exploration drilling, and air photo interpretation.

The main value of these maps is providing an understanding of the stratigraphy, structure and geological history of the region, i.e. the broad geological environment in which the project is to be located. As set out in Chapter 3, this understanding can provide a useful insight into the range of geotechnical conditions which might be expected at the project sites. The maps are also useful during studies associated with access routes, material sources and reservoir and catchment areas.

Local deposits of alluvial, colluvial or residual weathered materials are usually ignored or briefly mentioned on regional geological maps. Because of this the maps generally do not answer detailed questions related to dam projects.

Geological reports prepared for different purposes, e.g. for mineral exploration, can provide some useful information.

Regional maps showing distribution of soil groups, classified for agricultural purposes, may also be available. These maps may be a useful supplement to the regional geology map, particularly in the search for construction materials.

Land capability maps, based on the regional soil surveys with particular attention to potential landsliding and erosion, may be available for limited areas.

5.4.2   Geotechnical mapping for the project

5.4.2.1   Regional mapping

When published regional geological maps are available they are usually able to provide the regional geological understanding required for a dam project. For major dam projects, and for all projects where these maps are to be used in access, material or reservoir studies, they should be checked on the ground and on air photos. Where necessary they should be updated by the addition of data of engineering importance. Such data will include lineaments, landslides, scarps, swamps, springs and areas of problem soils.

If no satisfactory regional geological map is available, and the proposed dam is of major size and high hazard, then it would be prudent to prepare a regional map specifically for the project.

However produced, the regional geological plan as used and included in the project reports should show the location of the proposed works and the outline of the proposed storage.

5.4.2.2   *Geotechnical mapping*

Geotechnical mapping at and near the sites of the proposed works is the key to the success of the site investigation. The mapping involves the identification and location of all surface features relevant to the establishment of geotechnical models at the sites. The plans and sections produced by the mapping form the initial geotechnical models which are the basis for the subsurface exploratory work, aimed at checking the models, filling in gaps and answering specific questions raised by the indicated geological environments (see Chapters 3 and 4).

The geotechnical maps are usually produced at an intermediate scale (1 : 5000 or 1 : 2500) covering the general works sites, and at 1:1000 or 1:500 covering the immediate area of the sites. The maps show the following types of factual information as shown on Figure 5.6.

- ground surface contours;
- geomorphic features, e.g. slope changes, areas of hummocky ground;
- geological surface features, e.g. areas of rock outcrops, scree, boulders and soil;
- features of *in situ* rock, e.g. rock types and their boundaries, attitudes of bedding and foliation, the nature, location and orientation of important geological defects such as sheared or crushed zones;
- groundwater features, e.g. springs, seepage, areas of swamp and vegetation indicating moist or wet ground;

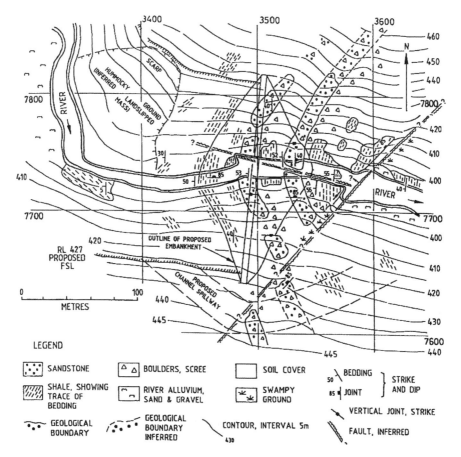

Figure 5.6.   Plan showing some of the features presented on a large scale geotechnical plan of a site for an embankment dam.

- the location of tracks, roads, test pits and trenches, with summary logs of the soils and rocks exposed;
- the position of drillholes and geophysical traverse lines;
- the proposed works in outline including the full supply level of the proposed storage.

Figure 5.6 is an example of the first stage of compilation of a geotechnical plan at the site for a concrete faced rockfill dam about 35 m high. At this stage the plan shows a factual record of surface geological and geomorphological features. Three relatively important features are inferred – the landslide upstream from the site and the two faults. Other important features might also be suspected, for example bedding-surface faults at the boundaries for the sandstone beds, because no actual shale-sandstone boundaries are exposed in outcrop, even at river level.

It should be noted that such a plan would normally be prepared on 1:500 or even 1:200 scale on an A2 or A1 size sheet. For the sake of legibility for publication at its present size the plan has been simplified greatly, i.e. it contains much less geological detail and wider contour spacing than would usually be present.

Position identification is important during the mapping. In sparsely vegetated areas the combination of enlarged air photos and contour plans at the same scale may enable positioning to an acceptable level of accuracy. Where required, more accurate control can be achieved by ground survey methods or by use of GPS.

5.5   GEOPHYSICAL METHODS, SURFACE AND DOWNHOLE

Geophysical methods have been extensively used in dam investigations, both on dam construction projects and in the assessment of the condition of existing dam structures. There are many different methods; the selection of the method(s) relevant to a particular problem will usually depend on the regional and local site conditions, the nature of the problem, timing and cost.

The *advantages* of the use of geophysical methods include:

- They are non-invasive and can be carried out from the surface or from existing boreholes;
- They can provide information on site conditions between data points e.g. boreholes;
- They may be able to identify local areas of concern which have no surface expression e.g. cavities;
- The surveys can usually be performed quickly and cover a relatively large area;
- Recent development of computer analysis and presentation of results (tomography) has assisted interpretation.

The *disadvantages* of the use of geophysical methods include:

- Each method measures a particular physical subsurface property which may or may not be relevant to the problem under examination;
- Some borehole surveys are affected by the presence of steel casing;
- Each method requires expensive equipment and skilled operators;
- The results involve the recording and analysis of a great amount of numerical data;
- The mathematical analysis is based on an assumed subsurface model which may be different from the actual geological situation;
- Much of the data involves the averaging of information;
- The accuracy of the results may be lower than required;
- The interpreted subsurface profile usually needs to be confirmed by drilling;
- Establishment of the equipment and operators is expensive and also stand-by time waiting for other operations, e.g. drilling.

Areas of investigation where the *correct use* of geophysical methods have provided valuable information include:

– Delineation of boundaries between the underlying *in situ* rock and transported materials such as alluvium, colluvium, glacial debris and landslide debris;
– Delineation of boundaries between residual soil, weathered rock and fresh rock;
– Delineation of boundaries between sandy and clayey soils;
– Location of anomalous foundation features e.g. igneous dykes, cavities, deeply weathered zones, fault zones, buried river channels;
– Assessment of rippability, depth of foundation excavation, depth of cutoff excavation, liquefaction potential;
– Location in existing structures of seepage paths, low density zones, cavities.

It is the authors' experience that no surface geophysical method has been able to contribute to the location of thin, weak seams in the rock mass. This information is often an issue of great importance especially for concrete dams.

The different geophysical methods and their application to dam engineering are briefly discussed below with reference to papers containing detailed method descriptions. Further discussion is included in Whiteley (1983, 1988), Stapledon (1988b), Fell (1988a), Fell et al. (2000) and Joyce et al. (1997).

5.5.1   Surface geophysical methods

The application and the limitations of the different geophysical methods which can be carried out from the surface without the aid of boreholes (other than to assist interpretation) are summarised in Table 5.1.

5.5.1.1   Seismic refraction

This method utilises the fact that seismic waves travel at different velocities in different materials; in rock and soil masses the velocity increases with increase in substance strength and compactness. Whiteley (1988) provides details of the method and interpretation of the results.

Profiles of apparent seismic "P" wave velocity are produced as shown on Figure 5.7, a section along the line of Sugarloaf Reservoir Inlet Tunnel. Air photo interpretation had shown two lineaments crossing the line of the tunnel. Seismic refraction traverses in combination with trenching and core drilling indicated that the lineaments are the surface expressions of a deeply weathered rock unit and a minor fault. This interpretation was confirmed during the driving of the tunnel.

Seismic refraction using the "P" waves is the method most commonly used for delineation of boundaries between soil and weathered rock, and within weathered rock profiles. It is the authors' experience that, in a residual weathered profile, the base of the lowest velocity layer is usually a reasonable approximation of the probable general foundation stripping level for an embankment dam, and the base of the second layer is a reasonable approximation of the cutoff excavation level.

When combined with geological information about the rock mass, "P" wave velocities can be used to estimate rock rippability (MacGregor et al., 1994). Seismic velocities alone should not be used for predicting rippability.

Some limitations and requirements of the method are:

– The method cannot distinguish between sandy and clayey soils, and between soil and weathered rock where the boundary is gradational;
– The accuracy of the method is affected by poor velocity contrast between "layers" and where there are "cliffs" in the rock profile;

Table 5.1.   Application of surface based geophysical methods (adapted from Fell et al., 2000 and McGuffey et al., 1996).

Type of survey	Applications	Limitations
Seismic refraction	Determines depth to strata and their characteristic "P" wave seismic velocities	May be unreliable unless strata are thicker than a minimum thickness, velocities increase with depth and boundaries are regular Information represents average values
Self-potential (SP) profiling	Locates seepage areas and low density materials. The only technique that directly indicates subsurface water flow	Accuracy affected by presence of ferrous objects. Needs to be accompanied by a magnetic survey. Not available "off–the–shelf" No "textbook" manuals
Electrical resistivity	Locates boundaries between clean granular and clay strata; assessment of degree of saturation, position of water table and soil-rock interface. Resistivity imaging may locate cavities	Difficult to interpret and subject to correctness of the assumed subsurface conditions; does not provide any indication of engineering strength properties. Accuracy affected by presence of ferrous objects
Electromagnetic conductivity profiling	Locates boundaries between clean granular and clay strata; assessment of degree of saturation, position of water table and soil-rock interface. More rapid reconnaissance than electrical resistivity	Difficult to interpret and subject to correctness of the assumed subsurface conditions; does not provide any indication of engineering strength properties. Accuracy affected by presence of ferrous objects
Magnetic	Locates ferrous objects. Used in combination with self potential and resistivity surveys	Does not locate other metals
Microgravity	May locate small volumes of low density materials and larger cavities	Use of expensive and sensitive instruments in rugged terrain may be impractical Requires precise levelling.
Ground-penetrating radar	Provides a subsurface profile, locates cavities, buried objects, boulders and soil-rock interface	Has limited penetration particularly in clay materials and laterites

- The geophysical "model" assumes sub-horizontal layering or boundaries. Results may be misleading where this is not so;
- The method cannot detect velocity inversions i.e. high velocity material above low velocity material (unless crosshole or downhole techniques are used, see Section 5.5.3);
- Large energy input (usually explosives) is required in loose material;
- Accurate ground survey is essential for good results.

A recent development in seismic methods has been Through Dam Tomography (TDT) where the condition of Bennett Dam in British Columbia was assessed by placing the seismic energy source on the surface of one side of the dam with the geophones on the other side (Garran, 1999).

5.5.1.2   Self potential

The self potential or SP method is based on the measurement of the natural potential difference which exists between any two points on the ground. The method has little application for new dam construction projects but has been extensively used to locate areas of leakage from existing dams and reservoirs (Ogilvy et al., 1969; Bogoslovsky and Ogilvy, 1972; Corwin, 1999).

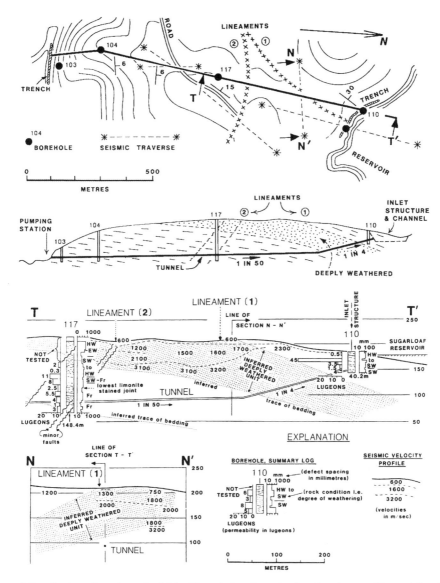

Figure 5.7.   Section along Sugarloaf Inlet Tunnel showing seismic refraction profiles and deeply weathered unit (Stapledon, 1988b).

5.5.1.3   *Electrical resistivity*

The electrical resistivity method measures the resistance of the ground to induced electrical current. Interpretation assumes a horizontally layered model. Whiteley (1983) and Corwin (1999) provide details of the method.

The method has been used to locate fault zones, zones of deep weathering and cavities (McCann et al., 1987; Smith and Randazzo, 1987). It can also be used in the exploration of alluvial deposits where permeable gravel and sand beds can be distinguished from low permeability clays or rock. This capability has been applied in searches for construction materials beneath alluvial terraces and for foundation materials at dam sites where significant alluvial deposits occur.

At dam sites in valleys the results of resistivity surveys are affected by the irregular terrain and by changes in the electrical properties between dry materials on the abutments and the wet material beneath the valley floor. In situations where the valley floor width is less than five times the thickness of alluvial material there are significant side effects that produce misleading results.

These limitations and the generally poor depth resolution (about 10% of depth) indicate that the resistivity method has limited application in dam site investigations. It has however been successfully used in the review of some existing embankment structures (Corwin, 1999).

5.5.1.4   *Electromagnetic conductivity*

EM surveys have been widely used in North America (Corwin, 1999) in the assessment of existing embankments. The development of commercial software packages for the interpretation of EM data has considerably assisted the analysis of the data. Care has to be taken to select the correct frequency and to ensure that the geophysical interpretation matches the likely geotechnical model. EM is affected by above ground metal objects and by AC noise sources, e.g. power lines.

The transient electromagnetic (SIROTEM) method produces resistivity-depth profiles (Whiteley, 1983). The method is quicker and requires less space than the resistivity method, but may not be suited to the shallow depths normally associated with dam foundations.

5.5.1.5   *Magnetic*

Surveys using a sensitive magnetometer do not provide significant detail on the subsurface profile but can indicate the presence of buried ferrous objects, e.g. pipelines that can have a major effect on EM and resistivity traverses.

5.5.1.6   *Microgravity*

Accurate measurement of the earth's gravitational field, together with tomographic processing of the data, can indicate the presence of anomalous subsurface structures e.g. dykes, cavities.

The survey procedure is simple but requires precise location and level of each measuring point, which can make the method tedious, particularly in rough terrain.

5.5.1.7   *Ground penetrating radar*

Ground penetrating radar (GPR) uses the ability of VHF electromagnetic (radio) signals to penetrate through soils and rocks. Usually the waves are radiated from an antenna mounted on a wheeled sled. Reflections are received from features such as fractured zones and cavities in rock, and from distinct layers in soils. The method has also been used downhole.

The method has been claimed achieve to some success in the location of sinkholes and cavities in karst areas and this appears to be its main potential in dam engineering. The results of surface surveys in karst areas are described by Wilson and Beck (1988). Cooper and Ballard (1988) also describe the location of cavities at the El Cajon dam site in Honduras, using a downhole technique.

Limitations of the method include the need for a near regular surface to permit passage of the sled and very limited depth penetration, particularly through clayey soils.

Siggins (1990) and Morey (1974) describe details of the method.

5.5.2   *Downhole logging of boreholes*

Table 5.2 summarizes the available methods for downhole logging of boreholes.

Table 5.2.   Applications and limitations of borehole logging methods (adapted from Fell et al., 2000 and McGuffey et al., 1996).

Method	Parameter measured	Applications	Limitations
Ultra-sonic log	Determination of sonic velocities, attenuation of shear velocity in shear zones, or imaging the sides of the borehole wall	Shows fractures, other discontinuities	Requires uncased hole. Image less clear than borehole camera
Thermal profile	Temperature	Zones of inflow (lower/ higher temperature water)	Open hole not necessarily required
Caliper log	Borehole diameter	Infers lithology by borehole enlargement by erosion during drilling by drill fluid	Requires an uncased hole
Induction log	Electrical conductivity	Infers lithology by clay content, permeability, degree of fracturing	Lower resolution than resistivity log, but can evaluate unsaturated zone and PVC cased boreholes
Resistivity	Electrical resistivity	Infers lithology, (particularly between sand and clay)	Applicable only in saturated zone; requires open hole
Self potential	Electrical potential from mineral reaction and groundwater flow	Infers lithology, oxidation/ reduction zones, subsurface flows	Applicable only in saturated zone; requires open hole. Data difficult to interpret.
Natural gamma	Natural gamma radiation	Infers presence of clay and shale	Mud coating on borehole wall may affect results
Gamma gamma	Natural density	Provides log of density, from which lithology may be inferred	Provides only density. Health and safety issues.
Neutron neutron	Moisture content (above water table). Porosity (below water table)	Log of moisture content, from which lithology or wetter zones may be inferred	Provides only water contents. Health and safety issues
Borehole camera or video e.g. RAAX	Visual image of fractures and structure	Can identify structure. May be able to interpret dip of fractures and opening of joints, and bedding partings	Requires uncased hole. Images affected by water quality, smearing of sides of hole
Acoustic reflectivity	Acoustic reflections	Can identify lithology and structure. May be able to interpret dip of fractures and orient cores	Requires uncased hole

Profiles of electrical resistivity, self potential, gamma-ray emission and neutron absorption are routinely obtained during exploration for oil and coal, where most boreholes are uncored. Geophysical logging of each hole produces graphical plots from which the characteristic features of the different lithological formations can be recognised (Whiteley, 1983; Whiteley et al., 1990a). The development of commercial software packages for the interpretation of the data has considerably assisted its analysis.

The logging commonly requires uncased and water or mud-filled holes. The equipment is expensive and surveys have to be coordinated with the drilling programme.

Dam site investigations involve many shallow holes and nearly all holes are sampled or cored. Geophysical logging can assist in correlation between boreholes, the location of potential leakage zones and in the assessment of the properties of valley floor sediments. In the opinion of the authors in many cases the additional cost of retaining the expensive equipment on site may not be justified by the benefit of the additional information obtained. However in some situations some techniques are valuable e.g. borehole imaging can provide invaluable data on joint and bedding parting openings greatly complementing the logging of core, and water pressure testing.

5.5.3   *Crosshole and Uphole seismic*

The use of seismic waves in combination with boreholes has provided a range of methods to identify subsurface conditions both in the dam foundation and within the dam structures. In crosshole seismic a hydrophone array is lowered into one borehole and an energy source – a specially developed drop hammer – into a second borehole. By changing the location of the source and also the position of the hydrophone array the arrival times of the direct and refracted seismic waves can be computed into a tomographic image of the section between the two boreholes.

An alternative method using a single borehole involves the placement of the hydrophone array in the borehole, with the energy source at different points on the surface. The results can be computed to provide information on a cone of material up to 100 m deep with a radius up to twice the depth of the hydrophones.

The methods are described by Joyce et al. (1997), Whiteley (1998) and Van Aller and Rodriguez (1999).

5.6   TEST PITS AND TRENCHES

5.6.1   *Test pits*

Test pits excavated by a rubber-tyred back-hoe or tracked excavator are effective in providing information on subsurface conditions in dam foundations or existing dams, for the following reasons:

– They are relatively cheap and quick;
– The subsurface profile is clearly visible and can be logged and photographed;
– Material types, the nature and shape of their boundaries and structure can be observed and recorded in three dimensions;
– The absence or presence of groundwater is indicated and the sources of inflows can usually be observed and their flow rates recorded;
– Undisturbed samples can be collected;
– *In situ* tests can be carried out;
– The resistance to excavation provides some indication of excavation conditions likely to be met during construction.

It may be possible to leave pits open for inspection by design engineers and representatives of contractors. In this case fencing is usually required. The presence of fissures in clays can be assessed by leaving the pit walls to dry for some time e.g. overnight. Fissures can be classified using the systems shown in Figure 6.28.

Figure 5.8 shows a log of a test pit which includes a sketch of one wall with descriptions of each material type, traces of their boundaries, and traces of defects. Some limitations of test pits are:

– Pit depth is limited by the reach of the machine and its ability to dig the material. The maximum depth of a pit dug by backhoe is 4 m but the machine commonly refuses on very weak rock. Larger excavators can reach to depths of more than 6 m and, depending on the nature of the rock defects, can excavate into weak rock;

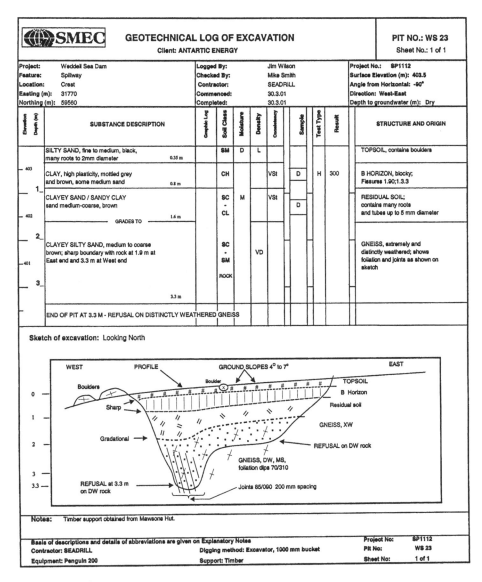

Figure 5.8.   Log of test pit.

– Local laws on support requirements of excavations must be followed. Collapse of pit walls occurs commonly, particularly in wet ground. The effective use of steel or timber support (Figure 5.9) can enable safe access to pits for logging and sampling;
– Groundwater inflows may cause excessive collapse and limit effective excavation, even with supports;
– Test pits disturb the local environment, but if carefully backfilled and the site revegetated their long term impact is minor.

5.6.2   Trenches

A logical extension of the use of test pits is the excavation of trenches, using either a tracked excavator or bulldozer and ripper. Such trenches provide virtually continuous exposures

Figure 5.9.   Timber supported trench/pit in Happy Valley Dam, Adelaide, South Australia (courtesy SMEC and SA Water).

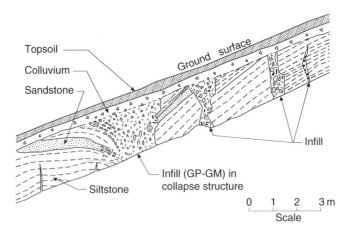

Figure 5.10.   Part of log of trench, Sugarloaf Dam showing infill features resulting from slope movement (Regan, 1980).

of the subsurface materials at sites where there is little natural outcrop. Careful cleanup by small backhoe or hand tools is usually required to expose defects, particularly in rocky materials. Trench exposures are logged in a similar manner to test pits as shown in Figure 5.8 and Figure 5.10.

Following the slope movements which caused disruption of construction of Tooma Dam (as described in Chapter 2, Section 2.10.3.2), clear evidence of past slope movements were exposed as infilled extension features in cut faces of access tracks. It was concluded that this evidence would have been found, and probably understood, if the site investigation had included bulldozer trenches.

Following that example the sites for most dams built in Australia have been explored extensively by bulldozer trenches which have been designed to provide the answer to geotechical questions and also access for drilling rigs.

These trenches have totalled more than 3 km in length at each of the following large embankment dams – Talbingo, Dartmouth, Sugarloaf and Thompson.

The effectiveness of well prepared trenches in providing an understanding of dipslope stability at the Sugarloaf site is described in Chapter 2, Section 2.10.3.4.

A major benefit of trenching is the experience gained from the use of plant of similar size and type which may be used during construction. It is good practice to record machine and ripper types, material types and excavation rates.

At sites for large structures in geologically complex areas with little natural exposure, trenching can be the most practical method of providing the answer to vital feasibility and design questions. A disadvantage of trenching is that it causes much disturbance of the site but, as with test pits, trenches can be backfilled and revegetated if construction work does not proceed.

5.7   SLUICING

In some situations during site investigation the bulk removal of soil by slicing using high-pressure water jets can expose the rock surface in the dam foundation and thus extend the information obtained from natural exposures on rock condition and structure.

Sluicing is an effective option on steep, rocky slopes where the soil is relatively thin and sandy. Stiff to hard clay is difficult to sluice. Sluicing requires a powerful pump and careful control of the operation to limit environmental effects. The movement of the mud slurry produced by the sluicing has to be controlled and contained using sedimentation basins.

Sluicing has been used with success at the Little Para Dam (South Australia), Gordon Dam (Tasmania) and Upper Ramu diversion dam (Papua New Guinea).

5.8   ADITS AND SHAFTS

In investigations for a major dam it may prove necessary to investigate a part or parts of the site, e.g. an area underlain by cavernous limestone or disturbed by a landslide, in more detail than can be achieved by the combination of surface excavations and drilling.

In such situations the excavation of adits or shafts into the area of concern provides the opportunity for:

– direct observation of the ground conditions;
– measurement of orientation of defects and comparison with surface measurements;
– *in situ* testing;
– underground investigation drilling.

Exposures in adits and shafts should be logged and photographed in detail using the same general approach as for test pits and trenches. All exposed faces, including the floor if practicable, should be logged.

Adits and shafts are expensive and slow to excavate. They are likely to require support, ventilation and drainage.

The requirements for, and location of, adits or shafts should be carefully considered in relation to all the available design and geotechnical information. It is necessary to justify the expense by the clarification of site conditions which they may provide. Wherever possible exploratory adits should be incorporated in the design of the dam, e.g. as drainage or grouting galleries.

Shafts using 1000 mm or larger diameter augers or clam-shell excavators are sometimes used to explore alluvial and glacial deposits.

5.9   DRILL HOLES

5.9.1   *Drilling objectives*

The main objectives of drilling are to extend the knowledge obtained from surface mapping, test pits and trenches below the depth limitations of these methods and to:

- provide control for the interpretation of any geophysical investigations;
- provide samples from these greater depths;
- provide access for test equipment e.g. for measurement of water levels, pore pressures and permeability etc.

The major advantage of drilling is that, subject to the choice of a suitable rig, there is little restriction to the depth to which the investigation can be taken. In the assessment of the subsurface profile the properties of the lowest rock or soil unit which could affect the structure usually have to be determined. In many cases drilling provides the most practical method.

Drilling also has little effect on the environment. Maximum hole size is usually less than 200 mm and holes can be easily covered, backfilled or neatly preserved. Surface disruption is commonly restricted to the preparation of a drilling pad on sloping ground.

In site investigation drilling it is imperative to recover as much information as possible on the subsurface profile. The drilling and sampling method should be selected to maximise the recovery of low strength material. Zones of crushed rock, gouge or clay seams are particularly important but some core losses can hardly be avoided. The recording of drilling parameters e.g. the drill penetration rate, drill fluid pressure, can assist in the interpretation of subsurface conditions.

5.9.2   *Drilling techniques and their application*

Table 5.3 summarises the different drilling techniques available, their applicability, advantages and disadvantages.

In the selection of the drilling method care should be taken not to disturb the existing site conditions. In the investigation of landslides the use of drilling fluid could lead to increased pore pressures, softening and hydraulic fracture. For the same reason the investigation of the clay core of existing embankment dams should always use dry drilling techniques, e.g. hollow flight auger, auger, dry cable tool.

Drilling has the disadvantage that information obtained is almost always indirect – either from the observation of resistance to rig penetration, by the measurement of *in situ* properties with equipment lowered down the hole or by the logging of samples recovered by the drilling (e.g. Thin Wall Tube samples). Direct observation of the ground is restricted to the use of mirrors, down-the-hole camera, television or fiberscope. Large diameter holes drilled by augers or the calyx method can, in some cases, be entered and logged.

Drilling rates depend on the machine and the material type and typically are 4 to 5 auger holes to 10 m in soil per 8-hour shift or 15 m of core drilling and water pressure testing per 8-hour shift. Rigs are often truck mounted – which makes them mobile but restricts access in sloping ground. Skid mounted rigs require considerably more time for establishment at each site.

Table 5.3. Drilling techniques and their applicability.

Drilling techniques	Hole diameter	Support	Applicability	Advantages	Disadvantages and precautions	Approx. cost & Productivity (2001) $Australian
Auger drilling solid flight	Usually 100 mm	Self-support i.e. the hole must support itself	Clayey soils, moist sand above the water table	Ease of setting up, rapid drilling, continuous disturbed sample recovered	Limited depth – 30 m normal maximum. Not applicable for sandy soils below water table or soft clays because hole needs support	$40–50/m, 30 m/day with stopping for sampling
Auger drilling hollow flight	Usually 100 or 150 mm, up to 300 mm	Self-support above water table if augers used as for solid flight augers. Augers used to support below water table	As above, and for sandy soils and soft clays below the water table by using the augers for support and sampling through centre of augers	As above, can be used below water table if flights filled with water or mud. Easy sampling Can be used in dam cores	Hollow flights are not normally filled with water or mud so "blowing" or disturbance at base of hole may occur in granular soils and soft clays when withdrawing the plug in the base of the augers for sampling or testing	$50–70/m, 20 m to 30 m/day for 100 mm diameter
Rotary drilling – Non coring in soil & rock (wash boring)	75–100 mm common	Self-support in clayey soils. Casing and/or drilling mud in granular soils and soft clays Usually no support in rock, but need mud if rock is weak or highly fractured	All soil and rock types	Allows full range of sampling and testing techniques. Rapid drilling in rock	Poor identification of soil types in drill cuttings Virtually impossible to identify thin beds of sand in clay. Proper identification requires tube or SPT samples Unable to penetrate gravels efficiently. Must case or fill hole with water or mud to prevent blowing in loose granular soils and disturbance of soft clays. Note that even stiff clays can fail by shear at 30 m depth if not supported by mud. Rock samples difficult	$50–60/m soil, $60–70/m rock plus $50/m for river gravel Good investigation rotary drill rig with hydraulic top drive and 3 m stroke Variable speed head for augering/coring. Truck mounted with 2 or 3 winches to facilitate rapid removal of drill rods for testing. Hire (continued)

Table 5.3.   (Continued).

Drilling techniques	Hole diameter	Support	Applicability	Advantages	Disadvantages and precautions	Approx. cost & Productivity (2001) $Australian
					to identify in cuttings. May induce hydraulic fracture in dam cores	rate is approx. $150–170/hr
Rotary drilling. Coring in soil & rock	"N" hole 76 mm, "H" hole 96 mm. Use "H" in soil and poor rock	Hole is supported by casing through soil overlying rock. No support in rock. Water, mud or foam used as drill fluid	All rock types. With care good recovery can be achieved in soils with any cohesion.	Continuous core of rock. Continuous "undisturbed" sample allows identification of stratification in alluvial soils.	Poor recovery in weak, friable, and closely jointed rock. No recovery in clean sand and silt. Soil and weathered rock samples usually disturbed by stress relief and swell due to water used from drilling. "H" size core gives better recovery but is heavier and more expensive.	NMLC or NQ3 coring $80–90/m, HMLC or HQ3 coring $90–100/m. 10–15 m/day coring. To ensure best practicable core recovery drilling on hourly hire is recommended. Rig hire $150–200/hr. Add $5/m for bit wear
Reverse circulation (rotary)	Usually BQ Duo-tube	Outer casing of reverse equipment	Sands, silty sand, sand with fine gravel	Rapid drilling. Complete recovery of soil drilled (disturbed). Useful for investigation of sand sources	Less commonly available. More expensive equipment. Mixes strata. No testing or undisturbed sampling possible	Rig hire $150–170/hr plus compressor hire. Up to 30 m/hr
Cable tool percussion bit and bailer	Up to 750 mm	Casing which may or may not be recovered	Sands, gravel, clay, rock (fractured)	Can penetrate gravels. Can drill large diameter hole. Simple equipment.	Slow drilling. Results in strata mixing, breakage of rock fragments and loss of fines which can give misleading particle size distribution. Very heavy tools for transport and handling	$80–100/hr. 2 m/hr down to 0.10 m/hr
Cable tool tube	Up to 200 mm	Casing usually recovered	All soils and weak rocks	Provides continuous samples and useful groundwater data.	Slow drilling. Very heavy tools for transport and handling.	Rig Hire $80–100/hr. 2 m/hr down to 0.10 m/hr

Method	Hole diameter	Casing	Suitable for	Comments	Limitations	Cost
Hammer drill (impact drill)	Usually 50–100 mm	Self supporting in rock. Needs casing in soils	Gravel, rockfill, boulders, rock (to just drill a hole)	Very rapid penetration rate (e.g. 200 m in 24 hours). Can penetrate rockfill and soil with gravel and cobbles	Samples limited to rock chips. May jam and lose equipment in loose fractured rock, rockfill or boulders. Penetration in clay is limited due to blockage of air passages	$150–170/hr plus compressor hire
Bucket auger, casing and auger, casing and clam shell	900–1200 mm	Casing which may or may not be recovered	Sands and gravels but applicable to all soils	Drills through gravely soils without breaking up into smaller sizes. Suitable for investigation of sand/gravel filter sources	Costly. Sub-sampling and testing also expensive because of large sample size	$300/m for holes to 10 m. 10 m/day
Rotary drilling with rig on allterrain vehicle	As for other rotary drilling	As for other rotary drilling	Drilling in swamps, water up to 0.5 m deep	Allows access where truck mounted rig would not enter because of soft ground	Smaller rig mounted. Applicable to 30 m depth.	Bombardier $350 per day in addition to ordinary drilling cost
Rotary drilling with rig on small jack-up barge	As for other rotary drilling	As for other rotary drilling	Drilling in water up to 6 m deep	Float rig into position and set up on bottom, unaffected by tide or waves	Costly to establish. Smaller rig mounted. Applicable to 30 m depth	$800 per day in addition to ordinary drilling cost. May need hire of work-boat to service rig

Other costs: U50 tubes including sampling $45; U75 tubes including sampling $60; SPT $35 each to 30 m; Core boxes $7/metre; Water cart $130/day; Compressor $200–300/day; Installation of rotary drill casing 10/m; establishment costs (and accommodation if applicable).

Figure 5.11.   Large diameter hollow auger drill rig (courtesy CMEC).

It is not good practice to specify drilling programmes in detail before geotechnical mapping and logging of test pits and/or trenches is well advanced or completed. By this time, understanding of the geological situation should be sufficient to allow boreholes to be aimed at answering specific geological and/or engineering questions.

5.9.3   *Auger drilling*

The most common type of drilling in cohesive soils uses a spiral flight auger to penetrate and remove the material below the surface. The simplest form is the hand auger which is usually restricted by the physical effort involved to about 3 m.

Most augers are machine driven – and range from portable to large truck-mounted hydraulic drill rigs (Figure 5.11).

The common auger rig equipped with either 100 mm or 150 mm diameter solid or hollow flight augers, can reach up to 30 m in soil strength materials. A steel blade "V" bit will penetrate most fine-grained soils and extremely weak rocks but usually refuses on coarse gravel or weak strength rock (Table 2.4). A tungsten-carbide "TC" bit will grind slowly through weak and medium strong rock.

Auger drilling allows the logging of disturbed material collected from the flights during drilling. By removal of the augers it is practical to regularly recover tube samples and carry out *in situ* testing of the material properties.

Auger drilling is suited to the investigation of areas with thick soil deposits which extend beyond the practical limit of pits or trenches. It does not provide the same amount of data on soil structure but can supplement other information. In many cases it is used as

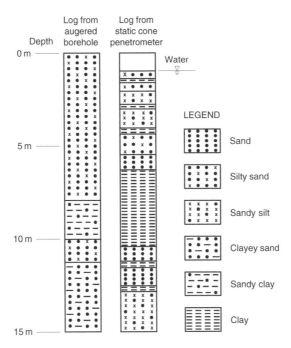

Figure 5.12.   Log of sandy and clayey soils below the water table from an augered borehole and from static cone penetrometer.

a rapid method of establishing the depth and general properties of the material overlying rock which will be investigated in more detail by some other method.

A major difficulty in auger drilling in cohesionless soils or soft clays is the stability of the sides of the drill hole particularly below water. Figure 5.12 shows a log prepared from an augered drill hole in sandy soils below the water table, compared with the log of a nearby static cone penetrometer probe. Mixing and collapse of the hole have led to gross inaccuracies in the auger hole log. Auger drilling in non-cohesive and weak soils is therefore restricted to above the water table.

Hollow flight augers have the advantage of providing support for the hole in these conditions and allow sampling through the augers. However the action of removing the plug at the end of the auger will often loosen and disturb the soil below the auger and hollow flight augers are not recommended for drilling in cohesionless soils, particularly if they are loose, or in weak clays.

The development of large diameter hollow auger drilling has enabled the recovery of continuous core with effectively undisturbed soil samples up to 300 mm in diameter (Figure 5.11 and Figure 5.13). This equipment is used to drill up to 80 m deep in dam cores in the USA.

Hollow flight augers are an acceptable technique for drilling in the core of existing dams because no drilling fluid is used which can cause hydraulic fracture. However the holes must be carefully backfilled in stages with cement/bentonite grout to avoid hydraulic fracture.

5.9.4   Percussion drilling

There are three main types of percussion drilling as follows:

(a) *Cable tool (bit and bailer) drilling* (Figure 5.14a) involves the successive dropping of a heavy chisel type bit to the bottom of the hole. The fragments are recovered using a

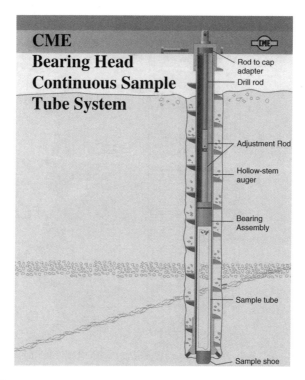

Figure 5.13.   Hollow auger continuous sample tube system (courtesy CMEC).

bailer. If the hole is dry, water is added to aid the bailing. As the hole is advanced steel casing is driven to preserve the hole. The method is slow, but is effective in penetrating gravels. It should be appreciated that the penetration process involves a reduction in particle size and may mix material from several layers. The grading of samples recovered from cable tool drilling may be different from that of the natural material.

(b) *Cable-tool Tube (also called Shell and Auger) drilling.* This method described by Wilson and Hancock (1970) uses a cable-tool rig as in Figure 5.14a but the hole is advanced and sampled by driving 100 mm diameter sampling tubes (Figure 5.14b and Figure 5.15). After each 300 mm long tube is driven the sample is extruded into a plastic sleeve, labelled and placed into a core tray. Before the next sample is taken the hole is reamed out using a slightly belled cutting shoe. When required, due to hole collapse or to seal off groundwater flow, the hole is reamed further and casing is driven. No drilling fluid is required. The method provides a continuous geological record in most soil types, a record of moisture content changes and also of the position and source level of any groundwater inflows (Figure 5.15). The tube samples are too disturbed for triaxial or similar testing but the fabric and defects in soils are preserved and can be seen when the samples are split. If rock, coarse gravel or boulders are met, these materials can be penetrated by the bit and bailer method.

Cable-tool drilling can be used for investigating the cores of existing dams as it is a dry method and will not induce hydraulic fracture. However it is slow and expensive compared to auger drilling. Completion of such holes should include backfilling with grout in stages designed to avoid hydraulic fracture.

(c) *Hammer (or Impact) Drilling.* Machines for this type of drilling range from the very small "air-tracks" through larger versions to large truck-mounted downhole hammers with

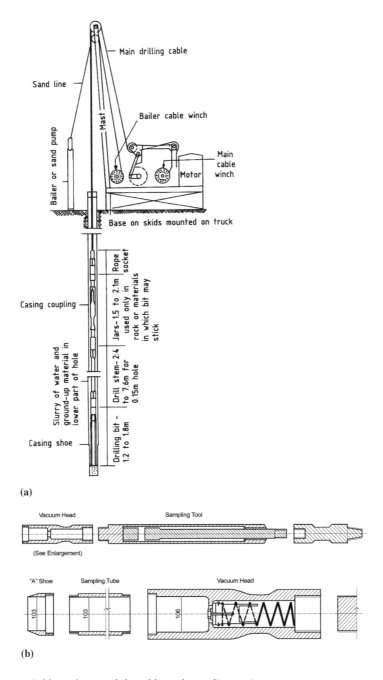

(a)

(b)

Figure 5.14.   (a) Cable tool rig, and (b) cable-tool sampling equipment.

reverse air circulation. These methods provide rapid penetration in medium to very strong rock but the rock quality must be assessed from penetration rate measurements and logging of sand and gravel sized rock fragments which provide no data on rock structure. The efficiency of the drilling is affected by excessive groundwater and drilling is slowed if the bit vents become blocked with clay.

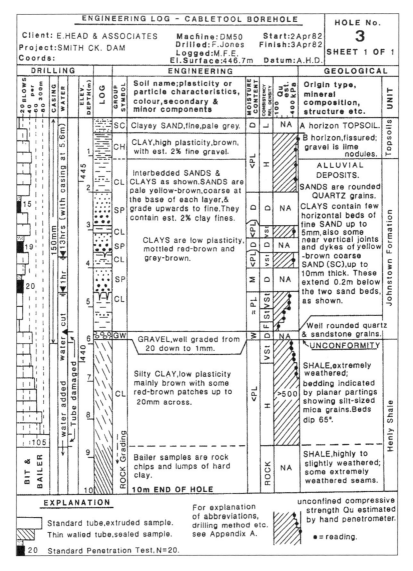

Figure 5.15.   Engineering log of cable tool drill hole (Courtesy South Australian Department of Mines).

It is considered that hammer drilling has little or no application in dam investigations other than through rockfill of existing dams for which it is particularly suited. It can be useful as a supplement to core drilling in the investigation of overburden depths and rock quality at sites for spillways or quarries.

5.9.5   Rotary drilling

Drilling using a rotary machine (Figure 5.16, Figure 5.17 and Figure 5.18) in soil and rock can involve either core or non-core methods.

(a) *Non-core drilling*, sometimes called wash boring, involves the use of a solid roller, button or drag bit at the end of drill rods. The bits break up or grind the full face of

Figure 5.16.   Rotary drill rig (a) Side view. Mast can be tilted to drill angled holes. (b) Rear view showing top drive hydraulic motor. (c) Dismantled core barrel showing inner tube, core catcher
and drill bit (courtesy Jeffery and Katauskas).

the bottom of the hole. The soil and rock fragments are removed by the circulating
fluid – commonly water. Samples from non-core rotary drilling through soil and rock
are usually a slurry of water and fine-grained material. This leads to poor identification of the materials.

It is usual to collect thin-walled tube samples of cohesive soils for more accurate
material description and laboratory testing and to carry out *in situ* testing in the
borehole.

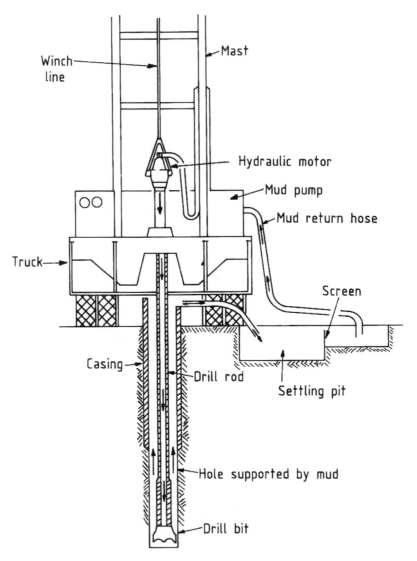

Figure 5.17.   Rotary drilling rig-non coring, also known as wash boring.

Non-coring bits can penetrate rock and care must be taken to change to coring methods as soon as resistance to penetration is encountered.

When drilling in sandy soils above the water table and in most soils below the water table, drilling mud is used to support the hole sides and assist in the recovery of drill cuttings. Either bentonite or more commonly chemical mud is used to form a cake on the sides of the hole (Figure 5.19). By maintaining a head of mud above the water table the excess pressure supports the sides of the hole. Bentonite mud must remain dispersed to be efficient. In saline ground-water the bentonite may flocculate and reduce the ability of the mud to support the sides of the hole.

Some chemical muds can be designed to maintain their viscous nature for a limited time and revert to the viscosity of water after a few hours. These muds have the advantage of not affecting the permeability of the material surrounding the hole and are suited to holes where permeability testing is required.

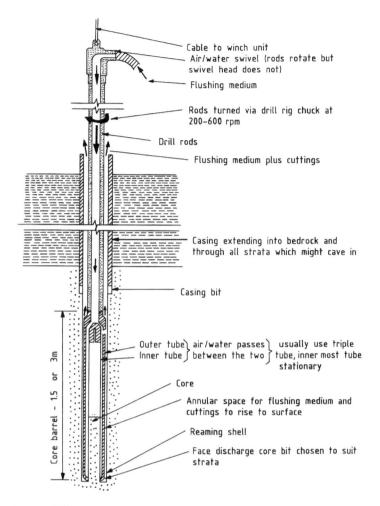

Figure 5.18.   Rotary drilling rig-coring.

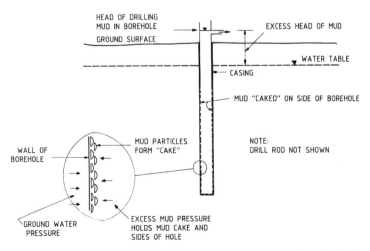

Figure 5.19.   Principle involved in the use of drilling mud to support the sides of a borehole.

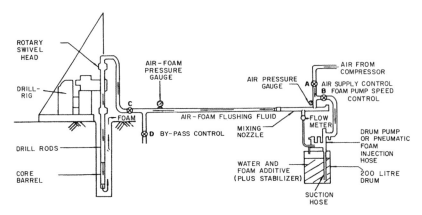

Figure 5.20.   Schematic layout of air-foam and flushing system (Brand and Phillipson, 1984).

Despite the "reverting" nature of the mud, the authors' experience is that permeabilities obtained by water pressure testing in holes drilled with mud are lower than
those obtained in holes drilled with water.

Chemical drilling muds can be used to improve recovery when coring in weak, erodible rock or stiff to hard soils. Core recovery can also be improved by using chemical
foam instead of water or mud. This is described in Brand and Phillipson (1984), and
illustrated by Figure 5.20.

(b) *Core Drilling*, an annulus of rock is removed using a hollow bit with a leading edge
impregnated with fragments of diamond or tungsten carbide, leaving a cylinder, or
core, of rock which can be removed by the core barrel. In conventional core drilling
(Figure 5.18) the barrel is connected to the end of drilling rods which convey the circulating fluid to the bit. Each time the core barrel is filled the rods have to be
removed.

In wireline drilling (Figure 5.21) the use of thin-walled rods and a special core barrel
enables the core to be raised inside the rods and significantly reduces the time involved in
drilling deep holes. The core recovered by wireline is slightly smaller in diameter than for
conventional drilling (NMLC core 52 mm, NQ3 core 45 mm).

The objective of site investigation drilling is to obtain the maximum amount of information on subsurface conditions. Every effort is required to recover as much core as possible. In low strength, fractured rock this may involve many short runs with low thrust
and drilling water pressures. Triple tube, stationary inner tube, core barrels reduce core
disturbance, improve core recovery and should be used in dam investigations.

In the analysis of foundation conditions, particularly in relation to slope movement and
potential leakage, the nature of the low strength material is most important. A drilling
programme which recovers all the strong rock and loses all the weak material is ineffective and its results may be misleading. As a matter of course drillers should be required to
comment on the probable reasons for every section where core has not been recovered.

Drilling contracts should be worded to ensure effective core recovery rather than rapid
drilling progress. The loss of 1 m of core in the upper part of the hole may be much more
important than the recovery of 10 m of fresh core at the base of the hole.

Placement of the core in boxes and logging of the core should emphasise sections where
core was not recovered as these may represent crucial zones in the foundation.

The choice of the core size to be used has to be considered in relation to the ability to
provide full core recovery. Larger diameter core is more expensive to drill, but is likely to
produce a higher core recovery in low strength and fractured rock.

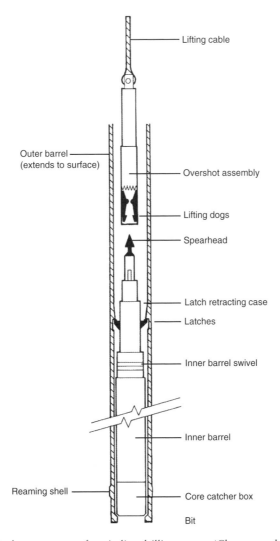

Outer barrel
(extends to surface)

Lifting cable

Overshot assembly

Lifting dogs

Spearhead

Latch retracting case

Latches

Inner barrel swivel

Inner barrel

Reaming shell

Core catcher box

Bit

Figure 5.21.   Principal components of a wireline drilling system (Clayton et al., 1995).

The ratio between natural fragment size in the rock to be sampled and core diameter is important. It is almost impossible to core conglomerate with 50 mm pebbles with N size equipment (76 mm hole, 52 mm NMLC core, or 45 mm NQ3 core). As a general rule coarse-grained fractured rock requires large diameter core. N size drilling is commonly used for high-strength rocks, but H size (96 mm hole, 63 mm HMLC core, or 61 mm HQ3 core) is significantly more effective in lower strength materials and only costs about 10% more than N size. On many dam sites defects which control strength and permeability in the dam foundation are close to vertical. Vertical drill holes are unsuitable for sampling these features and in the assessment of their effect on the foundation permeability.

The ability to drill angled holes at selected azimuths enables the choice of hole direction to provide maximum information including the orientation of rock defects and to interact defects which control the rock mass permeability. Angled holes are commonly slower to drill and slightly more expensive (contract rates are usually 10% to 15% higher than for vertical drilling). Drill holes at angles of less than 45° to the horizontal require anchoring of the rig to provide thrust.

5.10   SAMPLING

5.10.1   *Soil samples*

Soil samples recovered from the subsurface site investigation are described as either disturbed or undisturbed.

Disturbed samples are collected from pits, trenches and from auger flights as representative of different material types or units. They are identified by location and depth and stored in sealed containers (usually plastic bags) for laboratory testing. For cohesive soils about 3 kg of sample is required for classification testing and 30 kg for compaction tests to evaluate probable performance as engineered fill in an embankment.

Undisturbed samples consist of material which is extracted from the site and transported to the laboratory with a minimum of disturbance. The ideal sample is a cube of approximately 0.3 m sides, hand cut from a test pit, carefully packed and sealed on site and transported to the laboratory without delay.

It is more usual, due to economic factors, and the limitations of depth of test pits, to use thin-walled steel tubes ("Shelby" tubes) to obtain samples of cohesive soils from boreholes. The tubes are pushed into the soil using an adaptor connected to the drill rods. Care should be taken to ensure that the drill hole is cleaned out before sampling and that the drilling water or mud level is maintained during sampling to avoid "blowing" of material into the hole. Choice of sample diameter depends on the hole size but, in general, the larger the sample diameter the less the disturbance. Common thin walled tube sizes are 50 mm, 63 mm and 75 mm. The wall thickness and cutting edge shape are defined by codes to limit sample disturbance. Samples should be identified and sealed against moisture loss using either a sample tube sealing device or several layers of molten wax as soon as the sample is recovered from the hole. Before sealing, loose/disturbed soil at the top of the tube should be removed, so moisture does not migrate from that area to the less disturbed part of the sample. Undisturbed samples should preferably be tested within two weeks of sampling as they rust into the tube, or despite all efforts, dry out. On extrusion in the laboratory a proportion of "undisturbed" samples often prove to have been partly disturbed by the sampling process and it is prudent to take enough samples to allow for this.

In any case, even apparently undisturbed samples are affected by stress changes during sampling, and this needs to be recognised when analysing the results.

5.10.2   *Rock samples*

Samples of rock exposed in the sides/floor of pits/trenches can be taken for testing of substance strength and mineralogy. Samples should be individually numbered and located. Storage in plastic bags prevents loss of field moisture.

The most common form of rock sample is core. A drill hole with full core recovery should present a complete linear profile through the rock mass below the ground surface. It is important that the core is systematically stored, properly logged, photographed and sampled as soon as practicable after drilling.

In most cases the core needs to be kept for the following purposes:

– To enable the site investigator to make an accurate, clear and concise log of those characteristics of the core which are significant to the project and to use this log in the compilation of the geotechnical model;
– To provide samples for testing;
– For inspection by designers and those preparing bids to build the dam;

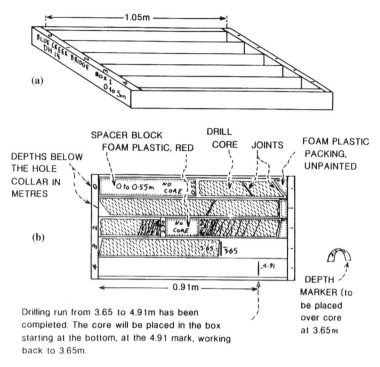

Figure 5.22.   Systematic boxing of drill core. (a) Empty core box labelled ready for use (b) Partly filled box during drilling.

– For inspection by those reviewing dam safety during the operating life of the dam. To allow this, core should be stored indefinitely.

The characteristics of most significance to the project will usually be:

– the substance strength;
– the lengths of the individual pieces of core;
– the lengths and positions of core losses;
– the length, position and engineering character of important defects such as sheared or crushed zones
– the engineering character of joint surfaces;
– the depths at which significant changes in rock substance type or strength occur.

The systematic method of boxing core is illustrated by Figure 5.22 and Table 5.4. This method "reconstructs" the core in the core box, to scale, packed with all joint or broken faces fitting, without any spacers apart from sections of core loss. The box is constructed in compartments of convenient lengths – usually either 1 m or 1.5 m – which become the "scale" for use by the driller and others who examine the core.

Changes in moisture content can produce significant changes in substance strength. Representative samples should be sealed, as wrapping in the thin plastic film used for food is only effective in the short term.

Core can deteriorate rapidly on exposure and should be stored inside in covered boxes. This is a particular problem for some shales and siltstones which can slake to soil within several days of sampling.

Table 5.4.   Good practice in boxing drill core.

Item	Description
Tray length	1.05 m or 1.10 m internal to hold 1.0 m of core and allow for increase in length due to bulking or oblique breakage.
Boxing scale	Core is boxed to scale in the ground; each row in the box represents 1 m of hole length. Any core loss is filled by a block of timber or equivalent, painted red, the length of the core loss and positioned where the core is considered to have been lost. If the position is uncertain the block is placed at the top of the run.
Boxing layout	Core is stored in the box like the lines of writing on a book.
Box marking	The left hand external end of the box is identified with the following information: Project, hole no., box no., depths. The same information should be recorded inside the box. The drilled depth in metres from the surface is marked on the left hand end of each row. Detachable core tray lids should not be marked.
Core marking	The bottom end of the core in each run is marked by paint or a "permanent" felt pen with the depth of the hole and a labelled line at that depth is marked on the bottom of the tray. A depth marker (Figure 5.22) is placed over the core at the bottom of each run with the depth recorded on the marker. Runs should not be separated by spacers. Where the core tends to break along oblique layering, or where core is very broken, the full 1.05 m or 1.10 m may be required to store the 1 m of core drilled. Where the core has few joints and can be broken normal to the axis the extra length allowed for bulking may not be needed and can be filled with white plastic foam to keep the core tightly packed and avoid disturbance.
Core samples	Core samples are removed only by authorised persons and are replaced immediately by yellow blocks of equivalent length marked with the name of the person who removed the core, the date of removal and reason for removal.

5.11   *IN SITU* TESTING

In many cases it is preferable to measure the properties of soil and rock in dam foundations using *in situ* tests, rather than taking samples and testing in the laboratory. In some cases (e.g. estimation of the relative density of sands) *in situ* testing is the only method available.

The topic is wide and rapidly developing and this section is restricted to directing readers to the available literature.

5.11.1   In situ *testing in soils*

The most commonly used *in situ* tests in soils are:

- Standard penetration test;
- Static cone penetrometer and piezocone;
- Vane shear;
- Pressuremeter;
- Dilatometer;
- Plate bearing test.

Jamiolkowski et al. (1985) and Wroth (1984) give overviews of the topic, and more recent conferences, e.g. Penetration Testing 1988, ISOPT-1 (de Ruiter, 1988), and Pressuremeter Testing (ICE, 1989) contain state of the art papers on the individual methods.

The main methods and their applicability to dam engineering are:

(a) *Standard penetration test (SPT)*. Refer to Decourt et al. (1988), Skempton (1986) and Nixon (1982) for details.

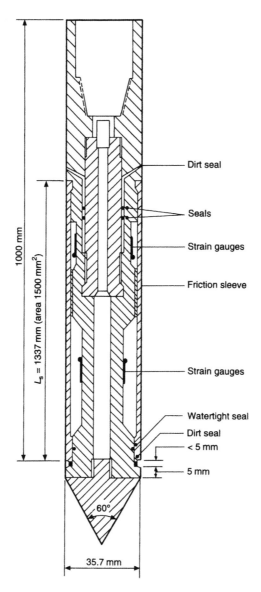

Figure 5.23.   Electric cone penetrometer.

The test is useful to obtain estimates of the relative density (density index), effective friction angle and deformation modulus (E) of cohesionless soils, and to assess the liquefaction potential of saturated sands and silty sands. The test is widely available but non-standardised and inherently approximate.

(b) *Static cone penetrometer (CPT) and piezocone (CPTU)*. Refer to De Beer et al. (1988), Campanella and Robertson (1988), Jamiolkowski et al. (1988) and Kulhawy and Mayne (1989) for details. Figure 5.23 shows details of an electric cone penetrometer.

The tests are useful to obtain estimates of the relative density, effective friction angle, drained Youngs modulus (E) of cohesionless soils, and the undrained shear strength of soft cohesive soils. There are many methods of interpretation of cone penetrometer

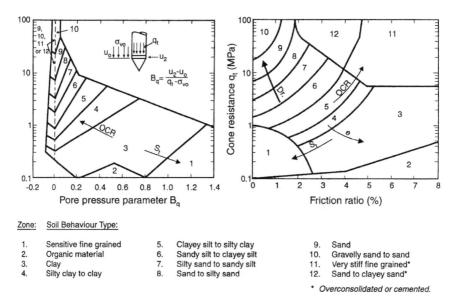

Zone:	Soil Behaviour Type:				
1.	Sensitive fine grained	5.	Clayey silt to silty clay	9.	Sand
2.	Organic material	6.	Sandy silt to clayey silt	10.	Gravelly sand to sand
3.	Clay	7.	Silty sand to sandy silt	11.	Very stiff fine grained*
4.	Silty clay to clay	8.	Sand to silty sand	12.	Sand to clayey sand*

* *Overconsolidated or cemented.*

Figure 5.24.   Campanella and Robertson (1988) method of soil classification using piezocone.

results. The authors consider that the soil classification using a piezocone proposed by Campanella and Robertson (1988) provides the most reliable interpretation (Figure 5.24). The cone resistance q_t is obtained from the CPTU q_c value adjusting for cone geometry. For CPT, Figure 5.24 can be used using q_c instead of q_t. The friction ratio ((sleeve friction/q_c) × 100%) is an output of CPT and CPTU tests. B_q is obtained from the CPTU dynamic pore pressure response. That graph is only used as a check on the other for CPTU tests.

The CPT and CPTU are particularly useful in alluvial foundations where sandy soils are interlayered with clayey soils as the instrument is able to detect the layering better than most drilling and sampling techniques. This test can also be used to assess the liquefaction potential of saturated cohesionless soils. The CPT has been used very successfully to locate softened/weakened zones in earth dams, including the core. The hole formed by the cone should be backfilled with cement/bentonite grout, placed by passing the grout down a tube in the hole under gravity, staged if necessary to avoid hydraulic fracture.

The CPT test is widely available but non-standard. The CPTU test is less widely available and requires some corrections to allow for the different designs of equipment. Both are inherently approximate but are generally regarded as more precise than the SPT test.

(c) *Vane shear.* Refer to Bjerrum (1973), Aas et al. (1986), Walker (1984), Azzouz et al. (1983) and Wroth (1984) for details.

This test can only measure the undrained shear strength of very soft to firm clays (maximum undrained shear strength of about 70 kPa), and has little use in dam engineering unless there are poorly compacted, saturated weak clay fills, or the foundations are weak clays. The need for correction of the field vane strength to give design strengths is well documented in Bjerrum (1973), Azzouz et al. (1983) and Aas et al. (1986).

(d) *Pressuremeter.* Refer to Ervin (1983), Jamiolkowski et al. (1985), Wroth (1984), Campanella et al. (1990), Powell (1990) and Clarke & Smith (1990) for details.

Both the self-boring (SBPM) (Figure 5.25) and Menard type pressuremeters have application in dam engineering, particularly in the assessment of the modulus of soil and weathered rock foundations. These pressuremeters give the most accurate estimate of modulus provided that the test method and interpretation are correct.

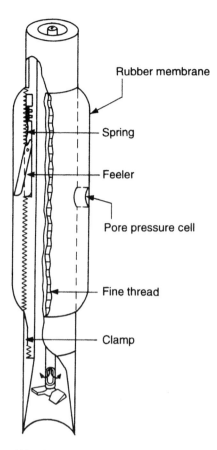

Figure 5.25.   Features of the self-boring pressuremeter.

The instruments can be used to estimate the effective friction angle of cohesionless soils and he undrained shear strength of cohesive soils. However disturbance on insertion gives an over-estimate of the undrained strength, so the results should be viewed with caution. SBPM test results can also be used to estimate the *in situ* horizontal stress in the ground, an important factor in modelling deformations in dam foundations using finite element methods, and to establish *in situ* stress conditions for dynamic analysis of liquefaction. The tests are relatively expensive (in comparison to CPT and SPT) but affordable on most large dam site investigations.

(e) *Dilatometer*. Refer to Kulhawy and Mayne (1989), Lacasse and Lunne (1988). Powell and Uglo (1988) for details.

The (Marchetti) dilatometer can be used in a similar way to the CPT. Its proponents claim advantages over the CPT but these are not apparent to the authors at this time.

(f) Plate bearing tests. Refer to Pells (1983), and Powell and Quarterman (1988) for details.

Plate bearing tests on the surface, in pits or downhole can be useful in assessing the deformation modulus of soil and weathered rock in a dam foundation. However as the tests are usually restricted to near the ground surface they are of limited value in a situation with a thick compressible layer.

Table 5.5 reproduced from Lunne et al. (1997) summarises the applicability of the different *in situ* tests. The authors broadly agree with the table.

Table 5.5. The applicability and usefulness of *in situ* tests (Lunne et al., 1997).

Group	Device	Soil type	Profile	u	*φ'	S_u	I_D	m_v	c_v	k	G_o	σ_b	OCR	σ-	Hard rock	Soft rock	Gravel	Sand	Silt	Clay	Peat
Penetrometers	Dynamic	C	B	–	C	C	C	–	–	–	C	–	C	–	–	C	B	A	B	B	B
	Mechanical	B	A/B	–	–	–	–	–	–	–	–	–	–	C	A	A	A	A	A	A	A
	Electric (CPT)	B	A	–	C	C	B	C	–	–	C	C	C	C	–	C	C	A	A	A	A
	Piezocone (CPTU)	B	A	A	C	B	A/B	C	A/B	B	B	B/C	B	–	–	C	C	A	A	A	A
	Seismic (SCPT/SCPTU)	A	A	A	B	B	A/B	B	A/B	B	A	B/C	B	–	–	C	–	A	A	A	A
	Flat dilatometer (DMT)	A	A	C	B	A/B	A/B	B	–	–	A	B	B	C	C	C	–	A	A	A	A
	Standard penetration test (SPT)	B	A	C	B	B	C	B	–	–	B	B	B	B	–	C	B	A	A	A	A
	Resistivity probe	A	B	–	–	–	A	–	–	–	–	B	B	–	C	A	B	B	B	A	B
Pressuremeters	Pre-bored (PBP)	B	B	–	C	C	C	B	C	–	B	C	C	C	B	B	–	B	B	A	B
	Self boring (SBP)	B	B	A¹	B	B	B	B	A¹	B	A²	A/B	B	A/B²	–	C	–	B	B	A	B
	Full displacement (FDP)	B	B	–	C	B	C	C	C	–	A²	C	C	C	–	–	–	–	–	A	B
Others	Vane	B	C	–	–	A	–	–	–	–	–	–	B/C	–	B	A	B	B	A	A	A
	Plate load	C	–	–	C	B	B	B	C	C	A	C	B	B	B	–	–	A	A	A	A
	Screw plate	C	C	–	C	B	B	B	C	C	A	C	B	B	A	A	A	A	A	A	B
	Borehole permeability	C	–	A	–	–	–	–	B	A	–	–	–	–	B	B	–	–	C	A	C
	Hydraulic fracture	–	–	B	–	–	–	–	–	C	–	B	–	B	A	A	A	A	A	A	A
	Crosshole/down-hole/surface seismic	C	C	–	–	–	–	–	–	–	A	–	B	–	A	A	A	A	A	A	A

Applicability: A = high; B = moderate; C = low; – = none.; *φ' = Will depend on soil type; ¹ = Only when displacement sensor fitted; ² = Only when pore pressure sensor fitted. Soil parameter definitions: u = *in situ* static pore pressure; φ' = effective internal friction angle; S_u = undrained shear strength; m_v = constrained modulus; c_v = coefficient of consolidation; k = coefficient of permeability; G_o = shear modulus at small strains; σ_b = horizontal stress; OCR = overconsolidation ratio; σ- = stress-strain relationship; I_D = density index.

5.11.2   In situ *testing of rock*

The most common used *in situ* tests in rock are:

- water pressure permeability test;
- pressuremeter;
- plate bearing;
- borehole orientation;
- borehole impression.

The water pressure test is described in Section 5.14. The references given in Section 5.11.1 for pressuremeter and plate bearing tests cover testing in rock as well as in soil.

Borehole orientation. Refer to Hoek and Bray (1981) and Sullivan et al. (1992) for details.

The logging of samples of soil and rock recovered from drill holes can identify the nature and spacing of defects and orientation in relation to the axis of the hole (more commonly expressed with reference to the plane normal to the core axis), but the absolute orientation is required for analysis of the pattern of defects which affect the project.

It is the authors' experience that rock mass defects in the drill core can be effectively oriented provided that they can be related to distinctive features (e.g. bedding), which have been identified and oriented on the surface or in underground exposures.

A recent development has been the use of remanent magnetism to orient drill core from Coal Measure Rocks (Schmidt, 1991). This technique involves the establishment of the regional orientation of remanent magnetism. Specific samples can then be oriented in relation to this direction.

It is possible to orient core from an angled drill hole using several different methods.

The simplest method applies when a hole is angled through rock in which the orientation of distinct bedding, foliation or cleavage is known, and is believed to remain constant throughout the depth of the hole. A length of core is simply held in the same orientation as the borehole, and then rotated until the bedding, foliation or cleavage lies in its known orientation. The orientation of joints or other defects can then be measured by compass and clinometer.

Other methods require down-the-hole instruments such as the Craelius core orientation device shown in Figure 5.26.

Analysis of the results of these measurements should consider the inherent problems involved. Accuracy of better than 20° in azimuth is difficult.

Borehole impression packer. This instrument records defects on the sides of the drill hole (Figure 5.27). It involves a tube with an expandable rubber packer, within foam-covered split metal leaves and a sleeve of thermoplastic film. When the packer is inflated – by water or gas – an impression is retained on the film of the defects which are present on the side of the hole.

If the impression packer is oriented (using marked drill rods), an impression survey should provide orientation data on defects from both vertical and angled drill holes.

In practice it has been found that successful use of the impression packer has been limited to medium strong to strong rocks with widely spaced open joints. Closed defects may not form a sufficient indentation in the wall of the drill hole to show on the sleeve. Tearing of the thermoplastic film is common in fractured rock.

5.12   GROUNDWATER

In all aspects of the investigation for embankment dams the groundwater situation should be continually monitored. This involves the location and monitoring of seepage points

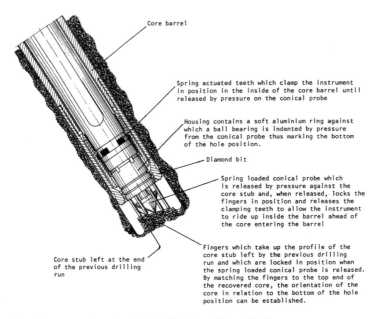

Core barrel

Spring actuated teeth which clamp the instrument in position in the inside of the core barrel until released by pressure on the conical probe

Housing contains a soft aluminium ring against which a ball bearing is indented by pressure from the conical probe thus marking the bottom of the hole position.

Diamond bit

Spring loaded conical probe which is released by pressure against the core stub and, when released, locks the fingers in position and releases the clamping teeth to allow the instrument to ride up inside the barrel ahead of the core entering the barrel

Core stub left at the end of the previous drilling run

Fingers which take up the profile of the core stub left by the previous drilling run and which are locked in position when the spring loaded conical probe is released. By matching the fingers to the top end of the recovered core, the orientation of the core in relation to the bottom of the hole position can be established.

Figure 5.26.   Craelius core orientation device (Hoek and Bray 1981).

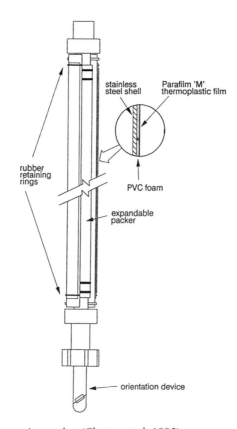

stainless steel shell

Parafilm 'M' thermoplastic film

rubber retaining rings

PVC foam

expandable packer

orientation device

Figure 5.27.   Borehole impression packer (Clayton et al. 1995).

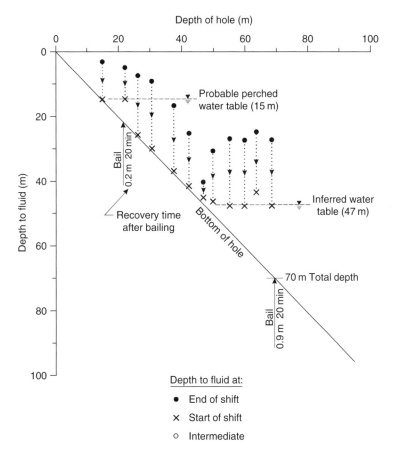

Figure 5.28.   Method for estimating perched and main water table levels from monitoring during drilling.

during surface mapping, recording of groundwater inflows into pits and trenches, and the monitoring of groundwater levels during the drilling investigation. Piezometers (with sealed tips and not just open slotted PVC "wells") should be installed in drill-holes (see Chapter 20) and regularly measured throughout the project. The response of groundwater levels to rainfall can provide a useful indication of mass permeability.

Measurement of seepage and flow rates, recording of rainfall, evaporation, and water levels installed in exploration drill holes gives a broad indication of the proportion of rainfall which infiltrates, the response of the groundwater to storms, and areas where high groundwater flow rates may be expected. This programme provides a relatively low cost general picture of the regional hydrogeology which can be refined by an investigation of specific permeability values at individual sites.

Claims related to unforeseen groundwater conditions form a significant proportion of contractual disputes. Many of these claims originate from a failure to record adequate groundwater information during site investigations and during the period between the completion of investigations and the start of construction.

Figure 5.28 shows how a perched and main water table can be inferred from the water levels measured in a borehole during drilling using the DVD (depth of hole versus depth of water) plot.

5.13   *IN SITU* PERMEABILITY TESTS ON SOIL

The permeability of soil in a dam foundation may be of importance if the cutoff is founded in the soil, as in dams on alluvial, colluvial and glacial deposits and on some deeply weathered residual or lateritised soils.

In general the structure of the soil controls water flow, for example:

- sandy layers in alluvial soils;
- root holes and fissures in residual and alluvial soils;
- worm burrows in alluvial and residual soils;
- leached zones, infilled root holes and relict joints in lateritised soil and extremely weathered rocks.

Hence it is seldom possible to obtain a realistic estimate of soil mass permeability from laboratory testing. Some of the difficulties associated with laboratory testing are discussed in Chapter 6.

Because the soil structure tends to be blocked or smeared by the drilling action, pump-in type permeability tests in boreholes can also give quite misleading results, with measured permeabilities one or two orders of magnitude lower than the actual soil mass permeability.

Where possible pump out tests should be conducted with a pump well and observation wells. Figure 5.29 shows a typical arrangement.

In a pump out test the inflow of water removes the soil blocking or smearing the hole side and realistic estimates of permeability can be obtained.

The design of the pump well, observation wells and the test program analysis need to recognise the effect of different vertical and horizontal permeabilities in the soil. The results are often difficult to assess and it is recommended that a groundwater hydrologist be employed in the planning, execution and analysis of such tests.

In the event that the groundwater level is below the depth of interest, it is necessary to carry out pump-in tests. Even when considerable care is taken to clean the hole, it should be understood that the results of such tests give lower bound estimates and that the actual permeability may be up to one or two orders of magnitude greater than indicated by the test result.

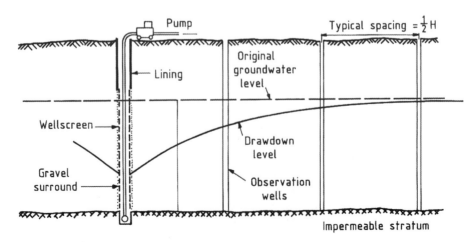

Figure 5.29.   Typical arrangement for a well pumping test with observation wells.

5.14   *IN SITU* PERMEABILITY TESTS IN ROCK

5.14.1   *Lugeon value and equivalent rock mass permeability*

The permeability of the rock mass can be determined by either constant head tests or falling head tests. A section of a drill hole is isolated using a sealing packer and water is added to maintain a constant head, or the rate of fall in water level is measured after a slug of water is added to the hole. Both methods suffer from the potential effects of smear and clogging of defects, but by careful flushing of the hole before testing, reasonable values can be obtained. In falling head tests the additional pressure which can be added to the test section is limited by the level of the test section and the practicability of extending the pipe imposing the head above ground level.

The most common and effective method of measuring rock mass permeability is the water pressure test (also known as the Lugeon or "packer" test). The test consists of isolating a section of drill hole and pumping water under pressure into that section until the flow rate for any given pressure is constant (i.e. it is a constant head test). The use of successive rising and falling test pressures establishes the relationship between the volume of water accepted into the section and the pressure, to provide an estimation of permeability, and indicate water flow mechanisms.

As rock substance is generally almost impermeable, the permeability determined in this test represents an indication of the number, continuity and opening of the rock defects which intersect the wall of the borehole in the test section.

Results are expressed in Lugeon (uL) units. A Lugeon is defined as the water loss of 1 litre/minute per metre length of test section at an effective pressure of 1 MPa.

Indicative rock permeabilities are:

Lugeon	Range	Condition
<1	Low	Joints tight
1–5	Low/Mod	Small joint openings
5–50	Mod/High	Some open joints
>50	High	Many open joints

There is no unique relationship between Lugeon value and equivalent rock mass permeability (ke). Moye (1967) recommended use of the equation:

$$ke = \frac{QC}{LH} \tag{5.1}$$

where ke = the equivalent coefficient of permeability (m/sec)
Q = the flow rate (m^3/sec)
L = the length of the test section (m)
H = Net head above the static water table at the centre of the test section (m)

$$C = \frac{1 + \ln(L/2r)}{2\pi} \tag{5.2}$$

r = the radius of the hole (m).

This is based on the assumption of radial laminar flow in a homogeneous isotropic rock mass, a condition seldom, if ever, achieved.

Hoek and Bray (1981) suggest the use of the equation:

$$ke = \frac{Q\ln(2mL/D)}{2\pi LH} \tag{5.3}$$

where m = $(ke/kp)^{\frac{1}{2}}$
kp = equivalent permeability parallel to the hole
ke = equivalent permeability normal to the hole
D = diameter of hole.

They suggest that for most applications ke/kp is about 10^6. This implies no fractures parallel to the hole and in most rocks would not be a reasonable approximation.

If it is assumed that ke/kp = 10, and water pressure testing in 5 m lengths of NMLC hole (75 mm diameter), then 1 Lugeon is equivalent to ke = 1.6×10^{-7} m/sec.

For ke/kp = 1, i.e. homogeneous, isotropic conditions, m = 1 and 1 Lugeon is equivalent to ke = 1.3×10^{-7} m/sec. The Moye (1967) formula gives similar results to this homogeneous isotropic case.

5.14.2 Test methods

There are two common methods of water pressure testing in a drill hole (Figure 5.30). The "down-stage" (or "single-packer") method is recommended and involves isolating and testing successively the bottom sections of the drill hole. This method enables progressive assessment of permeability and allows later stabilisation of the wall of the hole by casing or grouting if caving occurs. A disadvantage of down-stage packer testing is that it disrupts drilling progress, but this is far outweighed by its advantages over the alternative method discussed below.

The alternative method is to complete the drilling of the hole and water test in sections by sealing the hole above and below the test area (the "double-packer" method). This method has the advantage of convenience in that all water testing is carried out at one time but results can be affected by:

– Damage to the sides of the hole by drill rods and casing;
– Possible leakage from the test section past the lower packer, which cannot be detected, and
– Sections of the hole which have been stabilised by cement or by casing to enable deeper drilling cannot be tested. Commonly these sections will be highly permeable.

The purpose of the test is to estimate the potential of water to pass through rock defects. The use of drilling mud to stabilise the hole can block these defects and make the results of water pressure testing meaningless. For best results the drilling fluid used in holes where permeability testing is required should be water. If necessary a small amount of soluble oil appears to improve drilling efficiency without affecting permeability.

There are several proprietary chemical drilling muds such as "Revert" which are reputed to break down and dissolve when treated with "Fastbreak". Testing experience indicates that this procedure is only partly effective and that holes drilled with Revert give lower indicated permeabilities than holes in the same situation drilled with water.

5.14.3 Selection of test section

To maximise the information on rock mass permeability the investigation drill hole should be oriented to intersect as many joints as practicable (see Figure 5.31).

Every effort should be made to test the total length of hole in rock. It is preferable to overlap sections and thus have two tests over a short length than to miss some length of hole.

The upper limit which can be tested is the highest level at which a packer can be satisfactorily sealed, often in distinctly weathered rock. Location of the packer within casing above the rock does not seal the hole as water may leak past the casing.

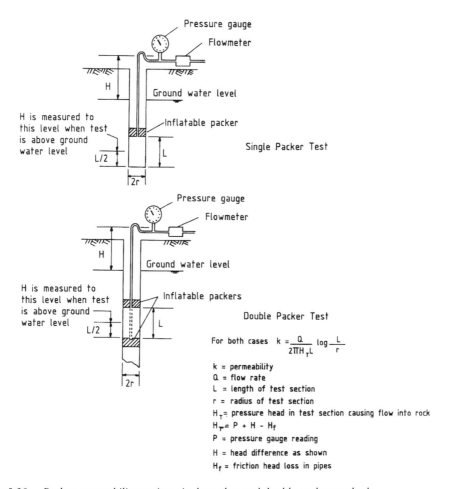

Figure 5.30.   Packer permeability testing; single packer and double packer methods.

The length of test section depends on the nature of the rock defects and the type of structure under investigation. Examination of the drill core usually indicates the presence of typical fracture spacing which represents probable background permeability and anomalous structural features which may be associated with zones of higher permeability.

Test sections should be selected to provide an indication of the relative width and permeability of these zones. In many cases water flow may be concentrated through a few fractures. The water loss is averaged over the length of the test section and, if the section is too long, the presence of a high permeability zone may not be recognised.

Usual test section lengths range from 3 m to 6 m but the length may be increased in essentially unfractured rock. When a particularly high water loss is recorded it is good practice to repeat the test over a shorter section of the hole to further define the zone of high permeability.

5.14.4   Test equipment

The equipment required for water pressure testing includes:

– a packer to seal the test section;
– a water line from the packer to the supply pump;

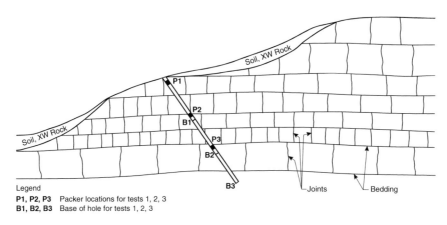

Figure 5.31.   Borehole inclined to intersect joints in the rock.

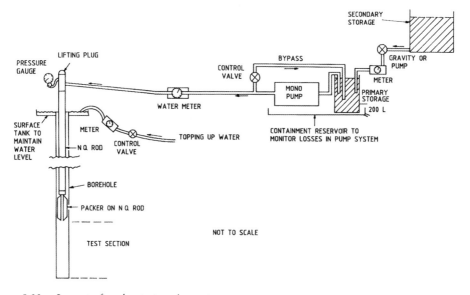

Figure 5.32.   Layout of packer test equipment.

– a pump to supply water under pressure;
– a bypass to control the pressure;
– a surface tank at the borehole to maintain the water level outside the water line;
– water storages;
– a pressure gauge;
– a water meter.

A layout used for water pressure testing is shown on Figure 5.32.

5.14.4.1   Packers

There are several types of packer used:

– A *hydraulic packer* (Figure 5.33) consists of a double tube with rubber sleeve. When the packer is in the "down" position water can be pumped into the sleeve to inflate the packer and seal the hole. In the "up" position water flows into the test section;

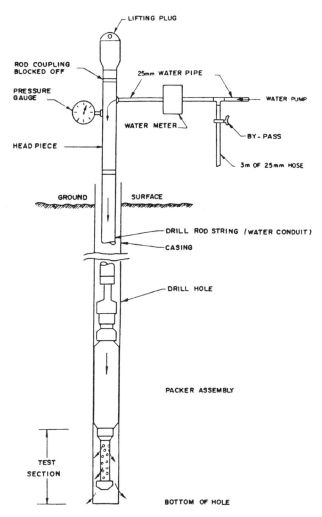

Figure 5.33.   Hydraulic packer equipment (Coffey Geosciences).

Hydraulic packers have proved reliable under most conditions. It is necessary to ensure that the hole outside the water line is kept full, as outflow from this area indicates possible leakage past the packer, and also it is necessary to equalise the pressures and allow effective inflation and deflation of the sleeve;

–   In a *pneumatic packer* the sleeve is inflated by air from a compressed air bottle using a separate air line to the surface. This method is effective in shallow holes, but with deeper holes fracturing of the air line has caused problems. Water levels do not affect inflation or deflation;

–   *Wireline packers* have been developed for testing holes drilled with wireline equipment. These enable water pressure testing without the withdrawal of the drill rods. The packer incorporates two sealing sleeves, the upper seal within the drill rods and the lower seal in the hole below the drill bit;

–   The *mechanical packer* seals by the expansion of two or more rubber rings when compressed (Figure 5.34). The test section length is controlled by the insertion of a selected length of perforated rod below the packer. Sealing is achieved by downward pressure on the drill string by the drill chuck. This pressure must be maintained throughout the test.

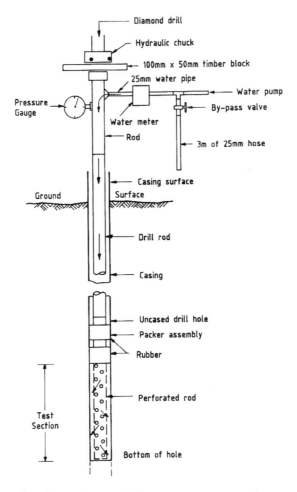

Figure 5.34. Mechanical packer equipment (Coffey Partners International).

The capability of the rubber rings to expand is limited. With a sealing length of about 200 mm the formation of an effective seal in closely fractured rock is difficult. Alteration of the test section length involves removal of the whole drill string and the addition or removal of rods below the packer. This operation is time consuming in deep holes. Mechanical packers do however have the advantage of fewer operational problems.

A double mechanical packer requires complete withdrawal and addition or removal of rods below the bottom packer between each test.

It is considered that mechanical packers should only be used where other packers are not available and results are not critical. Double packers should be avoided if at all possible.

5.14.4.2 *Water supply system*

The line connecting the packer to the surface should be watertight with minimum restriction to water flow. Flush coupled rods or casing are commonly used. The test water should be clean and supplied by a pump with bypass to enable control of pressure. Dirty water will result in clogging of fractures and lower Lugeon values than are correct for the rock.

A centrifugal pump or reciprocating pump with surge chamber is necessary to ensure constant pressure. Pressure and water flow is measured by meters which should be recently calibrated. Before testing the water supply rods and packer should be calibrated for friction losses at different flow rates.

5.14.4.3 *Selection of test pressures*

The object of the test is to:

– measure the natural permeability;
– indicate the probable water flows under the expected hydraulic loading by the proposed structure.

The pressures applied during the test should not be sufficient to produce hydraulic fracturing (jacking) of the rock around the test section. In weak rocks near the ground surface this fracturing does take place at relatively low pressures and is usually indicated by a unexpected increase in water loss following a raising of test pressures. It is recommended that, to avoid potential "jacking", maximum effective test pressures be limited below overburden pressure (approximately 22 kPa/m). In low strength weathered rocks, lower values will probably be necessary.

5.14.5 *Test procedure*

The test involves:

– measurement of ground water level;
– washing out of drill hole. Circulation of drilling water should be continued for at least 15 minutes after the water appears clear;
– installation of packer at the selected level;
– connection of the water supply system;
– application of the test pressures and measurement of water loss;
– removal of equipment.

The testing is carried out in several stages with different pressures. Commonly at least three test pressures are used (five are desirable). Pressures are applied in an increasing and then decreasing sequence. For example with three pressures a, b, c, the water loss is measured at stages with pressure successively at a, b c, b, a. Each stage should be continued until a constant rate of water loss (within 10%) for a 5 minute period is recorded.

5.14.6 *Presentation and interpretation of results*

The results are best plotted with the effective test pressure at the centre of the test section against the flow rate. The effective test pressure is the gauge pressure corrected for the elevation difference between gauge and water table and for friction losses in the system. (i.e. H_T in Figure 5.30). Where the water level is not known, or suspected to be locally elevated due to the effects of the testing, the results are plotted using gauge pressure only. Typical test results (Table 5.6) plotted in this way are shown in Figure 5.35. Note that the data from the increasing pressures do not plot identically to those on the decreasing pressures.

Table 5.6.   Sample results of water pressure test.
Hole No.: JB24; Test Section: 20.05 m to 26.50 m; Section Length: 6.45 m

Duration of test mins	Gauge pressure kPa	Water loss litres	Litres per min	Litres per min/m
5	20	20.0	4.0	0.62
5	20	19.0	3.8	0.59
5	60	28.5	5.7	0.89
5	60	28.5	5.3	0.82
5	110	36.0	7.2	1.12
5	110	35.0	7.0	1.08
5	170	48.5	9.7	1.50
5	170	46.5	9.3	1.44
5	110	34.0	6.8	1.05
5	110	32.0	6.4	0.99
5	60	25.0	5.0	0.78
5	60	25.5	5.1	0.79
5	20	15.5	3.1	0.48
5	20	16.0	3.2	0.50

Interpreted result = 6 lugeons.

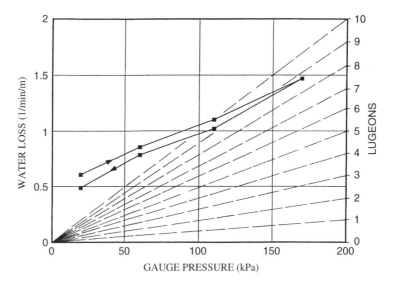

Figure 5.35.   Typical results of packer test.

The plot is judged to be parallel to the 6 Lugeon line. A range of different results is possible, indicative of the following possible mechanisms:

– Laminar flow and no change in permeability during the test;
– The occurrence of turbulent rather than laminar flow;
– Scour of joint infill, weathered or crushed rock;
– Leakage past the packer;
– Sealing of joints by fines from the water;
– Hydraulic fracture of the ground;
– Inaccuracies in measurement.

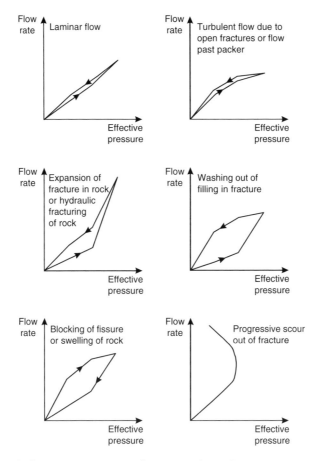

Figure 5.36.   Typical effective pressure versus flow curves for packer tests.

Figure 5.36 shows examples of plots indicating these types of behaviour.

5.15   USE OF SURFACE SURVEY AND BOREHOLE INCLINOMETERS

5.15.1   *Surface survey*

In the investigation of dam sites and the assessment of existing dams the precise measurement of location and elevation of the ground surface provides information on:

– Existing mass movement – e.g. landsliding;
– Foundation settlement;
– Deformation or displacement of structures.

A range of surface survey methods is included in Table 5.7.

The development of computer assisted survey instruments has enabled the determination of the position of many surface points to be carried out rapidly and with great accuracy. A vital component of any dam investigation should be the establishment and regular measurement of a network of surface stations. Survey marks should be founded below the level of reactive soil and away from traffic. The points should be clearly marked and numbered.

Table 5.7.　Instrumentation and methods for the measurement of surface and subsurface displacement (adapted from Dunnicliff, 1995 and Fell et al., 2000).

Parameter	Instruments/Methods
Surface deformation	Surveying methods, including GPS Crack gauges/surface extensometers Tiltmeters Multi-point liquid level gauges Photogrammetry Satellite images Remote video
Subsurface deformation	Inclinometers Simple borehole deformation measurements Fixed borehole extensometers Slope extensometers Shear pin indicators In-place inclinometers Multiple deflectometers Acoustic emission monitoring Time domain reflectometry (coaxial cables) Pendulum

The results of repeat surveys of these marks should be recorded as amount of differential movement and plotted as vectors of vertical and horizontal displacement. These plots enable a quick assessment of the amount and consistency of the movements that have been recorded.

5.15.2　Borehole inclinometers

Subsurface movement can be measured using several different instruments as listed in Table 5.7 and discussed in Chapter 20. These instruments are also used in the study of landslide movements and are described by Dunnicliff (1995).

The most common method uses a borehole inclinometer, the principles of which are shown in Figure 20.27. Successive readings on a borehole inclinometer provide information on the amount and direction of deflection and the location of the point of rupture (Figure 5.37). A borehole extensometer uses magnets which can be installed outside the PVC casing in soil or rock (Figure 20.23). Sliding couplings in the casing allow relative movement which can be measured using a probe lowered down the borehole.

The installation of electrical shear strips within a slope or embankment when monitored by a readout unit can indicate the location of an area of slope movement but not the amount of movement.

5.16　COMMON ERRORS AND DEFICIENCIES IN GEOTECHNICAL INVESTIGATION

Many problems during design and construction of dam structures are caused by poor quality or inadequate geotechnical investigation. This is often due to the investigation not following good engineering practice. Many of the problems are known to competent practitioners. In some situations lack of available finance can force the adoption of poor

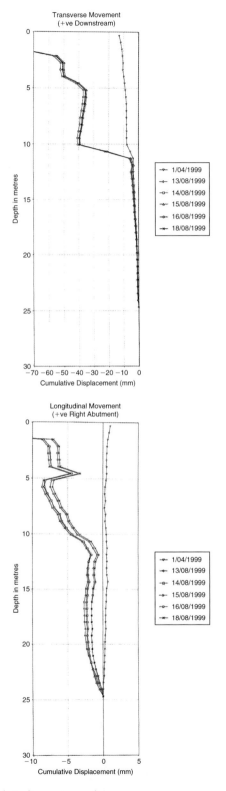

Figure 5.37.   Plot of borehole inclinometer readings.

Table 5.8.   Deficiencies in drilling and sampling.

Problem	Consequences	Remedy
Only drilling vertical holes where jointing is near vertical	Joint spacing and rock mass permeability incorrectly assessed	Angle holes to intersect joints, bedding and other features
Poor identification of orientation of joints in boreholes	Lack of knowledge of actual joint orientation affecting slope stability	Careful orientation in boreholes and mapping in trenches
Water table not measured, not measured often enough or data not recorded	Dewatering problems with resultant contractual claims; lack of information for design, e.g. pore pressures for slope stability	Routine measurement and or recording. Install casing and or piezometers
Drill water inflows or outflows not recorded	Poorer understanding of reasons for high/low water tables	Employ good drillers and supervise full time
Poor identification of layering in soil deposits particularly when using augers below water table, or wash boring without sampling, e.g. see Figure 5.12 which compares log from augered borehole and static cone penetration test	Soil strata are mixed together, hole may collapse causing greater mixing. Leads to under/over-estimation of horizontal/vertical permeability, potential overestimation of strength by mixing clay layers	Use rotary drilling with mud and/or casing, sample systematically with thin wall tubes and SPT. Use static cone penetrometers
"Blowing" in boreholes in silty sand, sand and soft clay i.e. flow of material towards borehole	Low SPT values in silt and sand and disturbed samples (in clay) leads to over-estimation of settlement, underestimation of strength	Use drilling mud and excess head of mud in the borehole to prevent blowing
Only drilling holes, not test pits, in "structured" clays, e.g. fissured or lateritised soils, soils with root holes	Failure to recognise the structure usually leads to overestimation of strength and underestimation of permeability (often by orders of magnitude)	Dig backhole and excavator pits and have experienced personnel log them
Drilling in gravel and gravelly sands with percussion drill	Gravel is broken up to finer particles, mixed with sand to give the impression of uniform sandy gravel. Fines may be lost in the drilling process. Horizontal permeability is under-estimated, possibly by orders of magnitude (see Figure 6.55)	Recognise the problem. No real drilling solution. Test permeability with pumpout tests
Ground surface level and location of boreholes not surveyed	Errors in plotting sections, plans misinterpretation of conditions	Survey all investigation location

Table 5.9. Deficiencies in *in situ* testing.

Problem	Consequences	Remedy
Testing only at predetermined depths, e.g. etc., e.g. SPT, undisturbed tube sampling 1.5 m, 3 m, 4.5 m	Poor identification of strata, poor selection of strata to test and sample	Supervise full time and test at strata changes as well as at pre-determined depths
SPT (and CPT) test in gravelly soils affected by coarse particles	Overestimation of SPT "N" value, with resultant overestimation of relative density, underestimation of compressibility	No real remedy. Just recognise the problem or seek other ways of estimating the parameters
Not washing borehole carefully before water pressure testing	Joints remain clogged with drill cuttings, Lugeon value underestimated	Take care in washing hole. Use clean water for testing
Lack of *in situ* permeability tests in soils	Contractual claims because "conditions are worse than contractor assumed"; gross errors in estimation of permeability	Do appropriate *in situ* tests despite the costs involved
Estimation of permeability from particle size distribution	Underestimation of permeability because of mixing of finer layers. See Figure 6.55	Only use "Hazen" type formulae in uniform, clean fine-medium sand for which it was derived
Use of pump-in permeability tests in soil, particularly structured clay and in augered boreholes	Gross underestimation of permeability due to smearing and clogging of fissures, root holes, sandy layers, (by factor of 10 to 10^3).	Use pump-out tests where soils are below water table. Above water table use pits, with the sides carefully cleaned to remove smearing. Adopt "realistic" values for design regardless of results
Use of seismic refraction survey to estimate rippability	Incorrect prediction of rippability, contractual claims	Do seismic refraction correctly (see Whiteley, 1988). Couple with geological factors (see MacGregor et al., 1994 and Stapledon, 1988b), and recognise estimates are approximate in contractual arrangements
Installation of "wells" instead of properly constructed piezometers in boreholes	Measures phreatic surface, not pore pressure. May over/underestimate pore pressures (see Figure 20.12 and fell, 1987)	Install piezometers, properly sealed in borehole

Table 5.10.   Deficiencies in logging of boreholes, test trenches and pits.

Problem	Consequences	Remedy
Logging rock in soil description terms, e.g. "black silty clay" in holes drilled by auger, with the rock ground to soil consistency by the drilling bit	Contractual claims relating to difficulty of excavation; incorrect design assumptions	Use correct drilling techniques and/or log correctly and/or log in soil and rock terms
Logging joints, partings and drill breaks all together as "fractures"	Incorrect assessment of the joint spacing leading to incorrect assessment of ease of rippability and size of ripped rock and contractual claims; incorrect assessment of slope stability and grouting conditions	Log joints, bedding plane partings and drill breaks separately and present data clearly
Failure to log condition of joints, e.g. clay coating, iron stained, and continuity of points. Poor definition of weathering classification	Incorrect assessment of joint strengths in slope stability, and of likely flow rates of water, grouting conditions, rippability. Confusion on acceptable rock conditions during construction with resultant contractual claims	Use experienced personnel to log core (or at least check logging). Define weathering classification and/or define acceptable conditions accurately
Incorrect description of cemented soils, either failing to describe the cementing, e.g. in calcareous sands OR describing cemented soil as rock	Incorrect assessment of conditions for design and contractual claims for excavation or tunnel support conditions	Inspect exposures in large cuttings, dig pits, relate drilling results to the local geology and log accordingly
Failure to log soil structure, e.g. fissures, root holes, minor interbedding	Incorrect assessment of shear strength of fissured soils, underestimation of permeability	Log carefully and systematically in pits
Classifying soils in the dry state	Underestimation of clay content and plasticity leading to incorrect design specification and contractual claims	Moisten soil before classifying
Incomplete description of soil, e.g. omission of moisture conditions, consistency, colour	Incorrect assessment of conditions for design, specification and contractual claims based on unforeseen "wet soil" etc.	Log carefully and systematically
Inadequate description of "organic matter" and "fill"	Contractual claims and incorrect design	Describe in detail in the logs

Table 5.11.   Deficiencies in data presentation and interpretation.

Problem	Consequences	Remedy
Too much detail on irrelevant features, e.g. lengthy description of trench logs, or mineralogy of rock types in borelogs. Confusion of "facts" and interpretation	Important features lost in the mass of irrelevant data. Misinterpretation of geological conditions, potential for contractual claims	Proper planning and briefing of personnel by experienced geotechnical practitioners Clear distinction and definition of terms
Straight line interpolation between boreholes without regard for lack of data or geological conditions	Misinterpretation of geological conditions, overconfidence in interpolation, potential for contractual claim	Draw interpretive sections with due allowance for geology e.g. core stones in granite, buried land surfaces
Use of exaggerated scales in preparing sections	Misinterpretation of geological conditions, overconfidence in interpolation, potential for contractual claim	Use natural scale or provide both natural and exaggerated scale
Consideration of data on a hole to hole basis	Misinterpretation of geological conditions, incorrect assessment of design parameters and the range of values	Determine a proper geotechnical model of the site based on all data
Failure to recognise that it is the exception which sometimes causes the problems e.g. a thick bed of rock or a thin bed of high strength rock causes problems with rippability and size of ripped material	Contractual claims	Recognise the importance and include in reports and interpretive sections
Incorrect projection of information from boreholes, in particular altering levels	Incorrect interrelation of conditions	Project along contour and/or with consideration of geological controls

Table 5.12.   Deficiencies in earthfill borrow area investigations.

Problem	Consequences	Remedy
Use of boreholes to investigate earthfill and sand/gravel deposits	Failure to recognise variability of soils, water contents incorrectly determined, contractual claims	Use backhoe or excavator pits where possible ($<6\,m$)
Lack of water content profiles, or profiles taken in periods not typical of contract	Lack of knowledge of moisture condition requirements, contractual claims	Take profiles in period representative of construction
Lack of information on "bulking" factors in earthworks	Deficiency of material when "shrinkage" occurs leading to contractual claims	Do density in place tests and laboratory compactions in borrow area investigations
Variability of alluvial and non alluvial soil deposits	Difficulty in selecting materials during construction, contractual claims	Dig plenty of test pits

investigation practice. In many cases these inadequacies only become apparent during construction and lead to costly redesign and/or contractual disputes. Tables 5.8 to Table 5.12 list some specific problems that occur in geotechnical investigations, the consequences, and measures which can be taken to avoid the problems.

6

Shear strength, compressibility and permeability of embankment materials and soil foundations

6.1 SHEAR STRENGTH OF SOILS

6.1.1 *Drained strength – definitions*

When a soil is sheared slowly in a drained condition (so that there is sufficient time for dissipation of pore pressures induced by shearing), the stress displacement curves will take the general form shown in Figure 6.1.

The behaviour is dependent on whether the soil has a high or low clay fraction (finer than 0.002 mm) content and on whether it is normally consolidated (NC) or overconsolidated (OC), but has the following common features:

- A peak strength is obtained at a small displacement.
- A reduction of strength to the critical state or fully softened strength then occurs with further displacement. For overconsolidated soils this is due to increase in water content with dilation of the soil as it is sheared. The fully softened strength corresponds to the critical state (Skempton, 1985), i.e. when continuing displacement occurs without further change in volume or water content. Note that normally and overconsolidated

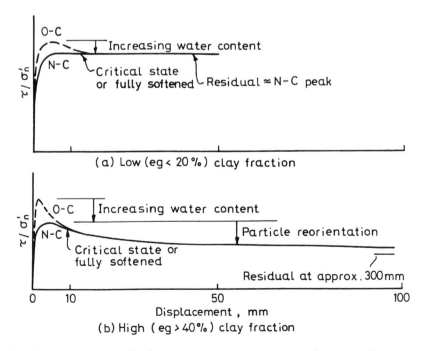

Figure 6.1. Diagrammatic stress displacement curves at constant normal stress σ'_n (Skempton, 1985).

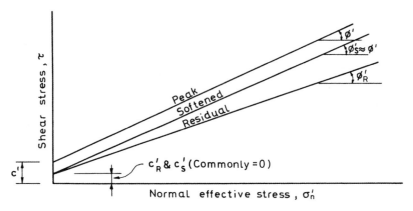

Figure 6.2.   Relationship between peak, softened and residual strengths.

samples of the same soil will tend to achieve the same critical state or fully softened condition.

- With continuing displacement of soils with a high clay fraction content, particle reorientation occurs, resulting in a further reduction in shear strength. The minimum value of shear strength achieved at large displacements is the "residual" strength.
- The residual and fully softened strengths are significantly different for high clay fraction content soils, but not for soils with low clay fraction content (see Figure 6.1). Sands behave similarly to the low clay fraction soils.
- Strength envelopes for these cases are defined by:
  - peak strength $c'\phi'$;
  - softened strength $c'_s\phi'_s$;
  - residual strength $c'_R\phi'_R$.

Note that often $c'_s \approx c'_R \approx 0$, and $\phi'_s \approx \phi'$. Figure 6.2 shows the relationship between these parameters.

Skempton (1985) adopts the use of the term "field residual" strength (ϕ'_R) as the strength of fully developed shear or slide surfaces in nature. As discussed in Section 6.1.4.4 this value differs from the laboratory residual strength depending on the testing method.

In practice, it may also reflect the roughness and waviness of the field failure surface compared to the planar surface in the laboratory sample.

An undisturbed sample of soil may behave in an overconsolidated manner at low normal stress and in a normally consolidated manner at high normal stresses (in excess of the preconsolidation pressure). This affects the pore pressure response of the soil during shear (see Section 6.1.2) and also the drained load-deformation behaviour as shown in Figure 6.3.

6.1.2   Development of drained residual strength ϕ'_R

There is a large amount of evidence that softening, with increased water content, and particle reorientation, occurs on slide planes and that these lead to a reduction in shear strength from the peak strength. Skempton (1985), Lupini et al. (1981), Mesri and Cepeda-Diaz (1986) and Hawkins and Privett (1985) give good summaries of these effects.

Lupini et al. (1981) carried out a series of ring shear tests on sand-bentonite mixtures, and suggested that there were different mechanisms of residual plane development depending on the clay fraction percentage present. The mechanisms are turbulent or rolling shear, where the presence of rounded silt size particles prevents alignment of clay particles on the shear surface;

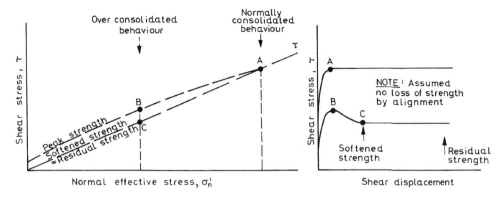

Figure 6.3.   Diagrammatic stress displacement behaviour.

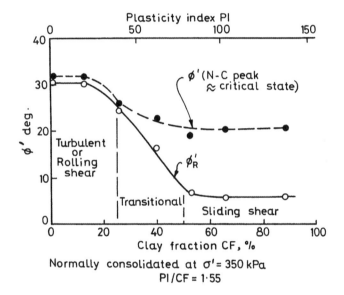

Figure 6.4.   Ring shear tests on sand-bentonite mixtures (from Skempton, 1985; after Lupini et al., 1981).

sliding shear when the effects of the silt is overridden by the predominance of clay particles; and a transition between these two conditions. These results were summarized by Skempton (1985) and are reproduced in Figure 6.4.

Skempton (1985) presented results of field residual and ring shear tests on a range of soils. These are reproduced in Figure 6.5.

These figures show that "sliding shear" with complete reorientation of the clay particles is likely to occur only where the clay fraction (finer than 0.002 mm) exceeds 50% of the total soil, and that for less than about 25% clay fraction, "turbulent" or "rolling" shear occurs without the influence of clay particle alignment. Note that Figure 6.5 only applies to clays with a Plasticity Index/Clay Fraction (PI/CF) ratio of 0.5 to 0.9.

Apart from the clay fraction, the mineralogy of the clay also has an effect on residual strength. This is particularly so when the clay fraction is large. This reflects the fact that the different clay minerals have different particle shape and different interparticle bonding. Most clay minerals, e.g. kaolinite, illite, chlorite and montmorillonite are platey structures, and are therefore subject to alignment when sheared. Montmorillonite has a particularly thin plate structure, and weak interplate bonds and leads to the lowest residual strength value with ϕ'_R as low as 5°. Kaolinite has a ϕ'_R of approximately 15° and illite approximately 10°.

Some clay minerals do not have a plate structure, e.g. halloysite has a tubular structure, attapulgite a needlelike structure and some amorphous clay sized minerals such as gibbsite, haematite, bauxite have essentially granular structures. This leads to much higher residual friction angles, commonly greater than 25° (Skempton, 1985).

The pore water chemistry may have an influence on the residual shear strength. Di Maio (1996a,b) showed that, with several clays from southern Italy and a commercial bentonite, there is a change in ϕ'_R from about 14° when prepared with saturated NaCl solution to about 6° when prepared with distilled water. Figure 6.6 shows this change for the bentonite. From a practical point of view, few natural soils are likely to be as affected as bentonite, but the residual strength parameters should be evaluated with pore water with a chemistry close to the one existing *in situ* or anticipated during the life of the dam.

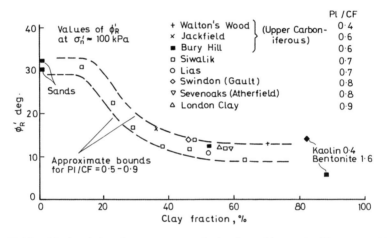

Figure 6.5.   Field residual and ring shear tests on sands, kaolin and bentonite (Skempton, 1985).

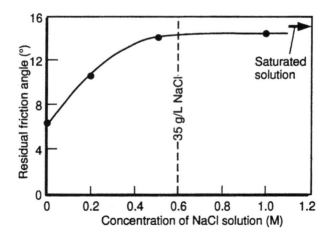

Figure 6.6.   Residual friction angle of Ponza bentonite under normal stresses of 250–400 kPa for various concentrations of NaCl solution (Leroueil, 2000; after Di Maio, 1996b).

6.1.3 *Undrained strength conditions*

Undrained strength conditions occur when the loading time and the properties of the soil in the slope are such that there is insufficient time for pore pressures generated during the loading period to dissipate. Duncan (1996) suggests this can be assessed from the dimensionless time factor, T which is expressed as:

$$T = \frac{C_v t}{D^2} \tag{6.1}$$

where C_v = coefficient of consolidation (m²/year); t = loading time (years); D = length of the drainage path (metres).

Duncan indicates that, if the value of T exceeds 3.0, it is reasonable to treat the material as drained, and, if T is less than 0.01, undrained. If the value of T is between 3.0 and 0.01, both undrained and drained strengths should be considered. The undrained strength of a soil may be less than or greater than the drained strength depending on whether the soil is contractive or dilatent when it shears.

It is important to note that the strength available in the case of undrained shear of the contractive soil (Su) is less than that for drained loading and less than assumed for conventional effective stress analysis, because of the positive pore pressures generated during shearing. As discussed below, this can happen in coarse grained as well as fine grained soils.

Figure 6.7 shows stress strain and stress paths of saturated loose cohesionless sand loaded in triaxial compression under undrained conditions at a constant rate of strain, undrained creep, and an initial shear stress and cycle loading. In all cases the large strain or ultimate strength, also known as the steady state strength or residual undrained strength, is reached at large strains. In an e–log p' diagram (Figure 6.8a), where e is the void ratio and p' is the mean effective stress, such a test would have its initial point at I_1, and the ultimate critical or steady state at point U_1 on the steady state line. For initial condition at point I_2, below the steady state line, the undrained behaviour would be dilatant with a steady state at point U_2.

Whether a soil will be contractive or dilatant depends on its initial state condition, i.e. whether it is above or below the steady state line, but also the stress path. This is discussed further in Leroueil (2000) and Fell et al. (2000).

In some contractive soils, e.g. loose sands, loaded in saturated, undrained conditions, the triaxial compression stress path is more complex, as shown in Figure 6.9. The soil reaches a peak undrained strength at stresses less than the peak strength (at A).

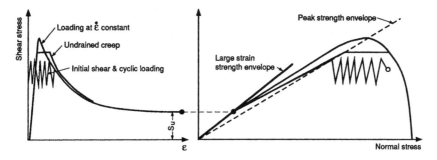

Figure 6.7.   Typical undrained response of loose sands in monotonic, creep and cyclic loading (Leroueil et al., 1996).

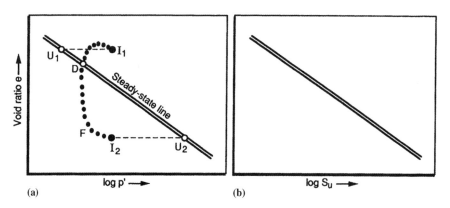

Figure 6.8.   Steady state or critical state concept (Leroueil et al., 1996).

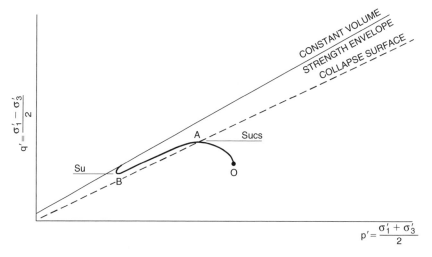

Figure 6.9.   Stress path for saturated loose sand which displays collapse behaviour.

The line through these points is termed the collapse surface (Sladen et al., 1985; Dawson et al., 1998), or critical stress ratio (Vaid and Eliadorani, 1998) or the instability line (Lade, 1992). At these stress conditions the soil wishes to contract as it shears but, being constrained by the no volume change and conditions of saturated undrained conditions, develops positive pore pressures resulting in weakening to the ultimate strength S_u at B.

The important point here is that the collapse surface strength may be lower than the peak, or critical state (constant volume) strength, which is commonly used in stability analysis, so failure may occur at stresses lower than expected (e.g. Dawson et al., 1998). As pointed out by Sladen et al. (1985), for very loose sands whose state lies on the collapse surface, only minimal excess pore pressure is necessary to trigger collapse. Loading may be essentially drained up to the point of reaching the collapse surface.

The degree of strain weakening has been characterized in terms of the Brittleness Index I_B (Bishop, 1971):

$$I_B = \frac{\tau_p - \tau_r}{\tau_p} \times 100\% \qquad (6.2)$$

where τ_p and τ_r are the peak and residual strengths defined under the same effective normal stress.

However, as indicated by Vaughan & Hamza (1977) and by Chandler (1984) the brittleness index alone is not sufficient to characterise the susceptibility of a soil to progressive failure and the rate at which the strength decreases from peak strength to ultimate strength is also important. D'Elia et al. (1998) propose a generalised Brittleness index, I_{BG}, defined as follows:

$$I_{BG} = \frac{\tau_p - \tau_{mob}}{\tau_p} \times 100\%$$ (6.3)

in which τ_{mob} is the mobilised shear stress at the considered strain or displacement.

I_{BG} thus varies with strain or displacement from 0 at the peak to a value equal to I_B at large displacements. I_{BG}, is associated to the stress paths that are representative of those followed *in situ*, and must thus not be seen as a fundamental characteristic of a soil. With this extended definition, not only overconsolidated clays, clay shales, sensitive clays, residual soils and loess, but also cohesionless soils such as loose sands, may appear brittle in undrained conditions.

Contractant conditions may occur in mine tailings, end-tipped mine waste dumps, uncompacted or poorly compacted road or railway fills, normally and lightly over-consolidated clays, poorly compacted clay fills (e.g. old dams which were compacted by horses hooves or rolled with light rollers in thick layers), puddle cores, and in the core of large dams where the confining stresses exceed the pre-consolidation effects of compaction, and potentially in loose dumped dirty rockfill in dams (although the authors are not aware of this happening).

Cooper et al. (1997) discuss an example where the use of conventional effective stress analysis over-estimated by about 50% the factor of safety of an earth dam constructed with poorly compacted clay and with concrete core wall.

As has been recognized for a long time, e.g. Jamiolkowski et al. (1985), Ladd (1991) and Kulhawy (1992), the undrained shear strength used in the analysis should take account of the failure mechanism. As shown in Figure 6.10 the loading conditions under an embankment are best simulated by triaxial compression, while beyond the toe, they are best simulated by triaxial extension and direct simple shear. Centrally within an embankment they are probably represented by triaxial compression, and at the upstream and downstream toe by triaxial extension.

Figure 6.11 shows the mean normalized undrained strength ratios for the major laboratory tests.

For this figure the reference strength ratio is given by the modified Cam Clay model, which Kulhawy (1992) indicates is reasonable for relatively unstructured soils. For sensitive, cemented and other structured fine grained soils, the reference strength is a lower bound. It can be seen that to use Ko triaxial compression tests is unconservative for those parts of the failure surface best modelled by triaxial extension and direct simple shear tests.

The behaviour of contractant granular soils is a complex, developing science, and will not be further discussed here. Reference should be made to Fell et al. (2000) and the references given therein and to the recent literature.

6.1.4   *Laboratory testing for drained strength parameters, and common errors*

The following discussion on shear strength of soils briefly outlines the test procedures and some of the problems which arise in testing and interpretation of results. It is the authors' experience that many organisations, at least in the past, failed to follow correct test procedures despite the fact that they have been long established.

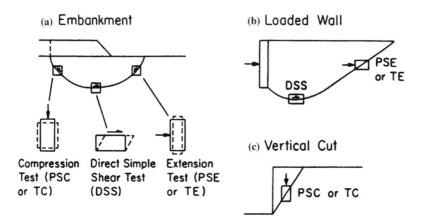

Note: Plane strain tests (PSC/PSE) used for long features
      Triaxial tests (TC/TE) used for near symmetrical features
      Direct shear (DS) normally substituted for DSS to evaluate $\bar{\phi}$

Figure 6.10.   Relevance of laboratory shear tests to shear strength in the field (Kulhawy, 1992, reproduced with permission of ASCE).

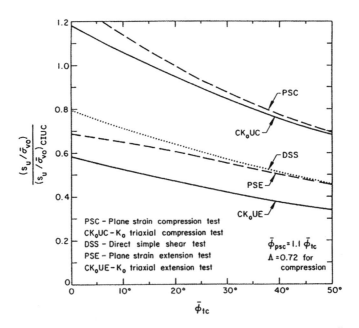

Figure 6.11.   Mean normalised undrained strength ratio for major laboratory tests (Kulhawy, 1992, reproduced with permission of ASCE).

6.1.4.1   *Triaxial test*

Triaxial tests are the most common means of obtaining the peak shear strength parameters c', ϕ' and, provided constant volume conditions are reached within the strain capability of the test, the fully softened strength. The triaxial test cannot be used to obtain residual strengths because it is not possible to subject the soil to sufficient displacement on the shear surface to reach these values.

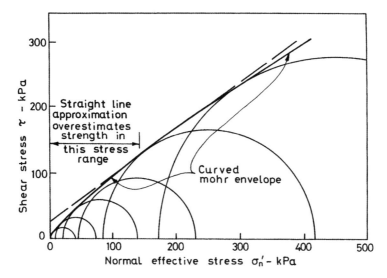

Figure 6.12.   Typical curved Mohr's envelope plot.

Most triaxial testing is triaxial compression, carried out at a nominally constant rate of axial strain. Tests are usually saturated consolidated undrained with pore pressure measurement (CUDPP) or saturated consolidated drained (CD). For practical purposes these yield the same strength (Duncan, 1994) provided the tests are performed correctly. Most triaxial testing is CUDPP because the time for testing is less than for CD. However, in testing some soils, the pore pressure response during testing is such that the CUDPP circles fall close together and CD tests may be needed to obtain a wide enough range of Mohr circle plots to draw an accurate envelope. Staged testing, where one sample is tested for three consolidating stresses, is sometimes used because it involves less sample preparation, but it can lead to errors as described below.

Details of testing equipment and test procedures are given in Head (1985), Bowles (1978) and Bishop and Henkel (1971) and the relevant ASTM (D2850-95, D4767-95), British (BS1377-8:1990) or other standards (e.g. Australian Standard AS1289 6.4.1 and 6.4.2, 1998). Lade (1986), Saada and Townsend (1981) and Baldi et al. (1988), Germaine and Ladd (1988), Lacasse and Berre 1988) give summaries of triaxial testing.

There are several common sources of error in triaxial testing, which often lead to an overestimation of effective cohesion, usually with a reduction of effective friction angle. This may lead to an overestimation of strength in the working stress range. The sources of error are:

(i) *Testing at a too high stress range.* The Mohr's circle envelope for most over consolidated soils is curved in the manner shown in Figure 6.12. If laboratory tests are carried out at higher stresses than will be encountered in the field, this will result in an overestimation of shear strength in the field. If the field stresses were between 0 and 100 kPa and testing had been carried out at confining stresses between 100 and 400 kPa, the strengths would be overestimated by 10% to 300%. This potential problem can be overcome by specifying the correct stress range for the triaxial tests and recognising possible curvature effects when selecting design parameters from the p'-q plot. It is also important to plot the individual Mohr's circle plots for a sample, so the curvature can be identified.

(ii) *Not saturating the sample adequately.* Soil samples (even if taken from below the water table) may not be saturated before testing in the laboratory because:
    – Fissures in the soil open up due to sampling and unloading;
    – Pore pressure redistributes in non homogeneous soils on unloading;

- Air intrusion may occur at the surface of the sample;
- Dissolved gas may come out of solution;
- Air may be trapped between the sample and the membrane in the triaxial test;
- Compacted soils even if compacted wet of optimum water content, as is commonly required in dam construction, are usually only 90% to 95% saturated.

The effects of partial saturation are:

- Pore pressure changes during shearing are less than for a saturated sample. However, the trend (whether +ve or −ve changes) is not affected (Lade, 1986);
- Because pore pressure changes are less than for a saturated soil, the undrained strength will be affected. For over-consolidated clays, Δu is −ve, and lower than for a saturated soil, so the undrained strength is reduced; for normally consolidated clays, Δu is +ve and lower than for a saturated soil, so the undrained strength is increased;
- The magnitude of actually measured pore pressures may be incorrect, usually underestimated, because of the presence of air in the pore system.

Provided the pore pressures are measured correctly, the effective strength envelope will not be greatly affected even though the total stress envelope is markedly affected.

To avoid problems of partial saturation, the laboratory samples should be saturated by percolation followed by back pressure saturation. Percolation by itself will not achieve saturation even in permeable soils (e.g. sands). Back pressure saturation is needed to reduce the volume of the air bubbles in the pore water (Boyle's Law) and to drive air into solution (Henry's Law). The degree of saturation should be checked by monitoring the increase in pore pressure for an increment in cell pressure in undrained conditions:

$$\Delta u = B \, \Delta\sigma_3 \qquad\qquad (6.4)$$

where B is defined by Skempton (1954) as:

$$B = \frac{1}{1 + \dfrac{nC_v}{C_{sk}}} \qquad\qquad (6.5)$$

where n = porosity; C_v = compressibility of pore fluid; C_{sk} = compressibility of the soil skeleton; For a saturated soil $C_v = C_{water}$ and C_v/C_{sk} is very small, hence B = 1. However, for very stiff soils, and very weak rocks, C_{sk} is also small and it is difficult to achieve B = 1 as shown in Figure 6.13. In these circumstances it is sufficient to check B for several increments of cell pressure and, provided B is constant, the correct effective strength parameters will be obtained (Lade 1986). For very stiff soils and very weak rocks, the stiffness of the testing system becomes significant and may lead to low measured B values.

Typically back pressure saturation will be carried out at 300 kPa to 1000 kPa, but may be as low as 200 kPa for soft soils. It should be noted that dissolving the air in the water takes time (often many hours). Black and Lee (1973) give a method to calculate this time.

(iii) *Testing at too high a strain rate.* CD and CUDPP triaxial tests must be sheared at a sufficiently slow rate to allow dissipation of pore pressure in the CD test and equalisation of pore pressure throughout the sample in the CUDPP test.

The effect of testing at too high a strain is illustrated in Figure 6.14 for an over-consolidated clay.

If a CUDPP test is sheared too quickly, the pore pressure changes measured at the ends of the sample are less than the actual changes at the centre of the sample where

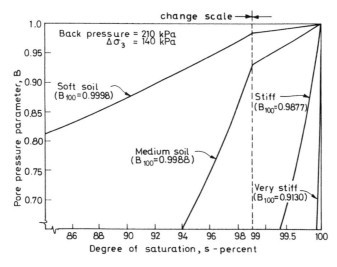

Figure 6.13. Variation of B-value with degree of saturation for four classes of soil (Black & Lee, 1973, reproduced with permission of ASCE).

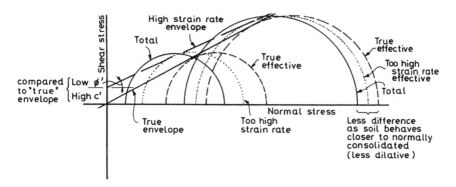

Figure 6.14. The effect of high strain rate on CUDPP triaxial tests.

shearing is occurring, because of the restraint effects of the top and bottom of the sample. Since there is a reduction in pore pressure during shearing of an overconsolidated clay, the Mohr circle plots based on measured pore pressures are displaced to the left of where they should be.

The negative pore pressure response to shearing is less for the sample at higher confining stress where the soil behaves closer to a normally consolidated soil, so the displacement of the Mohr's circle for the higher confining stress is less than that for the lower confining stress.

This yields overall an increase in c', and a decrease in ϕ'. The effect is similar for CD tests, where negative pore pressure will remain in the centre of the sample if the strain rate is too high, giving an apparent increase in failure stress (σ'_1), which again will be greater for samples at lower confining stresses than for higher confining stresses.

It is generally accepted (e.g. Germaine and Ladd, 1988; Bishop and Henkel, 1957, 1971; Lade, 1986) that 95% dissipation (for drained tests) or 95% equalisation (for CUDPP tests) gives adequately accurate results.

Bishop and Henkel (1957, 1971) used consolidation theory to develop the following approximate equations to estimate the time for failure:

(a) Drained tests (CD)

$$t_f = 6.7H^2/C_a \tag{6.6}$$

for top and bottom drainage

$$t_f = 0.5H^2/C_a \tag{6.7}$$

for top, bottom and side drainage

(b) Undrained tests with pore pressure measurement (CUDPP)

$$t_f = 1.7H^2/C_a \tag{6.8}$$

without drainage

$$t_f = 0.07H^2/C_a \tag{6.9}$$

with top, bottom and side drainage. where t_f = time for failure; H = half height of sample = diameter; C_a = coefficient of consolidation.

To determine the strain rate, the strains to failure must be estimated. This may be up to 20% for normally consolidated soils, and only 1 to 2% for heavily over consolidated soils.

C_a should be determined from the consolidation phase of the test, but it should be noted it will vary with degree of over consolidation (Germaine and Ladd, 1988; Leroueil et al., 1988), so should be measured at the end of the consolidation.

Blight (1963), Lade (1986), Germaine and Ladd (1988), Leroueil et al. (1988) all discuss the problem of efficiency of the side drains (usually strips of filter paper placed between the soil sample and the rubber membrane in the triaxial test). Leroueil et al. (1988) point out that, in the over-consolidated range, the drains are usually able to cope with the required water flow, but in the normally consolidated range, they may not have adequate discharge capacity. Germaine and Ladd (1988) indicate this is a problem for soils with a permeability greater than 10^{-10} m/sec (i.e. virtually all soils) Lade (1986), recommends the use of Figure 6.15 based on the theory of Bishop and Henkel and experimental results.

Akroyd (1957, 1975), Head (1985) and Bowles (1978) all give relationships for estimating time for failure based on monitoring the consolidation stage of the test.

Lade (1986) and Germaine and Ladd (1988) indicate that, by using lubricated end plates, pore pressures are more uniform and CUDPP tests can be carried out more quickly. However no guidelines are given to determine an acceptable strain rate. Lubricated end plates do not reduce the required time for failure in CD tests since pore pressure dissipation, not equalisation, is required.

(iv) *Staged testing.* In staged testing, the one soil sample is usually saturated, consolidated and sheared at the lowest selected confining stress, then consolidated to the second confining stress and sheared, then further consolidated to the third confining stress and sheared. This reduces the amount of sample preparation time, and hence the cost of testing. The procedure often gives acceptable results, but may tend to give lower strengths for the second and, particularly third stage, when testing strain weakening or compacted dilative soils due to sample dilation and softening, and loss of strength

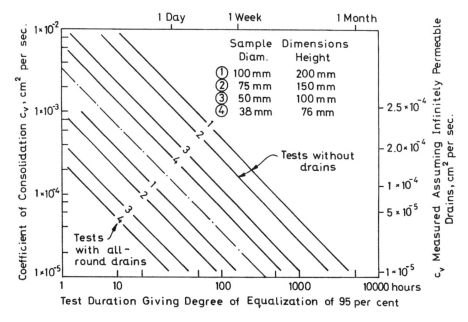

Figure 6.15.   Chart for finding durations of drained and undrained tests for 95% dissipation at failure (Blight, 1963).

due to displacement on the shear plane. Where staged testing gives high effective cohesions, the third (and second) stage testing should be checked with tests on samples consolidated to the third stage confining pressure without earlier staged testing. The author's recent experience is that staged testing almost always gives excessively high c′ for compacted soils, and recommend it not be used. Staged testing should not be used for sensitive or cemented soils.

The shear strength c′, ϕ′ of a compacted earthfill will be somewhat dependent on the degree of compaction, with higher compaction, leading to minor increase in c′ and ϕ′. When testing earthfill for dam construction, the laboratory tests should be carried out on samples compacted at compaction density and moisture content consistent with what will be specified for the dam construction. Where there are fabric features present (e.g. fissures), staged testing is desirable so the complete test is carried out on the one feature.

6.1.4.2   Direct shear test

Direct shear tests are used to obtain the peak and fully softened or critical state (constant volume) strengths of granular soils (sand, silt, gravel) and cohesive soils. They are also used to determine the residual strength of cohesive soils, by using the multiple reversal technique which is described below, and the strength on existing surfaces of weakness in soil, e.g. shear surfaces, slide surfaces and fissures and in the opinion of the author's should not be.

These tests are all done in drained conditions. Those for cohesive soils are saturated prior to testing. The test is seldom used to determine undrained strengths and in the opinion of the authors should not be.

Details of testing equipment and test procedures are given in Head (1980, 1981, 1985), ASTM D3080-98, BS1377:1990 and AS1289, 6.2.2-1998.

It is the author's experience that when direct shear and triaxial test data are available, the peak strengths derived from direct shear tests are sometimes higher than from the triaxial tests. This is mainly due to the inability to saturate the soil properly, along with the problem that the direct shear forces failure on a pre-determined plane, which is not coincident with the principal stress direction at failure (Saada and Townsend, 1981). This means that care should be taken in using the direct shear results and the test should not be used as the primary means to obtain peak strengths for cohesive soils. The direct shear is a good way to obtain the residual strength.

True laboratory residual strength of clay soils only develops with significant displacement. Skempton (1985) gives the "typical displacement values" reproduced in Table 6.1.

Most laboratory shear box equipment is 75 mm × 75 mm or 60 mm × 60 mm, and the maximum practical displacement is of the order of 6 mm to 10 mm, sufficient only to measure the peak and possibly the softened strength.

Residual strength is therefore obtained with a direct shear machine by repeatedly shearing the sample until the strength is not further reduced by further shearing (the "multiple reversal" method). Typical load displacement curves for "turbulent" and "sliding" shear are shown in Figure 6.16 for tests which are progressing well. Most tests are more irregular.

Table 6.1.   Typical displacements at various stages of shear in clays having clay fraction[3] >30% (Skempton, 1985).

| Stage | Displacement mm[2] | |
	Overconsolidated	Normally consolidated
Peak	0.5–3	3–6
Rate of volume change approx zero[1]		4–10
At residual +1° ($\phi'_R + 1°$)		30–200
At residual (ϕ'_R)		100–500

Notes: [1] i.e. fully softened strength.
[2] For σ'_n <600 kPa.
[3] Clay fraction = % finer than 0.002 mm.

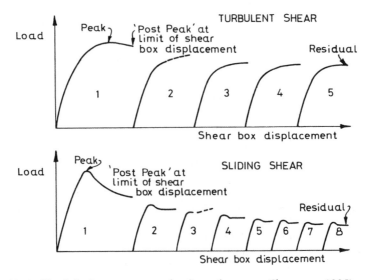

Figure 6.16.   Typical load displacement curves for direct shear tests (Skempton, 1985).

The repeated shearing is achieved in several ways:

- Shear to limit of travel, unload (partly), rewind rapidly to start, then reload and reshear etc. until a minimum is reached, or
- Shear to peak, then unload partly, and wind back and forth rapidly until a total travel of say 50 mm is achieved, reload and reshear etc. until a minimum is achieved, or
- Shear back and forth under load at a slow strain rate until a minimum is achieved.

Further acceleration of achievement of the residual strength is sometimes accomplished by cutting a slide plane into the sample prior to testing. The cut surfaces are sometimes polished on a glass plate prior to testing.

Problems which arise in the use of the direct shear to assess peak strength include:

- It is difficult to ensure that the sample is properly saturated. Saturation is achieved by soaking under a confining stress until there is no further volume change. Unlike triaxial testing, pore pressure cannot be measured, so it is necessary to rely on monitoring the consolidation deformation of the sample during saturation and consolidation. The best approach is to saturate under a low confining pressure until swell (or consolidation) ceases, then load to the first test confining pressure, and monitor until settlement ceases.
- Testing at a too high strain rate leads to overestimation of shear strength in a similar way to triaxial testing, i.e. c' is overestimated, ϕ' underestimated. This can be overcome by monitoring the consolidation to obtain t_{50}, t_{90} or C_v value, and then using formulae developed by Akroyd (1975), Bowles (1978) or Head (1985) to estimate the strain rate. Skempton (1985) showed however that the residual strength is not greatly sensitive to strain rate provided reasonably low strain rates are used. This fact may be used to speed up test rates, provided that when a constant shear stress has been reached the strain rate is reduced, and the resulting shear stress is not significantly lower than the previous minimum (see Figure 6.17).
- Testing at a too high confining stress range, and staged testing can lead to an overestimation of c', underestimation of ϕ' in the same way as for triaxial testing.

Problems which arise in the use of the direct shear to assess residual strength include:

- Repeated reversal of the direction of shearing is necessary, and this may partially destroy the alignment of particles on the shear plane, preventing a true residual strength being achieved;
- Repeated reversal of the shear box often leads to soil squeezing out between the two halves of the box. If this happens, or if the box tilts, interference can occur between the two halves of the box leading to erroneous results.

The authors' experience is that one of the biggest problems is that laboratory personnel (and the engineers supervising them) do not understand the definition of residual strength. Figure 6.18 shows some of the all too common misconceptions.

6.1.4.3   Ring shear test

To overcome the problems of multiple reversals of a shear box and to obtain sufficient displacement on the slide plane to achieve residual strength, Bishop et al. (1971) developed a ring shear apparatus.

In the ring shear, an annular ring shaped specimen (Figure 6.19) is subject to a constant normal stress σ'_n confined laterally, and caused to shear on a plane of relative rotary motion at a constant rate of rotation.

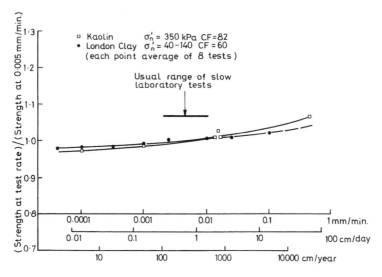

Figure 6.17.   Variation in residual strength of clays at slow rates of displacement (Skempton, 1985).

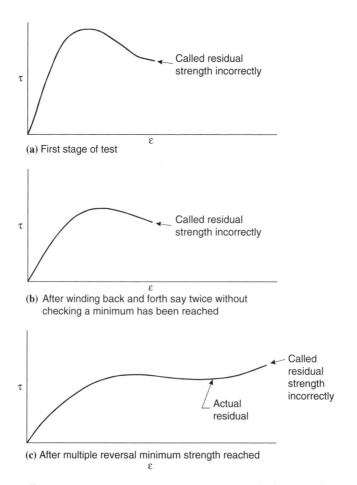

Figure 6.18.   Some all too common misconceptions in interpreting residual strength from direct shear tests.

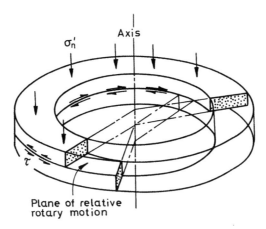

Figure 6.19.   Ring shear test sample.

For the Bishop et al. (1971) machine (manufactured by Wykham Farrance) the sample is 150 mm outside diameter, 100 mm inside diameter, and 19 mm thickness. Remoulded samples are usually tested, but undisturbed samples can be placed in the equipment. The machine is mechanically complex and expensive, putting it largely in the research category. Bromhead (1979) and Bromhead and Curtis (1983) describe a simpler ring shear device. The test procedure is detailed in BS1377-7:1990, and ASTM D6467-99. Bromhead (1986) and Clayton et al. (1995) give some practical details on how to operate the equipment. The sample in the Bromhead machine is outside diameter 100 mm, inside diameter 70 mm and only 5 mm thick.

The ring shear devices have essentially unlimited "strain" availability, so overcoming the major objection to the reversing direct shear test. However, there are limitations:

- only remoulded samples can be tested (in the Bromhead machine);
- only the residual strength can be obtained. The peak strength is affected by remoulding and non uniform stress conditions;
- the sample may tend to squeeze out the sides of the ring. The Bishop machine is constructed to overcome this and can be used for softer clays than the Bromhead machine;
- saturation and testing at a strain rate appropriate to give the required drained conditions is necessary as for the direct shear test.

6.1.4.4   Comparison of field residual with laboratory residual strength obtained from direct shear and ring shear

There have been a number of papers published comparing field residual strengths obtained by back-analysis of landslides, and laboratory direct shear testing of the undisturbed slide plane, with residual strengths obtained from multiple reversal direct shear, and ring shear tests. These include Bishop et al. (1971), Townsend and Gilbert (1973), Bromhead (1979) and Bromhead and Curtis (1983) and Skempton (1985).

In summary the literature indicates that:

- Direct shear tests on the slide plane or bedding plane shear are the most reliable indicator of field residual strength;
- Ring shear devices either underestimate by 1° to 2° or approximate the field residual;
- Multiple reversal direct shear on clays will probably overestimate field residual by 1° or 2°;

– The difference between multiple reversal direct shear and field residual is only small in weak rocks such as shales and claystones.

The authors' experience in testing clays and very weak claystone is that the difference may be significantly greater (up to 6°) and direct shear tests on the slide surface or bedding surface shear should always be done if practicable.

The authors' experience in testing soil from several landslides where soils were derived from sedimentary rocks, tuff and basalt is that the Bromhead ring shear is very susceptible to the presence of particles of sand. These "catch" on the shear surface and lead to higher residual strengths than obtained from multiple reversal direct shear or tests on slide planes. Comparable results were only obtained by drying and sieving the soil on a 0.150 mm sieve to remove any particles of sand before re-wetting it to the plastic limit for testing.

It can be seen that ideally tests should be carried out on the low strength planes if already present, or that ring shear tests should be carried out provided that the soil is free of sand particles. Unfortunately, ring shear machines are not widely available and reversing direct shear tests have to be used. In these cases, the design strength should be selected at the lower bound of results unless exhaustive testing shows this to be conservative. This is valid since few (if any) of the reversing tests may have achieved a proper alignment of particles, and the low results (provided these were not testing errors) are probably the best indicator of residual strength. The validity of the results should also be checked by comparison with published data, e.g. that in Figure 6.5. The notable exceptions to the data on that figure are allophane and halloysite clays which do not have classical platey structure, so give higher values (e.g. $\phi'_R = 25°$ to $35°$).

In many cases of slope instability, where stabilization measures are being designed, the strength can be determined by back analysis and stabilization works designed on the basis of increasing the factor of safety by no less than 30%. This obviates some of the need for accurate knowledge of the shear strength from laboratory tests.

A method which has been used by the authors to determine the "lower bound" residual strength of a soil, is to separate the clay fraction from the silt (and sand) of the soil by sedimentation. The clay fraction is then tested by overconsolidating it in the direct shear machine, and then determining the residual strength by multiple reversal direct shear testing. This procedure was suggested by Pells (1982) and has given reasonable values when used. More recently we routinely use the ring shear on the minus 0.150 mm sieve soil as described above.

In nature, it is not uncommon to observe a clay rich zone 0.2 mm to 1 mm thick at the slide surface, and in these circumstances the procedures outlined above can be expected to give a good representation of the field condition.

6.1.5   Laboratory testing for undrained strength

The undrained strength of cohesive soils can be obtained from several laboratory tests:

(a) Unconfined compression (UC);
(b) Unconsolidated undrained (UU);
(c) Consolidiated undrained (CU).

It must be recognised (Germaine and Ladd, 1988; Jamiolkowski et al., 1985) that the undrained strength is affected by:

– The degree of disturbance of the soil sample. Disturbance can occur because of stress relief on sampling; the sampling technique (disturbance as the thin wall tube or other samples is pushed into the soil); handling (e.g. drying or disturbance during transportation, extrusion of the sample from the sampling tube, or trimming in the laboratory);

- The water content of testing e.g. soils allowed to dry from their field water content will be stronger than those tested at the field water content. Thin-walled tubes should be sealed as soon as they are taken to avoid drying;
- The stress path, in particular whether tested in compression, extension, or direct simple shear (see Figures 6.9 and 6.10);
- The test procedure e.g. UC and UU tests do not simulate the in-situ test stresses.

Sample disturbance for soft clays, such as may be present in foundations of small dams, puddled cores, or in saturated, poorly compacted clay fill; and the sensitivity of undrained strength to the water content for stiff clays, make UC and UU tests totally unreliable, and they should not be used for design purposes. In particular for soft clays they may seriously underestimate the real strength – but you never know by how much.

Tests should therefore be done by CU methods, and they should, so far as practicable, simulate the in-situ stress conditions, and the stress path.

For testing of soft clays, the most commonly used methods are the recompression test, and the SHANSEP technique. These are discussed in Jamiolkowski et al. (1985), Germaine and Ladd (1988) and Lacasse and Berre (1988).

The recompression technique was developed by the Norwegian Geotechnical Institute (NGI), and is described in Bjerrum (1973) and Jamiolkowski et al. (1985). Samples are consolidated to the same effective stress condition as experienced in-situ, before being sheared undrained, either in compression or extension. That is, the samples are consolidated under σ'_{vo} and σ'_{ho}. Ideally a constant Ko ($= \sigma'_{ho}/\sigma'_{vo}$) is maintained during consolidation, but NGI have shown that the simpler procedure of consolidating isotropically under σ'_{ho}, then increasing σ'_v to σ'_{vo} and consolidating further, gives satisfactory results. Lacasse and Berre (1988) indicate that for soils with an OCR less than 1.5, they recommend a two stage application of the cell pressure, and application of the deviatoric stress in 4 to 6 steps. For other soils, a single step consolidation is acceptable.

The SHANSEP – "Stress History And Normalised Engineering Properties" – technique was developed by Ladd and others. It is described in Ladd and Foot (1974), Ladd et al. (1977) and Jamiolkowski et al. (1985).

Note that for both SHANSEP and the recompression techniques, S_u should be determined in compression or extension, whichever is appropriate. Both techniques are claimed to give high quality strength data, which compares well with strengths determined by back analysis of failures. The recompression technique is simpler to use, requiring less laboratory testing, but can give an overestimate of strength in normally and lightly over-consolidated clays. SHANSEP involves consolidation of samples beyond their in-situ stresses, and may lead to destruction of soil structure and cementing, giving an underestimate of strength.

In many projects the SHANSEP technique would be too expensive, particularly since a S_u/σ'_{vo} vs. Over Consolidated Ratio (OCR) relation has to be developed for each soil type on the site. It is also dependent on being able to determine the OCR of the soil accurately, and this is often difficult because the pre-consolidation pressure of the soil as determined in oedometer tests is affected by sample disturbance. The recompression technique on the other hand is relatively simple and is recommended. It is emphasised that a lesser number of samples tested by recompression or SHANSEP techniques is far preferable to a large number of cheaper UC or UU tests.

The recompression and SHANSEP methods require a knowledge of the at rest earth pressure coefficient, Ko, in situ. This can be determined by:

- Self boring pressuremeter (SBPM) tests;
- Use of published relationships between Ko and effective friction angle ϕ', plasticity index I_p, and over-consolidation ratio OCR.

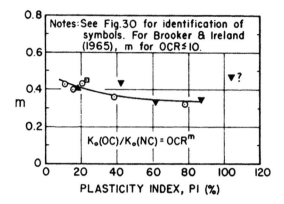

Figure 6.20.   Coefficient m relating Ko and OCR vs. plasticity index (Ladd et al., 1977).

The SBPM method is preferred as it gives a direct measure of Ko, but is relatively expensive and in many cases use of published relationships is sufficient. In any case SBPM results should be checked against the other methods such as:

(a) *For normally consolidated clays*
    Jaky (1944) derived the theoretical relationship

$$Ko_{NC} = \left(1 + \frac{2}{3}\sin\phi'\right)\left(\frac{1 - \sin\phi'}{1 + \sin\phi'}\right) \tag{6.10}$$

where ϕ' = effective friction angle.
    An approximation which is more commonly used is:

$$Ko_{NC} = 1 - \sin\phi' \tag{6.11}$$

Ladd et al. (1977) showed that this latter relationship was accurate within ±0.05
(b) *For overconsolidated clays*
    Measure the preconsolidation pressure (σ'_{vp}) in oedometer tests and calculate the OCR ($= \sigma'_{vp}/\sigma'_{vo}$). Then calculate Ko_{oc} from equation 6.12 (Ladd et al. (1977))

$$Ko_{OC}/Ko_{NC} = OCR^m \tag{6.12}$$

where Ko_{NC} is estimated from the Jaky formula, m is determined from the Figure 6.20.

Wroth (1984) shows that Ko varies in loading and unloading emphasising that the Ko_{OC}/Ko_{NC} relationship is only applicable for reloading. On this basis the graphs would not be applicable to soils subject to over-consolidation by desiccation. For these soils, Ko should be determined by SBPM tests for important projects.
    The undrained shear strength is dependent on the rate of shearing. Figure 6.21 shows the general effect, with the strength measured increasing with the strain rate.
    Germaine and Ladd (1988) indicate that most laboratories (in their experience) use a strain rate of 0.5% to 1.0% of the sample height per hour, for soft cohesive soils and that this is reasonable based on case histories of undrained failures.
    Note this is a lot slower than typically used in many laboratories. A strain rate of 1% per minute, often used for undrained testing, will over-estimate the undrained strength by about 20 ± 10% for low OCR clays, and up to 50% for high OCR clays. In view of this, it is recommended that a strain rate of 1% per hour be specified.

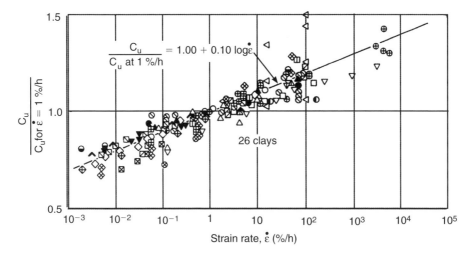

Figure 6.21.   Influence of strain rate on the undrained shear strength ($C_u \equiv S_u$) measured in triaxial compression (Kulhawy and Mayne, 1990).

6.1.6   Estimation of the undrained strength from the over-consolidation ratio (OCR), at rest earth pressure coefficient Ko, and effective stress strengths

6.1.6.1   Estimation of undrained strength from OCR

Jamiolkowski et al. (1985) using data from Koutsoftas and Ladd (1984) showed that the normalised shear strength S_u/σ'_{vc} versus OCR graphs for some soils were close to linear on a log-log plot. Hence:

$$S_u/\sigma'_{vc} = S(OCR)^m \qquad (6.13)$$

where σ'_{vc} = effective vertical stress in the test; $S = S_u/\sigma'_{vc}$ for normally consolidated clay

$$\text{or} \quad \frac{(S_u/\sigma'_{vc})_{OC}}{(S_u/\sigma'_{vc})_{NC}} = OCR^m \qquad (6.14)$$

Jamiolkowski et al. (1985) show that for non cemented clays with a plasticity index I_P less than 60%, $m \approx 0.8$ and $(S_u/\sigma'_{vc})_{NC} = 0.23 \pm 0.04$. So:

$$(S_u/\sigma'_{vc})_{OC} = (0.23 \pm 0.4)\ OCR^{0.8} \qquad (6.15)$$

This method can be used as an initial estimate of strength, or as a check on other methods.

6.1.6.2   Estimation of undrained strength from effective stress shear parameters

Wroth and Houlsby (1985) give for normally consolidated soils:

$$\left(S_u/\sigma'_{vo}\right)_{NC} = 0.5743\ \frac{3\sin\phi'_{tc}}{3 - \sin\phi'_{tc}} \qquad (6.16)$$

where subscript tc indicates triaxial compression tests, with isotropic consolidation.

Davis in Thorne (1984) gave the relationship:

$$Su_{tc} = \frac{c' \cos \phi' - U_o \sin \phi'}{1 - \left(1 - 2A_f\right) \sin \phi'} \tag{6.17}$$

where $U_o = -(1 + 2K_o)\, \sigma_{vo}'/3 = -(\sigma_{vo}' + 2\sigma_{Ho}')/3$; A_f = Skempton "A" at failure; c' = effective cohesion; ϕ' = effective friction angle; K_o = earth pressure coefficient at rest.

This relationship is applicable to both normally consolidated and over-consolidated clay, with A_f and K_o dependent on the degree of over consolidation.

6.1.7 *Estimation of the undrained strength of cohesive soils from* in-situ *tests*

The undrained strength of cohesive soils in a dam foundation or within the embankment can be determined by in-situ test methods. These include:

– Cone Penetration Test (CPT) and Piezocone Test (CPTU);
– Vane Shear;
– Self boring Pressuremeter.

6.1.7.1 *Cone penetration and piezocone tests*

The undrained strength of clays can be estimated from the cone resistance q_c of the static cone penetrometer test (CPT).

The undrained strength S_u is related to q_c by

$$S_u = \frac{q_c - \sigma_{vo}}{N_K} \tag{6.18}$$

where q_c = cone resistance; σ_{vo} = in-situ total overburden stress and N_K = empirical cone factor.

Lunne and Kleven (1981) presented graphs showing that for normally consolidated clays, N_K varies depending on the plasticity index of the soil. When uncorrected vane shear strengths are used as the base for correlation the values shown in Figure 6.22 are obtained. Note that in Figure 6.22 Po = in-situ total vertical stress ($=\sigma_{vo}$)

Aas et al. (1986) showed that when the vane shear strengths were corrected using Bjerrum's (1973) correction factor μ, where $S_u = \mu S_{uv}$, the cone correction factor N_K values were less dependent on the plasticity index of the soil, as shown in Figure 6.23.

As summarised in Jamiolokowski et al. (1985) and in Aas et al. (1986) water pressure acting on the cone modifies the cone resistance by an amount which depends on the cone geometry and the pore pressure generated as the cone is inserted. The corrected cone resistance q_r is calculated from:

$$q_T = q_c + u\left(1 - a\right) \tag{6.19}$$

where q_T = corrected cone resistance; q_c = measured cone resistance; u = pore pressure at cone and a = area ratio = A_N/A_T (see Figure 6.24).

$$S_u = \frac{q_T - \sigma_{vo}}{N_{KT}} \tag{6.20}$$

where N_{KT} is the empirical cone factor related to q_T.

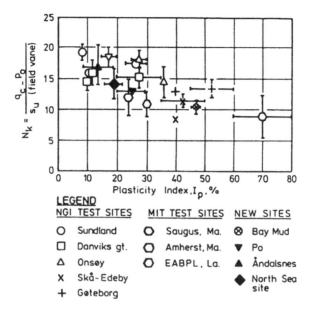

Figure 6.22.   Empirical cone correlation factor N_K – based on uncorrected vane shear strengths – Lunne and Kleven (1981), reproduced with permission of ASCE.

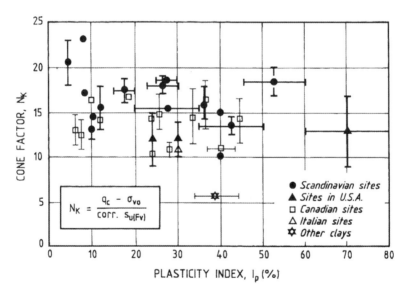

Figure 6.23.   Empirical cone correction factor N_K based on corrected vane shear strengths – Aas et al. (1986), reproduced with permission of ASCE.

Aas et al. (1986) point out that the N_K values proposed by Lunne and Kleven (1981) and others, are not corrected for the effect of pore pressure, and since they use different cones with varying geometry, and the pore pressure generated depends on clay type, much of the scatter in N_K values can be explained by these effects. They recommend that future correlations should be based on piezocones, where pore pressure is measured, and corrected cone resistance q_T can be estimated, i.e. S_u should be estimated from:

Aas et al. (1986) present cone factors N_{KT} for several sites in Norway. These are reproduced in Figure 6.25.

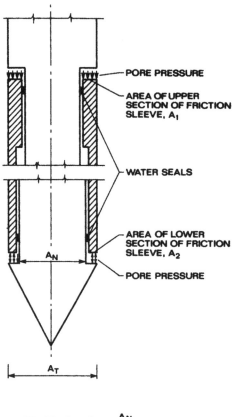

PORE PRESSURE
AREA OF UPPER SECTION OF FRICTION SLEEVE, A1
WATER SEALS
AREA OF LOWER SECTION OF FRICTION SLEEVE, A2
PORE PRESSURE
AN
AT
NET AREA RATIO, a = AN/AT
FRICTION SLEEVE UNEQUAL END AREA,
A1 ≠ A2

Figure 6.24. Vertical components of pressures acting on a cone penetrometer – Campanella and Robertson (1988).

It can be seen that N_{KT} varies from about 12 to 20 when correlated with the laboratory undrained strength (the average of TC, TE and DSS tests), and 12–22 when correlated to corrected field vane strength; the latter larger variation due probably in part to scatter in the vane shear strengths.

N_K and N_{KT} values show a much larger range for overconsolidated clays (e.g. Figure 6.25). N_K values of 15–20 are commonly adopted for overconsolidated clays.

It can be seen from the above discussion that the static cone penetrometer or piezocone can be used to estimate undrained strength, but the value of the correlation factor N_K or N_{KT} is uncertain, and strengths cannot be estimated with a high degree of accuracy. The potential error is higher in overconsolidated clays.

In practice the static cone should only be used for assessment of undrained strength in slope stability problems in clay when it has been calibrated on the site by comparison with good quality laboratory undrained strengths.

The published correlations are too variable for the cone to be used alone for undrained strength assessment in slope stability other than for preliminary design with an appropriate factor of safety. In foundation design, where factors of safety adopted are usually greater than in slope stability analysis, cones may be used alone, but conservative N_K factors should be adopted.

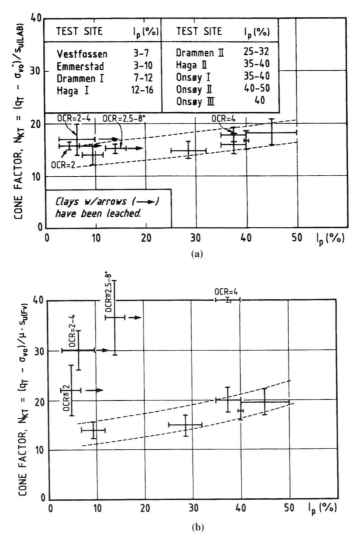

Figure 6.25.   Cone factor N_{KT} for normally consolidated and overconsolidated clays (a) Cone factor vs laboratory undrained strength (Piezocone), (b) Cone factor vs corrected vane shear strength (Piezocone). – Aas et al. (1986), reproduced with permission of ASCE.

Compacted earthfill in dams will be over-consolidated, so care should be exercised in selecting the N_K or N_{KT} values. It should also be noted that the correlations between undrained strength and core resistance are developed for saturated soil conditions. When the CPT or CPTU is used in dams, the soils may be partially saturated above the phreatic surface. In view of these problems, where the undrained strength is needed with some accuracy, the CPT or CPTU should be calibrated against laboratory tests on samples from the dam.

The CPT and CPTU have the advantage of providing a continuous profile of q_c and hence a continuous profile of undrained strength. The cone resistance is however affected by the soil up to 10 to 20 diameters ahead of the cone, and may therefore under or over-estimate the q_c (or q_T) depending on the situation.

Based on a simplified elastic solution, Vreugdenhil et al. (1994) provided a method to correct cone data for thin layers. They showed that the error in the measured cone resistance within thin stiff layers is a function of the thickness of the layer as well as the stiffness of the layer relative to that of the surrounding soil. The relative stiffness of the layers is reflected by the change in cone resistance from the soft surrounding soil (q_{c1}) to that in the stiff soil layer (q_{c2}). Based on this work, a corrected cone resistance (q_{c*}) is derived as:

$$q_{c*} = K_c \cdot q_{c2} \tag{6.21}$$

where K_c = a correction factor for the cone resistance as a function of the layer thickness.

The corrections recommended by Vreugdenhil et al. (1994) were considered by Lunne et al. (1997) to be rather large, so they recommended a more conservative correction which is shown in Figure 6.26 and given by the following expression:

$$K_c = 0.5\left(\frac{H}{1000} - 1.45\right)^2 + 1.0 \tag{6.22}$$

where H = layer thickness, in mm; q_{c2} = cone resistance in the layer; q_{c1} = cone resistance in the soil surrounding the layer.

The CPT or CPTU may not be able to penetrate well compacted earthfill. They will refuse on gravelly soils.

CPT or CPTU probes through dam cores should be backfilled with cement – bentonite grout inserted by pushing the rods without the cone on them or a steel pipe down the hole, and injecting the grout at low pressures while withdrawing. Care should be taken to avoid hydraulic fracture and in some situations it may be judged too risky to probe through the core, or preferable to leave the probe hole ungrouted, rather than risk hydraulic fracture.

6.1.7.2   Vane shear

The vane shear may be used to measure the undrained strength of very soft to firm (S_u less than about 60 kPa) cohesive soils. The equipment and procedures are described in many texts (e.g. Clayton et al., 1995; Walker, 1983).

It is normal practice to correct the strength measured by the vane to account for anisotropy and rate of shearing. Bjerrum (1973) developed a method for doing this using

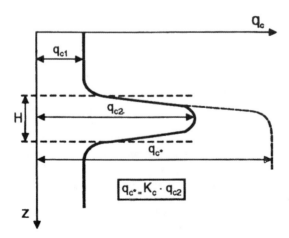

Figure 6.26.   Suggested correction to CPT cone resistance in thin sand layers (Lunne et al., 1997; based on Vreugdenhil et al., 1994).

back-analysis of case studies of embankment failures. Azzouz et al. (1983) reviewed the same failures as Bjerrum and some additional failures. They also considered 3 Dimensional (3D) effects. Azzouz et al. (1983) concluded that the correction factors obtained when 3-D effects were included gave better estimates of the actual strength as measured by the SHANSEP technique, and recommended use of the "average correction curve including end effects" for conventional 2-Dimensional analysis shown in Figure 6.27.

On balance it seems wise to apply the Azzouz et al. (1983) correction using the equation $S_u = \mu S_{uv}$ where S_u = design undrained shear strength; S_{uv} = undrained strength measured by the vane; μ = correction factor from Figure 6.27, but to also calibrate the vane with good quality laboratory testing on samples from the site. For design of embankments on soft clays it is recommended the Ferkh and Fell (1994) method be adopted, but with more conservative factors of safety than recommended in the paper as detailed in Section 11.3.2.

6.1.7.3 *Self boring pressuremeter*

The undrained strength of cohesive soils can be measured using the Self Boring Pressuremeter (SBPM).

However the measured strength is affected by disturbance, being larger the greater the disturbance, and the extent of disturbance is unpredictable. In view of this use of the SBPM to estimate Su is not recommended.

6.1.8 *Shear strength of fissured soils*

6.1.8.1 *The nature of fissuring, and how to assess the shear strength*

Fissures are discontinuities in soils which have similar implications to the strength and permeability that joints have in rock masses – they reduce the soil mass strength, and (usually)

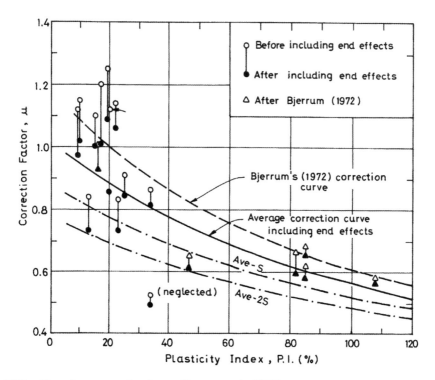

Figure 6.27.   Vane shear correction factors (Azzouz et al., 1983).

increase the soil mass permeability. Fissured soils are sometimes present in the foundations of dams – either left in place by design, or because they were not removed by those who built the dam, not realising the effect on the shear strength.

Fissures in soil have a wide range of characteristics relating to the properties of the soil mass, the depositional and weathering history and climate (when the fissures were formed and now). It is useful to log, and classify the fissures systematically.

Walker et al. (1987) provide a fissure classification system which is simple and practical to use. The main components are shown in Figure 6.28. Figures 6.29 and 6.30 show examples of fissured soils.

Fissures are caused by one or more of the following:

(a) Stress relief due to erosion or retreat of glaciation;
(b) Freeze-thaw;
(c) Shrink-swell (desiccation) due to seasonal moisture content changes;
(d) Differential consolidation and settlement in sedimentary soils.

It is important to recognise that the fissures are often formed in a different geological and climatic condition from the present – e.g. in periods of glacial retreat, or when the sea level was lower than now. The fissured soils may in some cases be overlain by more recent sediments which are not fissured (e.g. Thorne, 1984).

The shear strength of fissured soils is dependent on:

−   The peak, fully softened and residual strength of the soil substance;
−   The continuity, orientation, shape and spacing of the fissures and the nature of the fissure surfaces;
−   These are in turn related to the origin of the fissures.

For example, smooth, planar fissures may only reduce the soil mass strength to the fully softened strength, but polished and slickensided, planar or undulose, continuous fissures, will reduce the strength in the direction of the fissures to near the residual strength.

Number	1st Number Continuity	2nd Number Orientation	3rd Number Shape	4th Number Spacing	5th Number Surface
1	Individual fissures continuous across sample	Actual measured angle(s) to horizontal	Planar in two directions with maximum amplitude across 75mm sample of 2mm	Fissures less than 5mm apart	Fissure surfaces grooved, striated or slickensided
2	Individual fissures not continuous but make up potential continuous failure plane because of many intersections		Undulose	Fissures 5 to 10mm apart	Fissure surfaces polished
3	Individual fissures not continuous, some intersections		Conchoidal	Fissures 10 to 40mm apart	Fissure surfaces smooth
4	Individual fissures not continuous, very few intersections		Irregular	Fissures more than 40mm apart	Fissure surfaces rough

Figure 6.28.   Fissure classification explanation sheet (Walker et al., 1987, after Coffey Geosciences).

It has been common practice to adopt the fully softened strength to approximate the mass strength of the soil. This is based on back analyses of slides in English clays (Skempton, 1977; Chandler, 1984). The authors' experience is that in some soils, e.g. fissured residual basalt and tuff (MacGregor et al., 1990; Moon, 1992) and fissured marine clays (Thorne,

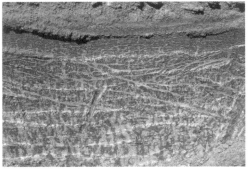

(a) View of side of test pit. Pit is about 1.5 m deep

(b) View of pieces of the fissured soil. The large piece is about 300 mm across

Figure 6.29.   Weathered Tertiary clay, which exhibits shrink-swell (desiccation) induced fissures, the major sets being inclined at about 20° to the horizontal, consistent with passive failure on swelling. (a) view of side of test pit, (b) view of pieces of the fissured soil.

Figure 6.30.   Fissures in pieces of weathered volcanic tuff. The fissures are induced by desiccation when the ground surface was exposed above sea level. Largest pieces are about 450 mm across.

1984), there was evidence from back analysis of failures, or from laboratory testing and mapping of the fissures, that the mass strength could be considerably lower than the fully softened. Table 6.2 summarises the data. Raby Bay and Plantes Hill both had failures which could be back-analysed (albeit with rather poor information on pore pressures at the time of failure).

Stark and Eid (1997) present data from 14 cases which show back-analysed mobilised shear strengths, for fissured clays as being about the average of the fully softened and residual strengths for soils with a liquid limit between 50–130%. They explain this in terms of progressive failure, rather than the nature of the fissures controlling the shear strength, which is considered more likely by the authors.

It is recommended that the following procedure be followed to assess the shear strength of fissured clays.

(a) Map the fissures for orientation, continuity and surface characteristics as detailed in Figure 6.28;
(b) Analyse the orientation using stereo projection methods as one does for rock joints. MacGregor et al. (1990) give an example;
(c) Carry out laboratory tests to determine the peak, fully softened and residual strengths and, if possible, the strength along fissures (this procedure is described below);
(d) Assess the strength of the soil mass parallel to the fissure surfaces accounting for the continuity and nature of the fissure surfaces eg. polished slickensided fissures would be assigned residual strength, and if they were 50% continuous, the strength determined as approximating to the average of residual and fully softened strength. Normal to the fissures, and for rough fissure surfaces, fully softened strength would apply.

This procedure is highly judgemental, and should where practicable, be backed up with back analyses of failures in the same geological materials.

The undrained strength is also affected by the presence of fissures. This is best assessed by using the procedures described in Section 6.1.6 to estimate the strength from the effective stress strengths (which have been assessed allowing for the fissures).

CPT, CPTU and Vane Shear Tests do not allow the soil to fail along the fissures, so they will generally measure the soil substance strength, not the fissured mass strength, and hence will over-estimate the strength.

6.1.8.2 Triaxial testing of fissured soils

The shear strength of a fissured soil mass can be determined by triaxial testing, provided the sample is sufficiently large to contain a representative selection of the fissures, and the

Table 6.2. Shear strengths of fissured soils from three projects in Australia.

Project	References	Peak c'	Peak ϕ'	Residual c'_R	Residual ϕ'_R	Adopted c'	Adopted ϕ'
Raby Bay[1]	Moon (1992)	5	24	2	8	0	13
Plantes Hill[2]	MacGregor et al. (1990)	10	25	0	11	5	18
Botany Bay[3]	Thorne (1984)	25	24	0	13	5	15

Notes: [1] Residual soils derived from tuff. Strengths are those used while the first author was involved in project. One failure was back analysed to give $c' = 0\,kPa$, $\phi' = 13°$.
[2] Residual soil derived from basalt.
[3] Marine clay fissured by desiccation and differential consolidation.

fissures are oriented in such a way that failure will occur along the fissures, rather than the soil substance.

In practice this means (Rowe, 1972) that large diameter samples – at least 100 mm, and preferably 250 mm diameter are needed. This is usually impractical because even if the samples can be obtained by block sampling, the test duration would be very long.

Hence it is usually more practicable to determine the fissure strength, either by judgement, from the nature of the fissures, and the peak, fully softened and residual strength, as described above, or by triaxial testing, and then use the orientation and continuity of the fissures to assess the mass strength. The strength of the fissure surfaces can be determined by the procedure described by Thorne (1984). This can be applied to block samples or thin walled tube samples and consists of:

– Test 75 mm diameter samples, being the maximum size which can readily be obtained with thin wall samples tubes;
– Carefully examine the samples before and after testing to determine whether shearing has occurred on a fissure surface or through the soil substance, and note the orientation of the failure surface;
– Determine the shear stress on the fissure surface from the Mohr's circle using the construction shown on Figure 6.31;
– Plot the strengths on the fissure surfaces in this manner to determine the design fissure strength (as a p′–q plot of points).

Thorne (1984) shows that whether a sample fails on a fissure depends on the orientation of the fissure and the magnitude of the stresses. Figure 6.32 illustrates this.

For Figure 6.32a 60° fissure: a, b and c all fail on fissure.
For Figure 6.32b 25° fissure: Circle a fails through substance.
For fissured soils which can be sampled by block sampling from test pits, shafts or other excavations, it is often practical to test the actual fissure surface in a direct shear test to obtain the fissure strength.

On a practical note, it is very unusual to be able to see the fissures in soil samples prior to testing, so it is usually necessary to test a large number of samples to obtain sufficient samples which fail on fissure surfaces not intersecting the end plates. Some fissured samples tend to fall to pieces on ejection from a sample tube. This can be overcome to a degree by cutting down the sides of the sample tube, so the sample can be lifted out, rather than forced out of the end in the normal manner.

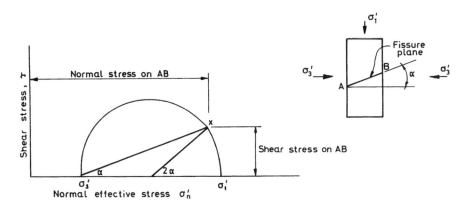

Figure 6.31.   Resolution of stresses on failure surface.

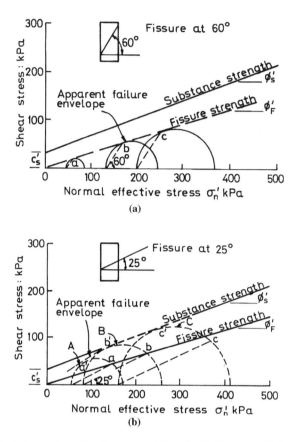

Figure 6.32.   Mohr's circle envelopes for fissures at 60° and 25° (Thorne, 1984).

6.1.8.3   *Methods usually adopted*

The shear strength (ϕ') of silts, sands, and gravels in the dam foundation or within the embankment can be assessed in three ways:

(a) In-situ testing, such as Standard Penetration Test (SPT), Cone Penetration Test (CPT) or Piezocone (CPTU);
(b) Sampling, and laboratory testing (direct shear or triaxial);
(c) Estimation, based on the particle size distribution, and compaction layer thickness, number of passes of a roller and/or relative density (density index) of compaction;

Method (a) is commonly used for assessing the strength of foundation soils, and occasionally embankment materials e.g. in hydraulic fill, or upstream construction tailings dams.

Method (b) is seldom used in view of the fact that the peak effective friction angle (ϕ') is dependent on the relative density, so laboratory testing requires a knowledge of the in-place relative density of the soil which, for dam foundations, can only be found by *in-situ* testing (SPT, CPT, CPTU). The normal practice therefore is to estimate the shear strength of those soils directly from the in-situ test, rather than doing laboratory tests. For very large embankment dams, it was common to do laboratory shear strength tests on the filters, but in reality this was not necessary, as they seldom influence stability very much. For outer zones ("shells") of embankments constructed of silt-sand-gravels, it would be good practice to test in the laboratory, at the relative density to be specified for construction.

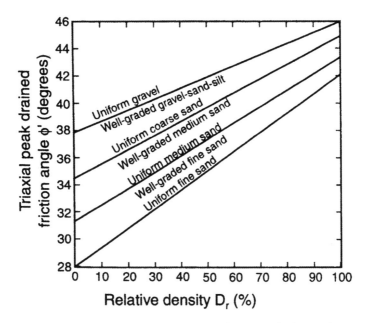

Figure 6.33.   Peak effective friction angle of sands as a function of relative density and grain size (Schmertmann, 1978).

Method (c) is commonly used for estimating the strength of filter zones and is sufficiently accurate in most cases.

6.1.9   Estimation of the effective friction angle of granular soils

6.1.9.1   In-situ tests

6.1.9.1.1   Standard Penetration Test (SPT)

The effective friction angle, ϕ' can be estimated directly from the SPT "N" values (e.g. using the methods of Mitchell and Kalti, 1981; Peck et al., 1974) or by using the SPT to estimate the relative density and then estimating the strength from the relative density and general grain size characteristics. Schmertmann (1978) developed such a method which is shown in Figure 6.33. Lunne et al. (1997) indicate the laboratory tests on which Figure 6.33 is based were at a confining stress of 150 kPa, and that lower values of ϕ' should be used at high confining stresses.

The relative density is best estimated using the method suggested by Skempton (1986). He reviewed the available information and proposed a method where the SPT value is corrected for the energy of the hammer, and to an effective overburden pressure of 100 kPa by:

$$N_{60} = Nm\ ERm \qquad (6.23)$$

where Nm = measured SPT "N" value; N_{60} = "N" value corrected to a hammer of 60% efficiency and ERm = Energy Rating of the SPT hammer.
and

$$(N_1)_{60} = C_N\ N_{60} \qquad (6.24)$$

Table 6.3.   Skempton (1986) method for estimating relative density (D_R) from $(N_1)_{60}$ value.

Relative Density (%)	Condition	$(N_1)_{60}$
0		0
	Very loose	
15		3
	Loose	
35		8
	Medium dense	
65		25
	Dense	
85		42
	Very dense	
100		58

where C_N = factor to correct N_{60} to 100 kPa effective overburden stress

$$= \frac{200}{100 + \sigma'_{vo}} \text{ for fine, normally consolidated sand} \qquad (6.25)$$

$$= \frac{300}{200 + \sigma'_{vo}} \text{ for coarse, normally consolidated sand} \qquad (6.26)$$

$$= \frac{170}{70 + \sigma'_{vo}} \text{ for overconsolidated sand} \qquad (6.27)$$

where σ'_{vo} = effective overburden stress at the level of the SPT test (in kN/m^2)
Skempton (1986) then uses the $(N_1)_{60}$ value to estimate relative density from Table 6.3. For $D_R \geq 35\%$, this can be approximated by

$$(N_1)_{60}/D_R^2 = 60 \qquad (6.28)$$

This method applies to naturally occurring sands, of other than very recent deposition.
Skempton (1986) also considers the effect of the age of the deposit on the SPT values. Discussing this, Jamiolkowski et al. (1988) produced Figure 6.34 which indicates that $(N_1)_{60}/D_R^2$ is dependent on the age of the deposit and will be less for recently deposited soil than for the more common aged soil.
This also means that for recently dredged fills, or deposits of mine tailings, a $(N_1)_{60}$ value of say 10, will imply a relative density of $\approx 50\%$, compared to 40% for an aged natural deposit.

6.1.9.1.2   Cone Penetration (CPT) and Piezocone (CPTU) Tests
The effective friction angle ϕ' can be determined directly from the CPT and CPTU using charts which have been developed by several authors e.g. Robertson and Campanella (1983a,b), Marchetti (1985, Kulhawy and Mayne (1988), Jamiolkowski et al. (1985),and Olsen and Fan (1986). Lunne et al. (1997) reviewed the methods and concluded that the

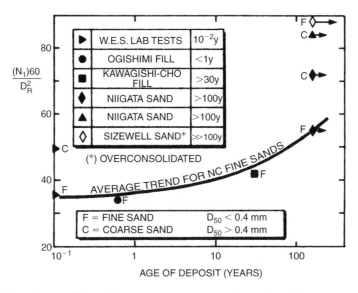

Figure 6.34. Effect of aging of the soil on SPT resistance (Jamiolkowski et al., 1988).

Robertson and Campanella (1984) and two other methods were to be preferred. The Robertson and Campanella (1984) method is simple to use and is shown in Figure 6.35.
   They indicate that:

– The method applies to normally consolidated, uncemented, moderately incompressible (grains), predominantly quartz sands but can be used for silty sand, sandy silt, sand and gravelly sand;
– For highly over-consolidated sands ϕ' may be up to 2° lower than predicted from Figure 6.35;
– For highly compressible sand (e.g. sand with many shells, or mica pieces) the method predicts conservatively low friction angles. For the sands included in Robertson and Campanella's analysis the effect could be up to 3°.

   The in-situ methods described give peak strength (not large strain, steady state strength) and do not appear to account for any curvature of the strength envelope. Hence if it is intended to construct a large dam on such a foundation, it may be necessary to account for this by using somewhat lower strengths. The Jamiolkowski et al. (1985) method allows for this.

6.1.9.2   *Laboratory tests*
Laboratory tests to determine the effective friction angle of granular materials are described in ASTM and British standards and in Head (1985).
   The most difficult aspect is to be able to prepare the sample at the correct relative density. Coarse grained materials (containing gravel) may have to be re-graded so they can be tested. The normal procedure is to prepare soils with a particle size distribution finer than and parallel to the actual.

6.1.9.3   *Empirical estimation*
A good guide to the strength of compacted sand and sandy gravel is to use the relative density specified (or expected), and particle size and use Figure 6.33.

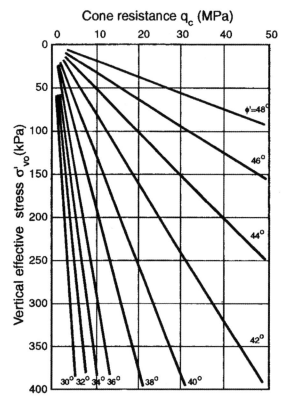

Figure 6.35.   Robertson and Campanella's (1984) method for estimating ϕ' for cohesionless soils from the cone penetration test (CPT).

Normal compaction of granular filters in dams e.g. 4 passes of a 10 tonne smooth drum vibratory roller, in layers 500 mm compacted thickness, should give a relative density of at least 70%, so the effective friction angle should be around 38° for fine sand and 40° for well graded filter sands and sandy gravels.

In reality at low confining stresses e.g. less than 200 kPa, ϕ' is likely to be 2° to 5° higher than these figures. If the figures become critical to the stability analysis, they should be checked by laboratory testing.

6.1.10   *Shear strength of partially saturated soils*

See Section 11.2.4 for a discussion on this topic.

6.1.11   *Shear strength of rockfill*

The shear strength of rockfill is dependent on the effective normal stress, and to a lesser extent on the bulk dry density, void ratio or degree of compaction; the unconfined compressive strength of the rock, the uniformity coefficient of the grading, maximum grain size, fines content, and particle shape (Marsal, 1973; Douglas, 2003). Table 6.4 summarises the effect of these variables. Figure 6.36 shows the methods for representing the shear strength.

Douglas (2003) carried out an extensive analysis of the factors affecting the shear strength of rockfill. This involved establishing a large database from published and unpublished triaxial testing, and statistical analysis.

Table 6.4.   Factors affecting the shear strength of rockfill.

Variable	Description of effect on shear strength	Importance of variable	References
Confining pressure	Mohr-Coulomb shear strength envelope is curved, with lower friction angles at high confining stress	Major – see Figures 6.36, 6.37	Douglas (2003), Indraratna et al. (1993, 1998), Charles and Watts (1980), Marsal (1973)
Density or void ratio	Shear strength increases with increased density, lower void ratio. The effect is greatest at low confining stress, with little effect at high confining stress	Medium to minor	Douglas (2003), Marsal (1973), Chiu (1994), Nakayama et al. (1982)
Unconfined compressive strength of the rock	Shear strength increases with increased rock strength below UCS 100 MPa but little effect for stronger rock. Little effect at low confining stress, medium effect at high confining stress	Minor to medium	Douglas (2003), Anagnosti and Popovic (1982)
Uniformity coefficient	Uniformly graded rockfill exhibits more curvature in the Mohr Coulomb strength envelope than well graded. No clear effect on strength (Douglas)	Minor	Saras and Popovic (1985), Chiu (1994), Douglas (2003)
Maximum particle size	Most accept that shear strength increases with particle size (Chiu, Douglas), but some claim no effect (Charles et al.), or the opposite effect (Anagnosti and Popovic)	Minor	Chiu (1994), Charles et al. (1980), Anagnosti and Popovic (1981)
Finer particles content	Shear strength reduced significantly with more than about 35–50% silty or clayey sand passing 2 mm size (USBR). Rockfill with >20% fines passing 0.075 mm had lower strength (Douglas)	Medium	USBR (1966), Douglas (2003)
Particle angularity	Angular particles have higher strength than sub angular and rounded particles at low confining pressures, little difference at high confining pressures	Minor	Douglas (2003), Sarac and Popovic (1985), Bertacchi and Berlotti (1970)

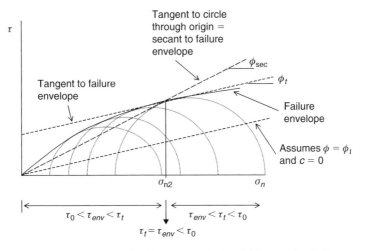

Figure 6.36.   Methods for representing the shear strength of rockfill (Douglas 2003).

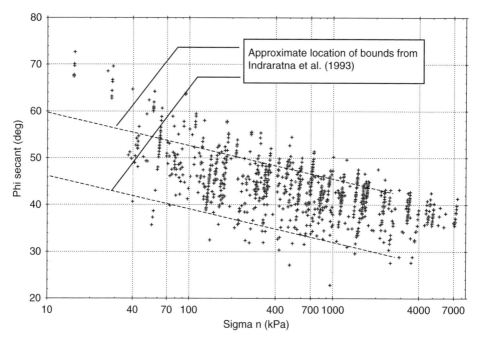

Figure 6.37. Variation of secant friction angle ϕ'_{sec} with normal stress σ_n (Douglas, 2003).

Figure 6.37 shows the variation of secant friction angle ϕ'_{sec} with normal stress σ_n for dense rockfill. The strong dependence of ϕ'_{sec} on σ_n is evident, as is the wide variation in ϕ'_{sec}.

Douglas (2003) carried out non-linear statistical anlaysis on the database. The estimation method used was least squares. An analysis of the form shown below was found to be the most effective for relating ϕ'_{sec} to σ_n.

$$\phi'_{sec} = a + b\sigma_n'^c \qquad (6.29)$$

The analysis resulted in the following constants:

$$a = 36.43 - 0.267\text{ANG} - 0.172\text{FINES} + 0.756(c_c - 2) + 0.0459(\text{UCS} - 150) \qquad (6.30)$$

$$
\begin{aligned}
b = {} & 69.51 + 10.27\text{ANG} + 0.549\text{FINES} - 5.105(c_c - 2) \\
& - 0.408(\text{UCS} - 150) - 0.408
\end{aligned}
\qquad (6.31)
$$

$$c = -0.3974 \qquad (6.32)$$

where ANG = (angularity rating -5) for angularity rating >5.5; otherwise 0. rounded particles have an angularity rating of 0, crushed angular rock a rating of 8. The rating is subjective (Douglas, 2003).

FINES = percentage of fines passing 0.075 mm (%). (6.30a)

$$c_c = \text{coefficient of curvature} = \frac{d_{30}^2}{d_{10}d_{60}} \qquad (6.30b)$$

UCS = unconfined compressive strength of the rock substance (MPa).

This function resulted in a variance explained of 61.7%. The addition of the uniformity coefficient, c_u and the maximum and minimum particle sizes, d_{max} and d_{min}, did not result in a better fit to the data.

Douglas (2003) also carried out an analysis of the database using the principal stresses σ'_1 and σ'_3. This showed an equation of the form shown in Equation 6.33 was found to be the most effective for relating σ'_1 to σ'_3.

$$\sigma'_1 = RFI\, \sigma'^{\alpha}_3 \tag{6.33}$$

The analysis resulted in the following equations (based on 869 data sets and with a variance explained = 98.8%):

$$\alpha = 0.8726 \tag{6.34}$$

$$RFI = 6.3491RFI_e + 0.48763RFI_{ANG} - 0.0027RFI_{d_{max}} - 1.1568RFI_{FINES} \\ + 0.30598RFI_{UCS} \tag{6.35}$$

$$RFI_e = \frac{1}{1 + e_i} \tag{6.36}$$

$$RFI_{ANG} = 1 \text{ if angular, otherwise} = 0 \tag{6.37}$$

$$RFI_{d_{max}} = d_{max}(mm) \tag{6.38}$$

$$RFI_{FINES} = \frac{e^{(Fines-20)}}{1 + e^{(Fines-20)}} \text{ where fincs is in \%} \tag{6.39}$$

$$RFI_{UCS} = \frac{e^{(UCS-110)}}{1 + e^{(UCS-110)}} \text{ where UCS is in MPa} \tag{6.40}$$

For many analysis of stability it will be sufficient to use the data in Figure 6.37 and equations 6.29 to 6.32 to select the strength of well compacted rockfills. A bi-linear strength envelop should be used to better represent the curved Mohr-Coulomb envelope. This can be plotted by calculating ϕ'_{sec} from equations 6.29 to 6.32 at varying σ_n applicable to the stability analysis.

Loose dumped rockfill at low σ_n will have ϕ'_{sec} about 4 degress lower than calculated using this approach. At high σ_n the difference will be negligible. For more important projects, equations 6.33 to 6.40 should be used, to give the relationship between σ'_1 and σ'_3. To convert this to the more familiar c', ϕ', use the method described in Section 16.3.2.5, Equations 16.16 to 16.18.

If the rockfill has high percentages of finer particles (say greater than 25–30%, passing 2 mm), it would be wise to carry out triaxial tests, using scaled coarse fraction particle size distribution, but retaining the same percentage passing 2 mm.

For very large dams, it may be necessary to carry out triaxial tests on scaled gradings of the rockfill to confirm the strength, but only if the strength is critical (which in reality it seldom is).

6.2   COMPRESSIBILITY OF SOILS AND EMBANKMENT MATERIALS

6.2.1   *General principles*

6.2.1.1   *Within the foundation*

The compression of soils in the foundation of a dam, as load from the dam is applied during construction, has three components:

- Initial "elastic" compression;
- Primary consolidation, as the pore pressures induced by the weight of the dam are dissipated;
- Secondary consolidation, or creep, which takes place with no further change in pore pressures.

The theory and principles involved are covered in soil mechanics text books and will not be discussed here. There are some points to keep in mind:

(a) The in-situ vertical and horizontal stresses, σ'_{vo} and σ'_{HO} are affected by overconsolidation ratio (OCR = $\sigma'_{vp}/\sigma'_{vo}$, where σ'_{vp} = preconsolidation pressure) of the foundation soils. These are influenced by periods of erosion of the site, lower groundwater levels than at present, desiccation from drying of soils exposed to the sun (or by freezing) and, in some areas, glaciation.

(b) The coefficient of consolidation, along with the drainage path distance determines the time for primary consolidation:

$$t = \frac{TH^2}{C_v} \tag{6.41}$$

where t = time for primary consolidation (years); T = time factor; H = drainage path distance (metres) and C_v = coefficient of consolidation (m²/year).

The coefficient of consolidation is dependent on the OCR (it is larger in the over-consolidated range of stresses, than in the normally consolidated range) and by fabric in the soil (e.g. fissures, fine layers of silt or sand interbedded with clay). Care must be taken to test sufficiently large samples to include the fabric e.g. by using triaxial consolidation. However, in these tests, the side drains may become a control on the rate of flow of pore water from the soil.

The drainage path H, is dependent on the thickness of the clay layer, and the presence of more permeable sandy, or silty layers within the clay. These can often best be located by cone penetration (CPT) or piezocone (CPTU) testing.

(c) Consolidation within the over-consolidated stress range (O-A in Figure 6.38) results in relatively small changes in void ratio (and hence consolidation) as the soil follows the unload-reload index C_r. Consolidation within the primary or normally consolidated range (A–B in Figure 6.38) gives larger changes in void ratio as the soil follows the compression index C_c. Hence, to estimate settlements, it is important to clearly define the pre-consolidation pressure σ'_{vp} in oedometer testing.

It is recommended that the Schmertmann (1955) correction to C_c be applied to allow for sample disturbance. This involves selecting point 0 from the existing effective over burden

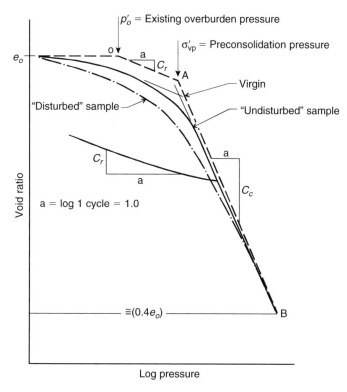

Figure 6.38. Void ratio versus log effective stress plot for a typical overconsolidated soil with the Schmertmann (1955) method shown to correct C_c for sample disturbance.

pressure σ'_{vo} from which the sample was taken, and the in-situ void ratio e_o, drawing OA parallel to the unloading curve with A on the preconsolidation stress σ'_{vp}; and joining A to point B, which is on the consolidation plot at $0.4e_o$.

The preconsolidation pressure measured in the laboratory should be checked against the history of the soil deposit, and indirectly checked using the relationship between the undrained strength of clays in the deposit, and the OCR (and hence σ'_{vp}) using Equation 6.14.

6.2.1.2 *Within the embankment*

As for the foundation, the materials within the embankment have three components to their compression – initial "elastic", primary consolidation, and creep. Considering each type of material in the embankment:

6.2.1.2.1 Earthfill

Soils in modern embankment dams are compacted to a degree of saturation which is typically 90–95%, so their early behaviour is that of partially saturated soils. Much of the consolidation process consists of the gas in the voids being compressed, and forced into solution, rather than flow of water from the system. It therefore occurs relatively quickly and, as discussed in Section 6.2.2, most of the settlement in dams occurs during construction. It is incorrect to try to predict the behaviour of these zones by carrying out conventional oedometer consolidation tests (which use saturated soils), and classical Terzaghi consolidation theory, because these do not properly model the partially saturated behaviour, and impose zero lateral strain (K_o) conditions, which are often not correct, particularly if the earth core of the dam is narrow and the rockfill compressible giving little lateral constraint.

The compaction process typically overconsolidates the soil to at least 200 kPa, but this depends on the compaction equipment, soil and the compaction water content. Hunter (2003) and Hunter and Fell (2003d) determined from internal settlement data that for earthfill cores placed generally 0.5 to 1% dry of standard optimum moisture content, the preconsolidation pressure was from 200 to 400 kPa for clayey earthfills, and 700 to 1,000 kPa for silty gravel, silty sand and clayey sand.

The theory and practice of partially saturated soil behaviour is rapidly developing and readers should seek the latest information from the literature. Some useful references include Alonso et al. (1990), Bardin and Sides (1970), Khalili and Khabbaz (1998) and Loret and Khalili (2002).

6.2.1.2.2 Filters and sand-gravel zones

These will, if well compacted, be stiffer (have a higher modulus E) than earthfill and are usually stiffer than rockfill. The modulus will depend on the grading and degree of compaction. Because they are stiff, they may attract load by arching of the core. Some data on the modulus of compacted "gravel" fill in concrete face rockfill dams is given in Section 15.2.4 and this can be used to estimate the modulus of filters.

6.2.1.2.3 Rockfill

The rockfill compressibility is dependent on the grading, degree of compaction (layer thickness, roller weight and number of passes), rock substance strength and the effect of wetting on the substance strength. The modulus is best estimated from the performance of similar dams and is discussed in detail in Section 15.2.4.

The rockfill upstream of the core of the dam is subject to wetting by the reservoir and to cyclic loading as the reservoir levels fluctuate. Some rockfills are subject to "collapse settlement" due to softening of the rock substance as they are wetted. Rockfill is likely to display a lower equivalent modulus than on previous drawdowns of the reservoir if the reservoir is drawn down below the previous lowest level, that is, the rockfill displays both over and normally consolidated behaviour.

6.2.2 *Methods of estimating the compressibility of earthfill, filters and rockfill*

6.2.2.1 *Using data from the performance of other dams – earthfill*

Generally speaking, it will be sufficient to estimate the settlement of the embankment earthfill during and after construction, using the results of monitoring of other dams. G. Hunter (2003) and Hunter and Fell (2003d), has gathered data from a large number of dams, where the core has been generally compacted to a minimum density of 98% of standard maximum dry density, with a water content in the range 2% dry to 1% wet of standard optimum water content. Most of the cores are "dry placed", with a moisture content 0.5 to 1% dry of standard optimum moisture content.

(a) *Settlement during construction*

Figure 6.39 shows total settlements of the core during construction of earth and earth and rockfill embankments. These have been obtained from monitoring of cross arm settlement gauges embedded in the dam core near the dam centreline.

Tables 6.5 and 6.6 give the equations of best fit for the data.

(b) *Strains and equivalent moduli during construction*

Figure 6.40 and Table 6.7 shows the vertical strains versus effective vertical stress (ignoring suction effects) and estimated confined secant moduli in dry placed earthfill cores at the end of construction. The confined secant moduli (M) can be converted to Youngs moduli using equation 6.30. The Poissons ratio are likely to be in the range 0.25 to 0.35 for dry placed earthfill.

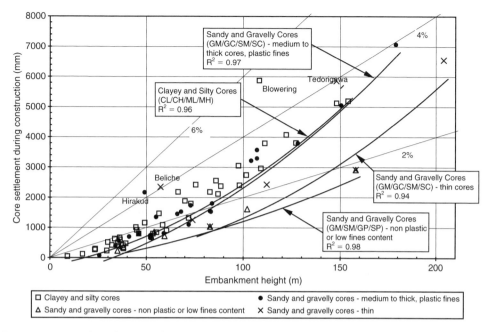

Figure 6.39.   Total settlements during construction of earth and earth and rockfill embankments (Hunter 2003; Hunter and Fell, 2003d).

Table 6.5.   Equations of best fit for core settlement versus embankment height during construction (Hunter, 2003; Hunter and Fell, 2003d).

Core material/ shape type	No. cases	Equation for settlement[*1]	R^2 [*2]	Std. err. of settlement[*3] (mm)
Clay cores – all sizes	42	$H(0.152\,H + 12.60)$	0.96	275
Sandy and gravelly cores:				
Medium to thick,				
Plastic	25	$H(0.183\,H + 7.461)$	0.97	290
Thin	5	$H(0.136\,H + 2.620)$	0.94	635
Non-plastic	7	$H(0.0635\,H + 8.57)$	0.98	130

Notes: [*1] Settlement in millimetres, embankment height H in metres.
　　　　[*2] R^2 = regression coefficient.
　　　　[*3] Std. err. = standard error of the settlement.

(c) *Post construction settlement*
　　　Figure 6.41 shows post construction crest settlements 10 years after the end of construction. Figure 6.42 shows long term post construction crest settlement rates for zoned earthfill and earth and rockfill embankments.
　　　Table 6.8 shows typical ranges of post construction settlement and long term settlement rates for different core materials, core widths and compaction moisture content. These exclude "outliers".
　　　The core width is classified as:
Thin – width < 0.5 dam height.

Medium – width ⩾ 0.5 and ⩽ 1.0 dam height.
Wide – width > 1.0 dam height.

Table 6.6.   Equations of best fit for core settlement (as a percentage of embankment height) versus embankment height during construction (Hunter, 2003; Hunter and Fell, 2003d).

Core material/ shape type	No. cases	Equation for core settlement[1] (as a percentage of embankment height)	R^2 [2]	Std. err. of settlement[3] (%)
Clay cores – all sizes	42	$0.179\,H^{0.60}$	0.77	0.41
Sandy and gravelly cores: Medium to thick,				
Plastic	25	$0.079\,H^{0.76}$	0.81	0.39
Thin	5	$0.063\,H^{0.72}$	0.70	0.47
Non-plastic	7	$0.187\,H^{0.46}$	0.71	0.24

Notes: [1] Settlement as a percentage of the embankment height, embankment height H in metres.
        [2] R^2 = regression coefficient.
        [3] Std. err. = standard error of the settlement (percent of embankment height).

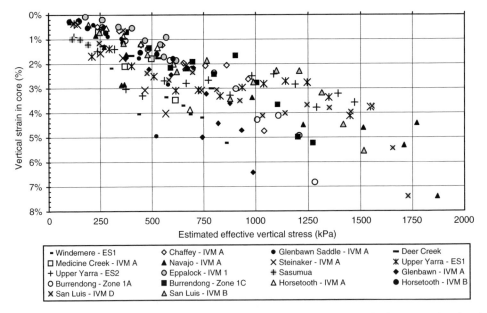

Figure 6.40.   Vertical strain versus effective vertical stress in the core at end of construction for dry placed dominantly sandy and gravelly earthfills (Hunter, 2003; Hunter and Fell, 2003d).

Hunter (2003) and Hunter and Fell (2003d) conclude:
For the crest region, the data has been sorted based on core width, core material type and placement moisture content, and indicates that:

– The post construction crest settlements are generally much smaller than the core settlement during construction;
– Nearly all dams experience less than 1% crest settlement post construction for periods up to 20 to 25 years and longer after construction;
– Most experience less than 0.5% in the first 3 years and less than 0.75% after 20 to 25 years;
– Smaller magnitude settlements are observed for dry placed clayey sands to clayey gravels and dry to wet placed silty sands to silty gravels;

Table 6.7.   Confined secant moduli during construction for well-compacted, dry placed earthfills (Hunter, 2003; Hunter and Fell, 2003d).

| Core material type | No. cases | Confined secant modulus (MPa)[1] | | |
| | | Effective vertical stress range (kPa) | | |
		500 to 700[2]	1000	1500 to 2000
Clayey soil types (CL) – sandy clays and gravelly clays	20	15 to 45 (27)	15 to 50 (30)	25 to 42
Silty soils (ML), data from Gould	3	30 to 60	–	–
Clayey sands to clayey gravels (SC/GC), plastic fines, >20% fines	11	20 to 65 (33)	30 to 65 (40)	25 to 65
Silty sands to silty gravels (SM/GM), plastic fines, >20% fines	9	35 to 65 (46)	45 to 65 (50)	35 to 65
Sandy and gravelly soils – non-plastic fines or <20% plastic fines	6	35 to 80 (60)	35 to 90 (60)	40 to 90

Notes:[1] Values represent range of confined secant moduli, values in brackets represent average.
[2] Includes Gould (1953, 1954) data.

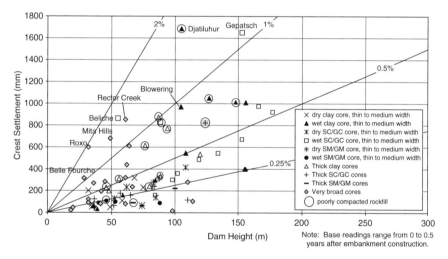

Figure 6.41.   Post construction crest settlement at 10 years after end of construction (Hunter, 2003; Hunter and Fell, 2003d).

- A broader range of settlement magnitude is shown for clay cores, wet placed clayey sand to clayey gravel cores, and embankments with very broad core widths;
- For zoned earth and rockfill dams, poor compaction of the rockfill is over-represented for case studies at the larger end of the range of crest settlement.

The data for the shoulder regions indicates:

- Settlements in the order of 1 to 2% are observed for poorly compacted rockfills. Greater settlements are observed for the dry placed, poorly compacted rockfills;
- For reasonably compacted rockfills the range of settlement is quite broad, from 0.1% up to 1.0%, but the number of cases is limited. Settlements toward the upper range are

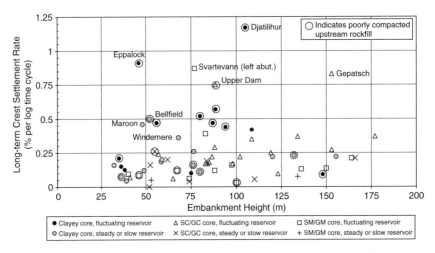

Figure 6.42.   Long-term post construction crest settlement rate for zoned earthfill and earth-rockfill embankments of thin to thick core widths (Hunter, 2003; Hunter and Fell, 2003d).

Table 6.8.   Embankment crest region, typical range of post construction settlement and long-term settlement rate (Hunter, 2003; Hunter and Fell, 2003d).

Core properties			Crest settlement (%) [*1, *2]			Long-term settlement rate [*1, *3]	
Classification	Core width	Moisture content	3 years	10 years	20 to 25 years	Steady/slow reservoir	Fluctuating reservoir
CL/CH	Thin to medium	dry	0.05 to 0.55	0.10 to 0.65	0.20 to 0.95	0.04 to 0.50	0.09 to 0.57
		wet	0.04 to 0.75	0.08 to 0.95	0.20 to 1.10	(most <0.26)	
		all	0.02 to 0.75	0.10 to 1.0	0.5 to 1.0		
	Thick	(most dry)					
SC/GC	Thin to medium	dry	0.10 to 0.25	0.10 to 0.40	<0.5	0 to 0.26	0.06 to 0.37
		wet	0.15 to 0.80	0.20 to 1.10	<1.1		
		all	0.05 to 0.20	0.10 to 0.35	0.10 to 0.45		
	Thick	(most dry)					
SM/GM	Thin to thick	all	0.06 to 0.30	0.10 to 0.65	<0.5 to 0.7	<0.10	0.03 to 0.21
Very broad earthfill cores – most CL and dry placed			0.0 to 0.60	0.0 to 0.80	0.05 to 0.76	0.08 & 0.44	0.07 to 0.70 (most <0.35)

Notes: [*1] Excludes possible outliers.
 [*2] Crest settlement as a percentage of the embankment height.
 [*3] Long-term settlement rate in units of % settlement per log cycle of time (settlement as a percentage of dam height).

observed for dry placed and/or weathered rockfills, where settlements due to collapse compression are likely to be significant;
– Much lower settlements, generally less than 0.5 to 0.7% at ten years after construction, are observed for well and reasonably to well compacted rockfills, and compacted earthfills;
– Very low settlements (less than 0.25% at 10 years) are observed for embankments with gravel shoulders.

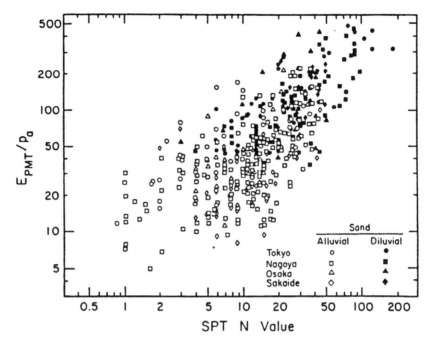

Figure 6.43.   Drained Youngs Modulus E' for sands versus SPT "N" value (Ohya et al., 1982; from Kulhawy and Mayne, 1988).

6.2.2.2   *Using data from the performance of other dams – rockfill*
See Section 15.2.4.

6.2.2.3   In-situ *testing*
The drained modulus E' for free draining soils, such as sands, silty sands and fine gravelly sands, and the undrained modulus Eu for clayey soils can be estimated using Standard Penetration and Cone Penetration Tests.

Figures 6.43 and 6.44 show plots of E' and Eu respectively versus SPT "N" value. The moduli were obtained by pressuremeter testing. It will be seen that there is a considerable uncertainty in the relationship between the moduli and SPT "N" value. Note that the moduli are in bars, where 10 bars = 1 MPa.

Figure 6.45 presents a chart to estimate the secant Young's modulus (E'_s) for an average axial strain of 0.1% for a range of stress histories and aging. This level of strain is reasonably representative for many well-designed foundations. In Figure 6.45 a = 100 kN/m² and σ'_{mo} is the mean stress level = $(\sigma'_{vo} + 2\sigma'_{Ho})/3$. The stiffness of normally consolidated aged sands (>1,000 years) appears to fall between that of very recent normally consolidated sands and over-consolidated sands.

Recent sands would include those deposited by dredging, or in historic time by for example scour and deposition in river channels.

Kulhawy and Mayne (1988) produced Figures 6.46 and 6.47 showing the relationship between M/q_c ($=\alpha_n$) and relative density for normally consolidated (NC) and over-consolidated (OC) sands. As can be seen there is a trend for decreasing α_n with increase in relative density, but a large scatter of test data. Note that:

$$M' = \frac{E'(1 - v)}{(1 + v)(1 - 2v)} \qquad (6.42)$$

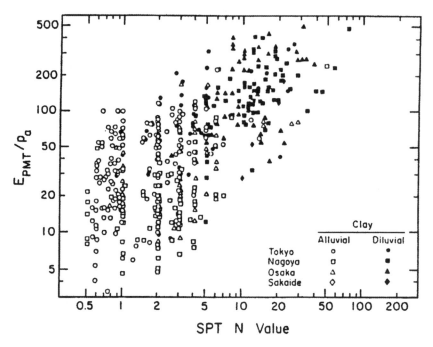

Figure 6.44.   Undrained Youngs Modulus E_u for clays versus SPT "N" value (Ohya et al., 1982; from Kulhawy and Mayne, 1988).

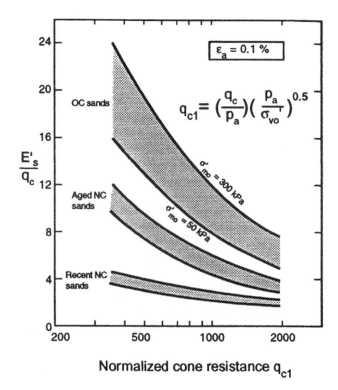

Figure 6.45.   Drained Youngs modulus E' of sands versus Static Cone Penetrometer resistance (Bellotti et al., 1989; from Lunne et al., 1997).

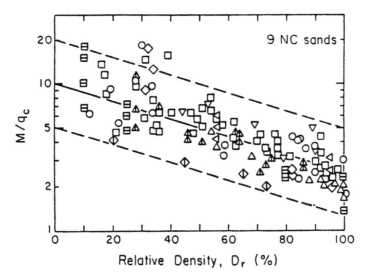

Figure 6.46.   Variation of α_n ($=M/q_c$) with relative density for normally consolidated sand (Kulhawy and Mayne, 1988).

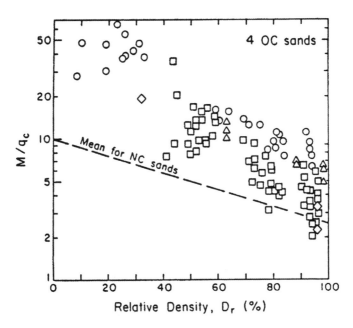

Figure 6.47.   Variation of α_n ($=M'/q_c$) with relative density for overconsolidated sand (Kulhawy and Mayne, 1988).

where E' = Youngs modulus, M' = constrained drained modulus, v = Poissons ratio, q_c = cone resistance.

Kulhawy and Mayne (1988) also discuss Poisson ratio and suggest that for drained loading:

$$v' = 0.1 + 0.3\,\phi'_{rel}$$

(6.43)

where

$$\phi'_{rel} = \frac{\phi' - 25°}{45° - 25°} \tag{6.44}$$

with $0 \leq \phi'_{rel} \leq 1$ where ϕ' is the friction angle of soil at its relative density.

They suggest the use of Figure 6.48 (based on work by Wroth, 1975) to estimate v' for lightly over-consolidated soils. For over-consolidated soils, Figure 6.49 may be used to

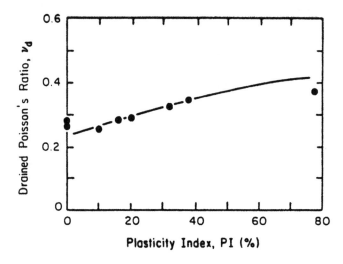

Figure 6.48.   Drained Poisson Ratio v' versus plasticity index for some lightly over-consolidated soils (Kulhawy and Mayne, 1988; Wroth, 1975).

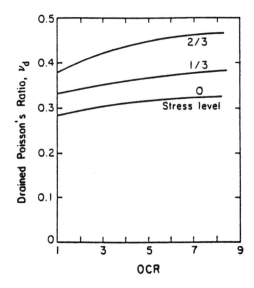

Figure 6.49.   Drained Poisson ratio v' versus OCR and stress level for clay (Kulhawy and Mayne, 1988 and Poulos, 1978).

adjust the values from Figure 6.48 (based on Poulos, 1978). In Figure 6.49, the stress level is the ratio of actual stress to failure stress.

In reality it is often very difficult to assess the degree of over-consolidation of sands, unless they are overlain by clays for which the OCR can be determined by laboratory oedometer tests, or the erosion/depositional history of the site can be established. In other cases it is usually necessary to assume the soil is normally consolidated which may give conservatively low moduli.

The undrained Youngs Modulus E_u for cohesive soils cannot be estimated with any degree of accuracy directly from q_c values. E_u can be estimated indirectly from CPT or CPTU, by first estimating S_u, and then relating E_u to S_u by using Figure 6.50.

The Self Boring Pressuremeter (SBPM) can be used to determine the drained modulus in sandy soils and undrained modulus in clayey soils. Details are given in Clayton et al. (1995).

The SBPM test gives the most reliable means of determining the modulus, but is relatively expensive. Usually it would only be used on important projects, and then often as a means of calibrating CPT tests.

The Menard Type Pressuremeter is inserted into a pre-drilled borehole. It is very dependent on the drill hole staying open and is affected by disturbance, particularly in sands and weak clays. More recent developments of the Menard Pressuremeter include a push-in pressuremeter and a self boring type. These should have a similar accuracy to the SBPM.

6.2.2.4   Laboratory testing

The use of laboratory triaxial tests to determine the modulus of a soil is discussed in Baldi et al. (1988), Jardine et al. (1984), Atkinson and Evans (1984). They point out that conventional triaxial equipment is not able to measure the modulus and, if used, will generally significantly underestimate the modulus. Generally it will be better to rely on estimated moduli, or values determined by *in-situ* testing as described above.

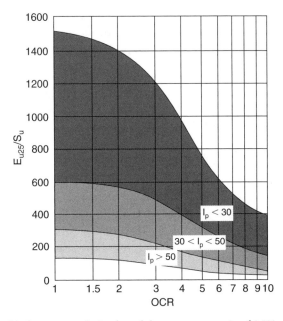

Figure 6.50.   Relationship between undrained modulus at a stress ratio of 25% of the ultimate strength (Eu_{25}) versus over-consolidation ratio (OCR), and plasticity index (PI) (Lunne et al., 1997; after Duncan and Buchignani, 1976).

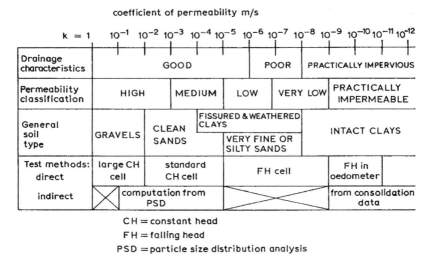

Figure 6.51.   Permeability and laboratory testing method for the main soil types (Head, 1985).

6.3   PERMEABILITY OF SOILS

6.3.1   *General principles*

The "permeability", or more correctly "permeability coefficient" or "hydraulic conduc-tivity" of the soil in a dam embankment or foundation, is not a fundamental property of the soil but depends on a number of factors. Head (1985) outlines these as: particle size distribution; particle shape and texture; mineralogical composition; void ratio; degree of saturation; soil fabric; nature of fluid; type of flow and temperature.

In embankment dam engineering these factors have varying degrees of influence:

Particle size distribution. The permeability is dependent on the particle size distribution of the soil with fine grained clay soils having permeabilities several orders of magnitude lower than that of coarser soils, i.e. sands and gravels. Figure 6.51 shows in general terms the range of permeabilities which can be encountered.

It has been recognised that the finer particles in a soil largely determine its permeabil-ity, and granular soils are often compared by their "effective grain size", D_{10}, which is the particle size for which 10% of the soil is finer.

Particle shape and texture. This affects permeability to a lesser extent. Elongated ("platey") particles tend to have a lower permeability than rounded, and rougher textured particles a lower permeability than smooth.

Mineralogical composition. This is a factor in clay soils. Montmorillonite clays, for example, are finer grained and have a greater tendency to adsorb water (and hence a lower permeability) than say a kaolin clay. The mineralogy of sand and gravel has little effect except where it results in elongated particles.

Void ratio. The void ratio of a soil has an important effect on permeability. Cohesive soils which are compacted to a high density ratio (e.g. 98% of standard maximum dry density) will have lower permeability than those compacted to a low density ratio (e.g. 90%). The difference may be orders of magnitude. Figure 6.52 shows the effect of void ratio for several clays.

The void ratio also has an effect on the permeability of granular soils, with soils com-pacted to a small void ratio (dense) having lower permeability than those with a high void ratio (loose). This effect is allowed for in the Kozeny-Carman formula discussed below.

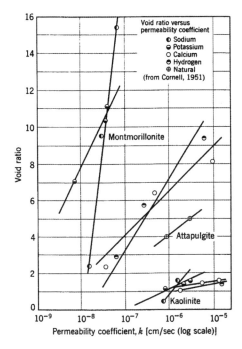

Figure 6.52.   Effect of void ratio and clay mineralogy on permeability (Lambe and Whitman, 1981). Reprinted by permission of John Wiley and Sons Inc.

Degree of saturation. When a soil becomes partially saturated, the permeability is reduced. This occurs due to a reduction in the total cross section of pores filled with water and surface tension effects. These water filled pores tend to be the finer pores, because water is most readily removed from the larger pores which have low suction potential. The remaining finer water filled pores have naturally lower permeability. The partially saturated permeability may be an order of magnitude lower than the saturated permeability. Jackson et al. (1965), Gillham et al. (1976), Richards (1974) and Hillel (1979) discuss the relationship between saturated and unsaturated permeability for granular and cohesive soils.

Soil fabric. The fabric of a soil can have a major effect on the permeability. This is particularly important for soil masses in-situ, where stratification or layering of different soil types, e.g. sand, silt, clay can lead to markedly different permeability along the strata to that across the strata. Ratios of horizontal to vertical permeability (k_H/k_V) of 100 or more are not uncommon.

Relict jointing, fissures, root holes, worm holes, lateritisation etc. all influence the soil fabric and lead to much higher permeability for the soil mass than the soil substance permeability.

When testing samples from such soil masses in the laboratory, it is essential that the presence of fabric is allowed for, e.g. in orienting samples along or across the strata, and in taking sufficiently large samples to include representative fissures. Rowe (1972) suggests that samples of up to 250 mm diameter may be needed to correctly sample fabric. Generally speaking if good quality *in-situ* testing can be carried out, it will be more reliable than laboratory testing.

The permeability of recompacted soil is also affected by the soil fabric, which is dependent on the method of sample preparation. Lambe and Whitman (1981) describe this as micro-structure or micro-fabric effects. Figure 6.53 shows the effect of mixing of the soil before compaction, and of adding a dispersant (0.1% polyphosphate) ("complete mixing")

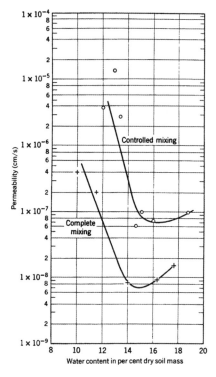

Figure 6.53.   Effect of structure on permeability (Lambe and Whitman, 1981). Reprinted by permission of John Wiley and Sons Inc.

to the soil. In the latter the dispersant breaks up the flocculated structure, resulting in a lower permeability.

The water content at which a clay is compacted affects its permeability. In particular a cohesive soil compacted dry of optimum results in a less oriented structure, with potentially aggregates of soil separated by voids, leading to relatively high permeability. An example of this behaviour is given in Figure 6.54 where soils are compacted with the same compactive effect at different water contents.

It is virtually impossible to simulate in the laboratory the effect of different degrees of compaction at the top and bottom of layers and other factors, such as desiccation at the surface of layers during construction. In the dam, the ratio of k_H/k_V is likely to be high, possibly as high as 100, with the laboratory permeability probably best representing the vertical permeability k_V.

Nature of the permeating fluid. The permeability (coefficient of permeability k) is dependent on the properties of the permeating fluid. It is related to the absolute (also known as intrinsic, or specific) permeability, K, by:

$$k = \frac{K\rho g}{\mu} \tag{6.45}$$

where ρ = fluid density; g = acceleration due to gravity; μ = fluid viscosity.

For most dam projects water is the permeating fluid and the small variations in viscosity and density, resulting from differences in temperature and dissolved salts content, are not significant compared to other variables. However the chemical composition of the

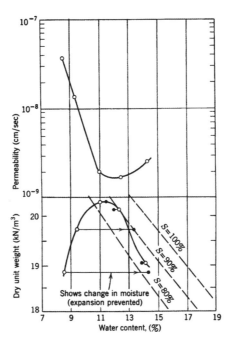

Figure 6.54.   Effect of compaction water content on permeability (Lambe and Whitman, 1981). Reprinted by permission of John Wiley and Sons Inc.

permeating water can affect soil dispersivity (see Chapter 7) and to a certain extent the permeability.

Type of flow. The basic assumption in calculations using Darcy's Law q = kiA is that the soil is saturated, and the flow is laminar. This is generally the case for flow through and beneath dams but flow in medium and coarse gravels may become turbulent in some cases, e.g. around dewatering wells.

6.3.2   Laboratory test methods

6.3.2.1   Permeameters

Most permeability testing of soils is carried out using constant head or falling head permeability apparatus. As shown in Figure 6.55 the constant head test is best suited to high to medium permeability soils, sand and gravel, and the falling head test to low to very low permeability soils, silts, fine sands and some clays.

For very low permeability soils, the flow rates are too low to carry out testing in a conventional permeameter, and it would be more usual to do the testing in an oedometer cell (e.g. a Rowe cell), or to determine the permeability indirectly from consolidation test data.

Many texts in soil mechanics include details of falling head and constant head tests so they are not repeated here. Head (1985) gives detailed instructions on laboratory procedures, as do Standards Association of Australia (1980) and American Society for Testing and Materials D2434. The USBR (1980) also gives detailed instructions for the test in their "*Earth Manual*".

Some common errors and problems which can arise with permeability testing include:

(a) leakage past the sample in the test mould;
(b) variability in test results for compacted soils due to difficulty in maintaining uniform water content and density in sample preparation;

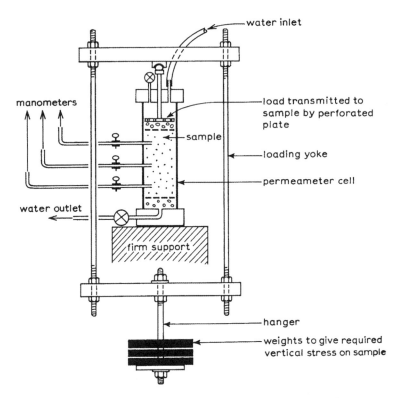

Figure 6.55.   Suitable assembly for constant head permeability apparatus (Head, 1985).

(c) failure to saturate sample;
(d) failure to adequately test the fabric of the soil;
(e) testing under little or no confining stress. For dams work in particular it is important to use apparatus such as that shown in Figure 6.55 which allows simulation of the overburden pressure on the soil and in so doing allows testing at the correct void ratio;
(f) excessive flow rates in the apparatus (when testing in coarse sand or gravel) leading to unaccounted losses in the pipework or porous plates at the ends of the sample and an underestimation of permeability.

From the above, (a) and (c) can be eliminated by careful test procedures; (b) is to be expected and requires several tests so that variability can be identified; and (d) can be overcome by correct sampling, using larger samples and/or using in-situ tests. Problems with high flow rates (f) can be readily checked by testing the apparatus without soil.

6.3.3   *Indirect test methods*

6.3.3.1   *Oedometer and triaxial consolidation test*
For "very low" to "semi impermeable" clays, it is often more accurate to estimate the permeability from the results of oedometer consolidation tests, rather than to carry out falling head tests. From the theory of consolidation:

$$k = \gamma_w c_v m_v \qquad (6.46)$$

where γ_w = unit weight of water; c_v = coefficient of consolidation; m_v = coefficient of volume change.

In any oedometer tests c_v and m_v commonly depend on the pressure applied to the soil, so the permeability k also varies depending on the applied pressure. If a Rowe consolidation cell is used with radial drainage (rather than vertical), the horizontal coefficient of consolidation and hence horizontal permeability will be calculated. Details are given in Head (1985).

Similarly, the consolidation phase of a triaxial test can be used to estimate m_v and c_v and hence the permeability. If radial drainage is promoted by surrounding the sample with filter paper, the horizontal coefficients of consolidation, volume change and permeability will be determined.

An advantage of using triaxial testing is that saturation of the sample can be assured by back pressure saturation. However the permeability of the filter paper drains may control the flow from the sample (Germaine and Ladd, 1988), so care must be taken in using them.

6.3.3.2   *Estimation of permeability of sands from particle size distribution*

An approximate estimate of the permeability of sands and sandy gravels can be obtained from the particle size distribution of the soil. The most commonly used of the available formulae are:

(i) *Hazen's formula.* The most commonly quoted form of the equation is

$$k = C(D_{10})^2 \tag{6.47}$$

where C = factor, usually taken as = 0.01; D_{10} = effective grain size in mm; k = permeability in m/sec.

It should be noted that Hazen's formula was developed for clean sands (less than 5% passing 0.075 mm) with D_{10} sizes between 0.1 mm and 3.0 mm. Even within these constraints C is quoted to vary from 0.004 to 0.015 (Holtz and Kovacs, 1981 and Head, 1985). Lambe and Whitman (1979) show values of C varying from 0.0001 and 0.0042 with an average of 0.0016 for a range of soils from silt to coarse sand. Sherard et al. (1984) suggest C = 0.0035 for graded filter materials.

It is recommended that if the formula is used, an appropriate degree of conservatism be applied to the value of C selected. Depending on the particular circumstances, conservatism may involve selection of a low or high value of C. Hazen's formula is NOT applicable to clays, or to coarse gravels.

(ii) *Kozeny-Carman formula.* Head (1985) discusses the application of the Kozeny-Carman formula and suggests the form

$$k = \frac{2}{fS^2}\left(\frac{e^3}{1+e}\right) \text{ m/sec} \tag{6.48}$$

where S = specific surface = $\dfrac{6}{\sqrt{d_1/d_2}}$ $\tag{6.49}$

and d_1, d_2 = maximum and minimum sized particles in mm; e = void ratio; f = angularity factor = 1.1 for rounded grains; for 1.25 for subrounded grains; 1.4 for angular grains.

This equation attempts to account for the effect of particle size, void ratio and particle shape or permeability.

It is unlikely that the Kozeny-Carman formula is significantly more accurate than the Hazen formula and its use should be limited to initial estimates of permeability of sand in non critical projects.

6.3.4   Effects of poor sampling on estimated permeability in the laboratory

Many alluvial soils are stratified in-situ, e.g. a sandy gravel/gravelly sand deposit in a river bed will, in fact, consist of separate beds of fine sand, coarse sand, and gravel. When sampled in boreholes or test pits, particularly below the water table, there is often a tendency to mix the strata, in the drilling process and/or in the sampling. If these mixed samples are used for laboratory permeability tests, or their permeability is inferred from their particle size distribution, it is likely that the permeability for flow along the strata, i.e. essentially horizontal permeability, will be significantly underestimated.

Figure 6.56 gives an example where soils A, B and C are present as distinct layers. Their approximate individual permeability estimated using Hazen's formula are as shown for horizontal flow. Soil C (gravel) dominates the conditions, and the equivalent horizontal permeability is of the order of 250×10^{-4} m/sec.

If in the drilling and sampling process the soils are mixed in proportion to their thicknesses, the combined soil particle size distribution would be as shown. For the combined soil the permeability is controlled by the finer particles, resulting in an estimated permeability of 1.5×10^{-4} m/sec, or only 0.6% of the actual field horizontal permeability. This

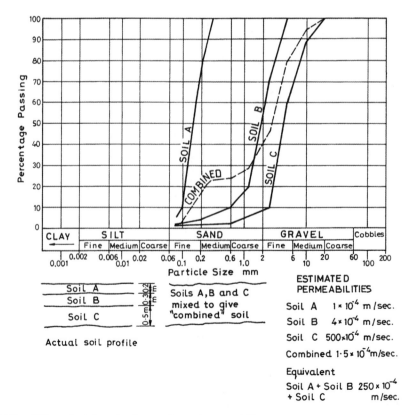

Figure 6.56.   Effect of mixing of soil on estimated permeability.

can only be overcome by very careful drilling and sampling, or preferably by in-situ permeability testing.

6.3.5   In-situ *testing methods*

In-situ permeability testing in soil is described in Section 5.13. In general pump-out in-situ tests, are the preferred method of estimating permeability, because the effects of the soil structure and fabric are properly accounted for. Pump-in in-situ tests are potentially inaccurate as described in Section 5.13.

7

Clay mineralogy, soil properties, and dispersive soils

7.1 INTRODUCTION

The properties of soils are determined by the properties of the constituent soil particles, the nature and quantity of water in the soil, the past consolidation history of the soil and soil structure. In this chapter the influence of the mineralogy of clays present in fine grained soils is discussed, with a particular emphasis on dispersive soils, and their use in the construction of embankment dams.

Dispersive soils are those which by the nature of their mineralogy, and the chemistry of the water in the soil, are susceptible to separation of the individual clay particles and subsequent erosion of these very small particles through even fine fissures or cracks in the soil under seepage flows.

This is distinct from erodible soils, such as silt and sand, which erode by physical action of the water flowing through or over the soil.

For many years it has been recognised (e.g. Aitchison, Ingles and Wood, 1963; Aitchison and Wood, 1965) that the presence of dispersive soils either in the soil used to construct a dam, or in the dam foundation, greatly increases the risk of failure of the dam by "piping failure", i.e. development of erosion to the extent that a hole develops through the embankment, with rapid loss of water from the storage. Figure 7.1 shows such a case.

An understanding of the basic concepts of clay mineralogy is essential to the understanding of identification and treatment of dispersive soils in dam engineering. The

Figure 7.1.   Examples of piping failure of a dam due to the presence of dispersive soils (Soil Conservation Service of NSW).

following gives an outline of the subject. For more details the reader is directed to Mitchell (1976, 1993) and Grim (1968). Holtz and Kovacs (1981) give a useful summary of the topic.

7.2   CLAY MINERALS AND THEIR STRUCTURE

7.2.1   *Clay minerals*

The basic "building blocks" of clay minerals are silica tetrahedra and aluminium (Al) or magnesium (Mg) octohedra. These give sheet like structures as shown in Figure 7.2. The alumina octohedra are known as gibbsite, the magnesium octohedra are brucite.

These in turn combine to give the clay minerals. Figure 7.3 shows the structure of montmorillonite and kaolin.

Some silicate clay minerals do not have a crystalline structure, even though they are fine grained and display claylike engineering properties. These are known as allophane and are present in most soils. Mitchell (1976) indicates they are particularly common in some soils formed from volcanic ash because of the abundance of "glass" particles.

Oxides also occur widely in soils and weathered rock as fine-grained particles which exhibit claylike properties.

Examples are:

– Gibbsite, boehmite, haematite and magnetite (oxides of Al, Fe and Si which occur as gels, precipitates or cementing agents);

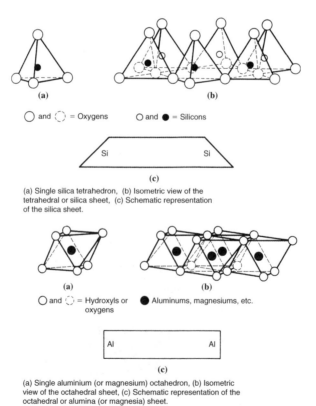

○ and ◌ = Oxygens   ○ and ● = Silicons

(a) Single silica tetrahedron, (b) Isometric view of the tetrahedral or silica sheet, (c) Schematic representation of the silica sheet.

○ and ◌ = Hydroxyls or oxygens   ● Aluminums, magnesiums, etc.

(a) Single aluminium (or magnesium) octahedron, (b) Isometric view of the octahedral sheet, (c) Schematic representation of the octahedral or alumina (or magnesia) sheet.

Figure 7.2.   Silica tetrahedra, and aluminium and magnesium octohedra (Holtz and Kovacs, 1981, reproduced with permission of Pearson Education).

– Limonite: amorphous iron hydroxide;
– Bauxite: amorphous aluminium hydroxide.

They are common in tropical residual soils and soils derived from volcanic ash.

7.2.2   Bonding of clay minerals

The properties of clays, particularly in a soil-water environment, are determined by the nature of the atomic bonding between the atoms which make up the sheets of clay mineral, and the bonding between the sheets of clay mineral. The bonding mechanisms which are present includes primary bonds and secondary bonds.

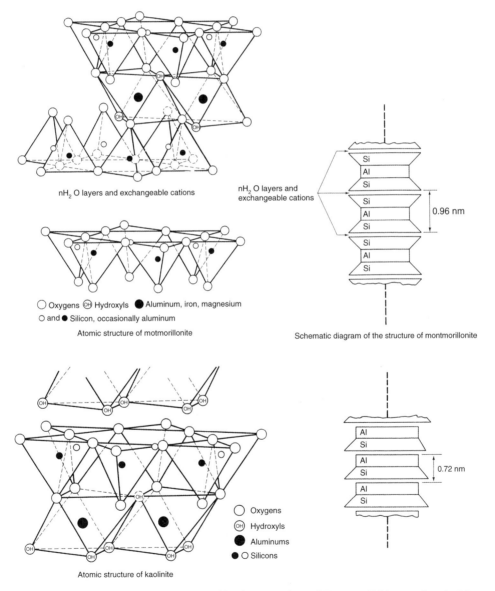

Figure 7.3.   Structure of montmorillonite and kaolinite (Holtz and Kovacs, 1981, reproduced with permission of Pearson Education).

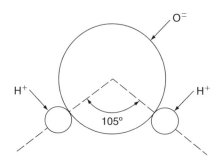

Figure 7.4.   Schematic diagram of a water molecule showing its dipole nature (Holtz and Kovacs, 1981, reproduced with permission of Pearson Education).

7.2.2.1   *Primary bonds*
These are either:

(i) *Covalent bonds*, where two atomic nuclei share one or more electrons, e.g. H• + •H = H:H, or the hydrogen molecule;

(ii) *Ionic bonds*, which result from the electrostatic attraction between positive and negative ions (ions are free atoms which have gained or lost an electron). Positive charged ions are known as cations, negatively charged as anions;

When an ionic bond forms, the centres of negative and positive charge are separated and form a dipole. The dipole then has a positive and negatively charged "end", even though it is neutral overall;

$$Na^+ + Cl^- = NaCl \tag{7.1}$$

Water is an example of a molecule which forms a dipole as shown in Figure 7.4. This dipolar nature has an important effect on the behaviour of water in clays;

Mitchell (1976) indicates that purely ionic or covalent bonds are not common in soils, and a combination is more likely. Silica (SiO_2) is partly ionic, partly covalent;

(iii) *Metallic bonds*, which are of little importance in the study of soils.

7.2.2.2   *Secondary bonds*

(i) *Hydrogen bonds*. Attraction will occur between the oppositely charged ends of permanent dipoles. When hydrogen is the positive end of the dipole, the resulting bond is known as hydrogen bonding;

(ii) *Van der Waal's bonds*. As the electrons rotate around the nucleus of an atom there will be times when there are more electrons on one side of the atom than the other, giving rise to a weak instantaneous dipole. This will not be permanent in one direction. The attraction of the positive and negatively charged ends of the fluctuating dipole causes a weak bond known as Van der Waal's bond. These bonds are additive between atoms, and although weak compared to hydrogen bonds, they decrease less with distance from the nucleus, than the latter.

7.2.3   *Bonding between layers of clay minerals*

Bonding between layers of clay minerals occurs in five ways:

- *Van der Waal's forces* – these occur in most clay minerals;
- *hydrogen bonding*, e.g. between the positive end of the OH at the base of the Al octahedra and the negative end of the O at the top of the silica tetrahedra in kaolinite (Figure 7.3);

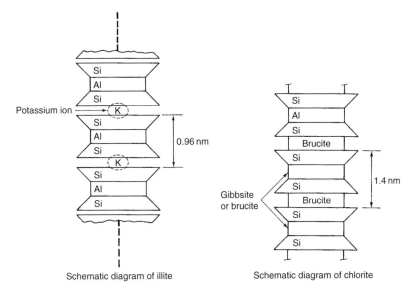

Figure 7.5.   Schematic diagram of the structure of illite and chlorite molecule (Holtz and Kovacs, 1981, reproduced with permission of Pearson Education).

Table 7.1.   Schematic structures of the common clay minerals. (Adapted from Lambe and Whitman 1981).

MINERAL	STRUCTURE SYMBOL	MINERAL	STRUCTURE SYMBOL
SERPENTINE		MUSCOVITE	
KAOLINITE		VERMICULITE	
HALLOYSITE (4H$_2$O)		ILLITE	
HALLOYSITE (2H$_2$O)		MONTMORILLONITE	
TALC		NONTRONITE	
PYROPHYLLITE		CHLORITE	

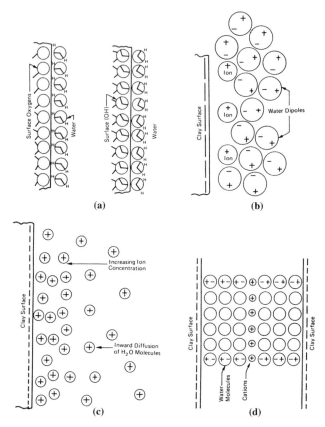

Figure 7.6.   Possible mechanisms of water adsorption by clay surfaces: (a) hydrogen bonding; (b) ion hydration; (c) attraction by osmosis; (d) dipole attraction (Mitchell, 1976).

- *hydrogen bonding with polar water* – as shown in Figure 7.6;
- *exchangeable cation bonding* – cations, eg. Ca^{++}, Na^+ act to bond between the negatively charged surfaces of such clay minerals as montmorillonite. The negative charge is brought about by substitution of cations within the sheet structure e.g. Mg^{++} for Al^{+++}. This is known as isomorphous substitution and is permanent. The interlayer cations are not permanent and may be substituted by other cations;
- *interlayer cation bonding* – cations such as K^+, Mg^{++}, Fe^{++} which fit in the space between layers giving a strong bond. Examples are micas and chlorites. These cations are not affected by the presence of water. Figure 7.5 shows the structure of illite and chlorite.

Table 7.1 lists the predominant bonding for the common clay minerals.

7.3   INTERACTION BETWEEN WATER AND CLAY MINERALS

7.3.1   *Adsorbed water*

There is much evidence that water is attracted to clay minerals, e.g.:

- Dry soils take up water from the atmosphere, even at low relative humidity;

- Soils wet up and swell when given access to water;
- Temperatures above 100°C are required to drive off all water from soils.

Much of the water in soil is adsorbed i.e. the electrical charge of the water dipoles and the electrical charge associated with the surface of the clay minerals attract each other.
  There are several explanations for this as shown in Figure 7.6:

- *hydrogen bonding,* where the positive H side of the water dipole is attracted to surface oxygens, or the negative O side of the water dipole is attracted to surface hydroxyls;
- *ion hydration,* where water dipoles are attracted with cations, which are in turn attracted by the negative charge of the surface of the clay mineral;
- *osmosis*: as discussed below, there is a greater concentration of cations close to the surface of the clay than further away. Water molecules tend to diffuse towards the surface in an attempt to equalise concentrations;
- *dipole attraction* – water dipoles orienting themselves as shown in Figure 7.6(d), with a central cation to equalise charges.

Some clay minerals, particularly those of the smectite, or montmorillonite group, have undergone significant isomorphous substitution giving a large net negative charge on the surface. They also tend to have particles which are very thin giving a large exposed surface area. As a result, these clays tend to attract water more than those which do not have a large negative surface charge and are thicker.
  The surface area of clay minerals is measured by the specific surface, where

$$\text{specific surface} = \frac{\text{surface area}}{\text{volume}} \qquad (7.2)$$

$$\text{or} \qquad = \frac{\text{surface area}}{\text{mass}} \qquad (7.3)$$

Figure 7.7 shows the typical dimensions and specific surface for four common clay minerals.

7.3.2 *Cation exchange*

As outlined above, one of the bonding mechanisms present in the montmorillonite (or smectite) group of clay minerals is exchangeable cation bonding. The cations, which are held in the space between sheets of clay minerals, are able to be exchanged for other cations, and, in the process, the properties of the clay can be altered. This process is known as cation exchange.
  This source of cation exchange is the most important one in clay minerals except for kaolinite. Other sources are:

- Broken bonds: electrostatic charges on the edges of particles and particle surfaces. This is the major cause of exchange in kaolinite;
- Replacement of the hydrogen of an exposed hydroxyl.

The most commonly present cations are Ca^{++}, Mg^{++}, Na^+ and K^+. The ease with which the cations replace each other is dependent on:

- Valence – divalent ($^{++}$) substitute readily for univalent ($^+$), e.g. Ca^{++} for Na^+;
- Relative abundance or concentration of the ion type;
- Ion size, or charge density – smaller high charge cations will displace large, low charge cations.

Edge view	Typical thickness (nm)	Typical diameter (nm)	Specific surface (km²/kg)
Montmorillonite	3	100–1000	0.8
Illite	30	10,000	0.08
Chlorite	30	10,000	0.08
kaolinite	50–2000	300–4000	0.015

Figure 7.7.   Average values of the relative sizes, thicknesses and specific surfaces of the common clay minerals (Holtz and Kovacs, 1981, reproduced with permission of Pearson Education).

A typical replacement series is:

$$Na^+ < Li^+ < K^+ < Mg^{++} < Ca^{++} < Al^{+++} < Fe^{+++} \qquad (7.4)$$

However, Ca^{++} can be replaced by Na^+ if the concentration of Na^+ is sufficiently high. The rate at which cation exchange occurs depends on:

- Clay type – for example, exchange will be almost instantaneous in kaolinite, a few hours in illite, and longer in montmorillonite;
- Concentration of cations in the soil water;
- Temperature.

The quantity of exchangeable cations required to balance the charge deficiency of a clay is known as the cation exchange capacity (CEC) and is usually expressed as milliequivalents per 100 grams of dry clay.

7.3.3   Formation of diffuse double layer

In a dry soil, the exchangeable cations are held tightly to the negatively charged clay surface. Excess cations and their associated anions precipitate as salts.

When placed in water, the salts go into solution and because of the high concentration of adsorbed cations near the clay surface these diffuse from the clay surface as they repel each other. However, this is counteracted by the negative charge on the clay surface. The anions in solution are repelled by the negative charge surface of the clay (but attracted to the cations). The overall effect is to result in a distribution of ions shown in Figure 7.8.

The negative surface and the distributed ions adjacent are together known as the diffuse double layer. Details are discussed in Mitchell (1976, 1993).

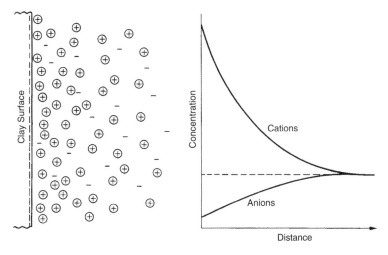

Figure 7.8.   Distribution of ions adjacent to a clay surface according to the concept of the diffuse double layer (Mitchell, 1976).

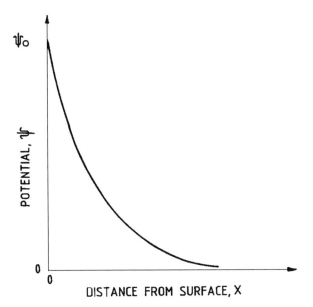

Figure 7.9.   Variation in electrical potential with distance from the clay surface (Adapted from Mitchell, 1976).

The resulting electrical potential varies with distance from the clay surface as shown in Figure 7.9. The potential is negative, reflecting the large negative charge on the clay surface.

7.3.4   Mechanism of dispersion

When two clay particles come near each other, the potential fields overlap, leading to repulsion if the particles are close enough. These repulsive forces are counteracted by Van der Waal's attractive forces as shown in Figure 7.10. If the repulsive forces are greater

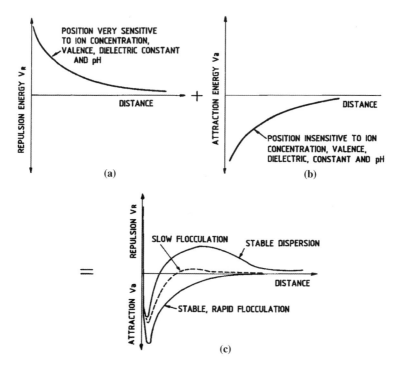

Figure 7.10. Interaction of (a) repulsive and (b) Van der Waals attractive forces to give (c) curves of net energy of repulsion or attraction (Adapted from Mitchell, 1976).

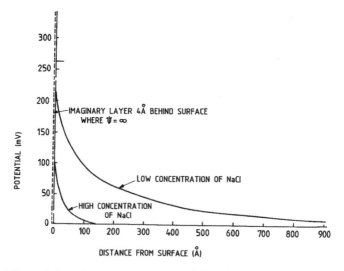

Figure 7.11. Effect of electrolyte concentration on diffuse double layer potential for montmorillonite (Adapted from Mitchell, 1976).

than the Van der Waal's forces the soil will disperse. In cases where the repulsive forces are small, the Van der Waals attractive forces dominate and flocculation results.

The repulsive forces in the diffuse double layer are affected by several factors:

(a) *Electrolyte concentration*: As shown in Figure 7.11, a high concentration of dissolved salt in the soil water leads to a smaller diffuse double layer (as the greater concentration of cations (Na^+) more readily overcomes the negative charge on the clay surface). Hence the repulsive forces are lower;

(b) *Cation valence*: Exchange of Na^+ cations by Ca^{++} cations leads to a smaller, higher charge density diffuse double layer and hence lower repulsive forces.

Other factors which affect the diffuse double layer include:

– dielectric constant of the electrolyte;
– temperature.

More details are given in Mitchell (1976, 1993).

7.4   IDENTIFICATION OF CLAY MINERALS

When required, the identification of clay mineral(s) present in a soil is usually carried out using at least two of the following techniques. It should be noted that most soils have clay sized particles mixed with silt and sand sized particles, the latter usually being quartz or other rock mineral. The results of the clay mineralogy must therefore be considered allowing for that proportion of the soil they represent in total.

7.4.1   *X-ray diffraction*

A sample of the powdered soil is subject to x-rays, which are diffracted depending on the mineral crystals present. Each clay mineral has a characteristic pattern which is known. Mitchell (1976, 1993) discusses the technique.

7.4.2   *Differential thermal analysis (DTA)*

A sample of the soil, and an inert substance are heated at a controlled rate (10°C/minute) up to 1000°C, and the differences in temperature in the soil and the inert substance are recorded.

Endothermic reactions (ie. heat being absorbed) are usually related to the driving off of:

– Adsorbed water;
– Water of hydration.

Exothermic reactions are usually related to recrystallisation and oxidation. Each clay mineral has a characteristic thermogram allowing its identification. Mitchell (1976, 1993) indicates that the method can be calibrated for quantitative analysis, accurate to the determined value plus or minus 5%.

7.4.3   *Electron microscopy*

Scanning electron microscopy can be used to assist in identification of clay minerals. Magnifications from X20 to X150,000 can be used and the micrographs compared with those for known clay minerals. The main clay minerals have distinctly different micrographs.

In practice the presence of several clay minerals, and silt and sand complicates identification but, coupled with x-ray diffraction or differential thermal analysis, the method can be useful.

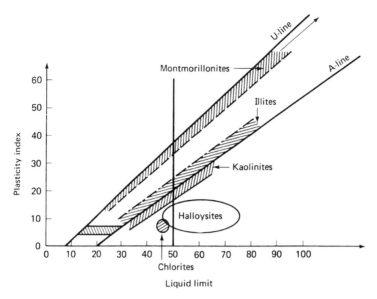

Figure 7.12.   Location of common clay minerals on Casagrande plasticity chart (Holtz and Kovacs, 1981, reproduced with permission of Pearson Education).

The above methods tend to be qualitative rather than quantitative. Mitchell (1976) indicates that an quantitative analysis can be made using DTA, glycol adsorption, cation exchange capacity and K_2O content. A semi quantitative analysis can be done by examining the silt and sand fractions under a microscope to ascertain the proportion of non clay minerals. The clay size % (finer than 0.002 mm) can be obtained by grain size analysis, and the proportion of the clay minerals in the clay size obtained from x-ray diffraction. The authors' experience is that few organisations are willing to quantify the amount of different clay minerals present.

7.4.4   Atterberg limits

Apart from the above methods a rough idea of which clay minerals are present can be obtained from the Atterberg limits, see Section 7.4.4.

As shown in Figure 7.12 the position of the soil on the Casagrande plasticity chart can give an indication of which minerals are present. It should be remembered that most soils contain several clay minerals, so the Atterberg limits may not fall exactly in the shaded zones.

7.4.5   The activity of the soil

$$\text{Activity} = \frac{\text{plasticity index}}{\text{clay fraction}} \tag{7.5}$$

where clay fraction = % finer than 0.002 mm. Table 7.2 shows typical values. Soils with high activity are likely to contain montmorillonite and illite, while those with lower activities are likely to contain kaolinite. However the plasticity index is carried out on that part of the soil which includes clay, silt and fine sand, so its value relates also to the amount of clay present.

Table 7.2.   Activities of various minerals (Holtz and Kovacs, 1981).

Mineral	Activity
Na-montmorillonite	4–7
Ca-montmorillonite	1.5
Illite	0.5–1.3
Kaolinite	0.3–0.5
Halloysite (dehydrated)	0.5
Halloysite (hydrated)	0.1
Attapulgite	0.5–1.2
Allophane	0.5–1.2
Mica (muscovite)	0.2
Calcite	0.2
Quartz	0

Note: Activity = (plasticity index)/(clay fraction).

7.5   ENGINEERING PROPERTIES OF CLAY SOILS RELATED TO THE TYPES OF CLAY MINERALS PRESENT

The engineering properties of clay soils depend on compositional factors (Mitchell 1976, 1993):

- type of clay minerals present;
- amount of each mineral;
- type of adsorbed cations (and anions);
- organic content;
- shape and size distribution of particles;
- pore water composition.

and on environmental factors:

- water content;
- density;
- confining pressure;
- fabric;
- availability of water;
- temperature.

Virtually all clay soils in nature are mixtures of clay and silt size particles (and sometimes sand), not just clay size particles. The silt and sand size particles are usually rounded or sub-rounded and are derived from the parent rock. The most abundant mineral present is usually quartz, followed by feldspar and mica.

7.5.1   *Dispersivity*

Soils in which the clay particles will detach from each other and from the soil structure without a flow of water, and go into suspension are termed dispersive clays.

The dispersivity of a soil is directly related to its clay mineralogy. In particular soils with a high exchangeable sodium percentage such as montmorillonite present, tend to be dispersive, while kaolinite and related minerals (halloysite) are non dispersive. Soils with illite present tend to be moderately dispersive.

Table 7.3.   Swelling index values for several minerals (Olson and Mesri, 1970).

Mineral	Pore fluid, adsorbed cations electrolyte concentration, in gram equivalent weights per litre	Void ratio at effective consolidation pressure of 100 psf	Swelling index
Kaolinite	Water, sodium, 1	0.95	0.08
	Water, sodium, 1×10^{-4}	1.05	0.08
	Water, calcium, 1	0.94	0.07
	Water, calcium, 1×10^{-4}	0.98	0.07
	Ethyl alcohol	1.10	0.06
	Carbon tetrachloride	1.10	0.05
	Dry air	1.36	0.04
Illite	Water, sodium, 1	1.77	0.37
	Water, sodium, 1×10^{-3}	2.50	0.65
	Water, calcium, 1	1.51	0.28
	Water, calcium, 1×10^{-3}	1.59	0.31
	Ethyl alcohol	1.48	0.19
	Carbon tetrachloride	1.14	0.04
	Dry air	1.46	0.04
Smectite (montmorillonite)	Water, sodium, 1×10^{-1}	5.40	1.53
	Water, sodium, 5×10^{-4}	11.15	3.60
	Water, calcium, 1	1.84	0.26
	Water, calcium, 1×10^{-3}	2.18	0.34
	Ethyl alcohol	1.49	0.10
	Carbon tetrachloride	1.21	0.03
Muscovite	Water	2.19	0.42
	Carbon tetrachloride	1.98	0.35
	Dry air	2.29	0.41
Sand			0.01–0.03

The dispersivity depends also on the pore water chemistry since, as discussed above, this affects the diffuse double layer geometry and electrical charge. In particular low electrolyte (pore water) salt concentrations lead to a large diffuse double layer and greater dispersivity. Hence percolation of a saline soil with fresh water can lead to dispersion.

Cation exchange of say Ca^{++} for Na^{+} leads to a smaller diffuse double layer, and lower dispersivity. Hence addition of lime (CaO or $Ca(OH)_2$), or gypsum ($CaSO_4$) leads to cation exchange and reduced dispersivity.

There have been many failures of dams constructed of dispersive clays. Mitchell (1993) indicates these have been mostly low to medium plasticity (CL and CL-CH) clays that contained montmorillonite. Bell and Maud (1994) concur with this view. The authors' experience is similar, that is the most susceptible soils are not those with a high plasticity, but those with limited clay size fractions sufficient only to give low to medium plasticity.

Sherard et al. (1976a, b) however tested some soils as dispersive in the pinhole test with % passing 0.005 mm greater than 50%. The better performance of the higher plasticity clays probably relates to greater resistance to erosion and greater likelihood for cracks to close as the soil swells.

Sherard et al. (1976a, b) indicate that based on their tests, soils with less than 10% finer than 0.005 mm may not have enough clay to support dispersive piping.

7.5.2   Shrink and swell characteristics

The shrink and swell characteristics of a soil can be related to clay mineralogy. Table 7.3 shows swelling index ($de/d(\log p')$, where e = void ratio, p' = effective stress on the soil)

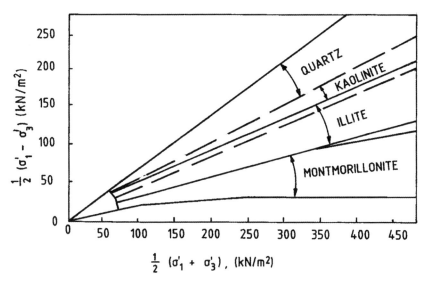

Figure 7.13.   Ranges in effective stress failure envelopes for pure clay minerals and quartz (Olson, 1974).

values for several clay minerals, with varying electrolyte concentrations and for different cations in the electrolyte.

It can be seen that:

– At the same pore fluid adsorbed cations concentration, montmorillonite swells to a greater void ratio and has a higher swelling index than illite, which in turn swells more than kaolinite;
– Exchange of Na^+ by Ca^{++} in montmorillonite (smectite) significantly reduces swell potential, but has only a minor effect on kaolinite;
– Increase of concentration of Na^+ causes a marked decrease in the swelling index of montmorillonite, but no change in the swelling index of kaolinite.

These phenomena are to be expected given the large surface area, and higher negative charge on the surface of montmorillonite, compared to illite and kaolinite.

7.5.3   Shear strength

The shear strength, parameters c', ϕ' (peak strength) and C'_R, ϕ'_R (residual strength) are dependent on the clay fraction percentage as shown in Figures 6.4 and 6.5. Clays with a high clay fraction percentage exhibit lower shear strengths and a greater reduction from peak to residual strength. As shown in Figure 6.5 and Figure 7.13 the shear strength is also dependent on the types of clay minerals present, with lowest strength for pure montmorillonites, highest for kaolinite. However for normal soils, the behaviour is also influenced by the mixture of clay minerals present, and the presence of silt and clay sized particles.

7.6   IDENTIFICATION OF DISPERSIVE SOILS

7.6.1   Laboratory tests

There are several laboratory tests which can be used to determine the dispersivity of a soil.

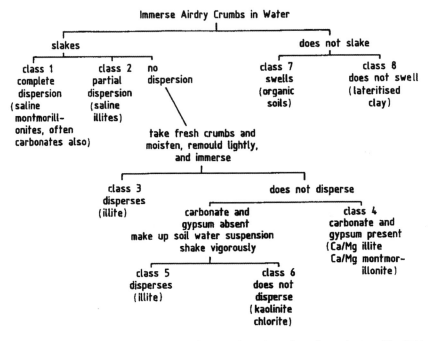

Figure 7.14.   Determination of the Emerson class number of a soil (Ingles and Metcalf, 1972).

7.6.1.1   *Emerson class number*

In this method, the soil is sieved through a 4.75 mm sieve and collected on a 2.36 mm sieve and tested as outlined in Figure 7.14. The test procedure is detailed in Standards Australia (1997), test AS 1289, 3.81 and USBR (1979).

The test is carried out in distilled water, but may be repeated in water from the dam, or groundwater. This often gives significantly different results due to the presence of dissolved salts in the water (higher salt content gives less dispersive results). The soils are graded according to class, with Class 1 being the highly dispersive, Class 8 non dispersive. Soils with Emerson Class 1 to 4 need to be treated with extra caution in dam construction.

Sherard et al. (1976b) indicate that if a soil tests as dispersive in the crumb test, it will also test as dispersive in the pinhole test, but 40% of soils testing as dispersive in the pinhole test, test as non dispersive in the crumb test (note that Sherard et al., 1976b, use different classes to the Australian standard).

The Emerson test allows identification of dispersive soils, but does not provide a measure of their erodibility. It is inexpensive, and a useful first check on dispersivity.

7.6.1.2   *Soil conservation service test*

Soil Conservation Service test, also known as the double hydrometer test, or percent dispersion test (Standards Australia, 1997, test AS1289, 3.8.2; and ASTM 2001 test D42291-99).

This involves two hydrometer tests on soil sieved through a 2.36 mm sieve. The hydrometer tests are carried out with dispersant and without. The percent dispersion is:

$$\frac{P}{Q} \times 100 \qquad (7.6)$$

where P = percentage of soil finer than 0.005 mm for the test without dispersant and Q = percentage of soil finer than 0.005 mm for the test with dispersant.

Sherard et al. (1976a) indicate that soils with a percent dispersion greater than 50% are susceptible to dispersion and piping failure in dams, and those with a percent dispersion less than 15% are not susceptible. A percent dispersion of less than 30% is unlikely to test dispersive in the pinhole test. Bell and Maud (1994) indicate that soils with >50% dispersion are regarded as highly dispersive, 30–50% moderately dispersive, 30% to 15% slightly dispersive, and <15%, non dispersive.

7.6.1.3 Pinhole dispersion classification

Pinhole dispersion classification, also known as the pinhole test, or Sherard pinhole test (Standards Australia 1997, test AS1289, 3.8.3 and ASTM 1998, test D4647-93).

This test was developed by Sherard et al. (1976b). A 1.0 mm diameter hole is preformed in soil to be tested, and water passed through the hole under varying heads and for varying durations. The soil is sieved through a 2.36 mm sieve and compacted at approximately the plastic limit to a density ratio of 95% (to simulate conditions in a dam embankment with a crack or hole in the soil).

Soils which tested as D1 and D2 were found by Sherard et al. (1976a) to have suffered piping failure in earth dams, and severe erosion damage by rainfall in embankments and natural deposits while those with ND1 and ND2 classification had not.

As for the Emerson class number, the results are dependent on the chemistry of the water used for the test (the standard test uses distilled water). The method is relatively simple with moderate cost, and has the advantage that "it identifies soil erodibility directly, rather than indirectly". All dispersive soils are erodible to some degree – usually highly – but erodibility depends on other factors than the clay fraction. Craft and Acciardi (1984) describe some modifications to the Sherard et al. (1976b) method. These are incorporated in Standards Australia (1997) and ASTM (1998).

Gerber and Harmse (1987) found that some dispersive soils with free salts in solution in the pore water were not identified as dispersive by the pinhole, SCS or crumb tests.

7.6.1.4 Chemical tests

Based on correlation with many dam failures and soil from dams which have leaked continuously (without any filters to control erosion) and not failed (Sherard et al., 1976a) proposed Figure 7.15 to determine the dispersivity of soil.

In this figure,

$$\text{percent sodium} = \frac{Na^+}{\text{Total dissolved salts}} \times 100 \tag{7.7}$$

$$\text{or} \qquad = \frac{Na^+}{Ca^{++} + Mg^{++} + Na^+ + K^+} \times 100 \tag{7.7a}$$

$$\text{Sodium absorption ratio (SAR)} = \frac{Na^+}{\left[\frac{1}{2} (Ca^{++} + Mg^{++}) \right]^{1/2}} \tag{7.8}$$

in which Na^+, Ca^{++}, and Mg^{++} are measured in milliequivalents per litre of saturation extract.

Sherard et al. (1976a) indicate that the case histories and pinhole tests confirm that the main factor governing dispersibility is the relative content of pore-water sodium, and

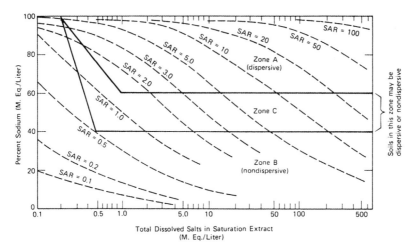

Figure 7.15. Relationship between dispersivity and dissolved pore water salts content (Sherard et al., 1976a, reproduced with permission of ASCE).

claimed a general relationship, valid for most soils, has now been established with considerable confidence as shown in Figure 7.15. They indicate that in this figure:

Zone A – much experience shows that damaged and failed dams all over the world have been constructed of these dispersive soils. Almost all soils are dispersive in the pinhole test.

Zone B – the great majority of these soils are nondispersive. These are the soils generally considered "ordinary erosion resistant clays", but include silts of low plasticity (ML), also nondispersive. A small percentage of exceptional soils in Zone B erode in the pinhole test in exactly the same fashion as soils of Zone A, and some of these can be identified only by the pinhole test.

Zone C – soils in this group may range from dispersive to nondispersive. This group contains a few soils which give intermediate reaction in pinhole tests, with apparently colloidal erosion but at a very slow rate compared with soils of Zone A.

These relationships are only valid for relatively pure eroding waters (such as TDS less than 0.5 meq/1 or about 300 ppm). This is because they are based on results of pinhole tests using distilled water and case histories of erosion in nature caused mainly by rainwater and relatively pure reservoir waters.

While Sherard et al. (1976a) found good agreement with this test and field behaviour, Bell and Maud (1994) indicate the method was unreliable in South African conditions. They quote Craft and Acciardi (1984) as concluding that the extraction of the soil water is difficult because of the low volumes used. They indicate it may be used where prior testing correlates well with the pinhole test.

Sherard et al. (1976a) did have small number of soils which tested as erodible in the pinhole test falling in Zone B. These soils were not from failed dams or areas showing erosion.

Another terms which are useful for assessing the likely dispersivity of a soil are:

$$\text{Exchangeable sodium percentage (ESP)} = \frac{\text{exchangeable sodium}}{\text{cation exchange capacity}} \times 100 \qquad (7.9)$$

$$\text{or} \qquad = \frac{Na^+}{Na^+ + K^+ + Ca^{++} + Mg^{++}} \times 100, \qquad (7.9a)$$

where the units are milliequivalents/100 grams of dry soil.

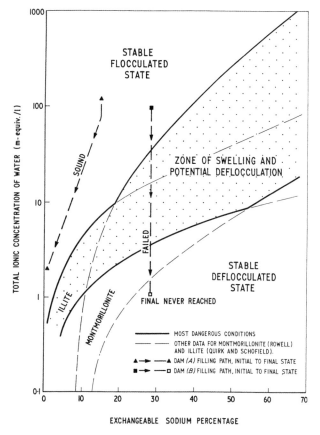

Figure 7.16.   Estimation of dispersivity from exchangeable sodium percentage of the pore water, and total ionic concentration of the seepage water (Ingles and Metcalf, 1972).

ICOLD (1990) indicate that soils with ESP of 10% or above, which are subject to having free salts leached by seepage of relatively pure water are classified as dispersive. They indicate that Australian research showed that soils are dispersive if SAR exceeded 2.

ESP and SAR are determined by chemical analysis of the soil pore water extracted by vacuum from saturated samples.

Ingles and Metcalf (1972) suggest that to assess dispersivity potential, the exchangeable sodium percentage and total ionic concentration of the water e.g. the water to be stored in the dam, should be determined. Then for montmorillonite and illite clays, Figure 7.16 is used to determine whether the soil is in a flocculated, or dispersed (deflocculated) state, and whether a change in pore water chemistry during filling of the reservoir can lead to dispersion.

In Figure 7.16 filling of the storage for Dam A maintains a stable flocculated state, but for Dam B, dispersion would occur. This can be readily related to the diffuse double layer concept, where a change in concentration of salts in the soil water can lead to a larger diffuse double layer, and a tendency to disperse (Dam B), but exchange of sodium cations by say calcium would retain a stable flocculated structure even though the total concentration was reduced (Dam A). McDonald, Stone and Ingles (1981) produced Figure 7.17 to include some work by Moriwaki and Mitchell (1977).

This indicated that the Ingles and Metcalf (1972) boundaries were reasonable except that dispersion could occur in montmorillonites at higher ESP than indicated by the earlier

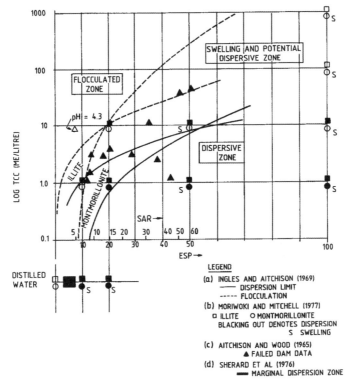

Figure 7.17. Estimation of dispersivity from exchangeable sodium percentage (McDonald, Stone and Ingles, 1981).

work. McDonald et al. (1981) suggested the use of the pinhole test for situations where water would be flowing (eg. a crack in a dam) and Emerson test for quiescent conditions, e.g. reservoir. They also point out that the ESP can vary a lot for soils in any dam and make use of cheaper tests such as Emerson and pinhole tests more attractive. It is also pointed out that where soils which are rich in bicarbonates are allowed to dry out, precipitation of relatively unsoluble $CaCO_3$ may occur, resulting in a lower Ca^{++} ion concentration and hence an increased potential for dispersion.

It should be noted that Sherard et al. (1976a) believe that such transformation of a stable, flocculated soil into a dispersive soil by the seepage water is unlikely, citing that most failures occur on first filling, pinhole tests eroding at the start of a test, or not at all (even after long periods of flow), and that there is no reason why the ESP should reduce (their experience being the contrary).

7.6.1.5 Recommended approach

Most authors consider that it is necessary to use more than one test to ascertain the dispersivity of a soil. Sherard and Decker (1977) suggest that four tests should be used: Soil Conservation Service, pinhole, Emerson and chemical test. They were of the opinion that the pinhole test was best. Moore et al. (1985) indicate that in their experience:

– similar results were obtained using the pinhole and Emerson tests;
– the pinhole test was not particularly sensitive to compaction water content;
– only approximate agreement was found between the SCS test and the pinhole and Emerson tests;

Figure 7.18.   Deep erosion gullying in dispersive soils (Soil Conservation Service of NSW).

– the Sherard et al. (1976a) chemical test shown in Figure 7.15 showed a soil to be dispersive when it was known not to be (by its field behaviour and other tests).

Bell and Maud (1994) suggested using the crumb test, SCS dispersion test, ESP/CED, SAR and PH, weighting the results, taking most account of the ESP/CEC, followed by the crumb and SCS dispersion test.

The authors have used the Emerson class number and pinhole tests: usually in a ratio of 2 or 3 crumb tests, to 1 pinhole test, reflecting the greater cost of the pinhole test. If one accepts that the objective is only to identify potential problems this would seem an adequate approach, but better supplemented by some SCS dispersion tests or chemical tests. The chemical approach in Figure 7.17 has attractions in that it allows a prediction of behaviour with altered water quality and can be related to clay mineralogy. However, the Emerson class and pinhole tests can also be related to changes in water quality by carrying out these tests in water other than distilled water.

It should be recognised that the tests identify the degree of dispersiveness of soils. They do not measure quantitatively the rate of erosion, or critical hydraulic shear stress at which erosion begins. Wan and Fell (2002, 2004) have developed the Hole Erosion Test and Slot Erosion Test which allow this quantification.

7.6.2   Field identification and other factors

While laboratory tests are a useful way of identifying dispersive soils, much can be determined by observing the behaviour of the soils in the field, e.g.:

– The presence of deep erosion gullies and piping failure in existing small farm dams usually indicates the presence of dispersive soils – see Figures 7.18 and 7.1;
– Erosion of road cuttings, tunnel erosion along gully lines and erosion of weathered or clay infilled rock joints may indicate potentially dispersive soils;

Table 7.4.   Clay mineral identification from the environment (Ingles and Metcalf, 1972).

Observation	Dominant clay component [1]
Turbid water of strong yellow-brown to red-brown colour	Montmorillonites, illite plus soil salinity
Clear waters	Calcium, magnesium or ironrich soil, highly acid soil sands
Clear waters with a bluish cast	Non-saline kaolins
Erosion gullies and/or field tunnelling in the natural soil	Saline clays, usually montmorillonites
As above, mild erosion	Kaolinites
Landslips	Kaolinites, chlorites, montmorillonites, illites
Surface microrelief (gilgai)	Montmorillonites
Country rock type granitic	Kaolinites, micas
Country rock type basaltic, poorly drained topography	Montmorillonites
Country rock type basaltic, well drained topography	Kaolinites
Country rock type sandstones	Kaolinites
Country rock type mudstones and shales	Montmorillonites or illite, often soil salinity
Country rock type limestone	Alkaline montmorillonites and of very variable properties
Country rock type recent pyroclastics	Allophanes

Note: [1] The dominant component of the soil may be silt or sand sized eg. for granites, sandstones.

– The presence of cloudy water in farm dams and puddles of water after rain indicates dispersive soils.

One can infer the clay mineralogy from such observational techniques. Some guidelines are given in Tables 7.4 and 7.5 reproduced from Ingles and Metcalf (1972).

The geology of the area can also be a guide to dispersivity.

Sherard and Decker (1977) indicate that:

– Many dispersive clays are of alluvial origin. (The authors' experience is similar but there are many non dispersive alluvial clays. Some slopewash clays are also dispersive);
– Some soils derived from shales and claystones laid down in a marine environment are also dispersive;
– Soils derived from weathering of igneous and metamorphic rocks are almost all non dispersive (but may be erodible, e.g. silty sand derived from grandiorite);
– Soils with a high organic content are unlikely to be dispersive (this needs to be treated with caution, since many "black cotton" soils are dispersive).

7.7   USE OF DISPERSIVE SOILS IN EMBANKMENT DAMS

7.7.1   *Problems with dispersive soils*

Dispersive soils are a major contributing factor to piping failure of embankment dams, particularly for small homogeneous dams constructed without filters and often with poor construction supervision. This view is widely held e.g. Sherard et al. (1976a, b, 1985). However, failures do occur in structures which are reasonably well engineered. The main contributory factors are:

– The presence of dispersive soils in the embankment or foundation;
– Poor compaction of the soil, i.e. to a low density ratio and/or dry of optimum water content;
– Poor compaction of soil around pipes or conduits which pass through the embankment;

Table 7.5.   Clay mineral identification from the soil profile (Ingles and Metcalf, 1972).

Inferences from the profile Observation	Dominant clay component
Mottled clays, red-orange-white mottle	Kaolinites
Mottled clays, yellow-orange-gray mottle	Montmorillonites
Medium to dark gray and black clays	Montmorillonites
Brown and red-brown clays	Appreciable illite, some montmorillonite
White and light gray clays	Kaolinites and bauxites
Discrete microparticles of high light reflectance (micas)	Micaceous soils
Discrete microcrystals, easily crushed	Gypsum-rich soils, or (rarer) zeolites
Soft nodules, acid-soluble, disseminated	Carbonates
Hard nodules, red-brown	Ironstones, laterite
Extensive cracking, wide, deep and closely spaced at 5 to 6 cm or less	Calcium-rich illites and montmorillonites
Up to intervals of 30 cm and more	Illites
Open-textured friable loamy soils with appreciable clay content	Usually associated with carbonate, allophane or kaoline, but never montmorillonite and seldom illite
Open-textured friable loamy soils with appreciable clay content, black	Organic soils, peats
Open-textured friable loamy soils of low clay content	Carbonate, silts and sands
Wormy appearance on exposed pre-existing weathered profile	Montmorillonites, plus soil salinity
Relatively thin, strongly bleached horizon near the soil surface (up to 60 cm from the top)	Above the bleach, fine silt: below the bleach, dispersive clay. Probably a seasonal perched water table at the bleach level

- The presence of soil structure, e.g. root holes, fissures or cracks, in the soil in the dam foundation, and with no adequate cutoff or erosion control measures;
- Erosion of dispersive or erodible embankment soils into open fractures in the rock in the sides of the cutoff trench;
- Poor cleanup of loose soil, grass etc from the cutoff foundation prior to placing earthfill;
- Cracking due to differential settlement or desiccation;
- Rapid filling of storages, giving insufficient time for cracks in the soil in the embankment to be sealed by the soil swelling or being wetted. The cracks may be due to desiccation during or after construction, differential settlement or hydraulic fracturing.

As indicated in ICOLD (1990) the majority of piping failure in dams constructed of dispersive soil occur on first filling, including cases when the reservoir has been raised after being at a given elevation for a period of time. The reasons for this include the presence of cracks and high permeability zones being present and "found out" on first filling, and the cracks not having time to close as the soil around swells on saturation.

Wan and Fell (2002, 2004) have also shown that the rate of erosion of saturated clay soils is lower than partially saturated.

Piping failure can also be caused by introduction of water with low ionic concentration (i.e. "fresh" water with a low salt content) into soils which were naturally saline, e.g. as shown in Figure 7.17. Ingles and Wood (1964) describe such a case. ICOLD (1990) indicates that virtually all failures occur in the embankment, not through soil in the foundation. However the authors have observed failures through the foundation in small farm dams. Examples of piping failure of dams constructed using dispersive soils are given in Aitchison, Ingles and Wood (1963), Aitchison and Wood (1965), Cole and Lewis (1960), Sherard and Decker (1977), Phillips (1977), Cole (1977), Rallings (1960), Wagener et al. (1981), Bell and Maud (1994).

The authors' experience has been that piping failures usually occur in farm dams which have not been compacted other than by tracking of a bulldozer, but that even apparently well engineered structures have failed.

Examination of these cases shows failure to be related to the presence of dispersive soils (Emerson Class 1 or 2) compacted poorly (density ratio as low as 90% of standard compaction, as much as 2% to 4% below optimum water content, around outlet pipes through the embankment. In one case failure on first filling occurred despite the downstream slope of the dam being very flat (15H:1V). Sherard (1985a) describes a near piping incident of Wister Dam, where the hydraulic gradient was 1 in 50, with piping occurring over 200 metres from upstream to downstream.

A contributory factor also has been use of rough compacted surfaces on the concrete surrounding the outlet pipe, precluding proper compaction of the soil. In another case, cracking of soil by drying during construction around the trench into which the outlet pipe was placed and concreted was a contributory factor (see Figure 13.9).

It should however be emphasised that soils other than dispersive soils are subject to piping. Foster et al. (2000a) record that 9 out of 51 piping failures through embankment of large dams were recorded in dispersive soils, but note that the % may have been greater with the dispersive soils not necessarily having been recorded. Wan and Fell (2002, 2004) show dispension class is only one of several factors which influence the erosion rate for a soil.

7.7.2 Construction with dispersive soils

As indicated in ICOLD (1990), safe dams can be built with dispersive clay provided certain precautions are taken. ICOLD (1990) indicates that the concern about the problems of dispersive soils and attention to precautions increase with dam size. While the authors understand the sentiment that failure of a large dam may be more important than a smaller one, they would argue that in reality it is easier to build in the necessary precautions to larger dams than smaller. In most cases, normal good practice for high consequence of failure dams will be all that is needed, e.g. well-designed filters, proper foundation preparation and good construction control. In small dams such measures may be regarded as uneconomic.

The following outlines the main precautions which should be adopted. These are based on ICOLD (1990) and the authors' own experience.

7.7.2.1 Provide properly designed and constructed filters

Sherard et al. (1984a, b) have proven conclusively that erosion of dispersive soils can be controlled by properly designed filters. If the guidelines outlined in Chapter 9 are followed, and filters are provided:

- Within the embankment to control erosion of the Zone 1 earthfill;
- On the foundation, if it is dispersive or erodible soil, either as a horizontal drain or as a filter layer between the foundation and rockfill;
- Around outlet pipes, as shown in Figure 13.10.

Then there should not be problems with piping failure.

As discussed in Chapter 9 if the soils are dispersive, more conservative criteria for selecting the grain size of the filter should be used, than if the soil is non erodible.

7.7.2.2 Proper compaction of the soil

Dispersive soils, particularly if being placed around outlet pipes, or at the contact between the earthfill and concrete structures, or if no filters are being provided (e.g. in a small low hazard dam), must be properly compacted or there is a high probability of piping failure.

ICOLD (1990) recommend compaction at a water content above optimum water content so as to avoid a flocculated soil structure, and to avoid brittleness which will promote formation of cracks. They suggest that a permeability of lower than 10^{-7} m/sec is required.

The authors' opinion is that a water content between optimum and optimum plus 2%, and a density ratio of greater than 98% (standard compaction) is desirable, and that the water content should not be below optimum -1%. If the soil was at optimum $+2\%$ or even say optimum $+3\%$, one might relax the density ratio requirement to 97%, in would probably be of low permeability and less likely to crack than if compacted at optimum -1%.

This will necessitate use of thin layers, particularly adjacent outlet pipes, where rollers may not be used in some cases. Supervision and testing need to be very thorough. Care must be taken to avoid drying of the surface of layers of earthfill, which could result in cracking, and a preferred path for piping failure to initiate.

7.7.2.3   Careful detailing of pipes or conduits through the embankment

Pipes through embankments should be avoided if possible since it is very difficult to ensure good compaction around the pipe, and differential settlement around the pipe can also lead to cracking of the soil. If pipes must be placed through an embankment then support the pipe on a concrete footing and use filters to surround the downstream end of the pipe as shown in Figures 13.16, 13.17 and 13.18.

It is considered that reliance only on good compaction with or without lime modification is a relatively high risk option and is not recommended.

7.7.2.4   Lime or gypsum modification of the soil

Most dispersive soils can be rendered non dispersive by addition of a small quantity of lime ($Ca(OH)_2$ or CaO) or gypsum ($CaSO_4$ or $CaSO_4(2H_2O)$).

This process is one of cation exchange, with the Ca^{++} ions exchangeable for Na^{++} ions. Laboratory tests can be carried out to determine the required amount of lime or gypsum, e.g. Sherard pinhole tests using soil with different amounts of lime or gypsum. Commonly one would require 2% to 3% lime, and would allow a margin of 1% or 2% above that indicated by the laboratory tests to allow for difficulty in mixing the lime with the soil. The lime should be mixed with the soil using a pulveriser. This breaks up the soil so that 80% to 90% of the particles are less than 25 mm diameter and facilitates good mixing of the lime.

7.7.2.5   Sealing of cracks in the abutment and cutoff trench

If the soil used for construction is dispersive, particular care must be taken to seal cracks in the cutoff foundation, and the sides of the cutoff trench so the soil will not erode into the cracks. Extensive use of slush concrete or shotcrete is likely. If the cutoff trench is in soil, it may be necessary to provide a filter on the downstream side as shown in Figure 10.8.

7.7.3   Turbidity of reservoir water

The presence of dispersive soil in the reservoir area, or in the catchment, can lead to turbidity of the water in the reservoirs. Grant et al. (1977) describe how turbidity in the water for Cardinia Creek Dam, Melbourne, was prevented by adding 6500 tonnes of gypsum to the water on first filling, at a concentration of 40 mg/litre of water. Again the process is one of Ca^{++} displacing the Na^+ in the soil and makes it less dispersive. Gypsum was added to water as it entered the storage. Cardinia Creek Dam is an off river storage reservoir.

McDonald, Stone and Ingles (1981) describe how turbidity in Ben Boyd Dam, Eden NSW, was controlled by adding 1% by weight of gypsum of the soil depth in the reservoir area (27 tonnes per hectare). Gypsum was mixed in to 150 mm depth by disc harrow and compacted with a sheep's foot roller and gave initially clear water, but flood rains washed soil from the catchment and led to turbidity. The storage water was subsequently flocculated and clarified by adding 40 mg/l of alum ($Al_2(SO_4)_3$) and 20 mg/litre NaOH.

8

Embankment dams, their zoning and selection

8.1   HISTORIC PERFORMANCE OF EMBANKMENT DAMS AND THE LESSONS TO BE LEARNED

Much of the history of development of embankment dams can be related to failures and accidents.

ICOLD (1974) define failures accidents and incidents as follows:

Accident: Three categories of accidents are listed by ICOLD:

Accident Type 1 (A1): An accident to a dam which has been in use for some time but which has been prevented from becoming a failure by immediate remedial measures, including possibly drawing down the water.

Accident Type 2 (A2): An accident to a dam which has been observed during the initial filling of the reservoir and which has been prevented from becoming a failure by immediate remedial measures, including possibly drawing down the water.

Accident Type 3 (A3): An accident to a dam during construction, i.e. settlement of foundations, slumping of side slopes etc., which have been noted before any water was impounded and where the essential remedial measures have been carried out, and the reservoir safety filled thereafter.

Failure: Collapse or movement of part of a dam or its foundation, so that the dam cannot retain water. In general, a failure results in the release of large quantities of water, imposing risks on the people or property downstream (ICOLD, 1995). Incidents to dams during construction are only considered as failures when large amounts of water were involuntarily released downstream.

Two categories of failures are listed by ICOLD (1974):

Failure Type 1 (F1): A major failure involving the complete abandonment of the dam.

Failure Type 2 (F2): A failure which at the time may have been severe, but yet has permitted the extent of damage to be successfully repaired, and the dam brought into use again.

Incident: Either a failure or accident, requiring major repair.

ICOLD (1974, 1995), carried out extensive surveys on dam incidents, and from this developed some overall statistics.

Foster, Fell and Spannagle (1998, 2000a, 2000b), used the ICOLD data, and information on the dams which had experienced incidents to develop historic performance data for dams constructed up to 1986.

They showed that about 1 in 25 large embankment dams constructed before 1950, and about 1 in 200 of those constructed after 1950 failed. Table 8.1 summarises the statistics of failure of embankment dams during operation (i.e. excluding failures during construction, showing that after overtopping, internal erosion and piping is the most important mode. Slope instability accounts for only 5.5% of failures and only 1.6% for dams built after 1950. Their data shows that for dams in Australia, USA, Canada and New Zealand designed and constructed after 1930, about 90% of failures were related to internal erosion and piping.

If 2 failures of hydraulic fill dams, and two of unknown zoning are excluded, only 1 in 5000 dams have failed by instability. However 1 in 200 have experienced an instability accident. This reflects the fact that it is possible to reliably assess the factor of safety of a

Table 8.1. Statistics of failures of large embankment dams up to 1986 (Foster et al., 1998, 2000a).

Mode of failure	% Total failures[1]	% Failures pre 1950[2]	% Failures post 1950[3]
Overtopping	34.2	36.2	32.2
Spillway/gate (appurtenant works)	12.8	17.2	8.5
Piping through embankment	32.5	29.3	35.5
Piping from embankment into foundation	1.7	0	3.4
Piping through foundation	15.4	15.5	15.3
Downstream slide	3.4	6.9	0
Upstream slide	0.9	0	1.7
Earthquake	1.7	0	3.4
Totals[3]	102.6	105.1	100
Total overtopping and appurtenant works	48.4	53.4	40.7
Total piping	46.9	43.1	54.2
Total slides	5.5	6.9	1.6
Total no. of embankment dam failures	124	61	63

[1] Percentages based on the % of cases where the mode of failure is known.
[2] Percentages are for failures of embankment dams in operation only, i.e. excluding failures during construction.
[3] Percentages do not necessarily sum to 100% as some dams were classified as multiple modes of failure.

Table 8.2. Statistics of failures and accidents by internal erosion and piping of embankment dams up to 1986 (Adapted from Foster et al., 1998).

	Average probability during life of dam ($\times 10^{-3}$)			
	Piping in embankment		Piping in foundation	
Zoning category	Failures	Accidents	Failures	Accidents
Homogeneous earthfill	16.0	9.2	3.0	11.2
Earthfill with horizontal filter	1.5	0.6	–	3.9
Earthfill with rock toe	8.9	8.0	7.0	3.9
Zoned earthfill	1.2	2.4	0.4	4.6
Zoned earth and rockfill	1.2	7.3	–	7.6
Central core earth and rockfill	<1.1	22.0	–	9.8
Concrete face earthfill	5.3	2.4	10.4	5.8
Concrete face rockfill	<1	3.5	–	–
Puddle core earthfill	9.3	20.7	–	–
Concrete core wall earthfill	<1	8.1	11.8	4.9
Concrete core wall rockfill	<1	21.6	–	–
Hydraulic fill	<1	32.4	15.7	91.8
Zoning unknown	–	–	–	–
All embankment types	3.5	26.8	1.7	6.2

dam, that modern criteria for acceptable factors of safety are reasonably conservative, and that most dams will show excessive deformation if they are marginally stable, allowing remedial works or lowering of the reservoir water level before failure occurs.

Tables 8.2, 8.3 and 8.4 present additional statistics on internal erosion and piping failures and accidents.

In Table 8.4, the annual probabilities have been calculated allowing for the estimated average life of the dams up to 1986.

Table 8.3.   Timing of internal erosion and piping failures in relation to the age of the dam.

Time of failure	Piping in embankment (%)	Piping in foundation (%)
During first filling	49	25
During first 5 years operation	16	50
After 5 years operation	35	25

There are a number of points which can be seen from this data:

(a) About 1 in 300 embankment dams have failed by piping through the embankment. The most likely to fail are homogeneous earthfill, earthfill with a rock toe and all dams with little or no filters or seepage control zoning;

(b) About 1 in 40 embankment dams have experienced an accident by piping through the embankment. The most likely to experience an accident are central core earth and rock-fill, puddle core earthfill, concrete corewall rockfill and hydraulic fill;

(c) Homogeneous dams are more likely to fail than experience an accident, which almost certainly relates to the absence of filters or any zoning to control erosion and piping once it begins;

(d) By contrast, while there have been a number of accidents in central core earth and rock-fill dams, only one failed (and it had coarse rockfill adjacent to the core). This reflects the fact that most central core earth and rockfill dams have filters or transition zones and high discharge capacity in the downstream rockfill, so while internal erosion and piping may initiate and progress to form a leak, failure is unlikely;

(e) About 1 in 600 embankment dams have failed, and 1 in 300 had an accident by piping through the foundation. Dams with narrow cut-offs, such as concrete face earthfill and concrete core wall rockfill, are over-represented in the failures. Accidents are spread over the dam zoning types;

(f) Most failures by internal erosion and piping occur on first filling or in the first 5 years of operation. This may relate to the weaknesses in the dam and foundation being "found out" in this early filling, but there is evidence (Wan and Fell, 2002, 2003) that soils are more erodible when partially saturated than when saturated and this may explain some of the improved performance after 5 years.

It is this history of failure and accidents due to internal erosion and piping which has led to the incorporation of filters and high permeability zones in dams beginning in the 1950s and more particularly in the 1970s. However filters and free draining zones are often costly, so there may be a wish to use simpler designs, such as homogeneous earthfill or zoned earthfill, particularly where the consequences of failure are small. If this is done, those involved should appreciate that there will be significant chance of failure, particularly for homogeneous dams and if there are other unfavourable circumstances such as dispersive soils, conduits through the dam and/or poor compaction. Foster et al. (1998, 2000a, 2000b) give a method for assessing how these factors can affect the historic probabilities in Table 8.4.

8.2   TYPES OF EMBANKMENT DAMS, THEIR ADVANTAGES AND LIMITATIONS

8.2.1   *The main types of embankment dams and zoning*

Figures 1.1, 1.2 and 1.3 show the main types of embankment dams. Table 1.1 describes the main zones and their function.

Table 8.4. Average probabilities of failure of embankment dams by internal erosion and piping (Foster et al., 1998, 2000a).

Zoning category	Embankment			Foundation			Embankment into foundation		
	Average P_{Te} ($\times10^{-3}$)	Average annual P_e ($\times10^{-6}$)		Average P_{Tf} ($\times10^{-3}$)	Average annual P_f ($\times10^{-6}$)		Average P_{Tef} ($\times10^{-3}$)	Average annual P_{ef} ($\times10^{-6}$)	
		First 5 years operation	After 5 years operation		First 5 years operation	After 5 years operation		First 5 years operation	After 5 years operation
Homogeneous earthfill	16	2080	190						
Earthfill with filter	1.5	190	37						
Earthfill with rock toe	8.9	1160	160						
Zoned earthfill	1.2	160	25						
Zoned earth and rockfill	1.2	150	24						
Central core earth and rockfill	(<1.1)	(<140)	(<34)	1.7	255	19	0.18	19	4
Concrete face earthfill	5.3	690	75						
Concrete face earthfill	(<1)	(<130)	(<17)						
Puddle core rockfill	9.3	1200	38						
Earthfill with corewall	(<1)	(<130)	(<8)						
Rockfill with corewall	(<1)	(<130)	(<13)						
Hydraulic fill	(<1)	(<130)	(<5)						
All dams	3.5	450	56	1.7	255	19	0.18	19	4

1. P_{Te}, P_{Tf} and P_{Tef} are the average probabilities of failure over the life of the dam.
2. P_e, P_f and P_{ef} are the average annual probabilities of failure.

8.2.2   The general principles of control of seepage pore pressures and internal erosion and piping

As will be apparent from Section 8.1, much of what differentiates between the likelihood of failure of the different types of dam is the degree of control over internal erosion and piping in the dam and foundation. Second, but still important, is the degree of control over pore pressures in the dam and the foundation so the factor of safety against slope instability can be reliably assessed. When considering internal erosion and piping, and slope instability, the confidence with which seepage pore pressures can be assessed is critical. This in turn is controlled by the permeability of the zones in the dam and the foundation, whether the dam is cracked by differential settlement, desiccation or earthquake, or whether there are high permeability layers in the earthfill due to poor compaction or desiccation of layers of earthfill during construction. In fact, it is not necessary to have poor compaction to give a high horizontal to vertical permeability ratio, k_H/k_V, which can lead to high pore pressures in the dam. This is shown schematically in Figure 8.1, which shows the normal layering of compacted earthfill.

Monitoring of pore pressures in dams shows that k_H/k_V may be anything from 1.0 to 100.

Figure 8.2 shows schematically the effect of the k_H/k_V ratio in the earthfill on pore pressures in an earthfill dam with a horizontal drain and an earthfill dam with horizontal and vertical drains.

For the earthfill dam with horizontal filter drain, if k_H/k_V is low (say 1), the seepage through the embankment will flow towards the horizontal drain. If it is designed as a filter to the earthfill, and has sufficient discharge capacity, internal erosion and piping is controlled. Pore pressures in the downstream slope would be low so the factor of safety would be high. If however k_H/k_V is high (say 15), the flow net is quite different, and some seepage is likely to emerge on the downstream slope without a filter to protect against internal erosion and piping. The pore pressures on a potential slide surface are much higher (U_H compared to U_L), so the factor of safety is lower.

For the earthfill dam with horizontal and vertical filter drains, the flow net is similarly affected by the k_H/k_V ratio, but the implications are far less, because the vertical filter drain intercepts the seepage regardless of k_H/k_V ratio, so controlling internal erosion and piping, and the pore pressures affecting slope stability are predictably low.

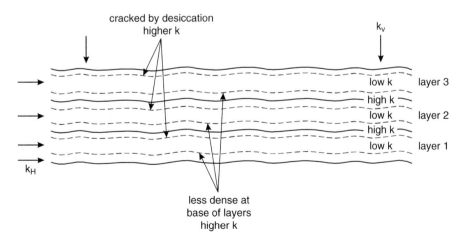

Figure 8.1.   Schematic diagram of the effects of compaction layers, and desiccation during construction, on k_H/k_V in compacted earthfill.

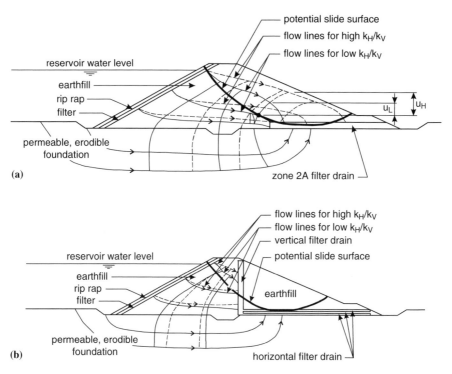

Figure 8.2.   Effect of horizontal to vertical permeability ratio k_H/k_V on pore pressures within an earthfill dam (a) with horizontal filter drain, (b) with horizontal and vertical filter drain.

It is important to recognize that it is not possible to predict k_H/k_V, or simulate the effects leading to high k_H/k_V in the laboratory. The way to make the pore pressures more predictable, and to control internal erosion and piping, is to provide filters/drains for the full height of the dam, such as with an earthfill with horizontal and vertical drains, or a central core earth and rockfill dam. The other important matter to recognize is that regardless of k_H/k_V ratio, if an embankment core is cracked by differential settlement, hydraulic fracture or by desiccation, seepage will flow preferentially in the cracks. This will occur quickly (in hours) when the reservoir water reaches the level of the cracking. The rate is not related to the permeability of the compacted earthfill, only the hydraulics of the crack.

While the conditions for cracking and hydraulic fracture are understood in qualitative terms (see Section 10.7.1), it is not possible to predict them quantitatively. Good embankment design recognizes this and provides filters and drains to cope with the flows from such cracking.

If the foundation of a dam is soil, or erodible (e.g. completely weathered) rock, there is potential for internal erosion and piping in the foundation. Good design requires that the seepage is, so far as practical, allowed to flow into a filter drain placed on the foundation as in Figure 8.2, or intercepted by pressure relief wells with screens designed to prevent erosion.

8.2.3   *Taking account of the likelihood and consequences of failure in selecting the type of embankment*

It is apparent from Section 8.1 that some types of embankment are much more likely to fail by internal erosion and piping than others. Logically this should be taken into account during the selection process.

Table 8.5.   ANCOLD (1999) consequences of failure classification.

| Population at risk | Severity of damage and loss | | | |
	Negligible	Minor	Medium	Major
0	Very low	Very low	Low	Significant
1 to 10	Low	Significant	Significant	High
	Note 4	Note 5	Note 5	
11 to 100	Note 1	Note 2	High	High
101 to 1000			High	High
			Note 3	
More than 1000			Note 3	Extreme

Note 1: With a PAR exceeding 5 people, it is unlikely that the severity of damage and loss will be "Negligible".
Note 2: "Minor" damage and loss would be unlikely when the PAR exceeds 10.
Note 3: "Medium" damage and loss is unlikely (but could be possible) when PAR exceeds 100.
Note 4: Change to *Significant* where it is possible that one life could be lost.
Note 5: Change to *High* where it is possible for more than one life could be lost.
Note 6: ANCOLD (1999) use the term "Hazard" rather than "consequences of failure".

Also, logically, the consequences of failure should also be considered, with more reliable types used where the consequences of failure are high. This can be done within a formal risk framework or, in a less formal way, using the population at risk and severity of damage and losses. Organisations such as ICOLD and ANCOLD have classification schemes for consequences of failure, usually called "Hazard" classification. For the purposes of this book we will use the ANCOLD (1999) classification. This is shown in Table 8.5.

The severity of damage and loss is assessed from estimated costs, loss of services and business, social and economic and natural environmental. These are shown in Table 8.6. The highest classification of damage and loss assessed from Table 8.6 is used when assessing the consequence of failure classification in Table 8.5.

8.2.4   Types of embankment dams, their advantages, limitations and applicability

Table 8.7 summarises the types of embankment dams, their degree of control of internal erosion and piping and pore pressures for slope instability, and suitability in relation to the consequences of failure classification.

This table is for new dams and assumes the dams are well designed and constructed so filters meet no-erosion filter criteria and drainage zones have sufficient capacity to discharge seepage through the dam and the foundation.

There are many existing homogeneous, earthfill with toe drains, zoned earthfill and earthfill with horizontal drain dams which are in significant and high consequences of failure locations. Whether these dams are suitably safe depends on their performance, monitoring and surveillance, and other detailed factors. These can be assessed as detailed in Chapters 9, 10 and 11.

8.3   ZONING OF EMBANKMENT DAMS AND TYPICAL CONSTRUCTION MATERIALS

8.3.1   General principles

The following discussion describes "typical" zoning and construction materials for the most common types of embankment dams. Each dam should be designed to satisfy the particular

Table 8.6.   ANCOLD (1999) sample guidelines on the selection of the severity of damage and loss – Part A.

Type	Negligible	Minor	Medium	Major
1. Estimated costs				
Residential	Not expected	Damage to one house	Destroy two houses or damage a number	About 10 houses destroyed
Infrastructure	<$10 000	$10 000 to $1M	$1M to $10M	More than $10M
Commercial	<$10 000	$10 000 to $1M	$1M to $10M	More than $10M
Dam repair or replacement cost	<$100 000	$100 000 to $10M	$10M to $100M	More than $100M
Provision of temporary services to replace those provided by dam	<$10 000	$10 000 to $1M	$1M to $10M	More than $10M
Cleanup within the flood affected zone	<$10 000	$10 000 to $1M	$1M to $10M	More than $10M
2. Service and business relating to the dam				
Importance to the business	No effect	Restrictions needed during dry periods	Restrictions needed during peak days and peak hours	Essential to main supply
Effect on services provided by the owner	Services can easily be replaced	Minor difficulties in replacing services	Reduced services are possible with reasonable restrictions. About 80% of full supply for some months	Services supplied by the dam are essential and cannot be provided from another source. Severe restrictions would be applied for at least one year
Practicality of replace the dam	No impediment	Some impediment	Replace at a high cost to the community	Impossible to replace
Community resistance to replacement, effect on continuing credibility, and political implications	None expected	Some reaction but shortlived	Severe widespread reaction	Extreme discontent, high media coverage, long term distrust
Impact on financial viability	None expected	Able to absorb in one financial year	Significant with considerable impact in the long term	Severe to crippling in the long term
Value of water in the storage	<$10 000	$10 000 to $1M	$1M to $10M	More than $10M

3. Social and economic

Loss of service to the community	None expected	<10% of the community affected	10% to 30% of the community affected	>30% of the community affected
Public health adversely affected	No effect	<100 people affected for one month	100 to 1000 people affected for one month	>1000 people affected for one month
Cost of emergency management	None expected	<10 person days	10 to 100 person days	>100 person days
Dislocation of people	No effect	<5 families	5–100 families	>100 families
Dislocation of business	No effect	<5 businesses	5–20 businesses	>20 businesses
Employment affected	No jobs lost	<10 jobs lost	10 to 100 jobs lost	>100 jobs lost
Production affected	None expected	Regional output affected by <10%	Regional output affected by 10% to 30%	Regional output affected by >30%
Post disaster trauma and stress	None expected	<100 person months	100 to 500 person months	>500 person months
Injured and hospitalized	Nil	<10	10 to 50	>50
Loss of heritage	None expected		Assessment based on community consultation	
Loss of recreation facility	None expected	Local facility	Regional developed facility	National facility

4. Natural environmental

Area of environmental impact	<0.1 km^2	0.1 km^2 to 1 km^2	1 km^2 to 10 km^2	>10 km^2
Duration of environmental impact	<1 month	1 month to 1 year	1 year to 10 years	>10 years
Significant factors	None expected	Forested land may contain environmental factors	Diverse species are likely to exist	Endangered species identified
Ecological value	None expected	Likely to alter, say, fish habitat	Significant habitat may exist	Impact on an important value, say fish breeding area
Habitat units	None expected	Minor effect on wetland or forest	Medium effect	Major effect

Table 8.7. Types of embankment dams, their degree of control of internal erosion and piping, and pore pressures for stability, and suitability in relation to consequences of failure classification.

Embankment type	Description of the zones	Degree of filter control of internal erosion and piping	Degree of control of pore pressures for stability	Consequences of failure classifications to which suited – new dams
Homogeneous earthfill	Zone 1 earthfill	None. Seepage in earthfill and from cracks is likely to emerge on the downstream face regardless of k_H/k_V	Poor. Pore pressures not predictable. Depend on k_H/k_V in earthfill	Very low and low
Earthfill with toe drain	Zone 1 earthfill, Zone 4 rockfill, may have Zone 2A/2B filter drain	Poor. Seepage in earthfill and from cracks is likely to emerge on the downstream face for high k_H/k_V. Poor into rockfill if no filters provided	Poor. Pore pressures not predictable. Depend on k_H/k_V in earthfill	Very low and low
Zoned earthfill	Zone 1 earthfill. Zones 1–3 of borrow pit run alluvial silt/sand/gravel; or weathered and low strength rock, compacted to form silt/sand/gravel	Moderate (poor to good). All seepage will be intercepted by Zones 1–3. Depends on particle size distributions of Zones 1–3 to act as a filter to Zone 1	Good provided Zones 1–3 is much higher permeability than Zone 1	Very low to significant, depending on material particle size distributions and construction control
Earthfill with horizontal drain	Zone 1 earthfill. Zone 2A filter drain	Poor. Seepage in earthfill and from cracks is likely to emerge on the downstream face for high k_H/k_V	Poor. Pore pressures not predictable. Depend on k_H/k_V in earthfill	Very low and low. Significant if no population at risk

Earthfill with vertical and horizontal drain	Zone 1 earthfill, Zones 2A, 2B filter drain	Very good to good. Seepage in earthfill and cracks is intercepted by the vertical drain, and provided the drains are designed as filter and have sufficient discharge capacity	Good, provided the filter drains have sufficient capacity to discharge seepage from through the dam and the foundation	Very low to high
Earth and rockfill, central core	Zone 1 earthfill, Zones 2A, 2B filters, Zones 3A and 3B rockfill	Very good. Seepage in earthfill and from cracks is intercepted by the filters and discharged in the rockfill	Very good, provided the rockfill is free draining	Significant to extreme. Likely to be too complicated and costly for dams less than about 20 m high
Earth and rockfill, sloping upstream core	Zone 1 earthfill, Zones 2A, 2B filters, Zones 3A and 3B rockfill. Earthfill quantity reduced compared to central core earth and rockfill	Very good. Seepage in earthfill and from cracks is intercepted by the filters and discharged in the rockfill	Very good, provided the rockfill is free draining	Significant to extremes. Likely to be too complicated and costly for dams less than about 20 m high. Suitable for staged construction
Concrete face rockfill	Zones 3A and 3B rockfill, with upstream Zones 2D and 2E to limit leakage in the event the face slab leaks	Very good. Provided Zones 2D and 2E have particle size distributions to act as filters and be internally unstable	Very good, provided the face slab Zones 2D and 2E are effective, and the rockfill is free draining	Significant to extreme. The cost of plinth and foundation preparation makes CFRD unsuitable for dams less than about 30 m high. Suited for staged construction. Usually not suited for other than rock foundations

topographic and foundation conditions at the site and to use available construction materials, so there really are no "typical" or "standard" designs. To highlight this, examples are given of the various types of dams. Many of these are drawn from Australian practice as detailed in the ANCOLD (Australian National Committee on Large Dams) Bulletins and from the authors' own experience. Figures 1.1 and 1.2 show typical cross sections for the most common types of embankment dams. Table 1.1 describes the zoning numbering system used in Figures 1.2 and 1.3 (and throughout this book) and the function of the zones. Table 1.2 describes terms used relating to foundation treatment.

Table 8.8 describes typical construction materials used for the different zones in embankment dams. It is emphasised that good dam engineering involves use of the materials available at the site rather than to look for materials with a preconceived ideas about the material properties needed. However this rule is not followed in the search for Zones

Table 8.8.   Embankment dam typical construction materials.

Zone	Description	Construction material
1	Earthfill	Clay, sandy clay, clayey sand, silty sand, possibly with some gravel. Usually greater than 15% passing 0.075 mm, preferably more. Note that weathered siltstone, shale and sandstone can sometimes be compacted in thin layers to give sufficiently fine material
2A	Fine filter	Sand or gravelly sand, with less than 5% (preferably less than 2%) fines passing 0.075 mm. Fines should be non plastic. Manufactured by crushing, washing, screening and recombining sand-gravel deposits and/or quarried rock
2B	Coarse filter	Gravelly sand or sandy gravel, manufactured as for Zone 2A. Zones 2A and 2B are required to be dense, hard durable aggregates with similar requirements to that specified for concrete aggregates. They are designed to strict particle size grading limits to act as filters
2C	Upstream filter and filter under rip rap	Sand gravel/gravelly sand, well graded e.g. 100% passing 75 mm, not greater than 8% passing 0.075 mm fines non plastic. Usually obtained as crusher run or gravel pit run with a minimum of washing, screening and regarding. Relaxed durability and filter design requirements compared to Zones 2A and 2B
2D	Fine cushion layer	Silty sandy gravel well graded, preferably with 2–12% passing 0.075 mm to reduce permeability. Obtained by crushing and screening rock or naturally occurring gravels or as crusher run. Larger particles up to 200 mm are allowed by some authorities, but segregation and internal instability is likely to result
2E	Coarse cushion layer	Fine rockfill placed in 500 mm layers to result in a well graded sand/gravel/cobbles mix which satisfies filter grading requirements compared to Zone 2D
3A	Rockfill	Quarry run rockfill, possibly with oversize removed in quarry or on dam. Preferably dense, strong, free draining after compaction, but lesser properties are often accepted. Compacted in 0.5–1 m layers with maximum particle size equal to compacted layer thickness
3B	Coarse rockfill	Quarry run rockfill. Preferably dense, strong, free draining after compaction, but lesser properties are often accepted. Compacted in 1.5–2.0 m layers with maximum particle size equal to compacted layer thickness
4	Rip rap	Selected dense durable rockfill sized to prevent erosion by wave action. In earth and rockfill dams often constructed by sorting larger rocks from adjacent 3A and 3B Zones. In earthfill dams either selected rockfill or a wider zone of quarry run rockfill may be used

2A and 2B filters where invariably one might seek materials satisfying carefully specified particle size grading limits and dense hard durable materials.

8.3.2   *Examples of embankment designs*

8.3.2.1   *Zoned earthfill dams*

Figures 8.3 and 8.4 show examples of zoned earthfill design. Both were designed for low or very low consequence of failure classification sites. The Tahmoor damsite was underlain by sandstone and there were limited sources of non-dispersive earthfill. The sandstone rock breaks down to a silty sand when compacted in thin layers, yielding an acceptable transition filter material. The design (which was done by the first author) would be improved if Zone 2 was taken to the top of Zone 1 to provide filter protection under flood conditions.

The Blair Athol Dam was built over alluvium, and over an existing dam, which had been built without conventional design or construction control. A cutoff trench was excavated to the weathered sandstone and siltstone foundation. Zoned earthfill construction

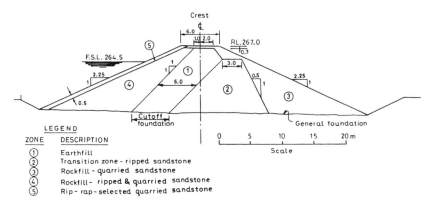

Figure 8.3.   Mine water supply Tahmoor Dam (courtesy of Coffey Geosciences).

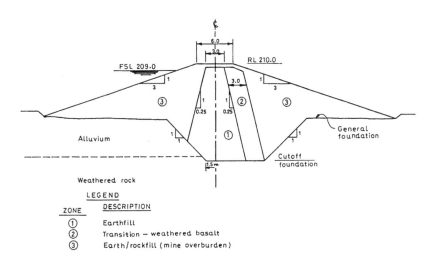

Figure 8.4.   Water storage dam, Blair Athol Mine (courtesy of Coffey Geosciences).

was practical, using a transition zone of weathered basalt rock compacted in 150 mm thick layers. The "earthfill" was sandstone from mine overburden compacted in 150 mm thick layers.

8.3.2.2   Earthfill dams with horizontal and vertical drains

The mine runoff water dam at Drayton Mine (Figure 8.5) is constructed on a foundation of interbedded siltstone and sandstone. It is a low consequence of failure site, allowing adoption of a design without vertical drain. Zone 3 was compacted mine overburden (mainly sandstone), which results in a low permeability earth/rockfill with some degree of internal erosion and piping control to Zone 1. A horizontal drain was provided to collect seepage through the foundation in a controlled manner.

The Ash Pond Dam for Loy Yang Power Station (Figure 8.6) is an off river storage founded on Quarternary sediments overlying Tertiary Coal Measures.

The sediments include widely interspersed beds of overconsolidated fissured silty and sandy clays, and clayey and silty sands. The Coal Measures consist of inferior coal and coal seams separated by interseam sediments composed of sandy clays, silts and silty sands which form a series of aquifers. Increased piezometric levels within these aquifers following filling of the ash storage, required the installation of a series of relief wells at the downstream toe of the dam. The dam has a vertical chimney drain and horizontal drain, to control seepage through the dam and its foundation. Internal sand drains were provided to control construction pore pressures.

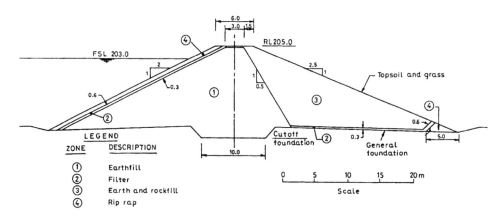

Figure 8.5.   Mine runoff water dam, Drayton Mine (courtesy of Coffey Geosciences).

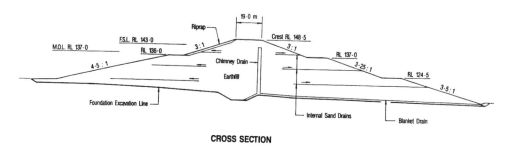

Figure 8.6.   Ash Pond Dam, Loy Yang Power Station (courtesy State Electricity Commission of Victoria).

Plashett Dam (Figure 8.7) is founded on coal measure rocks including sandstone, silt-stone and coal. The design incorporates a cutoff through alluvial soils in the foundation, a vertical drain, and horizontal drainage provided by strips of filter material. (The authors do not favour such strips on erodible foundations as it leaves gaps in the control of internal erosion and piping in the foundation. In this case the drains and cutoff extend to rock so the problem is not so great). A vertical drain is provided through the alluvium to intercept seepage which bypasses the cutoff.

8.3.2.3   *Central core earth and rockfill dams*

The Bjelke-Petersen Dam (Figure 8.8) is largely founded on a very complex sequence of andesitic volcanic rock and limestones, which have been intensely sheared and folded. These sequences alternate with metamorphosed phyllites, with steeply dipping fault zones occurring at the contacts between sequences. The limestone exhibited solution channels and sink holes, often infilled with stiff, high plasticity clay. These were less frequent and less permeable below the water table.

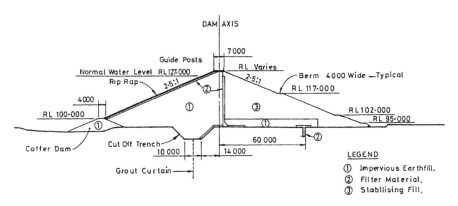

Figure 8.7.   Plashett Dam (ANCOLD, 1990).

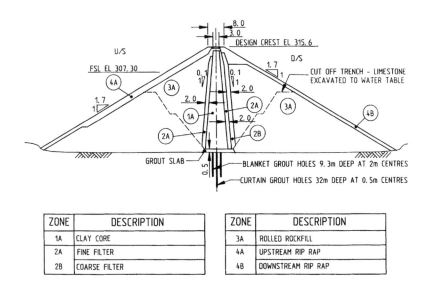

ZONE	DESCRIPTION
1A	CLAY CORE
2A	FINE FILTER
2B	COARSE FILTER

ZONE	DESCRIPTION
3A	ROLLED ROCKFILL
4A	UPSTREAM RIP RAP
4B	DOWNSTREAM RIP RAP

Figure 8.8.   Bjelke-Petersen Dam (courtesy of Water Resources Commission of Queensland).

The cutoff trench was excavated through the limestone to the water table and a reinforced concrete grout slab formed. A high-pressure grout curtain, with close hole spacing, was adopted. The core was kept narrow to minimize the cost of excavation and foundation treatment. McMahon (1986) gives more details of the project. See also Chapter 3, Section 3.7.2.

The Lungga Dam (Figure 8.9), in the Solomon Islands, was to have been constructed on 60 m of alluvial sand and gravel. A partially penetrating diaphragm wall was to assist in reducing seepage flow in the foundation, but substantial seepage was still anticipated. The available rockfill was relatively low permeability, so a substantial horizontal drain and protective filters was incorporated into the design to cope with the very large anticipated under seepage.

Blue Rock Dam (Figure 8.10) is a fairly conventional central core earth and rockfill dam, constructed on a foundation of mainly Silurian mudstones with thin interbedded sandstones. The rockfill was zoned to accommodate poorer quality rock in the random rockfill zone. An unusual feature of the embankment is the use of reinforced shotcrete on the upstream face for wave protection, as the available rock was of inadequate size and quality.

Hinze Dam (Figure 8.11) is a central core earth and rockfill dam which is founded on greywacke, greenstone and chert. It is unusual (for a central core dam) in that it has been designed to be constructed in three stages. This leads to a modified form of geometry, and necessitates placing some rockfill under water.

Blue Rock and Hinze Dams, would be improved if the filters were taken higher to control internal erosion and piping under flood conditions higher than anticipated by the designers.

Thomson Dam is 166 m high with a saddle dam on the right abutment. Figures 8.12 and 8.13 show sections through the dam and saddle dam.

Each of the two dams is a rockfill dam with a central earthfill core. The cores are flanked by processed chimney filters, a fine and a coarse filter on the downstream side, a fine filter on the upstream side below 7 m below the crest and a coarse filter on the upstream side in the top 11 m of the dam. The main dam has a partial processed two-filter blanket under its downstream shoulder; the saddle dam has a full filter blanket under its downstream shoulder. Both dams have riprap zones for wave protection on their upstream faces. Both dams have multi-row grout curtains up to 80 m deep in places. The grout curtain extends beyond the left end of the main dam and right end of the saddle dam and between the dams to form a continuous seepage barrier.

A stabilizing fill up to 50 m high was placed upstream of the dam and saddle dam. This is discussed in Section 2.10.3.5.

The sources of the various zones were:

Zone 1	"Impervious" cores.	Decomposed granite slope wash and residual soil from borrow pits.
Zones 3A, 3C and 4	All rockfill except a small amount of initial main dam stabilizing fill.	Sandstones and siltstones from quarry on left bank. Initial placement in stabilizing fill came from sedimentary rock right bank quarry.
Zone 3B	Upstream face wave protection (riprap).	Fresh granitic rock from right bank quarry.
Zones 2A and 2B	Filters – process materials.	Mainly indurated sandstones from quarry on right bank.

8.3.2.4   *Sloping upstream core earth and rockfill dam*

An example of a sloping upstream core dam is given in Figure 19.25. The section has been designed to facilitate staged construction.

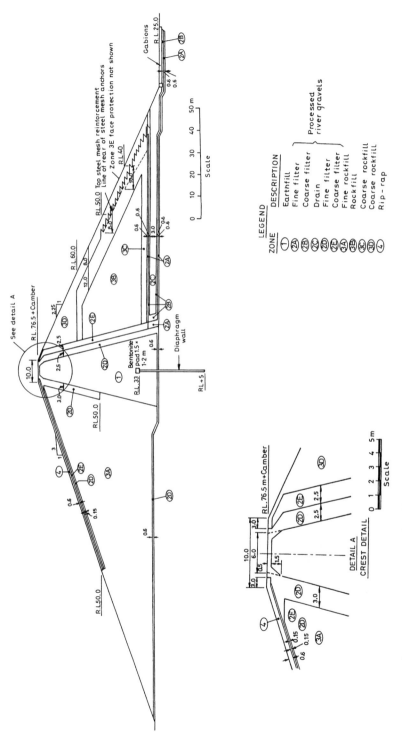

Figure 8.9.   Proposed Lungga Dam (courtesy of Coffey Partners International).

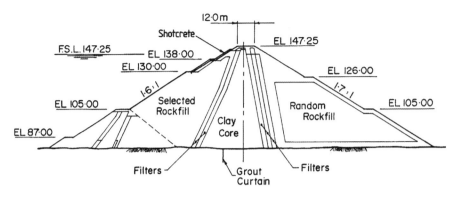

Figure 8.10.   Blue Rock Dam (ANCOLD, 1990).

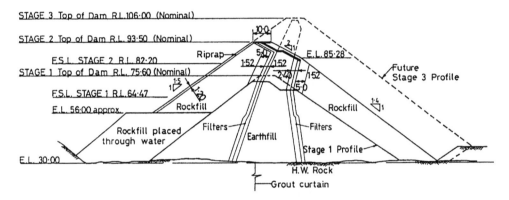

Figure 8.11.   Hinze Dam (ANCOLD, 1990).

8.3.2.5   Concrete face rockfill dams
Several examples of a concrete face rockfill dams (CFRD) are given in Chapter 15.

8.4   SELECTION OF EMBANKMENT TYPE

The selection of the type of dam embankment to be used at a particular site is affected by many factors, some of which are outlined below. The dam engineer's task is to consider these factors and adopt a suitable design. The overriding consideration in most cases will be to construct an adequately safe structure for the lowest total cost. Hence, preparation of alternative designs and estimates of cost for those alternatives will be a normal part of the design procedure. Usually the most economic design will be that which uses a construction materials source close to the dam, without excessive modification from the "borrow pit run" or "quarry run" material.

The following outlines some of the factors involved and their effect on embankment type.

8.4.1   Availability of construction materials

8.4.1.1   Earthfill
Clearly the availability of suitable earthfill within economic haul distance is critical in the selection of the embankment type. If there is no earthfill available – for example in an area

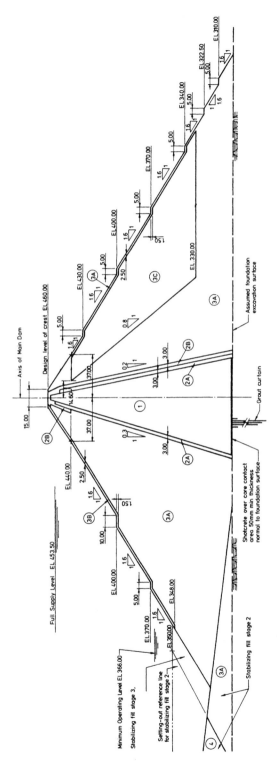

Figure 8.12.   Thomson Dam – maximum section of the main dam (Courtesy of Melbourne Water).

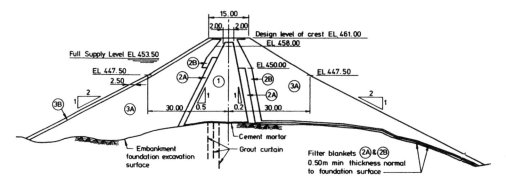

Figure 8.13.   Thomson Dam – section through the saddle dam. (Courtesy of Melbourne Water)

underlain by sandstone, which weathers to give only a shallow cover of sandy soil, it will normally be appropriate to construct a rockfill dam with concrete (or other) impervious membrane, or a concrete gravity dam.

The uniformity of the available earthfill will also influence design and the method of construction. If the borrow areas produce two different types of earthfill, the earthfill may be zoned into two parts, e.g. for an earthfill dam with vertical and horizontal drains, the earthfill with lower proportion of fines and more variable properties would be best placed downstream of the vertical drain. Alternatively the earthfill zone may be separated into Zones 1a and 1b with the coarser soil in Zone 1b adjacent the Zone 2A filter.

Alluvial clayey soils are often more variable than residual soils derived from weathering of underlying rocks and, hence, it may be necessary to provide additional zoning as described above. Alternatively, the soils may be mixed by borrowing from a vertical face with a shovel and truck operation (rather than use of scrapers). Blending of soils on the dam embankment is generally avoided as it leads to increased cost and difficulty in quality control.

If cobbles and boulders are present in clayey soil deposits, these will have to be removed prior to compaction either by passing the earthfill through a "grizzly" or by grader or hand labour on the embankment. This is necessary to prevent the oversize particles affecting compaction.

Relatively permeable soils can be used for earthfill in many dam projects. The permeability of most dam foundations is between 1 Lugeon and 10 Lugeons (10^{-7} m/sec to 10^{-6} m/sec) so for most dams an earthfill core with permeability, say 10^{-9} m/sec, seepage through the foundations will far exceed that through the dam. Even if silty sand is used for the earthfill, a permeability of 10^{-6} m/sec should be achieved, i.e. not higher than the foundation permeability, and from a seepage viewpoint such a high permeability soil would be acceptable for many dams. Weathered or even relatively unweathered siltstone and sandstone with a clayey matrix, may break down sufficiently when compacted in thin layers to achieve a satisfactory core material. This will usually require field trials to observe the actual properties. MacKenzie and McDonald (1985) and ANCOLD (1985) describe trials and actual performance of compacted siltstone and sandstone, during construction of Mangrove Creek concrete face rockfill dam.

8.4.1.2   Rockfill

Over the last 30 years the unit rate for construction of rockfill has reduced compared to that for earthfill and as usually less rockfill is required (because the side slopes can be steeper), the availability of rock which can be quarried to yield free draining rockfill often leads to economic dam design.

Most igneous and many metamorphic rocks (even low grade metamorphics), e.g. granodorite, diorite, granite, basalt, rhyolite, andesite, marble, greywacke quartzite and some indurated siltstones and sandstones will, when fresh, yield free draining rockfill.

Some metamorphic rocks, e.g. phyllite, schist, gneiss and slate, may break down under compaction to yield a poorly draining rockfill even though it may be dense with a high modulus. The amount of breakdown depends upon the degree to which foliation or cleavage is developed in the rocks.

Most highly weathered, and many moderately weathered igneous and metamorphic rocks, will not yield totally free draining rockfill.

The spacing of the bedding and joint planes influences the size and grading of rockfill obtained from a quarry. Blasting may be varied to yield the required sized product, but this is not always practicable.

Thick beds of sandstone within a sequence of thinner bedded siltstone and sandstone are likely to yield oversize rock, which would require either secondary breaking in the quarry or sorting and disposal on the embankment.

Often a substantial amount of the rockfill for a dam will come from "required excavations", i.e. from the spillway, foundation for the dam, inlet and outlet works etc. This is in principle desirable as the effective cost of the rockfill could be only the cost of placement (and any additional haulage costs). However, the rock quality from these excavations may not be ideal (due to rock type, weathering, method of excavation) necessitating changes to the embankment zoning to accommodate the material. Also the timing of production of rock from required excavations may not be ideal from the viewpoint of dam construction and in the event, may not be as scheduled. Some flexibility in zoning is desirable to allow for such circumstances. Such flexibility is also useful to allow for different quality of rock being obtained from required excavations from that anticipated at the time of designing the embankment.

Most sedimentary rocks, e.g. sandstone, siltstone, shale and mudstone generally tend to break down under compaction, even when fresh, and yield poorly draining rockfill. In these circumstances it may be necessary to incorporate zones of free draining rockfill to ensure the embankment rockfill as a whole is capable of remaining free draining, e.g. Figures 8.9 and 8.10.

The use of so called "random" rockfill zones in a dam section also facilitates use of rock, which does not yield free draining rockfill.

8.4.1.3 *Filters and filter drains*

A source of high quality sand and gravel is necessary for construction of filters. It may be necessary to obtain these materials from many kilometres' distance (50 km is not unusual), despite the haulage costs.

Filter aggregates may be obtained from alluvial sand and gravel deposits, or from quarries. Generally suitable aggregates are of igneous and, less commonly, metamorphic origin. It is unusual to manufacture filters from sedimentary rocks, as these rocks are usually not sufficiently durable and often have poor shape (as measured by flakiness index).

For large dams it is usually necessary to establish a separate crushing and screening plant for manufacture of filter and concrete aggregates. It is sometimes necessary to let a separate early contract to begin the manufacture and stockpiling of aggregates, as their production rate controls dam construction progress.

Where sources of filter aggregates are far from the dam and for this or other reasons filters are expensive, the width of filter zones may be reduced by using spreader boxes (see Chapter 9).

8.4.2 *Foundation conditions*

The strength, permeability and compressibility of the dam foundation have a major influence on the embankment type, e.g.:

- A soil foundation will have a relatively low strength, which may determine the embankment stability, and will require relatively flat embankment slopes. Such a formation is likely to favour construction of earthfill dams, i.e. earthfill with horizontal and vertical drains, rather than earth and rockfill;
- A permeable soil foundation will be susceptible to leakage and erosion, requiring construction of some form of cutoff and a filter drain under the downstream slope of the dam (see Chapter 10);
- A strong low permeability rock foundation is suited to any type of dam construction, but may favour construction of a concrete face, concrete gravity or, in particular cases, concrete arch dam.
- In earthquake zone areas, the presence of loose to medium dense saturated sandy soils in the foundation will be important, as liquefaction may occur during earthquakes. This may necessitate removal or densification of the sandy soil and/or the provision of weighting berms;
- Dams on karst limestone foundations are a special case, where extensive grouting and other work may be needed to limit leakage to acceptable levels. Such situations favour adoption of a design, which allows for grouting to continue during embankment construction or after it is completed. This may tend to favour concrete face rockfill or earth and rockfill with a sloping upstream core;
- In some sedimentary rocks, particularly interbedded weak claystone and mudstone and strong sandstone, which have been subject to folding and/or faulting, bedding plane shears may exist, resulting in low effective friction angles (Hutchinson,1988, Casinader, 1982). In these cases, flat slopes may be required on the embankment, favouring earthfill with vertical and horizontal drains, or earth and rockfill with random rockfill zones. Figure 8.14 shows the cross section for Maroon Dam, where such conditions required very flat slopes;

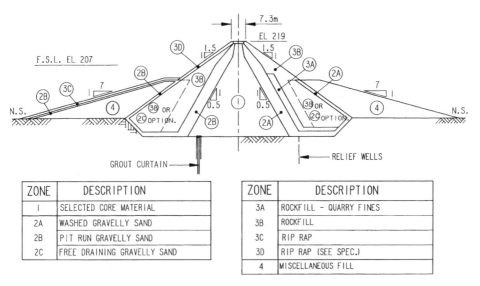

ZONE	DESCRIPTION
I	SELECTED CORE MATERIAL
2A	WASHED GRAVELLY SAND
2B	PIT RUN GRAVELLY SAND
2C	FREE DRAINING GRAVELLY SAND

ZONE	DESCRIPTION
3A	ROCKFILL – QUARRY FINES
3B	ROCKFILL
3C	RIP RAP
3D	RIP RAP (SEE SPEC.)
4	MISCELLANEOUS FILL

Figure 8.14. Maroon Dam (courtesy of Water Resources Commission of Queensland. Note: Drawing is not to scale).

- In some areas, often but not always tropical, the rock is deeply weathered and some-times with a lateritic profile which may lead to a high permeability, soil strength foundation, favouring embankments with flatter slopes and good under drainage, e.g. earthfill with vertical and horizontal drain;
- Embankments constructed on deep soils, e.g. deep alluvium in or adjacent to the river bed, may be subject to a large amount of settlement, leading to differential movement and cracking. In such dams it is particularly important to provide good filters to control seepage and prevent internal erosion.

8.4.3   *Climate*

It is difficult (and in many cases impossible) to construct earthfill embankments during wet weather, or in freezing temperatures. This is particularly critical when the rain is relatively continuous without high evaporation (and not so critical when the rain is in short storms, followed by hot sunshine).

In these circumstances, it is often advantageous to adopt concrete face rockfill or sloping upstream core construction, so that the rockfill can continue to be placed in the wet weather, and the face slab or core constructed when the weather is favourable.

In very arid areas there may be a shortage of water for construction, thus favouring concrete face rockfill rather than earthfill.

8.4.4   *Topography and relation to other structures*

The selection of embankment type and overall economics of a project is determined with consideration of all components of the project, i.e. embankment, spillway, river diversion outlet works etc. These components are interrelated, e.g.:

- The diversion tunnel will be longer for an earthfill dam (with relatively flat side slopes) than for a rockfill dam;
- The spillway may generate rockfill (and possibly earthfill and random fill) for use in the embankments. The size of spillway can be varied by storing more floodwater in the reservoir, necessitating a higher embankment but smaller spillway so the optimisation of total project cost may influence embankment zoning and size;
- It is common practice in Australia to allow large floods during construction to pass over the embankment, rather than providing a larger diversion tunnel and coffer dam. This necessitates incorporation of some rockfill in the downstream toe of the dam with steel mesh reinforcement (see Chapter 13).

The topography of the site, i.e. valley cross section or slope, curve of the river in plan or the presence of "saddles" in the abutments, can have a significant effect on embankment selection, e.g.:

- In narrow steep sided valleys there is restricted room for construction vehicles and haul roads, favouring embankments with simple zoning, e.g. concrete face rockfill;
- The curve of the river in plan, and changes in valley cross section, may favour adoption of an upstream sloping core rather than central core (or vice versa) to reduce the quantities of earthfill;
- Local changes in slope of the abutments may lead to differential settlement and cracking, necessitating more extensive filter drains, or favouring concrete face rockfill construction.

8.4.5   *Saddle dam*

Directly related to the question of topography is a group of dams called saddle dams.

Experience suggests that saddle dams have been often treated as "distant" to the main dam in projects. This type of approach is very dangerous and could lead to major expenditure by a dam owner. A saddle dam must be treated in the same way and to the same detail as a main dam. At first glance, the loss of a saddle dam might seem minor insofar as downstream consequences or even in loss in storage. The recent experience of a fuse plug failure in the U.S. in which erosion went a long way below the plug's foundation should serve as a warning to an owner assuming the loss of a saddle dam would have little intact on his business.

A particular issue that can cause significant problems to the designer of a saddle dam is the dam's foundations. Tops of ridges often present different geological conditions than one would expect to find in a valley. An experienced engineering geologist should be asked to investigate any significant saddle damsite in a dam project.

8.4.6   Staged construction

It is often economic to construct a dam in two or more stages, e.g.:

- In water supply, irrigation or hydropower projects, demand in the early years can often be met with a lower dam and smaller storage;
- In mine tailings dams, the storage required increases progressively as the tailings are deposited in the dam.

If staging is planned, this favours adoption of concrete face rockfill, earth and rockfill with sloping upstream core, or possibly earthfill with vertical (or sloping) drain and horizontal drain. Figures 8.11, 8.15 and 19.25, show examples of dams designed for staged construction.

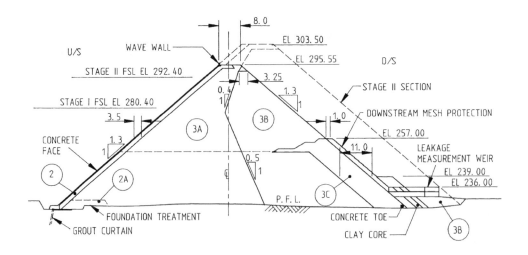

ZONE	DESCRIPTION
2	FACE MATERIAL
2A	FILTER MATERIAL

ZONE	DESCRIPTION
3A	ROCKFILL–UP TO 600 mm DIA.
3B	ROCKFILL–UP TO 900 mm DIA.
3C	AS PER 3B WITH NO MORE THAN 5% PASSING THROUGH A 100 mm APERTURE

Figure 8.15.   Boondooma Dam (courtesy Water Resources Commission of Queensland).

8.4.7 *Time for construction*

The time available for construction may influence the selection of dam type, particularly if considered in relation to other factors such as the climate, e.g. in a climate of well defined wet and dry seasons it may be practicable to construct an earth and rockfill or earthfill dam, but only over two dry seasons. A concrete face rockfill dam may be constructed in lesser time by continuing to place rockfill in the wet season.

Foundation treatment and zoning details may also be influenced by the time available for construction. For example, if constructing a dam on a permeable soil foundation, cutoff may be achieved by a cutoff wall at the upstream toe rather than a rolled earth cutoff under the central core, so that the cutoff wall can be constructed at the same time as the rest of the embankment.

9

Design, specification and construction of filters

9.1   GENERAL REQUIREMENTS FOR DESIGN AND THE FUNCTION OF FILTERS

9.1.1   *Functional requirements*

Filters in embankment dams and their foundations are required to perform two basic functions:

(a) Prevent erosion of soil particles from the soil they are protecting;
(b) Allow drainage of seepage water.

Filters are usually specified in terms of their particle size distribution. They are required to be sufficiently fine, relative to the particle size of the soil they are protecting (the "base soil"), to achieve function (a), while being sufficiently coarse to achieve function (b).
To achieve these functions the ideal filter or filter zone will (ICOLD 1994):

– Not segregate during processing, handling, placing, spreading or compaction;
– Not change in gradation (by degradation or break down) during processing, handling, placing and/or compaction, or degrade with time e.g. by freeze-thaw or wetting and drying by seepage flow;
– Not have any apparent or real cohesion, or ability to cement as a result of chemical, physical or biological action, so the filter will not allow a crack in the soil it is protecting to persist through the filter;
– Be internally stable, that is the fines particles in the filter should not erode from the filter under seepage flows;
– Have sufficient permeability (and, if a drain, thickness) to discharge the seepage flows without excessive build-up of head;
– Have the ability to control and seal the erosion which may have initiated by a concentrated leak, backward erosion, or suffusion (internal instability) in the base soil.

9.1.2   *Flow conditions acting on filters*

Figure 9.1 illustrates the basic flow conditions that can occur between a filter and base soil. These are:

N_1 Flow normal to the base soil – filter interface, with potentially high gradient conditions; e.g. at the downstream face of the earthfill core, the contact between the horizontal drain and the foundations, and within the foundation where seepage is across bedding.
N_2 Flow normal to the base soil – filter interface, with low gradient conditions; e.g. at the upstream face of the earthfill core under reservoir drawdown conditions, or into the upper Zone 2A filter for the horizontal drain.
P   Flow parallel to the interface, e.g. at the base of rip-rap layers, or in the foundation where seepage is along bedding.

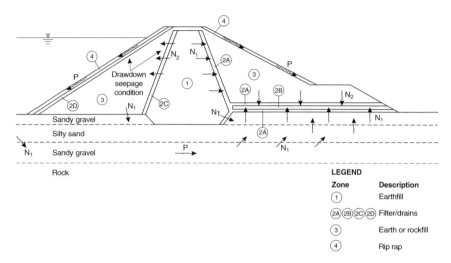

Figure 9.1.   Flow conditions acting on filters. P = flow parallel to interface; N_1 = flow normal to interface, high gradient conditions; N_2 = flow normal to interface, low gradient conditions.

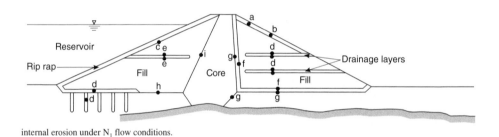

internal erosion under N_1 flow conditions.

Figure 9.2.   Filter functions (adapted from ICOLD, 1986).

As will be described in Section 9.2, the erosive stresses are greatest for case N_1, and less for N_2 because for N_2 the flow is draining from the base soil under gravity, not under reservoir water head.

The erosive action for case P is different and also less severe than for N_1. As a result less conservative (and therefore coarser) filter may be used for cases N_2 and P than for N_1.

9.1.3   Critical and non critical filters

Figure 9.2 and Table 9.1 show a number of applications of filters in dams. Some filters are critical to the control of internal erosion in the dam and, if they fail, give an increased likelihood of piping progressing and potentially breaching the dam. These are termed critical filters, and they should be designed and constructed to meet stringent, no-erosion filter criteria. Examples are filters "g" in Figure 9.2. Some filters are non-critical, in that, if some erosion occurs, it can be repaired (e.g. beneath rip-rap, locations "a", "b" and "c" in Figure 9.2) or it will cause problems only during construction ("d" and "e", in Figure 9.2).

Most critical filters are in an N_1 flow condition (Section 9.1.2) and non-critical filters in an N_2 or P flow condition.

Filters "f" and "h" are critical to the performance of the dam, but are in N_2 flow conditions, so may be designed and constructed to lesser standards. Filters "d" and "I" may be critical or non-critical, depending on whether or not erosion of the embankment fill into

Table 9.1.   Critical and non-critical filters (adapted from ICOLD, 1986).

Filter location	Purpose of filter	Type of flow or loading	Significance of filter	Access for repair
a. Downstream slope protection	Control of erosion by rainfall	P – Occasional surface flow.	Non-critical	Easy
b. Downstream surface drains	Removal of surface seepage	P – Continuous or occasional local seepage	Non-critical. Local wet areas may reappear	Easy, possible
c. Upstream slope protection	Control of erosion by wave action and by outward flow during drawdown	P – Cyclic flow during wave action. N_2 – Small flow during drawdown	Usually non-critical	Possible, but may be difficult
d. Temporary internal drainage during construction	Dissipation of excess pore pressure during construction of wet fills	N_2 – Temporary flow, limited quantity. Some migration of fines allowable if drains not blocked	Non-critical. Failure may lead to instability during construction, or delays	None
e. Upstream internal fill boundary	Prevention of unacceptable migration of fines in upstream direction	N_2 – Transient and small flows during drawdown	Non-critical. Only significant if migration is large and continuous	None
f. Downstream internal interface	Prevention of unacceptable migration of fines into filter-drains	N_2 – Flow only due to infiltration of rainfall, not from reservoir or foundation	Critical, but only if erosion is large and continuous	None
g. Downstream interface, e.g. downstream core boundary or foundation inter-face near core	Prevention of internal erosion of core including effects of concentrated flow in cracks, etc.	N_1 – Continuous flow from reservoir, potentially large and increasing if erosion occurs	Critical	None
h. Upstream interface between embankment and foundation	Prevention of internal erosion of core into foundation	N_1 – Continuous flow from reservoir, potentially large and increasing if erosion occurs	Critical	None
i. Upstream internal interface	Prevention of unacceptable migration of fines from core into upstream fill	N_2 – usually approaching N_1 for pumped storage dams	Critical, but only if erosion is large and continuous	None

the foundation can occur. If it can, the filters are critical. As shown, "d" is a construction drainage system, so would probably be non-critical.

Much of the discussion in this chapter is about critical filters designed to control internal erosion under N_1 flow conditions.

9.1.4   *Filter design notation and concepts*

9.1.4.1   *Notation*

The notation used in this book is described in the following examples:

D_{15F} Particle size of the filter material for which 15% by weight is finer.

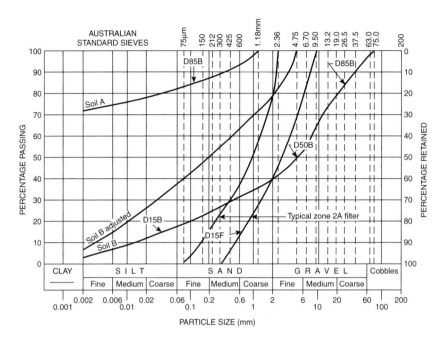

Figure 9.3.   Filter design notation and adjustment of the particle size distribution for gravelly base soils.

D_{85B} Particle size of the base material for which 85% by weight is finer.
pp% 0.075 mm   Percent finer than a particle size of 0.075 mm.
Fines content   pp% 0.075 mm.

Figure 9.3 shows particle size distributions for two base soils, A and B, and a Zone 2A filter.

For broadly graded soils, such as soil B, which have some medium and coarse gravel, the USBR (1977 and 1987), USA-SCS (1994), Sherard and Dunnigan (1989) and other methods use an adjusted particle size distribution. As shown in Figure 9.3, this involves taking the particle size distribution and adjusting it to what it would be if only the fraction passing the 4.75 mm sieve were used.

9.1.4.2   *Filtering concepts*

Figures 9.4(a) and (b) show the interface between a filter and base soil. The basic concept of filter design is to design the particle size distribution of the filter so that the voids in the filter are sufficiently small to prevent erosion of the base soil.

The void size in the filter is controlled by the finer particles and, for design purposes, the D_{15F} is usually used to define this void size. Sherard et al. (1984a) showed that for granular soils the void size between the soil particles, known as the opening size, is given by $O_E = D_{15F}/9$ (see Section 9.2.1.5). Testing by Foster (1999) confirmed this.

A further basic concept, inherent in filter design, is that the base soil will generally provide a degree of "self-filtering". Hence in Figure 9.4(a), in a well graded base soil, the coarser particles in the base soil are prevented from eroding into the filter and they in turn prevent the medium sized particles in the base soil from eroding and the medium sized particles in the base soil prevent the fine particles in the base soil from eroding.

If the base soil is gap-graded or graded concave upwards (Figure 9.5), there is a deficiency of medium sized particles, as shown in Figure 9.4(b). The self filtering does not

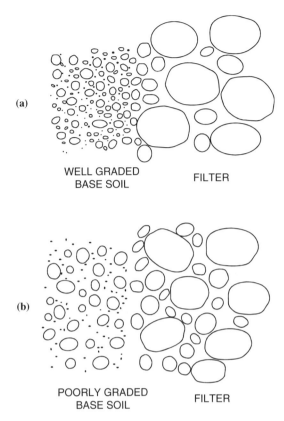

Figure 9.4.   Filtering and self filtering concepts, (a) well graded base soil, (b) gap graded, or concave graded soil, deficient in medium sized particles.

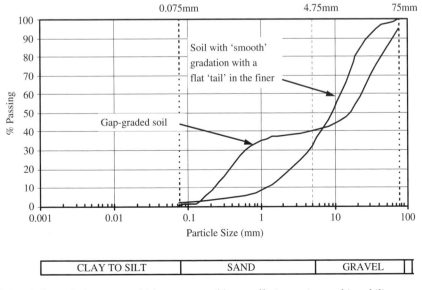

Figure 9.5.   Soils gradation types which are susceptible to suffusion or internal instability.

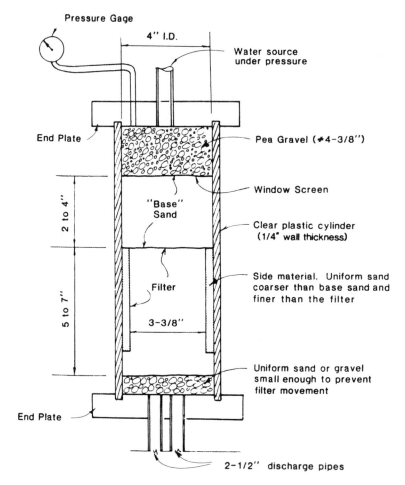

Figure 9.6.   USSCS filter test apparatus details for fine to coarse sand base soils (Sherard et al., 1984a, reproduced with permission of ASCE).

occur and the fine particles in the base soil will erode through the coarse particles, a process called suffusion, or internal instability. In these situations, for the filter to be successful in controlling erosion, it must be able to control the erosion of these finer particles.

In most filter design methods, the base soil particle size is characterised by the D_{85B} size, although some use D_{95B} and D_{50B}. For the USBR (1977, 1987), Sherard and Dunnigan (1989) and USDA-SCS (1994) methods, and as recommended by the authors, the D_{85B} for gravelly base soils is based on the particle size distribution adjusted to what it would be if all the gravel and cobble sized particles greater than 4.75 mm size were removed as shown in Figure 9.3.

9.1.4.3   Laboratory test equipment

The particle size distribution of critical filters is determined using methods which are based on laboratory testing, where base soils are placed against a filter and water is passed through the sample under pressure. Figures 9.6 and 9.7 show the test apparatus used by the US Soil Conservation Service which is reported in Sherard et al. (1984a and b) and Sherard and Dunnigan (1985).

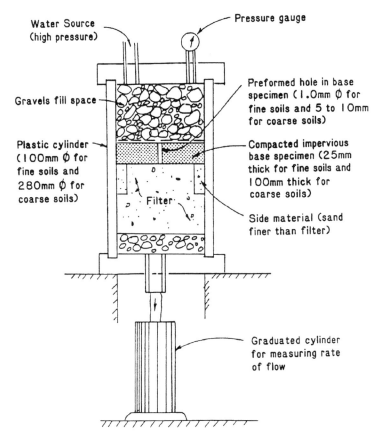

Figure 9.7. USSCS "no erosion" filter test apparatus (schematic, no scale) (Sherard and Dunnigan, 1985, reproduced with permission of ASCE).

Sherard et al. (1984a and b) also used a test set up with the base soil as a slurry and slot tests.

The test shown in Figure 9.7 is intended to model the situation where a concentrated leak has formed through the dam core, so that high hydraulic heads and gradients can occur at the interface between the base soil and the filter (Figure 9.8).

The tests carried out by Sherard et al. (1984a and b) were directed towards finding the filter particle size distribution which would give no-erosion (or at least very minor erosion) conditions.

Figure 9.9 shows the type of equipment used by Bakker et al. (1990) to test for parallel flow conditions.

9.2 DESIGN OF CRITICAL AND NON-CRITICAL FILTERS

9.2.1 Review of available methods for designing filters with flow normal to the filter

ICOLD (1994) describes the evolution of filter design practice. These have basically evolved from the concepts of Terzaghi (1926) who proposed that $D_{15F}/D_{85B} \leqslant 4$ to control erosion and $D_{15F}/D_{15B} \geqslant 4$ to ensure the filter was sufficiently permeable.

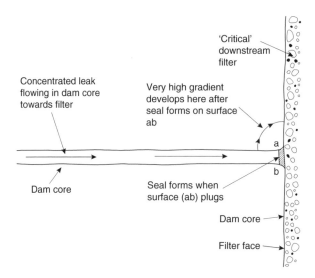

Figure 9.8.   Sketch showing concentrated leak through dam core discharging into downstream filter (no scale) (Sherard et al., 1984b, reproduced with permission of ASCE).

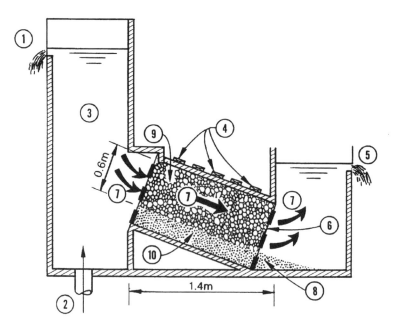

Figure 9.9.   Delft Hydraulics Laboratory filter-box, with parallel flow conditions (ICOLD, 1994, from Bakker et al., 1990). Note: (1) Overflow, (2) water supply, (3) inflow, (4) ballast, (5) overflow, (6) screen, (7) direction of flood, (8) sand trap, (9) filter, (10) Base soil (sand).

Extensive laboratory test programes were carried out by Bertram (1940), who worked under the direction of Terzaghi and Casagrande and then by United States Bureau of Reclamation (USBR 1955) and the US Corps of Engineers (US Corps of Engineers 1941).

The USBR method was widely adopted for design, so many existing dams were designed using this method.

9.2.1.1   Original USBR method

The original USBR method as described in USBR (1977) was:

(i) (a) $D_{15F}/D_{15B} = 5$ to 40, provided that;
    (b) the filter does not contain more than 5% fines passing 0.075 mm, and the fines should be cohesionless.
(ii) $D_{15F}/D_{85B} \leqslant 5$;
(iii) The grain size curve of the filter should be roughly parallel to that of the base material;
(iv) Maximum size particles in filter = 75 mm to prevent segregation during placement;
(v) For base materials which include gravel particles, the base material D_{15B} and D_{85B} etc. should be analysed on the basis of the gradation of the soil finer than 4.7 mm.

Criterion (i), i.e. $D_{15F}/D_{15B} \geqslant 5$ ensures that the filter is more permeable than the base soil. Criteria (ii), (iii) and (ib) are designed to ensure that the filter is sufficiently fine to control erosion of the base soil.

The authors' experience with applying this method was that, while it could be successfully applied to base soils which are clayey sands or sandy clays, it cannot be rigorously applied to base soils with a high clay and silt content. In particular:

– Rule (iii) cannot be followed (this is also the experience of Sherard and Dunnigan, 1985);
– Rule (ib) will almost always overrule (ia), often to the concern of inexperienced engineers, who feel that the "mathematic rule", i.e. (ia), should stand.

The authors' approach has been to be comfortable with rule (ib) overriding rule (ia), to ignore rule (iii) and substitute in its place the US Corps of Engineers (1941) requirement that the uniformity coefficient $D_{60F}/D_{10F} \leqslant 20$.

However we would not now use this method, nor does the USBR.

9.2.1.2   Sherard and Dunnigan method

The USSCS (United States Soil Conservation Service) (Sherard et al., 1984a, 1984b and Sherard and Dunnigan, 1985) carried out extensive laboratory testing to check filter criteria. They used several different test apparatus to simulate a concentrated leak in a dam.

These included:

– Tests using the equipment shown in Figure 9.6 for base soils in the fine to coarse sand ($D_{15B} = 0.075$ mm to 2.36 mm) range;
– Tests using slot and slurry tests on silt and clay base soils, in which a small amount of erosion was accepted as a successful filter;
– Tests using the "no-erosion" filter test apparatus shown in Figure 9.7. For a filter to be successful it was required that there should be no visible increase in the size of the hole in the base soil, and only "very slight" erosion of the base soil;
– Sherard and Dunnigan claimed to have tested dispersive and non-dispersive silt and clay soils, but in fact they only tested two dispersive soils (Foster, 1999, Foster and Fell, 1999a).

Based on these tests, Sherard and Dunnigan (1985, 1989) recommended the following:

1. For all soils with a gravel component (except Group 3 below), the filters should be designed on the grading of that part of the soil finer than 4.75 mm;
2. *Impervious Soil Group 1* (fine silts and clays): For fine silts and clays that have more than 85% by weight of particles finer than the 0.075 mm sieve, the allowable filter for design should have $D_{15F} \leqslant 9D_{85B}$;

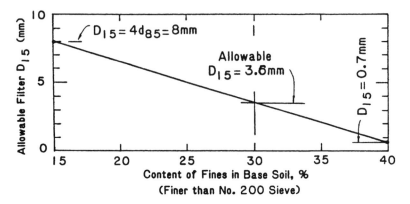

Figure 9.10.   Determination of allowable D_{15F} for filters for impervious Soil Group 4 (having between 15% and 40% finer than 0.075 mm sieve). (Sherard and Dunnigan, 1985, reproduced with permission of ASCE).

3. *Impervious Soil Group 2* (sandy silts and clays and silty and clayey sands): For sandy (and gravelly) impervious soils with 40% to 85% by weight (of the portion finer than the 4.76 mm sieve) finer than the 0.075 mm sieve, the allowable filter for design should have $D_{15F} \leqslant 0.7$ mm;

4. *Impervious Soil Group 3* (sands and sandy gravels with small content of fines): For silty and clayey sands and gravels with 15% or less by weight (of the portion finer than the 4.76 mm sieve) finer than the 0.075 mm sieve the allowable filter for design should have $D_{15F} \leqslant 4D_{85B}$ where D_{85B} can be the 85% finer size of the entire material including gravels;

5. *Impervious Soil Group 4*: For coarse impervious soils intermediate between Groups 2 and 3 above, with 15% to 40% passing the 0.075 mm sieve, the allowable filter for design is intermediate, inversely related linearly with the fines content and can be computed by straight line interpolation. As an example (Figure 9.10) for an impervious sandy soil with 30% of silty or clayey fines and $D_{85B = 2}$ mm, the allowable filter for design is in between the value of $D_{15F = 0.7}$ mm (for soils of Group 2) and $D_{15F} = 4 \times (2) = 8$ mm (for soils of Group 3), and is calculated as follows:

$$D_{15F} = \frac{40 - 30}{40 - 15} (8 - 0.7) + 0.7 = 3.6 \text{ mm}$$

6. As well as having D_{15F} sizes as set out above, the filters for Soil Groups 1 and 2 must be composed wholly of sand or gravelly sand in which $\not> 60\%$ is coarser (that is $< 40\%$ finer) than 4.76 mm and the maximum particle size is 50 mm;

7. The above criteria can be applied for all soils in Groups 1 and 2 regardless of the shape of the particle size distribution curve. For soils of Groups 3 and 4 the criteria apply to reasonably well-graded soils. For soils in Groups 3 and 4 which are highly gap-graded it is desirable to provide a filter for the finer portion of the gap-graded soil, or carry out No Erosion Filter Tests in the laboratory to select the appropriate filter.

Sherard and Dunnigan (1985) indicate that these criteria already incorporate an adequate factor of safety.

The criteria listed above do not include specific reference to limiting the "fines" content, i.e. silt and clay passing 0.075 mm sieve, except that (6) indicates filters for Soil Groups 1 and 2 should be composed "wholly of sand and gravel". Some limitation on fines content and nature such as that required by the USBR (1977) seems to be implied i.e. filters should contain not more than 5% fines passing 0.075 mm, and the fines should be "cohesionless".

It should be noted that for Soil Group 1 hydrometer particle size analysis will be required to define the particle size below 0.075 mm.

The Sherard and Dunnigan (1985, 1989) method has become widely adopted including by USBR (1987), USDA-SCS (1994) and throughout Australia.

9.2.1.3   *Kenney and Lau*

Kenney and co-workers at the University of Toronto, Canada, have carried out extensive testing on filters. Their work is reported in Kenney et al. (1985) and Kenney and Lau (1985).

(a) *Filters for cohesionless base soils*. Kenney et al. (1985) developed the concept of a controlling constriction size D_C^* which is a size characteristic of the void network in a granular filter and is equal to the diameter of the largest particle that can possibly be transported through the filter by seepage. They show that:

$$D_C^* \leq 0.25\, D_{5F} \tag{9.1a}$$

and

$$D_C^* \leq 0.20\, D_{15F} \tag{9.1b}$$

They found that D^* is primarily dependent on the size of the small particles in the filter, and not strongly dependent on the shapes of the particle size distribution curve for the filter as a whole.

The test set-up used by Kenney et al. (1985) is similar to the USSCS apparatus shown in Figure 9.6, but the seepage gradient was between 3 and 50. The cell was tapped lightly during the test to induce vibration. As might be expected, this was found to have a significant effect on the results, with vibration dislodging particles and facilitating their penetration into the filter.

Kenney et al. (1985) suggest that for filtering cohesionless base soils the following relationships should be applied:

$$D_{5F} < 4\, D_{50B} \tag{9.2a}$$

and

$$D_{15F} < 5\, D_{50B} \tag{9.2b}$$

This is based on the requirement that the cohesionless bases and filters have a uniformity coefficient $Cu \leq 6$, and the coarser of the filters given by the relationships should be adopted. If the base soil has a potential for "grading instability", i.e. it will not self filter, Kenney et al. suggest that consideration of allowable loss of fines may be necessary.

(b) *Filters for cohesive soils*. The work of Kenney et al. (1985) was largely on cohesionless base soils. They describe personal experience of two dam cores composed of widely graded glacial tills with particles from gravel to clay size, where silt and clay size particles (finer than 0.060 mm) were removed through the filters by seepage. They attribute this to low seepage velocities being unable to transport the coarser sand particles in the soil to form a "self filter". They suggest a rather arbitrary selection of particle size in the base soil (0.020 mm in the example given) which must not pass through the filter (i.e.

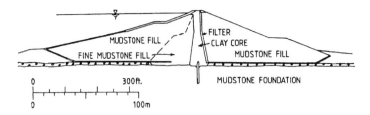

Balderhead Dam: Cross-Section

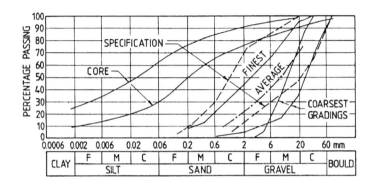

Balderhead Dam: Core and Filter Gradings

Figure 9.11.   Balderhead Dam core and filter gradings (Vaughan and Soares, 1982).

this becomes the controlling constriction size D_C^*) and application of $D_C^* \leqslant 0.25D_{5F}$ and $D_C^* \leqslant 0.20D_{15F}$.

They point out this gives finer filter requirements than those given by Sherard, Dunnigan and Talbot (1984b) and coarser than that of the method proposed by Vaughan and Soares (1982). They concluded that to design on use of precedents and experiments (essentially the method used by Sherard, Dunnigan and Talbot, 1984b) is acceptable, but that checking with some other approach may be prudent.

9.2.1.4   *Vaughan and Soares*

Vaughan and Soares (1982) provoked considerable discussion when, in a review of partial piping failure of Balderhead Dam in northern England, they claimed that in some base soils self filtering could not be relied upon and that it was necessary to design filters to prevent the passing of clay flocs into the filter.

The core for the Balderhead Dam had been constructed from well graded glacial till, which had a high resistance to erosion. Filters were crusher run hard limestone with grading limits shown on Figure 9.11.

Cracks occurred in the dam core by hydraulic fracture and Vaughan and Soares (1982) postulated that the partial piping failure which occurred more than a year after the dam was filled was due to erosion of the finer clay particles in the cracks. The flow velocity in the cracks was postulated to be too low to transport coarser particles from the core, so preventing self filtering within the core material and allowing the clay particles to erode into the filter. As discussed above, Kenney et al. (1985) observed similar behaviour in two dams constructed of widely graded glacial till.

To overcome this, Vaughan and Soares (1982) proposed the concept of a "perfect filter". In this approach, the particle size distribution of the clay core is obtained in a hydrometer

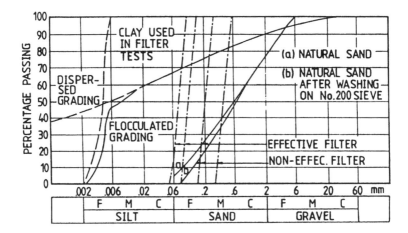

Filter Design for Cow Green Dam

Figure 9.12.   Filter designs for Cow Green Dam (Vaughan and Soares, 1982, reproduced with permission of ASCE).

test, but with no dispersant added. The water used for the test should be of the same chemistry as the water to be seeping through the dam since the floc size (degree of dispersion) is dependent on the water chemistry.

Having determined the particle size distribution, a sample of the floc sized sediment is prepared by sedimentation and used in filter experiments. In these experiments the flocs are introduced to the filter in a dilute suspension.

Tests carried out in this manner on soils from two dams indicated the need for a significant silt fraction in the filter and that even a fine to medium sand size filter may not be satisfactory, e.g. for Cow Green Dam shown in Figure 9.12.

The general tone of the discussion which followed the paper was that "based on experience with other laboratory experiments and performance of dams" the Vaughan and Soares method was too conservative. Vaughan and Soares in the closure discussion stated "the writers would stress that the need to meet these criteria completely is one for the designer of any particular dam to assess, along with other factors involved in the provision of effective filters".

The authors are not aware of any general acceptance of the "perfect filter" concept. Vaughan and Soares' case is weakened somewhat by the coarse grading of some of the filters as constructed for Balderhead Dam – these would not meet USBR (1977) or Sherard and Dunnigan (1985) criteria.

9.2.1.5   *Lafleur and co-workers*

Lafleur et al. (1989, 1993) prefer to work in terms of the opening size of the filter O_E and consider this in terms of the "indicative" or "self filtering" grain size, d_{SF} of the base soil. This has advantages where, as they do, one is considering conventional sand – gravel filter, and geotextiles. They indicate that if the retention ratio $R_R = O_E/d_{SF}$ is $>>1$, then continuous erosion will occur and that if $R_R <<1$, clogging (or "blinding") of the base soil/interface may occur, leading to a build up in pore pressure on the filter. Clogging is particularly a problem for internally unstable soils and the fines accumulate upstream of the filter, not within it.

Figure 9.13 summarizes the Lafluer et al. (1993) design approach. Note that it is not identical to Lafleur (1989) as reported in ICOLD (1994). In particular the retention

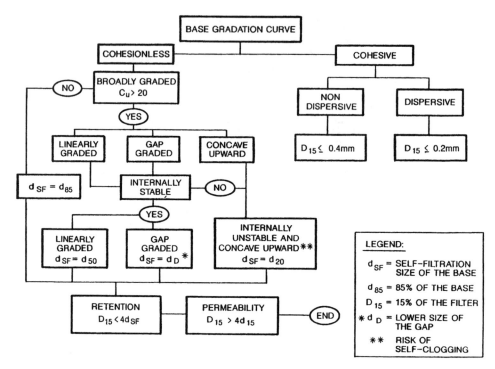

Figure 9.13.   Lafluer et al. (1993) method for design of conventional and geotextile filters.

criteria for aggregate filters is $D_{15}/5 < d_{SF}$, not $D_{15}/4 < d_{SF}$, so they have opted for coarser filters than in the ICOLD (1994) and Lafleur et al. (1989).

Figure 9.13 does not include design criteria for cohesive soils. Lafleur et al. (1989) recommend $D_{15F} \leq 0.4$ mm for non dispersive cohesive soils and $D_{15F} \leq 0.2$ mm for dispersive soils.

Sherard and Dunnigan (1985, 1989) recommend, for their Group 3 soils, that $D_{15F} \leq 4D_{85B}$, whereas Lafleur et al. 1993 recommend, for uniform granular soils, $D_{15F} < 5D_{85B}$. Given that Foster (1999), Foster and Fell (1999a) showed that erosion would continue for $D_{15F} > 9D_{95B}$ and would be excessive for $D_{15F} > 9D_{85B}$ (see Section 9.3), both criteria seem to have some margin of safety against large erosion losses.

9.2.1.6   Brauns and Witt

Brauns and Witt (1987), Schuler and Brauns (1993, 1997) and Witt (1993) propose that filter criteria should be based on a probabilistic analysis to allow for the variability in the particle size distributions of the base soil and the filter. Honjo and Vaneziano (1989) had a similar concept. The authors can see some merit in these approaches, but they are predicated implicitly that filters either "fail" *or "succeed". As discussed in Section 9.3, the reality is that such criteria apply to no-erosion acceptance criteria and that much coarser filters will eventually seal.

9.2.1.7   Foster and Fell

Foster (1999) and Foster and Fell (1999a) carried out extensive no-erosion tests using a test set-up similar to that of Sherard et al. (1984a,b), and reviewed the results of the USSCS tests as reported in internal reports by Sherard (Sherard, 1985a and b).

Table 9.2.   Summary of statistical analysis of the results of no-erosion filter tests, and proposed no-erosion filter criteria (Foster, 1999; Foster and Fell, 1999a, 2001).

Base soil group	Fines content (a)	Design criteria of Sherard and Dunnigan (1989)	Range of DF15 for no erosion boundary D_{85B}	Criteria for no erosion boundary
1	≥85%	$D_{15F} \leqslant 9D_{85B}$	$6.4–13.5D_{85B}$	$D_{15F} \leqslant 9D_{85B}$ (b)
2A	35–85%	$D_{15F} \leqslant 0.7\,\mathrm{mm}$	0.7–1.7 mm	$D_{15F} \leqslant 0.7\,\mathrm{mm}$ (b)
3	<15%	$D_{15F} \leqslant 4D_{85B}$	$6.8–10D_{85B}$	$D_{15F} \leqslant 7D_{85B}$
4A	15–35%	$D_{15F} \leqslant (40–pp\%$ $0.075\,\mathrm{mm}) \times (4D_{85B}–0.7)/$ $25 + 0.7\,\mathrm{mm}$	$1.6\ D_{15F}–2.5D_{15F}$ of Sherard and Dunnigan design criteria	$D_{15F} \leqslant 1.6D_{15F}\mathrm{d}$, where $D_{15F}\mathrm{d} =$ $(35–pp\%\ 0.075\,\mathrm{mm})$ $(4D_{85B}–0.7)/20 +$ $0.7\,\mathrm{mm}$

Notes: (a) The subdivision for Soil Group 2 and 4 was modified from 40% passing 0.075 mm, as recommended by Sherard and Dunnigan (1989), to 35% based on the analysis of the filter test data. The modified soil groups are termed Group 2A and 4A. The fines content is the % finer than 0.075 mm after the base soil is adjusted to a maximum particle size of 4.75 mm.
(b) For highly dispersive soils (Pinhole classification D1 or D2 or Emerson Class 1 or 2), it is recommended to use a lower D_{15F} for the no erosion boundary:
For Soil Group 1, use the lower limit of the experimental boundary, i.e. $D_{15F} \leqslant 6.4D_{85B}$.
For Soil Group 2A, use $D_{15F} \leqslant 0.5\,\mathrm{mm}$.

Table 9.2 summarizes the results. The proposed "criteria for no-erosion boundary" are intended for applying to the assessment of filter performance of existing dams and do not include any margin of safety.

This work showed that the division between Group 2 and 4 soils is better defined on a fines content of 35% than on the 40% used by Sherard et al. (1984a, b) and that the Sherard and Dunnigan (1989) criteria generally have a margin of safety but they are not sufficiently conservative to define no-erosion condition for dispersive soils.

9.2.2   Review of other factors affecting filter design and performance

9.2.2.1   Criteria to assess internal instability or suffusion

Numerous criteria exist for the assessment of suffusion including those by Sherard (1979), De Mello (1975), Kenney and Lau (1985) and Burenkova (1993). These criteria are based on the analysis of the grain size distribution of the soil. Schuler (1993) critically reviews these criteria and recommends the following:

– For gap-graded soils, use the method of splitting the grain size distribution at the point of inflection in the gap, Figure 9.14 (Sherard, 1979; De Mello, 1975);
   The soil is considered to be self-filtering if:

$$\frac{D_{15coarse}}{D_{85fine}} < 5 \tag{9.3}$$

where $D_{15coarse} = D_{15}$ of the coarse fraction, and $D_{85\,fine} = D_{85}$ of the fine fraction;
– For soils with smooth gradations with a tail of finer sizes use the criterion of Burenkova (1993), shown in Figure 9.15, or Kenney and Lau (1985, 1986) in Figure 9.16 and Figure 9.17.

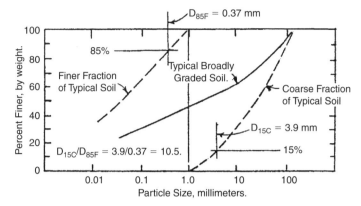

Figure 9.14.   Method of splitting the grain size distribution curve at an arbitrary point of separation for assessing internal instability, after Sherard (1979).

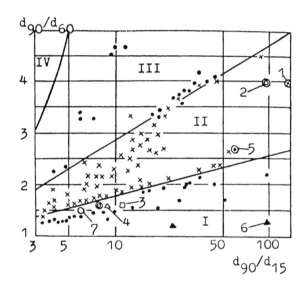

Key:	I	Zone of suffusive soils
	II	Zone of non-suffusive soils
	III	Zone of suffusive soils
	IV	Zone of artificial soils
	1-7	Non-suffusive soils by various other authors
	•	Suffusive soil (individual test)
	x	Non-suffusive soil (individual test)

Figure 9.15.   Method of assessing internal instability by Burenkova (1993).

9.2.2.2   *Segregation*

Characteristics which make sand-gravel filter segregate on placement include:

– A broad grading, particularly with maximum particle size >75 mm;
– A low percentage of sand and fine gravel sizes (<40% finer than 4.75 mm);
– Poor construction practices e.g. end dumping from trucks, high lift heights and poor control of stockpiling operations.

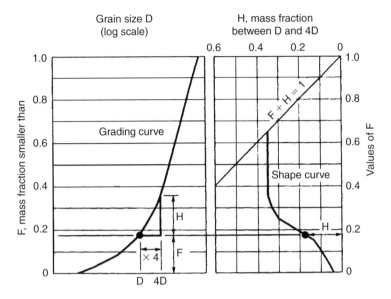

Figure 9.16.   Method of characterising the shape of a grading curve by Kenney and Lau (1985).

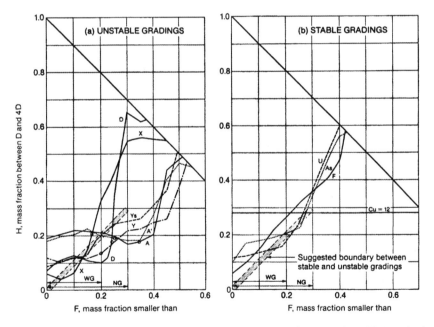

Figure 9.17.   Method of assessing internal instability by Kenney and Lau (1985, 1986), revised criteria from Kenney and Lau (1986).

Sherard and Dunnigan (1989) suggested issuing filters with a maximum size not greater than 50 mm and required no less than 40% sand and fine gravel passing 4.75 mm. Ripley (1986) recommended for filters in contact with the core:

– Maximum size 19 mm;
– Not less than 60% finer than 4.75 mm;
– Not greater than 2% finer than 0.075 mm.

These criteria were based on construction experience.

Kenney and Westland (1993) carried out laboratory tests and concluded that:

- All dry soils consisting of sands and gravels segregate in the same general way, independent of grain size and grain size distribution;
- Dry soils containing fines <0.075 mm segregate to a smaller extent than soils not containing fines;
- Water in sandy soils (mean size finer than 3 mm to 4 mm) inhibits segregation but has little influence on the segregation of gravels (mean size coarser than 10 mm to 12 mm).

The USDA-SCS (1994) and USBR (1987) have adopted a maximum size of 75 mm and limits on the minimum D_{10F} and maximum D_{90F} to limit segregation. These are described in Section 9.2.3. The authors believe that for narrow or thin filter zones, 75 mm may be too large and recommend the use of maximum size of 37 mm or 50 mm in these situations.

9.2.2.3   *Permeability*

The filter must be sufficiently permeable for the seepage flow to pass through it without significant build up of pressure. This has been taken into account by using the criteria $D_{15F}/D_{15B} > 4$ or 5, which ensures that the permeability of the filter is 15 to 20 times that of the soil. However just as important is to keep the fines content (silt and clay sized particles) to a minimum.

Figure 9.18 shows the influence on permeability of the type and amount of fines.

The authors' preference is to specify not greater than 2% fines and that the fines be non plastic. Some, e.g. USDA-SCS (1994), USBR (1987), allow 5% (non plastic) fines, but as is evident from Figure 9.18, this may reduce the permeability by one or two orders of magnitude compared to clean filter materials. The cost in washing to achieve not more than 2% fines is not high and generally worthwhile. The second advantage of low fines content is that the filters are unlikely to hold a crack.

Note that the discharge capacity of a filter drain system is a separate issue, discussed in Chapter 10.

9.2.3   *Review of available methods for designing filters with flow parallel to the filter*

ICOLD (1994) summarize testing at Delft Hydraulic Laboratory (Bakker, 1987) to test the condition where flow in the filter is along, or parallel to, the filter. Den-Adel et al. (1994) provide details of a method for designing such filters.

It is apparent that in these situations, erosion is less likely and considerably coarser filters than obtained from the criteria discussed in Sections 9.1 and 9.2.4 can be used.

Figure 9.19 summarizes some of the Delft Laboratory Testing.

In this figure the hydraulic gradient, i_{cr}, e.g. equal to the slope of the filter under rip rap under the upstream face of a dam, is plotted against D_{15F}/D_{50B} (D_{15}/d_{50} in Figure 9.19). Curves for sandy base soils with $D_{50} = 0.15$ mm and 0.82 mm are shown. Thus for a uniform sand base material with a d_{50} of 0.15 mm on a slope of 2.5H:IV ($i_{cr} = 0.4$), the ratio of D_{15F}/D_{50B} can be as high as about 8.

However great care should be taken in applying this approach, particularly to situations where, if the filter fails and erosion occurs, it is not practical to repair the damage. Readers should seek the latest literature and be willing to use the rather daunting detail outlined in Adel et al. (1994) if they are to use other than normal filter criteria.

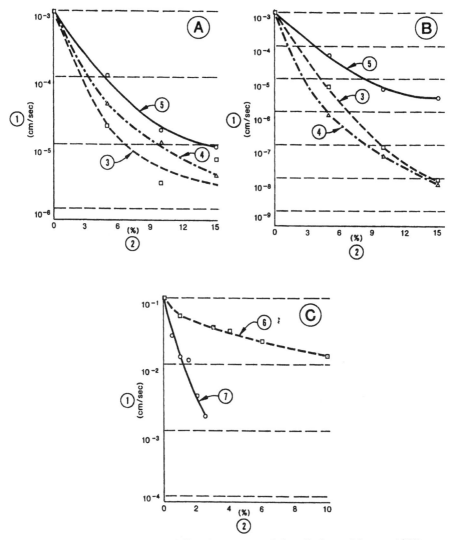

Notes: (A) Effect of fines on permeability of concrete sand (from Barber and Sawyer, 1952).
(B) Effect of fines on permeability of sand-gravel mixture (from Barber and Sawyer, 1952).
(C) Effect of fines on permeability of uniform fine sand (from Fenn, 1966).
(1) Coefficient of permeability, cm/s. (2) Percent of fines. (3) Silt. (4) Clay. (5) Limestone. (6) Kaolinite. (7) Calcium montmorillonite.

Figure 9.18.   Influence of the type and amount of fines on permeability of concrete sand, sand-gravel mixture and uniform fine sand (ICOLD, 1994, from US Corps of Engineers, 1986).

9.2.4   Recommended method for design of critical filters, with flow normal to the filter

The recommended method for design of critical filters, with flow normal to the filter (flow types N1 and N2), is based on Sherard and Dunnigan (1985, 1989). They however did not give explicit instructions on how to apply their criteria. This has been better detailed in USBR (1987) and USDA-SCS (1994).

The following is taken largely from the US Soil Conservations Service, USDA-SCS (1994). It has been modified to include the changes to the Base Soil Group boundaries and

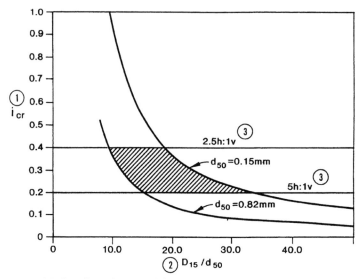

Notes: (1) critical gradient (2) ratio of the D_{15} size of the filter (D_{15F}) to the d_{50} size of the base (D_{50B}), (3) slope of the interface between the filter and base.

Figure 9.19. Critical gradient with steady flow parallel to the interface (ICOLD, 1994, based on Bakker, 1987).

for dispersive soils proposed by Foster (1999) and Foster and Fell (1999a). The base Soil Groups are re-named 1, 2A, 3A and 4 to distinguish them from those used by Sherard and Dunnigan (1985, 1989). The method is to determine filter gradation limits using the following steps:

Step 1: Plot the gradation curve (grain-size distribution) of the base soil materials. Use enough samples to define the range of grain sizes for the base soil or soils. Design the filter using the base soil that requires the smallest D_{15F} size for filtering purposes. Base the design for drainage purposes on the base soil that has a representative (say median) D_{15B} size.

Step 2: Proceed to Step 4 if the base soil contains no gravel (material larger than 4.75 mm) or if designing coarse filters, where the base soil is the fine filter.

Step 3: Prepare adjusted gradation curves for base soils that have particles larger than the 4.75 mm sieve.

– Obtain a correction factor by dividing 100 by the percent passing the 4.75 mm sieve;
– Multiply the percentage passing each sieve size of the base soil smaller than 4.75 mm sieve by the correction factor determined above;
– Plot these adjusted percentages to obtain a new gradation curve;
– Use the adjusted curve to determine the percentage passing the 0.075 mm sieve in Step 4.

Step 4: Place the base soil in a category determined by the percent passing the 0.075 mm sieve from the regraded gradation curve data according to Table 9.3.

Step 5: To satisfy filtration requirements, determine the maximum allowable D_{15F} size for the filter in accordance with the Table 9.4.

Step 6: To ensure the filter is sufficiently permeable, determine the minimum allowable D_{15F} in accordance with Table 9.5. The permeability requirement is determined from the D_{15} size of the base soil gradation before regrading.

Step 7: The width of the allowable filter design band must be kept relatively narrow to prevent the use of possibly gap-graded filters, but wide enough to allow manufacture. Adjust

Table 9.3.   Base soil categories.

Base soil category	% finer than 0.075 mm (after regrading, where applicable)	Base soil description
1	>85	Fine silts and clays
2A	35–85	Silty and clayey sands; sandy clays; and clay, silt, sand, gravel mixes
4A	15–35	Silty and clayey sands and gravel
3	<15	Sands and gravel

Table 9.4.   Filtering criteria for critical filters – maximum D_{15F}.

Base soil category	Filtering criteria
1	$\leq 9 \times D_{85B}$ but not less than 0.2 mm (a)
2A	≤ 0.7 mm (b)
3	$\leq 4 \times D_{85B}$ of base soil after regrading
4A	$(35 - A/35 - 15) [(4 \times D_{85B}) - 0.7$ mm] + 0.7 mm A = % passing 0.075 mm sieve after regrading (If $4 \times D_{85B}$ is less than 0.7 mm, use 0.7 mm)

Note: (a) $\leq 6D_{85B}$ for dispersive soils. (b) ≤ 0.5 mm for dispersive soils.
Dispersive soils are soils with pinhole classification D1 or D2, or Emerson class 1 or 2.

Table 9.5.   Permeability criteria.

Base soil category	Criteria
All categories	Minimum $D_{15F} \geq 4 \times D_{15B}$ of the base soil before regrading, but not less than 0.1 mm $\leq 2\%$ (or at most 5%) fines passing 0.075 mm sieve in the filter; fines non plastic

Table 9.6.   Other filter design criteria.

Design element	Criteria
To prevent gap-graded filters	The width of the designed filter band should be such that the ratio of the maximum diameter to the minimum diameter at any given percent passing value $\leq 60\%$ is ≤ 5
Filter band limits	Coarse and fine limits of a filter band should each have a coefficient of uniformity of 6 or less

the maximum and minimum D_{15F} sizes for the filter band determined in Steps 5 and 6 so that the ratio is 5 or less at any given percentage passing of 60 or less and adjust the limits of the design filter band so that the coarse and fine sides have a coefficient of uniformity D_{60}/D_{10} of 6 or less. Criteria are summarized in Table 9.6. This step is required to avoid the use of gap-graded filters. The use of a broad range of particle sizes to specify a filter grada-tion could result in allowing the use of gap-graded materials. Materials that have a broad range of particle sizes may also be susceptible to segregation during placement.

   Step 8: To minimize segregation during construction, use a maximum size of 75 mm in filter zones which are not less than 2 m wide or 0.5 m thick. For narrower and thinner fil-ter zones (particularly Zone 2A filters) use a maximum size of 37 mm or 50 mm. Consider the relationship between the maximum D_{90} and the minimum D_{10} of the filter. Calculate a preliminary D_{10F} size by dividing the minimum D_{15F} by 1.2. (This factor of 1.2 is based

Table 9.7. Segregation criteria

Base soil category	If D_{10F} (mm) is:	Then maximum D_{90F} (mm) is:
All categories	<0.5	20
	0.5–1.0	25
	1.0–2.0	30
	2.0–5.0	40
	5.0–10	50
	>10	60

on the assumption that the slope of the line connecting D_{15F} and D_{10F} should be on a coefficient of uniformity of about 6.) Determine the maximum D_{90F} using Table 9.7. For Zone ZB filters, use the coarse limit D_{10F} in Table 9.7.

Sand filters that have a D_{90F} less than about 20 mm generally do not require special adjustments for the broadness of the filter band. For coarser filters and gravel zones that serve both as filters and drains, the ratio of D_{90F}/D_{10F} should decrease rapidly with increasing D_{10F} sizes.

Step 9: Connect the control points to form a preliminary design for the fine and coarse sides of the filter band. Complete the design by extrapolating the coarse and fine curves to the 100 percent finer value. For purposes of writing specifications, select appropriate sieves and corresponding percent finer values that best reconstruct the design band and tabulate the values.

For situations where filters close to the allowable limits are being contemplated, or being checked in an existing dam, it is recommended that no-erosion tests as described by Sherard et al. (1984b) be carried out. This is particularly important for fine-grained soils, where the separation of the no-erosion and continuing erosion boundaries is small (see Section 9.3) and for dispersive soils, because of the limited amount of testing upon which the design criteria are based.

9.2.5 *Recommended method for design of less critical and non-critical filters*

9.2.5.1 *Filters upstream of the dam core*

The filter Zone 2C upstream of a central core earth and rockfill dam is not subject to continuous seepage exit gradients or to the risk of high exit gradients if the core cracks.

Zone 2C filters can therefore be designed less conservatively than Zone 2A and 2B filters. It is common only to require that "Zone 2C filters shall be constructed of well graded rockfill with a maximum size of 150 mm". This will result in most cases in a sand/gravel/cobble size mixture of rock fragments, which provides a transition between the earthfill core and the upstream rockfill. Since this zone will be won from the quarry (probably after passing through a "grizzly" to remove oversize rock) it is relatively inexpensive and it would be normal to place at least 4 m width, possibly 6 m on a larger dam. The added width is important so as to counter the effects of likely segregation.

However, care should be taken in adopting this approach. A more robust and defensible approach, which the authors would recommend, would be to accept a relaxed specification for these filters, but to ensure they are finer than that required to satisfy the excessive (or, at worst, continuing) erosion criteria. Be aware that using filter/transition materials with rocks up to 150 mm is a recipe for segregation problems and it would be better to limit the size to a maximum of 75 mm.

Even if a relaxed specification is adopted, the fines content (passing 0.075 mm) should not exceed 5% and the fines should be non plastic.

For any dam, including pumped storage dams, where the drawdown could be as high as several metres within hours, or on a daily or even twice daily cycle, it would be wise to use conventional filters designed according to Section 9.2.4, at least in the zone of daily drawdown.

9.2.5.2   *Filters under rip-rap*

There are two requirements for these filters:

– That they are coarse enough not to wash out of the rip rap;
– They are fine enough to prevent erosion of the soil beneath the filter.

Two layers of filters may be required to satisfy these requirements. If proper protection is required, no-erosion filter criteria should be satisfied. This is what is recommended by US Corps of Engineers (1984a).

It is fairly common to use more relaxed criteria in non-critical conditions where some damage can be tolerated. ICOLD (1993, 1994) suggest that, for bedding filter under rip rap, filters which are reasonably well graded, between a maximum of 80–100 mm and coarse sand sizes, are satisfactory for the great majority of dams, but warn that segregation will be a problem and that such filters will not protect against frequent surging from wave action.

The authors' view is that the design of these Zone 2D filters (Figure 13.5a and c) is not as critical as for Zone 2A filters in that, if damage does occur, it can usually be repaired, and so a relaxation of the strict no erosion rules may be appropriate. If there is a reasonably well-graded sandy gravel/ gravelly sand from 0.075 mm to 50 mm or 75 mm available either naturally or with a minimum of processing, this should be satisfactory in most cases. This is also the experience of Sherard et al. (1963). Filters which consist only of fine or medium (or even coarse) sand are unlikely to be satisfactory as they will not self-filter against the rip-rap.

The authors take the view that for thin layers of rip-rap, on important dams, with significant wave action, properly designed filters should be provided. Advantage may be taken of the knowledge of the work described above, or simply standard filters designed using Section 9.2.4.

For smaller dams, where the risk of deterioration and the need for repairs may be acceptable, geotextiles may be used under rip-rap. If so, the following points should be noted:

(a) Use non-woven needle punched heavy quality fabric e.g. Bidim A44;
(b) Use a maximum size of 150 mm cobbles or gravel-cobble mix directly onto the fabric to avoid developing folds in the fabric between the particles, below which soil could be eroded.

9.3   ASSESSING FILTERS AND TRANSITION ZONES IN EXISTING DAMS

9.3.1   *Some general issues*

When assessing the likelihood of internal erosion and piping failure of dams, it is necessary to consider the whole of the piping process – initiation, continuation, progression to form a pipe and breach (See Chapter 10). This section considers only the filters or transition zones, which mainly affect the continuation of erosion.

The starting point for assessing existing dams for internal erosion and piping is to determine, as reliably as is practical, what materials were used to build the dam and their properties. It is not sufficient to rely on a specification and assume those who built the dam followed the specification. Too often it has been found that materials with a higher fines content or coarser than specified have been used, particularly near the crest of the dam, as borrow areas became worked out. There seemed to be a lack of awareness in the 1950s and 1960s of the need for good filters near the crest of the dam, so the quality control was sometimes relaxed. Some dams even had the filters omitted near the crest (even though they were shown on drawings). Such "sins and omissions" were not restricted to small organisations – it happened in large organisations with good reputations as dam engineers.

The authors prefer to have available the following:

– Particle size distributions on samples taken from the dam during construction ("record testing");
– As-constructed drawings;
– Photographs of construction (illustrate whether, for example, filters were placed ahead of or after rockfill, or were placed in high lifts with subsequent risk of segregation);
– Test pits into the dam, to log and sample the relevant zones, with associated particle size distributions.

9.3.2   Continuing and excessive erosion criteria

Mark Foster (Foster, 1999; Foster and Fell, 1999a, 2001) used the test data from Sherard (1985a, b) and additional tests using no-erosion test equipment similar to that shown in Figure 9.3 to develop the concept of no erosion, some erosion, excessive erosion and continuing erosion shown in Figure 9.20.

The terms are defined as:

 (i) *No erosion*: filter seals with practically no erosion of the base material;
 (ii) *Some erosion*: filter seals after "some" erosion of the base material;
(iii) *Excessive erosion*: filter seals, but after "excessive" erosion of the base material;
(iv) *Continuing erosion*: the filter is too coarse to allow the eroded base materials to seal the filter.

The continuing erosion boundary was determined by carrying out additional filter tests, using a modified version of the NEF test, called Continuing Erosion Filter (CEF) tests, to

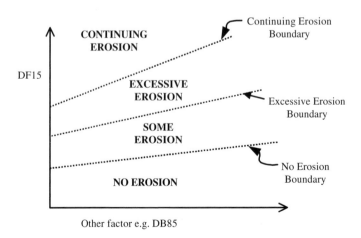

Figure 9.20.   Filter erosion boundaries (Foster, 1999; Foster and Fell, 1999a, 2001).

determine the continuing erosion boundary for soils with $D_{85B} > 0.1$ mm. The test procedures of the CEF tests were essentially the same as those of the NEF test, as described by Sherard and Dunnigan (1989), but with the following modifications:

– Water passing through the filter during the tests was collected and the eroded materials dried and weighed to determine the loss of base soil required to seal the filter;
– Progressively coarser filters were used until the filter was not sealed;
– Thicker base specimens were used to allow for greater erosion losses.

Details of the CEF test setup are shown in Figure 9.21.
The tests were carried out for until it was evident the filter was sealed or it was judged that the filter was not going to seal no matter how much erosion of the base soil occurred. The filters were judged to have sealed when all of the following conditions were reached:

– Full mains pressure was maintained in the space above the base specimen as measured on the pressure gauge;
– Water passing through the filter was clear;
– The flow rate of water passing through the filter had decreased substantially from the initial flow and was relatively constant.

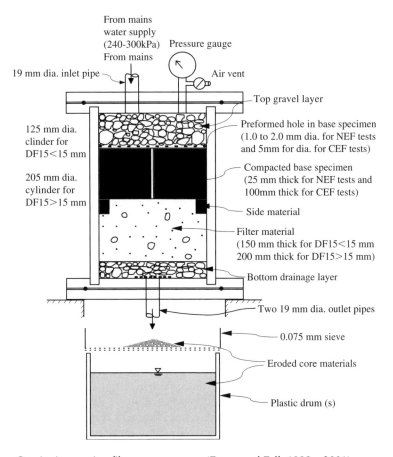

Figure 9.21.   Continuing erosion filter test apparatus (Foster and Fell, 1999a, 2001).

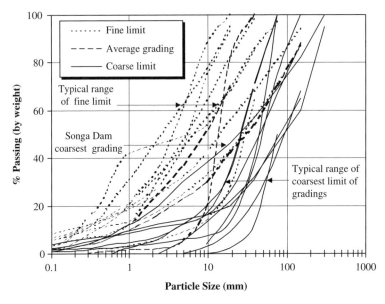

Figure 9.22.   Gradings of filters which have experienced poor filter performance.

Details of the tests and the results of testing are given in Foster (1999), Foster and Fell (1999a) and Foster and Fell (2001).

The excessive erosion boundary was determined from consideration of the performance of dams which had resulted in damage to the dam in the form of sinkholes and large leakages, but none had resulted in failure (i.e. breaching) of the dam (Foster et al., 1998).

Figure 9.22 presents the gradations of the filters of some of the dams with poor filter performance. The case histories generally involved piping of core materials into coarse or segregated downstream filters in zoned earthfill or central core earth and rockfill dams. The dams were generally constructed in the 1960s to 1970s which coincides with a period when there was a trend away from the use of uniformly graded multiple filters and towards the use of a single filter of substantial width and broad gradation (Response by Ripley in ICOLD, 1994). The filter gradings shown in Figure 9.22 have wide gradings and low proportions of sand sizes which would tend to make them susceptible to segregation during construction and also potentially make them internally unstable.

There are also several reported cases of concentrated leaks that have developed through the cores of dams but which have evidently sealed due to the effectiveness of the downstream filter, as evidenced by observations of near hydrostatic piezometer levels in the downstream section of the core and "wet seams" in the core (Sherard, 1985a). Peck (1990) also describes several examples from the literature of dams, which have shown evidence that some form of filtering action has taken place at the core-filter interface.

Only two dams, Rowallan Dam and Whitemans Dam, were found in the literature which have experienced poor filter performance involving piping of fine grained core materials with $D_{95B} < 2$ mm. In both cases, the finest core material and coarsest filter combination fall into the continuing erosion category as defined by the laboratory tests, i.e. $D_{15F}/9 > D_{95B}$. At Rowallan Dam where very fine abutment contact earthfill was protected by the filter provided for the bulk of the earthfill, the filter with the coarsest grading has a filter opening size $(D_{15F}/9)$ of $11/9 = 1.2$ mm. This is larger than the D_{95B} of the finest grading of contact clay soil of 0.9 mm. At Whitemans Dam, core materials were eroded into the

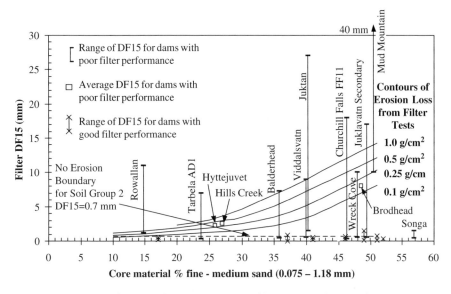

Figure 9.23. Comparison of erosion losses measured in filter tests to dams with poor and good filter performance.

downstream gravel zone; the filter opening size of this coarsest gravel zone material was 1.0/9 − 0.1 mm and the D_{95B} of the finest core grading is 0.075 mm.

The other dams with poor filter performance generally have broadly graded core materials which fall into Soil Groups 2 and 4 (fines content 15–85%) and have $D_{95B} > 2$ mm, which places them in the soil types where a continuing erosion boundary could not be identified by the CEF tests. Figure 9.23 shows the range of D_{15F} of the filter plotted against the average percentage of fine-medium sand sizes (% 0.075 mm to 1.18 mm) of the core material for the dams which have had poor and good filter performance. The contours of equal erosion losses from the CEF tests and the no erosion boundary for Soil Group 2 soils ($D_{15F} = 0.7$mm) are shown on the plot. The percentage fine-medium sand has been taken off the grading curves of the core materials after adjustment to a maximum particle size of 4.75 mm.

Dams with good filter performance generally have filters with an average $D_{15F} \leqslant 0.5$ mm, which is finer than the Sherard and Dunnigan design criterion for Soil Group 2 base soils ($D_{15F} \leqslant 0.7$ mm). The coarsest gradings are only slightly coarser than this (D_{15F} up to 1.5 mm).

Dams with poor filter performance have filters with an average $D_{15F} > 1.0$ mm and generally with D_{15F} greater than or equal to about the 0.25 g/cm^2 contour of erosion loss. Where a range of filter gradations is given, the coarsest grading is significantly coarser than the design criteria. Balderhead Dam has the finest coarse limit grading of the filter ($D_{15F} = 7$ mm) and this is 10 times coarser than the recommended design criteria. The D_{15F} for the coarsest gradings is typically greater than or equal to about the 1.0 g/cm^2 contour.

One notable exception is Songa Dam, which has a range of D_{15F} of 0.4–1.5 mm. This is considerably lower than the other dams with poor filter performance. However, the gradings of the filter for Songa Dam (Figure 9.22) have a wide grading and low proportion of sand sizes which, as discussed later on, would have made the filters particularly susceptible to segregation during placement. Therefore it is likely that the actual gradings of the filter in this dam are probably locally much coarser than that shown.

Table 9.8. Criteria for the excessive and continuing erosion boundaries (Foster, 1999; Foster and Fell, 1999a, 2001).

Base soil	Proposed criteria for excessive erosion boundary	Proposed criteria for continuing erosion boundary
Soils with $D_{95B} < 0.3$ mm	$D_{15F} > 9\,D_{95B}$	
Soils with $0.3 < D_{95B} < 2$ mm	$D_{15F} > 9\,D_{90B}$	
Soils with $D_{95B} > 2$ mm and fines content $>35\%$	Average $D_{15F} > D_{15F}$ which gives an erosion loss of $0.25\,g/cm^2$ in the CEF test ($0.25\,g/cm^2$ contour line in Figure 9.23) or Coarse limit $D_{15F} > D_{15F}$ which gives an erosion loss of $1.0\,g/cm^2$ in the CEF test ($1.0\,g/cm^2$ contour line in Figure 9.23)	For all soils: $D_{15F} > 9D_{95B}$
Soils with $D_{95B} > 2$ mm and fines content $<15\%$	$D_{15F} > 9\,D_{85B}$	
Soils with $D_{95B} > 2$ mm and fines content 15–35%	$D_{15F} > 2.5\,D_{15F}$ design, where D_{15F} design is given by: D_{15F} design = $(35\text{-pp}\%\ 0.075\,\text{mm})(4D_{85B}\text{–}0.7)/20 + 0.7$	

Table 9.8 shows the criteria for the excessive and continuing erosion boundaries. The criteria for the excessive erosion boundary are selected from the case studies and the laboratory testing, and dams which experience erosion to this limit may have large piping flow discharges – up to say 1 to 2 m^3/sec. Whether a dam can withstand such flows without breaching, depends on the discharge capacity of the downstream zone and whether unravelling or slope instability may occur. It is also likely that, if a dam experiences a piping event which does eventually seal, it will experience another later, as the erosion process moves laterally.

The criteria listed in Table 9.8 should be used with caution and, for final decision making, should be supported by laboratory tests using the filter/transition and core materials from the dams. The results should be tempered with sound dam engineering judgement.

The procedure has been used in a limited number of dams and has proven to be a valuable aid to decision making. Bell et al. (2001) describe the assessment of internal erosion and piping for Eucumbene Dam. Figure 9.24 shows the particle size distributions of the core, the particle size required for a filter designed according to Sherard and Dunnigan (1989) and filter/transition materials tested in the continuing erosion tests. These distributions are truncated at the coarse end to facilitate laboratory testing.

Tests on these materials all eventually sealed, although considerable erosion occurred.

There are some broader implications to these studies in that it is apparent that, for most soils, there is a considerable margin between design no-erosion and continuing erosion criteria. This assists in explaining why, despite the statistical variability in the particle size distributions of the base soil and filters, (which might on first consideration imply that there was a significant potential for piping to occur), this is unlikely because the no-erosion and continuous erosion boundary criteria are not exceeded. Figure 9.25(a) shows this conceptually.

However it is important to recognise that for fine ground base soils, the separation between the design limit, no-erosion and continuing erosion boundaries is much less – Figure 9.25(b) and Tables 9.2 and 9.8. Hence if the filter design is close to the limiting D_{15F}/D_{85B}, very close control on construction will be necessary.

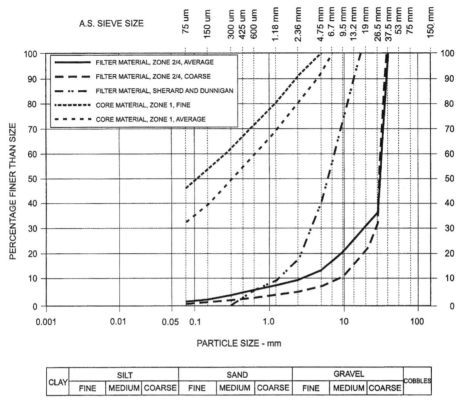

Figure 9.24. Particle size distribution for core and filters for continuing erosion tests, Eucumbene Dam (Bell, et al., 2001).

It is important to recognize the limitation of the laboratory test apparatus used to assess the performance of a filter. If the filter is a non-erosion filter or perhaps a filter with a slightly coarser grading, the ultimate seal may only affect a very small area of the soil/filter interface in the test. On the other hand, for a very coarse filter, such as the in-place filter for Eucumbene dam, one must be cautious when deciding if this test indicates "ultimate" sealing or whether it shows significant localized weakening of the soil with the high likelihood of the leak moving sideways. Care must also be taken in the laboratory test to prevent erosion up the sides of the sample.

9.3.3 Excessive fines content in filters or transition zones

It is unfortunately not uncommon to be presented with information showing that the filters or transition zone in an existing dam were constructed with fines content (% passing 0.075 mm) greater than the 5% normally accepted as an upper limit and/or that the fines are, contrary to accepted practice, plastic.

The question which arises is whether the filter or transition will "hold a crack", and not perform its filter function. Vaughan (1982) and ICOLD (1994) describe a simple test attributed to Vaughan for assessing this.

"A simple test, suitable for use in a field laboratory, has been devised to examine filter cohesion. It consists of forming a cylindrical or conical sample of moist compacted filter, either in a compaction mould, or in a small bucket such as is used by a child on a beach;

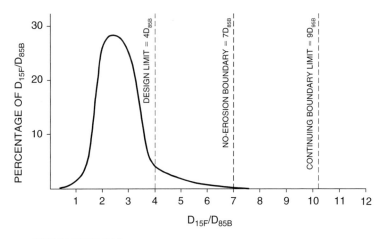

(a) SANDY BASE SOIL

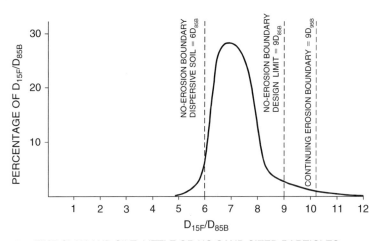

(b)   FINE CLAY AND SILT, LITTLE OR NO SAND SIZED PARTICLES

Figure 9.25.   Conceptual distribution of D_{15F}/D_{85B} for (a) clean sandy base soil, (b) base soil fine clay and silt with little or no sand sized particles, with design, no erosion and continuing erosion criteria.

standing the sample in a shallow tray (if a bucket is used the operation is exactly as building a child's sand castle) and carefully flooding the tray with water. If the sample then collapses to its true angle of repose as the water rises and destroys the capillary suctions in the filter, then the filter is non-cohesive. Samples can be stored for varying periods to see if cohesive bonds form with time. This test is, in effect, a compression test performed at zero effective confining pressure and a very small shear stress, and it is a very sensitive detector of a small degree of cohesion."

The authors' judgement is that it is the combination of fines content, the plasticity of the fines and degree of compaction which is important. One is more likely to accept that a filter with excessive and/or plastic fines will collapse and not hold a crack, if it is not well compacted. Each case should be considered on its merits, taking into account whether zones downstream of the filter may meet continuing erosion criteria, so even if the filter holds a crack, erosion will eventually seal.

Table 9.9.   Criteria for filters adjacent to perforated collector pipes.

Situation	Criteria
Non critical drains where surging or gradient reversal is not anticipated	$D_{85F} \geqslant$ perforation size
Critical drains where surging or gradient reversal is anticipated	$D_{15F} \geqslant$ perforation size

9.4   DESIGN CRITERIA FOR PIPE DRAINS AND PRESSURE RELIEF WELL SCREENS

9.4.1   *Pipe drains*

The authors do not favor the use of pipe drains to increase the discharge capacity of filter/drains, because they are susceptible to corrosion and blocking and joints may open with differential movement in the dam and the foundation, allowing the surrounding filter material to enter the pipe and block it. This view is supported by the USBR (1987) among others.

For the assessment of existing installations, the USDA-SCS (1994) requirements as shown in Table 9.9 are recommended.

The USBR (1987) use rather more relaxed criteria, with $D_{85F} \geqslant 2 \times$ (perforation size).

9.4.2   *Pressure relief well screens*

ICOLD (1994) quote Hadj-Hamou et al. (1990) requirements for well screen sizing that, for minimum wash-in of filter material during well development, the well screen slot size should be equal to D_{10F} of the filter surrounding the screen.

9.5   SPECIFICATION OF PARTICLE SIZE AND DURABILITY OF FILTERS

9.5.1   *Particle size distribution*

The design rules outlined in Section 9.2.4 do not in themselves entirely allow selection of the particle size distribution of filters. The emphasis in those rules is on "spot" limitations within the grading curve. For filters to be manufactured and placed in a dam the full particle size distribution limits have to be designed as for example in Figure 9.26.

Some factors which should be taken into account are:

– The wider the particle size grading allowed and the more compatible this is with the available sand and gravel, the lower the cost of manufacture. If natural sand and gravel deposits are available it may be possible to match the filter grading requirements to the grading of those deposits.
– Breakdown of particles (particularly coarser particles) occurs during placement and compaction. Specifications should apply to the filter after placement in the dam. This means that the contractor may have to manufacture the filters coarser than the specifications.
– On most large new dam projects concrete aggregates and filters are produced in the same crushing and screening plant, so sieve sizes on which specifications are based should be the same for both.

Charleton and Crane (1980) and Magnusson (1980) discuss some of the practical problems associated with filter manufacture and design.

An example of design of Zone 2A and 2B filters is given in Figures 9.26 and 9.27. The "steps" are those detailed in Section 9.2.4.

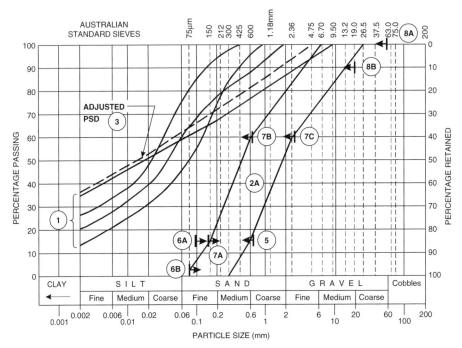

Notes: Step 1 – Plot particle size distributions (PSD); Step 3 – Adjust PSD for gravelly soils; Step 4 – Select base soil category for finest soil – Type 2; Step 5 – Determine D_{15Fmax} from Table 9.4 = 0.7 mm; Step 6 – Determine permeability criterion; $D_{15Fmin} = 0.1$ mm, 6(A), less than 2% passing 0.075 mm (6B); Step 7 – D_{XFmax}/D_{XFmin} less than 5 e.g. $D_{15Fmax} = 0.7$ mm; $D_{15Fmin} = 0.14$mm (7A); D_{60}/D_{10} less than 6, say $D_{10Fmin} = 0.12$; $D_{60Fmin} = 0.72$ (7B); $D_{10Fmax} = 0.45$, D_{60Fmax} less than $5D_{10Fmax} = 2.7$mm (7C); Step 8A – D_{100F} less than 50 mm or 75 mm, say 50 mm (8A); Step 8 B – $D_{90Fmax} = 20$ mm (Table 9.7).

Figure 9.26.   Example of design of a Zone 2A filter.

Most experienced dam designers will design the Zone 2A filter for the finest soil to be used in the dam, or which has been identified by investigations of an existing dam. It is not, in the view of the authors, defensible to design on an average grading – the filters are required to stop all soils in the dam core or foundation eroding, not just half of them!

If one can be confident that finer soils in the proposed borrow area for a new dam will be mixed with coarser soils by the loading, dumping and spreading of the soil, the Zone 2A filter may be designed for other than the finest soil. However this is difficult to control.

In this example the D_{85B} is less than 0.075 mm, and must be obtained from hydrometer particle size analysis. This should be done using the standard approach, in which dispersant is added to the soil as part of the test procedure.

For Zone 2B, a maximum particle size of 75 mm has been adopted. Larger size particles may be used in Zone 2B filters but at the risk of segregation in stock piles and during placement and the authors would not recommend this be done.

9.5.2   Durability

Filters are required to be constructed of sand and gravel which is sufficiently durable not to break down excessively during the mechanical action of placement in the dam, under the chemical action of seepage water or under wetting and drying within the dam.

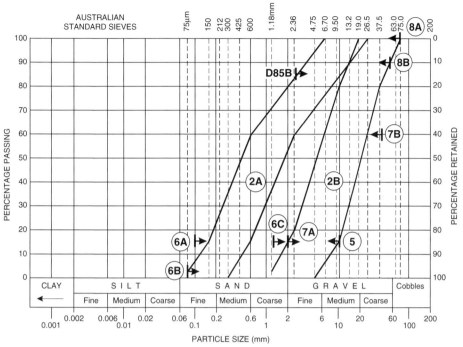

Notes: Step 1 – Particle size distribution for Zone 2A becomes the "Base Soil" – the fine side boundary is critical. $D_{85B} = 2.5$ mm. Go to Step 4. Step 4 – Soil is type 3 (<15% passing 0.075 mm). Step 5 – Determine D_{15Fmax} from Table 9.4. D_{15Fmax} less than $4D_{85B}$ = less than 4×2.5 mm = 10 mm (5). Step 6 – Determine permeability criterion, $D_{15Fmin} = 0.1$ mm, (6A), < 2% passing 0.075 mm (6B), $D_{15Fmin} = 4D_{15B} = 4 \times 0.7$ mm or = 2.8 mm. However for average $D_{15} = 4 \times 0.3$ mm = 1.2 mm accept this (6C). Step 7 – D_{XFmax}/D_{XFmin} less than 5 e.g. $D_{15Fmax} = 10$ mm, therefore $D_{15Fmin} = 2$ mm (7A), D_{60}/D_{10} less than 6, so say $D_{10Fmax} = 6.7$ mm, $D_{60Fmax} = 40$ mm (7B); Follow step 7 at 60% Passing when finalising grading. Step 8 – $D_{10F} = 7$ mm, so max $D_{90F} = 50$ mm from Table 9.7 (8B).

Figure 9.27.   Example of design of a Zone 2B filter.

9.5.2.1   *Standard tests for durability*

Table 9.10 shows the specified durability requirements for several dams constructed in Australia. Also shown for comparative purposes are the requirements for fine and coarse concrete aggregates as specified by Standards Association of Australia (1985).

These specifications can be used as a guide to selecting reasonable limits. The following comments are offered:

(a) The Los Angeles abrasion and wet strength, wet/dry strength variation tests largely assess the susceptibility to breakdown under the mechanical action of placement and rolling in the dam. It can be seen that dam specifications are generally less stringent than AS2758.1. If materials test marginally within the specification it would be advisable to carry out field compaction trials of the filters, to observe directly the degree of breakdown under rolling and the size of the "broken down" product. It may be practicable to make design changes, e.g. wider or thicker filters, to accommodate the breakdown.

(b) The Sodium Sulphate soundness test assesses the susceptibility to breakdown under wetting and drying, with the added involvement of sodium sulphate in the solution

Table 9.10. Durability requirements for zone 2A and 2B filters, and for concrete aggregates.

Authority and dam	Zone	General statement	Los Angeles abrasion value	Sodium sulphate soundness	Wet strength and wet/dry strength variation	Other requirements
Water resources Commission of Queensland, Peter Faust Dam (1989)	2A	Uncrushed, non plastic, free draining, clean, hard, durable sand free of organic material				
	2B	Hard, durable dense, fresh to slightly weathered (≤30% slightly weathered in +4.75 mm size)		≤15% on minus 4.75 mm wet/dry	≥80 kN wet strength, ≤35% variation	
Water authority of Western Australia, Harris Dam (1989)	2A	Hard, durable fresh rock	≤45% at 500 revolutions	≤14% on minus 0.3 mm	Not specified	Satisfy organic content in AS141 section 34
Melbourne and metropolitan Board of works, Cardinia Creek Dam (1970)	2A and 2B	Hard, durable, non plastic	≤40% at 500 revolutions	≤14% on minus 0.3 mm	Not specified	
Snowy mountains Hydroelectric authority, Talbingo Dam (1967)	2A	Hard, durable fresh rock	≤40% at 500 revolutions	≤14% on minus 0.3 mm		
	2B transition	Moderately weathered to fresh rock unaffected by chemical alteration (rhyolite, weathered tuff or porphry)				Zone 2B is wide and not strictly comparable to Zone 2B as discussed in this book

AS2758.1, 1985. Aggregates and rock for engineering purposes		Test either Los Angeles and sulphate soundness OR wet strength variation				Experience shows fine grained naturally occurring aggregates have generally undergone substantial weathering in their natural environment and rarely require durability testing. Also specify particle density, shape, water absorption
	Fine aggregates (<5 mm)			≤15% protected and moderate conditions, ≤12% severe conditions		
	Coarse aggregate		≤35% CG, ≤25% FG, moderate conditions; ≤30% CG, FG not specified, severe conditions	≤12% protected, ≤9% moderate, ≤6% severe	≤80 kN wet strength, ≤35% wet/dry variation moderate conditions, ≥100 kN wet strength, ≤25% wet/dry variation severe conditions	

CG = coarse grained rock; FG = fine grained rock.

used for soaking. The sodium sulphate penetrates fine cracks in the aggregates and expands on crystalization in the drying phase, breaking apart the aggregate particles.

The test is particularly severe for most dam applications where high salt contents are not present. This is reflected in the dam specifications in Table 9.10, where the requirements are less severe than for coarse concrete aggregates (similar for fine aggregates).

If materials are testing as marginal for this test, it is important to inspect the product of breakdown. In some rocks the product will be a fine silt and sand, which is likely to affect filter permeability markedly. In others it is simply the breakup of coarse particles into two or three smaller particles, which does not affect permeability greatly.

It is quite common to have different performance in these tests depending on the size fraction being tested. This is particularly the case for naturally occurring aggregates from a river which drains a mixed geological environment where the mineralogy of the individual particles may vary substantially.

The authors' opinion is that for important tailings dams which are storing water with a high acidity or high salts content, it may be necessary to impose more severe requirements on sulphate soundness than normally accepted for water dams. It may also be necessary to test with the more severe magnesium sulphate if the water in the tailings dam has a high magnesium sulphate content.

9.5.2.2 *Possible effects if carbonate rocks are used as filter materials*

Rocks containing significant proportions of carbonate minerals (e.g. limestone, dolomite or marble) may perform satisfactorily in the tests in 9.5.2.1, but could be unsuitable for use in filter zones, because any of the following effects may be possible, during the lifetime of the dam.

(a) Change of grading, due to dissolution;
(b) Partial dissolution and recemention;
(c) Interlocking of grains due to pressure-solution.

Effect (b) or (c) could cause a filter zone to become cohesive and be capable of sustaining an open crack or erosion tunnel. All 3 possible effects are discussed in more detail in 3.7.7, 3.7.7.1, 3.7.7.2 and 2.9.4.

The authors consider that carbonate materials should not be adopted for use in a filter zone, until the risk of malfunction due to any of the above, during the lifetime of the dam, has been assessed by experienced geochemical specialist or team.

9.5.2.3 *Effects if rocks containing sulphide minerals are used as filter materials*

Rocks containing small amounts (1% or less) of sulphide minerals may also perform satisfactorily in the tests in 9.5.2.1, but in service, the sulphides will oxidise and produce sulphuric acid and sulphate minerals. Effects can include clogging or cementation of the filter, by sulphate minerals (or iron hydroxides?) and acid drainage problems (see 2.9.4 and 3.7.8). The effects may also include weakening of the filter rock and acid attack on other embankment materials.

The authors recommend that rocks containing sulphide minerals should be avoided, whenever possible, when selecting filter zone materials.

9.5.2.4 *Other investigations for filter materials*

In the light of all of the above, the authors recommend that as well as the laboratory tests discussed in 9.5.2.1, the following should always be undertaken:

(i) If the proposed material is bedrock, quarried and crushed, or alluvial sand/gravel, the mineral composition of the particles should be determined.

(ii) If material is found to contain even small amounts (e.g. 1% to 5%) of carbonates, gypsum, or much less than 1% of sulphide minerals, then it should be subjected to further laboratory and field studies to assess its likely performance in the proposed filter zone.

In the field the studies are primarily of the observational type. For all types of materials a search should be made for existing (preferably very old) deposits of the materials e.g. old fills, pavements or natural screes. In the case of alluvial materials being considered for use, the actual deposit in question is also studied.

The deposits are carefully exposed by trenching or pitting. Evidence of any cohesive behaviour is looked for during excavation and the exposed faces and spoil are examined for any particles which have become cemented together. Evidence of weathering or solution of particles is also looked for.

When judging the suitability or otherwise of a material, from the results of such observational tests, differences between the environments (particularly moisture) in the old deposit and the proposed filter zone must be taken into account.

(iii) Volcanic ash materials should also be checked carefully for long term chemical stability and for any tendency to develop cohesion when compacted. As well as the laboratory and field tests described in (ii) above, field compaction trials are advisable for sand-sized volcanic ash materials.

9.6   DIMENSIONS, PLACEMENT AND COMPACTION OF FILTERS

9.6.1   *Dimensions and method of placement of filters*

9.6.1.1   *Some general principles*

The following factors should be taken into account when selecting the dimensions of zones of filters.

(a) For filters upstream and downstream of the dam core;
  – The theoretical width to achieve filtering action (see below);
  – The discharge capacity required along the filter zone e.g. for chimney drains (see Section 10.3);
  – The particle size distribution of the filter and the potential for segregation. Wider filter zones should be used if segregation is likely and narrower zones may be used for more uniformly graded materials;
  – Potential displacement under major earthquake – filters need to be wide enough to remain effective even if displacements occur in an earthquake;
  – The size of the dam – there is a tendency to use minimum width of 2.5 m or 3 m for very large (>100 m to 150 m high) dams, to allow for differential movement during construction and simply for conservatism;
  – The method of placement, as detailed below.
(b) For horizontal, or near horizontal filters e.g. in horizontal drains, on the surface of erodible foundations;
  – The theoretical thickness to achieve filtering action (see below);
  – The discharge capacity required along the filter zone e.g. for a horizontal drain (see Section 10.3);
  – The particle size distribution of the filter, the potential for segregation and maximum particle size. As a guide, the horizontal filter layer thickness should be not less than 20 times the maximum particle size, to ensure that, given good segregation control, there are very unlikely to be continuous coarse zones through the filter layer;
  – The method of placement, as detailed below.

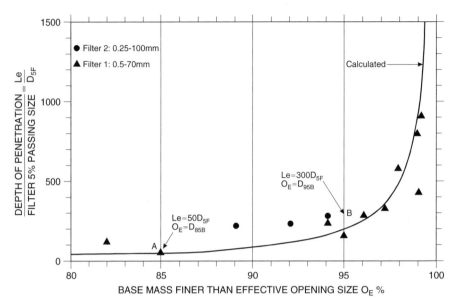

Figure 9.28.   Depth of penetration of base soil into filter material (Le) versus effective opening size (O_E) (Adapted from Witt, 1993).

The theoretical minimum width or thickness for filters designed according to no-erosion criteria, is very small, and does not control the dimensions of filters. Witt (1986, 1993) carried out some calculations and experiments which demonstrated that the depth of penetration of the base soil into the filter is small, even if the filter is somewhat coarser than required by the design criteria described above.

Figure 9.28 shows the depth of penetration versus effective opening size O_E

The graph shows:

At A, $O_E = D_{85B}$,   Depth Penetration = $50D_{5F}$
and at B, $O_E = D_{95B}$,   Depth Penetration = $300D_{5F}$

Given that $O_E = D_{15F}/9$ (Sherard et al. 1984a) this means that for $D_{15F} = 9D_{85B}$, Depth penetration = $50D_{5F}$.

Sherard and Dunnigan (1985, 1989) design criteria range from $D_{15F} \leq 4D_{85B}$ for Group 3 soils, to $D_{15F} \leq 9D_{85B}$ for Group 1 soils, so the depth of penetration will be less than $50D_{5F}$ for filters designed according to Sherard and Dunnigan criteria. Commonly, Zone 2A filters would have a D_{5F} less than 0.5 mm, so the depth of penetration would be less than 25 mm.

Larger depths of penetrations, and hence larger minimum thicknesses would be calculated using Witt (1986, 1993) approach for Zone 2A filters penetrating Zone 2B, but it is not clear if his work would apply to this case.

For assessing the filters in existing dams, the degree of compliance with these requirements would be taken into account.

9.6.1.2   Placement methods

Given that the width needed to achieve a proper filter against the protected soil is small, the width of filter zones is often determined by construction requirements.

The selection of the economic dam zoning should account for filter placement requirements and the cost of manufacture of the filter materials. In most dam projects the cost of

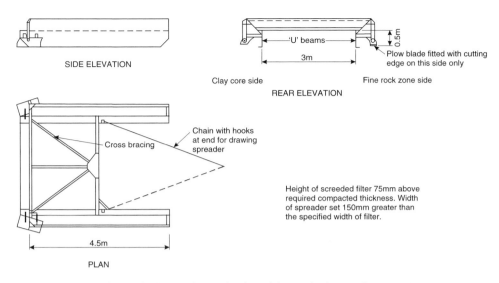

Figure 9.29.   Typical spreader box and screed (adapted from Charlton and Crane, 1980).

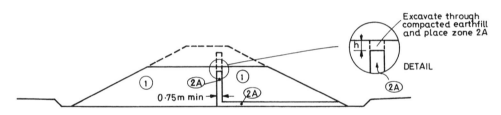

Figure 9.30.   Construction of vertical chimney drain by excavation through earthfill.

manufacture of filters is high and there is a need to keep widths to a practical minimum. The possible exception is the rare occasion when naturally occurring sand and gravel deposits satisfy filter design requirements and are therefore relatively inexpensive.

The following is offered as a guide to practical minimum widths:

(a) Filters upstream or downstream of an earth core, when constructed by end-dumping off a truck should be at least 2.5 m and preferably 3 m wide.
(b) If a spreader box such as that shown in Figure 9.29 is used, a minimum width of 1.5 m is practicable. The filter material is dumped off the truck into the spreader box, which spreads the filter out of its base as it is pulled along by a small bulldozer.
(c) If filter materials are very scarce or high cost, formwork can be used to contain bands of filters as narrow as one metre. Sherard et al. (1963) show an example of such placement. This is very unusual and would only be contemplated in exceptional circumstances.
(d) For homogeneous or zoned earthfill dams with a vertical chimney, a relatively narrow filter (as narrow as 0.75 m or 1.0 m) can be constructed by placing the earthfill for up to 2 m over the filter layer, and then excavating through the earthfill with a backhoe or excavator to expose the filter, as shown in Figure 9.30. Careful cleanup of the surface of the exposed filter is necessary and the filter is compacted with small vibrating sleds or other compaction equipment. The depth "h" is best limited to say 1.5 m to reduce the risk of collapsing of the trench and allow access of men into the trench. Contamination of the

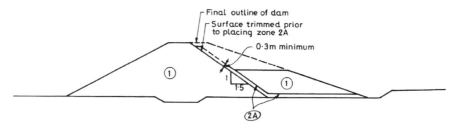

Figure 9.31.   Construction of inclined chimney drain by placement on downstream slope of earthfill.

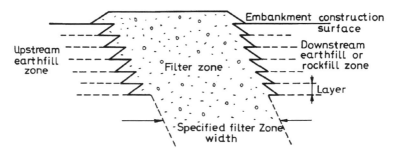

Figure 9.32.   Filter zone placement ahead of other zones which is generally desirable.

filter during placement can be reduced by spreading it from a movable steel plate placed adjacent to the trench (Charlton and Crane, 1980). This also reduces the risk of collapse of the trench under the surcharge load of the filter material.

(e) For smaller dams, an inclined chimney drain can be constructed by dumping the filter on the trimmed downstream slope of the earthfill core as shown in Figure 9.31. The filter can be compacted by rolling up the slope or by running rubber tyred equipment up against the slope. The downstream earthfill (or rockfill) is then placed in layers adjacent to the filter. In this way thin layers of filter (say as thin as 0.3 m normal to the slope, 0.5 m horizontal) can be placed.

Horizontal filters can be placed in layers as thin as 150 mm or as thick as about 500 mm after compaction (the upper limit being determined by compaction requirements). However, for thick layers of filters (e.g. 400 mm), it is considered preferable to place 2×200 mm layers rather than 1×400 mm layer since, if segregation occurs during placement, it is less likely that coarse zones will coincide with each other in the two layers.

9.6.2   Sequence of placement of filters and control of placement width and thickness

Filters generally should be placed ahead of the adjacent earthfill or rockfill zones as shown in Figure 9.32.

This is desirable because it allows good control of the width of the filter zone compared to the specified width and reduces the risk of contamination of the filter zone with materials from the adjacent zones and from water eroding off adjacent areas generally. Figure 9.33 shows the alternative which is not recommended by the authors or such organisations as USBR (1987). However, both systems are used successfully by experienced contractors.

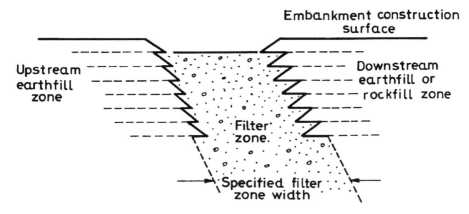

Figure 9.33.   Filter zone following construction of other zones which is generally undesirable.

Care must also be taken to avoid contamination where haul roads cross the filter zones. This can be facilitated by covering the filter with a geotextile or geomembrane before building the haul road. If practicable, it is wise to shape the surface of the embankment fill, so that surface water flows away from the filter.

9.6.3   Compaction of filters

The general objectives for the compaction of filter zones are:

(a) Filters upstream and downstream of the dam core;
  – Sufficient compaction to ensure the filter shear strength does not influence the stability of the embankment;
  – For saturated filters, compaction should be to a relative density (density index) such that flow liquefaction under earthquake is not possible, i.e. a relative density ≈70% is required;
  – The deformation modulus of the filter should be compatible with the adjacent core material and rockfill;
  – Avoid excessive breakdown of the filter materials by the compaction equipment.
(b) Horizontal or near horizontal filters;
  – The compaction should be to a relative density (density index) such that flow liquefaction under earthquake is not possible, i.e. a relative density of ≈70% is required;
  – Avoid excessive breakdown of the filter materials by the compaction equipment.

Filters should be compacted in layers using a vibratory smooth steel drum roller. Filters are usually well graded granular materials and are readily compacted to a dense condition. Hence for the majority of dams, a "methods" type specification for compaction is the most practicable, i.e. a maximum layer thickness, coupled with a number of passes of a vibratory roller of a specified static weight and centrifugal force. Table 9.11 gives some examples of "standards" type specifications, where a density index (relative density) is specified. The latter is not the preferred basis for the routine (daily) control of placement on the dam because of the cost involved in carrying out the tests, the delay in the results and the difficulty of carrying out the testing required, particularly in Zone 2B filters. Zone 2B filters will commonly have gravel up to 50 mm or 75 mm size, requiring a density in place test hole at least 200 mm or 300 mm diameter and laboratory compactions in non standard 300 mm or 500 mm diameter cylinders.

Table 9.11. Typical filter compaction specifications.

Authority and dam	Zone	Standards specification	Methods specification
Water Resources Commission of Queensland, Peter Faust Dam (1989)	2A and 2B	AS1289. Density index between 60% and 70%	At least one pass of a suitable smooth drum vibratory roller to give standards specification Maximum layer thickness 350 mm after compaction
Water Authority of Western Australia, Harris Dam (1989)	2A chimney drain	AS1289. Density index >70%	Not given. Maximum thickness 500 mm before compaction
	2A horizontal drain and 2B	Not specified	4 passes of a vibratory roller with static mass between 8 and 12 tonnes, and centrifugal force not less than 240 kN. Maximum thickness 600 mm before compaction
Melbourne and Metropolitan Board of Works, Cardinia Creek Dam (1970)	2A and 2B	Not specified	2 passes of a vibratory roller with static weight not less than 8 tonnes and a centrifugal force not less than 160 kN Maximum layer thickness 450 mm after compaction
Snowy Mountains Hydroelectric Authority, Talbingo Dam (1967)	2A and 2B	Not specified	4 passes of a vibratory roller with static weight not less than 10 tonnes static weight and a centrifugal force not less than 350 kN. Maximum layer thickness 450 mm after compaction

If a minimum density index is specified, there should be trials early in the construction to define a method which will give the required compaction and with daily control by the method specification. In this case a requirement to compact to a density index of say 70% or 80% would be appropriate. It should be noted that requiring a density index greater than say 80% is likely to result in excessive breakdown of filters under the compactive effort and will achieve little or no benefit. There are some arguments, e.g. Sherard (1985a), that over-compaction can lead to excessively high moduli for the filters, resulting in their settling less than the dam core during construction and first filling and thereby reducing vertical stresses and horizontal stresses in the core. This can foster the development of hydraulic fracture. Some therefore favour minimal compaction of these filters – e.g. 1 or 2 passes of a 10 tonnes steel drum roller, using layer thickness of 500 mm. It is arguable whether it is practicable to set an upper limit on density index in dam construction, such as for Peter Faust dam in Table 9.11, other than as a guide to contractors. It would seem difficult to insist on removal of over-compacted material without raising the issue of contractual claims.

9.7   USE OF GEOTEXTILES AS FILTERS IN DAMS

9.7.1   *Types and properties of geotextiles*

Koerner (1986) defines geotextiles as: "Geotextile: a permeable textile material (usually synthetic) used with soil, rock or any other geotechnical engineering-related material to enhance the performance or cost of a human-made product, structure or system."

A wide range of different geotextiles is available. These differ in the type of polymer used for manufacture, the type of fibre construction and how this fibre is manufactured into a fabric.

Giroud and Bonaparte (1993) describe the different types of geotextiles as follows:

Woven geotextiles are composed of two sets of parallel yarns systematically interlaced to form a planar structure. Generally, the two sets of yarns are perpendicular. The manner in which the two sets of yarns are interlaced determines the weave pattern. By using various combinations of three basic weave patterns, i.e. plain, twill, and satin, it is possible to produce an almost unlimited variety of fabric constructions. Woven geotextiles can be classified according to the type of yarn, as discussed below:

– Multifilament woven geotextiles are made with yarns that comprise many filaments (typically 100 or more);
– Monofilament woven geotextiles are made with yarns that are single filament;
– Slit film woven geotextiles are obtained by weaving small tapes produced by slitting a plastic film with blades;
– Fibrillated woven geotextiles are obtained by weaving yarns that are bundles of tape-like fibers still partially attached to each other. Such bundles are film strips that have been nicked and broken up into fibrous strands.

Knitted geotextiles include two types: the classical knits and the insertion knits.

– Classical knits are formed by interlocking a series of loops of one or more yarns to form a planar structure. The way the loops are interlocked identifies the type of knit, such as jersey. The knitted geotextiles of the classical knit type are highly deformable materials;
– Insertion knits are produced by inserting yarns into a knit that is being made. Using this manufacturing process, it is possible to make geotextiles that combine properties of wovens, fabrics, and classical knits.

Nonwoven geotextiles are formed from continuous filaments or from short fibers (also called staple fibers), arranged in all directions (but not necessarily at random) and bonded together into a planar structure. Synthetic filaments are produced by extrusion and are, therefore, continuous, whereas synthetic fibers are obtained by cutting filaments into lengths of 100–150 mm (4–6 inches). Natural fibers are generally short. For nonwoven geotextiles, the continuous filaments or short fibers are first arranged into a loose web, then bonded together using one, or a combination, of the following processes: chemical bonding, thermal bonding and mechanical bonding by needle-punching.

Chemically bonded nonwoven geotextiles are obtained by bonding the filaments or fibers together using a cementing medium such as glue, rubber, latex, cellulose derivative or, more frequently, synthetic resin.

Heatbonded nonwoven geotextiles are obtained by bonding the filaments or fibers together using heat. When heat is provided by heated rollers that press the fabric, the resulting heatbonded nonwoven geotextile is relatively stiff and thin, typically 0.5–1 mm.

Needle-punched non-woven geotextiles are obtained by entangling the filaments or fibers together using thousands of small barbed needles, which are set into a board and punched through the fabric and withdrawn many times as the fabric passes under the board. Needle-punched non-woven geotextiles are relatively thick, typically 1–5 mm.

The three bonding processes described above can be combined. For example, needle-punched non-woven geotextiles are typically compliant, but, if they are exposed to controlled heat on one side, they become stiffer on that side while remaining compliant on the other side.

Multilayer geotextiles are produced by bonding together several layers of fabrics. Two techniques for manufacturing multiplayer geotextiles are stitch bonding and needle-punching. Stitch bonding consists of stitching together several layers of fabric. This is often used to produce high strength geotextile from slit film woven fabrics. The needle-punching process (described above for non-woven geotextile manufacturing) is used, for example, to associate woven and needle-punched non-woven fabrics.

Yarns, filaments, fibers, and tapes are usually made from polypropylene or polyester. Other polymers used include polyamide (commercially known as Nylon), polyethylene and polyaramide (commercially known as Kevlar). Some natural fibers such as jute and coconut fibers are also used.

There are also other geotextile related products such as geomats, geonets, geogrids and geocells. These are described in Giroud and Bonaparte (1993).

When considering the use of geotextiles as filters in dam construction, the important properties are the "particle size" that affects its ability to act as a filter and its "permeability", or ability to allow water to pass through. The "particle size" of the geotextile is usually measured by sieving a standard soil or single sized glass spheres on the geotextile for a fixed period and observing the quantity and particle size of the soil which passes through. As described in Bertacchi and Cazzuffi (1985), ICOLD (1986), Heerten (1993), Fischer et al. (1992) and Christopher et al. (1993) there is no one accepted standard and different methods use dry or wet sieving and consider the geotextile size as that allowing zero, 2% or 5% of the soil passing.

Koerner (1986) and most authors suggest the use of the apparent opening size AOS (also known as equivalent opening size EOS). The test uses known size glass beads and determines by sieving using successively finer beads, that size of bead for which 5% or less pass the fabric. The AOS or EOS is the US standard sieve number of this size bead. It is also, more commonly, quoted as the equivalent sieve opening in millimetres or 95% opening size O_{95}. This is the form used for most filter design criteria as discussed below. Many manufacturers provide data on the 95% opening size in their product literature and it appears to have become the industry standard, at least for the time being. However for important designs, given the dependence of the D_{95} (or O_{90}, O_{50}) on the test method, caution should be exercised in using the manufacturers' values, unless the test method is compatible with the design method.

Some typical values are given in Table 9.12.

In most cases, the ability of the geotextile to transmit water across the fabric is also important.

This is assessed as the permittivity

$$\psi = \frac{kn}{t} \tag{9.4}$$

Table 9.12.   Some typical values for geotextiles.

Product name	Structure woven (W)	EOS (mm)	Permeability cm/sec
Terrafirma s/2100	W	0.06–0.07	0.03
Polyfelt TS500	NW Needle P	0.21	0.5
Polyfelt TS800	NW Needle P	0.13	0.4
Bidim U14	NW Needle P	0.1–0.12	N/A
Bidim U64	NW Needle P	0.06–0.075	
Polytrac	W	0.07	

where
ψ = permittivity
kn= cross plane permeability coefficient
t = geotextile thickness at the normal pressure on the geotextile.
In Darcy's Law:

$$q = ki\, A \tag{9.5}$$

$$= kn\frac{\Delta h}{t} \cdot A \tag{9.6}$$

So

$$\frac{kn}{t} = \psi = \frac{q}{(\Delta h)A} \tag{9.7}$$

where
q = flow rate
Δh = head loss across the geotextile
A = area of flow (i.e. of geotextile).

As pointed out by Bertacchi and Cazzuffi (1985) and others, the permittivity is dependent on the normal stress applied to the geotextile i.e. the ability to transmit water is reduced as the geotextile is placed under stress which compresses the geotextile and the values adopted for design should account for this. The permittivity is also affected by clogging of the geotextile by fine soil. This is discussed further below.

In some applications the geotextile may be used to transmit water along its plane i.e. it is in itself performing a drainage function. This is measured by transmissivity θ and is given by:

$$\theta = k_p t = \frac{q\, L}{\Delta h\, W} \tag{9.8}$$

where
θ = transmissivity
k_p = permeability in the plane of the fabric
t   = thickness of the fabric
q   = flow rate in the plane of the fabric
L   = length of the fabric
W   = width of fabric
Δh = head lost.

As shown in Figure 9.34, transmissivity is also affected by the applied normal stress.

9.7.2   Geotextile filter design criteria

9.7.2.1   General requirements

The requirements for the design of geotextile filters are as for conventional filters:

– Prevent erosion of soil particles from the base soil they are protecting;
– Allow drainage of seepage water;
– Be sufficiently durable to resist damage during construction and in service.

The first is achieved in a manner similar to that for conventional filters, but using the equivalent opening size (EOS) of the geotextile, instead of D_{15F}.

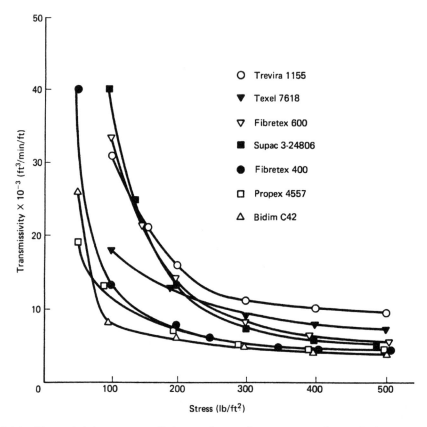

Figure 9.34.   Transmissivity versus applied normal stress for various needle punched non woven geotextiles (Koerner, 1986, reproduced with permission of Pearson Education).

The second is achieved through having sufficient transmitting (or permeability) across the plane of the geotextile, but in addition, there is a requirement to control the likelihood of clogging of the geotextile.

9.7.2.2   *Filtering requirement*

There are many different filter design criteria for geotextiles. Summaries are given in Bertacchi and Cazzuffi (1985), ICOLD (1986) and Christopher et al. (1993). Critical reviews are given in Giroud (1996) and Heerten (1993).

Table 9.13 summarizes some of the methods.

Palmeira and Fannin (2002) give a more recent review but they do not recommend a particular method.

The authors are not in a position to give clear guidance on what criteria should be used. In principle, we would favour a method which is based on the D_{85B}, (or D_{90B}, D_{95B}) of the base soil, rather than D_{50B}, because in conventional filters, Sherard et al. (1984a) and Foster (1999) have demonstrated D_{50B} is a poor guide to filter characteristics. This would favour the use of Loudiere et al. (1982) (upon which ICOLD 1986, also seem to have relied), and Heerten (1993). Giroud (1982, 1996) (along with Loudiere et al., 1982) specifically discuss the application to dams.

The authors' advice would be to try these three methods and, if necessary, seek guidance from the more recent literature and/or manufactures of geotextiles.

Table 9.13.   Methods for designing geotextiles as filters.

LOUDIERE et al. (1982)
For non cohesive soils:
   Uniformity coefficient $(D_{60B}/D_{10B}) > 4$
      $O_{95} < D_{95B}$
   Uniformity coefficient < 4
      $O_{95} < 0.8D_{50B}$
For cohesive soils:
   Uniformity coefficient > 4
      $O_{95} < D_{85B}$ and
      $O_{95} \geq 0.05$ mm
   Uniformity coefficient < 4
      $O_{95} < 0.8D_{50B}$ and
      $O_{95} \geq 0.05$ mm

AASHTO (FROM KOERNER, 1986)
For coarse grained soils, $\leq 50\%$, *passing 0.075 mm*
      $O_{95} \leq 0.6$ mm
For fine grained soils, $> 50\%$, *passing 0.075 mm*
      $O_{95} \leq 0.3$ mm
(Note: Koerner quotes $\geq$ not $\leq$ for these relationships, which appears to be in error)

CARROLL (1983)
      $O_{95} < (2 \text{ or } 3) D_{85B}$

GIROUD (1982) (*for needle punched and woven geotextiles*)

Relative density	$1 < C_u < 3$	$C_u > 3$
Loose (RD < 35%)	$O_{95} < (C_u) D_{50B}$	$O_{95} < 9D_{50B}/C_u$
Medium (RD 35% to 65%)	$O_{95} < 1.5 (C_u) D_{50B}$	$O_{95} < 13.5D_{50B}/C_u$
Dense (RD > 65%)	$O_{95} < 2 (C_u) D_{50B}$	$O_{95} < 18D_{50B}/C_u$
For woven and heat bonded non wovens		
	$1 < C_u < 3$	$O_{95} < (C_u) D_{50B}$
	$C_u < 3$	$O_{95} < 9D_{50B}/C_u$

where C_u = uniformity coefficient = D_{60B}/D_{10B}
GIROUD (1996) indicates Giroud (1982) still applicable

HEERTEN (1984)
For cohesionless soils:
   Static load conditions, $C_u \geq 5$
      $O_{98} < 10D_{50B}$, and
      $O_{98} \leq D_{90B}$
   Static load conditions, $C_u < 5$
      $O_{98} < 2.5D_{50B}$, and
      $O_{98} \leq D_{90B}$
   Dynamic load conditions (high turbulence, wave action or "pumping")
      $O_{98} \leq D_{50B}$
For cohesive soils:
      $O_{98} < 10D_{50B}$, and
      $O_{98} \leq D_{90B}$, and
      $O_{98} \leq 0.1$ mm

HEERTEN (1993)
      $O_{90, w} < D_{90B}$, and
      $O_{90, w} > D_{50B}$ where $O_{90, w}$ is obtained by wet sieving

FISCHER, CHRISTOPHER AND HOLTZ (1990)
      $O_{50} \leq 0.8 D_{85B}$ ⎫ depending on the geotextile pore size
      $O_{50} \leq 1.8$ to $7.0D_{15B}$ ⎬ distribution, and C_u of soil.
      $O_{50} \leq 0.8$ to $2.0D_{50B}$ ⎭

9.7.2.3   *Permeability requirement*

There are several views on the selection of geotextile permeability including:
  Giroud (1996) suggests

$$k_G > i_s \, k_s \tag{9.9}$$

where
k_G = permeability of the geotextile
i_s = hydraulic gradient in the soil adjacent the geotextile (which may be greater than 1)
k_s = permeability of the soil.
  Giroud recommends a large safety factor (e.g. 10) should be used to allow for uncertainties in the permeabilities.
  Christopher et al. (1993) suggest

$$k_G > 10k_s \tag{9.10}$$

  Heerten (1993) suggests the permeability will be satisfactory if other requirements are met. Heerten (1984) suggested

$$k_{Gclogged} > k_s \tag{9.11}$$

  Loudiere et al. (1982), reflecting the views of the French Committee on Geotextiles and Geomembranes suggests $k_G > 100k_s$, where k_G is the uncompressed geotextile permeability.
  From these data, it would seem wise to require an uncompressed geotextile permeability of at least $10k_s$ and preferably $100k_s$.
  The further requirement to maintain sufficient cross geotextile flow capacity is to control clogging, which occurs when the fine soil particles either penetrate into the geotextile, blocking off the fine (as well as the coarse) pore channels, or are deposited on the upstream surface of the geotextile. Heerten (1993) indicates that clogging is more likely where the base soil is cohesionless sand and silt, gap graded (internally unstable) cohesionless soils and high hydraulic gradients. Clogging is more likely if the geotextile is much finer than it needs to be to meet filtering requirements and if it is thin, with little pore capacity.
  Christoper et al. (1993) list several criteria to reduce the likelihood of clogging. These are reproduced in Table 9.14.
  Heerten (1993) suggests the use of the second of his criteria in Table 9.13 ($O_{90,w} > D_{50B}$), to control the minimum pore size in the geotextile, and a thickness of geotextile, t_G, from $30 \, O_{90,w} < t_G < 50 \, O_{90,w}$. He quotes several German organisations as suggesting $4.5 \, mm < t_G < 6.0 \, mm$. Thick needle-punched non-woven geotextiles are more likely to meet these criteria than woven geotextiles.
  Christopher et al. (1993) suggest that for critical/severe applications, filtration tests on the soil and geotextile should be carried out.
  The authors are aware of cases where limonite depositing from seepage water has clogged geotextile filters. Palmeira and Fannin (2002) also warn of chemical and biological clogging.

9.7.2.4   *Durability requirement*

There are two aspects to consider:

(a) Durability during installation – geotextiles are prone to tearing as they are being placed, or by materials being compacted on top of or adjacent to them;

Table 9.14.   Clogging criteria (Christopher, et al., 1994).

A   *Critical/severe applications*
     Perform soil/geotextile filtration tests (e.g. Calhoun, 1972; Haliburton, et al., 1982a and b;
     Giroud, 1982; Carroll, 1983; Christopher and Holtz, 1985 and 1989; Koerner, 1990).

B   *Less critical/non-severe applications*
     1. Perform soil/geotextile filtration tests;
     2. Minimum pore sizes alternatives for soils containing fines, especially in a non-continuous
        matrix:
        a. $O_{95} \geq 3D_{15B}$ for $C_u \geq 3$ (Christopher and Holtz, 1985 and modified 1989);
        b. $O_f \geq 4D_{15B}$ (French Committee on Geotextiles and Geomembranes, 1986);
        c. $O_{15}/D_{15B} \geq 0.8$ to 1.2;
           $O_{50}/D_{50B} \geq 0.2$ to 1 (Fischer, et al., 1992).
     3. For $C_u \leq 3$, geotextile with maximum opening size from retention criteria should be specified;
     4. Apparent open area qualifiers;
        Woven geotextiles: present open area $\geq 4\%$ to 6% (Calhoun, 1972; Koerner, 1990);
        Nonwoven geotextiles: porosity $\geq 30\%$ to 40% (Christopher and Holtz, 1985; Koerner,
        1990).

(b) Durability during operation, both from a viewpoint of mechanical issues, e.g. ability
    to cope with differential deformations, and chemical durability.

Heerten (1993) suggests using a geotextile with a mass of not less than 300 grams/m^2.
This was based on tests where sites were exhumed after construction;
    Christopher et al. (1993) suggest trialling the geotextile in the application proposed and
exhuming trial sections to show the geotextile is sufficiently robust;
    Giroud and Bonaparte (1993) give some advice on construction practice:

– If the soil is placed or excavated before geotextile placement, its surface should be as
  smooth as possible. This requirement is particularly important if the geotextile filter is
  part of a stiff geocomposite. If the soil is placed after the geotextile, it should be compacted to eliminate voids in the vicinity of the geotextile;
– The geotextile should be installed to be not too taut, but also not too loose (wrinkles must
  be avoided). If the geotextile is placed on a previously placed or excavated soil, all efforts
  should be made to follow the shape of the soil. If proper contact cannot be ensured,
  because the geotextile is part of a stiff geocomposite, sand should be placed between the
  soil and the geotextile. The sand particle size must be compatible with the soil particle size
  and the geotextile opening size;
– The aggregate, if any, used in contact with the geotextile must be small enough to
  apply a uniform stress on the geotextile in order to properly confine the soil located
  on the other side of the geotextile. A maximum aggregate size of 20 mm is recommended.

The long term chemical durability of geotextiles has been and remains a significant concern. Giroud and Bonaparte (1993) indicate examples of successful operation for up to 35
years. However concerns on long term durability remain e.g. ICOLD (1986) (and by implication ICOLD, 1994).

9.7.2.5   *Use of geotextile filters in dams*

Use of geotextiles to perform a permanent function, probably in lieu of a sand/gravel filter, does require consideration of the effectiveness of the geotextile as a "filter" and
"drain" as has been discussed in preceding Sections. It also requires consideration of long-
term function without damage due to deformation or clogging and durability, particularly

if the geotextile is performing a critical function and is not accessible for checking its condition or for repair.

ICOLD (1986) outlines possible uses of geotextiles in dam construction. They conclude that:

- Considerable caution is required in using geotextiles to prevent erosion at interfaces which are subject to seepage from the reservoir (i.e. location "g" in Figure 9.2);
- Successful use in non critical locations, e.g. under slope protection (as in "c" in Figure 9.2) should not be used as justification for use in critical areas;
- Geotextiles are subject to damage during construction, particularly from angular coarse-grained materials. This may have little effect on their use as a separator of different soils during construction but continuity is necessary for effectiveness as a filter;
- Geotextiles, once buried, have similar durability to other man made materials. However they are subject to rapid deterioration if left exposed to sunlight before, during or after construction. (The authors add that if polyester geotextiles are placed next to concrete, particularly where the concrete may be heated by sunlight, the polyester can degrade under the high pH conditions).

ICOLD (1994) quote Roth and Schneider (1991) as concluding:

"The availability of geosynthetics presents the dam engineering community with both opportunities and challenges – opportunities to design economical, safe dams and challenges to avoid inadvertently committing an unfortunate error leading to a dam failure. Blanket avoidance of geosynthetics is inappropriate, since they may permit safe, more economical designs in some cases. On the other hand, the potential for catastrophic damage and loss of life in the event of a major dam failure suggests a cautious approach and prudent use of geosynthetics in dams".

The authors offer the following suggested broad guidelines for the use of geosynthetics in dams in the hope of generating open and productive discussion by the dam engineering community on the appropriate uses of geosynthetics in modern dam design:

(1) Geosynthetics should not be used in a configuration where they serve as the sole defense against dam failure. For example, if a geomembrane is used as the sole water barrier within an embankment, the remainder of the embankment design should be checked to assure that the embankment would be completely safe if portions of the geomembrane were suddenly and completely removed.
(2) Geosynthetics should primarily be used where they can be readily exposed, repaired, or replaced. For example, a geomembrane used on the face of a concrete dam can be examined and repaired or replaced if required.
(3) We should expand our use of geosynthetics when they prove to be economically attractive in non-critical, redundant or superficial applications. In this way we can expand our knowledge of the behaviour of geosynthetics in dams and gain additional, valuable experience on the long-term performance of geosynthetics in service conditions.

While not endorsing these views directly, ICOLD (1994) appear to do so indirectly.

The current authors' views mirror those of Roth and Schneider (1991) and ICOLD (1986, 1994). We do not recommend the use of geotextiles in critical locations within dams. We are concerned at the high likelihood of damage during installation or by differential movements of the dam during and after construction, clogging, and hence loss of ability to pass the seepage flow through the geotextile and long term chemical durability.

10

Control of seepage, internal erosion and piping for embankment dams

10.1 GENERAL REQUIREMENTS AND METHODS OF CONTROL

The objective in designing a new dam, or when assessing an existing dam, is to be satisfied that the dam has adequate measures to control seepage and potential internal erosion so that:

(a) Pore pressures in the dam and the foundation are such that there is an adequate margin of safety against slope instability;
(b) Internal erosion, which might progress to form a pipe and breach the dam, will not occur. This applies to piping in the embankment, in the soil or rock foundation, from the embankment to the foundation and the foundation into the downstream zones of the embankment.

These objects are achieved by zoning of the dam, provision of filters and high seepage discharge capacity zones and other embankment design and foundation treatment measures. Figure 10.1 shows some of the measures which are used.
The design features are:

A: Low permeability "core" of the dam, to limit seepage through the dam;
B: Chimney filter drain, which intercepts seepage through the dam and controls internal erosion of the core and pore pressures in the dam, if the downstream zone C is low permeability;

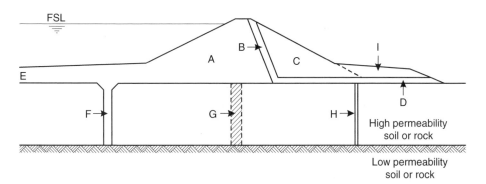

Figure 10.1.   Design features for controlling seepage and erosion through and beneath earth dams on highly permeable foundations A, low permeability core; B, chimney filter drain; C, downstream zones; D, horizontal filter drain; E, upstream low permeability blanket; F, slurry trench; G, grouted zone; H, relief wells;I, weighting berm.

C: Downstream zone, which maintains stability and may, if permeable compared to the core and filter, control pore pressures;

D: Horizontal filter drain – which controls exit gradients, erosion of the foundation and pore pressures (but may actually increase the amount of seepage because the gradients are increased);

E: Upstream low permeability blanket which increases the seepage path, reduces seepage and seepage exit gradients;

F: Slurry trench (or other types of cutoffs) which will reduce (or almost stop if fully penetrating to a low permeability base) seepage and seepage exit gradients (usually in soil or extremely weathered rock);

G: Grouting to reduce seepage and seepage exit gradients. Grouting is usually in rock, since grouting in soil is usually costly and ineffective;

H: Pressure relief wells – control exit gradients where confined aquifers in the foundation reduce the effectiveness of the horizontal drain;

I: Weighting berm – which improves downstream slope stability and overcomes potential liquefaction or "blow-up" of the foundation (with H).

Of course, not all these features are needed for every dam. For example, a central core earth and rockfill dam on a strong, non erodible rock foundation will usually have grouting, but will not need a horizontal filter or drain.

10.2   SOME PARTICULAR FEATURES OF ROCK AND SOIL FOUNDATIONS WHICH AFFECT SEEPAGE AND SEEPAGE CONTROL

Seepage occurs through all embankment dams and their foundations. The permeability of most compacted earthfill core materials is less than 10^{-8} or 10^{-9} m/sec. By comparison virtually all rock foundations have rock mass permeability greater than 1 lugeon (1 lugeon = 1 litre/second/metre of drill hole at a pressure of 1000 kPa) or approx. 10^{-7} m/sec, and most rocks have greater than 5 lugeons (approx. 5×10^{-7} m/sec). Foundations of alluvial or deeply weathered and lateritised rock, may have mass permeabilities as high as $10^{-3} - 10^{-5}$ m/sec. As a result most of the seepage is through the foundation, not the embankment, and significant rates of seepage may occur.

The following details some particular features of rock and soil foundations which influence seepage in the foundation and the design of seepage control measures.

(a) *Rock foundation.* As discussed in detail in Chapters 3 and 4, the permeability and seepage in rock foundations is controlled by joints, bedding, shears and faults and in some situations other features such as solution cavities.

The detailed characterisation and understanding of the hydrogeological model is vital to design of foundation grouting, assessing the likely magnitude of the under seepage and erodibility of the foundation and hence the need for filters, drains or other high permeability zones in the dam. It should be noted that clay infilling of joints and stress relief at depth can give marked stratification of permeabilities in weathered surfaces.

Table 10.1 shows the assessed rock mass permeability model for the Ben Lomond Tailings dam site which had steeply dipping meta-sediments in the foundations. Similar effects were present at the Ranger Tailings dam, giving a confined aquifer about 4 metres below the surface, below the clay infilled joints in the upper weathered rock. This necessitated construction of weighting berms to maintain adequate factors of safety against "blow-up", or erosion into rockfill in the dam if filters are not provided over the foundations.

Table 10.1.   Ben Lomond project – rock mass permeabilities (Coffey Partners, 1982).

Zone	Depth to base of zone (m) Avg.	Permeability (m/sec × 10⁻⁷) Avg.	Range
Weathered rock with clay infilled joints	0–12 (8)	1.5	0.1–54
Weathered rock (limonite stained and coated joints)	19–27 (22)	8.0	0.1–27
Rock in stress relief zones	20–36 (28)	108.0	4.8–>200
Average rock mass zone (fresh rock)	–	2.6	0.1–10
Sandstone bed – Unit 3 –	75		
Folded zone (western end of site)	–	43	0.5–>200

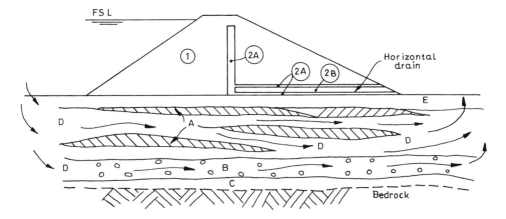

Figure 10.2.   Some common features of alluvial foundations.

Some rocks, such as basalt, inter-bedded sandstone and shale with stress relief joints, limestone and granite with stress relief effects, are particularly susceptible to having open joints. If these rocks are exposed in the cutoff foundation, the core material may erode into the joints. If the rock around the joints is completely weathered it may erode into the open joints under seepage from the reservoir.

(b) *Alluvial soils.* Some features of alluvial foundations are shown schematically in Figure 10.2. They include:

A: Lenses or layers of lower permeability sand, silty sand, or even clayey sand may occur, giving a very much reduced vertical permeability. These may be present in point bar deposits;

B: There is often a coarser gravel, or even boulder/gravel layer at the base of the alluvium (channel lag deposit), reflecting the time when the river was more active. This may be very permeable;

C: The upper part of the rock surface may be permeable because of the presence of open joints, potholing and destressing. The surface may also be very irregular;

D: The coarse alluvium – sand/gravel is in itself likely to be layered giving a high permeability ratio k_H/k_V, e.g. point bar deposits;

E: There is often a layer of silty sand/silty sandy gravel on the surface, giving a low permeability, e.g. in flood plain deposits.

In general these features combine to give a much greater horizontal than vertical permeability. Hence while water can enter into the foundation in the large area of exposed alluvium in the reservoir, it may be inhibited from flowing into the horizontal drain. Similar layering can occur in soils of colluvial origin.

Internal erosion may occur in the foundation e.g. by internal instability (suffusion) of gap-graded sandy gravels and by erosion of fine grained soils into adjacent coarse grained soils. As well as the particle size distribution of the soils, the continuity of the layers and whether they persist under the dam and exit in a way in which eroded material can be transported from the foundation is important, but often difficult to assess with confidence.

(c) *Lateritised soil and rock profiles.* Figure 10.3 shows some common features of lateritised soil and rock profiles which have developed by weathering of rock. The features include:

– Very high permeability upper zone of pisolithic gravel;

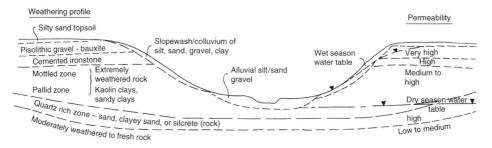

Figure 10.3.   Some common features of lateritic foundations.

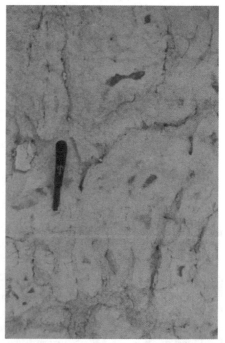

(a) overall profile, bauxite at top   (b) close-up showing sand infillied
overlying pallid, kaolinite zone.   root holes and fissures.

Figure 10.4.   Example of sand infilled roots and fissures – Weipa, North Queensland.

– Variably cemented and permeable ironstone zone;
– Mottled zone and underlying kaolin rich zone. This is commonly medium to high permeability due to relict joints and root holes as shown in Figure 10.4. Many of these features are near vertical and not readily intercepted by vertical boreholes. The mass permeability of the clay will commonly be of the order of 10^{-6} to 10^{-5} m/sec, i.e. the permeability of sand;
– The quartz rich zone overlying weathered bedrock is often a mixture of clean sand and clayey sand of high permeability. In some cases it may be cemented as silcrete, but will still be highly permeable;
– The upper bedrock is often fractured and has moderate permeability;
– The depth of weathering may be up to 30 m, e.g. in the Darling Ranges of Western Australia (Gordon, 1984) and in the subtropical monsoonal climate areas of Weipa, North Queensland and Ranger and Jabiluka, Northern Territory. The weathering profile usually does not follow the topography but rises more gradually in the abutments as shown in Figure 10.3. At Weipa, Ranger and Jabiluka, with low relief, it is the authors' experience that the depth of weathering is related to rock type rather than topography;
– The water table fluctuates markedly between end of wet season and end of dry season, often by as much as 10 m or 20 m.

10.3 DETAILS OF SOME MEASURES FOR PORE PRESSURE AND SEEPAGE FLOW CONTROL

10.3.1 *Horizontal and vertical drains*

Seepage beneath a dam on a permeable soil (or permeable weathered rock) foundation should be allowed to exit in a controlled manner into a horizontal drain. The horizontal drain may consist of a single layer of Zone 2A filter, or 3 layers of Zone 2A and 2B as shown in Figure 10.2. In both cases Zone 2A should be designed to act as a filter to control erosion of the soil from the foundation into the drain. The filter design criteria detailed in Chapter 9 should be used.

The 3 layer drain incorporates the layer of high permeability Zone 2B to ensure that the drain has sufficient discharge capacity.

It is good practice (Cedergren, 1972, USBR, 1987) to design the horizontal drain to have sufficient capacity to discharge the flow entering the drain from the dam foundation and from the vertical drain without the phreatic surface rising into the low permeability fill (see Figure 10.5).

If the horizontal drain has insufficient capacity, the phreatic surface will rise into the downstream low permeability fill as shown in Figure 10.6, reducing the stability

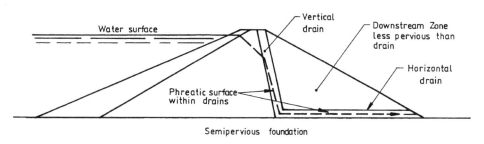

Figure 10.5.   Earth dam with internal drain designed to prevent the phreatic surface from rising above the top of the drain.

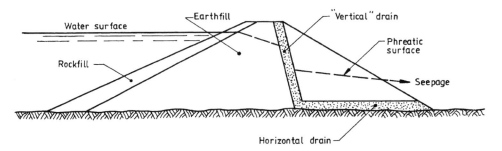

Figure 10.6.   Earth dam with internal drains of inadequate discharge capacity.

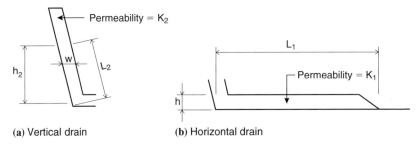

(a) Vertical drain   (b) Horizontal drain

Figure 10.7.   Design of drain dimensions for discharge capacity.

of the downstream slope and also potentially leading to piping failure in the down-stream fill.

Such lack of drain capacity can also occur for dams on permeable rock foundations where earthfill dams such as those shown on Figures 10.5 and 10.6 are used. It can also occur where dirty rockfill has been used as the downstream zone.

Cedergren (1972) gives a design method for estimating the discharge capacity of a horizontal drain without pressurization based on:

$$q = \frac{k_1 h^2}{2L_1} \tag{10.1}$$

where k_1 = permeability of the drain material – m/sec
 h = vertical thickness of the drain – m
 L_1 = length of the drain – m as shown in Figure 10.7b.
 q = discharge capacity per metre width of drain (width measured across river) – m³/m/sec.

The capacity of the vertical drain is seldom a critical issue because the quantity of seepage through the earthfill is small and the vertical drain width is dictated by construction factors. However its capacity should be checked by:

$$q_2 = \frac{k_2 h_2 w}{L_2} \tag{10.2}$$

where k_2 = permeability of vertical drain; h_2, L_2 are as shown in Figure 10.7a.
 w = width of drain.

If there is any doubt about the capacity, a two (2A, 2B) or 3(2A, 2B, 2A) layer filter drain system should be used.

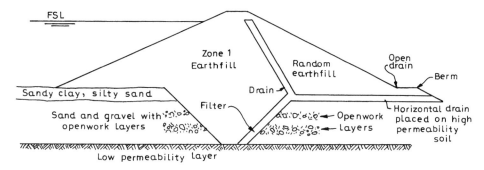

Figure 10.8.   Control of erosion of earthfill into permeable foundation using a filter.

When checking the design of the horizontal drain the following should be noted:

- Water from the abutments will flow towards the lowest part of the drain before flowing out the toe of the dam. Hence the required flow capacity per unit width in this area will be greater than the average flow per unit width under the dam;
- If a 3 layer filter drain is used, Zone 2B will dominate the discharge capacity;
- Conservative estimates of foundation permeability should be used for the design of the horizontal drain, since failure to provide adequate capacity can lead to failure of the dam, or the requirement for a stabilizing berm if the problem is recognised;
- Similarly, conservative estimates of the horizontal drain permeability should be used, and care taken to ensure the drain is not contaminated by soil from the foundation during construction. Generally it is good practice to provide a 3 layer drain, so there is a large discharge capacity;
- The downstream toe should be designed to ensure the drain outlet is not blocked by erosion of soil off the embankment or the abutment. It is wise to provide a berm at the toe of the dam above the drain to collect any material eroded off the downstream face of the dam. (as shown in Figure 10.1);
- Place the filter drain in contact with the higher permeability strata in the foundation. So, as shown in Figure 10.8, the lower permeability sandy clay-silty sand has been removed. Specify that the foundation not be rolled or trafficked by construction equipment which would reduce the permeability under the filter drain.

10.3.2   Treatment of the sides of the cutoff trench

Erosion of the earthfill core into the sides of the cutoff trench may occur where:

- There are open jointed rocks, with a joint opening large enough to allow the soil to erode into the joint e.g. in columnar basalt and other volcanics, pyroclastic rocks, open-jointed sandstone or other rock;
- Coarse gravels in alluvial or colluvial soils are present;
- There are fissures or holes in lateritic soils or weathered rock.

Figure 10.8 shows one way of controlling erosion into a soil foundation. The filter would have to satisfy critical filter criteria (see Chapter 9) in regard to the Zone 1 earthfill, and the "openwork" layer of alluvium. In open jointed rock it is normal to cover the open joints with shotcrete or other concrete, before placing the earthfill. This may also be done in cutoffs into soil and weathered rock.

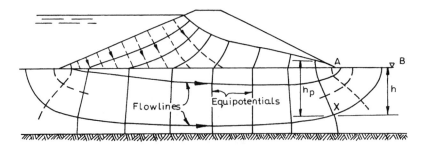

Figure 10.9.   Locations downstream of a homogeneous dam where blowup may occur.

10.3.3   Prevention of "blowup", "boiling" or "liquefaction" of the foundation

If seepage water emerges at too high a gradient from downstream of the toe of a dam on alluvial foundations, "blowup", "boiling" or "liquefaction" of the toe can occur with the possibility of:

– Loss of shear strength, leading to slope instability;
– Potential for development of piping failure.

Figure 10.9 shows the area (A–B) which is susceptible. Figure 11.13 shows another situation where "blow-up" is likely to occur – where a low permeability layer in the foundation confines the seepage flow.

The mechanism involved is one of the high pore pressure in the foundation leading to low effective stresses, with "liquefaction" or "blow-up" occurring when the effective stress becomes zero.

The factor of safety against this occurring can be calculated in two ways for Point X in Figure 10.9, the first being:
either

$$F_{UT} = \frac{\sigma_v}{u} \qquad (10.3)$$

Where σ_v = total vertical stress at any point in the foundation – kN/m^2
            u = pore pressure at the same point – kN/m^2
        or

$$F_{UT} = \frac{h \, \gamma_{sat}}{h_p \, \gamma_w} \qquad (10.4)$$

where γ_{sat} = unit weight of saturated foundation soil – kN/m^3
            γ_w = unit weight of water – kN/m^3
            h_p = piezometric head – metres
The factor of safety against blowup calculated in this manner should be compatible with the consequences of failure and degree of certainty of predicting pressures. However it should be at least 1.5, with pore pressures calculated conservatively (Cedergren, 1972).

The alternative method for estimating the factor of safety is to consider the gradient of the flownet. If the gradient approaches unity, liquefaction can be expected to occur. The

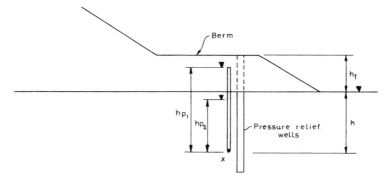

Figure 10.10.   Control of "blow-out" or liquefaction by pressure relief wells and weighting berm.

required factor of safety, i.e.

$$F_{UC} = \frac{\text{critical gradient}}{\text{actual gradient}}$$

$$= \frac{1}{\text{actual gradient}} \qquad (10.5)$$

should be at least 3 and preferably 5.

The authors' preference is to use the zero effective stress, F_{UT} definition, because this allows a better understanding of the ways in which the factor of safety can be improved.

In the event that pore pressures are calculated or known from piezometers to be too high, the factor of safety can be increased by adding a weighting berm (see Figure 10.10) or by providing pressure relief wells. The weighting berm is effective by adding to the vertical stress; the pressure relief drains by reducing the pore pressure.

If the original pore pressure at X was hp_1, the original factor of safety would have been

$$F_{UT} = \frac{\gamma_{sat}h}{\gamma_w hp_1} \qquad (10.6)$$

If the berm of height h_f and unit weight γ_f was constructed, then

$$F_{UB} = \frac{\gamma_{sat}h + \gamma_f h_f}{\gamma_w hp_1} \qquad (10.7)$$

If no berm was provided but instead pressure relief wells were constructed, resulting in a reduction of pore pressure to $\gamma_w hp_2$, then

$$F_{UW} = \frac{\gamma_{sat}h}{\gamma_w hp_2} \qquad (10.8)$$

If both the berm and pressure relief wells were constructed

$$F_{UWB} = \frac{\gamma_{sat}h + \gamma_f h_f}{\gamma_w hp_2} \qquad (10.9)$$

The following should be noted:

– It is difficult to predict pore pressures near the toe of a dam, because they are greatly affected by local variations in permeability (see for example Figures 11.14 and 11.15). Hence it is necessary to be conservative and to provide piezometers to check the assumptions;
– The berm will be most effective if it is free draining;
– The pressure relief wells will provide greatest benefit at each well and least benefit midway between wells. For conservatism the design should be based on the pore pressures midway between wells. Again, piezometers are required to check the effectiveness of the wells.

10.4   CONTROL OF FOUNDATION SEEPAGE BY CUTOFFS

10.4.1   *General effectiveness of cutoffs*

As shown in Figure 10.1, seepage through a permeable foundation can be reduced by constructing a low permeability cutoff through the permeable material. This may consist of:

– cutoff trench filled with earthfill;
– slurry trench;
– concrete diaphragm wall;
– contiguous or intersecting bored piles;
– sheet pile wall;
– grout curtain.

Such cutoffs will have varying degrees of effectiveness depending on their permeability and the depth to which they are taken. This is illustrated in Figures 10.11 and 10.12 which are reproduced from Cedergren (1972).

Figure 10.11 shows the effect on the line of seepage in the downstream shell of an embankment and the effect on exit gradients of cutoffs constructed to varying proportions of the total depth of a permeable foundation (note the example assumes the foundation and downstream fill zone have the same permeability).

It can be seen that there is not a significant improvement even with a cutoff which is 90% penetrating. Only where the cutoff is fully penetrating and properly connected into the low permeability zone will the seepage pressures be controlled. This is often difficult to achieve as will be discussed below.

Figure 10.12 shows the effect of fully penetrating grouted cutoffs on the line of seepage in the downstream zone. Unless the grouted zone (or cutoff constructed by another method) has a permeability much less than the foundation, there is little reduction in downstream pore pressures. This applies to soil and rock foundations.

The effectiveness of partially penetrating cutoffs will depend on layering of lower and higher permeability soils in the foundation. If there are continuous low permeability layers present, partially penetrating cutoffs can be effective. However it must be expected that, unless a cutoff is fully penetrating and of low permeability (say 10 to 100 times less than the permeable foundation), it will have little benefit in the reduction of leakage and exit gradients.

Figure 10.13 shows some important implications of constructing effective slurry trench or similar cutoffs. As shown in Figures 10.13(a) and (b), the seepage through the dam may flow towards the foundation downstream of the cutoff, and a filter may be required under the earthfill to prevent erosion. Figure 10.13(c) shows the problem of how far a cutoff should penetrate into the abutments, to prevent seepage flowing around the ends.

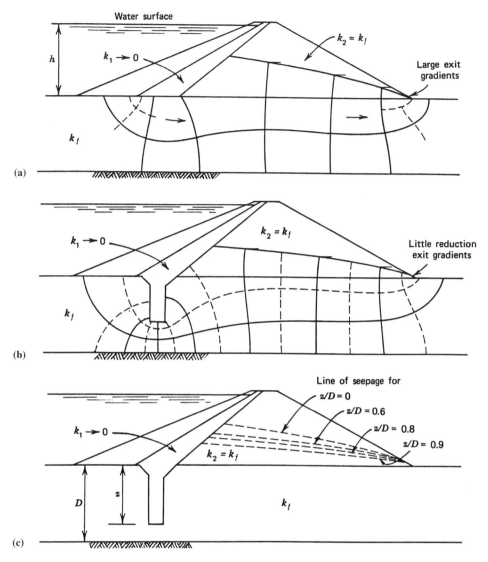

Figure 10.11.   Effect of partial cutoff on position of line of seepage: (a) flownets, $z/D = 0$. (b) flownet, $z/D = 0.6$. (c) position of line of seepage to various values of z/D (Cedergren, 1972).

10.4.2   *Cutoff trench*

When the depth of permeable soil is relatively small, an effective way of providing a cutoff is to excavate a trench through the permeable layer and backfill with Zone 1 earthfill as shown in Figure 10.8. If there are continuous relatively low permeability layers within the soil the cutoff trench may be stopped at such a layer rather than penetrating to rock.

The practicality and economics of constructing the cutoff this way depend on:

- Whether dewatering is necessary to construct the cutoff;
- The availability of equipment to construct cutoffs by other methods, e.g. slurry trench;
- The stability of the sides of the trench during construction (which is dependent on soil type, consistency and effectiveness of dewatering).

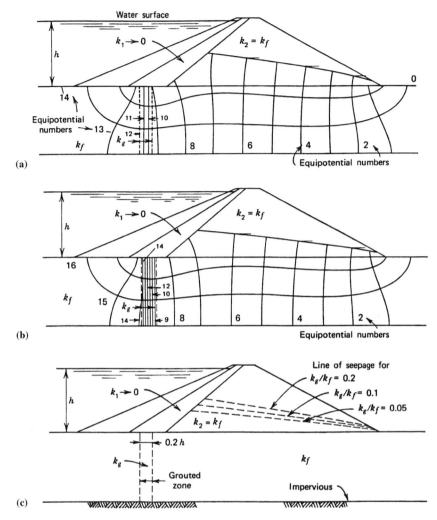

Figure 10.12.   Effect of grouted cutoffs on position of line of seepage: (a) flownet, $k_g = 0.2k_f$, (b) flownet, $k_g = 0.1\,k_f$, (c) position of line of seepage for various values of k_g/k_f (Cedergren, 1972).

In most cases the economic depth is likely to be less than 10 metres. Beyond this depth slurry trench and other wall cutoffs are likely to be more economic.

If dewatering of the permeable soil is necessary, it is likely that the dewatering has to continue while the trench is being backfilled with earthfill. Figure 10.14 shows a possible dewatering arrangement. However provision would have to be made for careful backfilling of the collector pipes with grout, and making the backfill around the pipes filter compatible with the Zone 1 and alluvium, or erosion may occur into the pipe or backfill.

As shown in Figure 10.8, a filter zone may have to be incorporated into the downstream side of the cutoff trench to prevent erosion of the earthfill into the foundation.

If a cutoff trench can be constructed it does provide a very good quality cutoff, with a low permeability. For compacted clayey earthfill it would not be unreasonable to expect a permeability of 10^{-8} to 10^{-9} m/sec. The contact with the foundation can be of high quality and, for rock foundation, grouting can be carried out if necessary from the base of the trench.

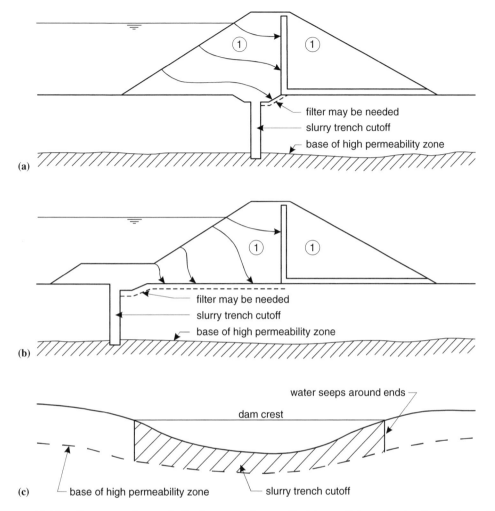

Figure 10.13. Slurry trenches and diaphragm walls cutoffs (a) cutoff beneath the dam (b) cutoff upstream of the dam (c) potential seepage around the ends of the cutoff.

10.4.3 *Slurry trench cutoff*

Slurry trench cutoffs consist of a continuous trench excavated by means of a backhoe, dragline or clamshell, or combination thereof, with the trench supported by bentonite slurry.

The trench is backfilled with sand or sand and gravel thoroughly mixed with the slurry. Often the sand and gravel will be from the excavation. Figures 10.15, 10.16 and 10.17 show the technique.

The slurry trench is constructed 1 m to 3 m wide. The maximum depth depends on the method of excavation. ICOLD (1985) and Xanthakos (1979) indicate that practical maximum depths are:

– backhoe (excavator)   10 m;
– dragline   25 m maximum, 20 m preferred maximum;
– dragline and clamshell   25 m for slurry trench, cutoffs.

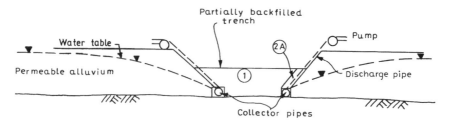

Figure 10.14.   Dewatering system for cutoff trench.

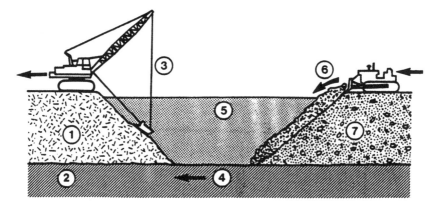

Figure 10.15.   Slurry trench technique (1) Permeable virgin ground, (2) Substratum, (3) Excavation, (4) Direction of progress, (5) Trench filled with bentonite slurry, (6) Filling with aggregate, (7) Finished trench (ICOLD, 1985).

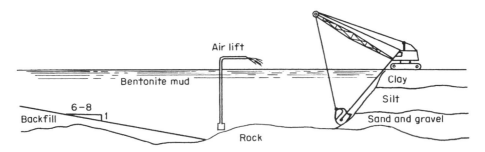

Figure 10.16.   Slurry trench excavation by dragline (Xanthakos, 1979, reproduced with permission of McGraw-Hill).

Note that it may be difficult to form a low permeability connection between the cutoff and the underlying low permeability stratum, particularly if the surface is irregular. An air lift or clamshell or scraper should be used as shown in Figures 10.16 and 10.17 to assist in cleanup of loosened debris at the base of the trench.

ICOLD (1985) indicates that the backfill will normally consist of bentonite slurry, with 5–15% bentonite (by weight), a marsh funnel viscosity of greater than 40 seconds, mixed with well-graded sand and gravel between 0.02 mm and 30 mm size. The mixture should have a standard concrete cone slump of 100–200 mm. Xanthakos (1979) indicates that naturally occurring clays from the site may be used, although they are unlikely to be suitable where their liquid limit is greater than 60% and may be difficult to mix if they remain

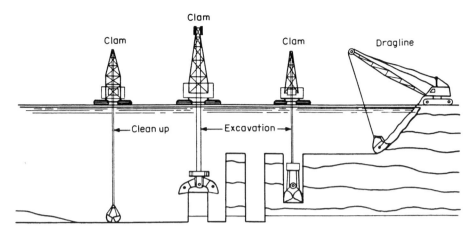

Figure 10.17.   Slurry trench excavation by dragline and clamshell (Xanthakos, 1979, reproduced with permission of McGraw-Hill).

Table 10.2.   Typical gradation limits for slurry trench backfills (adapted from Xanthakos, 1979).

Sieve size (mm)	Percentage passing	
	UK and Australian cutoffs	USA cutoffs
75	80–100	80–100
19	40–100	40–100
4.75	30–70	30–80
0.6	20–50	20–60
0.075	10–25	10–30

in lumps. Table 10.2 gives particle size distributions for the combined backfill material quoted by Xanthakos. Xanthakos indicates that at least 10% silt and clay fines passing 0.05 mm is needed to give adequately low permeability.

The backfill may be mixed by withrawing, dozing and blading the excavated soil to remove lumps of clay and silt or pockets of gravel. The soil is then mixed with the bentonite slurry, either by using the same earthmoving equipment or, preferably, by using a concrete batch plant type mixer to ensure a uniform product.

The backfill is placed into the trench by gradually pushing it in by bulldozer, so that the backfill slope is between 6 H to 1 V and 8 H to 1 V. To start backfilling, the initial slope should be formed by lowering the backfill using (say) a clamshell bucket, as dropping the backfill through the slurry would cause segregation. A gap of 15 m to 45 m is left between the excavation face and toe of backfill to ensure proper cleanup of the trench base can be carried out.

No attempt is made to densify or compact the backfill.

Xanthakos (1979) indicates that in some cases 0.6–0.9 m of concrete may be placed at the bottom of the trench to provide protection against erosion of the base under seepage flows. The top of the trench must be protected from drying by placing earthfill over the top. Settlement of the cutoff is quoted by Xanthakos (1979) as being between 25–150 mm for trenches 15 m to 25 m deep and 25 m wide. Less settlement is observed in narrower trenches. Settlement mostly takes place in the first 6 months.

10.4.4   Grout diaphragm wall

Grout diaphragm walls are excavated continuously in panels as shown in Figure 10.18, with the trench supported by a cement/bentonite slurry. This slurry is left in the trench and cures to give a low strength, low permeability compressible wall. The panels are excavated in the sequence shown in Figure 10.17, with the secondary panels being excavated before the slurry has hardened excessively in the primary panels but has hardened sufficiently to be self supporting. This obviates the need for end support for the panels as is used for cast in place diaphragm walls (see Section 10.4.5).

The trench may be between 0.5 m and 1.5 m wide (ICOLD, 1985) but will often be nearer the narrow limit for economy.

Excavation is carried out by grab buckets or clamshells. While ICOLD (1985) make no definitive statement on maximum practical depth, it can be inferred that grout diaphragm walls may be used to at least 50 m depth.

(ICOLD, 1985) indicate that the cement/bentonite grout will usually have the following composition per cubic metre of grout:

– 80–350 kg cement;
– 30–50 kg bentonite.

Xanthakos (1979) indicates that typical mixes will be:

– 15–20% cement;
– 2–4% bentonite;
– 5–10% sand and gravel.

The water cement ratio (by weight) will be between:

– 4:1 and 10:1 for ground granulated blast furnace cement (BLF);
– 3.3:1 and 5:1 for Portland cement (P).

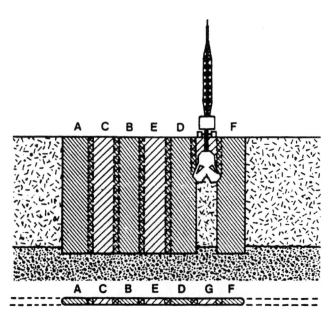

Figure 10.18.   Grout diaphragm wall, (1) Order of construction of panels A,B,C,D,E,F,G; (2) ABDF primary panels, CEG secondary panels (ICOLD, 1985).

The blast furnace cement has greater resistance to attack by aggressive groundwater (pore water) which dissolves the free lime in the cement and selenitic water. These result in formation of tricalcium sulpho-aluminates which destroys the hardened grout by expanding. Alternatively fly ash may be added in proportion 10% to 100% by weight of cement (ICOLD, 1985).

Retarders are added to the mix to control the curing process mainly to delay the initial set.

The addition of cement, which has free lime ($Ca(OH)_2$ and/or gypsum ($CaSO_4$) to the bentonite slurry makes it flocculate because of cation exchange of the Ca^{++} ions for Na^+ ions in the bentonite. The slurry remains stable (i.e. doesn't bleed) but does not form such an effective filter cake on the sides of the trench as bentonite and hence losses are greater. Xanthakos (1979) indicates losses may be up to 100% of the trench volume. This increases costs but gives a greater effective wall width. Admixtures can be used to reduce the flocculation effect and the use of BLF cement (with less free lime) instead of portland cement assists.

The cement bentonite grouts commonly have very low strength compared to concrete. ICOLD (1985) indicate an unconfined compressive strength of 100 kPa at 28 days, 150 kPa at 90 days. The strength is affected by water/cement ratio and cement type. The grout is able to withstand considerable plastic deformation to accommodate settlement due to embankment construction. It is best if the cement-bentonite grout has a Youngs Modulus just a little larger than the surrounding soil. In this way the load applied by the dam causes the wall to remain in compression.

10.4.5   Diaphragm wall using rigid or plastic concrete

Diaphragm walls are excavated in alternating panels as shown in Figure 10.19 with the panel supported by bentonite. The wall is constructed by tremie pipe placement of concrete or cement-bentonite concrete ("plastic concrete").

The ends of each panel are supported by a steel "stop-end" tube as shown in Figure 10.20. The pipe is removed after initial set of the concrete leaving a half round key which is used as a guide for the excavating tool, thus reducing potential misalignment of panels and leakage.

It should be noted that for dam cutoffs the walls are not usually reinforced with steel.

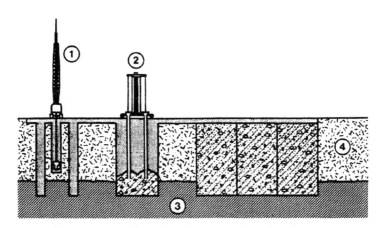

Figure 10.19.   Diaphragm wall cutoff, (1) excavation; (2) concreting; (3) substratum; (4) permeable layer (ICOLD, 1985).

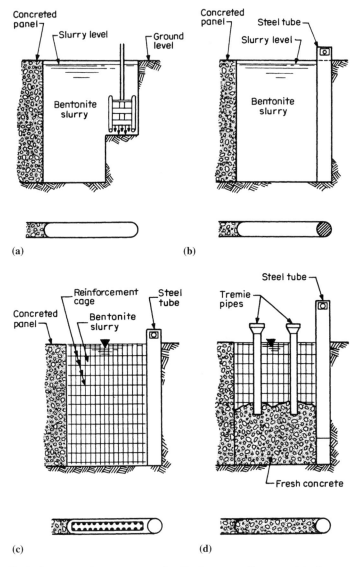

Figure 10.20.   Typical construction sequence of a diaphragm wall executed in four stages: (a) excavation; (b) insertion of steel tubing; (c) placement of reinforcement cage; (d) concrete placement (Xanthakos, 1979, reproduced with permission of McGraw-Hill).

The steel pipe should have a diameter equal to the trench width, so concrete does not leak past the tube. Figure 10.21 shows the effect of overbreak in a gravel layer, allowing concrete to surround the tube and make it difficult to remove the tube.

Other joint systems may be used to achieve a better contact between adjoining panels and improve water tightness. Some examples are given in Figure 10.22. Millet et al. 1992 give other examples.

The trench wall thickness is generally 0.6 m for walls up to about 30 m deep, increasing to 1 m or 1.2 m for deep walls e.g. 50 m. The added width is required to assist in maintaining overlap between adjacent panels. The usual specified tolerance for verticality is 1/100 or 1/200, with some instances of 1:500 being required (Xanthakos, 1979), Tamaro and Poletto (1992), Millet et al. (1992).

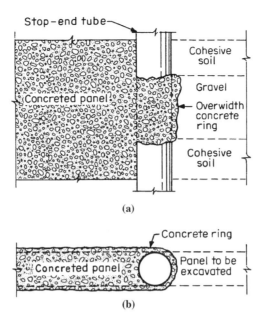

Figure 10.21.   Penetration of concrete beyond the stop-end tube due to overbreak (a) partial elevation; (b) partial section (Xanthakos, 1979, reproduced with permission of McGraw-Hill).

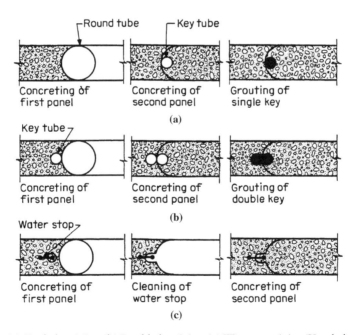

Figure 10.22.   (a) Single key joint; (b) Double key joint; (c) Water stop joint (Xanthakos, 1979, reproduced with permission of McGraw-Hill).

Excavation is carried out by clamshell, scraping bucket or rotary drilling equipment. Details are given in Section 10.4.6.

The walls may be constructed of conventional concrete and often are for building construction work. For dam applications, concrete is too rigid. As the soil mass surrounding the wall compresses under the weight of the embankment during its construction and the water load as the dam is filled, substantial loads will be shed onto the wall by negative skin friction. This can cause crushing of the wall and penetration of the wall into the dam fill.

For most dam applications it is preferable to use a plastic concrete backfill i.e. a concrete with bentonite added. ICOLD (1985) suggest that the following properties are desirable:

– The Youngs Modulus of the wall should relate to that of the soil, i.e. $E_{wall} \leqslant 5\ E_{soil}$;
– Failure strain should be high. Figure 10.23 shows some examples;
– Unconfined strength. A high strength is not important, in the order of 1–2 MPa (Xanthakos suggests an upper limit of 2 MPa).

ICOLD (1985) indicate that the composition of plastic concretes should be (per cubic metre):

– 400–500 litres – bentonite slurry;
– 100–200 kg – cement;
– 1300–1500 kg – well graded aggregates less than 30 mm size.

The water cement ratio will be between 3.3 to 1 and 10 to 1 according to the type of cement, the higher values being for BLF cement, the lower for portland cement.

The bentonite slurry is to keep the cement and aggregates in suspension during placement and assure plasticity and low permeability. The percentage (by weight) of bentonite to water varies from 2% to 12% according to its hydration. The marsh funnel viscosity should be 50 seconds.

If coarse aggregates are replaced by medium to fine sand the composition should be:

– 375–750 litres – bentonite slurry;
– 75–290 kg – cement;
– 500–100 kg – medium to fine sand.

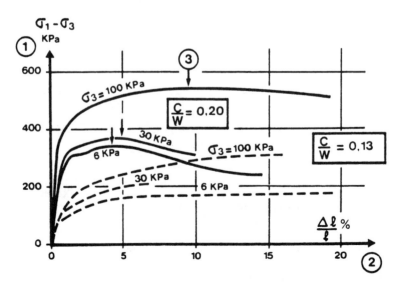

Figure 10.23.   Triaxial tests on plastic concretes (ICOLD, 1985). (1) Deviator stress, (2) strain, (3) failure.

Millet et al. (1992) emphasise the need to hydrate the bentonite before mixing in cement and aggregates.

The concrete (either conventional or plastic) must be placed in the wall by a tremie pipe to avoid segregation. Xanthakos (1979) indicates that it is desirable to complete the concrete pours in less than 4 hours to avoid any significant stiffening of the concrete. For panels up to 3.5–4.5 m long, a single tremie pipe is adequate but, for longer panels, two pipes may be needed.

Reinforced concrete guide walls are constructed ahead of the trenching operation, for diaphragm walls filled with grout or concrete. The guide walls:

– Control the line and grade in the trench;
– Support the sides of the trench from heavy construction loads;
– Protect the sides of the trench from turbulence and erosion;
– Can be braced to support the top of the trench;
– Act as a guide trench for the slurry.

Xanthakos (1979) gives details of guide wall design.

10.4.6   *Methods of excavation of diaphragm walls*

Diaphragm walls are usually excavated as a series of panels, each panel as a series of passes as shown in Figure 10.24.

This method can be applied to soft to medium hard, granular and cohesive soils, provided there are not boulders or other obstructions. The minimum "pass" is about 2 metres, the maximum about 6 m.

Excavation will usually be carried out using clamshell type bucket as shown in Figure 10.24. The clamshells are of two types:

(a) Cable suspended which:
      – Are easy to keep vertical;
      – Are manouverable;
      – Can work in limited headroom;

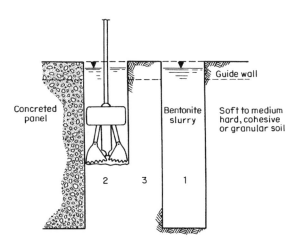

Figure 10.24.   Single stage panel excavation with clamshell bucket: Passes 1 and 2 spread of clamshell bucket; Pass 3 spread of clamshell bucket minus clearance for grab to embrace soil (Xanthakos, 1979, reproduced with permission of McGraw-Hill).

     – Depend on their self weight to close the grab. Pulleys or hydraulic systems are used
to assist in closing;

     – Can be used up to 75 m (cable operated), or 55 m hydraulic.

(b) Kelly bar which:

     – Are better suited to homogeneous soils;

     – Have a maximum depth of 40 m (single piece Kelly) telescoping up to 60 m;

     – Requires large headroom;

     – Are hydraulically or power operated, with the Kelly bar weight assisting to close
the bucket.

     Other types of excavation equipment which can be used include:

(c) Bucket scraper. This is attached to a Kelly bar as shown in Figure 10.25;

(d) Rotary drills

     In this technique the excavation is carried out by a drill which has several bits
mounted on a submersible frame. Figure 10.26 shows one type of such equipment.

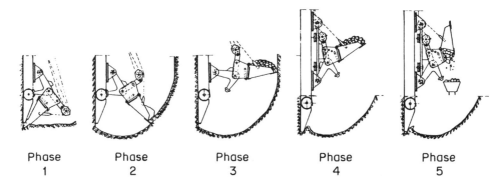

Phase 1   Phase 2   Phase 3   Phase 4   Phase 5

Figure 10.25.   Action of a scraping bucket excavator (Xanthakos, 1979, reproduced with permission of
McGraw-Hill).

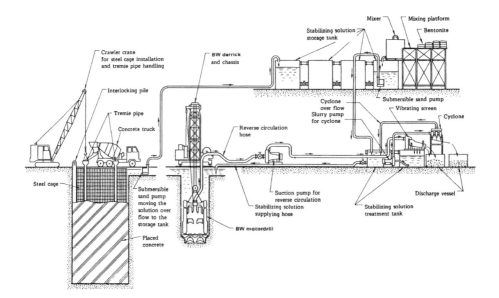

Figure 10.26.   Rotary drill technique (Xanthakos, 1979, reproduced with permission of McGraw-Hill).

The rotary drill bits can excavate into soil or soft rock. The cuttings are suspended in the slurry and flushed from the trench, usually by reverse circulation, to improve the lifting capacity. Screens and settling tanks are used to remove the cuttings.

(e) Percussion tools. Percussion tools and rock chisels are used where boulders, cobbles or other hard materials are encountered. The broken pieces are removed by clamshell bucket. Figure 10.27 shows examples of this equipment.

In some applications where boulders are present it may be economic to use a two stage operation (Figure 10.28), where the first stage consists of excavating pilot holes with percussion rotary drill and the second stage removing the material between the holes with a clamshell bucket. The first stage holes are spaced to leave an intermediate piece 0.3 m to 0.6 m narrower than the clamshell bucket grab.

(f) Hydrofraise – which is a down the hole reverse circulation rig which excavates a 2.8 m long panel. The use of this technique for Mud Mountain dam, is described in Davidson et al. (1992). They also describe the use of grouting to control hydraulic fracture in the dam core as the slurry wall was built.

Where bentonite mud is used to support an excavation, as required for virtually all the methods described above, it should be noted that the chemical composition of the groundwater will have an effect on the performance of the bentonite slurry.

In particular, groundwater with a high salt content (i.e. NaCl) or a high calcium ion concentration will cause the bentonite to flocculate. This is caused by contraction of the diffuse double layer around the clay particles as described in Chapter 8.

ICOLD (1985) indicate that:

– Fresh water should be used to make the bentonite slurry. Water with up to 5 g/litre of salt may be used, but the bentonite content will have to be increased;

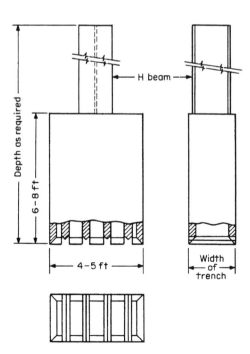

Figure 10.27.   Typical chisel details for breaking embedded boulders in site excavations (Xanthakos, 1979, reproduced with permission of McGraw-Hill).

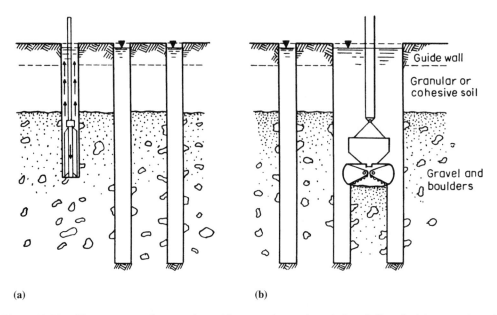

(a)   (b)

Figure 10.28.   Two stage panel excavation with percussive tools and clamshell grab: (a) excavation of
pilot holes with percussive tools; (b) panel excavation with clamshell bucket (Xanthakos,
1979, reproduced with permission of McGraw-Hill).

– "Colloid protectors" (called peptizers, or dispersing agents by Xanthakos, 1979) may
be used to prevent flocculation.

Xanthakos (1979) indicates that even if trenching into soil with groundwater of high
salts content, even of seawater quality, the slurry only takes up ionic salt and the flocculation can be controlled. If only saline water is available, attapulgite may be substituted
for bentonite.

More information on muds can be obtained from Xanthakos and from Chilingarian
and Vorabutr (1983).

10.4.7   Permeability of cutoff walls

ICOLD (1985) indicate that the permeabilities which can be achieved for cutoff
walls are:

– 10^{-7} to 10^{-8} m/sec – for grout walls;
– 10^{-8} to 10^{-9} m/sec – for plastic concrete walls;
– 10^{-9} to 10^{-10} m/sec – for concrete;
– compared to 10^{-6} m/sec for a grout curtain.

Millet et al. (1992) give similar values.
Powell and Morgenstern (1985) reviewed published case histories and concluded that
the range of permeabilities achieved was as shown in Figure 10.29.
The ICOLD (1985) figures are more likely to reflect the properties of the materials in
the cutoff wall, while the Powell and Morgenstern (1985) values reflect the equivalent
permeability of the constructed wall. The permeability of well constructed well constructed walls should approach that given by ICOLD (1985).

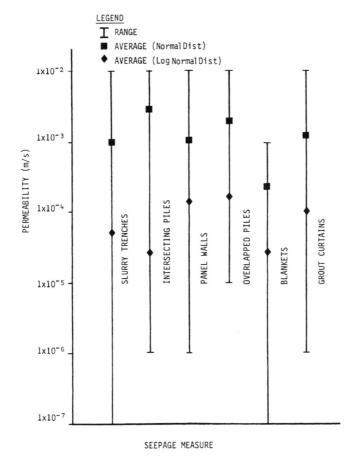

Figure 10.29.   Permeability of seepage control measures (Powell and Morgenstern, 1985, reproduced with permission of ASCE).

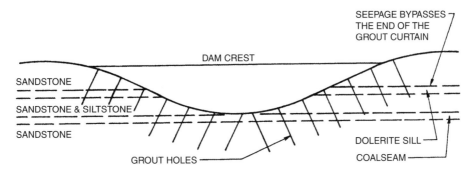

Figure 10.30.   Foundation for mine water dam, Drayton, NSW.

10.5   WE LIVE IN A THREE DIMENSIONAL WORLD!

We live in a three, not two, dimensional world. Figure 10.30 shows the foundation for a 20 m high dam for the Drayton Mine in the Hunter Valley for which the first author was responsible for the investigation and design. The grouting as shown was highly successful, closing to 1–3 lugeons. What was overlooked was that the dolerite sill daylighted in the storage, was highly permeable due to jointing and water simply went around the ends of the grout curtain. The water level in the storage dropped by ≈25 mm/day, and the water could be heard flowing in the dolerite sill in a borehole about 100 metres downstream!

Similar problems can occur with cutoffs into foundations of alluvial or lateritic soils (Figure 10.3) and it is often necessary to take the cutoff beyond where the crest of the dam meets the natural surface.

Another example of a 3 dimensional problem is seepage from tailings dams is given in Figure 19.33 which shows the plan of the tailings storage for the Boddington Mine, Western Australia. At the time of the first author's initial involvement, the emphasis on estimating the seepage from the storage had been on the seepage under the dam structures. The foundation rock was deeply weathered and lateritised and likely to have a high mass permeability. Hence seepage would occur through the hills to the north and east, not just under the dams. The groundwater table to the south was towards the dam. Being a tailings dam, the tailings themselves can control the seepage, except where the water on the tailings is in contact with the foundation.

10.6   THE MECHANICS OF INTERNAL EROSION AND PIPING

10.6.1   *Some general issues*

For internal erosion and piping to occur four conditions must exist (adapted from von Thun 1996):

1. There must be a seepage flow path and a source of water;
2. There must be erodible material within the flow path and this material must be carried by the seepage flow;
3. There must be an unprotected exit (open, unfiltered), from which the eroded material may escape;
4. For a pipe to form, the material being piped, or the material directly above, must be able to form and support "roof" for the pipe.

Piping may occur in the embankment, foundation and embankment to foundation.

10.6.2   *Piping in the embankment*

Piping in dam embankments initiates by one of three processes: backward erosion, concentrated leak and suffusion.

Backward erosion piping refers to the process in which erosion initiates at the exit point of seepage and progressive backward erosion results in the formation of a continuous passage or pipe.

Concentrated leak piping involves the formation of a crack or concentrated leak directly from the source of water to an exit point and erosion initiates along the walls of the concentrated leak.

Figure 10.31 shows conceptual models or the development of failure for backward erosion piping and concentrated leak piping. The sequence of events leading to failure by the two models is essentially the same, however the mechanisms involved in the initiation and progression stages are different.

Suffusion involves the washing out of fines from internally unstable soils. Soils which are gap-graded, or which have only a small quantity of fine soil in a mainly coarse sand or gravel are susceptible to suffusion.

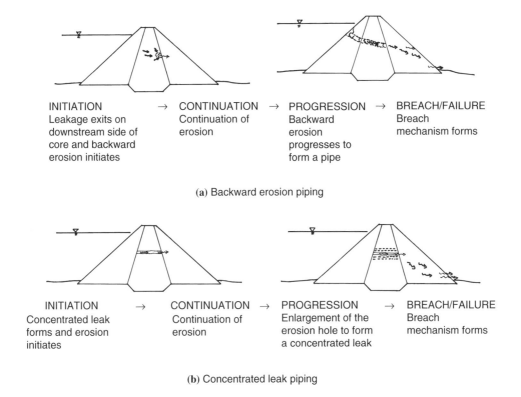

(a) Backward erosion piping

INITIATION → CONTINUATION → PROGRESSION → BREACH/FAILURE
Leakage exits on   Continuation of   Backward   Breach
downstream side of   erosion   erosion   mechanism forms
core and backward   progresses to
erosion initiates   form a pipe

(a) Backward erosion piping

INITIATION → CONTINUATION → PROGRESSION → BREACH/FAILURE
Concentrated leak   Continuation of   Enlargement of the   Breach
forms and erosion   erosion   erosion hole to form   mechanism forms
initiates   a concentrated leak

(b) Concentrated leak piping

Figure 10.31.   Model for development of failure by piping in the embankment (a) backward erosion, and (b) concentrated leak (Foster, 1999)

Potential breach mechanisms are:

– Gross enlargement of the pipe hole;
– Unravelling of the toe;
– Crest settlement, or sinkhole on the crest leading to overtopping;
– Instability of the downstream slope.

Figure 10.32 shows a failure path diagram illustrating the possible sequence of events leading to dam breaching.

Only marginal increases in the permeability of the core may be required to increase pore pressures in a low permeability downstream zone sufficiently to initiate downstream sliding, so it is assumed that the progression of piping to form a hole is not necessarily required for this failure mechanism. It is assumed that failure by gross enlargement of the pipe or unravelling of the toe requires continuing flow and therefore this is only possible if the pipe remains open.

10.6.3   *Piping through the foundation*

Piping in the foundation initiates by one of four processes: concentrated leak, backward erosion, suffusion, or blowout/heave followed by backward erosion.

As for piping through the embankment, it is possible to develop a single failure path diagram for the assessment of concentrated leak piping and backward erosion piping. Figures 10.33 and 10.34 show the failure path diagram for piping through the foundation.

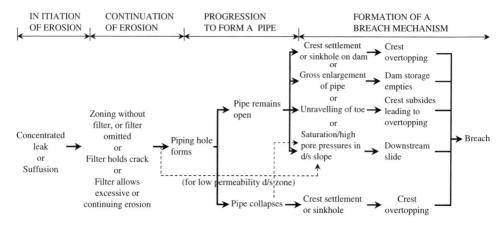

Figure 10.32.   Failure path diagram for failure by piping through the embankment (Foster, 1999).

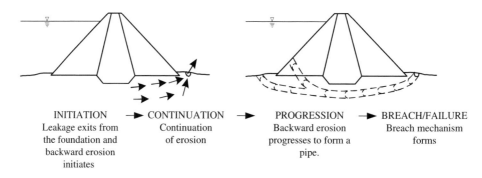

Figure 10.33.   Model for development of failure by piping in the foundation (Foster, 1999).

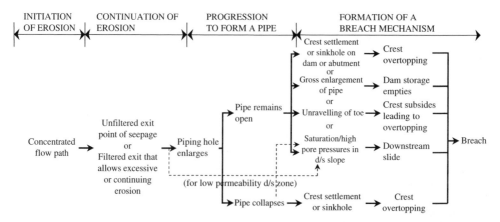

Figure 10.34.   Failure path diagram for failure by piping through the foundation – concentrated leak and backward erosion piping (Foster, 1999).

Failure can occur even if the pipe collapses, by either by crest settlement (sinkhole) leading to overtopping, or slope instability. Also it is assumed failure is possible by crest settlement, even if the pipe remains open due to the potential for settlements associated with crack filling action in the foundation.

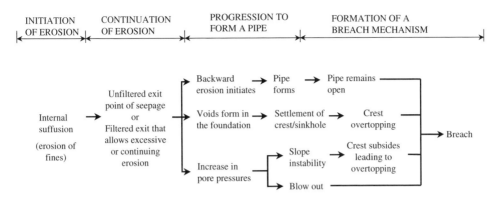

Figure 10.35.   Failure path diagram for failure by piping through the foundation – suffusion (Foster, 1999).

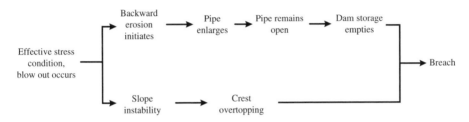

Figure 10.36.   Failure path diagram for failure by piping through the foundation – blowout followed by backward erosion (Foster, 1999).

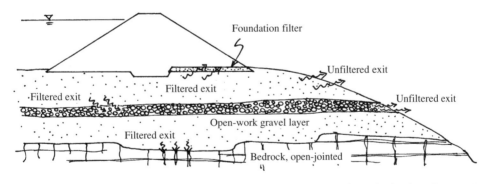

Figure 10.37.   Examples of filtered and free exit points for piping through the foundation (Foster, 1999).

Figures 10.35 and 10.36 show the failure path diagrams for failure by suffusion and blowout, followed by backward erosion.

Terzaghi and Peck (1967) and Von Thun (1996) believe if blowout were to occur it would be expected the first time the reservoir reached a critical elevation. However, if pore pressures increase with time due to internal erosion processes, or for example gradual deterioration of pressure relief wells, the risk of blowout may in fact increase with time. This has been observed at Mardi Dam (PWD of NSW, 1967).

Figure 10.37 gives an example of filtered and unfiltered exit points for piping through the foundation.

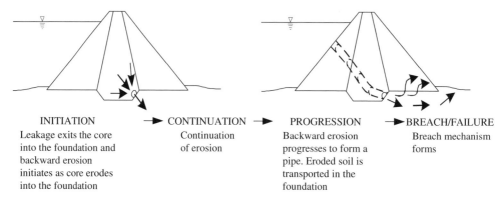

Figure 10.38. Model for development of failure by piping from the embankment into the foundation (Foster, 1999).

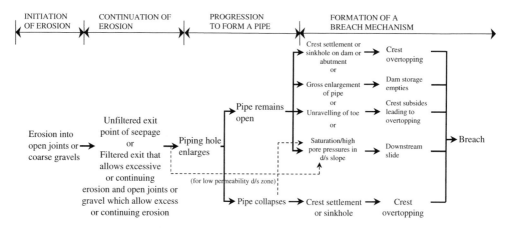

Figure 10.39. Failure path diagram for failure by piping from the embankment into the foundation – backward erosion piping (Foster, 1999).

10.6.4 *Piping from the embankment into the foundation*

Piping from the embankment to the foundation involves backward erosion, or suffusion initiated by erosion of the embankment soil into open joints or open gravels in the foundation. Figure 10.38 shows conceptually the model for development of failure. Figure 10.39 gives a failure path diagram.

10.7 FACTORS INFLUENCING THE LIKELIHOOD OF INTERNAL EROSION AND PIPING

Foster (1999), Foster and Fell (1999b, 2000) present a detailed discussion of the factors which influence the likelihood of initiation and continuation of erosion, progression to form a pipe and formation of a breach of the dam. The following is a summary of the information, presented mainly to assist those who are assessing existing dams, but the points are also a useful guide to good practice to reduce the likelihood of piping problems in new dams.

Table 10.3.   Effect of design and construction details on the likelihood of internal erosion and piping in the embankment (Foster and Fell, 2000).

Factor	Relative importance			
	Initiation	Continuation	Progression	Breach
Geometry				
General zoning	L	–	M	H
Core width	M/H	–	L	L
Core width/height	L/M	–	M	L
Crest width	–	–	–	L/M
Freeboard	–	–	–	M/H
Downstream zone properties	–	–	M/H	H
Filter/code	–	H	H	M
Compatibility dam core				
Classification	M	–	H	L
Erodibility/dispersivity	L	–	H	L
Compaction density ratio	M	–	M	L
Compaction water content	H	–	H	L
Permeability	M	–	M	L
Degree of saturation	M	–	H	L
Foundation				
Large scale irregularities	H	–	L	–
Small scale irregularities	M	–	L	–
Compressible soils	M	–	L	L
Conduits				
If present	H	–	H	L
Type/joint details	L	–	L	–
Settlement	L	–	L	–
Trench details	H	–	H	L
Walls abutting core				
If present	H	–	H	L
Slope	M	–	L	–
Collars/finish	L	–	L	–
Storage volume	–	–	–	M
Closure section	M	–	M	L

Notes: (1) Relative importance weightings are judgemental and will vary from dam to dam.
(2) – = not applicable, L = low, M = medium, H = high.

10.7.1   *Piping through the embankment*

Table 10.3 summarizes the effect of design and construction details on the likelihood of internal erosion and piping in the embankment. It is based on a detailed review of the literature (which generally only considers one aspect e.g. erodibility) and a review of case studies of dams which have failed or experienced piping accidents.

The relative importance weightings are judgemental and the table is only meant to highlight how the different factors influence the different stages of the development of piping, from initiation through to breach.

The following sections summarize the factors affecting the internal erosion and piping process. They are taken from Foster and Fell (2000).

10.7.1.1   *Formation of a concentrated leak*

A concentrated leak may form by transverse cracking, wetting induced collapse, hydraulic fracture, a high permeability zone or association with a conduit or wall. The assessment

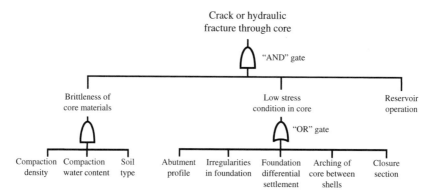

Figure 10.40.   Fault tree diagram for the formation of a crack or hydraulic fracture through the core (Foster, 1999).

Table 10.4.   Influence of factors on likelihood of cracking or wetting induced collapse-susceptibility of core materials (Foster and Fell, 2000).

| Factor | Influence on likelihood of cracking or collapse | | |
	More likely	Neutral	Less likely
Compaction density ratio[1]	Poorly compacted, <95% standard compaction density ratio[2]	95–98% standard compaction density ratio	Well compacted, ≥98% standard compaction density ratio
Compaction water content	Dry of standard optimum water content (approx. OWC-3%)	Approx OWC-1% to OWC-2%	Optimum or wet of standard optimum water content
Soil types[3]	Low plasticity clay fines	Medium plasticity clay fines	High plasticity clay fines Cohesionless silty fines

Notes: [1] For cracking, compaction density ratio is not a major factor. It is more important for wetting induced collapse.
[2] <93% Standard compaction, dry of OWC, much more likely.
[3] Soil type is not as important as compaction density and water content.

of likelihood of hydraulic fracture or crack through the core can be represented in a fault tree structure, as illustrated in Figure 10.40.

Cracking or hydraulic fracturing of the core is more likely if the soil is brittle and there is a low stress condition present. Therefore the "AND" gate is used to link these two issues in the fault tree. The factors contributing to the presence of a low stress condition in the core are linked using an "OR" gate as only one of these factors needs to be present.

Tables 10.4 to 10.8 and Figures 10.41 to 10.45 summarize the factors which influence the likelihood of a concentrated leak. These are based on the literature, including Lambe (1958), Leonards and Narain (1963), Sherard et al. (1963), Sherard et al. (1972a and b), Sherard (1973, 1985), Truscott (1977), Jaworski et al. (1981), Gillon and Newton (1988), Lo and Kanairu (1990), Lawton et al. (1992), Charles (1997), Høeg et al. (1998) and a review of case studies Foster (1999), Foster and Fell (1999b), Foster et al. (1998).

Table 10.9 summarises the factors influencing the likelihood of suffusion occurring in the core if the core or the foundation is cohesionless material. The susceptibility of soils

Table 10.5.   Influence of factors in the likelihood of cracking or hydraulic fracturing – features giving low stress conditions (Foster and Fell, 2000).

| Factor | Influence on likelihood of cracking or hydraulic fracture | | |
	More likely	Neutral	Less likely
Overall abutment profile	Deep and narrow valley. Abrupt changes in abutment profile, continuous across core. Near vertical abutment slopes	Reasonably uniform slopes and moderate steepness, e.g. 0.25 H:1 V to 0.5 H:1 V	Uniform abutment profile, or large scale slope modification. Flat abutment slopes (>0.5 H:1 V)
Small scale irregularities in abutment profile	Steps, benches, depressions in rock foundation, particularly if continuous across width of core (examples: haul road, grouting platforms during construction, river channel)	Irregularities present, but not continuous across width of the core	Careful slope modification or smooth profile
Differential foundation settlement	Deep soil foundation adjacent to rock abutments. Variable depth of foundation soils Variation in compressibility of foundation soils	Soil foundation, gradual variation in depth	Low compressibility soil foundation. No soil in foundation
Core characteristics	Narrow core, H/W > 2, particularly core with vertical sides	Average core width, 2 < H/W < 1	Wide core H/W < 1
	Core material less stiff than shell material Central core	Core and shell materials equivalent stiffness	Core material stiffer than shell material Upstream sloping core
Closure section (during construction)	River diversion through closure section in dam, or new fill placed a long time after original construction		No closure section (river diversion through outlet conduit or tunnel)

to suffusion depends on the particle size distribution. This can be assessed based on the general shape of the grading curve or using the techniques of Sherard (1979), Kenney and Lau (1985) or Burenkova (1993) to analyse the grading curve. Recent research still under way at UNSW indicates that the Kenney and Lau (1985) and Sherard (1979) methods may be conservative for soils with a large proportion of silty fines.

We are of the opinion that suffusion cannot occur in cohesive soils. Except possibly in low clay size content (<10% passing 0.075 mm) and low plasticity (plasticity index <7%) soils.

10.7.1.2   Continuation of erosion

Whether erosion will continue is dependent on the zoning of the dam, in particular, whether there are filters or transition zones which will eventually cause the erosion which has initiated to cease.

It is recommended that the assessment of the likely filter performance be considered by the compatibility of the core and filter (or downstream zone) materials in terms of the filter test erosion boundaries as described in Foster (1999) and Foster and Fell (1999a and b, 2000) and in Chapter 9.

These show that modern filter design criteria e.g. Sherard and Dunnigan (1989) are based essentially on no erosion of the dam core and that coarser filters may eventually seal with some erosion, which may only give limited leakage rates.

Table 10.6.   Influence of factors on the likelihood of a concentrated leak – high permeability zone (Foster and Fell, 2000).

| Factor | Influence on likelihood of a high permeability zone | | |
	More likely	Neutral	Less likely
Compaction density ratio	Poorly compacted, <95% standard compaction density ratio[1]	95–98% standard compaction density ratio	Well compacted, ≥98% standard compaction density ratio
Compaction water content	Dry of standard optimum water content (approx. OWC-3%)	Approx OWC-1% to OWC-2%	Optimum or wet of standard optimum water content
General quality of construction	Poor clean up after wet, dry or frozen periods during construction. No engineering supervision of construction		Removal of dried, wet or frozen layers before resuming construction. Good engineering supervision
Instrumentation details	Poor compaction around instrumentation, particularly if pass through the core		No instrumentation in the core
Characteristics of core materials	Large variability of materials in borrow area, moisture content, conditioning and grain size. Core materials susceptible to shrinkage cracks due to drying. Widely graded core materials susceptible to segregation		Low variability of materials in borrow areas. Low shrinkage potential. Narrow grading

Notes: [1] <93% Standard, dry of OWC, much more likely.

Table 10.7.   Influence of factors on the likelihood of a concentrated leak associated with a conduit (Foster and Fell, 2000).

| Factor | Influence on likelihood of concentrated leak | | | |
	Much more likely	More likely	Neutral	Less likely
Conduit type[1]	Masonry, brick. Corrugated steel	Steel, cast iron, not encased	Cast iron, concrete encased. Concrete precast	Concrete encased steel. Concrete cast in-situ
Conduit joints[1]	Open joints, or cracks signs of erosion	Open joints	High quality joints, "open" up to 5 mm but with waterstops	High quality joints, no openings, waterstops
Pipe corrosion[1]	Old, corroded cast iron or steel	Old cast iron, steel		New steel with corrosion protection
Conduit details[1]	Significant settlement or deep compressible foundation soils. Junction with shaft in embankment	Some settlement, shallow compressible foundation soils		Little or no settlement or rock foundation
Conduit trench details[2]	Narrow, deep, near vertical sides. Vertical sides, trench in soil (backfilled with concrete)	Medium depth, width, slope. Excavated through dam	Wide, side slopes flatter than 1 H:1 V	Trench totally in rock, backfilled with concrete

Notes: [1] Conduit type, joints, corrosion and details mostly affect piping into the conduit.
[2] Conduit trench details mostly affect piping along the conduit.

Table 10.8.   Influence of factors on the likelihood of a concentrated leak associated with a spillway wall (Foster and Fell, 2000).

	Influence on likelihood of a concentrated leak		
Factor	More likely	Neutral	Less likely
Slope of wall	Overhanging	Vertical	Sloping $\geqslant 0.1$ H:1 V
Founding material	Soil, wall subject to settlement and rotation	Soil or weathered rock, no rotation	Rock
Finish on wall	Rough, irregular		Smooth, planar
Concrete collars, buttresses, overhangs	Present, particularly if shape makes compaction of core difficult		Not present
Special compaction	No special compaction adjacent to wall		Careful compaction adjacent to wall

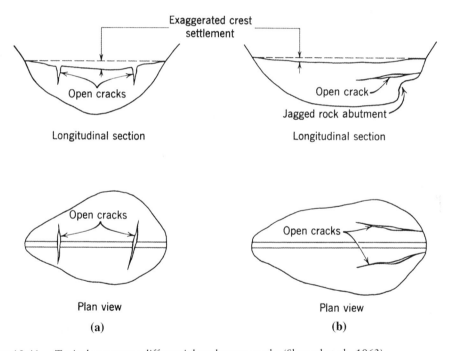

Figure 10.41.   Typical transverse differential settlement cracks (Sherard et al., 1963).

The likelihood of continuation of erosion will be dependent on the compatibility of the filters and the dam core: If they will:

(a) Seal with no erosion – rapid sealing of the concentrated leak with no potential for damage and no or only minor increases in leakage, continuation will be very unlikely;
(b) *Seal with some erosion* – sealing of the concentrated leak but with the potential for some damage and minor or moderate increases in leakage, continuation will be likely;
(c) *Excessive or continuing erosion* – slow or no sealing of the concentrated leak with the potential for large or continuing erosion losses and large increases in leakage, continuation will be highly likely.

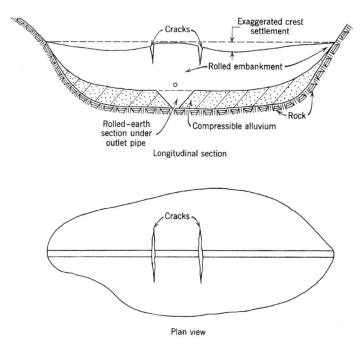

Figure 10.42.   Transverse cracking due to differential settlements over embankment discontinuities (Sherard et al., 1963).

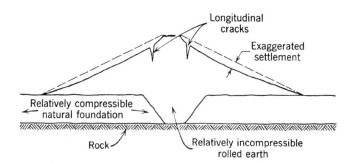

Figure 10.43.   Longitudinal cracking due to differential settlements over foundation discontinuities (Sherard et al., 1963).

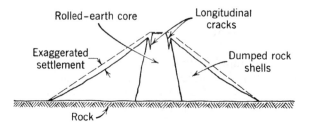

Figure 10.44.   Longitudinal cracking due to differential settlements between embankment zones (Sherard et al., 1963).

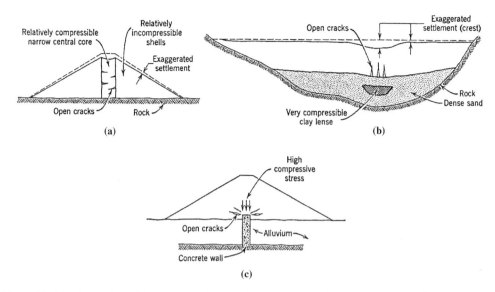

Figure 10.45.   Internal cracking in embankments due to differential settlements (Sherard et al., 1963).

Table 10.9.   Influence of factors on the likelihood of suffusion of cohension less soils (Foster and Fell, 2000).

Factor	Influence on likelihood of suffusion		
	More likely	Neutral	Less likely
Particle size distribution: – General – Gap-graded soils (Sherard, 1979) – Smooth gradations with a tail of fines based on Kenney and Lau (1985) or Burenkova (1993)	Gap-graded Flat tail in finer sizes d15c/d15f > 5[2] Potentially unstable		Uniform gradation, well graded d15c/d15f < 5[2] Stable
Compaction density	Poorly compacted, <95% standard compaction density ratio [1]	95–98% standard Compaction density ratio	Well compacted, ≥98% standard compaction density ratio
Permeability	High	Moderate	Low

Notes: [1] <93% Standard compaction, dry of OWC, much more likely.
       [2] d15c = particle size on coarse side of the distribution for which 15% is finer, d15f = particle size on the fine side of the distribution for which 15% is finer.

Continuation of erosion is highly likely if there is no downstream filter e.g. in homogeneous dams.

The presence of well designed filters to control the continuation and progression of erosion in the embankment and foundation can virtually eliminate the chances of piping. However this is dependent on good detailing, with the filters intercepting all potential seepage flow. "Blind Spots" are sometimes present e.g.:

– Filters only taken to full supply level, leaving the dam susceptible to internal erosion under flood conditions or if the dam crest is cracked and settles under earthquake loading. The authors have seen engineers modelling the rate of wetting of the soil related to the permeability of the core, arguing the flood loading is too short to cause saturation. It must be recognised that the crest of the dam is susceptible to cracking due to differential settlements, desiccation, and earthquake and piping can develop very quickly through the cracks. The permeability of the compacted soil is irrelevant in this situation.
– Filters not being taken around conduits which pass through the dam. Figure 10.46 shows Zoeknog dam which failed by piping. The filter stops half way up the conduit and fails to protect the areas adjacent to the lower part of the conduit where compaction in the narrow (1 m wide) zone would have been difficult and arching could have led to low vertical stresses, making hydraulic fracture likely.

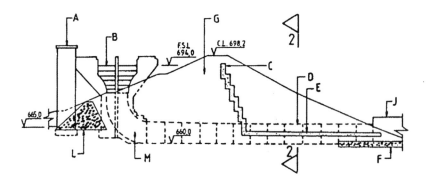

(a) Cross-section of embankment along the conduit.

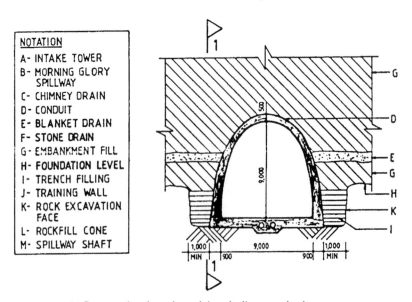

NOTATION
A- INTAKE TOWER
B- MORNING GLORY
     SPILLWAY
C- CHIMNEY DRAIN
D- CONDUIT
E- BLANKET DRAIN
F- STONE DRAIN
G- EMBANKMENT FILL
H- FOUNDATION LEVEL
I- TRENCH FILLING
J- TRAINING WALL
K- ROCK EXCAVATION
     FACE
L- ROCKFILL CONE
M- SPILLWAY SHAFT

(b) Cross-section through conduit and adjacent embankment.

Figure 10.46.   Zoeknog Dam – Details of filter near the conduit (Keller, 1994).

10.7.1.3   *Progression to formation and enlargement of a pipe*

There are two issues affecting the progression of piping through the embankment:

– The ability of the soil to support a roof of the pipe, i.e. will the pipe remain open or collapse;
– Enlargement of the hole i.e. will the pipe enlarge and how quickly?

   Tables 10.10, 10.11 and 10.12 summarize the factors which influence this. These are based on the literature, including Sherard (1953), Terzaghi and Peck (1967), Sherard et al. (1972a and b), Arulanandan et al. (1975), Vaughan and Soares (1982), Arulanandan and Perry (1983), Sherard and Dunnigan (1985), Chapius and Gatien (1986), Peck (1990), Bickel and Kuesel (1992), Fell et al. (1992), Hanson (1992) Hanson and Robinson (1993), Charles et al. (1995), Von Thun (1996), and a review of case studies (Foster, 1999, Foster and Fell, 1999b). The review of case studies showed that fines content was the critical factor in whether a roof could be supported. Soils with fines content ≥15% were found to be able to hold a roof, even if the fines were cohesionless.

10.7.1.4   *Formation of a breach mechanism*

Table 10.13 summarizes the mechanics of the breach for failures involving piping through the embankment (Foster, 1999, Foster and Fell, 1999b).

Table 10.10.   Influence of factors on the likelihood of progression of erosion – ability to support a roof (Foster and Fell, 2000).

| Factor | Influence on likelihood of fill materials supporting a roof of a pipe | | |
	More likely	Neutral	Less likely
Fines content (% finer than 0.075 mm)	Fines content >15%	Fines content <15% and >5%	No fines or fines content <5%
Degree of saturation	Partially saturated (first filling)		Saturated

Table 10.11.   Influence of factors on the progression of erosion – enlargement of the pipe (limitation of flows) (Foster and Fell, 2000).

| Factor | Influence on likelihood of pipe enlargement | | |
	More likely	Neutral	Less likely
Action of filter downstream of core	Considered in assessment of filter performance.		
Filling of cracks by washing in of material from upstream	Homogeneous zoning. Upstream zone of cohesive material		Zone upstream of core capable of crack filling (cohesionless soil)
Restriction of flow by upstream zones or concrete element in dam	Homogeneous zoning. Very high permeability zone upstream of core	Medium to high permeability zone upstream of core	In zoned dam, medium to low permeability granular zone upstream of core. Central concrete corewall and concrete face rockfill dam

Table 10.12.   Influence of factors on the progression of erosion – likelihood of pipe enlargement (erodibility) (Foster and Fell, 2000).

| Factor | Influence on likelihood of pipe enlargement | | |
	More likely	Neutral	Less likely
Soil type	Very uniform, fine cohesionless sand. (PI < 6) Well graded cohesionless soil. (PI < 6)	Well graded material with clay binder (6 < PI < 15)	Plastic clay (PI > 15)
Pinhole	Dispersive soils	Potentially dispersive soils	Non-dispersive soils
Dispersion test[3] Critical shear stress (Arulanandan and Perry, 1983)	Pinhole D1, D2 Soils with τ_c < 4 dyne/cm^2	Pinhole PD1, PD2 Soils with $4 < \tau_c < 9$ dyne/cm^2	Pinhole ND1, ND2 Soils with τ_c > 9 dyne/cm^2
Compaction density ratio	Poorly compacted, <95% standard compaction density ratio[1]	95–98% standard compaction density ratio	Well compacted, ≥98% standard compaction density ratio
Compaction water content	Dry of standard optimum water content (approx. OWC-3% or less)	Approx standard OWC-1% to OWC-2%	Standard optimum or wet of standard optimum water content
Hydraulic gradient across core[2]	High	Average	Low

Notes: [1] <93% Standard, dry of OWC, much more likely.
[2] Even dams with very low gradients, e.g. 0.05, can experience piping failure. (3) PI = Plasticity index.
[3] Using Sherard Pinhole test.

Table 10.13.   Mechanism of breach formation, failures by piping through the embankment.

Mechanism of breach formation	No. of failure cases
Gross enlargement of pipe	22
Crest settlement/sinkhole	3
Unravelling or sloughing	1
Instability	1
Not known	24
Total	51

Table 10.14 summarizes the influence of factors on these mechanisms. These are largely based on a review of case studies, (Foster, 1999; Foster and Fell, 1999b) with some data on unravelling and sloughing failures from Leps (1973), Olivier (1967) and Solvick (1991).

As listed in Table 10.14, the nature of the downstream zone has a major influence on possible breach mechanisms. If the downstream zone is free draining rockfill, large leakage flows can be sustained without breaching of the dam.

Table 10.15 lists some dams experiencing large flows without or before unravelling.

Free draining downstream zones are also very unlikely to experience slope instability which may occur in low permeability materials even under modest seepage flows. The ability of rockfill to experience large through flows can be assessed using the methods described in Olivier (1967), Leps (1973) and Solvick (1991).

Table 10.14.   Influence of factors on the likelihood of breaching (Foster and Fell, 2000).

Factor	Influence on likelihood of breaching		
	More likely	Neutral	Less likely
(a) Gross enlargement			
Zoning	Homogeneous type zoning	Zoned type dam, downstream zone of sand or gravel with fines	Zoned type dam with a downstream zone of gravel or rockfill
	Zoned type dam with a downstream zone able to support a roof		
Storage volume	Large storage volume		Small storage volume
(b) Crest Settlement/Sinkhole			
Freeboard at time of incident[1]	<2 m freeboard	≈3 m freeboard	≥4 m freeboard
Crest width	Narrow crest	Average crest width	Wide crest
Downstream zone[2]	Fine grained, erodible	Fine grained, non-erodible gravel	Rockfill
(c) Unravelling or sloughing			
Downstream zone	Silty sand (SM) Silty gravel (GM)	Fine grained rockfill. Gravel with some fines	Coarse grained rockfill Cohesive soil (CL, CH)
Flow-through capacity (Qc) of downstream zone	Qc ≪ estimated flow-through due to piping (Qp)	Qc ≈ Qp	Qc ≫ Qp
(d) Slope Instability			
(d1) Initiation of slide			
Existing stability	Analysis and/or evidence indicates existing marginal stability		Analysis and evidence indicates significant margin of safety against instability
Downstream zone	Cohesive soils (CL, CH)	Clayey/silty sands (SC, SM)	Free draining rockfill or gravel
(d2) Loss of freeboard and overtopping			
Freeboard at time of incident	<2 m freeboard	≈3 m freeboard	>4 m freeboard
Crest Width	Narrow crest	Average crest width	Wide crest
(d3) Breaching given overtopping occurs[2]	Fine grained, erodible	Fine grained, non-erodible Gravel	Rockfill
Downstream zone			

Notes: [1] Much more likely if ≤1 m, very unlikely >5 m. [2] Minor influence.

By contrast, in the piping failure of Teton dam (Independent Panel, 1976) and the major piping incident at Fontenelle Dam (Bellport, 1967), relatively small initial leakages of 300–400 litres/sec eroded the downstream zones of sandy gravel. Particles size distributions for these are shown in Figure 10.47.

The relatively fine grading, and low permeability of these zones would have been quite critical to the behaviour of the dams.

Table 10.15.   Dams with rockfill downstream zones which experienced large leakage flows without or before unravelling (Foster, 1999).

Dam	Estimated flows (m³/s)	Comment	Reference
Alto Anchicaya	1.8	Leakage through upstream concrete face	Regalado et al. (1982)
Bullileo	8	Piping incident, flow emerging on abutment	Appendix A, Foster (1999).
Churchill Falls FF11, Juklavatn Secondary, Songa and Suorva East	0.1	Piping through central core	Foster and Fell (1999)
Courtright	1.3	Leakage through upstream concrete face	USCOLD (1975)
Dix River	85	Dumped rockfill, flood passed through partly completed dam	Leps (1973)
Guadalupe	4	Leakage through upstream concrete face	Marsal (1982)
Hell hole	<540	Dumped rockfill, dominant rock size of 0.2–0.3 m, toe rocks 1.5–3 m in size. Unravelling initiated at 540 m³/s flow through leading to breach	Leps (1973)
Hrinova	0.18	Piping through central core	Verfel (1979)
Martin Gonzalo	1.0	Piping of upstream membrane bedding layer into rockfill zone	Justo (1988)
Omai Tailings	50	Piping incident, waste water lost through downstream rockfill zone	Vick (1996)
Salt springs	0.85	Leakage through upstream concrete face	USCOLD (1988)
Scofield	1.4–5.6	Piping incident	Sherard (1953)

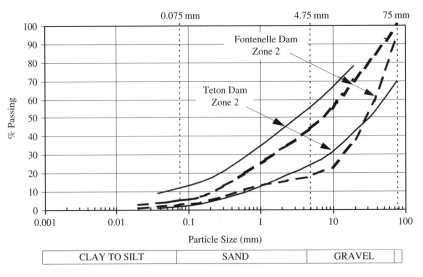

Figure 10.47.   Particle size distributions of downstream zones of Teton and Fontenelle Dams (Foster, 1999).

10.7.2 Piping through the foundation

10.7.2.1 Initiation of erosion

(a) Formation of a concentrated seepage path. A concentrated seepage path in a dam foundation is most likely related to a high permeability geological feature e.g. coarse gravels clay-infilled, or open joints in erodible or weathered rock foundations, or lateritic features. Table 10.16 summarizes the factors.

Stress conditions leading to cracking and hydraulic fracture are far less likely to occur in a foundation than in the embankment. However where there are potentially large differential settlements as shown in Figure 10.48 hydraulic fracture could occur.

(b) Suffusion. The factors influencing the likelihood of suffusion in the foundation are basically the same as for suffusion in the embankment; except that compaction density is replaced by in-situ relative density, with loose soils more likely, and dense soils less likely to experience suffusion.

Table 10.16. Influence of factors on likelihood of a concentrated seepage path through the foundation (Foster 1999, Foster and Fell, 1999b).

| Factor | Influence on likelihood | | |
	More likely	Neutral	Less likely
Geological environments soil foundations	Glacial Colluvial Volcanic (ash) Lateritic profile	Residual Aeolian Lacustrine	Alluvial
rock foundations	Limestone, dolomite Gypsum Basalt, rhyolite Interbedded sandstone and shale		Shale Sandstone (only) Conglomerate Igneous (other than basalt and rhyolite) Metamorphic
Geological features	Open jointed rock Openwork gravel Buried river channels Solution features Weathered faults and dykes		Demonstrated absence of such features
Continuity of high permeability features	Continuous from upstream to downstream, perpendicular to axis	Discontinuous feature	Not continuous below dam, cutoff by trench

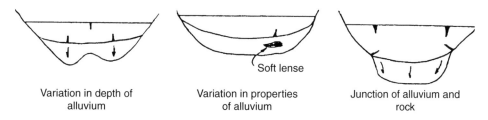

| Variation in depth of alluvium | Variation in properties of alluvium | Junction of alluvium and rock |

Figure 10.48. Sources of differential foundation settlements (Truscott 1977).

Table 10.17. Influence of factors on the likelihood of "blow out" (Foster, 1999; Foster and Fell, 1999b).

| Factor | Influence on likelihood | | |
	More likely	Neutral	Less likely
Foundation conditions at the downstream toe	Low permeability layer overlying high permeability layer		High or low permeability layer only
Observed behaviour	Sand boils at downstream toe "Quick sand" conditions		No sand boils
Factor of safety for zero effective stress condition $F_{UT} = \sigma_{vo}/u$	$F_{UT} < 1.2$	$F_{UT} \approx 1.5$	$F_{UT} > 2.0$

(c) "Blow-out" (or "heave", "liquefaction"). The likelihood of blow-out can be assessed by calculation as described in Section 10.3.3, Table 10.17 summarizes this and the other factors which influence the likelihood of blowout.

10.7.2.2 Continuation of erosion

A necessary requirement for piping in the foundation to occur is the presence of an unfiltered exit point for the seepage, which allows continuing removal of eroded materials (Von Thun, 1996). Figure 10.37 showed filtered and infiltered exit points.

Good dam design sets out to ensure that all seepage in the foundation is collected in filtered exit points such as:

– Horizontal filter drains;
– Toe drain;
– Relief wells.

Natural filtering can also happen due to stratification in the alluvium.

The approach should be to characterise the foundation geology, particle size distributions and other factors, and use the no erosion, excessive erosion and continuing erosion filter criteria (Section 9.3.2) to assess the likelihood of continuation of erosion.

10.7.2.3 Progression of erosion

The statistics of dam incidents, described in Foster et al. (1998, 2000a), suggests that piping initiating in the foundation is less likely to progress to failure than piping through the embankment. The progression of piping in the foundation is related to the ability of the foundation soils to support a roof and to the factors influencing pipe enlargement.

The formation of an open pipe through the foundation would be expected to be largely influenced by the foundation soil types, soil stratigraphy and by the presence of geological features such as cemented layers and infilled scour channels in cohesive soils. Homogeneous cohesionless materials are unlikely to maintain an unsupported roof and are therefore unlikely to be susceptible to piping unless they are overlain by an artificial roof, such as the base of a concrete spillway structure, or a cohesive material. (Terzaghi and Peck, 1967) (Sherard et al., 1972a). Well graded sandy gravels may be able to support a roof by arching action between the larger gravel particles.

Given that cohesive embankment materials can support open pipes, it is feasible that the base of the embankment dam could form a roof if piping developed along the

Table 10.18.   Factors influencing the enlargement of the pipe, piping through the foundation – flow limitation (Foster, 1999, Foster and Fell, 1999b).

Factor	Influence on likelihood of pipe enlargement		
	More likely	Neutral	Less likely
Hydraulic gradient[1]	High	Average	Low
Filling of "cracks" or voids by washing in of embankment or foundation materials	Homogeneous zoning or upstream zone of cohesive material Low permeability cohesionless foundation layer upstream of the dam	Cohesive layer upstream of the dam (may crack)	Zoned type dam with gravel or rockfill upstream shell High permeability layer upstream of dam
Restriction of flow path	Flow path unrestricted dimensions, or flow path restricted but large dimensions (e.g. large solution channels in limestone)		Flow path of small restricted width (e.g. piping through cracks in cutoff walls or narrow rock joints)

Note: [1] Even dams with very low overall gradients across the foundation, e.g. 0.05, can experience piping failure.

embankment/foundation interface. However, there appear to be no examples of this occurring in the foundation piping failures.

From consideration of case studies Foster (1999) concluded that to form a pipe foundation soils needed a high, more plastic fines content than embankment soils. The apparent difference in behaviour in the foundation and the embankment is possibly due to differences in the degree of saturation of the soil.

Foster (1999), Foster and Fell (1999b) concluded that the formation of an open pipe is more likely in the foundation if any of the following features are present:

(i) The erodible material is cohesive;
(ii) There is cohesive material overlying the erodible material. Examples are layers of clay, cemented soil or rock overlying erodible soil or interbedded cemented and non-cemented layers;
(iii) Solution features in rock, for example solution channels or cavities in limestone filled with erodible materials;
(iv) The erodible materials are below a rigid structure such as a concrete dam, concrete spillway structure or below an outlet conduit.

The factors influencing enlargement of a pipe in the foundation are summarised in Table 10.18 for factors related to mechanisms of flow limitation and Table 10.19 for factors influencing the erodibility of the foundation soils.

Zoning is an influencing factor for the "crack filling" mechanism in the foundation, as piping may progress back into the upstream zone of the dam. If the upstream zone is comprised of well-graded material (i.e. gravel or rockfill), washing in of the material may clog the pipe. Foundation soil layering upstream of the dam is also influential as it may limit flows into the area of piping.

The factors influencing erodibility are assumed to be essentially the same as those for piping through the embankment except the compaction characteristics are replaced with density (for cohesionless soils) and consistency (for cohesive soils).

Table 10.19.   Influence of factors on likelihood of pipe enlargement, piping through the foundation –
erodibility. (Foster 1999 and Foster and Fell, 1999b).

Factor	Influence on likelihood of pipe enlargement		
	More likely	Neutral	Less likely
Soil type	Very uniform, fine cohesionless sand (PI < 6) Well graded cohesionless soil (PI < 6)	Well graded material with clay binder (6 < PI < 15)	Plastic clay (PI > 15)
Pinhole dispersion test (AS1289.3.8.3)	Dispersive soils, Pinhole D1, D2	Potentially dispersive soils, Pinhole PD1, PD2	Non-dispersive soils, Pinhole ND1, ND2
Relative density	Loose	Medium dense	Dense
Consistency	Soft	Stiff	Very stiff

10.7.2.4   Formation of a breach

Most cases of failure through the foundation where the breach mechanism was known, failed by gross enlargement of the pipe, leading to emptying of the reservoir without breaching the crest. Two cases led to settlement of the crest and overtopping. The factors influencing the formation of a breach mechanism are generally similar to those as for piping through the embankment, described in Section 10.7.1.

Gross enlargement. The factors are similar to those for piping through the embankment shown in Table 10.14(a). However, dam zoning would only be important if the pipe exits through the downstream zone of the dam. Gross enlargement is likely if there is continuing enlargement of the pipe and the roof of the pipe can be supported along its full length and these factors are considered in the preceding branches of the event tree.

Sinkhole or crest settlement. The assessment of breach by sinkhole or crest settlement leading to overtopping is assumed to be similar to that for piping through the embankment (Table 10.14b). However, piping can also occur through the abutments of the dam and therefore the freeboard of the abutments would be important in this situation.

Unravelling. The likelihood of unravelling or sloughing initiating is related to the discharge capacity and characteristics of the soil at the exit point of seepage. It is assumed the assessment is similar to piping through the embankment (Table 10.14c), except that the characteristics of the soils at the exit point of seepage need to be considered. For piping through the foundation, seepage may exit through the dam, downstream of the dam or on the abutments.

Slope instability. The assessment is assumed to be similar to that for piping through the embankment, except the characteristics of the foundation soils (in terms of permeability and loss of strength on shearing) need to be considered as well as the embankment materials. Like unravelling, slope instability may also be an issue on the abutments of the dam.

10.7.3   Piping from the embankment to the foundation

10.7.3.1   Initiation of erosion

Factors which would influence the likelihood of initiation of erosion are:

- The presence of open joints or coarse soils in the foundation against which the core is placed;
- The particle size distribution of the core material compared to these joints or coarse soils;

– The treatment which has been applied to the foundation e.g. shotcrete or other concrete, or filters between the core and the foundation;
– Hydraulic gradients and stresses in the core material in the cutoff trench. High hydraulic gradients and narrow, deep cutoffs in which the core material may arch, giving low vertical stresses, are more likely to experience initiation.

10.7.3.2   Continuation of erosion

Continuation of erosion will be dependent on the "filtering" capability of the open joints, or coarse foundation soils, compared to the core material. This can be assessed as described in Chapter 9, considering no erosion, excessive erosion and continuing erosion criteria. The opening of joints can be considered as equivalent to the equivalent opening size, so therefore equal to $D_{15F}/9$. From the consideration of the penetration of cement grout into jointed rock, it seems unlikely that erosion will continue where the crack opening size is less than $3D_{100B}$ of the core material. (See Section 12.2.4.4). These two approaches could be compared, to assess whether they give consistent outcomes, and a judgement made. For new dams, a conservative approach should be taken.

10.7.3.3   Progression to form a pipe and breach

The factors controlling progression to form a pipe and breach will depend on the likely failure path and whether it is through the embankment or foundation. The factors described above would then apply. It is notable that in the case of Teton (Independent Pane, 1976) and Fontenelle (Bellport, 1967) dams, relatively small leakages of only 300–400 litres/sec, eroded and initiated unravelling and slope instability of the downstream slope. This was related to the fine material, with low discharge capacity, in the downstream zones of these dams. Figure 10.48 shows the particle size distributions.

10.7.4   How to apply this information

The authors have found the information in Section 10.7.1 to 10.7.3 useful in the assessment of existing dams, using an enhanced traditional engineering assessment approach e.g. Bell et al. (2001) and in a quantitative risk analysis framework e.g. Foster et al. (2002).

The rigour of breaking the internal erosion and piping process to initiation, continuation, progression and breach, for each of the areas of potential weakness of a dam (e.g. adjacent conduits, walls, over irregularities in the foundation, in the upper part of the dam where filters may have been omitted) leads to a better understanding of the dam. The tables allow a qualitative assessment of likelihoods – e.g. if all factors affecting initiation of erosion are "more likely", then the likelihood of initiation is high.

This can be converted to numbers (as in Foster et al., 2002) but that is not always necessary.

It is also useful to use this approach when considering new dams, where compromises on design or construction are being assessed and in the design of remedial works.

It should be noted that research at UNSW in 2002–2003 has developed better methods for quantifying the rate of erosion of soils, and the initial hydraulic shear stress which will case erosion to begin. This is described in Wan and Fell (2002, 2004a and b), this research is on-going.

11

Analysis of stability and deformations

11.1   METHODS OF ANALYSIS

The analysis of the stability of dams and other slopes is usually carried out using Limit Equilibrium Analysis (LEA). In most situations more detailed analysis using numerical methods, e.g. finite element analysis, is not necessary. However if it is necessary to model the deformations of the dam or the effects of strain weakening, as in progressive failure, numerical methods must be used. Comprehensive reviews of slope stability analysis have been written by Fredlund and Krahn (1977), Whitman and Bailey (1967), Fredlund (1984), Graham (1984), Mostyn and Small (1987), Duncan (1992, 1996a and b), Morgenstern (1992), and Hungr (1997).

Whether carrying out limit equilibrium or numerical analysis, it is important to formulate the problem correctly and, in particular, to establish whether undrained strengths (total stress analysis) or drained strengths (effective stress analysis) is used. Table 11.1 summarises the shear strengths, pore pressures and unit weights which should be used.

The reason for using total stress analysis (Su, ϕ_u) in low permeability zones for end of construction, rapid drawdown and staged construction condition is that it is difficult to predict the pore pressures accurately. Undrained strengths should also be used for contractive soils as described in Section 6.1.3.

If in doubt as to whether undrained or drained conditions will be most critical, it is wise to check both and adopt the results from the lowest strength case. Undrained conditions can occur even in relatively slow rates of change of reservoir levels. Cooper et al. (1997) describe a situation where annual cycles of rise of reservoir level were sufficient to induce undrained loading conditions in the downstream poorly compacted (contractive) clay zone of the concrete corewall Embankment No.1 at Hume Dam. This is described in Section 11.6.1.

Effective stress methods can be used for the end of construction and have the advantage that monitoring of pore pressures can assist in assessing the stability. Section 11.4.2 describes the methods for estimating the pore pressures in the partially saturated soil as it is compacted in the dam and loaded by the weight of the embankment as construction continues.

It is important to model the foundation of the dam in the analysis of stability, particularly where the dam is founded on soil or on rock with low strength surfaces within it, e.g. bedding surface shears. Most failures of dams have occurred where such weak zones in the foundation have not been recognised.

11.2   LIMIT EQUILIBRIUM ANALYSIS METHODS

11.2.1   *General characteristics*

Table 11.2 Summaries the characteristics of the methods of slope stability analysis.
These methods all employ the same definition of the factor of safety, F:

$$F = \frac{\text{Shear Strength of the Soil}}{\text{Shear Stress required for equilibrium}} \qquad (11.1)$$

Table 11.1. Shear strengths, pore pressures and unit weights for stability analysis (adapted from Duncan 1992).

	End of construction	Condition rapid draw-down and staged construction	Normal operating ("steady seepage")
Analysis procedure and shear strength for free draining zones – filters, rockfill, sand/gravel in foundations	Effective stress analysis using c', ϕ'	Effective stress analysis using c', ϕ'	Effective stress analysis using c', ϕ'
Analysis procedure and shear strength for low permeability zones	Total stress analysis using S_u and $\phi_u^{(1)}$ or effective stress analysis modelling partially saturated conditions	Total stress analysis using S_u and $\phi_u^{(1)}$ for the dam prior to draw down or construction of the second stage	Effective stress analysis using c', ϕ', unless soils are contractive[2] in which case use Su measured in the dam
Internal pore pressures	No internal pore pressures (u) for total stress analysis; set u equal to zero in these zones. Pore pressures determined from laboratory tests for effective stress analysis	No internal pore pressures (u) for total stress analysis; set u equal to zero in these zones. Pore pressures from seepage analysis for effective stress analysis	Pore pressures from seepage analysis and/or from piezometer readings for effective stress analysis
Reservoir water	Include (usually as a zone with c' = 0, ϕ' =0, γ = 9.8 kN/m³)	Include (usually as a zone with c' = 0, ϕ' = 0, γ = 9.8 kN/m³)	Include (usually as a zone with c' = 0, ϕ' = 0, γ = 9.8 kN/m³)
Unit weights[3]	Total	Total	Total

Notes: [1] S_u and ϕ_u describe the undrained strength envelope, so the variation in undrained strength, with increase in total stress, can be modelled in the analysis.
[2] Contractive soils include poorly compacted saturated clay fill, normally and lightly over-consolidated clays, and other situations described in Section 6.1.3. Effective stress analysis which ignores pore pressures generated on shearing over-estimate the factor of safety.
[3] For free draining zones use γ_{dry} or γ_{moist} for zones above water, γ_{sat} below. For low permeability zones, use γ_{sat} or γ_{moist}.

The analyses assume the factor of safety is uniform along the whole of the failure surface and cannot directly allow for localised strain weakening or progressive failure effects.

The number of equations of equilibrium available is smaller than the number of unknowns in limit equilibrium analysis, so assumptions are made to make the problem determinate. This is what differentiates most of the methods in Table 11.2. Of the methods listed, the ordinary (or Fellenius, 1927) method should not be used, because it generally underestimates the factor of safety. In force equilibrium methods the factor of safety is affected significantly by the assumed inclination of the side forces between slices and are therefore potentially not as accurate.

Most practitioners use the Bishop "modified" (or "simplified") method for circular analysis, which has been shown to be adequately accurate (Whitman and Bailey, 1967) and is stable computationally, but most methods give similar answers. For non-circular surfaces, the Morgenstern and Price (1965) and Spencer (1967) methods are widely used.

Hungr (1997) has demonstrated that for non-circular surfaces the methods may give different answers. The example is a simple bi-planar sliding block (Figure 11.1), defined

Table 11.2.   Characteristics of limit equilibrium methods of slope stability analysis (adapted from Duncan 1992; after Duncan and Wright, 1980).

Method	Characteristics
Slope Stability Charts (Janbu, 1968; Duncan et al., 1987)	Accurate enough for some purposes for initial estimates. Faster than detailed computer analyses
Ordinary Method of Slices (Fellenius, 1927)	Only for circular slip surfaces Satisfies moment equilibrium Does not satisfy horizontal or vertical force equilibrium Underestimates the factor of safety in most cases
Bishop's Modified Method (Bishop, 1955)	Only for circular slip surfaces Satisfies moment equilibrium Satisfies vertical force equilibrium Does not satisfy horizontal force equilibrium
Force Equilibrium Methods (e.g. Lowe and Karafiath, 1960, and U.S. Corps of Engineers, 1970)	Any shape of slip surfaces Do not satisfy moment equilibrium Satisfies both vertical and horizontal force equilibrium
Janbu's Generalised Procedure of Slices (Janbu, 1968)	Any shape of slip surfaces Satisfies all conditions of equilibrium Permits side force locations to be varied More frequent numerical problems than some other methods
Morgenstern and Price's Method (Morgenstern and Price, 1965)	Any shape of slip surfaces Satisfies all conditions of equilibrium Permits side force orientations to be varied
Spencer's Method (Spencer, 1967)	Any shape of slip surfaces Satisfies all conditions of equilibrium Side force are assumed to be parallel

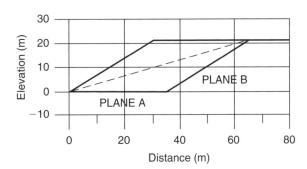

Figure 11.1.   A bi-linear rupture surface used in the parametric study to compare methods of stability analysis. The dashed line is the piezometric surface (Hungr, 1997).

by a flat basal surface (Plane A) and a steeper back-scarp (Plane B). Either of the two planes can be a weak surface. The second plane may be a stronger structural feature or it may be a rupture plane extending through intact material. A parametric study was carried out, varying the friction angle on Planes A and B between 10°, 20° and 30° with no cohesion. Both dry conditions and pore-pressures determined by the piezometric surface shown in Figure 11.1 were examined. The resulting Factors of Safety ranged between 0.5 and 2.5 for the range of conditions.

Figure 11.2 shows a comparison between the different methods of analysis. The abscissa in this figure is the ratio, R, between the friction coefficient on Plane A and that on Plane B: ϕ

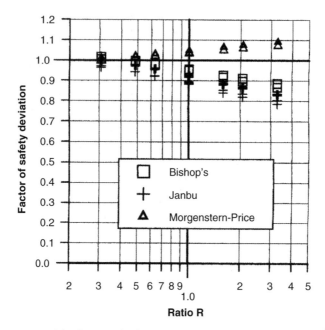

Figure 11.2.   Comparison of the factors of safety obtained by four methods of analysis, for the bi-linear surface shown in Figure 11.1. The factors are normalized with respect to the Spencer's Method. The abscissa is the ratio of strengths available on the two parts of the rupture surface (Hungr, 1997).

$$R = \frac{\tan \varphi'_A}{\tan \varphi'_B} \qquad (11.2)$$

The ordinate is a ratio between Factors of Safety determined by the various methods and that obtained by Spencer's Method, which is used as a reference. All the methods are shown to converge to the same value when R is low, i.e. when the flat basal plane is weaker than the back scarp. On the other hand, the methods diverge by as much as 30% in the case when the back scarp is weak and the sliding body is supported by the toe plane. Under these conditions there is a significant difference even between Spencer's and Morgenstern-Price methods.

Janbu's Simplified method produces a Factor of Safety which is consistently less than that of Bishop's (with the Fredlund and Krahn modification). The Janbu correction factor is required to compensate for this. All of these trends persist in equal measure for dry conditions and with pore-pressure.

In practice, cases where R is low are fortunately more frequent than the opposite, as Plane A often follows a thin weak layer in the stratigraphy, a bedding plane or a near-horizontal pre-sheared surface. Under such conditions there is not much difference between the four methods and the user is therefore justified in taking advantage of the high efficiency of Bishop's Simplified Method in two or three dimensions.

On the other hand, in cases where the strongest element of the sliding surface is at the toe, rigorous methods such as Morgenstern-Price should be used. Examples of this are cases where the back scarp (Plane B) follows steeply inclined bedding, a fault surface or similar and the sliding body is supported by a strong toe.

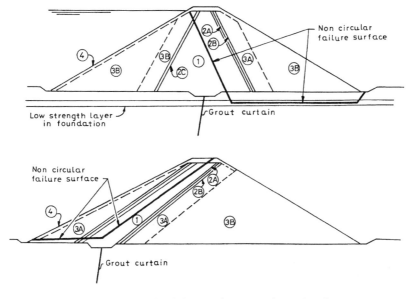

Figure 11.3.   Examples where non circular failure surfaces must be analysed.

11.2.2   *Some common problems*

Some common problems in LEA (which are often simply omissions and/or occur through lack of proper care) include:

- Use of circular rather than non circular analysis, where failure surfaces will clearly follow weak surfaces – e.g. bedding surface shears. Figure 11.3 shows examples where non circular analysis must be used;
- Omitting to model cracks in the ground surface and the water pressure which will often develop in the cracks due to rainfall or surface water flowing into the crack. This can have a significant effect on calculated factors of safety, particularly for smaller failure surfaces near a dam crest;
- Not modelling anisotropy of strengths, e.g. in fissured clays or stratified soil and rock;
- Not clearly showing on the drawing of the cross section and failure surface the properties used for that particular analysis. If the two are separated confusion is likely to occur.

These features can all be modelled with modern computer programs. If the program being used cannot model these features, consideration should be given to purchasing a program which can.

A more subtle problem, which is an extension of the discussion on the different methods of calculation described above, is where essentially translational sliding in rock is being modelled. Figure 11.4 shows such a slope, where the rock is relatively strong and with a few, widely spaced, near vertical joints AB, CD, EF (e.g. sandstone or conglomerate), with potential sliding on a weak layer (e.g. claystone) OBDF. The driving force for instability is the groundwater in the joints and the gravity forces along OBDF.

Use of a conventional computer program to analyse the stability of OAB, OCD, or OEF will overestimate the factor of safety substantially even if the failure surfaces are pre-defined,

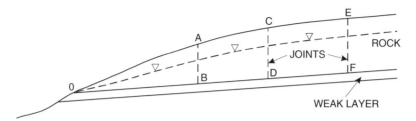

Figure 11.4.   Limit equilibrium analysis of a rock slope sliding on a weak layer.

because if one inserts a reasonable joint strength (say $c' = 0$, $\phi' = 40°$), the program assumes this applies to the rock and assumes slice slide forces will apply, equivalent to say a wedge of "rock" with $c' = 0$, $\phi' = 40°$ which will also drive the instability. This is difficult to overcome in computer programs other than modelling the vertical part of the failure surface (e.g. EF) as a crack. It is also possible this type of problem will lead to convergence problems in the non-circular analysis. It is better to analyse such cases using a wedge analysis (either by hand or a computer wedge analysis).

Other problems which arise are (Duncan 1992):

– Failure of the person performing the analysis to understand soil mechanics well enough to know how to define the water pressures, unit weights and shear strengths appropriate for the analysis;
– Failure of the person performing the analysis to understand the computer program well enough to define these quantities correctly in the input;
– Failure of the person performing the analysis and the person reviewing the results to check the results and properly evaluate their reasonableness.

11.2.3   Three dimensional analysis

The status of three-dimensional analysis is summarised in Duncan (1992), Morgenstern (1995) and Fell et al. (2000). The Bishop's Simplified, Janbu and Spencer's Methods have been extended into three dimensions (Hungr et al., 1989, Lam and Fredlund, 1993). However, none of the available algorithms can account for the internal stresses existing in a non-rotational and laterally asymmetric problem (Hungr, 1994). A true "rigorous method" that would satisfy all the available equilibrium conditions has not yet been formulated. As suggested by Hungr (1994), a rigorous three-dimensional method would require the definition of five spatially distributed inter-column force functions. The solution for the Factor of Safety and five inter-column force inclination ("i") coefficients would require six nested levels of iteration. While modern computers could easily handle the numerical computation, the correct selection of the inter-column force functions and possible convergence difficulties pose a daunting research problem.

As summarized by Fell et al. (2000) the use of "rules of thumb" such as a 10% increase to compensate for the neglect of 3D effects, as suggested in some textbooks, is not advisable, because although, as proven by Cavounidis (1987), the Factor of Safety of the critical 3D sliding surface always exceeds the critical 2D factor, the ratio between the two can vary within a range of 1.0 to as high as 1.4 (Morgenstern, 1992; Hungr et al., 1989). Potential problems lie in situations where the stability analysis is used to estimate the strength of certain materials through back analysis. Neglecting a strong 3D effect in the back-analysis could result in a serious overestimation of the back-calculated strength. However provided that the geometry of the failure surface remains the same for the back-analysis, and the "forward analysis", this is not a major practical problem.

The authors have on occasions resorted to, or have seen others resort to, reliance on three-dimensional effects to show that a factor of safety is adequate. In most cases this has been done simply by averaging the factors of safety of sections taken adjacent to each other. Usually, if the situation is so marginally stable, it is in reality in need of some remedial action to improve the factor of safety.

11.2.4   Shear strength of partially saturated soils

In recent years, there has been growing interest in the originally proposed effective stress approach for the determination of the shear strength of unsaturated soils (e.g. Fleureau et al., 1995; Oberg and Sallfors, 1995; Bolzon et al., 1996; Khalili and Khabbaz, 1996). In this approach the shear strength is determined on the basis of the effective strength parameters c' and ϕ' and a single stress variable defined as:

$$\sigma' = (\sigma - u_a) + \chi(u_a - u_w) \tag{11.3}$$

in which σ' is the effective stress and χ is the effective stress parameter, which has a value of 1 for saturated soils and a value of 0 for dry soils. The advantage of the effective stress approach is that the change in the shear strength with changes in total stress, pore water pressure and pore air pressure is related to a single stress variable. As a result a complete characterisation of the soil strength requires matching of a single stress history rather than two or three independent stress variables. Furthermore, the approach requires very limited testing of soils in an unsaturated state. A major difficulty of the effective stress approach has been the determination of χ.

Khalili and Khabbaz (1998) used experimental data to demonstrate that χ may be determined from the using the equation

$$\chi = \left[\frac{(u_a - u_w)}{(u_a - u_w)_b} \right]^{-0.55} \tag{11.4}$$

where $(u_a - u_w)_b$ is the air entry value and

$$\frac{u_a - u_w}{(u_a - u_w)_b} \quad \text{is termed the suction ratio.} \tag{11.5}$$

The air entry value, also called the bubbling pressure, corresponds to the matrix suction above which air penetrates into the soil pores and can be measured using the filter paper technique (ASTM, 1992; Swarbrick, 1992) or pressure plate technique (ASTM, 1968, 1992).

This approach was further assessed by Geiser (2000) who demonstrated an even stronger correlation between χ and the suction ratio than Khalili and Khabbaz (1998) had with their data.

In practice most stability analyses carried out ignore the effects of partial saturation, but in so doing, the factor of safety, particularly for the upper part of the dam, may be significantly underestimated. Hence either an approach such as that of Khalili and Khabbaz (1998) should be used to estimate the strength above the full supply level (or above the phreatic surface) or alternatively undrained strength estimated for existing dams by in-situ testing. However allowance should also be made for potential cracking or softening in areas affected by cracking as discussed in Section 11.3.2.

11.3   SELECTION OF SHEAR STRENGTH FOR DESIGN

11.3.1   *Drained, effective stress parameters*

11.3.1.1   *Peak, residual or fully softened strength in clay soils?*

Whether peak, softened or residual strengths are used in the analysis of slope stability depends on the presence or absence of existing slide "planes" (actually surfaces which may not be truly planar) and fissuring. The following guidelines are given:

- Where there is an existing slide plane, e.g. in the foundation of a dam or where a dam has failed by sliding, the field residual strength should be used for the slide plane. This applies regardless of how long it has been since sliding last occurred. The exception could be where there is definite evidence of recementing, e.g. in some slickensided joints in weathered rocks.
- Bedding surface shears formed by folding of rock strata have strengths approaching residual and, unless extensive tests on the surfaces show otherwise, residual strengths should be used.
- Compacted soils and soils which have no fissuring or cracks should be assigned peak strength parameters c' and ϕ' determined from saturated remoulded samples compacted to the same density ratio and moisture content the soil would be in the dam. The majority of dam embankment design is therefore based on peak strength parameters.
- Cuts in soils are often designed using fully softened strength to allow for the softening due to the unloading of the slope and some progressive failure mechanisms which may be present.
- Fissured soils have a strength between peak and residual strength depending on the nature of fissuring, orientation, continuity and spacing of the fissures. Such soils do occasionally occur in dam foundations, e.g. Ross River Dam (McConnel, 1987) and Prospect Dam (Landon-Jones, 1997). This is discussed in more detail in Section 6.1.8. Soils with relict joints, such as extremely weathered basalt, behave in a similar manner.
- Triaxial and direct shear tests do not properly simulate the actual plane strain stress conditions which exist in most slope stability problems. However, the uncertainty arising from this is not significant in practical slope stability problems and, provided the tests are carried out and interpreted correctly, the results from both tests can be adopted for design.

It should be noted that, as pointed out by Lade (1986) and Mitchell (1993), apart from cemented soils the effective cohesion (c') should be zero or very small. The value of c' adopted is particularly critical in the analysis of the stability of smaller dams, the upper parts of larger dams and in smaller landslides, because it has a major effect on calculated factors of safety when failure surfaces are shallow. It is recommended that unless there is definite evidence of higher values, the effective cohesion adopted for design should be between 0 and 10 kPa for peak strength and should be zero or say 1 kPa for residual and softened strengths, at least at low normal stresses. In some situations where the strength envelope is markedly curved a bi-linear envelope such as that shown in Figure 6.43 may be used.

11.3.1.2   *Selection of design parameters in clay soils*

When several triaxial tests have been carried out on the one soil, it is recommended that the design shear strength parameters are obtained from a p'-q plot of the test results, rather than by, say, averaging the individual c', ϕ' values from each test, or plotting all Mohr's circles on one diagram. It is also important to use results from the effective stress range applicable to the field problem. In many cases this is at low stress, e.g. less than 50 kPa to 100 kPa for slide surfaces at 5 m depth.

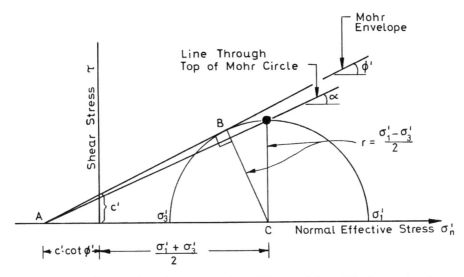

Figure 11.5.   Basis of p'-q plot where $p' = (\sigma'_1 + \sigma'_3)/2$, $q = (\sigma'_1 - \sigma'_3)/2$; it can be shown that $\sigma' = \sin^{-1}(\tan \alpha)$; $c' = a/\cos \phi'$; where a and α are obtained from the p'-q plot as shown on Figure 11.6.

The p'-q plot is a graph of the apex points of the Mohr's circles from the test results as shown in Figure 11.5.

When selecting the design parameters for design of dam embankments and landslide stabilizing works, it is common to bias towards the conservative by selecting a line with, say, 75% of the test points above and 25% below (i.e. a lower quartile line), but a line of best fit or a lower bound may be adopted depending on the circumstances.

If direct shear tests have been used to determine strength parameters the results should be plotted on a graph of normal stress vs shear stress and design parameters selected by a similar procedure to that outlined below for triaxial tests.

In any slope stability analysis it is good practice to check the calculated factor of safety for a range of strengths, e.g. lower quartile and lower bound, to determine the sensitivity of the factor of safety to the assumed strength.

It is essential to use judgement in selecting the design line, rather than using say a least squares regression analysis, as judgement enables allowance to be made for:

– Poor individual test results;
– The general trend for the second and particularly third stage of staged tests to give a lower strength than the true one because of excessive deformation and some loss of strength from the peak due to displacement on the shear plane, particularly in sensitive clays or cemented soils;
– General curvature of the Mohr's circle envelope;
– Adoption of low effective cohesion (c') which correctly models the behaviour of most soils.

The average of the c' and ϕ' values from which the p-q plot was derived, is also shown in Figure 11.6. It can be seen that using averages tends to give a larger c' and lower ϕ', than using the p'-q diagram. This is generally unconservative for smaller dams, shallow failure surfaces in larger dams and landslide stability, because the strength is overestimated in the working stress range.

Wroth and Houlsby (1985) have suggested a method based on the critical state parameters. This method is similar in principle to that proposed above using a p'-q diagram, but

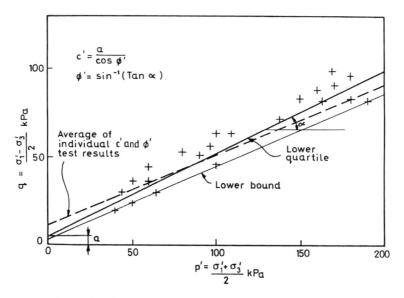

Figure 11.6.   Typical p'-q plot of triaxial test results.

takes account of the variation of c' with water content at failure. The critical P, Q values $P = (\sigma' + 2\sigma'_3)/3$ and $Q = (\sigma'_1 - \sigma'_3)/2$ at failure are normalised by dividing by the equivalent pressure at that water content (the pressure on the normally consolidated line at the same voids ratio as the specimen). Wroth and Houlsby claim that this gives a more rational way of selecting c' as a function of water content, particularly for fills. In practice this degree of sophistication may not be warranted.

11.3.1.3   *Selection of design parameters – granular soils and rockfill*

The strengths of granular soils and rockfill should be selected to account for the confining pressure, degree of compaction and particle size distribution and, for rockfill, the rock substance strength and finer particles content as described in Section 6.1.10.

   In most cases, particularly for rockfill, a bi-linear strength envelope should be used to model properly the curvature of the strength envelope. If this is not done, the factor of safety of smaller failure surfaces, particularly at the crest of the dam, will be significantly under-estimated.

   Depending on the confidence with which the strengths are able to be estimated (reflecting whether there are good records of how the dam was built or not and whether testing was carried out), either median or lower quartile strengths should be adopted.

11.3.2   *Undrained, total stress parameters*

11.3.2.1   *Triaxial compression, extension or direct simple shear strength*

As discussed in Section 6.1.3 and shown in Figures 6.10 and 6.12, it is important to allow for the stress path in selecting the undrained strength. The undrained strength estimated from triaxial compression tests only may significantly over-estimate the strength.

11.3.2.2   *Selection of design parameters*

The undrained strength of natural soft clay deposits usually shows a considerable scatter which is due to natural stratification of the deposit and measurement errors. Figures 11.7

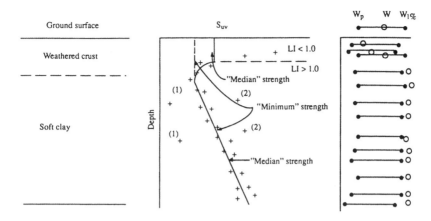

Figure 11.7.   Selection of foundation strength – Normal case. Notes: (1) Discard, probably disturbed. (2) Discard, probably affected by shells, roots etc. (Ferkh and Fell, 1994).

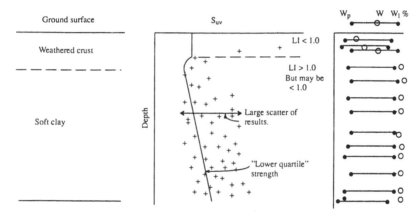

Figure 11.8.   Selection of foundation strength – influenced by shells, roots etc. and large scatter of data (Ferkh and Fell, 1994).

and 11.8 show typical soft ground strength profiles with an overconsolidated crust over the near normally consolidated soil.

In these circumstances it is recommended that where the data plots as in Figure 11.7 the median strength be used after discarding outlying results but, where there is considerable scatter as in Figure 11.8, a more conservative e.g. lower quartile strength be adopted.

In the overconsolidated crust Ferkh and Fell (1994) adopted an empirical approach, using the median strength line – which uses a strength equal to the median strength up to where the liquidity index – (water content – plasticity index) / (liquid limit – plastic limit) – is less than 1.0. These guidelines were developed for soft clay deposits, which are not usually found where dams are being built, but the principles are applicable. Rather than use the empirical approach detailed above it would be better to assess the fissured strength for the overconsolidated crust, as detailed in Section 6.1.8.

For embankments:

(a) The undrained strength varies within and between the layers of the compacted fill, reflecting varying degrees of softening as the soil wets up with seepage from the reservoir.

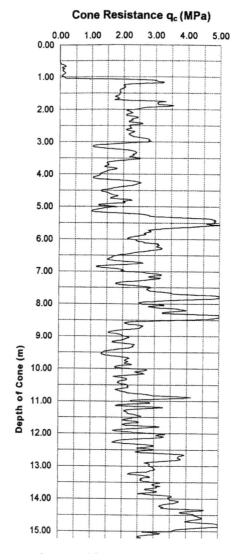

Figure 11.9.   Cone Resistance q_c determined from Cone Penetrating test, versus depth below dam crest for eppalock dam (courtesy Goulburn-Murray Water).

This is demonstrated in Figure 11.9 with the variation in CPT cone resistance. Note that $S_u = (q_c - \sigma_{vo})/N_K$ (see Section 6.1.7.1).

(b) The core has significant undrained strength at the surface, because the soil is compacted in a partially saturated condition. In Figure 11.9 the strength is somewhat less above Full Supply Level (the top 5 m approx.) due to the effects of cracking and softening as water has entered the cracks. This is not an uncommon feature particularly for central core earth and rockfill dams with poorly compacted rockfill and is due to spreading of the dam as it settles as shown in Figures 11.10 and 11.11.

The selection of undrained strength and stability analysis should allow for these features and may include:

– Allowance for cracks, or softened zones around old cracks with little or no shear strength. These are unlikely to persist below FSL;

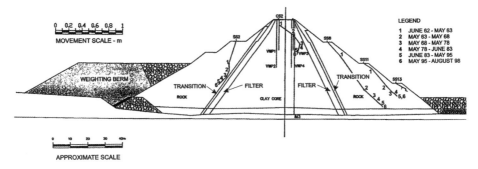

Figure 11.10.   Deformation of eppalock dam (courtesy of Goulburn-Murray Water).

Figure 11.11.   Cracking and softening due to settlement and lateral spreading at the crest of an embankment dam (courtesy Goulburn-Murray Water).

– Allowance for water pressures in these cracks;
– Averaging of the strength for failure surfaces which will pass through many layers of the compacted soil, i.e. steeply sloping failure surfaces;
– Adoption of the strength of lower strength layers for failure surfaces which are horizontal. However care needs to be taken in this, since it is not usually likely that weak layers will persist across the zone, unless they are due to desiccation or freezing in a prolonged shut-down during construction or a particularly poor period in compaction control.

11.3.3   Inherent soil variability

When assessing the strength of materials for stability analysis it should be recognised that there are inherent uncertainties in the strengths of the soil and there are potential errors in the measurement of the strength and the formulae used to transform the measured value (e.g. SPT "N" value or CPT q_c value) to strength. This is discussed in Kulhawy (1992).

454   GEOTECHNICAL ENGINEERING OF DAMS

Table 11.3. Coefficient of variation (COV) for available data on soil properties (Kulhawy, 1992 from Kulhawy et al., 1991).

	Property	No. of Studies	Mean COV without outliers (%)
Index	Natural water content, w_n	18	17.7
	Liquid limit, w_L	28	11.3
	Plastic limit, w_P	27	11.3
	Initial void ratio, e_i	14	19.8
	Unit weight, γ	12	7.1
Performance	Effective stress friction angle, ϕ'	20	12.6
	Tangent of ϕ'	7	11.3
	Undrained shear strength, S_u	38	33.8
	Compression index, C_c	8	37.0

Table 11.4. Estimates of in-situ test variability (Kulhawy, 1992; from Orchant et al., 1988).

Test[a]	COV[b] Equipment	COV (%) Procedure	COV (%) Random	COV[c] (%) Total	COV[d] (%) Range
SPT	5[e] to 75[f]	5[e] to 75[f]	12 to 15	14[e] to 100[f]	15 to 45
MCPT	5	10[g] to 15[h]	10[g] to 15[h]	15[g] to 22[h]	15 to 25
ECPT	3	5	5[g] to 10[h]	7[g] to 12[h]	5 to 15
VST	5	8	10	14	10 to 20
DMT	5	5	8	11	5 to 15
PMT	5	12	10	16	10 to 20[i]
SBPMT	8	15	8	19	15 to 25[i]

a – Test notation: SPT Standard Penetration Test; CPT Cone Penetration Test; VST Vane Shear Test; DMT Dilatometer Test; PMT Pressuremeter Test; SBPMT Self Boring Pressuremeter Test (M = mechanical, E = Electrical).
b – COV = standard deviation/mean.
c – COV(Total) = [COV(Equipment)^2 + COV(Procedure)^2 + COV(Random)^2]$^{1/2}$
d – Because of limited data and judgment involved in estimating COV, ranges represent probable magnitudes of test measurement error.
e – Best case scenario for SPT test conditions.
f – Worst case scenario for SPT test conditions.
g – Tip resistance, q_c, CPT measurements.
h – Side resistance, fs, CPT measurements.
i – Results may differ for P_o, P_f and P_L, but data are insufficient to clarify this issue.

Tables 11.3 and 11.4 present the coefficient of variation (COV = standard deviation/ mean) in soil properties and the variability in in-situ tests, excluding outliers in the data.

It should be noted that there is an inherent greater uncertainty in the undrained shear strength than in the effective shear strength. As discussed below, this should be reflected in the factor of safety used for undrained (total stress) analyses. The test variability shown in Table 11.4 shows a greater uncertainty in SPT tests, than in CPT tests.

However it must be remembered that there is considerable further model uncertainty in the assessment of strengths (and other properties) from the SPT, CPT and Vane Shear – see Sections 6.1.7 and 6.1.9. For example the shear strength of clays can be estimated from the CPT using the equation:

$$S_u = \frac{q_c - \sigma_{vo}}{N_k} \qquad (11.6)$$

where S_u = undrained shear strength; q_c = cone resistance; σ_{vo} = total overburden stress; N_k = Cone Factor.

N_k normally varies from around 10 to 20, but may be higher for overconsolidated soils. Hence the adoption of say $N_k = 15$, as is often done, may introduce an inherent bias of up to 30% or more. These potential errors can and should be reduced by using high quality sampling and laboratory testing to calibrate the in-situ test for the particular soil and conditions.

11.4   ESTIMATION OF PORE PRESSURES AND SELECTION OF STRENGTHS FOR STEADY STATE, CONSTRUCTION AND DRAWDOWN CONDITIONS

11.4.1   *Steady state seepage condition*

11.4.1.1   *Steady state pore pressures*
Steady state seepage conditions are usually assumed for the assessment of the long term stability of the downstream slope of the dam. These are usually based on the reservoir being at Full Supply Level.

In reality, it may take many years to reach steady state conditions in the dam core. LeBihan and Leroueil (2000) calculate that for typical central core earth and rockfill dams, full saturation will take a few years for a core saturated permeability of 10^{-6} m/sec, several decades for 10^{-7} m/sec and centuries for 10^{-8} m/sec. The authors' experience confirms that large dams may not have reached equilibrium pore pressures 20–30 years after construction, with construction pore pressures dominating in the lower part of the dam and fully saturated conditions not established in the upper parts. It is however good practice to design for the steady state conditions, making appropriately conservative assumptions to estimate the pore pressures.

Pore pressures for the steady state seepage condition are estimated by calculating the flownet for the embankment section either by graphical techniques or more commonly now by finite element methods. These techniques are described in detail in other references, e.g. Cedergren (1967, 1972), Bromhead (1988) and Desai (1975) and are not covered here.

The following issues are important when calculating the flownets for embankment dams.

(a) *Zoning of the embankment.* Embankment zoning clearly has a vital role in determining the pore pressures in the embankment. This is discussed in Chapter 8.
(b) *Anisotropic permeability of embankment earthfill.* Earthfill in dam embankments is compacted in layers and the action of rolling, possible drying and cracking of the surface of each layer and greater compaction of the upper part of the layers compared to the lower part will almost invariably lead to the horizontal permeability (k_H) being greater than the vertical permeability (k_V). It would not be unusual for $k_H/k_V = 15$ and even as high as $k_H/k_V \approx 100$. This has a marked effect in pore pressures in an embankment, particularly if no seepage control measures such as a vertical drain are incorporated in the design. Figure 11.12 shows pore pressures for a zoned earthfill dam for $k_H/k_V = 1, 16, 50$ and 100 in the core. The permeability of the downstream zone is 20 times the vertical permeability of the core. It can be seen that pore pressures on the downstream slope are affected greatly by the permeability ratio. The affect on pore pressures of using $k_H/k_V = 15$ and $= 1$ are also evident in Figure 11.13.
(c) *Foundation permeability.* The permeability of the foundation (compared to that of the embankment) has an important effect on the seepage flownet. No dam foundation is "impermeable" and in the majority of cases, even a rock foundation will have a permeability greater than that of compacted earthfill, e.g. most rocks have a permeability of between 1 and 100 lugeons (10^{-7} to 10^{-5} m/sec) compared to compacted clay earthfill with a $k_V \approx 10^{-7}$ to 10^{-10} m/sec. Soils in a dam foundation (which will usually

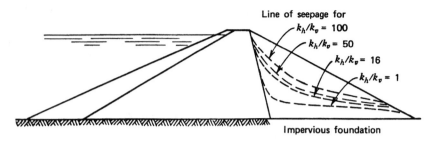

Figure 11.12.   Effect of k_H on pore pressures in a zoned earthfill embankment (Cedergren, 1972. Reprinted by permission of John Wiley & Sons Inc.).

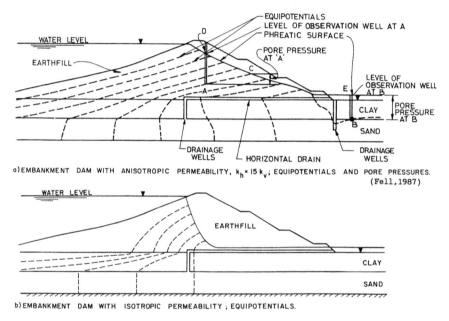

Figure 11.13.   Effect of foundation permeability on seepage flownet at Mardi Dam (PWD of NSW, 1985; Walker and Mohen, 1987).

be affected by fissures, root holes or layering during deposition) will almost invariably have a permeability far greater than compacted earthfill e.g. in the range 10^{-5} to 10^{-6} m/sec. Figure 11.13 shows the effect of such permeability contrasts on the seepage flownet for a earthfill embankment with a horizontal drain (Mardi Dam).

The following further points are made on Mardi Dam:

(i) Figure 11.13 shows the calculated seepage equipotentials for $k_H/k_V = 15$ in the embankment. These give pore pressures compatible with those observed in the embankment;

(ii) For both cases (a) and (b) most seepage from the reservoir is through the foundation – equipotentials in the embankment upstream of the centreline are nearer horizontal than vertical;

(iii) Even though k_H/k_V for the dam was ≈ 15 and seepage was emerging on the downstream face of the embankment, flow in the downstream part of the dam was towards the horizontal drain, yielding relatively low pore pressures. However the drain capacity was exceeded, so pore pressures built up in it.

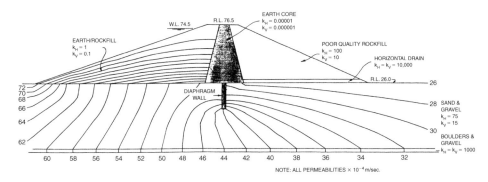

Figure 11.14.   Seepage equipotentials Lungga Dam, uniform foundation (Coffey and Partners, 1981b).

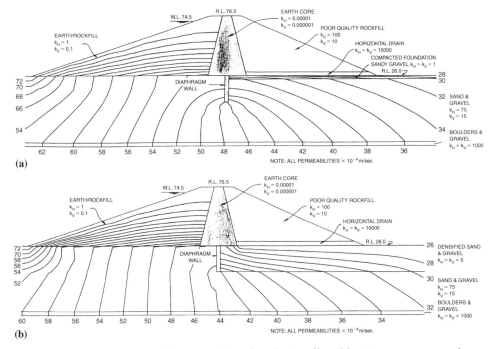

Figure 11.15.   Seepage equipotentials Lungga Dam foundation affected by (a) upper compacted zone; (b) densification by vibroflotation (Coffey and Partners, 1981b).

The upper foundation layer has a lower permeability than the lower layer and the confining effect led to pore pressures in the lower layer being above ground surface at the downstream toe of the embankment. To remedy this situation pressure relief wells were constructed which fully penetrated the lower (sand) layer. The original wells shown in Figure 11.13 were only partially penetrating and therefore not very effective in relieving pressures.

Figures 11.14 and 11.15 show the effect of foundation permeability on the seepage flownets for Lungga Dam, Solomon Islands, which was to be constructed on permeable alluvial soils.

The analysis for Lungga Dam includes the use of a diaphragm wall which penetrates part way through the highly permeable foundation sands and gravels. Being partially penetrating it reduces seepage only by a small amount (less than 5%) compared

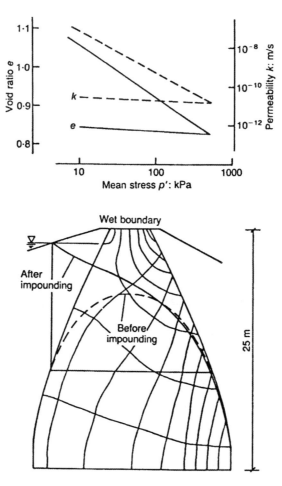

Figure 11.16.   Seepage through a wide dam core predicted by coupled finite element analysis in two dimensions: e ∝ log p′ and ln k ∝ e (Vaughan, 1994).

to no diaphragm cutoff. Seepage is largely through the foundation (>99% of total) and the seepage through the upstream part of the dam embankment passes directly to the foundation. Figure 11.15(a) shows the flownet when the upper 3 m of alluvium was compacted, resulting in a lower permeability. Figure 11.15(b) shows the effect of vibroflotation of the foundation which by densification and mixing yields a lower permeability. Lungga Dam was not constructed, due to the high overall cost of the project.

It should be noted that for both Mardi and Lungga Dams, the critical failure surface for slope stability passed through the foundation due to the high pore pressures.

(d) *Dependence of permeability on void ratio and effective stress.* The permeability of a soil is dependent on the void ratio (e), and hence the mean effective stress (p′). This is discussed at some length in Vaughan (1994) who shows that the effect of including this phenomenon in seepage analyses is to give higher seepage gradients at the downstream of the core as shown in Figure 11.16. It is apparent that this should be modelled if correct predictions of pore pressures are required.

(e) *Effects of partial saturation.* Le Bihan and Leroueil (2000) develop the concept of St. Arnaud (1995) and demonstrate that the high gradients of pore pressure which occur near the downstream of the core of some dams may be explained by the fact that the partially saturated permeability is lower than the saturated permeability. Hence as seepage begins to flow through the core of a dam, the flow regime is controlled by the contrasting permeability. It is hypothesised that the phenomenon is aided by dissolved gas in the seepage water coming out of solution in the lower furthermost downstream – part of the core where the pore pressures are lowest.

(f) *Clogging of the filters*. Peck (1990) suggests that fine particles from the core may migrate to the downstream filter, clog it and also lead to high pore pressure gradients on the downstream part of the core. This is more likely to be a problem if the soil in the core is internally unstable (see Chapter 9).

From the above discussion it will be apparent that it is difficult to predict steady seepage pore pressures. There are several mechanisms which can lead to pressures equivalent to those where a high k_H/k_V ratio is assumed, i.e. reservoir head persists through the core. The authors' advice is to design all embankments on the assumption that k_H/k_V is $\geqslant 15$. For larger, more critical dams, internal zoning should control seepage to such a degree that the stability is not greatly sensitive to k_H/k_V. The authors' own practice for an earth and rock-fill embankment with a vertical drain is to ignore the flownet effect, i.e. assume the flow-lines are horizontal. This is conservative as it implies $k_H/k_V = \infty$ but only affects the factor of safety marginally over that if $k_H/k_V = 9$ were used and allows one to be confident that pore pressures have been estimated conservatively. In some cases this allows adoption of a slightly lower factor of safety for the design.

For existing dams, there is often piezometer data on which to assess the pore pressures. The data need to be considered carefully, particularly in regard to whether equilibrium has been reached and whether there is sufficient instrumentation to define the pore pressures over the whole of the dam. The implication of the high k_H/k_V on pressures should be considered.

11.4.1.2   *Pore pressures under flood conditions*

Pore pressures in the low permeability zones in a dam under flood conditions are very difficult to predict. There are two broad issues:

– Will pore pressures in the body of the dam, for example in the core, rise with the flood water level?
– Will pore pressures in the vicinity of the dam crest above normal Full Supply Level rise with the flood water level?

Considering these in turn:

(a) *Pore pressures in the body of the dam*. There are some general observations:
  (i) Whether the pore pressures will respond rapidly to the reservoir rise depends on the degree of drainage which can be assessed as described in Section 11.4.3.3. If a zone is assessed to be free draining (Time Factor T > 3) or in the intermediate zone (Time Factor T 0.01 to 3), then it should be assumed the pore pressures will respond as the reservoir level rises in the flood;
  (ii) For non-draining zones (Time Factor T < 0.01) the pore pressures will respond slowly, so in most flood conditions there will be little or no rise in pore pressures;
  (iii) Where monitoring of the dam shows a rapid response to reservoir level, regardless of the assessments above, assume the pore pressures will respond to the reservoir level.

(b) *Pore pressures near the crest of the dam*. The pore pressures near the crest of the dam will depend on whether the zones are free draining or not, as detailed above, but also on whether the core is affected by cracking due to differential settlement and desiccation. In many dams the low permeability zones have not been taken to dam crest end and high permeability zones, e.g. rockfill, are used, so these will saturate under the flood.

The authors are reluctant to assume that low permeability zones will not develop pore pressure during floods due to the possibility of cracking and would normally check the case where the pore pressures above Full Supply Level rise with the flood. If the dam is known to have cracking in the core, this would become the base design case. If there is no evidence or no knowledge of cracking, it is recommended it be assumed the pore pressures above FSL may rise with the flood, but that a slightly lower factor of safety be accepted.

11.4.2   *Pore pressures during construction and analysis of stability at the end of construction*

11.4.2.1   *Some general principles*

When earthfill is compacted in a dam embankment the water content is usually controlled to be near standard optimum, so that compaction is facilitated. The soil will typically be compacted to a degree of saturation around 90% to 95% and will be heavily overconsolidated by rolling, with negative pore pressures in the partially saturated, compacted soil.

As additional fill is placed, the load imposed on the earthfill lower in the embankment will produce an increase in pore pressure, resulting eventually in positive pore pressures. The pore pressure will depend on the soil type, water content, roller used and magnitude of the applied stress (i.e. height of fill over the layer in question, and lateral restraint).

Larger dams constructed of earthfill with a high water content may develop sufficient pore pressure for this to control the stability of the upstream or downstream slope. If there is early closure of the diversion facility, the gradual rise in the storage may worsen the situation in so far as the upstream shoulder is concerned.

As the pore pressures will build up during construction and then (slowly) dissipate with time the critical condition may be during construction.

Another situation where construction pore pressures are important is where a dam is raised or when a berm is built to improve the stability. In these cases the core of the existing dam is usually saturated, so the pore pressure response will be that of a saturated soil and it is usual to analyse the stability using undrained strengths, which should be those existing in the dam, allowing for any swelling which may have occurred. This is discussed in Vaughan (1994).

The approaches which are available to consider the stability during construction are:

(a) Predict the pore pressures which will be developed using simplified methods such as Skempton (1954) and use effective stress stability analysis;
(b) Predict the pore pressures using more sophisticated partial saturated soil theory and use effective stress stability analysis or numerical analysis;
(c) Develop undrained strengths for the dam core and use total stress analysis;
(d) The following sections describe the Skempton (1954) approach, a variant on it by Vaughan reported in McKenna (1984), the status of partial saturated soil theory and the undrained strength approach.

11.4.2.2   *Estimation of construction pore pressures by Skempton (1954) method*

Construction pore pressures can be estimated by using Skempton's (1954) pore pressure parameters A and B. In this approach:

$$\Delta u = B[\Delta\sigma_3 + A(\Delta\sigma_1 - \Delta\sigma_3)] \tag{11.7}$$

Table 11.5.   Skempton "A_f" values for clays of varying degrees of overconsolidation (Skempton, 1954).

Type of clay	A_f
Highly senstive clays	+0.75 to +1.5
Normally consolidated clays	+0.50 to +1.0
Compacted sandy clays	+0.25 to +0.75
Lightly overconsolidated clays	0 to +0.50
Compacted clay gravels	−0.25 to +0.25
Heavily overconsolidated clays	−0.50 to 0

where Δ_u = change in pore pressure; $\Delta\sigma_1$ = change in major principal stress; $\Delta\sigma_3$ = change in minor principal stress.

A and B are pore pressure parameters, which are related to the compressibility of the soil as follows:

$$B = \frac{1}{1 + \dfrac{nC_v}{C_{sk}}}$$

(11.8)

where n = porosity; C_v = compressibility of the voids; C_{sk} = compressibility of the soil skeleton.

The pore pressure equation can be rewritten as:

$$\frac{\Delta u}{\Delta\sigma_1} = \bar{B} = B\left[\frac{\Delta\sigma_3}{\Delta\sigma_1} + A\left(1 - \frac{\Delta\sigma_3}{\Delta\sigma_1}\right)\right]$$

(11.9)

For saturated soils, B = 1 but, as shown in Figure 6.13, B varies considerably with degree of saturation and the degree of over-consolidation of the soil. "A" also varies with the degree of over-consolidation of the soil, the stress path followed and with the strain in the soil. Usually A at "failure" (i.e. at maximum principal stress ratio or maximum deviator stress) is used and is known as A_f. Some typical values are shown in Table 11.5.

The values of A will also be dependent on the lateral strain conditions within the dam embankment. It is normal to assume that K_o conditions apply, i.e. that there is zero lateral strain. This is a reasonable (and conservative) assumption for the bulk of the failure surface in an embankment. The use of A at failure is also slightly conservative in most cases.

A more general pore pressure equation was proposed by Henkel (1960) to take into account the effect of the intermediate principal stress. It is:

$$\Delta u = B(\Delta\sigma_{oct} + a\,\Delta\tau_{oct})$$

(11.10)

where

$$\sigma_{oct} = \frac{1}{3}(\sigma_1 + \sigma_2 + \sigma_3)$$

(11.11)

$$\tau_{oct} = \frac{1}{3}\sqrt{(\sigma_1 - \sigma_2)^2 + (\sigma_2 - \sigma_3)^2 + (\sigma_3 - \sigma_1)^2}$$

(11.12)

and a is the Henkel pore pressure parameter. The equivalent Skempton A from Henkel's a parameter for triaxial compression is:

$$A = \frac{1}{3} + a\frac{\sqrt{2}}{3} \qquad (11.13)$$

For triaxial extension conditions:

$$A = \frac{2}{3} + a\frac{\sqrt{2}}{3} \qquad (11.14)$$

The pore pressure parameters A and B can be estimated by laboratory tests on the soil to be used in the embankment. "B" is determined by placing soil compacted to the required water content and density ratio in a triaxial cell and observing the change in pore pressure Δu for changes in cell pressure under undrained conditions. Under these conditions $\Delta\sigma_1 = \Delta\sigma_3$ and the pore pressure equation becomes:

$$\Delta u = B[\Delta\sigma_3 + A(0)] = B\,\Delta\sigma_3 \qquad (11.15)$$

This relationship will not be linear and must be determined over a range of cell pressures. B will be larger for higher cell pressures than for low cell pressures.

To determine A, the soil is placed in the triaxial cell under undrained conditions and sheared. Knowing B from the earlier testing and observing Δu, $\Delta\sigma_1$ and $\Delta\sigma_3$, A can be determined from the pore pressure equation. As explained above, in most cases this will be done under K_o conditions. Head (1985) shows a suitable test set up and lateral strain indicators needed to control the test using conventional triaxial equipment.

More accurate estimation of the pore pressure parameters can be achieved by more closely following the stress history of the soil in the embankment using a Bishop-Wesley triaxial cell.

To apply this method one must determine:

– What the stresses in the embankment will be;
– The pore pressure parameters A and B relevant to the water content and density ratio at which the soil is placed and the stress conditions in the embankment.

To be conservative in the estimation of pore pressures, the laboratory tests to estimate A and B can be carried out at the upper limit of specified water content (e.g. optimum +2%) and/or degree of compaction (density ratio 98% to 100%, standard compaction). However it must be remembered that at optimum +2% water content the maximum achievable density ratio may be 97% to 99%, while at optimum water content a density ratio of 100% to 101% may be achievable and the laboratory testing should account for this (see Figure 11.17).

The stresses in the embankment due to construction of the dam can be estimated by finite element techniques. ICOLD (1986a) and Duncan (1992, 1996a) give descriptions of the methods and their limitations. It is concluded that:

– A two dimensional model can be used to estimate vertical stresses in a homogeneous dam on a rigid (rock) foundation;
– A three dimensional model is necessary to model cross valley stresses in an embankment on a compressible foundation;
– The initial stresses locked into the fill during compaction should be incorporated into the model;
– The models must be "built" in layers, being loaded progressively.

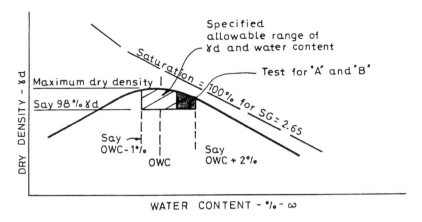

Figure 11.17.   Test water content and density ratio for determining A and B parameters.

In a practical sense it is difficult to carry out 3 dimensional modelling and the rolling stresses can only be crudely modelled (Naylor, 1975 and ICOLD, 1986a). Given the difficulty in predicting pore pressure parameters A and B with any degree of accuracy, it is not worthwhile spending too much effort on modelling the stresses. Two dimensional finite element models along the dam axis, as well as up and down river, might give some appreciation of three dimensional problems for narrow or irregular shaped valleys and compressible foundations.

11.4.2.3   *Estimation of construction pore pressures from drained and specified undrained strengths*

McKenna (1984) cites an approximate method for estimating construction pore pressures developed by P.R. Vaughan. This method was used at Yonki Dam in Papua New Guinea by the Snowy Mountains Engineering Corporation and is applicable when the control of the embankment soil construction is by undrained shear strength, rather than by specifying a minimum dry density ratio and water content relative to optimum. This approach is used where the field water contents are high relative to optimum and drying of the soil is impracticable, or the optimum water content and maximum dry density are difficult to define.

The method is based on the effective strength and total strength envelopes obtained from tests on the embankment soil. These are plotted as p'-q plots, yielding envelopes where for the effective strengths:

$$q = p' \tan \alpha' + a' \tag{11.16}$$

and for total strengths:

$$q = p \tan \alpha + a \tag{11.17}$$

where $\tan \alpha' = \sin \phi'$; $a' = c' \cos \phi'$, $\tan \alpha = \sin \phi_u$; $a = c_u \cos \phi_u$.

The displacement between the total and effective stress envelopes represents the pore pressures developed on shearing.

The pore pressure at failure is given by:

$$u_f = \frac{a(1 + \cot \alpha') - a' \cot \alpha'(1 + \tan \alpha')}{(1 - \cot \alpha' \tan \alpha')} + \frac{1 - \cot \alpha' \tan \alpha'}{1 + \tan \alpha'} \sigma_1$$

$$= \text{threshold pressure} + (\text{slope of curve}) \, \sigma_1 \tag{11.18}$$

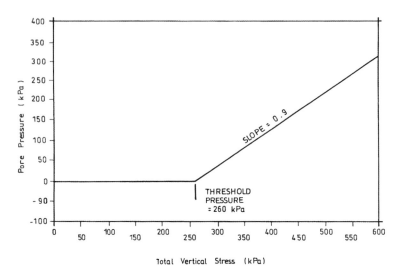

Figure 11.18.   Predicted construction pore pressures vs total vertical stress for Yonki Dam using Vaughan method.

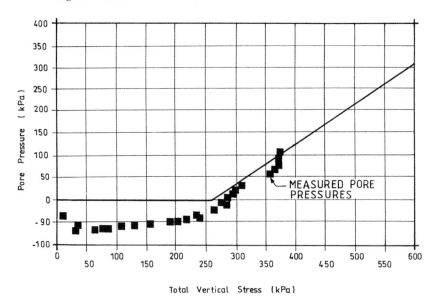

Figure 11.19.   Observed construction pore pressures for Yonki Dam – courtesy of the Snowy Mountains Engineering Corporation.

In this σ_1 is the total vertical stress in excess of the threshold pressure, e.g. for $c' = 0$, $\phi' = 30°$ and $\phi_u = 2°$; slope of curve = 0.9, "x" intercept = $3.23S_u$.

Hence if the dam is constructed so that Su = 80 kPa, the predicted pore pressures would be as shown in Figure 11.18.

In practice there are negative pore pressures in the embankment below the threshold pressure and the slope of the curve increases with increasing vertical stress. Figure 11.19 shows results from Yonki Dam.

11.4.2.4   *Estimation of pore pressures using advanced theory of partially saturated soil*

Fry et al. (1996) present a good overview of the state of art of application of partially saturated soil mechanics to embankment dams and slope stability up to 1996.

As they point out, there are two main approaches:

(a) Uncoupled analysis, where the first step is to model the pore pressures using partially saturated flow equations, followed by the use of limit equilibrium analysis. These analyses allow, with varying degrees of sophistication, for the dependence of the unsaturated permeability with the water content, stresses, degree of saturation and diffusivity (equivalent to the coefficient of consolidation in saturated soils). The many shortcomings of such analyses are detailed by Fry et al. (1996).
(b) Coupled analysis models (again with varying degrees of sophistication) the pore air and water pressures, volume change and stresses using finite element analyses and properties based on what are usually time consuming and sophisticated laboratory tests. Fry et al. (1996) provide a review of the approaches available at that time.

Now (2003) there is considerable research in progress on this topic which should provide better methods. In view of this and the reality that the science of partially saturated soils is well beyond most dam engineers (and the authors!), if it is wished to attempt a fully coupled approach, one of the handful of experts in this field should be engaged to assist.

11.4.2.5   *Undrained strength analysis*

In view of the difficulty in correctly estimating the pore pressures in partially saturated soils, for most dams it is sufficient to use the undrained strengths (for cohesive soils) and a total stress analysis.

The undrained strengths used are usually those estimated to apply at the placement condition, so any strength gained by (partially saturated) consolidation as the dam is constructed are ignored.

Alternatively, the undrained strengths can be determined by a series of laboratory tests which consolidate the samples to the effective stresses expected in the dam. These stresses should be estimated by numerical analysis and the testing should follow the stress path of the soil from compaction through to the conditions in the dam.

Just as there are uncertainties in estimating the pore pressures, there are uncertainties in these undrained strengths.

11.4.2.6   *Summing up*

It should be recognised that it is not practicable to estimate construction pore pressures accurately and that the best one can do is to get an indication of the likely pressures, and how they will build up as the embankment is constructed. It is necessary to monitor pore pressures with piezometers if it is considered they may be critical. Measures can be taken to reduce pore pressures, e.g. by using a lower water content. This was done at Dartmouth Dam (Maver et al., 1978). However by reducing the compaction water content the earthfill will be more brittle and more susceptible to cracking.

It is also important to monitor displacements of the dam during construction, so the onset of increasing rates of movement, possibly indicating construction instability, can be detected. Surface survey should always be possible. Bore hole inclinometers are valuable but some, including the authors, are reluctant to put too much instrumentation into dam cores because of the increased chances of piping problems along the instruments.

The discussion above, with the exception of Section 11.4.2.4, ignores the effects of dissipation of pore pressure with time in the embankment. Eisenstein and Naylor in ICOLD (1986a) discuss how this can be modelled by finite element methods. In most cases the earthfill will have a low permeability and coefficient of consolidation and drainage paths will be long which will result in long times for pore pressure dissipation. As the estimation of pore pressures is approximate, it would not be warranted to take account of pore pressure dissipation in most cases.

11.4.3   Drawdown pore pressures and the analysis of stability under drawdown conditions

11.4.3.1   Some general issues

When the reservoir level behind an embankment dam is lowered, the stabilizing influence of the water pressure on the upstream slope is lost. If the water level is dropped sufficiently quickly that the pore pressures in the slope do not have time to reach equilibrium with the new reservoir water level, the slope is less stable.

Many failures of dam slopes (and natural slopes around reservoirs) have occurred during drawdown. Duncan et al. (1990) give some examples.

It is often assumed, for design purposes, that the drawdown is rapid or even instantaneous. This assumption imposes severe loadings and it is often the controlling case for the design of the upstream slope.

Terzaghi and Peck (1967) summarized the problems in estimating pore pressures during drawdown in these words:

"... in order to determine the pore pressure conditions for the drawdown state, all the following factors need to be known: the location of the boundaries between materials with significantly different properties; the permeability and consolidation characteristics of each of these materials; and the anticipated maximum rate of drawdown. In addition, the pore pressures induced by the changes in the shearing stresses themselves ... need to be taken into consideration. In engineering practice, few of these factors can be reliably determined. The gaps in the available information must be filled by the most unfavorable assumptions compatible with the known facts."

In view of the difficulty in estimating the pore pressures, particularly those developed on shearing, in low permeability, clayey soils it is common practice to use an essentially total stress analysis to assess the stability under drawdown. The recommended procedure is detailed in Section 11.4.3.3. For free draining rockfill or gravels, the pore pressures will dissipate as the reservoir level drops, so there are no residual pore pressures.

For well compacted, medium permeability soils, e.g. silty sands, pore pressures can be estimated using flow nets. In these cases any shearing induced pore pressures will be negative, due to dilation of the dense soil, so they can be conservatively ignored. If the silty sands are loose and drawdown is in fact rapid, undrained strengths should be used.

Duncan (1992) gives guidance on how to assess whether drainage will occur. This is covered in Section 11.4.3.3.

11.4.3.2   Estimation of drawdown pore pressures, excluding the effects of shear-induced pore pressures

Pore pressures in the upstream slope of a dam under drawdown conditions are determined by the change in water level, geometry and zoning of the slope, the relative permeabilities of the zones and, importantly, layers within each zone and the relative permeability of the foundation.

Some of these points are illustrated in Figure 11.20 and discussed below:

(i) In most cases the assumption is made that drawdown is "instantaneous", and hence the flownet changes from the steady state case to the drawdown case instantaneously. This is somewhat conservative but sufficient for most analysis;

(ii) The pore pressures in the embankment are immediately reduced on drawdown due to the unloading effect of removing the water. Hence pore pressures on the surface of earthfill in Figures 11.20a to 11.20d are all zero;

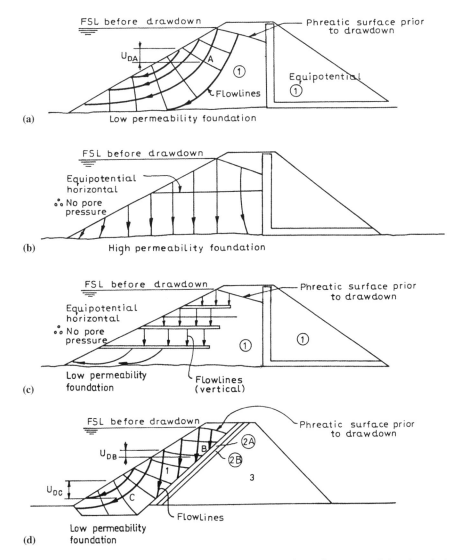

Figure 11.20.   Drawdown flownets (a) earthfill with chimney drain, low permeability foundation; (b) earthfill with chimney drain, high permeability foundation; (c) earthfill with horizontal drainage blankets; (d) sloping upstream core, low permeability foundation. Note that flownets assume isotropic permeability and $k_H = k_V$ and the reservoir is completely drawn down.

(iii) In Figure 11.20a the foundation permeability is low compared to that of the earthfill and flow will be towards the upstream face. This gives relatively high pore pressures compared to Figure 11.20b. In the latter case, the foundation is more permeable than the earthfill, so flow lines will be near vertical and equipotentials near horizontal, giving pore pressures which should be close to zero;

(iv) In Figure 11.20c horizontal drains are incorporated in the upstream slope. These cause flow lines to be near vertical on drawdown (except near the low permeability foundation) and should result in low pore pressures;

(v) In Figure 11.20d the flow lines are influenced by the presence of the filter zones and rock-fill, giving lower pore pressures than for 11.20a, at least in the upper part of the slope.

The diagrams in Figure 11.20 are sketched with the implicit assumption that $k_H/k_V = 1$ and permeability is isotropic (uniform). In reality, if there are lower permeability layers within the zones, perched water tables may result, completely altering the pore pressures. Hence great care must be taken in the analyses and it would be more common to assume that the pore pressures are those which exist prior to the drawdown, adjusted for the unloading effect of removing the water.

Drawdown flownets and pore pressures can be obtained by graphical methods or finite element analysis. Factors which must be considered in the analyses are:

- Permeability anisotropy in the earthfill, i.e. allowing for $k_H > k_V$ and low permeability layers;
- The relative permeability of the foundation and the embankment zones. In practice this means that the foundation should be modelled with the embankment;
- The effect of removal of buoyancy from the upstream rockfill zone (if present) should be allowed for. This will induce increased pore pressures.

The drawdown condition is the controlling case for design of the upstream slope of embankments such as those shown in Figure 11.20 and the analysis is sensitive to the assumptions made. For a central core earth and rockfill dam the analysis may control the slope but will not be so sensitive to assumptions.

11.4.3.3   *Methods for assessment of the stability under drawdown conditions*

Duncan et al. (1990) discuss several methods for analysing the stability of the upstream slope under drawdown conditions, including those developed by Bishop (1955), Morgensten (1963), US Corps of Engineers (1970) and Lowe and Karafiath (1960). These latter two methods have in common that they use a composite drained and undrained strength envelope. Duncan et al. (1990) favour these methods in preference to effective stress methods which rely on estimating the pore pressures generated by shearing. As discussed above this is very difficult.

They recommend a procedure which is a development of the method developed by Lowe and Karafiath (1960) and is somewhat complicated in that it is iterative and requires an estimate of the principal stress ratio at the base of each slice to assist in selecting the undrained strength. While it seems quite reasonable and logical, the complexity of its application makes it difficult to recommend its use. In view of this, it is recommended that the procedure given below, which is based on USBR (1987) and Duncan et al. (1990) be used.

The suggested approach is:

(a) Assess for each zone in the dam whether drainage will occur during drawdown, using the approach suggested by Duncan (1992). Estimate the dimensionless Time Factor T from:

$$T = \frac{C_v t}{D^2} \qquad (11.19)$$

where C_v = coefficient of consolidation (m²/year); t = time for drawdown (years); D = drainage path length (m).

If T > 3.0, the zone will be free draining, so effected stress parameters will apply.
If T < 0.01, the zone will be undrained, so undrained strengths should be used.
If T > 0.01 and <3, the zone will be partially drained. For drawdown analysis assume the zone will be undrained.

(b) For the zones which are below normal maximum reservoir operating level e.g. below Full Supply Level, develop effective stress and total stress strength envelope from conventional isotropically consolidated undrained (IC-U) triaxial tests, as shown in Figure 11.21.

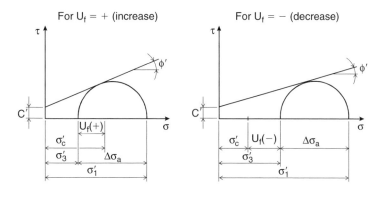

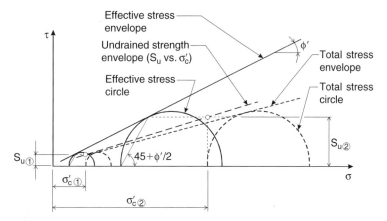

Figure 11.21.   Shear strength envelopes for isotropically consolidated undrained (IC-U) strength tests.

Note that the effective stress Mohr circle will be to the left or right of the total stress circle, depending on whether the pore pressures generated during shear are positive or negative.

(c) For these zones, develop the composite shear strength diagram shown in Figure 11.22, from the effective stress and undrained strength envelopes in Figure 11.21.

(d) For the zones which are above normal maximum operating level, e.g. above FSL, estimate the undrained strengths from the undrained strength envelope, allowing for any known cracking or softening.

(e) For undrained or partially drained zones below the normal maximum operating level, determine the strength from the composite envelope i.e. the minimum of effective stress and undrained strengths. The normal stress used to select the shear strength from the composite envelope should be the effective consolidation stress acting on the base of the slice after drawdown, i.e. the total normal stress minus the pore pressure taken from the drawdown flow net (or from measured pore pressures), after allowing for the factors discussed in Section 11.4.3.2.

This procedure has the following characteristics:

(i) It uses undrained strengths, where these are less than effective stress strengths, to allow for the effects of pore pressures developed during the undrained loading.

(ii) It uses effective stress strengths, where these are lower than undrained strengths, in areas normally submerged by the reservoir. This is as recommended by USBR (1987),

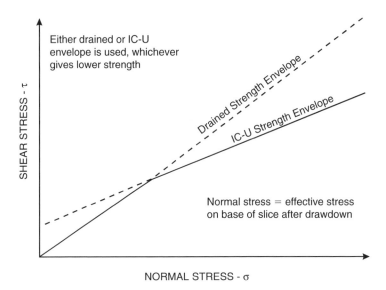

Figure 11.22.   Composite shear strength envelope.

US Corps of Engineers (1970), and Duncan et al. (1990), and is to allow for possible drainage (the drainage paths may be short in these low stressed areas) and softening under the low confining stresses. It will also allow for the cracking, dilation and softening of the earthfill which occurs because of settlements during construction and under repeated drawdowns.

(iii) It uses the effective stress after drawdown to estimate the strengths as recommended by Duncan et al. (1990). They demonstrated that the use of the effective stresses prior to drawdown as recommended by US Corps of Engineers (1970) and (USBR, 1987) is conservative. However the authors caution that these effective stresses should be estimated conservatively and the default position would be to use the effective stress prior to drawdown which is more readily estimated.

(iv) It uses undrained strengths above the normal maximum operating level (after allowing for cracking and softening) because effective stress strengths in this area, where confining stresses are low, greatly under-estimate the real strength of the partially saturated soil and, if used, could give misleadingly low strengths in the upper part of the dam. Dams which have experienced larger than normal settlements should be assumed to have cracking and softening present.

11.5   DESIGN ACCEPTANCE CRITERIA

11.5.1   *Acceptable factors of safety*

Most dams are designed and the safety of existing dams is assessed using factors of safety as acceptance criteria. The factors of safety adopted are reasonably universal, with for example similar values used by US Bureau of Reclamation (USBR, 1987), US Corps of Engineers (1970), Building Research Establishment Guide to the Safety of Embankment Dams in the United Kingdom (BRE, 1990) and Norwegian Geotechnical Institute (1992).

Table 11.6 lists what can be regarded as baseline minimum factors of safety applicable to design of new dams with high consequence of failure, with the qualifications listed in the notes.

Table 11.6.   Baseline recommended minimum acceptable factors of safety and load conditions.

Slope	Load condition	Reservoir characteristic	Minimum factor of safety
Upstream and downstream	End of construction	Reservoir empty	1.3
Downstream	Steady state seepage	Reservoir at normal maximum operating level (Full Supply Level)	1.5
Downstream	Maximum flood	Reservoir at maximum flood level	1.5, free draining crest zones, 1.3 otherwise
Upstream	Drawdown	Rapid drawdown to critical level	1.3

Note: These factors of safety apply to design of new high consequence of failure dams, on high strength foundations, with low permeability zones constructed of soil which is not strain weakening, using reasonably conservative shear strengths and pore pressures developed from extensive geotechnical investigations of borrow areas, laboratory testing and analysis of the results and using the methods of analysis detailed above. It is assumed there will be monitoring of deformations by surface settlement points during construction and during operation of the dams.

Table 11.7.   Factors which influence the selection of factor of safety, and their effect on the baseline minimum factor of safety.

Factor	Description	Recommended change to the baseline minimum factor of safety
Existing (vs new) Dam	A lower factor of safety may be adopted for an existing dam which is well monitored and performing well	0 to −0.1
Soil or weak rock foundation	A higher factor of safety may be needed to account for the greater uncertainty of the strength	0 to +0.2 for effective stress +0.1 to +0.3 for undrained strength analyses
Strain weakening soils in the embankment or foundation	A higher factor of safety may be needed to account for progressive failure, and greater displacements if failure occurs	0 to +0.2
Limited (little or no good quality) strength investigation and testing, particularly of soil and weak rock foundations	A higher factor of safety should be used to account for the lack of knowledge	+0.1 to +0.3 for effective stress analyses, +0.3 to +0.5 for undrained strength analyses
Contractive soils in the embankment or foundation	A higher factor of safety may be needed to account for the greater uncertainty in the undrained strength	+0.1 to +0.3 for undrained strength

Note: These figures are given for general guidance only. Experienced Geotechnical Professionals should use their own judgement, but note the principles involved in this table.

Table 11.7 details factors which should influence the selection of a minimum acceptable factor of safety and their extent of influence.

Tables 11.6 and 11.7 are based on the following:

(a) The strengths and pore pressures used in the analysis should be selected reasonably conservatively, e.g. use lower quartile shear strengths; high k_H/k_V in compacted earth-fill. This is always a matter of judgement and the authors' recent experience is that practitioners are sometimes overly conservative in selection of strengths, e.g. using lower bound strengths for earthfill, not allowing for the high friction angles of rockfill at low confining stresses.

(b) The lower minimum factors of safety for the construction condition is based on the consequences of failure usually being lower than for a dam under steady state seepage. However the uncertainties in the shear strength are often high, so unless there is good monitoring of pore pressures and displacements, high factors of safety should be used.

(c) The lower factors of safety for rapid drawdown are again predicated on the assumption that the dam is unlikely to breach under a drawdown failure but it should be recognized that, by adopting a lower factor of safety, there is a higher chance of failure.

(d) It should be recognised that by adopting lower factors of safety for the construction and drawdown conditions there is an acceptance of a higher probability of instability.

(e) The factors of safety apply to non-trivial failure surfaces. Failure surfaces passing only through free draining rockfill may acceptably have lower factors of safety, because provided the slope can be constructed, there is no possible change in conditions which will cause instability. If realistic (high) effective friction angles are used for rockfill at low confining stresses, this problem will generally not arise.

(f) The greater uncertainty in undrained strengths than in effective stress strengths means that higher factors of safety should be used where undrained strengths control stability.

11.5.2   Post failure deformation assessment

It is recommended that as a matter of routine the potential post failure deformation of the dam be assessed using the simplified method of Khalili et al. (1996).

The Khalili et al. (1996) model considers two modes of failure defining the upper and lower bounds of post-failure deformation:

- A "rapid model" based on the principle of conservation of energy, where the potential energy of the slide mass is resisted by the frictional forces along the surface of rupture, and approximated using the residual shear strength;
- A "slow model" based on equations of equilibrium between the driving force of the slide mass and resisting forces along the surface of rupture and approximated using the residual shear strength, i.e. assuming the rate of movement would be so slow that the effect of inertia forces of the sliding mass is negligible.

The failure model is shown in Figure 11.23 and is based on the assumptions of a circular slide surface, the failure is plane strain, the failing mass moves as a rigid body and energy losses during failure is only due to the frictional forces acting along the slip surface. Approximate solutions for the rapid model are calculated using Equation 11.20 and for the slow model using Equation 11.21, where $FS_{residual}$ is the factor of safety calculated using residual strengths along the surface of rupture, and θ_i and θ_f are the initial and final positions of the centre of gravity as defined in Figure 11.23.

$$\text{Rapid Model } \theta_f = 2\theta_i(FS_{residual} - 0.5) \tag{11.20}$$

$$\text{Slow Model } \sin\theta_f = FS_{residual}\sin\theta_i \tag{11.21}$$

Khalili et al. (1996) found that for most cases analysed the rapid model gave the best estimate of deformations.

In this manner, the consequences of the dam "failing" (reaching a factor of safety ≤ 1) can be determined. It will become apparent that in many cases, either the factor of safety using residual strengths is ≥ 1.0, in which case failure is virtually impossible, or it is only a little less than 1.0, because the materials in the dam are not significantly strain weakening.

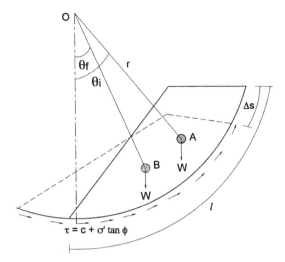

Figure 11.23.   Failure model for post-failure deformation (Khalili et al., 1996).

If this is the case, and if the assessment of the post failure deformation is such that free-board will not be lost, one might, after careful consideration, accept a lower factor of safety than usually adopted rather than carry out remedial works.

Situations which are likely to be non-brittle (low chance of large deformations) include dams with large rockfill or dense sand-gravel zones and/or earthfill with low clay content (less than ≈20% based on Figure 6.5).

Situations which are likely to be brittle are those with little or no rockfill and brittle, strain weakening soils, as described in Section 6.1.3.

11.6   EXAMPLES OF UNUSUAL ISSUES IN ANALYSIS OF STABILITY

The two case studies detailed below exemplify why conventional effective stress analysis has shortcomings in some situations.

11.6.1   Hume No.1 Embankment

Hume No.1 Embankment is a 39 m high, 1170 m long earthfill dam, with a concrete core-wall which is 0.6 m thick at the top and up to 1.8 m thick at the base. The corewall is lightly reinforced. Figure 11.24 shows the cross section and description of the zones. The embankment was constructed in the period 1921–1936 and rockfill was placed on the upstream slope in 1939 and 1941 following upstream slope failures on reservoir draw-down. The dam was raised 5.5 m in the period 1953–1965. The earthfill was placed by horse and dray with the horses' hooves providing the only compaction.

Monitoring of the dam (Mail et al., 1996) from 1966 showed downstream movements of the top of the corewall totalling 450 mm in one area, with 20 mm in 1993 alone. The movements only occurred at high reservoir levels.

Detailed investigations and analyses were carried out to determine the reason for the deformations. These are described in Cooper et al. (1997) and included drilling boreholes, installing piezometers, cone penetration (CPT) and piezocone (CPTU) testing, laboratory testing and limit equilibrium and numerical modelling of stability.

Figure 11.25 shows the geotechnical conditions in the bend area which deformed excessively.

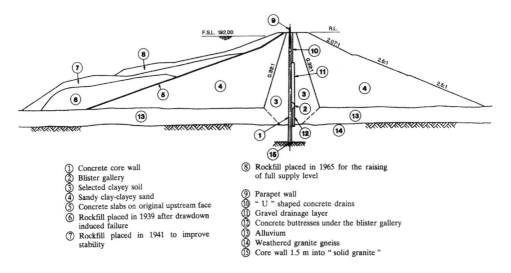

① Concrete core wall
② Blister gallery
③ Selected clayey soil
④ Sandy clay-clayey sand
⑤ Concrete slabs on original upstream face
⑥ Rockfill placed in 1939 after drawdown
   induced failure
⑦ Rockfill placed in 1941 to improve
   stability

⑧ Rockfill placed in 1965 for the raising
   of full supply level
⑨ Parapet wall
⑩ " U " shaped concrete drains
⑪ Gravel drainage layer
⑫ Concrete buttresses under the blister gallery
⑬ Alluvium
⑭ Weathered granite gneiss
⑮ Core wall 1.5 m into " solid granite "

Figure 11.24.   Hume No.1 Embankment cross section (Cooper et al., 1997).

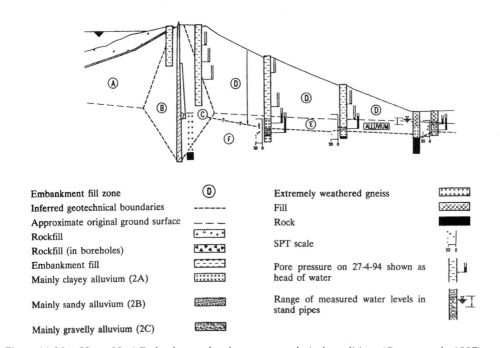

Embankment fill zone

Inferred geotechnical boundaries

Approximate original ground surface

Rockfill

Rockfill (in boreholes)

Embankment fill

Mainly clayey alluvium (2A)

Mainly sandy alluvium (2B)

Mainly gravelly alluvium (2C)

Ⓓ

－ － －

Extremely weathered gneiss

Fill

Rock

SPT scale

Pore pressure on 27-4-94 shown as
head of water

Range of measured water levels in
stand pipes

Figure 11.25.   Hume No.1 Embankment, bend area – geotechnical conditions (Cooper et al., 1997).

The earthfill downstream of the concrete corewall was generally unsaturated, with local perched water tables. The lower 2 m of the fill was saturated by seepage through the abutment or leakage through the corewall. The CPT showed the saturated zones had much lower undrained shear strength than the rest of the fill.

Limit equilibrium stability analyses were carried out using effective stress (peak and residual) and undrained strengths. The undrained strengths were estimated from the CPT, CPTU and, more particularly for the saturated zones, from the effective stress

Table 11.8.   Hume No.1 embankment results of limit equilibrium stability analysis (Cooper et al., 1997).

	Factors of safety	
Case	Effective stress	Undrained strength
Peak strengths	1.52 (1.52)*	1.23 (1.14)
Residual strength	1.06 (0.98)	0.97 (0.89)

Note: The figures in brackets assume no strength in the concrete core. For values marked "*", the critical failure surface does not pass through the wall.

strengths taking account of pore pressures generated during undrained shearing using the formula:

$$S_u = \frac{c'\cos\varphi' + P_o\sin\varphi'}{1 + 2A_f \sin\varphi'} \qquad (11.22)$$

$$\text{where } P_o = \frac{1 + K_o}{2} \, \sigma'_{vo} \qquad (11.23)$$

where c', φ' are the effective stress peak strengths, A_f = Skemptons pore pressure coefficient at failure, K_o = the rest earth pressure coefficient, σ'_{vo} = effective vertical stress. A_f was in the range 0.36 to 0.7 for the triaxial tests. A value of 0.7 was adopted.

The results of the analysis are given in Table 11.8.

The numerical analysis underestimated the actual displacements using the undrained strengths derived from peak strengths and overestimated them using the undrained strengths derived from the residual strengths. The analysis showed a yield surface developed in the 2 m of saturated fill just above the foundation downstream of the core wall and was able to replicate movements at other parts of the dam not affected by the saturation of the lower part of the fill.

From this it was concluded that:

– The behaviour of the dam was due to marginal slope stability;
– The use of effective stress analysis significantly over-estimates the factor of safety;
– Undrained strength analysis more correctly models the behaviour but allowance must also be made in this case for strain weakening.

It should be noted that the investigations had shown that the movements could not be explained by consolidation, collapse mechanisms or internal erosion and piping.

Subsequent investigations of an area where No.1 Embankment joins the concrete spillway section confirmed the need to use undrained strengths in stability analysis. In that area borehole inclinometers confirmed the development of shear surfaces in the saturated, softened fill. Both areas have had berms constructed to improve the stability.

11.6.2   Eppalock Dam

Eppalock Dam is a 47 m high central core earth and rockfill dam which was built between 1960 and 1962. Figure 11.10 shows a cross section through the dam.

The rockfill in the dam was placed in lifts 2 m to 4 m thick, spread with a light tractor. There was no rolling. The core was compacted by sheepsfoot roller, probably to > 98%

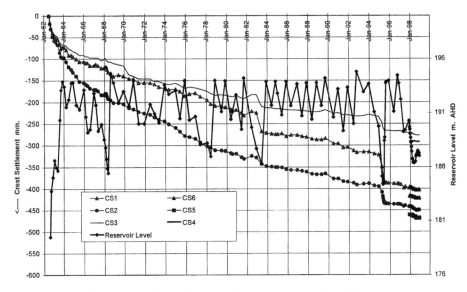

Figure 11.26.   Eppalock Dam – Observed crest settlement (courtesy of Goulburn Murray Water).

density ratio (standard compaction). A weighting berm was built on the upstream side from waste rock and soil to improve the stability relating to potential weak zones in the foundation. The berm material was probably not well compacted.

Since monitoring began in 1962, the dam has experienced relatively large deformations (vertically up to 850 mm, laterally 750 mm). Figure 11.10 shows typical movement vectors plotted on a time basis. It will be noted that the overall effect is one of spreading and that the displacement on the crest was downstream until about 1978, thence it has been upstream.

Figure 11.26 shows the observed settlements on the dam crest. The lateral movements were also measured. It will be noted that there is a general trend for on-going deformation which is reducing with time, but there have been rapid movements in 1968, 1983 and 1995. These relate to low reservoir levels and appear to coincide with rainfall events at the end of the reservoir drawdowns. The movements increased with each drawdown except that in 1998 the movement was less. This is probably because of a rigorous procedure of filling cracks in the dam crest as they formed, stopping rainfall infiltration.

In 1995 there was severe cracking and disruption of the dam crest and longitudinal cracks were observed in the other drawdown periods. Extensive investigation and analysis carried out by Snowy Mountains Engineering Corporation and Woodward Clyde showed that the factor of safety was low for the upstream slope above the first berm. However to model the observed behaviour of displacements at the surface and in an inclinometer which showed the development of a shear surface at a depth of about 11 m, it was necessary to model the cracks and softened zones which had formed in the core. These persisted at least down to full supply level. It was also necessary to model transient pore pressures which occur in the cracks/softened zones during rain events and pore pressures in the filters. Conventional effective stress analysis gave low factors of safety, but only for shallow surfaces through the rockfill, i.e. not modelling the real situations. The observed deformations could only be replicated by numerical modelling if very low moduli were used in the rockfill (in the order of 5 MPa) and weakening of the core.

During construction of remedial works, which included a rockfill berm on the upstream side and new filters and rockfill for the upper 10 m of the downstream side, a wedge of the core was displaced nearly 50 mm upstream. This displacement also sheared an inclinometer installed from the dam crest at depths of up to 16 metres. Calculations showed that failure surfaces encompassing this wedge had a limit equilibrium factor of safety of about 1.3. Numerical analysis was consistent with this. What appears to have happened is that the relatively stiff core has sheared locally particularly during drawdown events at a number of depths within the section due to lack of support from the poorly compacted rockfill.

11.6.3   *The lessons learnt*

The lessons learnt can be summarised as:

(a) Conventional effective stress analysis ignores pore pressures generated during shearing. This is conservative for over consolidated, dilative soils. As discussed in Section 6.1.3 Ladd (1991) shows that effective stress analysis of normally consolidated (and slightly over-consolidated) clays can very significantly over-estimate the factor of safety. The error can be as much as 2 times.
(b) What must be recognised is that even though the soil in modern dams is given a degree of over consolidation by rolling and by partial saturation suction effects, this is often only equivalent in over-consolidation to the stress due to ≈ 20 m to 30 m of dam height so, in the lower part of larger dams, the dam core is likely to be normally consolidated. If the dam is affected by transient loading, such as drawdown or high reservoir levels or high pore water pressures in cracks in the crest (see below), the loading is of short duration compared to the drainage path, so undrained loading occurs and to model this correctly, either the pore pressures generated in shearing should be included in the effective stress analysis or undrained strengths used.
(c) For poorly compacted fills such as those in Hume No.1 Embankment, which were wetted by seepage through the dam or foundation, the soils behave as normally consolidated clays even at shallow depths. In fact, the CPT data from adjacent to walls at Hume No.1 Embankment and at O'Shannassy Dam shows that the fill may "hang up" on the walls, so the effective vertical stress at depth is lower than that calculated by unit weight x depth to the surface and strengths are even lower than expected.
(d) For cases of marginal stability it is important to model the actual conditions carefully including using undrained strengths, cracks and softened zones in the partially saturated zone and pore pressures in cracks or softened zones during rain events. This of course means that site investigation methods capable of locating such cracks and transient pore pressures are needed, e.g. using trenching across the dam crest and using continuous reading or maximum pressure recording piezometers.
(e) One should be conscious of the possibility that large deformations of the core may occur where rockfill is poorly compacted, even though the factor of safety is well above 1.0.

11.7   ANALYSIS OF DEFORMATIONS

The analysis of deformations of embankment dams, either as an aid to the assessment of stability or to assess the likelihood of cracking or hydraulic fracture, has become possible within most dam engineering organisations, with the advent of powerful personal computers and relatively inexpensive finite element and finite difference analysis programs.

We do not propose to attempt to review these methods, and instead recommend that several overview and case-specific papers listed below be read. These include:

(a) *Overview – State of the art papers*:
    – Eisenstein and Naylor in ICOLD (1986a) Static analysis of embankment dams;

  – Duncan (1992, 1996a). State of the Art: Limit Equilibrium and Finite Element Analysis of Slopes. Duncan lists over 100 references of case studies;
  – Duncan (1994). The role of advanced constitutive relations in practical applications;
  – Vaughan (1994). Assumption, Prediction and Reality in geotechnical engineering.
(b) *Examples of strain softening*:
  – Progressive failure and other deformation analyses; Potts et al. (1990, 1997), Muir-Wood et al. (1995), Dounias et al. (1988, 1996), Kovacevic et al. (1997). Duncan (1996a) lists some of the uncertainties in numerical modelling of dams;
  – Degree of compaction of the embankment materials – the actual may be different to that specified;
  – Compaction water content – actual may be different to the specified;
  – No test data – e.g. it is not practicable to test rockfill zones;
  – Construction sequence and time;
  – Simplified approximation of stress–strain behaviour – all constitutive relations are approximation of real behaviour;
  – Reliability of field measurements.

He concludes: *"comparisons of the results of finite-element analyses with field measurements have shown that there is a tendency for calculated deformations to be larger than measured deformations. The reasons for this difference include: (1) soils in the field tend to be stiffer than soils at the same density and water content in the lab because of aging effects; (2) average field densities are higher than the specified minimum dry density, which is often used for preparing lab triaxial test specimens; (3) samples of inplace materials suffer disturbance during sampling, and are less stiff as a result; (4) many field conditions approximate plane strain, whereas triaxial tests are almost always used to evaluate stress–strain behaviour and strength; and (5) two-dimensional finite-element analyses over-estimate deformations of dams constructed in V-shaped valleys with steep valley walls.*

By the time of this writing there has been about 30 years of experience with using the finite-element method to estimate stresses and deformations in slopes and embankments. This experience has shown the considerable potential of the method for use in engineering practice and it has pointed up clearly the sources of uncertainty in the results of these analyses. These are related primarily to difficulties involved in being able to predict the actual densities and water contents of soils in the field and with being able to anticipate the sequence of operations that will be followed during construction. With due allowance for the uncertainties, finite-element analyses afford a powerful method of analysis that is applicable to a wide variety of engineering problems."

Duncan (1996a) concluded (as had Vaughan, 1994, who discusses several cases), that there is much to be gained from the analyses in the understanding of mechanisms, if not the absolute stresses and deformations.

The authors' experience in this area is limited to that of "user" rather than "doer" of the analyses. From this it is clear that it is difficult to model the deformation of dams with any degree of precision. The ability to numerically model generally is greater than the ability to provide accurate material properties.

There are however often more basic reasons for poor modelling, including not "constructing" (or numerically "building") the slope in steps correctly, assuming drained conditions will apply when in fact undrained behaviour is actually present and not paying enough attention to the stiffness of the layers (or in particular relative stiffnesses), so the stress distributions and deformations are poorly modelled.

Few model pore pressures accurately or allow for cracking, softened zones around cracks, crack water pressures or in many cases, strain weakening. Given that these factors often control the stresses and deformations in slopes, it is not surprising the modelling is not good.

The authors' view is that the overwhelming problem in too many cases is that those doing the analysis have neither the proper understanding of the mechanics of the problem and what are reasonable properties, nor a good, detailed understanding of the computer programs. The result is an unfortunate waste of time and money.

The techniques are very valuable, done by competent persons, who understand dams, material behaviour and numerical analysis.

11.8   PROBABILISTIC ANALYSIS OF THE STABILITY OF SLOPES

The use of probabilistic methods for assessing the stability of dam or other slopes, other than mine pit slopes, is not common. This is despite a large body of research and publication into the topic, which have brought the state of the art to a relatively high degree of sophistication.

Mostyn and Fell (1997) gives a thorough overview of the state of the art method. Other reviews are given in Christian et al. (1992) and Wu et al. (1996).

Given the low incidence of slope instability failures of dams and the fact that many accidents and failures can be attributed to weaknesses in the dam (e.g. by softening) or the foundation (e.g. the presence of bedding surface shear in the rock), which would be missed in probabilistic analysis just as they were in the deterministic, factor of safety approach, the authors do not see a great need to use probabilistic analysis. The exception could be for embankments built on soft clay foundations, where the method may help demonstrate that a considerable degree of conservation in selection of the factor of safety is required. See Duncan (2000) and Christian et al. (1992) for examples.

12

Design of embankment dams to withstand earthquakes

12.1 EFFECT OF EARTHQUAKE ON EMBANKMENT DAMS

Earthquakes impose additional loads on to embankment dams over those experienced under static conditions. The earthquake loading is of short duration, cyclic and involves motion in the horizontal and vertical directions. Earthquakes can affect embankment dams by causing any of the following:

- Settlement and cracking of the embankment, particularly near the crest of the dam;
- Instability of the upstream and downstream slopes of the dam;
- Reduction of freeboard due to settlement or instability which may, in the worst case, result in overtopping of the dam;
- Differential movement between the embankment, abutments and spillway structures leading to cracks;
- Internal erosion and piping which may develop in cracks;
- Liquefaction or loss of shear strength due to increase in pore pressures induced by the earthquake in the embankment and its foundations;
- Differential movements on faults passing through the dam foundation;
- Overtopping of the dam in the event of large tectonic movement in the reservoir basin, by seiches induced upstream;
- Overtopping of the dam by waves due to earthquake induced landslides into the reservoir from the valley sides;
- Damage to outlet works passing through the embankment leading to leakage and potential piping erosion of the embankment.

The potential for such problems depend on:

- The seismicity of the area in which the dam is sited and the assessed design earthquake;
- Foundation materials and topographic conditions at the damsite;
- The type and detailed construction of the dam;
- The water level in the dam at the time of the earthquake.

The amount of site investigation, design, and additional construction measures (over those needed for static conditions) will depend on these factors, the consequences of failure, and whether the dam is existing or new.

There are four main issues to consider:

- The general (or "defensive") design of the dam, particularly the provision of filters, to prevent or control internal erosion of the dam and the foundation, and provision of zones with good drainage capacity (e.g. free draining rockfill);
- The stability of the embankment during and immediately after the earthquake;
- Deformations induced by the earthquake (settlement, cracking) and dam freeboard;
- The potential for liquefaction of saturated sandy and silty soils and some gravels with a sand and silt matrix in the foundation, and possibly in the embankment, and how this affects stability and deformations during and immediately after the earthquake.

The aim in this chapter is to present an overview of the topic, rather than to give detailed methods, because these are far too complex and extensive to be covered here. The authors have drawn largely from several useful review papers including Seed (1979), ICOLD (1983), ICOLD (1986c), US National Research Council (1985), Martin (1988), Finn (1993), Youd and Idriss (1997), Stark et al. (1998), ANCOLD (1998), Youd et al. (2001). Much of the description of earthquakes and their characteristics is taken from ANCOLD (1998) and was written by G. Gibson.

12.2   EARTHQUAKES AND THEIR CHARACTERISTICS

12.2.1   *Earthquake mechanisms and ground motion*

Earthquakes result when the stresses within the earth build up over a long period of time until they exceed the strength of the rock, which then fails and displacement along a fault results.

The fault displacement in a particular earthquake may vary from centimetres up to a few metres. Once ruptured, the fault is a weakness which is more likely to fail in future earthquakes, so a large total displacement may build up from many earthquakes over a long period of time. This may eventually measure kilometres for thrust faults produced by compression, or hundreds of kilometres for strike-slip faults such as the San Andreas.

Figure 12.1 shows the types of fault and the effects on the ground near the fault.

The point on the fault surface where a displacement commences is called the focus or hypocenter, and the earthquake epicentre is the point on the ground surface vertically above the focus. The displacement usually propagates along the fault in one direction from the focus, but sometimes it propagates in both directions. The centre of energy release is near but not exactly at the focus.

As shown in Figure 12.2 the focal distance from an earthquake to a point on the earth's surface is the three dimensional slant distance from the focus to the point, while the epicentral distance is the horizontal distance from the epicentre to the point.

Earthquake ground vibration is recorded by a seismograph or a seismogram. Modern seismographs record three components of motion: east-west, north-south and vertical.

The rupture time for small earthquakes is a fraction of a second, for earthquakes of magnitude 5 it is about a second and for large earthquakes may be up to tens of seconds. However the radiated seismic waves travel at different velocities, and are reflected and refracted over many travel paths, so the total duration of vibrations at a site persists longer than the rupture time and shows an exponential decay.

Several types of seismic wave are radiated from an earthquake. Body waves travel in three dimensions through the earth, while surface waves travel over the two-dimensional surface like ripples on a pond. There are two types of body wave (P and S waves), and two types of surface wave (Rayleigh and Love waves).

Primary or P waves are compressional waves with particle motion parallel to the direction of propagation.

Secondary or S waves are shear waves, with particle motion at right angles to the direction of propagation. The amplitude of S waves from an earthquake is usually larger than that of the P waves.

P waves travel through rock faster than S waves, so they always arrive at a seismograph before the S waves.

The frequency content of earthquake ground motion covers a wide range of frequencies up to a few tens of hertz (cycles per second). Most engineering studies consider motion between about 0.2 and 25 Hz. For embankment dams, the lower frequency (0.5 to 5 hertz) motion corresponding to their natural frequency is most important.

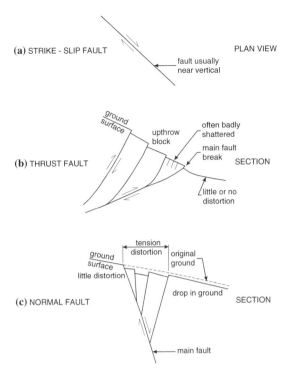

Figure 12.1.   Schematic diagram of the types of fault and the effects on the ground nearby (adapted from Sherard et al., 1974).

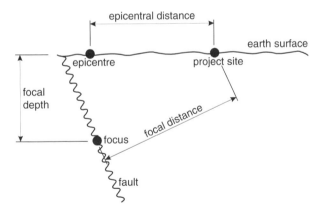

Figure 12.2.   Definition of earthquake terms.

The amplitude, duration and frequency content of earthquake ground motion at a site depend on many factors, including the magnitude of the earthquake, the distance from the earthquake to the site and local site conditions.

The larger the earthquake magnitude, the greater the amplitude (by definition a factor of ten for each magnitude unit), the longer the duration of motion and the greater the proportion of seismic energy at lower frequencies. A small earthquake has low amplitude (unless it is very close), short duration, and has only high frequencies.

The smaller the distance from an earthquake to the site, the higher the amplitude. The duration is not strongly affected by distance. High frequencies are attenuated by absorption within the ground more quickly than low frequencies, so at greater distances the proportion of seismic vibration energy at high frequencies will decrease.

12.2.2   Earthquake magnitude, and intensity

There are a number of different measures of the magnitude of an earthquake.

(a) *The Local or Richter magnitude*

$$M_L = \log_{10} A \tag{12.1}$$

where A = maximum seismic wave amplitude (in thousandths of a millimetre) recorded from a standard seismograph at a distance 100 km from the earthquake epicentre or

$$M_L = \log_{10} A - F(\Delta) + k \tag{12.2}$$

where $F(\Delta)$ = distance correction; k = scaling constant.
This allows calculation of M_L at different distances from the earthquake.
(b) *The Body Wave magnitude*

$$M_b = \log_{10} V + 2.3 \log \Delta \tag{12.3}$$

where V = maximum ground velocity in microns/sec recorded by the seismograph; Δ = distance from epicentre in kilometres.
(c) *The Moment Magnitude (M_W)* which assigns a magnitude to the earthquake in accordance with its seismic moment M_O, which is directly related to the energy released by the earthquake:

$$M_W = (\log_{10} M_O / 1.5) - 10.7 \tag{12.4}$$

where M_O is the seismic moment in dyn-cm.

Moment magnitude is the scale most commonly used for engineering applications. Figure 12.3 shows the relationship between the different scales.

Earthquakes with magnitude of less than 3 or 4 will usually not cause any felt effect, and earthquakes with a magnitude less than about 5 will usually not cause any damage unless they are shallow, near the structure and/or ground conditions cause amplification. The maximum recorded magnitude is approximately 8.9. The scale is not linear. Each step in the magnitude scale represents a thirty-fold increase in energy released by the earthquake e.g. a Magnitude 5 earthquake represents 900 times the energy of a Magnitude 3, and a Magnitude 8 represents about 10^6 times the energy of a Magnitude 4.

Another measure of earthquake size is the fault area, or the area of the fault surface which is ruptured. The fault area ruptured in an earthquake depends on the magnitude and stress drop in the earthquake. For a given magnitude, a higher stress drop will give a small rupture area. Typically, a Magnitude 4 earthquake ruptures a fault area of about 1 square kilometre, Magnitude 5 about 10 square kilometres, and Magnitude 6 about 100 square kilometres.

Earthquake intensity is a qualitative value based on the response of people and objects to the earthquake. The intensity depends on distance from the earthquake, ground conditions and topography, so there will be a range of intensity values for any earthquake. The

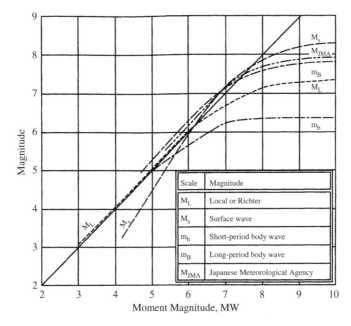

Figure 12.3.   Relationship between Moment, M_W and other magnitude scales (Youd and Idriss, 1997; after Heaton et al., 1982).

most commonly used scale is the modified Mercalli Scale which is reproduced in Table 12.1.

For the design of dams, the horizontal and vertical acceleration induced by the earthquake at the base of the dam is usually required. Information is best obtained from accelerograph measurements at the dam site, but in many cases will be obtained from records of sites with similar geological conditions. A typical accelerograph record is shown in Figure 12.4.

12.2.3   Attenuation and amplification of ground motion

Earthquake ground motion attenuates with increasing distance from the source due to radiation and hysteretic damping. High frequency motion is attenuated more quickly with distance than lower frequency motion.

For estimates of peak ground acceleration, attenuation is allowed for by using an attenuation function of the form:

$$a = b_1 e^{b_2 M} R^{-b_3} \qquad (12.5)$$

where a = acceleration; R = focal distance; M = Magnitude; b_1, b_2, b_3 are constants, which vary considerably over the world.

Some earthquake hazard studies use the Esteva and Rosenblueth (1969) attenuation functions, which give peak ground velocity (mm/s), peak ground acceleration (mm/s²) at an epicentral distance x kilometres from an earthquake at focal depth z kilometres with local magnitude M_L. The equations are:

$$R = \sqrt{x^2 + z^2 + 400} \qquad (12.6)$$

Table 12.1.   Modified Mercalli Scale, 1956 version (Richter, 1958; Hunt, 1984).

Intensity	Effects
I	Not felt. Marginal and long period effects of large earthquakes.
II	Felt by persons at rest, on upper floors, or favorably placed.
III	Felt indoors. Hanging objects swing. Vibration like passing of light trucks. Duration estimated. May not be recognized as an earthquake.
IV	Hanging objects swing. Vibration like passing of heavy trucks or sensation of a jolt like a heavy ball striking the walls. Standing motor cars rock. Windows, dishes, doors rattle. Glasses clink. Crockery clashes. In the upper range of IV wood walls and frames creak.
V	Felt outdoors, duration estimated. Sleepers wakened. Liquids disturbed, some spilled. Small unstable objects displaced or upset. Doors swings, close, open. Shutters, pictures move. Pendulum clocks stop, start, change rate.
VI	Felt by all. Many frightened and run outdoors. Persons walk unsteadily. Windows, dishes, glassware broken. Knickknacks, books, etc. off shelves. Pictures off walls. Furniture moved or overturned. Weak plaster and masonry D cracked. Small bells ring (church, school). Trees, bushes shaken (visibly, or heard to rustle – CFR).
VII	Difficult to stand. Noticed by drivers of motor cars. Hanging objects quiver. Furniture broken. Damage to masonry D, including cracks. Weak chimneys broken at roof line. Fall of plaster, loose bricks, stones, tiles, cornices (also unbraced parapets and architectural ornaments – CFR). Some cracks in masonry C. Waves on ponds; water turbid with mud. Small slides and caving in along sand or gravel banks. Large bells ring. Concrete irrigation ditches damaged.
VIII	Steering of motor cars affected. Damage to masonry C, partial collapse. Some damage to masonry B, none to masonry A. Fall of stucco and some masonry walls. Twisting, fall of chimneys, factory stacks, monuments, towers, elevated tanks. Frame houses moved on foundations if not bolted down; panel walls thrown out. Decayed piling broken off. Branches broken from trees. Changes in flow or temperature of springs and wells. Cracks in wet ground and on steep slopes.
IX	General panic. Masonry D destroyed; masonry C heavily damaged, sometimes with complete collapse; masonry B seriously damaged. (General damage to foundations – CFR). Frame structures, if not bolted, shifted off foundations. Frames cracked. Serious damage to reservoirs. Underground pipes broken. Conspicuous cracks in ground. In alluviated areas sand and mud ejected, earthquake fountains, sand craters.
X	Most masonry and frame structures destroyed with their foundations. Some well-built wooden structures and bridges destroyed. Serious damage to dams, dikes, embankments. Large landslides. Water thrown on banks of canals, rivers, lakes, etc. Sand and mud shifted horizontally on beachheads and flat land. Rails bent slightly.
XI	Rails bent greatly. Underground pipelines completely out of service.
XII	Damage nearly total. Large rock masses displaced. Lines of sight and level distorted. Objects thrown into the air.

Note: Masonry A, B, C, D. To avoid ambiguity of language, the quality of masonry, brick or otherwise, is specified by the following lettering (which has no connection with the conventional Class A, B, C construction).
– Masonry A: Good workmanship, mortar, and design; reinforced, especially laterally, and bound together by using steel, concrete, etc.; designed to resist lateral forces;
– Masonry B: Good workmanship and mortar; reinforced, but not designed to resist lateral forces;
– Masonry C: Ordinary workmanship and mortar; no extreme weaknesses such as non-tied-in corners, but masonry is neither reinforced nor designed against horizontal forces;
– Masonry D: Weak materials, such as adobe; poor mortar; low standards of workmanship; weak horizontally;
– CFR indicates additions to classification system by Richter (1958).

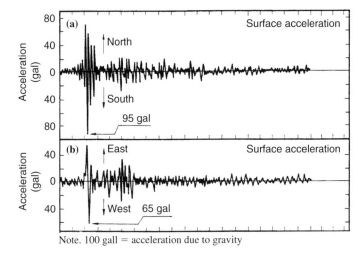

Figure 12.4.   Earthquake accelerograph (USNRC, 1985).

$$v_{peak} = 160 \; e^{1.0 M_L} R^{-1.7} \tag{12.7}$$

$$a_{peak} = 20000 \; e^{0.8 M_L} R^{-2.0} \tag{12.8}$$

Because of the "400" term in the expression for R, corresponding to a minimum R of 20 kilometres, these equations give low values of ground motion at distances closer than a few kilometres.

There have been many different attenuation relationships developed around the world. A seismologist should be engaged to advise on the relationship appropriate for the site.

The attenuation equations discussed above, allow estimation of peak bedrock acceleration at a site. As discussed by Martin (1988), sites underlain by stiff soil and thick cohesionless soil may experience similar peak accelerations to the "bedrock" values. However, in sites underlain by soft or medium soft clays, or shallow cohesionless soils, the maximum ground acceleration may be different from bedrock accelerations. Martin (1988) indicates that for ground accelerations less than 0.1 g amplification may occur, whereas for very high accelerations (e.g. >0.3 g) the peak accelerations may be attenuated (de-amplified) due to energy dissipation. He quotes the example of the 1985 Mexico earthquake where the bedrock accelerations were 0.03 g and the peak accelerations of 0.17 g (at a reduced frequency) were recorded at the surface of soft soil areas. The authors' experience is that, even for stiff and thick cohesionless soil, amplification may occur for small earthquakes. Amplification effects can be estimated using computer programs such as SHAKE or DESRA, using input ground motions from typical accelerograms.

Topographic effects are also important. An accelerograph sited on a ridge high on a dam abutment, is likely to record significantly higher accelerations than one sited at the river level, so care must be taken in siting such instruments. An appreciation of the extent of magnification due to topographic effects can be gained by comparison with the crest and base accelerations for a dam as discussed in Section 12.6.3.4.

12.2.4   Earthquakes induced by the reservoir

As discussed in ICOLD (1983) and ICOLD (1989c), there is documented evidence to prove that impounding of a reservoir sometimes results in an increase of earthquake activity at or near the reservoir.

ICOLD (1983) conclude that:

- Earthquakes of magnitude 5 to 6.5 were induced in 11 of 64 recorded events;
- The greatest seismic events have been associated with very large reservoirs (but there is insufficient data to show any definite correlation between reservoir size and depth and seismic activity);
- In view of the above, a study of possible induced seismic activity should be made at least in cases where the reservoir exceeds $10^9 \, m^3$ in volume, or 100 m in depth;
- The load of the reservoir is not the significant factor, rather it is the increased pore water pressure in faults, leading to a reduction in shear strength over already stressed faults.

ICOLD (1983) and (1989c) give more details and references on this issue.

ANCOLD (1998), as advised by G. Gibson, indicates that reservoir induced seismicity events usually occur initially at shallow depth under or immediately alongside a reservoir. As years pass after first filling and groundwater pore pressure increases permeate to greater depths and distances, the events may occur further from the reservoir. This occurs at a rate of something like one kilometre per year.

Reservoir induced seismicity is experienced under new reservoirs, usually starting within a few months or years of commencement of filling and usually not lasting for more than about twenty years. Once the stress field and the pore pressure fields under a reservoir have stablised, then the probability of future earthquakes reverts to a value similar to that which would have existed if the reservoir had not been built. Most of the earthquake energy does not come from the reservoir, but from normal tectonic processes. The reservoir simply acts as a trigger.

If there is a major fault near the reservoir, reservoir induced seismicity can produce earthquakes exceeding Magnitude 6.0 (Xinfengjiang, China, 1962, M6.1; Koyna, India, 1967, M6.3). Such events will occur only if the fault is already under high stress. A number of Australian reservoirs have triggered earthquakes equalling or exceeding Magnitude 5.0 (Eucumbene, 1959, M5.0; Warragamba, 1973, M5.0; Thomson, 1996, M5.2).

It is more common for a reservoir to trigger a large number of small shallow earthquakes, especially if the underlying rock consists of jointed crystalline rock like granite (Talbingo, 1973 to 1975; Thomson, 1986 to 1995). These events possibly occur on joints or local minor faults rather than major faults, so are limited in size and only give magnitudes up to 3 or 4. There is no hazard to the dam from such low magnitude reservoir induced earthquakes, even if they occur regularly. Their shallow depth means that they may often be felt or heard.

12.3   EVALUATION OF SEISMIC HAZARD

12.3.1   *Probabilistic approach*

This method is generally used where active faults cannot be identified, or to assess the hazard from earthquakes in areas away from active faults. The probabilistic estimation of ground motion requires the following seismicity information about the surrounding area:

- The rate of occurrence and magnitude of earthquakes;
- The relative proportion of small to large events (b value);
- The maximum earthquake size expected (maximum credible magnitude);
- The spatial distribution of earthquake epicenters including delineation of faults.

The seismicity can be evaluated by a modified Richter relation (Gibson, 1994)

$$\log_{10}(P) = -\log_{10}[10(10^{-bM} - 10^{-bM_{max}})] - \log_{10}(N_o) \qquad (12.9)$$

where P is the return period in years for an earthquake of magnitude M or greater and N_o is the rate of occurrence of earthquakes, given as the number of earthquakes of Magnitude zero or greater per year per unit volume or per unit area. An area of 100×100 kilometres is commonly used. This must be converted to per square kilometre or per cubic kilometre for ground motion calibrations.

"b" is the Richter "b" value, which gives the relative number of small earthquakes to large. It is the logarithm to the base 10 of the ratio of the number of events exceeding M + 1 to the number exceeding M. A value of 1.0 would correspond to ten times as many earthquakes exceeding magnitude M as would exceed magnitude M + 1.

M_{max} is the magnitude of the maximum credible earthquake for the area. Because of the low probability of very large earthquakes, M_{max} does not critically affect ground motion recurrence estimates for return periods up to hundreds of years, especially when the "b" value is high. However, it is more important for low AEP events such as may be important for design of high hazard dams. That maximum credible magnitude causes the magnitude recurrence plot to flatten out and asymptote to that value.

The seismicity parameters N_o and b are determined using available earthquake data. In most places there are insufficient earthquake data to determine M_{max} from historical records, but values can be estimated by considering the tectonic situation and local fault dimensions. The "b" value is much more critical than N_o, and a small adjustment to "b" will give a large change in predictions.

Hazard evaluation depends on the extrapolation of data from small earthquakes to larger magnitudes. The lower the "b" value, the greater the resulting hazard estimate. A catalogue with missing small earthquakes will give an invalid low "b" value, and an estimated hazard that will be too high. A catalogue with smaller quarry blasts incorrectly identified as earthquakes will give a high "b" value and an extrapolation that is non-conservative.

Figure 12.5 shows the output of a probabilistic assessment of seismic hazard expressed in peak ground (bedrock) acceleration – as would be required for analysis of liquefaction. Note that the contribution of earthquake magnitude is separated, because the magnitude, as well as peak ground acceleration is important.

12.3.2   *Seismic hazard from known active faults*

This method is used where faults in the vicinity of the dam can be identified. The procedure will usually involve:

(a) Identification of major faults within the vicinity of the dam. This may involve an area up to several hundred kilometres from the site. Figure 12.6 shows an example;
(b) Assessment of whether the faults are active or potentially active, by consideration of whether modern (including small) earthquakes have been recorded along the fault. This may also involve geomorphological studies, e.g. of displaced river terraces and/or trenching across faults to identify past displacements and determine their ages;
(c) Assessment of the Maximum Credible Earthquake magnitude on each identified fault. This will usually be determined by considering the length and/or area of the fault and the type of fault. The likely focal depth and, hence, focal distance are also estimated;
(d) For a deterministic approach, assess the peak ground acceleration (pga) at the project site resulting from the MCE at each of the faults and determine the most critical earthquake.

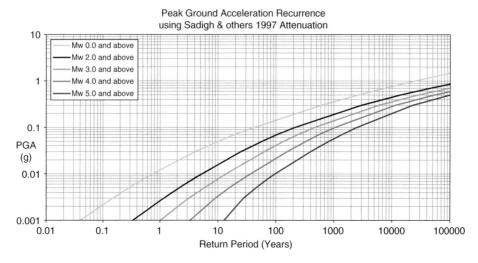

Figure 12.5.   Peak ground acceleration versus annual exceedance probability (AEP) and earthquake magnitude for a typical site in Australia.

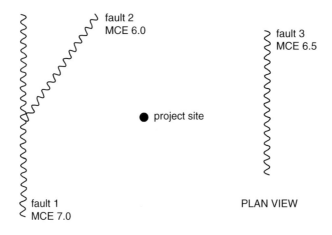

Figure 12.6.   Plan of a project site and potentially active faults for assessment of seismic hazard.

The duration of the earthquake and the period of oscillations are dependent on the magnitude of the earthquake and the effect on the dam is dependent on these factors as well as maximum accelerations, therefore more than one earthquake may have to be used in analysis;

(e) For a probabilistic approach, determine the annual probability versus earthquake magnitude for each fault. For example, for the situation in Figure 12.6.

Fault	Annual exceedance probability	Earthquake magnitude
1	0.01	5.5
	0.001	6.5
	0.0001	7.0(MCE)
2	0.01	5.0
	0.001	6.0(MCE)
3	0.01	5.5
	0.001	6.5(MCE)

From this estimate pga versus combined annual exceedance probability, allowing for the three faults and accounting for earthquake magnitude.

12.3.3   *Other forms of expression of seismic hazard*

The discussion so far has centred on assessing the hazard in peak ground acceleration (on bedrock) terms.

For more advanced dynamic analysis, it will be necessary to obtain accelerograms of earthquakes of a similar nature to that expected considering the source faults, scaled to the expected peak ground accelerations. In some cases, synthetic accelerograms may be used. More than one accelerogram will usually be needed.

For simplified analysis of concrete gravity dams and other structures response spectra may be used but with some limitations on the output of the analyses.

12.3.4   *Selection of design seismic loading*

There are two ways of selecting the design seismic loading:

(a) *Deterministic* – which requires the assessment of an operating basis earthquake OBE, and Maximum Design Earthquake MDE.

   ANCOLD (1998) adopted the following definitions for OBE and MDE:

   Operating Basis Earthquake – the OBE will produce a level of ground motion which will cause only minor and acceptable damage at the damsite. The dam, appurtenant structures and equipment should remain functional and damage from the occurrence of earthquake shaking not exceeding the OBE should be easily repairable.

   Maximum Design Earthquake – the MDE will produce the maximum level of ground motion for which the dam should be designed or analysed. It will be required at least that the impounding capacity of the dam be maintained when subjected to that seismic load.

   The OBE is often accepted as a loading which has a 10% chance of being exceeded in a 50 year period, or an annual probability of exceedance of 1 in 475; and the Maximum Design Earthquake (MDE), which may be taken as the maximum loading at the site from MCE on the known faults, or the load which has an annual probability of exceedance of say 1 in 10,000.

(b) *Probabilistic* – which assesses the effects on the dam and its foundations of a range of seismic loads, e.g. from 1 in 100 Annual Exceedance Probability (AEP) to 1 in 10,000 AEP, or 1 in 100,000 AEP, resulting from earthquakes on known faults, or estimated probabilistically as detailed in Section 12.3.1.

   The steps involved in this process are (ANCOLD, 1998):

   (i) determine the AEP of earthquake ground motion (P_E) over the range of earthquake events which may affect the dam. Table 12.2 gives an example for AEP vs ground acceleration;

   (ii) determine the conditional probability (P_{BC}) that for each of the ground motion ranges (e.g. 0.125 g to 0.175 g in Table 12.2) the dam will breach. In assessing this conditional probability all modes of failure should be considered and the probabilities combined, making allowance for interdependence and mutual exclusivity or otherwise (e.g. for embankment dams, slope instability, piping, liquefaction/ instability, and for concrete gravity dams, overturning, and sliding);

   (iii) assess the probability of failure for each range of ground motion by multiplying the AEP with P_{BC} i.e. $P_B = P_E \times P_{BC}$ – see Table 12.2;

   (iv) sum the probabilities to give the overall annual probability of failure due to earthquake.

Table 12.2. Example of assessing the probability of failure by earth-quake.

Acceleration	Annual probability	Conditional[1] probability (P_{BC})	P_B[2]
<0.075 g	0.874	0.0005	0.0004
0.075 g to 0.125 g	0.100	0.005	0.0005
0.125 g to 0.175 g	0.015	0.05	0.0007
0.175 g to 0.225 g	0.007	0.1	0.0007
0.225 g to 0.3 g	0.003	0.3	0.0009
>0.3 g	0.001	0.5	0.0005
Total	1.000		0.0037

[1] Given the earthquake occurs.
[2] $P_B = AEP \times P_{BC}$.

National codes, or guidelines to practice, will often specify which method should be used. The authors favour the probabilistic approach because it recognizes there is a chance a dam will fail at loads less than the MDE from the MCE.

Whichever approach is taken, the bedrock ground motions need to be adjusted where appropriate for amplification (or de-amplification) effects.

12.3.5   *The need to get good advice*

Seismic hazard should be assessed by a seismologist familiar with the seismic hazard in the project area, and experienced in assessing seismic hazard for dams.

12.4   LIQUEFACTION OF DAM EMBANKMENTS AND FOUNDATIONS

12.4.1   *Definitions and the mechanics of liquefaction*

12.4.1.1   *Definitions*

The following definitions are based on USNRC (1985), Robertson and Fear (1995), Robertson and Wride (1997), Lade and Yamamuro (1997) and Fell et al. (2000).

"*Liquefaction*". All phenomena giving rise to a reduction in shearing resistance and stiffness, and development of large strains as a result of increase in pore pressure under cyclic or monotonic (static) loading of contractive soils.

"*Initial liquefaction*" is the condition when effective stress is momentarily zero during cyclic loading.

"*Flow liquefaction*" is the condition where there is a strain weakening response in undrained loading and the in-situ shear stresses are greater than the steady state undrained shear strength.

"*Temporary liquefaction*" is the condition where there is a limited strain weakening response in undrained loading; at larger strain the behaviour is strain hardening. It is important to recognize that these phenomena apply to monotonic (static) as well as cyclic loading and are apparent in contractive soils, both cohesionless and those with some clay content.

Robertson and Fear (1995) and Robertson and Wride (1997) define two further terms which apply to cyclic loading.

"*Cyclic liquefaction*" is a form of temporary liquefaction, where the cyclic loading causes shear stress reversal and an initial liquefaction (zero effective stress) condition develops temporarily.

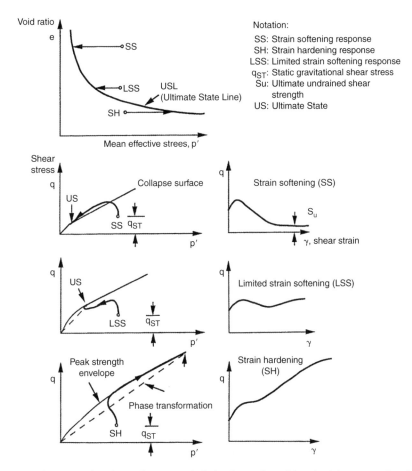

Figure 12.7.   Schematic of undrained monotonic behaviour of sand in triaxial compression (Robertson 1994).

"*Cyclic mobility*" is a form of temporary liquefaction where the shear stresses are always greater than zero.

These are discussed further below.

12.4.1.2   *Some consideration of the mechanics of undrained shear of granular soils, and of liquefaction*

12.4.1.2.1   Undrained shear of saturated granular soils

Figure 12.7 shows schematically the undrained behaviour of saturated sand in triaxial compression.

A soil which has an initial void ratio higher than the ultimate state (steady state, critical state/line) will tend to contract (densify) in drained loading, reaching a void ratio equal to that at the ultimate state (critical state) line at that mean effective stress. In saturated undrained loading the soil is unable to contract, so positive pore pressures are developed, which will give strain weakening behaviour, reaching the steady state undrained strength S_{us}. If the static gravitational shear stresses q_{st} are greater than S_{us}, flow liquefaction will result.

A soil which has an initial void ratio just above the ultimate state line may show limited strain softening, or temporary liquefaction, with the soil wanting to contract on initial

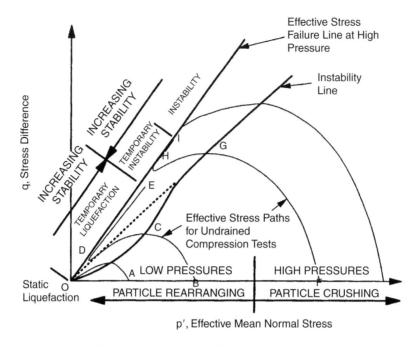

Figure 12.8.   Undrained effective stress behaviour of loose silty sands (Lade and Yamamuro, 1997).

shear and to dilate at larger strains, so developing positive, then negative pore pressures in undrained loading. A soil with an initial void ratio below the ultimate state line will tend to dilate, or develop negative pore pressures in undrained loading, giving a strain "hardening" response.

The susceptibility of a soil to static liquefaction is dependent on the particle size distribution, void ratio (or density index), the initial stress conditions and the stress path of loading, e.g. triaxal compression or extension.

Lade and Yamamuro (1997), and Yamamuro and Lade (1997), show that there are four different types of undrained stress paths after the instability line (also known as collapse surface). These are shown in Figure 12.8:

- *Static liquefaction* occurs at low stresses and is characterised by large pore pressure development, resulting in zero effective stresses at low axial strain levels (stress path AO). Increasing confining pressures result in increasing effective stress friction angles in this stress region;
- *Temporary liquefaction* occurs at higher stresses than the static liquefaction region and is characterized by stress path BCDE. Hence the behavior is contractant from C to D, then dilatant from D to E which occurs at large axial strains. The tendency to dilate (and less likely to contract) increases as the initial confining stresses increase;
- *Temporary instability* (stress path FGHI) is similar to temporary liquefaction except that the amount the stress difference increases beyond the initial peak (i.e. I vs G), is not as large as that exhibited by temporary liquefaction;
- *Instability* – at higher stresses, due partly to particle crushing.

It should be noted that a soil may reach the instability line (collapse surface) at stress conditions less than the effective stress failure line. Also, a soil can be in a stress condition just to the right (stable) of the instability line and cross over the line by change in

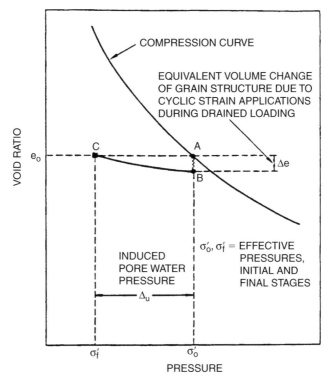

Figure 12.9.   Schematic illustration of mechanism of pore pressure generation during cyclic loading (Seed & Idriss 1982; USNRC, 1985, reproduced with permission of ASCE).

static stress conditions, e.g. a rise in pore pressures or strains induced for example by an earthquake.

12.4.1.2.2   The effects of cyclic loading

Cyclic loading, such as that from an earthquake, causes densification of dry granular soils by particle rearrangement due to the back and forth straining. If, however, the soil is saturated and not allowed to drain during cyclic loading, just as in monotonic loading the decreases in volume cannot occur and the tendency to decrease volume is counteracted by an increase in pore pressure and decrease in effective stress – see Figure 12.9.

Hence, soil starting at A and subject to cyclic loading, which would otherwise have ended at B, will in fact have stresses represented by C where total stress σ_o is taken by σ_f' and Δ_u. The pore pressures must increase to maintain equilibrium in this undrained condition.

The pore pressures build up gradually with the number of cycles of loading and only if the pore pressures build up to equal the total stress, does the "initial liquefaction" (effective stress $\sigma' = 0$) condition occur.

The stress path followed by the soil affects the liquefaction potential. Figure 12.10 shows the p'-q stress paths during a cyclic triaxial test, the effective stress path moving from cycle 1 to the left before reaching the failure envelope on the k_f line in cycle 21.

Once the failure line is reached, pore pressures and strain development accelerates. After $\sigma' = 0$ is reached the stress paths are up and down the failure lines passing through or near the origin twice in each cycle as shown. This is the cyclic liquefaction condition.

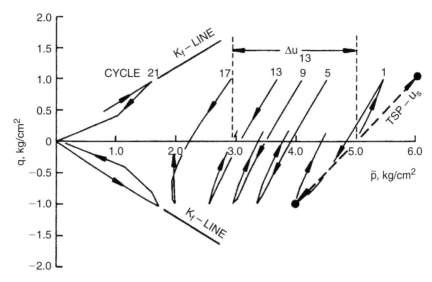

Figure 12.10.   Example of effective and total stress paths leading to cyclic liquefaction for isotropically consolidated fine sand in a cyclic triaxial test (Hedberg, 1977; USNRC, 1985).

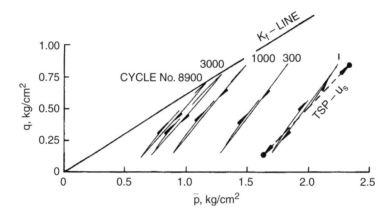

Figure 12.11.   Example of effective and total stress paths leading to cyclic mobility on anisotropically consolidated fine sand during cyclic triaxial compression test (Hedberg, 1977; USNRC, 1985).

If the sample is subjected to anisotropic consolidation, or to a constant shear stress over and above the cycled stress, the stress paths are altered significantly, as shown in Figure 12.11.

Here the cycling takes the sample from the failure line (compression side) away to a non-failure condition so, while the soil will continue to strain, the continuous $\sigma' = 0$ condition is not reached. This would be a cyclic mobility condition.

Laboratory tests show that the number of cycles to cause initial liquefaction is dependent on the cyclic stress ratio τ/σ'_0, the relative density (or void ratio related to the ultimate state void ratio) soil particle size and fabric, the stress conditions, and stress path. Figure 12.12 shows some tests on sand showing the effect of relative density and cyclic stress ratio.

It will be noted that for a given cyclic stress ratio, soils at lower relative density require fewer cycles of loading to achieve initial liquefaction.

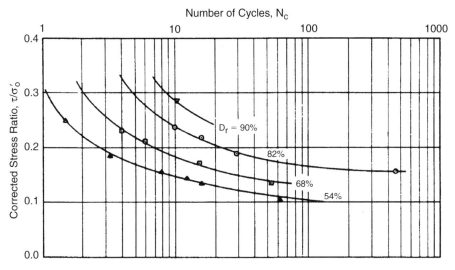

Figure 12.12.   Stress ratio τ/σ_o' versus number of cycles to initial liquefaction from tests on a shaking table (De Alba et al., 1976; USNRC, 1985, reproduced with permission of ASCE).

Table 12.3.   Representative number of cyles of ground motion versus earthquake magnitude (adapted from Seed and DeAlba, 1986).

Earthquake magnitude	Number of representative cycles at $0.65\ \tau_{max}$
8.5	26
7.5	15
6.75	10
6.0	5–6
5.25	2–3

Table 12.3 shows that the number of cycles of ground motion are related to earthquake magnitude, which can be related qualitatively to the effects shown in Figure 12.12 to show that only large magnitude earthquakes will have sufficient cycles or large enough cyclic stress ratio to liquefy medium dense to dense soils.

12.4.1.2.3   Post earthquake behaviour

As shown in Figure 12.13, the effect of cyclic loading is also to induce strain. If, as shown in Figure 12.13(a), the soil is strain weakening, the strain may be sufficient to take the soil to the condition that the static gravitational stresses τ_{ST} exceed the available strength, so straining will continue under gravity loads without further cycling by the earthquake. This is the flow liquefaction condition and large, relatively rapid deformations of the slope will occur. For a strain hardening soil, as shown in Figure 12.13(b) the cyclic loading causes deformations during the cycling, but the static gravitational stresses are less than the steady state undrained strength, S_{us}, so the slope will be stable after the cyclic loading ceases.

For slopes partly of strain softening and partly of strain hardening soil overall instability will occur after the cyclic stresses cease only if after stress redistribution in the strain softening soil, the remaining soil cannot support the gravitational shear stress. Robertson and Wride (1997) suggest a flow slide can only occur if a kinematically admissible mechanism can develop. Whether a flow slide would occur would also depend on the factor of safety

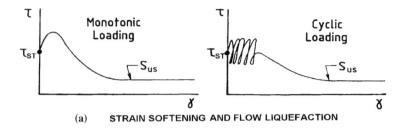

(a)   STRAIN SOFTENING AND FLOW LIQUEFACTION

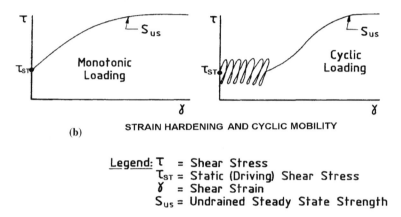

(b)   STRAIN HARDENING AND CYCLIC MOBILITY

Legend: τ = Shear Stress
τ_{ST} = Static (Driving) Shear Stress
γ = Shear Strain
S_{us} = Undrained Steady State Strength

Figure 12.13.   Contractive (a) and dilatent (b) soils in undrained static and cyclic loading.

using S_{us} in the liquefied soil. If it is significantly less than 1.0, large, rapid deformations will occur based simply on the momentum effects.

12.4.1.3   *Suggested flow chart for evaluation of soil liquefaction*

Robertson (1994) and Robertson and Wride (1997) suggest that the flow chart shown in Figure 12.14 be used for the evaluation of soil liquefaction. The following summary of the conditions for flow liquefaction, cyclic liquefaction and cyclic mobility is taken from those papers:

(a) *Flow liquefaction*
  – Applies to strain softening soils only;
  – Requires a strain softening response in undrained loading resulting in constant shear stress and effective stress;
  – Requires in-situ shear stresses greater than the ultimate or minimum undrained shear strength, as illustrated in Figure 12.13;
  – Either monotonic or cyclic loading can trigger flow liquefaction;
  – For failure of a soil structure to occur, such as a slope, a sufficient volume of material must strain soften. The resulting failure can be a slide or a flow depending on the material characteristics and ground geometry. The resulting movements are due to internal causes and can occur after the trigger mechanism occurs;
  – Can occur in any metastable saturated soil, such as very loose granular deposits, very sensitive clays, and loess (silt) deposits.

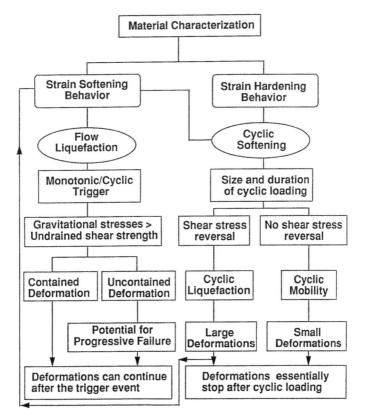

Figure 12.14.   Suggested flow chart for evaluation of soil liquefaction (Robertson, 1994; Robertson and Wride, 1997).

(b) *Cyclic liquefaction*
   - Requires undrained cyclic loading during which shear stress reversal occurs or zero shear stress can develop (i.e. occurs when in-situ static shear stresses are low compared to cyclic shear stresses);
   - Requires sufficient undrained cyclic loading to allow effective stresses to reach essentially zero.
   - At the point of zero effective stress no shear stress exists. When shear stress is applied, pore pressure drops as the material tends to dilate, but a very soft initial stress strain response can develop, resulting in large deformations;
   - Deformations during cyclic loading can accumulate to large values, but generally stabilize when cyclic loading stops. The resulting movements are due to external causes and occur only during the cyclic loading;
   - Can occur in almost all saturated sands provided that the cyclic loading is sufficiently large in magnitude and duration;
   - Clayey soils can experience cyclic liquefaction but deformations are generally small due to the cohesive strength at zero effective stress. Rate effects (creep) often control deformations in cohesive soils.
(c) *Cyclic mobility*
   - Requires undrained cyclic loading during which shear stresses are always greater than zero, i.e. no shear stress reversal develops;
   - Zero effective stress will not develop;

- Deformations during cyclic loading will stabilize, unless the soil is very loose and flow liquefaction is triggered. The resulting movements are due to external causes and occur only during the cyclic loading;
- Can occur in almost any saturated sand provided that the cyclic loading is sufficiently large in magnitude and duration, but no shear stress reversal occurs;
- Cohesive soils can experience cyclic mobility, but rate effects (creep) usually control deformations.

Note that strain softening soils also experience cyclic softening (cyclic liquefaction or cyclic mobility) depending on the ground geometry.

In Figure 12.14 the first step is to evaluate the material characteristics in terms of a strain softening or strain hardening response. If the soil is strain softening, flow liquefaction is possible if the soil can be triggered to collapse and if the gravitational shear stresses are larger than the ultimate or minimum strength. The trigger mechanism can be either monotonic or cyclic. Whether a slope or soil structure will fail and slide will depend on the amount of strain softening soil relative to strain hardening soil within the structure, the brittleness of the strain softening soil and the geometry of the ground. The resulting deformations of a soil structure with both strain softening and strain hardening soils will depend on many factors, such as distribution of soils, ground geometry, amount and type of trigger mechanism, brittleness of the strain softening soil and drainage conditions.

If the soil is strain hardening, flow liquefaction will generally not occur. However, cyclic softening can occur due to cyclic undrained loading, such as earthquake loading. The amount and extent of deformations during cyclic loading will depend on the density of the soil, the magnitude and duration of the cyclic loading and the extent to which shear stress reversal occurs. If extensive shear stress reversal occurs, it is possible for the effective stresses to reach zero and hence, cyclic liquefaction can take place. When the condition of essentially zero effective stress is achieved, large deformations can result. If cyclic loading continues, deformations can progressively increase. If shear stress reversal does not take place, it is generally not possible to reach the condition of zero effective stress and deformations will be smaller, i.e. cyclic mobility will occur.

Both flow liquefaction and cyclic liquefaction can cause very large deformations. Hence it can be very difficult to clearly identify the correct phenomenon based on observed deformations following earthquake loading. Earthquake-induced flow liquefaction movements tend to occur after the cyclic loading ceases, due to the progressive nature of the load redistribution. However, if the soil is sufficiently loose and the static shear stresses are sufficiently large, the earthquake loading may trigger essentially "flow liquefaction" within the first few cycles of loading. Also, if the soil is sufficiently loose, the ultimate undrained strength may be close to zero with an associated effective confining stress very close to zero (Ishihara, 1993). Cyclic liquefaction movements, on the other hand, tend to occur during the cyclic loading since it is the inertial forces that drive the phenomenon.

12.4.2 Soils susceptible to liquefaction

It has long been recognized that saturated sands, silty sands, and gravelly sands are susceptible to liquefaction. Figure 12.15 shows the boundaries suggested in 1985 by USNRC.

For mine tailings, USNRC recognized that soils with high silt and even clay size particles would be liquefiable as shown in Figure 12.16.

Hunter and Fell (2003a and 2003b) gathered data from case studies where flow liquefaction had occurred, mainly under static loading conditions. These would also apply to earthquake loading. Figure 12.17 shows the data separated into classes of slope, with a

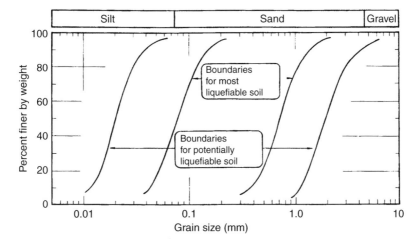

Figure 12.15.   Limits in the particle size gradation curves separating liquefiable and unliquefiable soils as suggested in 1985 by USNRC (USNRC, 1985).

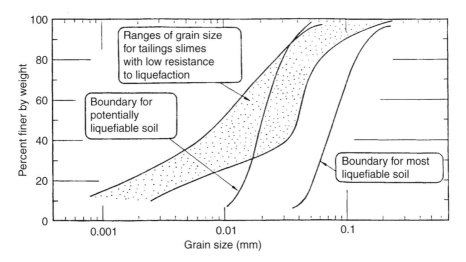

Figure 12.16.   Ranges of particle size gradation curves for mine tailings with low resistance to liquefaction as suggested in 1985 by USNRC (Ishihara, 1985; USNRC, 1985).

suggested range of soils susceptible to flow liquefaction. This data would be more reliable than Figure 12.5.

Seed and Idriss (1982), based on data on liquefaction in China produced by Wang (1979, 1981), recognised that clayey soils could liquefy provided their water content was high (relative to the liquid limit), the soils were not too plastic and not too much clay was present. They proposed that liquefaction can occur only if all three of the following criteria are met:

– The "clay" content (particles smaller than 0.005 mm) is ≤15% by weight;
– The liquid limit is ≤35%;
– The natural moisture content is ≥0.9 times the liquid limit.

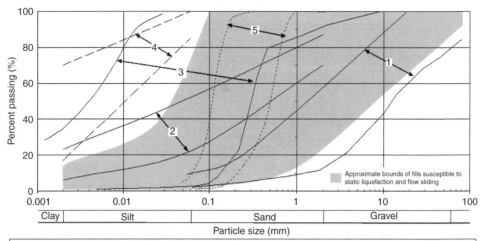

1- Coarse grained coal mine waste (Dawson et al., 1998; Taylor, 1984; Bishop et al., 1969; Hutchinson, 1986)
2- Loose silty sand fills, Hong Kong (upper and lower quartile of pre 1977 fills (HKIE, 1998))
3- Hydraulically placed mine tailings and fills in dam embankments (various published sources)
4- Sensitive clays (indicative limits from: Lefebvre, 1996; Bentley & Smalley, 1984; Mitchell & Markell, 1974; Hutchinson, 1961,1965)
5- Sub-aqueous slopes, natural and fill slopes (Koppejan et al., 1948; Kramer, 1988; Sladen & Hewitt 1989; Comforth et al., 1974)

Figure 12.17.   Particle size gradation of soils susceptible to flow liquefaction under static and earth-quake loading (Hunter and Fell, 2003a,b).

Collins and Tjoumas (2003) point out that the original Wang (1979, 1981) criteria were somewhat different, being that soils which satisfied the following criteria may liquefy:

1. $w_c \geqslant 0.9 w_L$,   where w_c is water content, w_L is liquid limit;
2. $I_L \geqslant 0.75$,   where I_L is Liquidity index, $(w_c - w_P)/I_p$;
3. $q_u \leqslant 50$–$70 \, kN/m^2$, where q_u is unconfined compressive strength;
4. SPT $N \leqslant 4$;
5. Sensitivity $S_t \geqslant 4$.

They also point out that the Seed and Idriss (1982) type criteria should only apply to shallow (less than 5 m) soils and, at higher stresses, the original criteria may be more meaningful. They quote experience from Saluda dam where silty (ML) and sandy (SM) soils outside the criteria were shown to be liquefiable in laboratory tests. They recommend the criteria not be applied to non-plastic ML or SM soils, and suggest that such soils, up to a plasticity index of 7, may be susceptible to liquefaction.

Youd et al. (2001), quoting Robertson and Wride (1998) suggest additional criteria if cone penetration tests are being used.

From the above data it is concluded that:

(a) Soils within the boundaries for flow liquefaction in Figure 12.17 are susceptible to flow liquefaction;
(b) Finer soils, particularly mine tailings, which may have clay size particles which are not clay minerals, and dredged fills may also be subject to liquefaction;
(c) Soils which meet all of the following criteria may be susceptible to liquefaction:
    – The "clay" content (particles smaller than 0.005 mm) is ≤15% by weight;
    – The liquid limit is ≤35%;
    – The natural moisture content is ≥0.9 times the liquid limit.
(d) ML and SM soils, which have a plasticity index less than 7%, may be susceptible to liquefaction, even if not all criteria in (c) are met;

(e) If soils are close to the boundaries of these criteria, and the consequences of liquefaction are great enough, err on the side of caution and/or carry out laboratory tests.

12.4.3   The "simplified procedure" for assessing liquefaction resistance of a soil

12.4.3.1   Background to the simplified method

The most widely accepted, simplest and most practical method of assessing whether there is a potential for liquefaction for horizontal ground conditions was developed originally by Professor H.B. Seed and his co-workers (Seed and Idriss, 1971; Seed, 1979b; Seed and Idriss, 1982; Seed et al., 1985b; Seed and De Alba, 1986). It was reviewed by a panel of experts and reported in USNRC (1985).

The method is semi-empirical and is based on the maximum acceleration induced by the earthquake a_{max}, the SPT "N" value corrected for the SPT hammer energy and for overburden pressure $(N_1)_{60}$, earthquake magnitude (M), and fines content of the soil (% passing 0.075 mm). It is based on recorded cases of liquefaction during earthquakes in USA, Japan and China.

Since then, there has been a gradual development of refinements of the method, including greater application of the cone penetration test, modification to earthquake magnitude corrections and to the corrections for high overburden stress, static shear stresses and age of the soil deposit.

In 1996 a NCEER workshop on the evaluation of Liquefaction Resistance of Soils was attended by 21 experts, who reached a consensus report on the state of the practice at the time. This is reported in NCEER (1997) and Youd et al. (2001).

While there have been some further developments since 1997, they have not significantly altered the understanding of the method, so what is produced here is a summary of the procedure as set out in Youd et al. (2001). Readers are urged to read that paper and to seek later papers to keep up with current practice.

It should be noted that the original method was applicable to level or gently sloping ground, underlain by Holocene age alluvial or fluvial sediments to a shallow depth (<15 m). NCEER (1997) and Youd et al. (2001) say their paper should be applied only to these conditions but give corrections factors for larger overburden stresses and for situations where there are static shear stresses, such as in dams and their foundations.

In reality, the method is used extensively for dams and their foundations, but readers need to be aware of the larger uncertainties which apply in these conditions than for level ground and should seek expert advice for important decision making, particularly in marginal cases.

12.4.3.2   The simplified method – outline

The method requires the calculation or estimation of two variables for evaluation of liquefaction resistance of soils:

(a) The Cyclic Stress Ratio, CSR, which is a measure of the cyclic load applied to the soil by the earthquake;
(b) The Cycle Resistance Ratio, CRR, which is the capacity of the soil to resist liquefaction.

The CRR is estimated from Standard Penetration Tests (SPT). Cone Penetration Tests (CPT) or, less frequently, the shear wave velocity.

If the CSR is greater than the CRR, liquefaction is likely to occur.

For most purposes a simple comparison of CSR with CRR with engineering judgement will be sufficient. However there are methods available which allow for the uncertainty in the boundary between liquefiable and non liquefiable soils. These include Liao, Veneziano and Whitman (1988) and Youd and Noble (1997). The latter indicate that the base curve in Figure 12.19 represents a probability of liquefaction of between 20% and 50% for SPT $(N_1)_{60}$ values between 5 and 25. That is, to have a CRR greater than CSR is not a guarantee of no liquefaction.

It is emphasised that the methods need to be used with reasonable engineering judgement and should not be regarded as giving precise outcomes.

12.4.3.3   Evaluation of Cyclic Stress Ratio (CSR)

The Cyclic Stress Ratio, CSR, is calculated from:

$$CSR = (\tau_{av}/\sigma'_{vo}) = 0.65(a_{max}/g)(\sigma_{vo}/\sigma'_{vo})r_d \qquad (12.10)$$

where a_{max} = peak horizontal acceleration at the ground surface generated by the earthquake; g = acceleration of gravity; σ_{vo} and σ'_{vo} are total and effective vertical overburden stresses, respectively and r_d = stress reduction coefficient, which can be calculated from:

$$r_d = 1.0 - 0.00765z \qquad \text{for } z \leq 9.15 \text{ m} \qquad (12.11a)$$

$$r_d = 1.174 - 0.0267z \qquad \text{for } 9.15m < z \leq 23m \qquad (12.11b)$$

where z = depth below ground surface in metres.

It should be recognized that there is some variability in r_d, as shown in Figure 12.18. This uncertainty is greater at depth. The peak horizontal acceleration at the ground surface should be calculated allowing for amplification or de-amplification of the peak horizontal bedrock acceleration estimated in the absence of increased pore pressure or the onset of liquefaction. The peak horizontal bedrock acceleration is the geometric mean of the two horizontal components (N-S and E-W) where they are available.

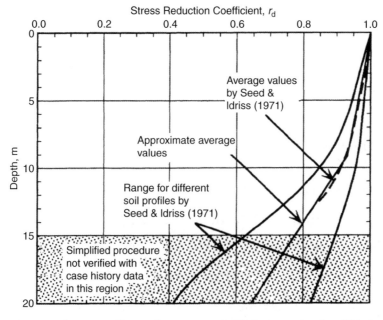

Figure 12.18.   Stress reduction coefficient factor r_d versus depth curves developed by Seed and Idriss (1971) with added mean-value lines plotted from Eq. 12.11 (Youd et al., 2001, reproduced with permission of ASCE).

To avoid bias in the analysis from short duration, high frequency acceleration estimated from near-field small magnitude earthquakes, the peak horizontal acceleration should be based on the influence of earthquakes Magnitude 5 or greater.

12.4.3.4   Evaluation of Cyclic Resistance Ratio from the Standard PenetrationTests

The CRR for M7.5 earthquakes can be estimated from the curves in Figure 12.19. It is recommended that the SPT Clean Sand Base Curve be used, with correction for fines content as detailed below.

In this figure, $(N_1)_{60}$ is the SPT blow count normalised to an effective overburden pressure of 100 kPa, a hammer energy of 60%, borehole diameter, rod length and sampling method. $(N_1)_{60}$ is given by:

$$(N_1)_{60} = N_m C_N C_E C_B C_R C_S \tag{12.12}$$

where N_m = measured standard penetration resistance; C_N = factor to normalize N_m to a common reference effective overburden stress; C_E = correction for hammer energy ratio

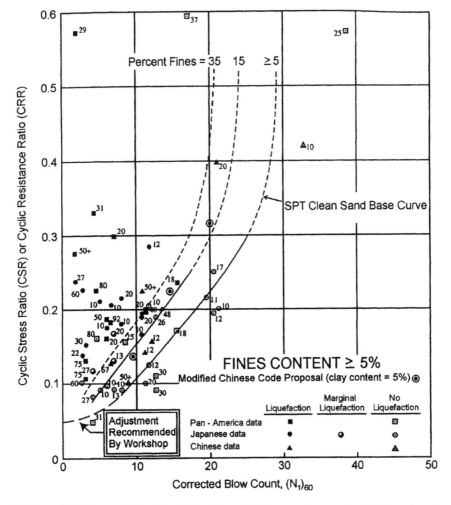

Figure 12.19.   SPT Clean-sand base Cyclic Resistance Ratio curve for magnitude 7.5 earthquakes with data from liquefaction case histories (Youd et al., 2001; modified from Seed et al., 1985b, reproduced with permission of ASCE).

(ER); C_B = correction factor for borehole diameter; C_R = correction factor for rod length; C_S = correction for samplers with or without liners.

C_N should be calculated from:

$$C_N = 2.2/(1.2 + \sigma'_{vo}/P_a) \qquad (12.13)$$

where σ'_{vo} = effective vertical stress at the time of doing the SPT test (kPa); P_a = atmospheric pressure (100 kPa). Youd et al. (2001) note that there is some uncertainty in C_N, and that equation 12.13 only applies for $\sigma'_{vo} < 300$ kPa.

The hammer energy ratio C_E varies between hammer types and for individual hammers and drill rig set ups. It varies between 0.5 and 1.0 for rope and pulley donut hammer, 0.7 and 1.2 for rope and pulley US Safety hammers and 0.8 and 1.3 for USA Safety hammers with automatic trip/free fall (Youd et al., 2001). Tests on Australian free fall SPT hammers at Hume Dam surprisingly gave low C_E of about 0.6. It is clear that C_E should be measured for each project. If it is not, conservative assumptions will have to be made which may prove costly in overly conservative assessments of liquefaction potential.

Correction for borehole diameter (C_B) is seldom needed. $C_B = 1.0$ for holes 65–125 mm diameter. $C_B = 1.05$ for holes 150 mm diameter and $C_B = 1.15$ for holes of 200 mm diameter. Correction for rod length, C_R is necessary for rod length <3 m, where $C_R = 0.75$. No other corrections are needed as they were not applied to develop the database.

SPT tests should be done with the split inner tubes in place, in which case $C_S = 1.0$. If the inner tube was left out, $C_S = 1.1$ to 1.3.

Correction for fines content, FC, (% passing 0.075 mm sieve) should be calculated from:

$$(N_1)_{60CS} = \alpha + \beta(N_1)_{60} \qquad (12.14)$$

where α and β = coefficients determined from the following relationships:

$$\alpha = 0 \qquad \text{for FC} \leq 5\% \qquad (12.15a)$$

$$\alpha = \exp[1.76 - 190/FC^2] \qquad \text{for } 5\% < \text{FC} < 35\% \qquad (12.15b)$$

$$\alpha = 5.0 \qquad \text{for FC} \geq 35\% \qquad (12.15c)$$

$$\beta = 1.0 \qquad \text{for FC} \leq 5\% \qquad (12.16a)$$

$$\beta = [0.99 + (FC^{1.5}/1{,}000)] \qquad \text{for } 5\% < \text{FC} < 35\% \qquad (12.16b)$$

$$\beta = 1.2 \qquad \text{for FC} \geq 35\% \qquad (12.16c)$$

It should be recognized that Figure 12.19 was originally developed by Seed et al. (1985b) from Figure 12.20, which shows three curves showing different cyclic strains τ_l. The 3% line is equivalent to the SPT clean sand base curve in Figure 12.19.

Sites which are just to the left of the 3% strain curve are likely to experience cyclic mobility or cyclic liquefaction; those to the left of the 20% are likely to experience flow liquefaction conditions.

Baziar and Dobry (1995), Ishihara (1993) and Cubrinovski and Ishihara (2000a, b) have further investigated the boundary of flow liquefaction conditions. Figure 12.21 summarizes the outcomes. Readers should read the original papers before applying this figure to serious decision making.

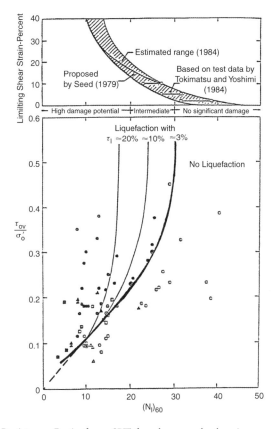

Figure 12.20.   Cyclic Resistance Ratio from SPT for clean sands showing curves of cyclic strain (Seed et al., 1985b, reproduced with permission of ASCE).

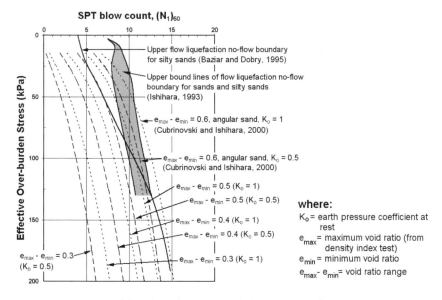

Figure 12.21.   Comparison of flow liquefaction boundaries in terms of SPT $(N_1)_{60}$ for sands and silty sands from monotonic laboratory undrained tests and earthquake triggered field case (Hunter and Fell, 2003a,b).

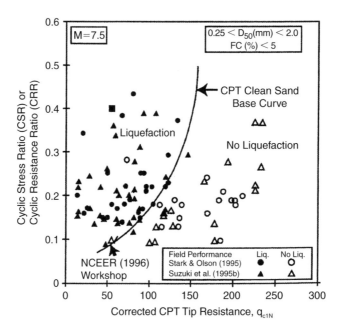

Figure 12.22.   CPT clean-sand base Cyclic Resistance Ratio curve for magnitude 7.5 earthquakes with data from compiled case histories (Youd et al., 2001, reproduced from Robertson and Wride, 1998, reproduced with permission of ASCE).

12.4.3.5   *Evaluation of the Cyclic Resistance Ratio for M7.5 earthquake (CRR7.5) from Cone Penetration Tests*

The CRR from M7.5 earthquakes can be estimated from the curve in Figure 12.22. In this figure q_{c1N} is the normalized dimensionless cone penetration resistance calculated from:

$$q_{c1N} = C_Q(q_c/P_a) \qquad (12.17a)$$

where

$$C_Q = (P_a/\sigma'_{vo})^n \qquad (12.17b)$$

where C_Q = normalizing factor for cone penetration resistance; σ'_{vo} = effective vertical stress at the time of doing the CPT test (kPa); P_a = atmospheric pressure (100 kPa); n = exponent that varies with soil type; and q_c = field cone penetration resistance measured at the tip in kPa.

The exponent n varies between 0.5 and 1.0 and can be estimated using the soil behaviour type index I_c as follows (Youd et al., 2001; from Robertson and Wride (1998):

(a) Assume $n = 1.0$ (characteristic of clays) and calculate the dimensionless tip resistance Q from:

$$Q = [(q_c - \sigma_{vo})/P_a][P_a/\sigma'_{vo}]^n = [(q_c - \sigma_{vo})/\sigma'_{vo}] \qquad (12.18)$$

and the Normalized Friction Ratio F, from

$$F = [fs/(q_c - \sigma_{vo})]100\% \qquad (12.19)$$

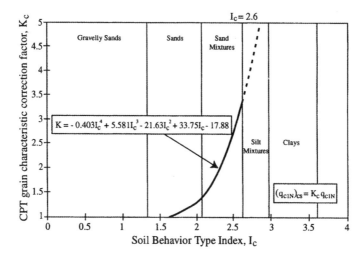

Figure 12.23.   Grain-characteristic correction factor K_c for determination of clean-sand equivalent CPT resistance (Youd et al., 2001; reproduced from Robertson and Wride, 1998, reproduced with permission of ASCE).

where fs = CPT sleeve resistance (kPa); σ_{vo} = total vertical stress at test level (kPa) and q_c is as defined above.

(b) Calculate I_c from Equation 12.20.

$$I_c = [(3.47 - \log Q)^2 + (1.22 + \log F^2)]^{0.5} \tag{12.20}$$

(c) If $I_c > 2.6$, the soil is classified as clayey, so is unlikely to liquefy. Check this using the criteria detailed in Section 12.4.2;

(d) If $I_c < 2.6$, the soil is likely to be granular, so recalculate c_Q, Q, and I_c using n = 0.5; If the recalculated $I_c < 2.6$, the soil is nonplastic and granular. This I_c is used to calculate liquefaction resistance.

   If the recalculated $I_c > 2.6$, the soil is likely to be silty, possible plastic, so q_{c1N} and I_c should be recalculated using n = 0.7, and this value of I_c used to calculate liquefaction resistance. Check if the soil is liquefiable as described in Section 12.4.2.

The correction to an equivalent clean sand value $(q_{c1N})_{cs}$ is done by:

$$(q_{c1N})_{cs} = K_c q_{c1N} \tag{12.21}$$

where K_c is the correction factor for grain characteristics, is defined by the following equations (Robertson and Wride, 1998):

$$\text{for } I_c \leq 1.64 \quad K_c = 1.0 \tag{12.22a}$$

$$\text{for } I_c > 1.64 \quad K_c = -0.403 \, I_c^4 + 5.581 \, I_c^3 - 21.63 \, I_c^2 + 33.75 \, I_c - 17.88 \tag{12.22b}$$

   The K_c curve defined by Equation 12.22b is plotted in Figure 12.23. For $I_c > 2.6$, the curve is shown as a dashed line, indicating that soils in this range of I_c are most likely too clay-rich or plastic to liquefy.

   With an appropriate I_c and K_c, calculate $(q_{c1N})_{cs}$ from Equation 12.21 and Figure 12.22 can then be used to calculate $CRR_{7.5}$.

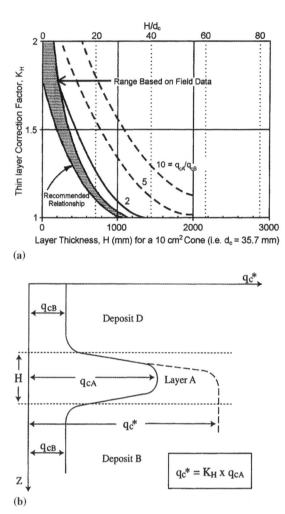

Figure 12.24.   Thin-layer correction factor K_H for determination of equivalent thick-layer CPT resistance (Youd et al., 2001; modified from Vreugdenhill et al., 1994; Robertson and Fear,
1995, reproduced with permission of ASCE).

Theoretical as well as laboratory studies indicate that CPT tip resistance is influenced
by softer soil layers above or below the cone tip. As a result, measured CPT tip resistance
is smaller in thin layers of granular soils sandwiched between softer layers than in thicker
layers of the same granular soil. The amount of the reduction of penetration resistance in
soft layers is a function of the thickness of the softer layer and the stiffness of the stiffer
layers.

Vreugdenhil et al. (1994) first studied this, and Robertson and Fear (1995) and Youd
et al. (2001) have modified that analysis to recommend the correction factor shown in
Figure 12.24.

In this figure, H = thickness of the interbedded layer in mm; q_{cA} and q_{cB} = cone resist-
ances of the stiff and soft layers, respectively; and d_c = diameter of the cone in mm.

As for the standard penetration test, the CPT clean sand base curve in Figure 12.22 is
for 3% cyclic strain. Figure 12.25 shows the curves for 3%, 10% and 20% cyclic strain.
For sites just to the left of the 3% curve, cyclic liquefaction or cyclic mobility are more
likely than flow liquefaction. Flow liquefaction is likely left of the 20% curve.

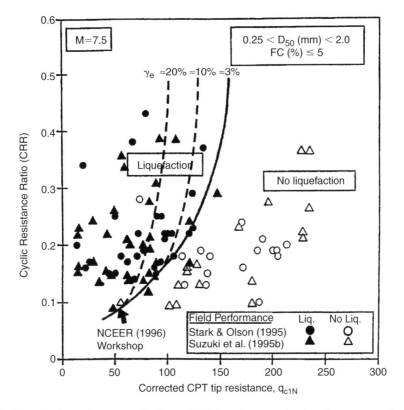

Figure 12.25.   Cyclic resistance ratio from CPT for clean sands showing curves of cyclic strain (Robertson and Wride, 1998).

12.4.3.6   Evaluation of Cyclic Resistance Ratio for M7.5 earthquake (CRR$_{7.5}$) from shear wave velocity

Shear wave velocity Vs, obtained from surface to downhole or by crosshole testing, can be used to assess the CRR. This is particularly useful where the potentially liquefiable soils are gravelly, and the gravel particles affect the SPT "N" value, and prevent the penetration of a CPT.

However the test may have difficulty locating thin, low versus strata. The authors' experience is also that there is often not clear discrimination using this test and it is recommended that it be used only in conjunction with SPT and/or CPT data. The CRR for M7.5 earthquake can be estimated from the curve in Figure 12.26.

In this figure Vs is corrected to an effective overburden stress of 100 kPa using:

$$V_{s1} = V_s \left(\frac{P_a}{\sigma'_{vo}} \right)^{0.25} \tag{12.23}$$

where V_{s1} = overburden-stress corrected shear wave velocity, m/sec; P_a = atmospheric pressure, 100 kPa; and σ'_{vo} = initial effective vertical stress in kPa.

The recommended lines in Figure 12.26 are dashed above CRR$_{7.5}$ of 0.35, to indicate field data is limited in this range, and do not extend below V_{s1} of 100 m/sec because there is a lack of data.

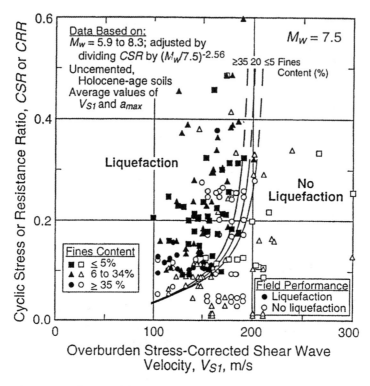

Figure 12.26.   Shear wave clean sand base cyclic resistance ratio curve for magnitude 7.5 earthquake with data from compiled case histories (Youd et al., 2001, from Andrus and Stokoe, 2000, reproduced with permission of ASCE).

12.4.3.7 *Earthquake magnitude scaling factors and factor of safety against liquefaction*

The procedures outlined in Sections 12.4.3.4 to 12.4.3.6 give $CRR_{7.5}$, Cyclic Resistance Ratios for M7.5 earthquakes.

Smaller magnitude earthquakes giving the same peak horizontal acceleration are less likely to initiate liquefaction, because the earthquake will have fewer cycles of motion. Larger magnitude earthquakes are more likely to initiate liquefaction. This is allowed for by using a magnitude scaling factor (MSF) in the equation for factor of safety (FS) against liquefaction:

$$FS = (CRR_{7.5}/CSR)MSF \qquad (12.24)$$

where CSR = calculated cyclic stress ratio generated by the earthquake shaking; and $CRR_{7.5}$ = Cyclic Resistance Ratio for magnitude 7.5 earthquakes.

Youd et al. (2001) present a discussion on MSF and conclude that the original Seed and Idriss (1982) factors are too conservative for M < 7.5. They recommend MSF in the range shown in Figure 12.27 for M > 7.5.

Liu et al. (2001) publish some additional information relating to MSF. Their studies show MSF lower than the NCEER workshop values for M < 7.5. Given this, it would seem prudent to use the lower bound of the NCEER workshop values in Figure 12.27, and, for critical studies, consider lower values after reviewing Liu et al. (2001) and the most recent literature.

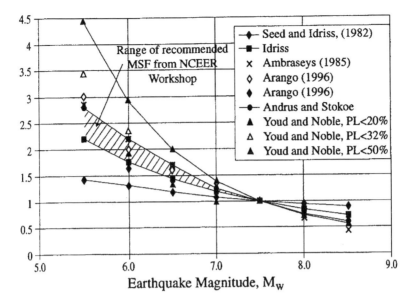

Figure 12.27.   Magnitude scaling factors (MSF). Earthquake use "range of recommended MSF from NCEER workshop" for M < 7.5; and Idriss for M > 7.5 (Youd et al., 2001; from Youd and Noble, 1997, reproduced with permission of ASCE).

It should be noted that there are no guidelines given in Youd et al. (2001) regarding an acceptable factor of safety against cyclic liquefaction using Equation 12.24.

12.4.3.8   *Corrections for high overburden stresses, static shear stresses and age of deposits*

The method of assessment of liquefaction potential described in Sections 12.4.3.2 to 12.4.3.7 is for horizontal or gently sloping ground and for depths less than about 15 m.

For assessments of liquefaction for embankment dams, the confining stresses may be higher than this and the dam will impose static shear stresses, which may alter the lique-faction potential.

Seed (1983) first proposed the factors K_σ and K_α to allow for the high overburden stresses and static shear stress respectively. They are applied by extending equation 12.24 to:

$$FS = (CRR_{7.5}/CSR).MSF.\,K_\sigma.\,K_\alpha \qquad (12.25)$$

K_σ and K_α can be estimated as follows (Youd et al., 2001):

(a) *Overburden stress factor K_σ* The overburden stress factor K_σ for $\sigma'_{vo} > \approx 100\,kPa$ in clean and silty sands, and sandy gravels/gravelly sands can be estimated from Figure 12.28.

Note that the vertical stress is normalised to atmospheric pressure, so $\sigma'_{vo} = 300\,kPa = 3$.

It should be noted that there is considerable scatter in the input data to derive these curves. They have been described as "minimal, or conservative estimates" by Hynes and Olsen (1999) and Youd et al. (2001), but considerable care should be taken in their application.

(b) *Sloping ground correction factor K_α* The NCEER workshop participants concluded that "although curves relating K_α to α have been published (Harder and Boulanger,

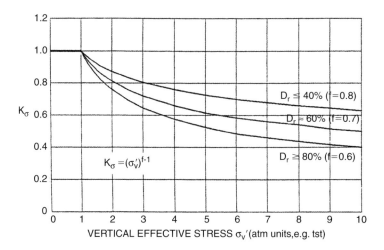

Figure 12.28.   Overburden stress factor K_σ versus vertical effective stress (Youd et al., 2001, reproduced with permission of ASCE).

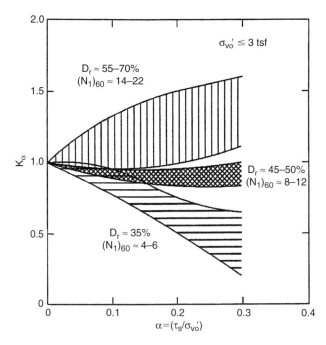

Figure 12.29.   Sloping ground correction factor K_α suggested by Harder and Boulanger (1997).

1997), these curves should not be used by nonspecialists in geotechnical earthquake engineering or in routine engineering practice".

Harder and Boulanger (1997) suggested the use of Figure 12.30 for soils with an effective confining stress less than 300 kPa. They point out that at significantly higher confining stresses, sandy soils will be more contractive and low K_α may be appropriate. For high risk and critical projects on sloping ground or where there are high initial static shear stresses, they recommend site-specific laboratory testing using high quality samples.

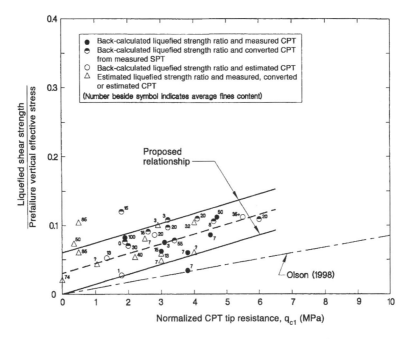

Figure 12.30.   Liquefied shear strength ratio $S_{u(LIQ)}/\sigma'_{vo}$ based on normalized cone penetration test q_{c1} (Olsen and Stark, 2002).

In Figure 12.29:

$$\alpha = \tau_s/\sigma'_{vo} \qquad\qquad (12.26)$$

where τ_s = static shear stress; σ'_{vo} = effective vertical stress.

It is clear there is considerable uncertainty in the estimation of K_α. It seems that K_α will be less than 1.0 (meaning liquefaction is more likely) for relative densities lower than about 50%, so, marginal cases should be assumed liquefiable if there are existing shear stresses, or laboratory tests and analysis of stresses should be carried out to assess the effects.

Idriss and Boulanger (2003) and Boulanger (2003a, 2003b) present later information and readers should review the latest literature.

12.5   POST LIQUEFACTION SHEAR STRENGTH AND STABILITY ANALYSIS

12.5.1   *Some general principles*

As shown in Figure 12.7, triaxial tests on saturated sands which display strain weakening behaviour, i.e. those susceptible to flow liquefaction, reach a steady state undrained strength, S_{us}. This is also referred to in the literature as the undrained steady state strength (Poulos et al., 1985) and undrained residual strength (Seed, 1987).

In field situations, such soils may experience some pore pressure redistribution and the equivalent term is "liquefied shear strength $S_{u(LIQ)}$", (NSFW, 1998; Olsen and Stark, 2002).

To check the stability of a dam after the earthquake motion has ceased, a limit equilibrium stability analysis is carried out with zones which have been assessed as liquefied under the earthquake loading and which are subject to flow liquefaction, assigned the steady state undrained strength, S_{us}, or liquefied shear strength, $S_{u(LIQ)}$, depending on how the strength is assessed.

Saturated zones of the dam or foundation which are subject to limited strain softening (temporary liquefaction), or even strain hardening, will experience pore pressure build up in the earthquake which should be modelled in this "post earthquake stability analysis".

If the factor of safety is greater than 1.0 in these conditions, the slope will have experienced deformations during the earthquake, but no further deformation should occur. If however the factor of safety is less than 1.0, a flow slide will occur, with large, rapid deformations.

In 1997, a National Science Foundation Workshop was held on "Shear Strength of Liquefied Soils" (NSFW, 1998). The main objectives of the workshop included seeking consensus on practice related issues concerning the shear strength of liquefied soils. Unfortunately, unlike the NCEER workshop the year before (reported in Youd et al., 2001), the participants did not reach consensus on important issues. The proceedings are however a valuable reference. For the purposes of this chapter, the authors have relied upon those proceedings, a later paper by Olsen and Stark (2002), Fell et al. (2000) and references on the topic referred to in that paper.

As discussed in Fell et al. (2000), the steady state undrained strength, S_{us}, of a soil is dependent on the void ratio, confining stresses, stress path of loading and soil fabric. The use of results from triaxial compression tests alone may be unconservative, because S_{us} in triaxial extension and direct simple shear may be lower.

Poulos et al. (1985) developed a laboratory procedure using triaxial compression tests on reconstituted specimens. However later studies, summarized in NSFW (1998), question the validity of this. NSFW (1998) recommend for high risk or high consequence of failure projects, laboratory testing using undisturbed samples obtained by ground freezing, but could not reach a consensus on test procedures.

There was a consensus that empirical corrections relating $S_{u(LIQ)}$ to back-analysed field failures represented an economical and reasonable means for estimating the liquefied shear strength. This has been the authors' experience and hence these procedures are given in some detail below.

If the outcomes of such analyses are marginal, and the consequences of failure are sufficiently high, it is recommended the services of an expert in the area be obtained to advise further, particularly if laboratory testing is being contemplated. It would seem that simplified laboratory procedures e.g. triaxial compression tests on reconstituted samples, have little merit.

12.5.2   Method for estimating $S_{u(LIQ)}$

The authors have for some time used the method of Stark and Mesri (1992) to assess the steady state undrained strength, S_{us}, for use in post earthquake analysis. This method uses the Standard Penetration Test. It requires adjustment of the SPT $(N_1)_{60}$ values for fines content (% passing 0.075 mm) to give $(N_1)_{60cs}$, or $(N_1)_{60}$ clean sand equivalent. Stark and Mesri (1992) back analysed failures, but also gathered a lot of laboratory test data and seem to have relied on the latter rather than the back analysis to reach their recommended S_{us}.

NSFW (1998) question the use of fines content corrections and recommend momentum effects be accounted for in the analysis of case studies.

Olsen and Stark (2003) have followed this procedure in the analysis of 33 case studies of flow liquefaction (excluding lateral spreading as recommended by NSFW, 1998). They demonstrate that it appears that the Stark and Mesri (1992) recommendation was generally conservative.

Given this, it is recommended that the Olsen and Stark (2003) method be adopted. This uses $(N_1)_{60}$ and q_{c1} values (for SPT and CPT), corrected to 100 kPa overburden stress, but

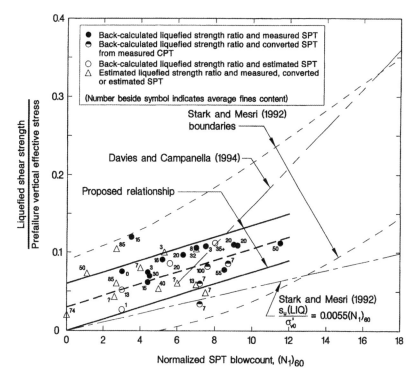

Figure 12.31.   Liquefied shear strength ratio $Su_{(LIQ)}/\sigma'_{vo}$ based on normalized standard penetration test $(N_1)_{60}$, (Olsen and Stark, 2002).

with no fines content correction. It is also based on the mean $(N_1)_{60}$ or q_{c1} for the liquefied zone, not the minimum.

Figures 12.30 and 12.31 show the back-analysed liquefied shear strength ratio $(S_{u(LIQ)}/\sigma'_{vo})$ where σ'_{vo} is the pre-failure vertical effective stress for CPT and SPT.

They indicate that the average trend lines which are described by:

$$S_{u(LIQ)}/\sigma'_{vo} = 0.03 + 0.0143\,(q_{ci}) \pm 0.03 \qquad \text{for } q_{c1} < 6.5 \text{ mPa} \qquad (12.27)$$

and

$$S_{u(LIQ)}/\sigma'_{vo} = 0.03 + 0.0075\,((N_1)_{60}) \pm 0.03 \qquad \text{for } (N_1)_{60} < 12 \qquad (12.28)$$

can be used. They prefer the use of CPT because of the continuous profile given by the test. The authors prefer to use a mix of CPT and SPT, as a check on each other. The SPT also allows recovery of a sample for classification.

NSFW (1998) caution against the use of $S_{u(LIQ)}/\sigma'_{vo}$ ratios, particularly for clean sands. This is particularly important if it is planned to use a berm to improve the post earthquake stability of a dam. Olsen and Starke (2003) indicate that, however in their view, at least for silty sands with $>12\%$ fines, it is reasonable to allow for the increase in $S_{u(LIQ)}$ which would be indicated by the increase in σ'_{vo} from the berm. For high risk projects they suggest laboratory consolidation tests be carried out to confirm that it is parallel to the steady state line which is implicit in the assumption that $S_{u(LIQ)}/\sigma'_{vo}$ is constant. Based on NSFW (1998) it would appear unwise to allow for the strength increase from the increased effective

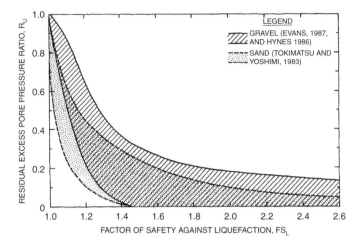

Figure 12.32.   Method for approximate estimation of the residual excess pore pressure R_u in non liquefied zones (Marcuson et al., 1990).

stress of the berm for clean sands, and the $S_{u(LIQ)}$ obtained without the berm should be
assumed to apply after the berm is built, unless sampling and laboratory testing indicates
otherwise.

Whether the mean lines or upper and lower lines are used for design, is for designers to
assess, based on the amount and quality of data available and the consequences of failure.
It would be wise to err somewhat on the conservative side given the uncertainty in the estimated strengths.

12.5.3   Estimation of pore pressure increase due to cyclic loading

Saturated soils which are potentially liquefiable, but which have not been subject to sufficient cyclic shear stress to reach a liquefaction condition, will develop positive pore pressures during the cyclic loading of the earthquake. The amount of positive pore pressure
will depend on many factors relating to the soil and the cyclic loading.

For assessment of the post earthquake factor of safety the pore pressure in these zones can
be estimated from Figure 12.32, which relates the residual excess pore pressure ratio R_u,
at the end of the earthquake loading to the factor of safety against liquefaction, FS calculated from Equation 12.24.

This procedure is used by USBR and US Corps of Engineers. For more critical projects,
carefully planned and executed cyclic laboratory tests should be used.

12.5.4   Post liquefaction limit equilibrium stability and deformation analysis

The post liquefaction stability is assessed as follows:

(a) Determine the zones which have liquefied (i.e. FS < 1.0 in Equation 12.24) under the
     earthquake loading using the methods in Section 12.4.3;
(b) Determine the liquefied shear strength $(S_{u(LIQ)})$ for these zones using the methods
     described in Sections 12.5.1 and 12.5.2;
(c) For potentially liquefiable soils with a factor of safety against liquefaction greater
     than 1.0, determine the residual excess pore pressure as detailed in Sections 12.5.3;
(d) For clay soil zones in the embankment and foundation assign a strength and pore pres-
     sure consistent with the soil's behaviour in static loading after being cracked and dis-

turbed by the earthquake. If the clay is contractive in nature, use undrained strengths. If it is dilative on shearing, use effective stress strengths c', ϕ'. Usually there will be some cracking and loosening, and if so fully softened strengths, would apply (e.g. $c' = 0$, $\phi' = \phi'_{peak}$). Some apply an arbitrary 10% or 15% loss of strength;

(e)   For well compacted free draining rockfill filters and dense sands and gravels adopt the effective stress strengths c', ϕ', with no change in the pore pressures. If large deformations are expected in the earthquake the dense granular materials may have loosened and will have a strength approaching the critical state strength, rather than the peak strength. In practice this can be accommodated by a small reduction from the expected peak strengths.

The analysis is done with conventional limit equilibrium analysis methods. No loading from the earthquake is applied, since this is a post earthquake analysis.

If the liquefied zone is subject to flow liquefaction and the post earthquake factor of safety is significantly less than 1.0, large, rapid deformations and flow sliding can be expected. If the factor of safety is only marginally less than 1.0, deformations may not be so large as to lose freeboard between the dam crest and the reservoir level. An approximate estimate of the deformations can be obtained using the Khalili et al. (1996) method which is detailed in Chapter 11. For more accurate methods of assessing the deformations numerical analysis using the post earthquake strengths can be used.

Indicative estimates of deformations can be obtained by performing a static deformation analysis which incorporates the earthquake induced pore pressures and the residual strength of the liquefied soils (Finn, 1993). The analysis is often performed in two stages. In the first stage, the numerical model is initialised to the pre-earthquake conditions of the dam by simulating the current in-situ stresses. Then, in the second stage, the earthquake induced pore pressures and residual strengths of the liquefied soils are incorporated into the model to simulate post-liquefaction conditions.

This type of analysis is also referred to as uncoupled deformation analysis and generally leads to conservative estimates of post liquefaction deformations, as it does not allow for dissipation of earthquake induced pore pressures with time. More accurate estimates of post liquefaction deformations can be obtained using fully and semi-coupled methods of analysis, as discussed in the following sections.

12.6   SEISMIC STABILITY ANALYSIS OF EMBANKMENTS

12.6.1   *Preamble*

The following discussion on seismic stability analysis of embankments has been adapted from ANCOLD (1998). The contribution of N. Khalili and other members of the ANCOLD working group is acknowledged.

The methods of analysis currently used in practice to evaluate seismic stability of embankment dams vary widely, ranging from simple limit equilibrium type analyses to highly sophisticated numerical modelling techniques. These include:

– Pseudo-static analysis;
– Simplified methods of deformation analysis;
– Numerical modelling techniques:
  (i) total stress;
  (ii) effective stress.

The simplified methods of analyses, including pseudo-static and post liquefaction, rely heavily on the lessons learnt from the performance of dams during past earthquakes. Major reviews of past performance have been conducted by Sherard (1967), Sherard et al.

(1974), Seed (1979b, 1983) and Seed et al. (1978, 1985b). These studies show that based on the data available at that time:

- Even at short distances from the epicentres, there have been no complete failures of embankments built of clay soils, but several dams have come close to failure;
- Well constructed dams of clay soils on clay or rock foundation, not susceptible to strain weakening, can withstand extremely strong shaking resulting from earthquakes of up to Magnitude 8.25 with peak ground accelerations ranging from 0.35 g to 0.8 g; (They suffer cracking but few, if any, have failed catastrophically.)
- Dams which have suffered complete failure as a result of earthquake shaking have been constructed primarily with saturated sandy materials or on saturated sand foundations. Liquefaction is a major contributory factor in these failures. The dams most susceptible to failure under earthquake loading are hydraulic fill dams and tailings dams constructed using upstream methods, because they are susceptible to liquefaction if the fill or tailings are granular and saturated;
- In dams constructed of saturated cohesionless soils the primary cause of damage or failure is the build up of pore water pressure (liquefaction) under the earthquake loading;
- There are very few cases of dam failures during the earthquake shaking. Most of the failures occur from a few minutes up to twenty-four hours after the earthquake. However, cracking and displacements do occur during the earthquake.

Based on the above and similar observations, the US Bureau of Reclamation (1989) classifies embankment dams into two main categories: (1) not susceptible to liquefaction and (2) susceptible to liquefaction. For the purposes of seismic stability assessment they also identify two types of analyses: deformation analysis and post earthquake liquefaction analysis. They recommend that deformation analysis be performed on dams not susceptible to liquefaction and post earthquake liquefaction analysis be performed on dams susceptible to liquefaction. The authors recommend that a similar approach be adopted but add the proviso that, where significant strain weaking of non liquefiable soils such as overconsolidated high clay content soils may occur due to displacements induced by the earthquake, post earthquake stability should be analysed taking account of the strain weakening. The authors also emphasise that the post earthquake piping risk must not be ignored as it may be a critical condition.

12.6.2   Pseudo-static analysis

Up until the 1970s, the pseudo-static analysis was the standard method of stability assessment for embankment dams under earthquake loading. The approach involved a conventional limit equilibrium stability analysis, incorporating a horizontal inertia force to represent the effects of earthquake loading. The inertia force was often expressed as a product of a seismic coefficient "k" and the weight of the sliding mass W. The larger the inertia force, the smaller the safety factor under the seismic conditions. In this approach a factor of safety (FOS) of <1 implies failure, whereas FOS >1 represents seismically safe conditions.

The seismic coefficients used in this approach were typically less than 0.2 and were related to the relative seismic activity in the areas to which they apply.

The US Corps of Engineers (1984b) used the basic pseudo-static method for dams not susceptible to liquefaction. They recommended use of a seismic coefficient equal to one-half of peak ground acceleration and the use of undrained conditions for cohesive soils and drained conditions for free draining granular materials, with a 20 percent strength reduction to allow for strain weakening during the earthquake loading. They required a factor of safety greater than 1.0. If a dam failed to satisfy this, they recommended more accurate and detailed analyses. Their approach has been calibrated against a large number of deformation analyses and they state that up to 1 m of deformation may occur.

The pseudo-static method of analysis, despite its earlier popularity, was based on a number of restrictive assumptions. For instance, it assumed that the seismic coefficient acting on the potential unstable mass is permanent and in one direction only. In reality, earthquake accelerations are cyclic, with direction reversals. Also, the concept of failure used in the approach was influenced by that used in static problems where a factor of safety of less than one cannot be permitted, as the stresses producing this state will exist until large deformations change the geometry of the structure. However, under seismic conditions, it may be possible to allow the FOS to drop below one, as this state exists only for a short time. During this time, earthquake induced inertia forces cause the potentially unstable masses to move down the slope. However before significant movement takes place, the direction of the earthquake loading is reversed and the movement of the soil masses stop as, once again, the FOS rises above one. In fact, experience shows that a slope may remain stable despite having a calculated FOS less than one and, on the other hand, it may fail at FOS >1, depending on the dynamic characteristics of the slope-forming material.

The authors have in the past recommended the US Corps of Engineers (1984) method as a screening method to select dams which need more detailed assessment. We are however now of the view that screening is best done using the simplified deformation analyses described below, and do not believe pseudo-static analyses are useful, because they do not model the actual load condition.

12.6.3   Simplified methods of deformation analysis for dams where liquefaction and significant strain weakening do not occur

There are a number of approaches for estimating the deformations of a dam which may occur during an earthquake. These include:

(a) Empirical methods based on recorded deformations, dam geometry and earthquake loading e.g. Swaisgood (1998), Pells and Fell (2002, 2003) extended this to include an empirical method to assess whether cracking would occur;
(b) Integration of the displacements which occur when the earthquake loading exceeds the available strength e.g. Newmark (1965) and developments of that approach using simplified numerical analyses programs such as SHAKE;
(c) Developments of the Newmark (1965) approach to allow for dynamic response of the embankment e.g. Makdisi and Seed (1978).

The following gives an outline of these methods. Readers will need to refer to the references given, or engage the services of a person experienced in these methods to apply all but the empirical methods.

12.6.3.1   Swaisgood (1998) empirical method for estimating crest settlements

Swaisgood (1998) gathered data on crest settlement, dam height, dam type, depth of alluvium in the foundation, earthquake magnitude and peak ground acceleration, and the focal distance of the dam to the earthquake.

Figure 12.34 shows relative crest settlement (settlement/{dam height + thickness of alluvium in the foundation}) versus peak ground acceleration (bedrock).

Swaisgood (1998) recommended the following equations to predict settlement:

$$CS = SEF \times RF \qquad (12.29)$$

where CS is the vertical crest settlement expressed as a percentage of the dam height plus the alluvium thickness. SEF is the seismic energy factor and RF is the resonance factor. These factors are calculated from:

$$SEF = e^{(0.72\,M+6.28\,PGA-9.1)} \qquad (12.30)$$

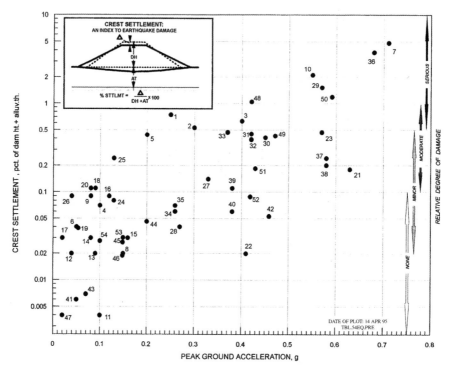

Figure 12.33.   Relative crest settlement versus peak ground acceleration (Swaisgood, 1998).

in which M is the magnitude of the earthquake (M_L or local magnitude below 6.5, and M_S or surface wave magnitude at 6.5 or above), and PGA is the peak horizontal ground acceleration at the damsite as a fraction of the acceleration due to gravity.

$$RF = 2.0\ D^{-0.35} \quad \text{for earthfill dams}$$
$$= 8.0\ D^{-0.35} \quad \text{for hydraulic fill dams}$$
$$= 0.12\ D^{0.61} \quad \text{for rockfill embankments}$$

in which D is the distance between seismic energy source and dam, in kilometres.

This method should only be used to give an approximate estimate of settlements, where there is no potential for liquefaction or significant strain weakening. Account should be taken of the scatter in the data as shown in Figure 12.33 and in practice it is probably best just to use Figure 12.33.

12.6.3.2   Pells and Fell empirical method for estimating settlement, damage and cracking

Pells and Fell (2002, 2003) gathered data from 305 dams, 95 of which reported cracking, and classified these for damage according to the system shown in Table 12.4.

Figures 12.34 and 12.35 show plots of damage contours versus earthquake magnitude and peak ground acceleration for earthfill and earth and rockfill dams.

In respect to cracking, Pells and Fell (2003) concluded:

– Earthquakes cause settlement, lateral spreading and cracking of embankment dams. Slope instability may occur but it is not common. Longitudinal cracks are more common than transverse cracks and are mostly in the upper part of the dam, more likely on the upstream face than the downstream.

Table 12.4.   Damage classification system for embankment dams under earthquake loading (Pells and Fell, 2002, 2003).

Damage class		Maximum longitudinal crack width[1] mm	Maximum relative crest settlement[2] (%)
Number	Description		
0	No or slight	<10 mm	<0.03
1	Minor	10–30	0.03–0.2
2	Moderate	30–80	0.2–0.5
3	Major	80–150	0.5–1.5
4	Severe	150–500	1.5–5
5	Collapse	>500	>5

[1] Maximum crack width is taken as the maximum width, in millimetres, of any longitudinal cracking that occurs.
[2] Maximum relative crest settlement is expressed as a percentage of the maximum dam height (from general foundation to dam crest).

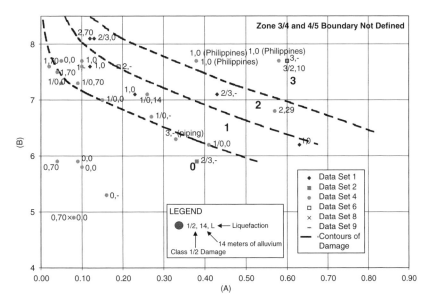

Figure 12.34.   Earthfill Dams – contours of damage class versus earthquake magnitude and peak ground acceleration. (A) Earthquake magnitude. (B) Foundation peak ground acceleration (as a fraction of acceleration due to gravity). Contours drawn without consideration for cases that had liquefaction (Pells and Fell, 2003).

– There are a number of mechanisms that can lead to formation of either of these types of cracks. Many of these mechanisms are common to both seismic loading and normal operating loading. There are seen to be more mechanisms that may lead to the formation of longitudinal cracks that transverse cracks, largely because lateral displacement is more readily achieved in the upstream-downstream directions.

– For visible longitudinal cracks to occur the dam needs to experience a Magnitude 6.5 or greater earthquake, and a peak ground acceleration greater than about 0.15 g for earthfill dams and 0.3 g for earth and rockfill dams. Alternatively it needs an earthquake of Magnitude 7.0 or greater, and a PGA of 0.05 g for earthfill and say 0.15 g for earth and rockfill dams. For hydraulic fill dams, visible cracking may occur for Magnitude 6 earthquakes, and 0.05 PGA.

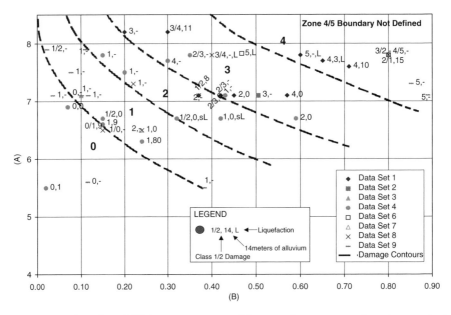

Figure 12.35.   Earth and Rockfill Dams – contours of damage class versus earthquake magnitude and peak ground acceleration. (A) Earthquake magnitude. (B) Foundation peak ground acceleration (as a fraction of acceleration due to gravity). Contours drawn without consideration for cases that had liquefaction. Boundaries shown assume rockfill is well compacted. For dumped or poorly compacted rockfill use Figure 12.34 to estimate the damage (Pells and Fell, 2003).

– Dams which experience damage of Class 2 (relative settlement of 0.2% to 0.5% and/or longitudinal cracks 30–80 mm wide) or greater are highly likely to experience transverse cracking. However, transverse cracking has been observed at under M6.0 to M6.5 earthquakes, at PGA as low as 0.1 to 0.15 g.
– There is evidence to show that higher seismic loads will result in the formation of larger and deeper cracks, and are more likely to cause transverse cracking due to the greater differential settlements across the valley.
– At low seismic loading, there is evidence to show that only one type of cracking is likely to develop. This may be either longitudinal (more common) or transverse, depending on factors within the embankment that make it predisposed to a particular form. It is the nature of the embankment, zoning and the foundation geometry and presence of compressible materials and not the seismic loading, that differentiate between which form of cracking develops, and in particular determine whether transverse cracking occurs at low seismic intensities.
– The susceptibility of an embankment to a particular type of crack could not be related to dam type, but a weak relationship has been developed with dam shape – dams in steeper valleys tend to be more susceptible to transverse cracking near the abutments.

Fong and Bennett (1995) give some limited information on maximum transverse crack depth as a function of crest settlement and crest length. This shows:

$$\text{Maximum crack depth} \approx (m)c \, (\text{settlement/dam crest length})$$

where "c" is usually about 1000 to 2500, but ranges from 125 to about 10,000.

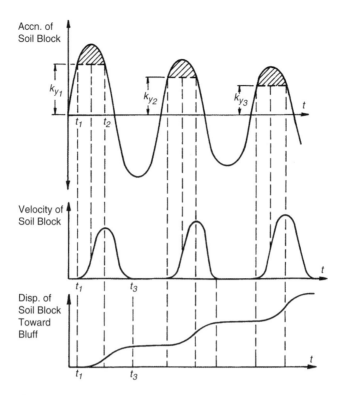

Figure 12.36.   Double integration method for determination of the permanent deformation period of an embankment (Newmark, 1965).

Pells and Fell (2003) had limited data to indicate the ratio of transverse crack depth to crack width at the surface is about 15 to 100, averaging about 40.

12.6.3.3   *The Newmark approach*

In 1965, Newmark introduced the basic elements of a procedure for evaluating the potential deformations of an embankment under earthquake loading (Newmark, 1965). In this contribution, sliding of a soil mass along a failure surface was likened to slipping of a block on an inclined plane. He envisaged that failure would initiate and movements would begin when the inertia forces exceed the yield resistance, and that movements would stop when the inertia forces were reversed. Thus, he proposed that once the yield acceleration and the acceleration time history of a slipping mass are determined the permanent displacements can be calculated by double integrating the acceleration history above the yield acceleration as shown in Figure 12.36.

According to this approach, the permanent deformation in a sliding mass is a function of:

– The amplitude of the average acceleration time history of the sliding mass, which in turn is a function of the base motion, the amplifying factor of the embankment and the location of the sliding mass within the embankment;
– The duration of the earthquake, which is a function of the magnitude of the earthquake;
– The acceleration of the potential sliding mass.

The validity of the basic principles of Newmark's approach has been demonstrated by many investigators (e.g. Goodman and Seed (1966), Ambraseys (1973), Sarma (1975)

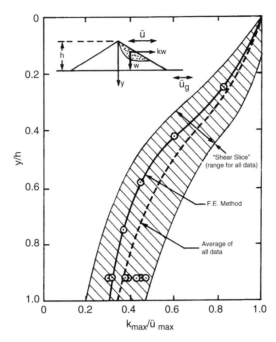

Figure 12.37.   Variation of seismic coefficient k_{max} with depth of the base of the potential sliding (Makdisi and Seed, 1978, reproduced with permission of ASCE).

and Makdisi and Seed (1978)). It is generally found that, provided the yield acceleration is accurately evaluated, the approach can estimate the permanent displacement of a soil mass in reasonably good agreement with those observed during past earthquakes.

It is also stressed that Newmark's approach is limited in application to compacted clayey embankments and dry or dense cohesionless soils that experience very little reduction in strength due to cyclic loading. The approach should not be applied where embankments or their foundations are susceptible to liquefaction or strain weakening because it will significantly underestimate displacements.

12.6.3.4   *Makdisi and Seed (1978) analysis*

The Makdisi and Seed (1978) approach is based on Newmark's method, but modified to allow for the dynamic response of the embankment as proposed by Seed and Martin (1966). The approach was developed from a series of deformation analyses performed on a large number of embankments subjected to earthquake loading.

The approach involves the following main steps:

(a) Determine y/h ratio for the potential sliding mass, where y is the depth to the base of the sliding mass and h is the embankment height;
(b) Calculate the yield acceleration k_y for the potential sliding mass;
(c) Determine the maximum crest acceleration $\ddot{u}_{max}$ and the predominant period of the embankment T_o (in seconds);
(d) Determine the maximum value of the acceleration history k_{max} using the normalized relationship given in Figure 12.37 and the values calculated in steps (a) and (c);
(e) Enter Figure 12.38 with the calculated values of k_{max} and T_o to determine the horizontal component of earthquake induced permanent displacement, U, in the potential sliding mass;

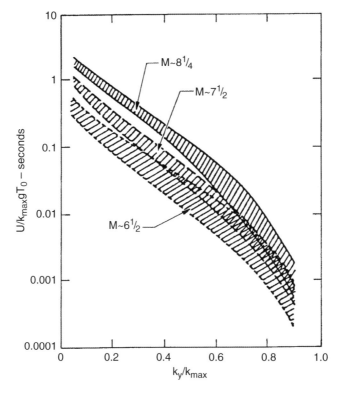

Figure 12.38.   Variation of yield acceleration with normalized permanent displacement (Makdisi and
            Seed, 1978, reproduced with permission of ASCE).

It should be noted that an important step in determination of k_{max} in step (d) is to establish the dynamic properties of the material forming the embankment and the foundation. This can be achieved by:

- Triaxial compression tests, simple shear tests or torsional shear tests conducted under cyclic loading conditions;
- Resonant column testing;
- Field measurement of shear wave velocities, either by downhole or crosshole techniques;
- Back calculation using finite element techniques, modelling measured responses to earthquake events;
- Empirical relationships, such as those given by Hardin and Drenerich (1972) and Seed et al. (1986).

Given that the laboratory and field based methods are expensive, it is recommended that the values of k_{max} be initially calculated using a range of G_{max} values obtained from the empirical relationships. Should the results of the analysis be marginal then a more elaborate program of laboratory and/or field testing may be warranted.

It should be noted that relating satisfactory dam performance to earthquake induced deformation is very subjective and generally depends on dam specific criteria about the allowable loss of freeboard or the tolerable extent of horizontal displacements.

The Makdisi and Seed approach is widely used and accepted among practicing engineers. However, like Newmark's approach, it is limited in application to dams not susceptible to liquefaction or strain weakening in the embankment or its foundations.

12.6.4   Numerical methods

Numerical modelling techniques such as the finite element method were first applied to the dynamic analysis of embankment dams by Clough and Chopra (1966). This was followed by major improvements by Ghaboussi (1967), Schnabel et al. (1972), Ghaboussi and Wilson (1973), Idriss et al. (1973), Martin et al. (1975), Finn et al. (1977), Lee and Finn (1978), White et al. (1979), Zienkiewicz and Shiomi (1984), Finn et al. (1986), Medina et al. (1990) and Li et al. (1992). Today, numerical methods are routinely used as both investigative and design tools in many geotechnical earthquake engineering problems.

The dynamic numerical codes used in practice may be divided into two main categories: total stress codes, and effective stress codes (Zienkiewicz et al., 1986; Finn, 1993). A brief discussion of some of the more frequently used codes within each category is provided in the following sub-sections.

12.6.4.1   Total stress codes

The total stress codes, as can be inferred from the classification, are based on the total stress concept and do not take account of pore pressures in the analysis. Therefore they are used in situations where the seismically induced pore pressures are negligible. The total stress codes may be divided into two main categories: (1) codes based on the equivalent linear (EQL) method of analysis, and (2) fully non-linear codes.

12.6.4.1.1   Equivalent linear analysis (EQL)

The earlier total stress codes were based on the EQL method of analysis developed by Seed and his colleagues in 1972. EQL is essentially an elastic analysis and was developed for approximating non-linear behaviour of soils under cyclic loading. Typical of the EQL codes used in practice are: SHAKE (Schnabel et al., 1972), QUAD-4 (Idriss et al., 1973) and FLUSH (Lysmer et al., 1975). SHAKE is a one dimensional wave propagation program and is used primarily for site response analysis. QUAD-4 and FLUSH are two-dimensional versions of SHAKE and are used for seismic response analysis of dams and embankments. Given the elastic nature of the EQL analysis, however, these codes cannot take account of material yielding and material degradation under cyclic loading. Therefore, they tend to predict a stronger response than actually occurs. Also, they cannot predict the permanent deformations directly. Indirect estimates of permanent deformations can however be obtained using the acceleration or stress data obtained from an EQL analysis and the semi-empirical methods proposed by Newmark (1965) and/or Seed et al. (1973).

12.6.4.1.2   Fully non linear analysis

More accurate and reliable predictions of permanent deformations can be obtained using the elasto-plastic nonlinear codes. Typical of the elasto-plastic non-linear codes used in the analysis of embankments are DIANA (Kawai, 1985), ANSYS (Swanson, 1992), FLAC (Cundull, 1993), etc. The constitutive models used in these codes vary from simple hysteretic non-linear models to more complex elasto-kinematic hardening plasticity models. Compared to the EQL codes, the elasto-plastic non-linear codes are more complex and put heavy demand on computing time. However, they provide more realistic analyses of embankments under earthquake loading, especially under strong shakings. Critical assessments of non-linear elasto-plastic codes can be found in Marcuson et al. (1992) and Finn (1993).

12.6.4.2   Effective stress codes

A major stimulus for the development of the effective stress codes has been the need for modeling pore pressure generation and dissipation in materials susceptible to liquefaction and thus to obtain better estimates of permanent deformations under seismic loading. The effective stress codes may be divided into three main categories: fully coupled, semi-coupled and uncoupled.

12.6.4.2.1   Fully coupled codes

In the fully coupled codes, the soil is treated as a two-phase medium, consisting of soil and water phases. Two types of pore pressures are considered, transient and residual. The transient pore pressures are related to recoverable (elastic) deformations and the residual pore pressures are related to non-recoverable (plastic) deformations. A major challenge in fully coupled codes is to predict residual pore pressures. The residual pore pressures, unlike the transient pore pressures, are persistent and cumulative and thus exert a major influence on the strength and stiffness of the soil skeleton. The transient pore pressures are cyclic in nature and their net effect within one loading cycle is often equal to zero. An accurate prediction of residual pore pressures requires an accurate prediction of plastic volumetric deformations. In the fully coupled codes this is often achieved by utilizing elasto-plastic models based on kinematic hardening theory of plasticity (utilizing multi-yield surfaces) or boundary surface theory with a hardening law. These models are very complex and put a heavy demand on computing time.

Generally speaking, fully coupled prediction of pore pressures under cyclic loading is very complex and difficult. To date, many of the numerical codes developed in this area are not fully validated and still are at their developmental stages. The validation studies performed on a number of these codes suggest that the quality of response predictions are strongly path dependent. When the loading paths are similar to the stress paths used in calibrating the models, the predictions are good. As the loading path deviates from the calibration path, the predictions become less reliable. Apart from the numerical difficulties, part of this unreliability is also due to the poor or less than satisfactory characterization of the soil properties required in the models. For instance, because of sampling problems, it is often very difficult to accurately determine volume change characteristics of loose sands as required by these models. In general, the accuracy of pore pressure predictions in fully coupled modes is highly dependent upon the quality of the input data. Typical of the fully coupled codes are DNAFLOW (Prevost, 1981), DYNARD (Moriwaki et al., 1988), SWANDYNE (Zienkiewicz, 1991) and SUMDES (Li et al., 1992).

12.6.4.2.2   Semi-coupled codes

Compared to the fully coupled codes, the semi-coupled codes are more robust and less susceptible to numerical difficulties. However they are theoretically less rigorous. In these codes empirical relationships, such as those proposed by Martin et al. (1975) and Seed (1983a), are used to relate cyclic shear strains/stresses to pore pressures. The empirical nature of the pore pressure generation in these codes generally puts less restriction on the type of plasticity models used in the codes. The semi-coupled codes are in general less complex and computationally demanding. Also, the parameters they require are often routinely obtained in the laboratory or in the field. There is extensive experience in using semi-coupled codes in practice.

Typical of the semi coupled codes are DESRA-2 (Lee and Finn, 1978), DSAGE (Roth, 1985), TARA-3 (Finn et al., 1986) and FLAC (Cundall, 1993).

12.6.4.2.3   Uncoupled codes

In the uncoupled analysis the pore pressures are estimated separately, using either a program of laboratory testing or an empirical relationship such as the one proposed by Seed (1983a). Then they are incorporated into an elasto-plastic non-linear code to obtain permanent deformations. The uncoupled analysis is widely used in practice and is generally believed to provide indicative estimates of the post liquefaction behaviour of earth dams. The permanent deformations obtained using this approach are often on the conservative side, as the analysis does not allow for dissipation with time of the estimated pore pressures.

12.6.4.3   *Recommended approach*

As a general guide, for most projects it will not be necessary to use other than simplified methods. Only for high consequence of failure projects, where stability and/or deformations show marginal safety, should it be necessary to use numerical methods and then it will usually be sufficient to model the post earthquake deformations.

More advanced dynamic methods are expensive and should only be done by very experienced persons who understand the limitations of the analysis and the need to use well considered properties.

There should not be an over-reliance on stability and deformation analysis at the expense of good engineering judgment and the consideration of the general design principles given in Section 12.7. At best, the analysis methods are approximate and controlled by the quality of data put into them, the limitations of the methods themselves and of those doing the analysis.

12.7   DEFENSIVE DESIGN PRINCIPLES FOR EMBANKMENT DAMS

The concept of "defensive design" of embankment dams for earthquake was developed by Sherard (1967) and Seed (1979a) and endorsed by Finn (1993), ICOLD (1986c, 1999) and ANCOLD (1998).

The general philosophy is to apply logical, commonsense measures to the design of the dam, to take account of the cracking, settlement and displacements which may occur as the result of an earthquake. These measures are at least as important (probably more so) as attempting to calculate accurately the stability during earthquake or the likely deformations. The most important measures which can be taken are:

(a) Provide ample freeboard, above normal operating levels, to allow for settlement or slumping or fault movements which displace the crest. For example, one might adopt a narrow spillway with large flood rise and large freeboard instead of a wide spillway with small flood rise and thus usually a lower freeboard, provided the costs were similar.

(b) Use well designed and constructed filters downstream of the earthfill core (and correctly graded rockfill zones downstream of a concrete face for concrete face rockfill dams) to control erosion if the core (or face) being cracked in the earthquake. Filters should be taken up to the dam crest level, so they will be effective in the event of large crest settlements, which are likely to be associated with transverse cracking. For larger dams, full width filters (2.5 m to 3 m) might be adopted instead of narrower (1.5 m say) filters placed by spreader boxes. This would give greater security in the event of large crest deformations.

(c) Provide ample drainage zones to allow for discharge of flow through possible cracks in the core. For example ensure that at least part of the downstream zone is free draining or that extra discharge capacity is provided in the vertical and horizontal drains for an earthfill dam with such drains. In this regard some embankment dam types are inherently more earthquake resistant than others. In general the following would be in order of decreasing resistance:
   – concrete face rockfill;
   – sloping upstream core earth and rockfill;
   – central core earth and rockfill;
   – earthfill with chimney and horizontal drains;
   – zoned earth-earth rockfill;
   – homogeneous earthfill.

(d) Avoid, densify, drain (to be non-saturated) or remove potentially liquefiable materials in the foundation or in the embankment. Figure 12.39 shows the principles for new

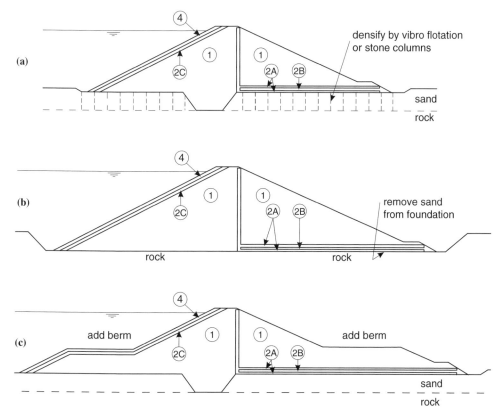

Figure 12.39.   Principles of ground improvement or treatment to reduce the likelihood of liquefaction in dam foundations. (a) Densification by vibroflotation or stone columns. (b) Remove potentially liquefiable soils. (c) Densification by vibroflotation or stone columns under a berm for an existing dam.

dams and for remediation of existing dams. Densification is usually carried out by vibro-flotation, dynamic consolidation, vibro piles or stone columns. Stone columns have the added role of draining excess pore pressures developed by the cyclic loading of the earthquake.

Filters, rockfill and other granular materials in the embankment should be well compacted if they are likely to become saturated, so they will be dilatant and not liquefy.

(e) Avoid founding the dam on the strain weakening clay soils, completely weathered rock or weak rock with the potential to strain weaken. (In many older, smaller dams, this was not done.) Post earthquake stability can be an issue if the earthquake causes even relatively minor movements which can take the foundation strength from peak to residual.

(f) The foundation under the core should be shaped to avoid sharp changes in profile across the valley. These are likely to make the core more susceptible to cracking due to differential settlement under earthquake (and normal) loading. Details of foundation slope modification are given in Section 17.6. ICOLD (1999) also suggest that it is desirable to keep the slope in the upstream-downstream direction across the core horizontal in the upper 30 m of the dam to reduce the likelihood of differential settlements.

There are a number of other less important measures which are listed by Sherard (1967), Seed (1979a) and Finn (1993). These include:

(g) Use a well-graded filter zone upstream of the core to act as a crack stopper, possibly only to be applied in the upper part of the dam. The concept is that, in the event that major cracking of the core occurs in an earthquake, this filter material will wash into the cracks, limiting flow and preventing enlargement of the crack. If well-designed filters are provided downstream, this upstream filter is of secondary importance.

(h) Provide crest details which will minimize erosion in the event of overtopping, e.g. by having a wide crest, and if the core is a highly dispersive soil, it could be modified with lime to make it non dispersive (in reality this is not likely to make a lot of difference). The common practice of using steeper slopes near the crest (usually to provide for camber of the crest) and the use of wave walls in concrete face rockfill dams will make the crest more susceptible to damage.

(i) Flare the embankment core at abutment contacts, where cracking can be expected, in order to provide longer seepage paths. Just as (or more) important is to consider the detailing of the contact with concrete walls and the provision of filters downstream of the contacts. This detailing is discussed in Section 13.6.

(j) Locate the core to minimize the degree of saturation of materials (e.g. use sloping upstream core). Finn (1993) also suggests positioning chimney drains near the central section of the embankment). These measures are intended to reduce to a minimum the extent of saturated zones which are more likely to reduce strength on cyclic loading. They are particularly relevant where sand, silty sand and sand-gravel soils are used in the dams as these are most susceptible to liquefaction. If all materials in the dam are well compacted these requirement is not important.

(k) Stabilize slopes around the reservoir rim (and appurtenant structures such as spillways) to prevent slides into the reservoir or onto the structures.

(l) Provide special details if there is likelihood of movement along faults or seams in the foundation.

(m) Site the dam on a rock foundation rather than soil foundation (particularly if it is potentially liquefiable) where the option is available.

(n) Use well graded (densely compacted) sand/gravel/fines or highly plastic clay for the core, rather than clay of low plasticity (if the option is available) (Sherard, 1967).

Seed et al. (1985) suggest that the slopes of concrete face rockfill dams should be flattened to limit displacements in earthquake. We are not convinced this is necessary for well-compacted rockfill.

When assessing an existing dam, the use of these "defensive design" measures is seldom practical (except in remedial works). However, it is useful to gauge the degree of security the existing dam presents by comparing it with this list. Where the dam fails to meet many or most of these features, particularly (a) to (e), this may be a better guide to the fact that the dam may not be very secure against earthquake than a lot of analysis.

Dams which have well designed and constructed filters, have adequate stability against normal loads, and do not have liquefiable or strain weakening zones or foundations, will be able to withstand the loading from very large earthquakes.

13

Embankment dam details

13.1 FREEBOARD

13.1.1 *Definition and overall requirements*

(a) *Freeboard.* The freeboard for a dam is the vertical distance between a specified reservoir water surface level and the crest of the dam, without allowance for camber of the crest of the dam;
(b) *Normal freeboard* is the vertical distance between the crest of the dam without allowance for camber, and the normal reservoir full supply level;
(c) *Minimum freeboard* is the vertical distance between the crest of the dam without allowance for camber, and the maximum reservoir water surface that results from routing the inflow design flood through the reservoir.

The objective of having freeboard is to provide assurance against overtopping resulting from (USBR, 1981):

– Wind setup;
– Wave runup;
– Landslide and seismic effects;
– Settlement;
– Malfunction of structures;
– Other uncertainties in design, construction and operation.

Other factors which may influence the selection of freeboard include (ANCOLD, 1986):

– Reliability of design flood estimates;
– Assumptions made in flood routing;
– Type of dam and susceptibility to erosion by overflow;
– Potential changes in design flood estimates, either through changes in flood estimation techniques or due to changed catchment conditions.

USBR (1992) have the following freeboard requirements for new embankment dams:

(1) *Freeboard at maximum reservoir water surface elevation.* The minimum freeboard should be the greater of (a) 0.9 m or (b) the sum of the wind set up and wave runup that would be generated by the average winds that would be expected to occur during large floods, as determined after seeking advice from local authorities and meteorologists. They point out that for large reservoirs and catchments the wind may be independent of the storm event that created the flood, and suggest a wind with a 10% exceedance probability (1 in 10 year) should be used.
(2) *Normal water surface freeboard.* The normal freeboard above full supply level should be the wind set up and wave runup for the highest sustained wind velocity "that could reasonably occur" e.g. 95–160 km/hr.

Table 13.1.   Freeboard requirements for preliminary studies (USBR, 1977).

Largest fetch (km)	Normal freeboard (m)	Minimum freeboard (m)
Less than 1.6	1.2	0.9
1.6	1.5	1.2
4	1.8	1.5
8	2.4	1.8
16	3.0	2.1

(3) *Intermediate water surface freeboard.* The intermediate freeboard requirement should be determined so that it has a remote probability (e.g. a 10^{-4}/annum probability) of being exceeded by any combination of wind generated waves, wind set up and reservoir surfaces occurring simultaneously. (Note that USBR, 1992, do not mention wind setup but logically it should apply.)

The adopted freeboard should be the largest of these three.

For existing embankment dams USBR (1992) require that, if the maximum water surface elevation is so close to the dam crest that wind generated waves and setup would wash over, or if the maximum water surface elevation is higher than the existing crest, the potential of this overtopping to cause failure of the embankment should be assessed.

They point out that freeboard requirements for a new dam may be different from that which may be accepted for an existing dam, because the cost of adding freeboard to a new dam is small, but to add to an existing dam may be very costly. They advocate the use of risk analysis to assess this. The authors would also advocate this analysis for existing dams. ANCOLD (2000) requirements are generally consistent with these and suggest that the freeboard at maximum water surface elevation should be below the top of the dam core. (In reality this means the dam core should be taken to the dam crest level.)

When considering water levels in the reservoir, consideration should be given to spillway gate reliability, since the water level may be very different if the gates fail to operate.

In some cases there may be a need to consider waves which could be generated by landslides failing into the reservoir and waves or seiches caused by earthquake faults in the reservoir.

It should be noted that camber is not part of the freeboard, since it is provided to allow for the long term settlement of the dam.

13.1.2   Estimation of wave runup freeboard for feasibility and preliminary design

For feasibility and preliminary design studies and for small dams, the USBR (1981) indicate that the method outlined in USBR (1977) may be adopted. This is based on a wind velocity of 160 km/hr (100 miles per hour) for determination of normal freeboard and 80 km/hr (50 miles per hour) for minimum freeboard. The effect of wind setup is ignored.

For rip-rapped slopes the freeboard requirements are as tabulated in Table 13.1. For dams with a smooth pavement or soil cement upstream slope, depending on the smoothness of the surface, freeboard of up to 1.5 times those shown in Table 13.1 should be used.

13.1.3   Estimation of wind setup and wave runup for detailed design

The following outlines the method described in USBR (1981). This is based on US Corps of Engineers (1976) and Saville et al. (1962). USBR (1992) up-dates this procedure, taking account of US Corps of Engineers (1984a), and favours a more probabilistic approach. The updated procedures are too lengthy to include here and for detailed design of important structures readers should refer to the original texts (USBR, 1992; US Corps of Engineers, 1984a).

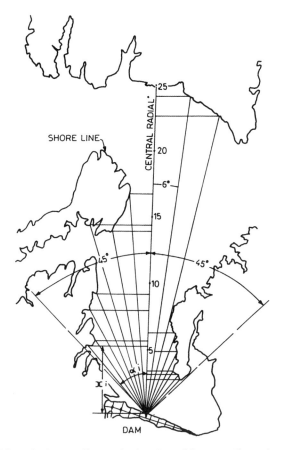

Figure 13.1.   Method for calculating effective fetch (adapted from Saville et al., 1962).

13.1.3.1   *Fetch*

In reservoirs, fetches are limited by the land surrounding the body of water. The shorelines are irregular and an effective fetch is calculated from:

$$F = \frac{\Sigma x_i \, \cos a_i}{\Sigma \cos a_i} \qquad (13.1)$$

where a_i = angle between the central radial from the dam and radial i; x_i = length of projection of radial i on the central radial.

A trial and error approach should be used to select the critical position on the dam and direction of the central radial to give the maximum effective fetch.

The radials spanning 45° on each side of the central radial should be used to compute the effective fetch. Figure 13.1 shows an example.

13.1.3.2   *Design wind*

Design wind estimates should be obtained from the Bureau of Meteorology or equivalent organisation. The critical wind duration depends on the fetch; longer durations are critical for longer fetches. The estimates should allow for local topographic effects.

USBR (1981) indicate that for maximum freeboard moderate winds should be used, i.e. winds that can reasonably be expected to occur concurrent with maximum reservoir flood

Table 13.2.   Wind relationships – water to land (USBR, 1981).

Effective fetch (Fe) (km)	0.8	1.6	3.2	4.8	6.4	8 (or more)
Wind velocity ratio $\frac{\text{over water}}{\text{over land}}$	1.08	1.13	1.21	1.26	1.28	1.30

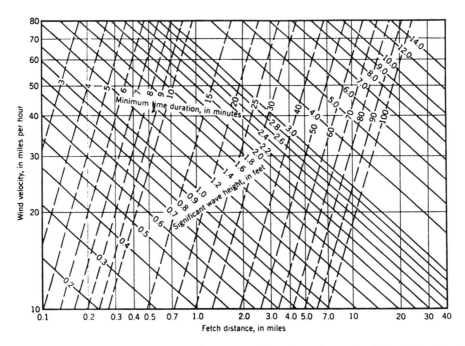

Figure 13.2.   Wave heights and minimum duration wind (Saville et al., 1962; USBR, 1981). Note 1 mile per hour = 1.6 km per hour, 1 foot = 0.3 metres.

level. If the rate of flood rise is rapid, the high wind velocity associated with the design storm may be applicable, but for a slower rate of flood rise a lower wind velocity would be appropriate. In many cases, a wind velocity of the order of 1 in 10 year event is appropriate.

For the normal freeboard calculation, maximum expected wind velocities, duration and direction should be used. The values should exceed 1 in 100 year winds. The estimates should be adjusted over land to over water wind velocity by applying the correction factors listed in Table 13.2.

The design wind velocity and duration are determined iteratively, by using Bureau of Meteorology estimates of wind speed vs duration and applying these to wind velocity vs fetch vs minimum duration shown in Figure 13.2.

13.1.3.3   Wave height

For the estimation of the minimum freeboard the significant wave height in metres (Hs), which is the average of the highest one-third of the waves in the wave spectrum, should be used. Hs can be estimated from Figure 13.2. The wave period (T) can be estimated from Figure 13.3.

The deepwater wave length in metres can then be computed from:

$$L = 1.56T^2 \tag{13.2}$$

in which T = wave period in seconds.

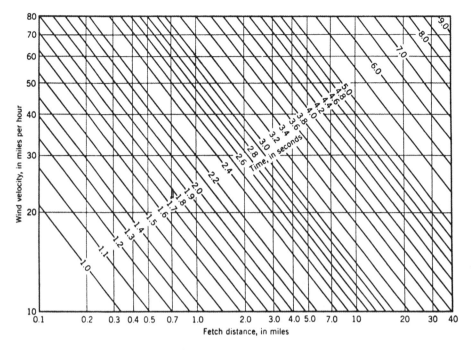

Figure 13.3.   Wave periods (Saville et al., 1962; USBR, 1981).

This equation is valid for most reservoirs where the reservoir is relatively deep compared to the wind generated wave length (i.e. water depth > 0.5L). For the normal freeboard computation, the runup should be calculated using the average of the highest 10% of waves, which is 1.27Hs. For shallow reservoirs it is suggested that US Corps of Engineers (1984a) or USBR (1992) be used.

13.1.3.4   Wave runup

Runup (Rs) from a significant wave on an embankment with rip-rap surface underlain by earthfill can be calculated from:

$$Rs = \frac{Hs}{0.4 + (Hs/L)^{0.5} \cot \theta} \tag{13.3}$$

where Rs = runup height in metres; Hs = significant wave height in metres; L = wavelength in metres; θ = angle in degrees of upstream face of the dam to the horizontal.

Rockfill dams act like permeable rubble mounds and have a different effect on energy dissipation from rip-rap placed on an "impervious" earthfill embankment. For these dams the "relatively permeable rubble mound" curves in Figure 13.4 should be used.

For smooth impermeable slopes of concrete or soil cement with a water depth at the dam greater than three times the wave height, the "smooth slope" curves in Figure 13.4 should be used. USBR (1981, 1992) indicate that these curves may underestimate runup by 15% to 20%.

For waves which are not normal to the dam, a correction should be applied to the computed runup. This is done by multiplying the runup by the cosine of the angle between the wave propagation direction and a line normal to the dam, provided the angle is less than about 50°.

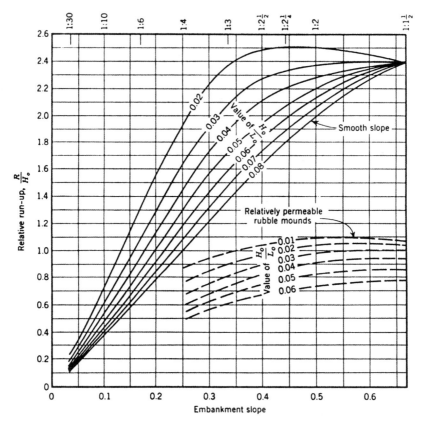

Figure 13.4.   Wave runup ratios vs wave steepness and embankment slopes (Saville et al., 1962). Note: R = runup height; Ho = wave height; Lo = wave length.

13.1.3.5   *Wind setup*

The wind setup (S) in metres is:

$$S = \frac{U^2F}{62000D} \qquad (13.4)$$

in which U = design wind velocity over water in km per hour; F = wind fetch in km, normally equals 2Fe; D = average water depth along the central radial in metres.

The wind setup should be added to the wave runup. It will be noted that, apart from long shallow reservoirs subject to high winds, the effect is very small.

13.2   SLOPE PROTECTION

13.2.1   *Upstream slope protection*

13.2.1.1   *General requirements*

The upstream slopes of earthfill on earth and rockfill dams need protection from erosion by wave action on the reservoir. Earlier dams were often protected by hand placed rock, but modern dams are generally protected by dumped rockfill, known as rip-rap.

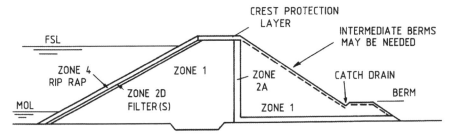

(a) SLOPE PROTECTION FOR EARTH FILL DAM

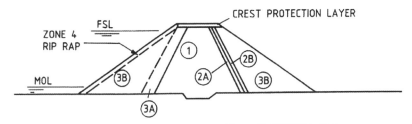

(b) SLOPE PROTECTION FOR EARTH AND ROCKFILL DAM

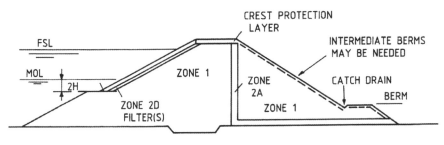

NOTE: H = DESIGN WAVE HEIGHT

(c) CASE WITH HIGH MINIMUM OPERATING LEVEL

NOTE   ALL SECTIONS DIAGRAMMATIC - NOT TO SCALE

Figure 13.5.   Upstream slope protection.

Rip-rap comprises quarried blocks of rock which have to be:

– Large enough to dissipate the energy of the waves without being displaced;
– Strong enough to do this without abrading or without breaking down to smaller sizes;
– Durable enough to withstand the effects of long term exposure to the weather and varying periods of inundation without becoming weaker and, hence, wearing or breaking down to smaller sizes.

For earthfill dams, the rip-rap is constructed as a separate layer and should be underlain by a filter to prevent erosion of the earthfill through the rip-rap as shown in Figure 13.5(a). Alternatively a wider zone of quarry run rockfill with the larger rocks pushed to the upstream slope may be adopted. For earth and rockfill dams the rip-rap is often obtained by pushing the larger rock from Zone 3B to the edge, although in some very large rockfill dams large rocks have been selectively placed by an excavation with "claws" (e.g. Thomson dam in Victoria where granite was used for the rip rap zone as distinct from sandstone/siltstone used tor the rest of the rockfill). Only in dams where severe wave action is anticipated and Zone 2B rockfill is too small or not sufficiently durable would a separate layer be placed.

Where the reservoir is operated so that the water level is maintained at a high level, i.e. there is a minimum operating level well above the base of the dam, it may be possible to provide lesser or no rip-rap protection on the lower part of the dam. Because wave action affects about two times the wave height below the water level, rip-rap should be provided to minimum operating level less two times the design wave height. A small berm should be provided at this level to support the rip-rap layer and prevent undermining of the rip-rap by erosion when the reservoir first fills as shown in Figure 13.5(c).

On some smaller dams, particularly those on mine sites where earth and rock moving equipment and waste rock are readily available, rip-rap may be designed in the knowledge that damage will occur due to larger waves or to breakdown of non durable rock, but that the damage can be readily repaired.

13.2.1.2   *Sizing and layer thickness*

The sizing of rock needed for rip-rap, the layer thickness required, and filter layer requirements are determined from the size of waves expected on the reservoir and the nature of the earthfill or rockfill under the rip-rap. The procedure suggested for sizing of rip-rap is that given in the US Corps of Engineers Shore Protection Manual (US Corps Engineers, 1984a). The steps involved are:

– Determine the design wave height H. This is taken on the average of the top 10% of the wave, which is 1.27 times the significant wave height determined as outlined in Section 13.1.3.
– Calculate the weight of the graded rock in the rip-rap from:

$$W_{50} = \frac{\gamma_r H^3}{K_{RR}(S_r - 1)^3 \cot \theta} \tag{13.5}$$

where W_{50} = weight in kilonewtons of the 50 percent size in the rip-rap; γ_r = unit weight of the rip-rap rock substance in kN/m^3; H = design wave height in metres; S_r = specific gravity of the rip-rap rock relative to the water in the dam ($S_r = \gamma_r/\gamma_w$); θ = angle of upstream slope of the dam measured from the horizontal in degrees; K_{RR} = stability coefficient = 2.5 for angular quarried rock and non breaking waves.

The maximum weight of graded rip-rap (W_{100}) is $4W_{50}$ and the minimum $0.125 W_{50}$. This is equivalent to the maximum size being 1.5 times the D_{50} size, and minimum size 0.5 times D_{50}. In practice if smaller minimum size rock is included it may be washed out under wave action. These values allow for less than 5% damage under the design wave. US Corps of Engineers (1984a) give factors which allow for greater damage.

For single size rip-rap, K_{RR} is replaced by K_D with K_D = 2.4 for smooth rounded rock and 4.0 for rough angular rock. These factors assume a rip-rap layer thickness allowing two layers of rock.

The equivalent sieve size of the rip-rap is approximately $1.15(W/\gamma_r)^{0.33}$. Where a thin rip-rap layer is being adopted, the average layer thickness can be calculated from

$$r = nK_\Delta (W/\gamma_r)^{0.33} \tag{13.6}$$

where r = average layer thickness in metres; n = number of sub layers of rip-rap weight W in kN in the layer; K_Δ = 1.02 for smooth rounded rock and 1.0 for angular quarried rock.

Table 13.3.   Rip-rap size vs wave height.

| Wave height (m) | Rip-rap size – metres | | | |
| | 3H:1V slope | | 2H:1V slope | |
	D_{50}	D_{100}	D_{50}	D_{100}
0.5	0.19	0.27	0.21	0.30
1.0	0.37	0.55	0.42	0.63
1.5	0.55	0.82	0.63	0.95
2.0	0.73	1.10	0.84	1.26
2.5	0.92	1.38	1.05	1.58

Table 13.4.   Damage to rip-rap in percent as a function of design wave height (US Corps of Engineers, 1984a).

H/H_{DO}	Damage (%)
1.00	0–5
1.08	5–10
1.19	10–15
1.27	15–20
1.37	20–30
1.47	30–40
1.56	40–50

H = actual wave height; H_{DO} = design wave height.

For rock with a specific gravity of 2.6 the estimated sizes of rockfill are as shown in Table 13.3. The wave height shown is the 10 percentile height. Values are given for upstream slopes of 3H:1V and 2H:1V.

Note that specifying rip-rap grading by linear dimension can sometimes lead to arguments about the acceptability of the rip-rap. This problem can be overcome by specifying the grading by mass of particles.

13.2.1.3   *Selection of design wind speed and acceptable damage*

On any project the decision has to be made on an acceptable degree of damage. This will influence selection of the design wind velocity and recurrence period. A useful guide can be obtained by considering alternative design wind velocity recurrence intervals, e.g. 1 in 10 year, 1 in 20 year, 1 in 50 year, 1 in 100 year, and assessing the effects of using a high recurrence design velocity using the damage estimates from Table 13.4. The figures in Table 13.4 are for quarried rock rip-rap.

The methods outlined in Thompson and Shuttler (1976), a CIRIA publication from the UK, are also valuable tools not only for designing the rip-rap but also for defining the expected damage.

What damage is regarded as acceptable will depend on the importance of the structure and the ease of access for repairs.

For existing dams, the authors' experience is that even rip-rap which is significantly smaller than a normal design has not suffered extensive damage after 30 years of operation. If the owner is willing to accept on-going maintenance, it is usually economic not to provide new rip-rap to modern design standards. Nevertheless it is important to recognize that inspection and maintenance are essential. If local failures are left without appropriate remedial works being done, the upstream edge of a crest could be seriously threatened, particularly in dams with continuously high storage level.

13.2.1.4   Rock quality and quarrying

The rock used for rip-rap must be able to withstand the repeated mechanical abrasion and wetting and drying action of the waves.

Rocks in the strong to extremely strong range are usually suitable for rip-rap if they can be quarried in intact blocks of sufficient size. Rock types which are commonly used when fresh include:

- Quartzite and sandstone;
- Limestone, dolomite and marble;
- Granite, diorite and gabbro;
- Basalt and andesite;
- Gneiss.

Most rocks containing siltstone, shale and claystone would be unsuitable for rip-rap because they would break down (slake) under repeated wetting and drying.

Ideally the rock should meet the durability requirements for concrete aggregates, but many rocks which do not meet these requirements have performed satisfactorily.

Laboratory tests which provide an indication of long term durability include:

- Apparent specific gravity and absorption;
- Petrographic examination;
- Methylene blue absorption;
- Accelerated weathering test such as wetting and drying, or sulphate soundness.

Fookes and Poole (1981) discuss the durability requirements for rip-rap.

Rip-rap is commonly obtained from rockfill quarries by stockpiling oversized rock from each shot. If a special quarry is required for rip-rap, selection of potential sites must bear in mind that the spacing of persistent joints in the rock mass should be appreciably greater than the dimensions of the required blocks. It is possible to get a good initial indication of the likely sizes and durability of blocks obtainable from an outcropping source by observing the joint spacing in the outcrops and the size and condition of surface boulders and/or scree derived from them. Experience suggests that the size of excavated block will rarely exceed the joint spacing.

At the site, or sites, selected for detailed exploration a special objective of the exploratory programs should be determination of the pattern of joints in the rock mass, i.e. the number of sets of joints and the orientation, persistence and spacing of joints in each set. In this regard, geotechnical mapping of surface rock outcrops and trench exposures (even if extremely weathered) is important, as it is usually difficult to determine the joint pattern at depth from drill cores.

During core logging, particular attention should be given to:

- The location of any zones in which the rock appears to be affected by chemical alteration (e.g. in granitic and basaltic rocks – see Chapter 2, Sections 2.7 and 2.9.3 and Chapter 3, Section 3.2.3). Such zones often contain minerals which weather rapidly on exposure;
- The distinction between minor joints which may extend for less than 500 mm, and major joints which extend for many metres;
- The mineral type(s) which form the joint cements, and the apparent strength of the joints. In general, only quartz-cemented joints are likely to be strong enough to resist parting during quarrying and later when blocks are in service.

Table 13.5. Minimum thicknesses of filters under rip-rap (from Sherard et al., 1963).

Wave height (m)	Minimum filter thickness (mm)
0–1.2	150
1.2–2.4	225
2.4–3.0	300

After logging, photography and sampling for strength and durability testing, it is useful to store the drill cores out in the open in a secure compound. Here they can be examined and photographed at regular intervals to record any deterioration. It is also possible to accelerate the "weathering" processes by a program of wetting and drying.

13.2.1.5 Design of filters under rip-rap

See Section 9.2.5.2 for a discussion on the particle size distribution requirements.

Sherard et al. (1963) recommend that the thickness be selected after consideration of:

– Size of wave;
– Gradation of the rip-rap. Rip-rap with less fines needs thicker filters;
– Plasticity and erodibility of the embankment earthfill. If the earthfill is well graded non erodible granular soil, it needs less protection than a fine silty sand or dispersive soil;
– The cost of the filter.

They recommend the minimum thicknesses shown in Table 13.5.

They indicate these should be absolute minimum thicknesses and that, if two layers of filters are needed, each should be at least 150 mm thick.

US Corps of Engineers (1984a) recommend that filter layers should be at least three 50 percent size stones thick, but not less than 230 mm.

The authors feel that the thicknesses in Table 13.5 seem reasonable for filters which satisfy no-erosion design rules, including limitations on maximum size particles. For filters which are quarry or pit run with particles up to 100 mm size, the minimum thickness would probably need to be at least 450 mm, so as to ensure that segregation of the larger particles does not leave areas with no finer particles to protect the soil under the filter.

13.2.1.6 Use of soil cement and shotcrete for upstream slope protection

In some locations rock suitable for rip-rap may not be available within economic haul distance and consideration may be given to the use of soil-cement as an upstream facing. Many dams in the USA have been constructed in this way.

ICOLD Bulletin 54 – Soil-Cement for Embankment Dams (ICOLD, 1986d) describes the use of soil-cement in some detail. The important features of the method are:

– The soil-cement is usually placed in horizontal layers as shown in Figure 13.6. The effective thickness normal to the slope has varied between 0.46 m and 0.76 m with 0.6 m being the most common. This gives a 2.4 m horizontal layer width for a 3H:1V upstream slope, allowing placement by trucks. Conveyor placement can be used to allow 0.46 m thickness, and on slopes 3H:1V or flatter, compaction up and down the slope can allow use of thinner covers. This may be acceptable for smaller dams;
– Portland cement is mixed with the available soil, conditioned to around optimum water content (standard compaction) and the soil-cement compacted in 150 mm to 300 mm thick layers to a density ratio of 98%. The soil cement is then cured under water spray;

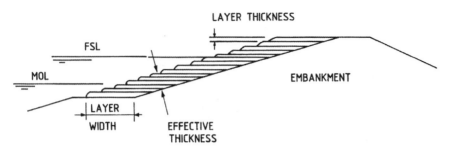

Figure 13.6.   Use of soil-cement for upstream slope protection.

– A wide range of soils can be used but usually clayey sand and/or silty sand with 10% to 25% passing 75 mm. The Plasticity Index is usually less than 8;
– The required cement content is determined by laboratory tests, which include wet-dry and freeze-thaw durability and unconfined compression strength. The requirements are detailed in ICOLD (1986d). Freeze-thaw testing would not seem appropriate in many countries with temperate or tropical climates. The USBR require an unconfined compressive strength of 4.1 MPa at 7 days and 6 MPa at 28 days. For field applications 2% cement (by weight) additional to that determined in the laboratory is used to allow for difficulties in mixing and compaction.

Soil-cement has proven to be a satisfactory slope protection over long periods – up to 30 years at least. Readers are directed to ICOLD (1986d) and more recent literature for additional details on the method.

13.2.2   Downstream slope protection

13.2.2.1   General requirements

The downstream slopes of earth and rockfill and concrete face rockfill dams are formed by rockfill. For these dams erosion of the face is not an issue and the requirement usually is simply to provide a uniform surface within the tolerance specified. Where no berms are provided access to survey markers or other instruments on the downstream face of rockfill can be difficult and consideration should be given to providing narrow berms at 30 metre maximum vertical intervals.

It is not unusual for contractors to go to some trouble to sort larger rocks on the downstream face to give a pleasing appearance to the dam.

For earthfill dams, the downstream face is potentially erodible and considerable care needs to be taken to prevent erosion. This is done by:

– Covering the surface with a layer of rockfill or by establishing grass cover;
– Providing berms to limit the vertical distance over which runoff can concentrate;
– Providing lined drains on the berms to catch the runoff and carry it to the abutments;
– Providing open lined drains at the contact between the abutment and the dam as shown in Figure 13.7(b), or the drains on the berms may be extended along the abutments as shown in Figure 13.7(a). In both cases erosion at the contact between the embankment and the abutment is being controlled.

The authors' experience is that berms should be provided at no greater than 10 m vertical intervals, although in arid climates berms may be unnecessary. A berm should be provided above the outlet of a horizontal drain as shown in Figures 13.5(a) and (c).

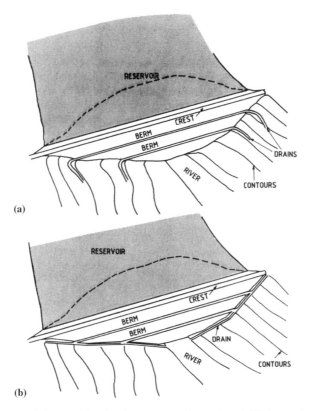

(a)

(b)

Figure 13.7.   Berms and drainage for the downstream slope of earthfill dams where grassing is used to
control erosion.

This is necessary to prevent blockage of the outlet to the drain by soil eroded off the
embankment.

When first constructed, drains will often block with eroded soil before the grass is well
established and must be inspected and cleaned out regularly. Even when grass is established inspection and maintenance of drains are necessary to ensure they continue to function properly.

13.2.2.2   *Grass and rockfill cover*

The type of grass to be used is dependent on local conditions, particularly the climate and
soil, and advice should be sought from local authorities such as the Soil Conservation
Service. It is common procedure to provide a layer of topsoil and then to seed the slope
using a bitumen hydromulch which provides initial protection against erosion before the
grass establishes. Low native bushes also have been successfully used. It must be expected
that the grass will need to be watered, at least until it is well established and that reseed-
ing and repairs will be necessary. Grass cover locally reinforced with a patented erosion-
resistant mat would be helpful at the dam/abutment intersection where run-off can
concentrate.

Where there is an ample supply of rockfill and/or climatic conditions preclude the use
of grass, dumped quarry run rockfill, placed directly on to the earthfill is usually a satisfactory way of controlling erosion on the downstream slope. USBR (1977) indicate that
0.3 m of rockfill usually is adequate although 0.6 m is usually easier to place.

13.3   EMBANKMENT CREST DETAILS

13.3.1   *Camber*

Embankment dams are subject to settlement after construction. In the case of earthfill this can be related to consolidation of the earthfill as pore pressures reach equilibrium but post construction settlement also occurs in rockfill, as with time high contact pressures between particles of rock are crushed.

Data on observed post construction settlement of earthfill and earth and rockfill dams are presented in Section 6.2.2, Table 6.8 and Figure 6.42.

From this it can be seen that most earth and earth and rockfill dams will experience less than 0.5% settlement in the first 10 years and a further 0.25% from 10 years to 100 years, i.e. a total of 0.75%.

Figure 13.8 shows the long term crest settlement rates for some concrete face rockfill dams, plotted as a function of dam height and the strength of the rock in the rockfill. The data is for well compacted rockfill.

It can be seen that, for high strength rockfill, an 80 m high dam could be expected to settle 0.3% in the 100 years after construction. This is consistent with Sherard and Cooke (1987) who suggest that settlements will generally be 0.15% to 0.3% in 100 years. They indicate that for dumped rockfill the crest settlements are 5 to 8 times that for compacted rockfill and the time dependent rate 10 to 20 times.

To maintain freeboard it is common to build the crest of the dam higher than the design level, i.e. to provide camber.

The camber relating to settlement of the embankment is best estimated by comparison with performance of other dams. Hence a fairly conservative approach would be to provide camber of about 0.5% to 1% for earthfill and earth and rockfill dams and about 0.2% to 0.4% for CFRD, depending on the height of the dam and strength of the rock in the rockfill. This is usually provided by oversteepening the upper slopes of the embankment, possibly by up to 0.25H:1V. The camber changes as a smooth curve or series of straight lines along the dam crest and is provided proportional to the height of the dam above the foundation.

The camber relating to settlements of the dam foundation is best calculated using conventional soil and rock mechanics principles.

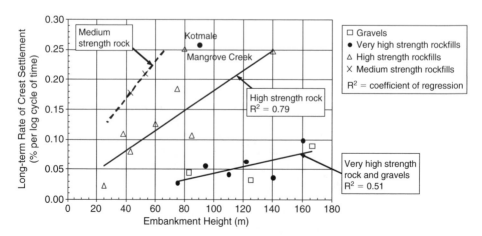

Figure 13.8.   Long-term crest settlement rates vs embankment height for compacted rockfills (Hunter 2003; Hunter & Fell 2003c).

13.3.2 *Crest width*

The crest width has no appreciable influence on the overall stability of a dam and is determined by the minimum practicable width for construction purposes, and possible roadway requirements. Unless a wide road is needed for traffic purposes a crest width greater than 6 or 8 metres is seldom required, even for large dams. For small dams, a width of 4 metres will often be adequate. To achieve this, it is necessary to carefully detail the zoning at the crest of the dam.

Figure 13.9 shows an example of a crest detail for a central core earth and rockfill dam. The principles here are:

(a) The filters are taken as close to the crest as possible, to provide internal erosion and piping control under extreme flood conditions. Many older dams did not have this detail and, with revised flood estimates, it is now likely the crests will be tested by flooding;
(b) The filter width can be narrowed at the crest. This involves special placement of the filter. Filters should not be left out. Leaving out the Zone 2B filter removes protection against piping of the Zone 2A into the rockfill;
(c) The rockfill (Zone 3A) on the downstream side should also extend to the crest to protect Zone 2B from eroding.

Figure 13.10 shows an example of crest detail for an earthfill dam with a vertical filter drain. In this case the width of the crest can be narrower than the earth and rockfill dam. It is assumed that the filter is constructed in the manner shown in Figure 9.30. i.e. by excavating through the previously placed earthfill. As for the earth and rockfill dam the filter must be taken as close to crest level as possible to provide protection from internal erosion and piping.

Any earthfill zone should be at least 3 m wide at the crest to allow compaction with normal rollers. No dam should have a crest width less than 3 m to allow vehicular access for maintenance purposes.

The crest is generally sloped towards the reservoir, to stop water ponding on it, and covered with a pavement to allow vehicles to traffic the crest. The authors' experience is that where no pavement has been provided, traffic causes wheel rutting and ponding of water on the surface which is undesirable. In this case it is wise to keep traffic off the crest by providing posts or a fence at each end. The pavement also serves the important function of reducing desiccation cracking of the core of the dam provided it is constructed of non-plastic materials. The Zone 2A filter may be taken across the top of the core in a 200 mm thick layer to provide a more positive control of desiccation.

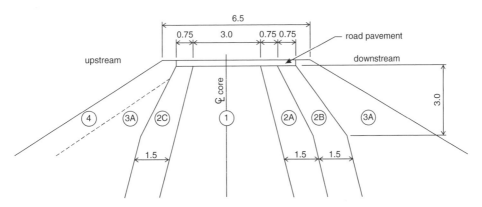

Figure 13.9.   Example of crest detail for a central core earth rockfill dam. All dimensions in metres. Steepening of the outer slopes for camber is not shown.

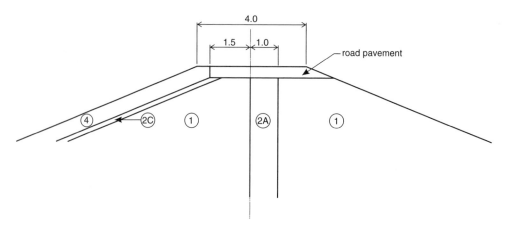

Figure 13.10.   Example of crest detail for an earthfill dam with a vertical filter drain.

13.3.3   *Curvature of crest in plan*

Some engineers have favoured curving the crest of earth and earth and rockfill dams in the upstream direction. As discussed by Wilson (1973), the concept is that, when the water load is imposed on a dam, the crest moves downstream as the water load is applied and, if the dam crest is curved upstream in plan, this can result in compressive strains along the axis of the dam. These compressive strains can counter tensile strains which are induced by settlement of the dam and hence reduce the likelihood of hydraulic fracture and leakage through the earthfill zones of the dam.

USBR (1977) suggest that for small dams the extra difficulty involved in constructing a curved axis is not warranted. Sherard (1973) indicates that the additional cost is very small, but the benefits of curvature are doubtful.

Sherard and Dunnigan (1985) consider that, with the greater confidence in the ability of well designed filters to control erosion, curvature of the dam axis is not necessary, even for high dams in steep valleys. The authors agree with this point of view.

13.4   EMBANKMENT DIMENSIONING AND TOLERANCES

13.4.1   *Dimensioning*

Embankment dimensioning is often done poorly by inexperienced engineers, because they fail to consider properly how the dam will be set out and constructed. Figure 13.11 shows some of the common errors. They are:

– Dimensioning height of dam rather than the reduced level of the crest including camber (RL b + camber). The height of dam varies across the valley and in any case is not known at any section prior to construction because the general foundation level is not known.
– Giving a reduced level at the base of the dam, RL t and/or depth of cutoff trench Z. These vary across the valley and are also not known before construction at any section. General foundation and cutoff foundation should be defined in geotechnical terms, not as levels and depths below ground surface or general excavation.
– Setting out the cutoff trench from the contact of earthfill zone with general excavation (i.e. point A) rather than as a fixed offset from the centreline. The position of A is not

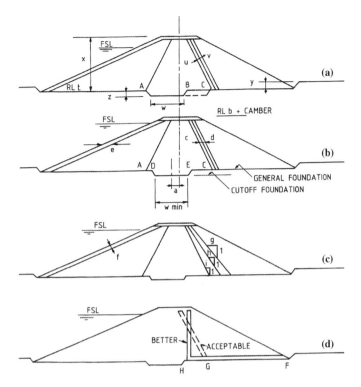

Figure 13.11.   Issues in dimensioning embankment dams. (a) and (c) are usually poor practice; (b) and (d) are usually correct practice.

known before construction and is a variable distance from the centreline giving a "meandering" location of the cutoff trench. The cutoff trench width is wrongly defined as shown, i.e. with width W at the cutoff foundation. In practice it is better to define the width D E which can be set after general excavation is completed, with an overriding dimension W_{min} to cover the situation where the depth of excavation to cutoff foundation is deeper than anticipated.

– Dimensioning filter and rip-rap layers as thicknesses normal to the slope (u, v and f) rather than as widths c, d and e. In most cases construction is in horizontal lifts so the widths are more appropriate.

– Dimensioning filter zones with varying widths using slopes g, h and i rather than as constant widths. There is no technical need for the filter width to increase with height; setting out is more complicated and volume of filters is increased by using the varying widths.

– Chimney drains should be vertical rather than inclined, as they are better constructed by digging back through compacted earthfill (see Figure 9.30) rather than placed concurrently with the fill.

13.4.2   Tolerances

Specifications for construction of embankment dams usually include tolerances on the dimensions of the embankment. These reflect the need to ensure that, for example, the thickness and the widths of filter zones are maintained even if the position of the filter in the embankment varies slightly.

The tolerances shown in Table 13.6 would be reasonable for larger earth and rockfill dams with filters, such as those shown in Figure 1.2.

Table 13.6.   Tolerances for embankment construction.

	Towards axis of dam	Away from axis of dam
(A) Tolerances on zone boundaries		
(a) Outside faces of dam embankment		
Crest	Zero	250 mm
Downstream slope	Zero	500 mm
Upstream slope	Zero	500 mm
(b) Division lines between zones upstream of dam axis		
Zone 1 and Zone 2A	Zero	1000 mm
Zones 3A and 3B	1500 mm	1500 mm
(c) Division lines between zones downstream of dam axis		
Zone 1 and Zone 2A	Zero	1000 mm
Zones 3A and 3B	1500 mm	1500 mm
(B) Tolerances on widths and thicknesses		
(a) Width of filter zones and rip-rap		
Zones 2A, 2B, 2C	Plus	250 mm or 500 mm
	Minus	Zero
Zone 4	Plus	500 mm or 1000 mm
	Minus	Zero
(b) Thickness of horizontal filter drains		
Zones 2A, 2B	Plus	250 mm
	Minus	Zero

13.5   CONDUITS THROUGH EMBANKMENTS

The placement of conduits or outlet pipes through earth dam embankments is a common cause of piping failure and accidents, particularly for small dams which do not have filters. Foster et al. (1998, 2000a) record that historically about half of piping failures can be attributed to the presence of conduits.

Figure 13.12 shows a dam which has lost all the reservoir water in the large pipe which has formed around the outlet conduit.

The traditional approach to overcome this problem has been to provide concrete cutoff collars around the conduit. These lengthen the seepage path and penetrate into the fill, lessening the likelihood of a stress induced crack penetrating outside the collar. Figures 13.13 and 13.14 show typical cutoff collars for an outlet works conduit.

USBR (1987) indicate the cutoff collars were generally 0.6 m to 0.9 m high, 0.3 m to 0.45 m wide and spaced from 7 to 10 times their height along the portion of the conduit within the core of the dam. For a conduit founded on soil, the collars should completely encircle the conduit. A bituminous or other joint filler is provided between the collar and the conduit to minimize stress concentration effects on the conduit.

While many dams have been built and operated successfully using this approach others have had problems.

Fell and Foster (1999) discuss the causes for this based on case studies, the literature and their experience. It is necessary to consider the mode of the piping (piping into the conduit, along and above the conduit or out of the conduit and the four phases of piping (initiation, continuation, progression and breach).

13.5.1   *Piping into the conduit*

(a) *Initiation* – the conduit allows erosion into the conduit if it is cracked, corroded, or joints have opened. This is most likely to occur if:
  – Settlement of the conduit has occurred due to compressible (soil) foundations;

Figure 13.12.   Piping failure of a dam around the outlet conduit.

- The conduit is joined to a "stiff" structure, e.g. outlet shaft or concrete core wall;
- Poor detailing of joints in design or construction;
- Water flows in the conduit under pressure fluctuate giving a surging effect;
- The conduit is old and corroded.

(b) *Continuation and progression* – the conduit facilitates continuation and progression by:
- Maintaining an open joint;
- Carrying away the soil which erodes into the conduit – so the seepage pressures do not have to erode the soil any great distance through the dam. However the progression of piping may be limited or slowed due to the limited width of the open joint or crack.

(c) *Breach* – the conduit does not in itself form a breach mechanism. Those failures which have involved erosion into the conduit have also involved erosion along the conduit. It is possible (probable?) that the initiation of erosion along the pipe was aided by erosion into the pipe. A breach mechanism can occur if the erosion progresses to form a large sinkhole which collapses at the crest leading to loss of freeboard.

Figure 13.13.   Typical cutoff collar on outlet works conduit (USBR, 1987).

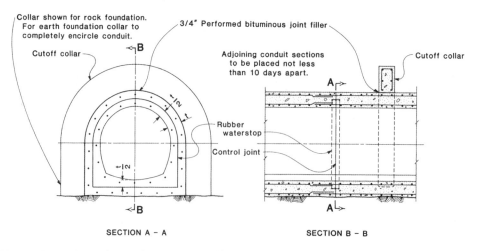

Figure 13.14.   Typical control joint and cutoff collars details (USBR, 1987).

13.5.2   *Piping along and above the conduit*

(a) *Initiation.* The conduit facilitates the initiation of piping by:
  – Causing stress distributions due to the stiff conduit and its surround which lead to
    low principal stresses and hydraulic fracture. This is discussed by Sherard et al.

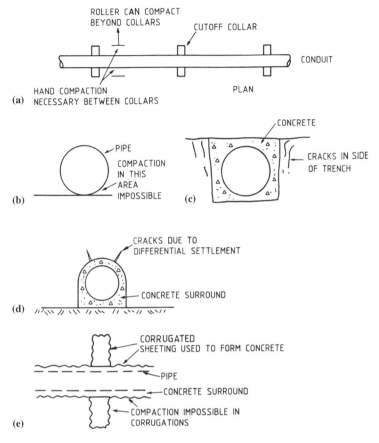

Figure 13.15.   Some causes of piping failure around conduits: (a) Inadequate compaction due to the presence of cutoff collars; (b) Inadequate compaction under pipe; (c) Cracking in soil or extremely weathered rock in the sides of a trench; (d) Cracks due to differential settlement; (e) Use of corrugations or other roughening of the surface of cutoff collars or concrete surround.

(1972) and Charles (1997). This can occur on the sides of culverts which are constructed in a trench. Melvill (1997) describes measurements on several South African dams. Knight (1990) measured very low horizontal stresses at Hinze dam in Queensland, Australia, sufficient to give virtually zero effective stress. Sherard et al. (1972b) point out that it can also occur where the concrete culvert, or concrete surround around a pipe, has a sharp corner. In this case piping can be expected above the culvert. Drying of the soil during construction can also cause cracks which allow initiation of piping. Sherard et al. (1972b) point out that, if the rate of filling is sufficiently slow, the soil swells, providing increased compressive stresses and preventing piping initiation.
– Making compaction of soil difficult, particularly if collars are provided at close intervals or the concrete is formed with corrugated steel sheet or other non-smooth formwork, preventing compaction of the soil adjacent to the conduit.
Figure 13.15 shows some of these features. There is also some evidence that excavating a trench through an existing dam to install a conduit can lead to conditions conducive to initiation of piping.

The likely effects of this are:
- Possible drying shrinkage cracks in the sides of the trench;
- Different moduli of the compacted backfill compared to the original dam leading to cracking and differential settlement;
- Potential for poor compaction of the backfill in confined spaces.

Note that it is the continuity of the potential defect caused by a conduit and the excavation through the dam which is so critical to the initiation of piping.

(b) *Continuation and progression.* The conduit facilitates this by providing a "side" or in the case of a circular outer shape, near its invert, a "roof" to the potential erosion hole which will not collapse. Hence the likelihood of erosion developing beyond the initiation is greater than without a conduit.

(c) *Breach.* The conduit itself does not cause a breach mechanism which is mostly controlled by such factors as overall zoning. Flow from the conduit may facilitate removal of eroded or slumped material and therefore contribute in a secondary way.

13.5.3 *Flow out of the conduit*

(a) *Initiation.* The conditions conducive to flow out of the conduit (if it is cracked, or has open joints) are as for piping into the conduit;

In addition, the conduit must flow under pressure, either because it fills and pressurises under some conditions, e.g. large flood flows for a spillway outlet, or if the conduit is an outlet with the control valves on the downstream side;

(b) *Continuation and progression.* The flow out of the conduit leads to erosion of the soil along the conduit which may then lead to a piping failure mechanism or to increased pore pressures in the downstream part of the dam, leading to slope instability;

(c) *Breach.* The breach mechanism may either be as for erosion along the pipe or by slope instability, leading to loss of freeboard.

13.5.4 *Conclusions*

Sherard (1973) concluded that where conduits must be placed through earth dams:

- *"It is particularly important that the embankment adjacent to the conduit be placed at a relatively high water content and not be a soil susceptible to piping;*
- *Even in small, homogeneous dams where no chimney drain is installed it is advisable to provide a drain and filter around the conduit at its downstream end for the purpose of intercepting concentrated leaks which follow the conduit;*
- *In cases where the soil foundation is thick and compressible, it is not desirable to excavate a trench under the conduit and fill it with compacted earth."*

Sherard and Dunnigan (1985) indicate that, because of concerns relating to compaction of soil adjacent to cutoff collars and the increased confidence in the use of filters to control erosion, they recommend that:

- No seepage cutoff collars be provided;
- Concrete surfaces surrounding conduits should be smooth and the sides should be sloped at 1H:8V or 1H:10V so that earthfill can be compacted directly against the concrete;
- A filter be provided to surround the downstream position of the conduit, i.e. underneath as well as on both sides and the top so that all potential leakage travelling along the concrete-earth core interface exits in a controlled manner.

At the ASDSO-FEMA 2000 workshop in Denver on issues, solutions and research needs related to seepage through dams, 29 participants, including senior dams engineers from the

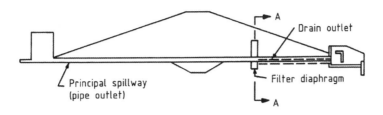

SECTION ALONG OUTLET PIPE

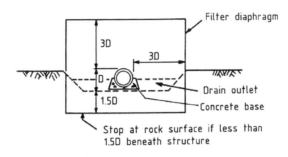

SECTION A-A

Figure 13.16.   Filter diaphragm for seepage and piping control around outlet pipe (Talbot & Ralston, 1985, reproduced with permission of ASCE).

USBR, Corps of Engineers, FERC, major USA Consulting Companies and the first author, were unanimous in agreeing that cutoff collars should not be provided because they made compaction more difficult.

Talbot and Ralston (1985) indicate that the Soil Conservation Service of the USA have replaced conduit cutoff collars with a filter diaphragm as shown in Figure 13.16.

They indicate that a single diaphragm is constructed around outlet conduits or other structures in the downstream section of the dam as shown in Figure 13.16. The diaphragm consists of a graded sand-gravel filter that projects outward a minimum of 3 times its diameter (for pipes) and extends 1.5 times its diameter below the pipe if the foundation is soil or erodible weathered rock. Where other embankment or foundation filter drainage systems are used, the diaphragm must tie into these systems to provide a continuous zone that intercepts all areas subject to cracking, poor compaction or other anomalies.

The "drain outlet", is a layer, or layers of sand/gravel designed to act as a filter against the soil in the embankment and foundation, and with sufficient discharge capacity to transmit seepage intercepted by the filter diaphragm.

13.5.5 Recommendations

The authors recommend the following:

(a) Where practicable, avoid placing conduits through the dam, or on soil and erodible rock foundations, by using outlet tunnels in the abutments (for larger dams) or placing the conduit in a trench excavated into non erodible rock and backfilled with concrete as shown in Figure 13.17(a). Avoid deep, partly backfilled trenches, with steep sides as shown in Figure 13.17(b) because these will become zones of potentially poor compaction and low stresses conducive to hydraulic fracture and initiation of piping.

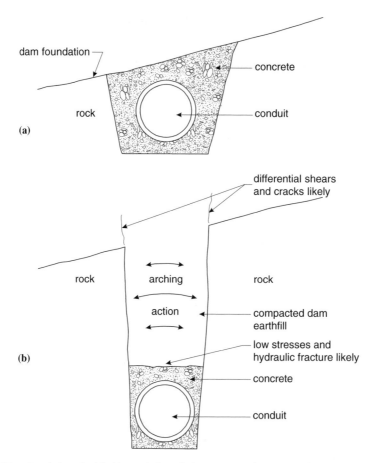

Figure 13.17.   Conduit embedded in rock foundation. (a) Good practice, (b) poor practice.

(b) Where conduits must past through the embankment, as will be the case for most smaller dams and many medium and large dams on soil foundations, the preferred detail is to provide filters for the dam core and to take these around the conduit. If the foundation is soil or erodible rock, the filter should be taken locally deeper into the foundation so as to surround the pipe in a similar way to the filter diaphragm in Figure 13.16.

Alternatively, for low consequence of failure dams, where filters are not being provided, or zoned earthfill construction is being used, and for flood control structures such as levee banks and dykes which have pipes or culverts through the levee, a filter diaphragm similar to that shown in Figure 13.16 should be provided.

(c) Concrete, steel, or corrugated steel pipes should be continuously supported on a concrete base similar to that shown in Figures 13.18(a) and (b). No cutoff collars should be provided. This is to allow compaction adjacent to the sides of the pipe, with rubber tyred rollers, rubber tyred construction equipment or hand held compaction equipment consistent with the strength of the pipe. The soil being compacted adjacent to the pipe should be wet of standard optimum water content to facilitate its being rolled into the irregular shape of the contact and to make it less likely to crack by differential settlement.

(d) Cast in-situ conduits should be constructed as shown in Figure 13.19 with the sides sloping no steeper than 1 Horizontal to 8 Vertical (1H:8V), so compaction equipment can be used to the side of the pipe. The trench in which they are placed should be wide

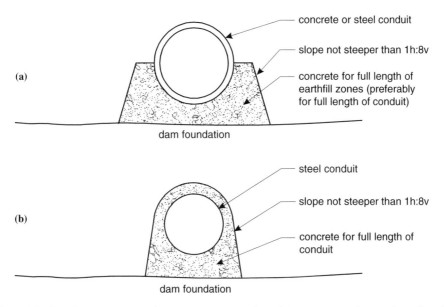

Figure 13.18.   Concrete support for (a) concrete or steel conduit strong enough to withstand earth pressure from the dam (b) steel conduit requiring concrete surround to withstand earth pressure from the dam.

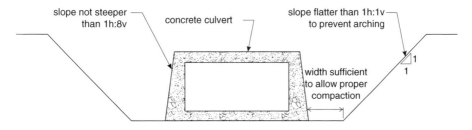

Figure 13.19.   Trench and culvert details for cast-in-situ or pre-cast concrete culvert.

enough to allow for compaction and the batter slopes should be no steeper than 1H:1V so arching does not develop and lead to low stresses in the trench, conducive to hydraulic fracture and cracking.

(e) Care should be taken to prevent desiccation cracking in the base and sides of any trench into which a conduit is placed and any cracked soil should be removed immediately prior to backfilling.

(f) Where the only earthfill available is dispersive (Pinhole D1 or D2, Emerson Class 1 or 2), it would be wise, even with all of the above factors, to add 2% or 3% lime (by weight) to the backfill soil, carefully mixed, to render it non dispersive.

(g) Care should be taken to provide joint details in the conduit and between conduits and other structures, such as intake works, which will accommodate the likely longitudinal and vertical deformations and prevent erosion of soil into the conduit or flow of water out of the conduit.

(h) Consideration should be given to the potential for corrosion of conduits and whether it is wise to rely on steel or corrugated steel pipes for long term operation.

13.6   INTERFACE BETWEEN EARTHFILL AND CONCRETE STRUCTURES

Just as the contact between earthfill and outlet conduits can be a potential zone of weakness in an embankment, the contacts between earthfill and concrete structures, e.g. the wall of a concrete spillway, can be potential sources of cracking and piping failure.

Earlier practice was to construct a cutoff wall at right angles to the wall of the structure to penetrate into the earthfill and increase the seepage path. Figure 13.20 shows an example of this and the difficulties that are created because rollers cannot be used to compact the soil adjacent the walls.

To overcome this Sherard and Dunnigan (1985) recommend that no cutoff wall be provided and that the contact be detailed as shown in Figure 13.21. This approach is recommended by the authors as it recognizes that the cutoff walls only hinder compaction and that seepage is inevitable leading to possible cracking. Such seepage should be controlled by well designed filters.

For this arrangement compaction of earthfill above standard optimum water content with rubber tyred equipment is possible right up to the wall. The wall is left smooth, off form concrete, not sandblasted, chipped, painted or treated in any way. The downstream filters may be widened. Sherard and Dunnigan (1985) suggest that, if the Zone 2A filter contains a significant quantity of gravel sized particles, there is a danger the gravels could segregate at the concrete-filter interface leaving pockets of gravel without sand in the voids. In this situation they suggest placing an additional filter consisting only of sand upstream of the Zone 2A filter (as well as widening the Zone 2A and 2B filters).

They suggest that the angle α at which the dam axis intersects the wall (angle α in Figure 13.21) may be made say 80° to increase the pressure on the contact as the dam water load is applied. They point out, however, that many dams have been successfully constructed with $\alpha = 90°$.

It is recommended that, if highly dispersive soils are being used for construction, consideration should be given to adding lime in the vicinity of the contact with the concrete structure (in addition to the features detailed above).

Where the dam generally has no filters, it is strongly recommended that filters are provided in the vicinity of the contact.

Figure 13.22 shows two situations where settlement of the dam fill during or after construction, or in an earthquake, can cause a gap to form between the embankment fill and

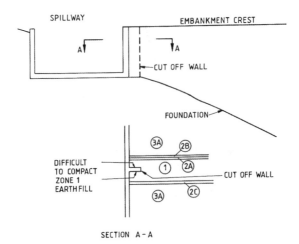

Figure 13.20.   Poorly detailed contact between earthfill and spillway wall with a cutoff wall provided.

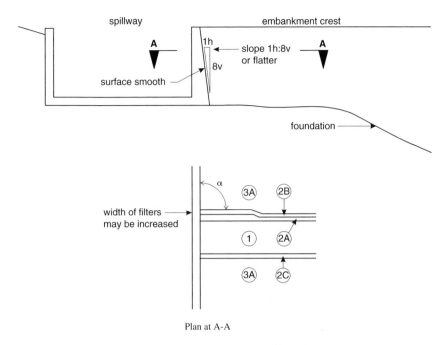

Figure 13.21.   Good details for the interface between earthfill and concrete structures.

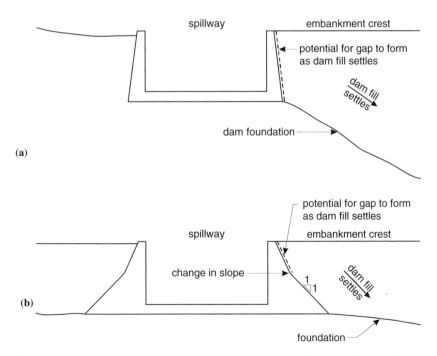

Figure 13.22.   Situations where a gap may form between the dam fill and spillway wall. (a) Steep foundation adjacent spillway wall; (b) change in slope of the retaining wall.

the spillway retaining wall. In Figure 13.22(a) the foundation slopes away at the base of the wall. In Figure 13.22(b) there is a change in slope of the back of the wall. Both situations are best avoided.

It should be noted that if a fine abutment contact soil is used against the wall (or a conduit in the previous section) the critical fine filter should be designed to suit this material as well as the general core material.

13.7   FLOOD CONTROL STRUCTURES

Flood control structures, e.g. flood storage basins and levee banks and dykes, have particular features which require consideration. These include:

– The embankments stand without water adjacent to them for long periods;
– Water levels rise quickly and usually fall within days or even hours;
– The consequences of failure may not be as critical as for other dams;
– The embankments may serve other purposes, e.g. as a parking area;
– The structures often have culverts or pipes passing through them;
– The foundations are usually soil and often these have sandy layers in them, so "heave" or "blow-up" and piping through fissures in the soil is an issue.

From a design viewpoint this has the following implications:

(a) The embankments may crack due to desiccation if built of clay soils;
(b) The rapid rate of rise of the water makes it more likely erosion will begin before cracks close by swelling of the soils adjacent to the cracks;
(c) There are potential points of weakness around the conduits as described in Section 13.5.

The design of such structures, particularly levee banks and dykes which are long, low structures, built with limited funds, is often a compromise between cost and the likelihood of failure. Many levees are built only to withstand a 1 in 100 AEP flood, so it is probably inappropriate to provide fully intercepting filters at great cost.

Figure 13.23 shows a possible detail for a levee bank where the consequences of failure are low and there is no "heave" or "blow-up" risk. The section is homogeneous earthfill. Cracking is controlled by providing a 50–100 mm thick sand layer which acts as a vapour barrier. A road surface on the crest would also assist in this. The cutoff is taken below the base of cracked and fissured soil.

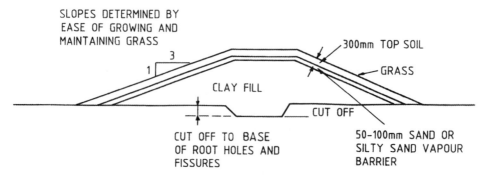

Figure 13.23.   An example of the design for a levee bank or dyke in a low consequence of failure situation.

If the soil is dispersive, it should be modified with lime or gypsum to render it non-dispersive.

Compaction control should be as rigorous as for any dam and all conduits, culverts, pipes and walls should be treated with a filter diaphragm as discussed in Section 13.5. This should give a likelihood of piping failure less than that of overtopping.

The embankments are unlikely to reach the steady state seepage condition (or instantaneous drawdown), so slopes can be steeper than a normal dam. Hence from slope stability considerations, embankments built of clay soils can stand at 1.5H to 1V. Silty sands, or sandy soil embankments will, however, saturate quickly and require flatter slopes.

From the erosion control viewpoint, slopes of 3H to 1V or flatter are preferable and this may override stability considerations.

The partial saturation effect may be more important for levee banks on the side of a river if it can be shown that the levee will not saturate. However, care must be taken as most levee bank failures occur on drawdown of the flood.

It should be recognised that use of such a cross section does have risks of failure higher than for well designed dams. Where practical and warranted from a consequences of failure viewpoint, it would be better to use a zoned earthfill cross section, or earthfill with vertical and horizontal drains.

For flood control structures in urban areas, the consequences of failure should be assessed and the embankment section selected accordingly. Where there is a potential for loss of life, the embankment should have proper internal erosion and seepage control.

13.8   DESIGN OF DAMS FOR OVERTOPPING DURING CONSTRUCTION

13.8.1   *General design concepts*

When embankment dams are constructed on a river it is necessary to divert the river from the river bed during construction (of the dam or a culvert along the lower part of one of the abutments). This is usually done by excavating a diversion tunnel through an abutment of the dam and constructing a coffer dam upstream of the embankment to divert the river into the diversion tunnel. This upstream coffer dam may be supplemented by a higher coffer dam which is incorporated into the main embankment as shown in Figure 13.24.

When designing coffer dams a balance is reached between the height (and therefore cost) of the coffer dam and the recurrence period of the flood which will overtop the dam. Also important are the consequences of overtopping in potential damage to the construction works, delay to construction schedules and potential for damage or loss of life downstream. Of particular concern is the situation if the coffer dam fails, giving a substantial flood wave.

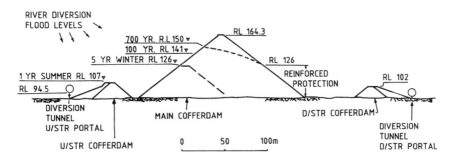

Figure 13.24.   Coffer dams for river diversion.

In the example shown in Figure 13.24, the upstream coffer dam has been designed to divert the 1 year summer (dry season) flood and the main coffer dam the 5 year winter (wet season) flood. There is therefore some risk that it will be overtopped.

It has been common practice in Australia since the late 1960s to provide protection to the coffer dam and/or to the main embankment to prevent failure in the event of overtopping. This is done by covering the downstream face of the coffer dam or the main embankment with steel mesh or rock filled gabions, which are anchored into the slope with steel reinforcing rods. If this reinforcement is provided on the main embankment it can be designed so that much greater return period floods can be accommodated, e.g. 1 in 100 year or 1 in 700 year floods. Whether these are acceptable should be determined in a risk analysis framework.

The use of such steel mesh reinforcement has allowed use of significantly smaller diversion tunnels and upstream coffer dams, with resultant savings in project cost. The design, construction and operation of steel mesh reinforcement is described in ANCOLD (1982) and ICOLD (1984). As detailed there, up to 1982 the method had been used on 50 dams; 41 in Australia, 9 in other countries.

The following outlines only the broad principles and for design purposes reference to the ANCOLD or ICOLD publications is recommended.

13.8.2   *Types of steel mesh reinforcement*

The technique has evolved over several years, being modified to take account of lessons learnt from failures and to reduce the costs involved. Some early systems failed during operation, due to:

- Damage of relatively lightweight steel meshing, by floating logs, or by boulders from the surface of the dam, e.g. Cethana Dam (HEC, 1969);
- Erosion of the contact between the steel meshing and the downstream abutment due to inadequate anchorage, e.g. Xonxa Dam (Pells, 1978).

A design which took account of these early problems, and successfully withstood 33 hours of overtopping, up to 2.5 m deep when the dam was 20 m high, was that employed at Googong Dam (Fokkema et al., 1977).

Figure 13.25 shows the meshing system for Googong Dam.

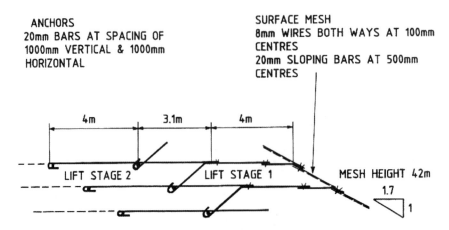

Figure 13.25.   Steel mesh reinforcement for Googong Dam (Fokkema et al., 197; ANCOLD, 1982).

There are several features of the design:

– The continuous concrete protection around the toe;
– The cranked anchor system for the avoidance of progressive failure;
– The combination of a medium duty surface mesh with small opening size and an over-
  lay of fairly closely spaced (500 mm centres) sloping bars to protect the mesh;
– The return of the sloping bar into the fill for tensioning to the anchor prior to welding
  the connection between the two;
– Infilling voids in the face rockfill with concrete to lessen the chances of individual rocks
  being plucked out (but not too much so as to lower the through flow capacity of the
  rockfill);
– Being able to secure the protection at the top of each layer as it is approaching completion, with the use of temporary gabions on the abutment section of each layer to direct
  the flow to the centre of the dam.

The Hydro-Electric Commission of Tasmania have carried out a significant amount of
developmental work in steel mesh reinforcement. Figures 13.26, 13.27 and 13.28 show systems they have developed to cope with their particular problems – regular overtopping, particularly of coffer dams, and heavily timbered catchments, which give potential for damage
during overtopping by floating logs.

These systems all use cylindrical gabions fabricated from 50 mm square mesh with
5 mm wires both ways (4 mm for Mackintosh Coffer Dam).

For the Mackintosh Coffer Dam (Figure 13.26), a concrete slab was placed over the
rockfill crest to secure the top row of gabions and prevent overflowing water from entering the downstream rockfill zone through the crest. Model tests were carried out to measure the pore pressures used in stability analysis which showed that anchor bars were not
required. The coffer dam was overtopped by small floods several times without any damage being sustained. A similar system has been used on other HEC dams.

For Mackintosh Dam the downstream face of the dam was protected to a height of 21 m.
As shown in Figure 13.27 this took the form of a vertical wall of gabions anchored to dowels
grouted into the rockfill. The protection was made vertical to keep the many floating logs that
were expected to be carried during floods clear of the face. The triangular wedge of rockfill
forming the toe of the dam was placed after the embankment was above the 100 year flood
level. The dam was overtopped once during construction but only to a depth of 300 mm.

The system was found to be costly mainly because of the very substantial anchors
required and the full strength bolted connections. Although this type of connection was

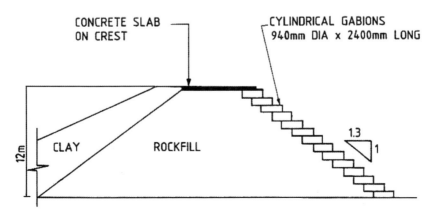

Figure 13.26.   Steel mesh reinforcement for Mackintosh Coffer Dam (ANCOLD, 1982).

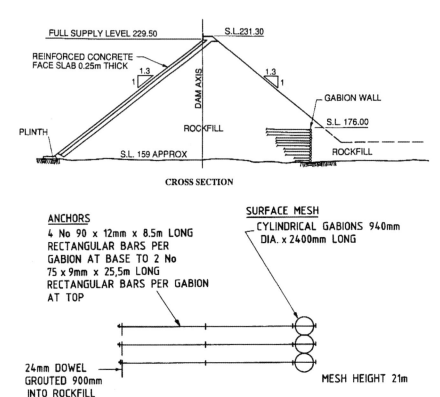

Figure 13.27.   Steel mesh reinforcement for Mackintosh Dam (ANCOLD, 1982).

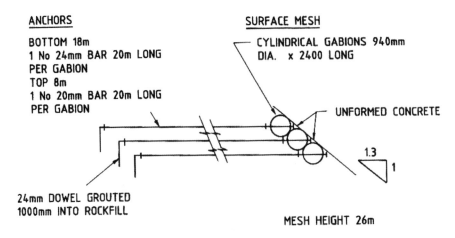

Figure 13.28.   Steel mesh reinforcement for Murchison Dam (ANCOLD, 1982).

used in preference to a welded connection to avoid delays in wet weather, the large number of anchors and connections slowed down the installation of the system and delayed embankment construction.

For Murchison Dam, the gabions were placed on the same slope as the face of the dam as shown in Figure 13.28. The cylindrical gabions could be installed rapidly and did not

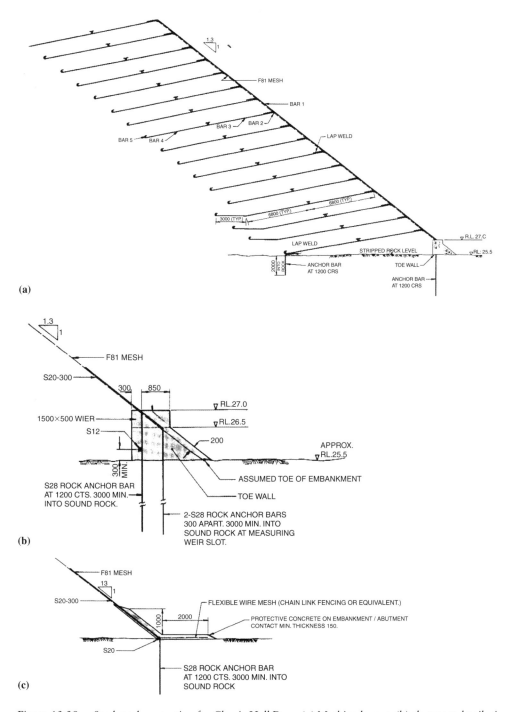

Figure 13.29. Steel mesh protection for Clarrie Hall Dam. (a) Meshing layout; (b) abutment detail, river section; (c) abutment detail, other than river sections (PWD of NSW, 1981).

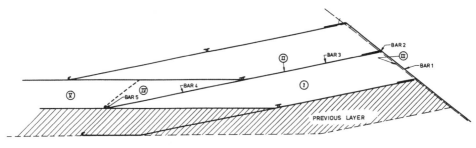

CONSTRUCTION SEQUENCE DIAGRAM

NOTES:

CONSTRUCTION SEQUENCE

1. BEFORE COMMENCEMENT OF ANY MESHING PREPARE AND HAVE ON HAND 24 GABIONS 2·0 m LONG × 1 m WIDE × 0·5 m DEEP.

2. CONSTRUCT   ROCKFILL WINDROW (I) COMPACT AND TRIM FACE.

3. CONSTRUCTION ON WINDROW SHALL NOT PROCEED MORE THAN ONE SHIFT IN ADVANCE OF MESH PLACEMENT.

4. PLACE AND TIE F81 FACE MESH (III) PLACE TIE BARS (II) (BARS 3 AND BAR 4 AND ANCHOR BAR 5).

5. PLACE ROCKFILL (IV) OVER BAR 4 AND PROCEED WITH PLACEMENT AND COMPACTION OF LAYER (V) PLACE FACE COVER BAR (BAR 1) AND HORIZONTAL BAR (BAR 2) AND WELD COVER BAR (BAR 1) TO PREVIOUS LAYERS.

6. WELD BAR 1 TO BAR 2.

7. BEND BAR 1 TO ALIGN WITH BAR 3. TENSION BAR 3 TO 5 kN AGAINST BAR 1 AND LAP WELD.

8. IN EVENT OF A THREATENED OVERTOPPING , DO NOT ADVANCE WINDROW FURTHER. COMPLETE TRIMMING AND MESHING INCLUDING TENSIONING AND WELDING OVER EXPOSED WINDROW. PLACE PROTECTIVE GABIONS ALONG WINDROW EDGE AND WELD TO MESH. TIE GABIONS TOGETHER.

Figure 13.30.   Steel mesh reinforcement for Clarrie Hall Dam – construction sequence (PWD of NSW, 1981).

control the rate of embankment construction. The exposed upper surface of the gabions on the dam face was protected with unformed concrete. There was no overtopping event during construction, although flow through the rockfill occurred in a flood which ponded above the upstream coffer dam when the dam was well above the mesh level. A similar system has been used on later HEC dams.

A more recent application of the Googong type of meshing system is shown in Figures 13.29, 13.30 and 13.31 for Clarrie Hall Dam. This incorporates a rather simpler system of anchorage than the cranked bars used at Googong and pays particular attention to preventing erosion at the abutment of the dam. Figure 13.29(b) shows the detail in the river, and Figure 13.29(c) the abutments.

A critical issue in steel mesh reinforcement for each layer is to keep the crest level as uniform as possible, and to be able to tie down the upper surface. Figure 13.31(b) shows gabions which were used in the event of imminent overtopping in lower parts of the embankment crest.

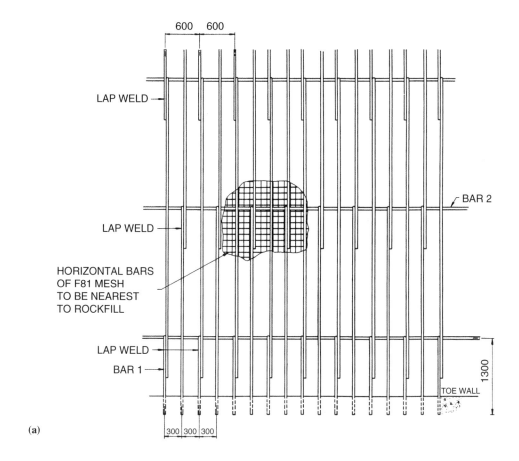

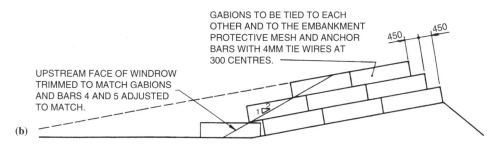

Figure 13.31.   Steel mesh reinforcement for Clarrie Hall Dam, (a) Dam detail; (b) crest gabions.

13.8.3   Design of steel reinforcement

The design principles are:

(a) Control ravelling of the rockfill by steel mesh or gabions;
(b) Protect the mesh from damage from floating debris by placing reinforcing bars up and down slope over the mesh; or construct the slope very steep with gabions so debris shoots out beyond the mesh/gabions; or cover (in part) with concrete;

(c) Prevent damage to the mesh from rockfill washing off the top of the dam by ensuring that the meshed section is always the highest point on the dam section. Rapid "closure" methods are needed if flooding is predicted;

(d) Anchoring the mesh into the dam in such a way that the anchors are always protected from erosion by the next layer of fill i.e. crank-shaped anchors (Googong); inclined anchors (Clarrie Hall); anchors fixed to grouted dowells (Mackintosh);

(e) Anchor length is determined by slip circle or wedge analysis, with anchor forces included. Pore pressures are estimated from design charts (see Fitzpatrick, 1977 and Lawson, 1987) or preferably by hydraulic model tests.

(f) The mesh is anchored into the abutment to prevent it being eroded out. Flow concentrates in this area during overtopping. Concrete protection may be provided (as in Clarrie Hall).

Detailing is vitally important and reference should be made to the papers referenced in ANCOLD (1982) and ICOLD (1984).

14

Specification and quality control of earthfill and rockfill

14.1 SPECIFICATION OF ROCKFILL

It is common to specify the following for rockfill for dam construction:

(a) *Type of rock, degree of weathering and source*
For example, "granodiorite, slightly weathered to fresh, from Quarry B" or "sandstone, with up to 20% siltstone, slightly weathered to fresh, from spillway excavation".

(b) *Layer thickness and maximum particle size*
The layer thickness varies depending on the rockfill zone, type of dam, and rockfill type, strength and degree of breakdown on rolling. Table 14.1 gives some examples. The maximum particle size can be the same as the layer thickness. Some e.g. USBR (1991) prefer a maximum particle size 90% of the layer thickness.

(c) *Rockfill grading and limits on finer particles*
The rockfill grading may be further specified depending on the properties required. Ideally rockfill is well graded, so it has a high density, strength and modulus after compaction and has a high permeability. However this is not always achievable if the rock available tends to break down on rolling to give a gap-graded material with a high percentage of sand. Many weakly cemented sandstones or weathered other rocks will do this. Even if the rock can be quarried and placed to meet the requirement of being well graded and with high permeability, care needs to be adopted in specifying such requirements, particularly for small dam projects as:
– It is costly and difficult to carry out the testing to determine the particle size in the embankment. For rock with a maximum size of 1 m, a hole about 4 m to 5 m diameter

Table 14.1. Examples of rockfill layer thickness.

Zone	Rock type, strength and breakdown	Layer thickness after compaction (m)
3A	Igneous or metamorphic, high strength, little or some breakdown on rolling.	0.9 or 1.0
3A	As above, adjacent Zone 2B[1].	0.45 or 0.5
3A	Metamorphic or sedimentary, or weathered igneous, medium or high strength, considerable breakdown on rolling.	0.45 to 0.6
3A	As above, adjacent Zone 2B[1].	0.45 to 0.6
3B	Igneous or metamorphic, high strength, little or no breakdown on rolling.	1.5 to 2.0
3B	Metamorphic or sedimentary or weathered igneous, medium or high strength, considerable breakdown on rolling[2].	0.9 to 1.5

Notes: [1] Thinner layers are to assist Zone 3A act as a filter to Zone 2B.
[2] Thinner layers would apply for concrete face rockfill dams where a high modulus is required or for upstream rockfill, with high fines content, or rock which weakens on saturation leading to collapse settlement.

Table 14.2.   Examples of rockfill grading requirements.

Dam	Dam type	Zone	Grading requirement		
			Layer thickness (m)	Size largest particle (mm)	% Other grading limits
Cardinia Creek	E & R	3A	1.8	>300	≤20% passing 25 mm
Harding	E & R	3A	1.0	>150	≤30% passing 20 mm and ≤10% passing 1.18 mm
Harris	E & R	3A	1.0	>150	≤30% passing 20 mm and ≤10% passing 1.18 mm
Thomson	E & R	3A	1.0	>150	≤30% passing 19 mm and ≤10% passing 1.18 mm
Winneke	CFRF	3A	0.7	Not specified	≤10% passing 0.075 mm
		3B random	0.4	Not specified	≤20% passing 0.075 mm
Boondooma	CFRF	3A	0.6	>75	≤15% passing 1.18 mm and ≤5% passing 0.075 mm
		3B	0.9	>75	≤15% passing 1.18 mm and ≤5% passing 0.075 mm

and to the depth of the layer would have to be dug to obtain a representative sample. Several tonnes of rockfill would be involved.

– The upper part of each layer is likely to be finer than the lower part, because of segregation in placing and breakdown of rockfill under the rolling action. For vertical seepage the permeability of this broken material will dominate. If, however, it is a normal situation where horizontal drainage is most important, such breakdown may not be important, provided the lower part of the layer remains permeable.

In any case the breakdown is a fact of life and the design must accommodate it. Table 14.2 gives some examples of rockfill grading requirements. These are for quite large dams for which it may have been practical to specify the grading to the extent done and to demonstrate that it was achievable by trial quarrying and fills. For most projects, for free draining high permeability rockfill, the authors favour an approach which requires that for the rockfill after compaction:

(a) The maximum particle size to be the compacted layer thickness;
(b) ≤10% or 15% passing 1.18 mm;
(c) ≤5% passing 0.075 mm, and
(d) Water not to pond on the surface of the compacted layer.

The grading would be checked by inspection and only measured if it became necessary from a contractual viewpoint to confirm the percentage of finer particles was excessive.

A guide to excessive fines can be if the fill moves excessively as it is being rolled, or trucks bog on the fill. This also indicates the fill does not have a high permeability.

(d) *Placement and spreading*

Usually the rockfill is dumped from trucks on the surface and spread by a bulldozer. In so doing the fines are moved onto the upper part of the layer, which creates a smoother working surface for the truck to place the next layer and the stratified permeability discussed above. Oversize rocks are often pushed into a specified zone (e.g. Zone 3B) in the outer part of the dam. To meet the filter requirements is important to ensure that the contact between Zones 2B and 3A does not have an accumulation of large rocks.

(e) *Roller type and number of passes*

It is normal to specify smooth steel drum vibrating rollers for compacting rockfill. The roller is usually specified as having a static mass of between 10 tonnes and 15 tonnes and a centrifugal force not less than say 240 kN "at the maximum frequency permitted by the manufacturer for the continuous operation of the roller" (PWD of WA, 1982). Alternatively (MMBW, 1980, for Thomson Dam) the static weight may be specified as a weight per metre of drum (50–55 kN/m), and a centrifugal force/m of drum (125 kN/m). The frequency was also specified (16–25 Hz) for Thomson Dam. Reference should be made to manufacturers' specifications for rollers.

Roller trials are often specified to determine the number of passes. These are discussed below. It is usual to require at least 4 passes of the roller.

(f) *Addition of water*

Water is often added to the rockfill to aid compaction and to weaken the rock-to-rock contact points in rock-types which are weakened by wetting. The water should be added by spraying on to the dumped rockfill before spreading. Commonly the amount of water is specified as a percentage of the volume of the rockfill e.g. 20%.

If the rock forming the fill is strong, produces few fines and is not greatly weakened by wetting, water may not be needed. Rolling trials can assist in assessing whether watering is needed and the amount of rolling required to achieve good compaction.

(g) *Durability requirements*

Rockfill is often required to be "hard" and "durable". The means of measuring this are seldom specified and in many dams such a requirement will be unobtainable, e.g. where siltstone or sandstone is being used, which may breakdown on repeated exposure to wetting and drying. It has been demonstrated (e.g. USBR, 1991 and Chapter 2, Section 2.9.1) that in some cases such breakdown only occurs on the surface of the rockfill, and is not detrimental to the shear strength or compressibility of the rockfill as a whole, provided the rockfill is well compacted and watered during compaction.

Hence, in general, there should be no requirements on durability for rockfill. The exception might be for some volcanic and altered granitic rocks which can deteriorate with time (see Sections 2.9.3 and 2.9.4.2). For rip-rap, or the outer layer of rockfill on the downstream slope, durability under wetting and drying is important and should be specified. The type of test will depend on the rock type proposed, but should include testing of the substance strength and assessment of the mineral composition, in particular the proportion of secondary minerals in the rock, and observation of blocks of rock left exposed to the weather.

(h) *Selection and placement of rip-rap*

Rip-rap on earth and rockfill dams is usually constructed by pushing the larger rocks from the adjacent rockfill zone to the face of the embankment and finishing the face by carefully positioning rocks with an excavator, sometimes fitted with "claws", to satisfy rock grading requirements for rip-rap, it will be normal to specify the grading more closely than rockfill, e.g. requiring that at least 50% of the rock should be greater than a certain size. Thin rip-rap layers for earthfill dams may be placed progressively as the embankment is built or on the completed face by dumping and spreading with a bulldozer.

(i) *Compacted density or void ratio*

Many designers in the past have also had a requirement for a compacted density for the fill, a common requirement being around 2.1–2.2 tonnes/m^3. Alternatively a void ratio (volume voids/volume solids) of 15–25% was required. The design is seldom sensitive to bulk density and, again for large rockfill, it is difficult and costly to measure density or void ratio because of the large volume which has to be sampled. Hence, it is unlikely (and unnecessary) that *insitu* density will be checked for smaller dams and relatively few tests will be done even for large dams.

14.2   SPECIFICATION OF EARTHFILL

It is common to specify the following for earthfill for dam construction.

(a) *The source and Unified Soil Classification of the earthfill*
For example, "clay and sandy clay, medium plasticity, from Borrow area A".
(b) *Maximum size and particle size distribution*
The maximum particle size (of gravel or rock fragments) in the earthfill is limited to ensure compaction is not affected. It will usually be specified as not greater than a size in the range 75–125 mm.
The particle size distribution is further defined to:

   (i) Ensure there is sufficient silt and clay fines passing 0.075 mm to give the required low permeability for the core. It is normal to require at least 15% passing 0.075 mm. For most clays, sandy clay and clayey sands this will be readily met.
   (ii) Be compatible with the grading requirements assumed for the design of the filters. This may be achieved by specifying the particle size grading envelope for the soil (based on laboratory tests from the borrow area, as shown in Figure 9.26). Alternatively for soils with no gravel content, the minimum D_{85B} i.e. the size for which 85% is finer used for the filter design could be specified. For gravelly soils, it will be necessary to express this in terms of the grading of the soil passing the 4.75 mm sieve, as is required for filter design.

   It is important to measure the particle size in the embankment after compaction to allow for breakdown of weak rock particles and mixing of fine and coarse soils from the borrow area. During site investigations, the effect of such breakdowns can be assessed by carrying out particle size distributions on samples which have been subjected to compaction.

(c) *The Atterberg limits*
Most authorities specify a minimum plasticity index and some also place a maximum on liquid limit. The latter may be due to the presence in the borrow area of particularly high plasticity clays, which may be difficult to compact, but in general there should be no need to place an upper bound on the liquid limit. In fact there is evidence that more plastic clays are likely to be less erodible than other soils so are advantageous to have in the core.

   The authors' preference is to specify an allowable range of liquid limit and plasticity index by relating to the "A" line on the plasticity chart. Figure 14.1 gives an example. The critical issue really is to specify limits which can be satisfied by the material you wish to use from the borrow area.

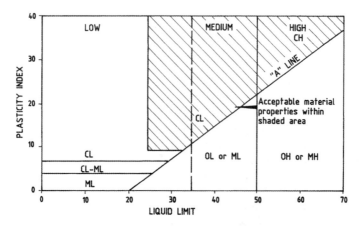

Figure 14.1.   An example of specification of Atterberg limits for earthfill.

The Atterberg limit requirements should be seen only as a way of allowing rejection of unsuitable material if a dispute arises with the contractor. Generally, it should be possible to accept or reject material visually in the borrow area, at least after the job has been under way for some time and the contractor and supervisors are familiar with the materials.

(d) *Density ratio, water content and layer thickness*

It is common practice to specify a density ratio ≥98% of standard maximum dry density, with a water content between OWC − 1% and OWC + 1%, or OWC and OWC + 2%, where OWC is standard compaction optimum water content. It should be noted that:

– Standard, not modified compaction should be used. This is to ensure moist compaction which leads to low permeability, flexible fills. Compaction at around modified optimum water content leads to high densities, but the soil structure is likely to be aggregated, leading to a higher permeability and more brittle fill. Recent studies (Wan and Fell, 2002, 2003) show dry compaction also is likely to result in clay soils which are more erodible than if they are compacted wet of optimum.

– The requirement for density ratio ≥98% is reasonable and compatible with the water content ranges shown. There is no advantage in specifying a higher density ratio and it may be detrimental in that the contractor will be forced to compact dry of optimum water content. For smaller dams, dams to be constructed in wet climates and soils which are difficult to compact, it would not be unreasonable to relax the compaction requirement to as low as 95% density ratio, provided that compaction is carried out above optimum water content. However, compaction to only 95% density ratio at say optimum −3% would lead to a permeable soil structure and would not normally be acceptable. Figure 14.2 shows some of these effects.

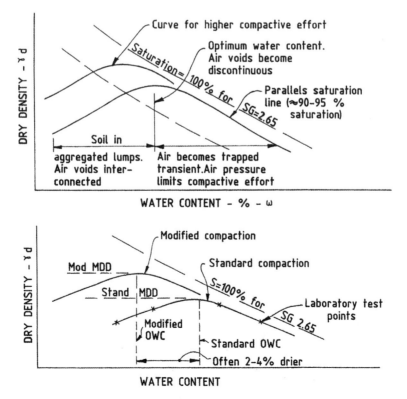

Figure 14.2.   Compaction water content and the effect on properties.

Table 14.3.   Practical maximum layer thickness (m) after compaction for different types of rollers at different applications (Forssblad, 1981).

Roller type static weight (drum module weights in brackets)	Embankment				Sub base	Base
	Rockfill (1)	Sand gravel	Silt	Clay		
Towed vibr. rollers						
6 tonnes	0.75	*0.60	*0.45	0.25	*0.40	*0.30
10 tonnes	*1.50	*1.00	*0.70	*0.35	*0.60	*0.40
15 tonnes	*2.00	*1.50	*1.00	*0.50	*0.80	–
6 tonnes padfoot	–	0.60	*0.45	*0.30	0.40	–
10 tonnes padfoot	–	1.00	*0.70	*0.40	*0.60	–
Self-propelled vibr. rollers						
7(3) tonnes	–	*0.40	*0.30	0.15	*0.30	*0.25
10(5) tonnes	*1.50	*1.00	*0.70	*0.50	*0.40	
8(4) tonnes padfoot	–	0.40	*0.30	*0.20	0.30	–
11(7) tonnes padfoot	–	0.60	*0.40	*0.30	0.40	–
15(10) tonne padfoot	–	1.00	*0.70	*0.40	0.60	–
Vibr. tandem rollers						
2 tonnes	–	0.30	0.20	0.10	0.20	*0.15
7 tonnes	–	*0.40	0.30	0.15	*0.30	*0.25
10 tonnes	–	*0.50	*0.35	0.20	*0.40	*0.30
13 tonnes	–	*0.60	*0.45	0.25	*0.45	*0.35
18 tonnes padfoot	–	0.90	*0.70	*0.40	0.60	–

Notes: (1) For rock fill only rollers especially designed for this purpose.
       (2) Most suitable applications marked*.

It has been common practice in Australian dam engineering (and elsewhere) to specify a layer thickness of 150 mm after compaction. USBR (1991) suggest 150–200 mm loose thickness for rolling with sheepsfoot rollers, and 225–300 mm loose thickness for rolling with 50 tonnes rubber typed rollers.

The 150 mm layer thickness after compaction is probably unnecessarily thin for modern heavy rolling equipment. Table 14.3 from Forssblad (1981) shows for different soil types the practical maximum layer thicknesses for a range of roller types and weights indicating that 6 tonnes to 10 tonnes pad foot rollers can compact layers up to 0.3 m and 0.4 m thick respectively. While these figures may represent an upper practical limit, they do show that a 150 mm layer is unnecessarily restrictive. It could be argued that, as a performance specification has been adopted for earthfill, i.e. water content and density ratio, there is no need for the "method or procedure specification" limitation on layer thickness. The authors' own preference is to specify layer thickness but to use more realistic figures, e.g. 200 mm or 250 mm for a 11 tonnes self propelled pad foot roller.

(e) *Placement and spreading*
Earthfill may be placed by scrapers or dumping from a truck and spreading with a grader or bulldozer. Oversize material should be removed before compaction.

The surface of the previously compacted layer should be scarified prior to placing the next layer of fill to ensure good bond. It may be necessary to add a small amount of water to the scarified surface prior to placing the fill. Care should be taken to ensure a uniform material. Segregation is likely to be a problem if gravel sized particles are allowed in the earthfill.

(f) *Roller type, weight, and number of passes*
It is normal to specify the roller type and weight. For earthfill it is most common to require a tamping foot (or "sheep's foot" or "pad foot") type roller because:
– The tamping foot action breaks up pieces of cemented soil or weathered rock in the fill;

- The rough surface of the layer left by a tamping foot layer allows better bond between layers;
- There is less tendency for formation of large shear surfaces beneath the roller when a tamping foot roller is used. This tendency is a particular problem in high plasticity, fine grained soils placed wet of optimum;
- The compaction water content is less critical than for rubber tyred rollers, i.e. the shape of the dry density-water content curve is broader for tamping feet type rollers.

The weight is best specified as a weight per metre length of roller, e.g. 6 tonnes/metre length as this better defines the contact pressure than total weight. Some authorities also specify the geometry and arrangement of the "feet" on the roller. This would seem to be unnecessarily restrictive for most dam construction. Sheepsfoot rollers are not readily available now, and are not so popular as they require more energy to pull them over the fill.

Sherard et al. (1963) report rolling trials carried out at the US Army Experiment Station, which showed that the foot size did affect the maximum dry density and optimum water content (as did the tyre pressure for rubber tyred rollers). Again, it could be argued that, provided the performance specification is defined, there is no need to also specify roller weight and type.

There are occasions where Zone 1 earthfill is better compacted with different roller types, e.g.:
- Where water contents are naturally high and compaction is required at well above optimum water content. In these cases a rubber tyred roller may be preferable as it will be less likely to become clogged with the wet, sticky clay. Knight et al. (1982) indicates that swamp dozers with low ground pressure were used to compact very high water content (80–120%) halloysite clays for the Monasavu dam in Fiji. Swamp dozer or similar techniques lead to compressible cores with high pore pressure, but they perform the primary function of low permeability. Sherard et al. (1964) and Kjaernsli et al. (1992) describe similar techniques used for wet, silty, sandy gravel glacial soils in Sweden and Norway. Penman (1983) also describes some recent dams where materials with high water content have been placed.
- Where the fill is a weathered rock which breaks down to a clayey silty sand/silty sand, steel drum rollers, compacting in thin layers (150 mm) may give better breakdown of the rock and compact it more readily. Best performance in these cases may be obtained without vibration.

The number of passes are best determined from field roller trials at the beginning of construction. Typically 6 to 8 passes is sufficient. Very little will be achieved by more than 8 passes and if more passes are required it is almost certain that control of layer thickness has been lost, or moisture conditioning is inadequate and should be rectified.

(g) *Water content adjustment*

Most specifications correctly require that water content adjustment be carried out in the borrow area, with only minor adjustment allowed on the embankment. This means that dry (or wet) soil in the borrow area is brought to a uniform water content by irrigating, harrowing, watering (or drying), and reworking before transportation to the embankment. It is impractical to do this on the embankment. Soils which are particularly dry or wet of the required water content may have to be conditioned for some days before use in the embankment. Failure to do this is the common cause of difficulty in achieving specified compaction requirement.

Sherard et al. (1963) give details of procedures which may be necessary to raise the water content of dry soils in the borrow area. They indicate that:
- Irrigation may be achieved by ponding of water or spray irrigation. Ponding is only suited to flat areas and can result in large evaporation losses;

  – The water content seldom becomes too high – most soils only take in water up to about the optimum;
  – Ripping to 0.6 m to 0.9 m will assist in allowing water to penetrate. Contour ploughing can assist in hilly borrow areas;
  – Water conditioning up to 4.5 m depth has been successfully achieved.

  Earthfill which dries out, or gets wet, on the embankment must be tyned, its moisture adjusted, then thoroughly reworked with a grader before recompaction. If the soil is judged too wet or dry to be adjusted on the embankment, it should be removed from the embankment.

  The risk of wetting earthfill on the embankment if rain is about to fall is usually reduced by requiring the surface to be "sealed" with a smooth drum roller, and contoured to allow runoff of surface water. The sealed surface must be tyned prior to placing the next layer.

(h) *Fill adjacent to the foundation*

  It is normal to specify that the earthfill which is to be placed adjacent (within say 0.6 m) of the dam foundation is to be composed of finer, more plastic soil available from the borrow area and is to be compacted at a higher water content (e.g. OWC + 2% or OWC + 3%) with rubber tyred construction equipment or rollers. This is to faciliate squeezing the soil into the irregularities in the foundation. Generally there is no "performance" specification (i.e. density ratio), because the layers are thin and testing is impracticable. On Thomson Dam, Snowy Mountains Engineering Corporation (SMEC), designing for the Melbourne and Metropolitan Board of Works (MMBW), required that the contact zone be compacted to a density 100% of that which would be achieved by compacting in the laboratory with standard compactive effort, at the field water content. This is a reasonable approach which allows for the high compaction water content, but should not be specified except for such large dams as Thomson (160 m high). For most dams, a methods specification with good water content control is sufficient.

(i) *Compaction of the edges of fill*

  Under normal operations, the outer 1 m to 1.5 m (measured horizontally) of an earthfill embankment will not be adequately compacted by rollers. It is necessary to specify either that the embankment is constructed oversize and trimmed back to the required lines by removing this poorly compacted soil, or to require rolling of the surface up and down the slope.

  If left in place, the poorly compacted soil will often soften and lead to surficial sliding, particularly on steeper slopes. While this will not in itself lead to failure of the embankment, subsequent erosion or sliding may cause problems.

14.3 SPECIFICATION OF FILTERS

It is common to specify the following for filters:

(a) *Type of material and source*

  For example "washed, graded sand and gravel alluvium from Borrow Area F" or "crushed, graded, slightly weathered to fresh granodiorite from Quarry G".

(b) *Particle size grading*

  This is specified as outlined in Section 9.5.1.

  It is important to make it clear in the specification for filters, that the particle size specification will be strictly adhered to, including the limitations on fines (passing 0.075 mm) content, and that the contractor should expect to have to wash, screen into stock piles of individual sizes and combine to obtain the required grading. It is unlikely that the required grading for fine filters in particular can be obtained directly from the borrow pit or by crushing alone.

The gradings should be specified as in place in the embankment unless trials or experience have shown that the material does not break down during compaction.

(c) *Durability*

This is discussed in Section 9.5.2.

(d) *Roller type, weight, number of passes, and required density index*

These are discussed in Section 9.6.3.

14.4   QUALITY CONTROL

14.4.1   *General*

The question of quality control for earth, earth and rockfill and rockfill dams was addressed by ICOLD and reported in Bulletin 56 (ICOLD, 1986b). The following discussion summarises the concepts in that bulletin, supplemented by some opinions of the authors and data from Australian practice.

14.4.2   *"Methods," and "performance" criteria*

There are two principal types of technical specifications:

(a) Method or procedure specifications which describe how the construction is to be carried out, in order to achieve the desired end product. The specifications stipulate to the contractor the materials to be used, the equipment and construction procedures, e.g. specification for earthfill includes:
– the source of materials;
– the water content;
– layer thickness;
– roller type and weight;
– number of passes.
This places the onus on the owner/engineer to have established procedures which will yield a satisfactory product.

(b) Performance or end product specifications which describe the end result to be achieved by the contractor. It is the contractor's responsibility to select materials, equipment and methods to obtain the specified end product, e.g. specification for earthfill includes:
– particle size gradation;
– Atterberg limits;
– water content;
– density ratio.

Many dam specifications are a mixture of these two alternatives. This is often unnecessary and can lead to inefficient construction procedures (e.g. requiring too thin a layer for compacting earthfill), unnecessary costly testing (as in requiring particle size and density of rockfill) and disputation between constructor and owner/engineers when the "methods" part of the specification fails to produce the required performance criteria.

In most projects, method's specifications become the routine quality control technique, even if a performance specification has been used for the contract documents. This is the only practical way of allowing work to proceed without unnecessary delay.

On some projects the owner/engineer will be responsible for the quality control, i.e. the inspection, testing and assessing whether the required quality standard has been achieved, and will have the authority to require additional work to achieve the standard or to reject the non conforming work. The owner/engineer in these projects will also be responsible for quality assurance, i.e. that the quality control standards are valid for the dam, that the specified tests are being implemented and correctly performed, that the quality control plan is working and that records and reports are verified and maintained.

On other projects the quality control aspect may be incorporated into the contract and be the contractor's responsibility. This can be quite satisfactory provided that the contract documents clearly define the level of inspection and testing required and the standards to be adopted and that the inspection and testing group are given authority by the contractor's management to enforce the standards. Often the testing will be subcontracted to a specialist geotechnical consultant. This assists in maintaining the independence of the inspection and testing group from the main contracting staff.

Whichever method is adopted, testing and QA procedures cannot replace experience visual inspection and commonsense changes to specifications, methods and control when they are needed.

14.4.3 Quality control

(a) *Inspection*

Inspection must always form a critical part of a quality control plan. The field and laboratory testing program should be seen first as establishing the methods required to achieve the required quality, then ensuring that the quality is being maintained and as a definitive, quantifiable means of rejecting substandard work. It is clearly impractical to test the whole of the completed product so one must rely on visual inspection to maintain overall quality.

It is important that the inspectors are properly trained and briefed on the implications of substandard work. It is also important to recognise that inspectors will often be needed in the borrow areas, as well as on the embankment, so that unsuitable material can be rejected before it reaches the embankment.

(b) *Testing*

Most testing which is carried out is for quality control, i.e. to ensure that the requirements of the specification are being met. The selection of areas or materials for testing may be done in either of two ways:

- Selecting those areas which are judged by the supervisor/inspector to be least likely to meet the specification. This can assist in reducing the quantity of testing and, if the testing shows acceptable performance, should ensure the overall adequacy of the construction work. It does, however, require independent and experienced supervisors;
- Selecting test areas at random, at the minimum recommended frequency. This will yield test results which better reflect the overall condition of the construction work, as the biased sampling of the first alternative is avoided.

The latter method is better suited to establishing statistical limits to the testing, allowing recognition of the fact that there is a statistical sampling error and that, within a large mass of earth and rockfill, the failure of a small proportion of the material to meet the basic specification criteria will not affect overall performance (provided the failures are not representative of poor compaction for example adjacent a conduit or wall).

Which of these two methods is adopted is also related to the specification limits used, and whether the supervisors are given any latitude in accepting material which falls below the specification.

For example, if earthfill is specified to have a density ratio in excess of 98%, with a water content optimum minus 1% to optimum plus 1%, it would be reasonable in most cases to accept an area which has tested at 97% density ratio, OWC + 2%, as this will still yield a low permeability fill. However, one would almost certainly reject an area which has tested at 95% density ratio, at OWC − 3%, as this will give a more permeable, erodible fill.

Similarly, if a relatively low compaction standard is set at (say) 95% density ratio, no "failures" should be accepted.

The authors' opinion is that it is desirable to establish such guidelines in the design stage of a project and to detail them in the specification either in a descriptive or a statistical form. Alternatively, a rigid specification limit may be set, but the site supervision staff should be advised as to the degree of flexibility which can be used in applying the specification.

The question of how many tests should be carried out on earth and rockfill is virtually impossible to answer, as it is interrelated to the specification standards, the competence of the contractor and the inspectors and the site conditions (e.g. variability of materials, climate).

ICOLD (1986b) gives some useful examples of minimum frequency of testing required for some large fill dams in USA, Austria, India, Canada and Italy. Tables 14.4, 14.5 and 14.6 summarize this information. Also shown in these tables is the testing specified for the random fill zone in Winneke Dam by the Melbourne and Metropolitan Board of Works and for the earthfill in the Ranger Mine Tailings dam (a zoned earth and rockfill dam).

(c) *Reporting*

It is important that a complete record should be kept of all construction operations. These are invaluable in the event that repairs or modifications are required, if the embankment is to be raised in the future and for surveillance during the life of the dam. Records are also important in respect to contractual and insurance claims. The reporting should include:

– Plans and specifications, including amendments and work as constructed;
– Final construction report written by the engineer;
– Monthly progress reports and reports on technical meetings;
– Reports from dam review panel if one is appointed;
– Laboratory test reports, including clear definition of location and level of samples tested, and differentiating between original tests and retests after failures;
– Daily reports by all supervisory personnel and inspectors in the form of diaries. These should concern adequacy of progress and comments on decisions;
– Photographs taken on a regular basis to show placement and compaction of earthfill, filters and rockfill.

The authors have been involved in a major dam foundation failure where daily reports and photographs proved invaluable.

14.4.4   *Influence of non technical factors on the quality of embankment dams*

In the Casagrande volume (Hirschfield and Poulos, 1972), Professor Ralph B. Peck highlighted some of the "facts of life" relating to the influence of non technical factors on construction of embankment dams. Professor Peck points out that many shortcomings in dam engineering relate to the attitudes and actions of the owner, designer, contractor and

Table 14.4.   Minimum required frequency of construction testing – Zone 1 earthfill.

Dam	Country	Fill volume	Particle size	Atterberg limits	Water content	Density ratio	Permeability	Shear strength
Oroville	USA	8,649,00 yd³	4,000 yd³	4,000 yd³	4,000 yd³	4,000 yd³	150,000 yd³	–
Culmback	USA	27,000 yd³	1,800 yd³	–	500 yd³	1,000 yd³	6,750 yd³	13,500 yd³
Beas	India	11,500,000 m³	56,000 m³	56,000 m³	1,150 m³	1,150 m³	–	56,000 m³
QA-8, La Grande	Canada	1,900,000 m³	4,000 m³	If required.	1,000 m³	4,000 m³	2 tests/season	800,000 m³
Brandy Ranch	USA	1,270,000 yd³ [1]	–	4,000 yd³	4,000 yd³	4,000 yd³	–	–
Bloomington	USA	1,750,000 yd³	5,000 yd³	5,000 yd³	5,000 yd³	5,000 yd³	–	15 tests
Winneke	Australia	–[2]	15,000 m³ or 2 per week	15,000 m³ or 2 per week	5,000 m³ or 1 per day	5,000 m³ or 1 per day	–	–
Ranger Mine Tailings Dam [3]	Australia	135,000 m³	3,000 m³	3,000 m³	1 per shift	500 to 1000 m³	Some	Some

Notes: [1] Earthfill and random fill.
[2] Random fill.
[3] Embankment is zoned earth and rockfill.

Table 14.5. Minimum required frequency of construction testing Zone 3A and 3B type rockfill.

Dam	Country	Dam type	Rockfill volume	Particle size	Compaction test	Density in place	Permeability	Shear strength
Oroville	USA	E & R	60,300,000 yd³	Each 24 hrs	100,000 yd³	100,000 yd³	–	–
Culmback	USA	E & R	130,000 yd³(1)	5,200 yd³	7,200 yd³	7,200 yd³	2 tests	2 tests
Finstertal	Austria	Asphalt core	2,790,000 m³	100,000 m³	Nil zone 3B 100,000 m³ zone 3A	2 tests per season	3 tests zone 3B 100,000 m³	2 tests/season 1 test/season zone 3A
Zirmsee	Austria	Asphalt face	550,000 m³	40,000 m³	–	2 tests	–	–
Beas	India	E & R	2,400,000 m³	–	–	15,000 m³	A number tests	A few tests
QA-8, La Grande	Canada	E & R	5,600,000 m³	4,000 m³	100,000 m³	5,000 m³	–	–
Bloomington	USA	E & R	3,010,000 yd³	100,000 yd³	–	100,000 yd³	–	–
Anapo upper and lower	Italy	Asphalt face	700,000 m³	15,000 m³	15,000 m³	45,000 m³	30,000 m³ (1.5 m dia) 100,000 m³ (0.6 m. dia)	–
Winneke	Australia	CFRF	–	50,000 m³ or 1 per week 25,000 m³ or 2 per week for random fill	–	50,000 m³ or 1 per week 25,000 m³ or 2 per week for random fill	N/A	N/A

Notes: (1) Gravel fill.
        (2) – indicates data not available.

Table 14.6.   Minimum required frequency of construction testing Zone 2A and 2B filters.

Dam	Country	Filter volume	Particle size	Relative density	Permeability	Shear strength
Oroville	USA	9,500,000 yd³	Each 24 hrs	4,000 yd³	150,000 yd³	–
Culmback	USA	36,000 yd³	1,400 yd³	1,400 yd³	2 tests	2 tests
Beas	India	1,670,000 m³	–	6,000 m³	Occasionally	–
QA-8, La Grande	Canadia	1,300,000 m³	4,000 m³	4,000 m³	–	–
Brandy Ranch	USA	22,000 yd³	1,500 yd³	1,500 yd³	–	–
Bloomington	USA	444,000 yd³	4,000 yd³	5,000 yd³	–	–
Winneke	Australia	Filter blanket	1,000 m³ or each layer	2,000 m³ or each alternate layer	N/A	N/A
Ranger Mine	Australia	297,000 m³[1]	1,000 m³ to 2,000 m³	Method specification	Some	
Tailings Dam						

Note: [1] Filter transition, combines function of Zones 2A & 2B.

technical consultants and these can outweigh the real technical issues. His comments are still relevant today and are recommended reading for all who are involved in dam engineering. Milligan (2003) also covers some of these issues.

14.5   TESTING OF ROCKFILL

14.5.1   *Particle size, density and permeability*

The testing of rockfill to determine the particle size compacted density and permeability is complicated by the large size of the rock particles.

Bertram (1972) describes field density tests in rockfill with a maximum size of 0.45 m, using a 1.8 m diameter density ring to define the density-in-place hole. The hole was dug to the full depth of the layer. ICOLD (1986b) give details of testing for Oroville Dam, where a 1.8 m diameter ring was used for rockfill up to 0.6 m size. Hence, a ring size 3 to 4 times the maximum particle size has been used to get representative samples. For equivalent samples of rock up to 1 metre maximum size, a ring and hole 3 metres diameter would be needed, giving about 15 tonnes of rockfill to be excavated for a 1 metre thick layer. This highlights the magnitude of the problem, and is the reason why it is impracticable to carry out density and particle size distribution tests in rockfill for smaller dams. Penman (1983) and Forssblad (1981) describe a compaction meter, which uses an accelerograph to monitor the response of a vibratory roller and assists in determining when the required degree of compaction has been achieved.

Test gradings on rip-rap are even more difficult to do. Spreading of a sample of rip-rap over the ground and individual measurement of rock particles within a defined strip has been used. On Prospect Dam remedial words (Sydney, 1996), one supplier went so far as weighing the individual particles.

Permeability of rockfill can be determined by ring infiltration tests *in situ*. If the rockfill is permeable the quantities of water involved would be huge and the results of doubtful value. For low permeability rockfill, tests through say 1.0 m to 1.8 m diameter rings, depending on particle size, could be carried out. In most cases a subjective assessment, by observing whether water will pond on the surface or in a hole dug through a layer, will be adequate to ascertain if the fill is "free draining" or not. Bertram (1972) and ICOLD (1986b) give references for field testing of rockfill.

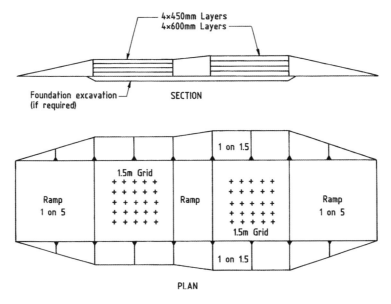

Figure 14.3.   Layout of rockfill roller trial (Bertram, 1972).

14.5.2   *Field rolling trials*

Field rolling trials are often carried out for larger dams:

(a) before construction, to ascertain the degree of breakdown of rock under rollers, the result-
     ing density and particle size distribution, permeability and modulus of compressibility;
(b) during construction, to confirm the number of passes of the roller required to achieve
     the required density of compaction.

Figure 14.3 from Bertram (1972) details a layout for roller trials on rockfill.
Bertram makes the following points:

1. An area sufficient to give 25–30 measuring points is necessary to overcome non uniformity in the rockfill;
2. The grids can be set out as shown (1.5 m) or 1.2 m × 1.3 m or 1.5 m × 2.1 m;
3. Measurements should not be taken less than 3 m from the edge;
4. Several layers (4 to 5 minimum) are required.

For construction, the specification usually requires that trials be carried out on the
embankment, requiring an area of about $200\,m^2$ for each of the test areas.

In the trials, the settlement of the surface of the rockfill is measured after each pass of
the roller and the results plotted as average settlement (or % settlement) vs number of
passes. Figures 14.4 and 14.5 show the results of roller trials at Murchison Dam, and at
Boondooma Dam.

The Boondooma Dam trials show the effect of watering the rockfill.

It can be seen that the additional compaction achieved after, say, 4 or 6 passes is relatively small. Since this additional compaction is often being achieved largely by breakdown of the upper part of the layer, it is common to limit the number of passes to 4 or 6,
seldom more than 8.

A key factor is any trial rolling is for equipment representative of what will be used on
the dam to be used in the trial. Too often, for example, the equipment is too small or not
enough water can be got to the trial rolling site, so the benefits of the trial are lost.

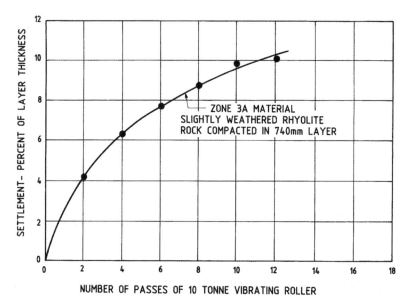

Figure 14.4.   Murchison Dam rockfill roller trials (Fitzpatrick et al., 1985).

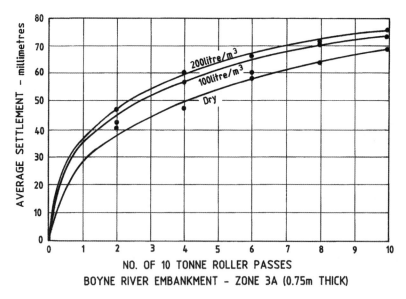

Figure 14.5.   Boondooma Dam rockfill roller trials (Rogers, 1985).

14.6   TESTING OF EARTHFILL

14.6.1   *Compaction-test methods*

The degree of compaction of cohesive earthfill is determined by the density ratio, where

$$\text{density ratio} = \frac{\text{dry density in place}}{\text{maximum dry density}} \qquad (14.1)$$

For earthfill in dams, the maximum dry density should be obtained using the standard compaction method (also known as standard Proctor) using the standard applied in the country, e.g.

- AS1289 E1.1 – Australia;
- ASTM D698-78 – USA;
- BS1377 4.1 Test 12 – United Kingdom.

The *in-situ* density should be determined using the sand replacement, rubber balloon or core cutter method

- AS1289 E3.1, E3.2 and E3.3;
- ASTM D1556-32, D2167-66 and D2937-71;
- BS1377 4.1 Tests 5, 15.

For cohesionless soil (sand, silt, sand/gravel) the degree of compaction is determined by density index, (or relative density), where

$$\text{density index} = \frac{e_{max} - e}{e_{max} - e_{min}} \times 100\% \qquad (14.2)$$

where

e = void ratio in place
e_{max} = void ratio in loosest state
e_{min} = void ratio in most compact state

$$\text{also density index} = \frac{\gamma_{d\,max}(\gamma_d - \gamma_{d\,min})}{\gamma_d(\gamma_{d\,max} - \gamma_{d\,min})} \times 100\% \qquad (14.3)$$

where

γ_d = dry density in place
γ_{dmax} = dry density in most compact state
γ_{dmin} = dry density in loosest state

The maximum and minimum dry densities are determined in the laboratory using the standard applying to the country, e.g.:

- AS1289 E5.1;
- ASTM D2049-69;
- BS1377 4.3 Test 14.

These methods all require the water content to be determined using an oven. As this usually takes about 24 hours at 110°C it may be unacceptably long for the construction condition.

For this reason it is common to specify that routine quality control will be carried out using the Hilf method which is described in detail in USBR (1985) and is a standard test (AS1289 E7.1). The method allows approximate determination of the density ratio within 1 hour of the density in place test, and accurate determination the next day when the water content is confirmed. Some specifications allow for disputes to be settled based on the measured water content.

Other methods which can be used to allow a more rapid determination of water content and, hence, density ratio, include drying the soil by:

- microwave oven;
- methylated spirits;
- high temperature oven;
- heating the soil on a hot plate with gas burners.

These methods are more approximate, and require calibration between them and water contents determined in the standard oven technique, to allow for adsorbed water being driven off by high temperature drying.

An alternative method of routine compaction control is the use of nuclear methods for determination of density and water content. In this method gamma rays are used to determine the density and neutrons to determine the water content. This may be done by direct transmission, backscatter or less commonly the air gap methods. The most common method is backscatter, because of the difficulty of penetrating the source into the compacted fill, particularly if the fill has gravel particles in it.

The method requires calibration against compacted materials of known density and water content from the site. Provided this is done, and the earthfill is not too variable or has too high a gravel content, the method can be adequate for routine control.

14.6.2   Compaction control – some common problems

Some common problems which arise in compaction control and which lead to disputation between contractor and engineer include:

(a) Specifying too high a compaction standard, e.g. 100% density ratio, standard compaction or 98% density ratio, modified compaction, for clay soils, or 100% density index for granular soils. These are virtually unobtainable, even with very heavy rolling equipment and, as discussed in Section 14.2, is undesirable for dam construction because it can only be achieved by compacting the clay fill dry, resulting in a brittle, permeable fill, or by overcompacting granular filter materials, giving excessive breakdown.

(b) Specifying unnecessarily restrictive water content range. Specification limits must be realistic to match the available materials. If the soil in the borrow area is, say, 4% wet of optimum in the borrow area, and the climate is wet, it is pointless requiring 98% density ratio at OWC $\pm$ 1%. The specification would be more realistically 95% density ratio at optimum to optimum +4%, or specified as detailed in Section 14.6.3.

(c) Carrying out insufficient laboratory compaction tests. The density ratio is obtained by comparing the density in place with the maximum dry density obtained in the laboratory. Ideally the soil for this laboratory compaction is sampled beside the density in place test. However, it is common practice to reduce the work involved by doing one laboratory compaction for every 2 to 4 or more density in place tests, assuming the soils are uniform. This is almost invariably not true and a small change in the maximum dry density can make the difference between acceptance and failure. Disputes often arise when such short cuts are used.

(d) Breakdown of materials during compaction. Soils which contain pieces of weathered rock, or gravels which break down under compaction, will often give a higher laboratory maximum dry density if the compaction test is carried out on soil dug next to the density in place test hole, i.e. on soil already compacted and broken down by the roller, than if the test is carried out on material sampled direct from the borrow area. The result of this is that the density ratio calculated is lower when the recompacted soil maximum dry density is used, often resulting in rejection of the fill. Since the rollers are dealing with the soil from the borrow area, not recompacted soil, this is unreasonable. There are two solutions for this problem – either lower the density ratio standard and use the recompacted maximum dry density, or ensure that laboratory compactions are done on representative uncompacted material. The latter may be difficult because of material variability.

(e) Property change on drying. Some soils change properties when dried in an oven or under lights. Halloysite clays are particularly prone to this, but most clays are affected. It is desirable not to dry the soil used for the laboratory compaction test completely, but to the water content needed for testing.

(f) Vibration from nearby construction equipment may affect the density obtained by sand replacement methods. This can be overcome by using the water balloon method, or by testing when equipment is not operating nearby.

(g) Inadequate curing of samples. Soils, particularly higher plasticity clays, need time for water added for laboratory compaction tests to evenly distribute throughout the sample. This is the reason why most standards require 12 hours to 24 hours "curing" of the soil before compaction. In a construction situation this may be regarded as impracticable and not adhered to. As a result, compaction results may be inconsistent and subject to error.

(h) Specification of standard soil tests for "gravelly" materials. The standard tests can be corrected for the presence of gravel particles up to a reasonable limit. However, if the soil to be tested is largely gravel size, the potential errors are too great and larger size compaction moulds must be used, or a methods type specification adopted. Inexperienced persons may even specify a density ratio rather than density index (relative density) for granular soils. One needs only to observe the loosening effect of a compaction hammer on sand in a compaction mould to appreciate that this cannot work.

14.6.3   Compaction control – some other methods

Some other approaches have been developed for routine compaction control of clay soils, particularly for fine grained clay soils which are compacted wet of optimum water content. These include:

– Specifying water content, and a density ratio based on wet density. The authors have used this approach for a dam constructed in Papua New Guinea, using halloysitic clay where previous experience had shown that it was very difficult to define maximum dry density and water content, i.e. the laboratory testing gave very variable results. The specification was written to require a water content between 44% and 50% (roughly OWC ± 3%) for the mean OWC and the required compaction dry density was 98% of the dry density achieved in the laboratory at the field water content. This wide range of water content was only practicable for the soil being used and would not normally be acceptable.

– Specifying undrained shear strength of the compacted earthfill. Knight (1990) describes the use of this technique, which is common in dams constructed in the United Kingdom. Undrained strengths used range from $\approx 40\,kPa$ to $110\,kPa$, with tests being carried out using triaxial tests on 100 mm diameter driven tube samples, or on remoulded samples. Lower bound and mean values were specified. Knight (1990) indicates that advantages of the method are that "measured strengths reflect design intention and testing is speedy". The authors' view is that the technique is useful where soils are to be compacted significantly wet of optimum, but that otherwise it is preferable to adopt a density ratio and water content specification. For most dams, the undrained shear strength of the core is not a critical issue, because stability is controlled largely by rockfill zones.

15

Concrete face rockfill dams

15.1 GENERAL ARRANGEMENT AND REASONS FOR SELECTING THIS TYPE OF DAM

15.1.1 *Historic development of concrete face rockfill dams*

The development of concrete face rockfill dams (CFRD) has been described by Galloway (1939), ICOLD (1989a), Cooke (1984, 1993, 1999, 2000) and Regan (1997) and is summarized in Table 15.1.

The first rockfill dam to have a concrete face was constructed in California in 1895. This followed on construction of timber faced dumped rockfill dams beginning in the 1850s.

These early CFRDs often had steep (0.54 H:1 V to 0.75 H:1 V) slopes, with a skin of hand placed rocks to stop the face from raveling. Then dams of greater height, up to 100 m high (Salt Springs Dam, California), were constructed with concrete faces and rockfill dumped in thick layers, often greater than 20 m or 35 m, and placed without compaction other than sluicing.

The design of faced rockfill dams was (and still is) mainly empirical, based on experience and judgement. The typical features of designs of CFRD up until the late 1950s are shown in Figure 15.1.

Many of these dams performed well. However several higher dams leaked excessively due to deformation of the concrete face, with resulting opening of joints and cracking. This could be attributed to the low modulus of the dumped rockfill, and to the detailing of joints, which allowed compression of horizontal and vertical joint fillers in the central part of the dam face (which is under compression) and resulted in increased opening of other joints, including the perimetral joint. The leakage did not endanger the dam stability but in some cases was unacceptably high for operating reasons.

Over the period 1955 to 1965 there was a general adoption of compaction of rockfill. This change was brought about by a realisation that dumping and sluicing of rockfill led to signifi-cant segregation, with the accumulation of larger rock at the base of the layer, leaving large voids and being particularly compressible. There was also a realisation that weaker rocks tended to lose strength on saturation leading to settlement if placed as dumped rockfill. Cooke (1984) and ICOLD (1989b) attribute the major change in approach to Terzaghi (1960a). Terzaghi (1960b) also introduced the change from a deep cutoff trench in rock as shown in Figure 15.1, to the adoption of the use of a plinth, or toe slab, on the grounds that excavation for the trench could loosen and fracture the rock, making it more permeable, and that the plinth could be an adequate cutoff if founded on suitable rock, grouted and anchored to the rock with steel bars.

15.1.2 *General arrangement – modern practice*

Figure 15.2 shows the modern practice for zoning of CFRD constructed of sound, free draining rockfill, on a strong rock foundation.

Table 15.1.   Historical summary of the trend in design and rockfill compaction in CFRD and Earth and Rockfill dams (Hunter 2003; Hunter and Fell 2002, based on Galloway 1939; Cooke 1984, 1993).

Approximate time period	Method of placement and characteristics of rockfill	Comments
Concrete faced rockfill dams		
Mid to late 1800s to early 1900s	Dumped rockfill with timber facing	Early embankments constructed with timber facing. Typically of very steep slopes (up to 0.5 to 0.75 H to 1 V). First usage of concrete facing in the 1890s. Height limited to about 25 m.
1920s–1930s	Dumped in high lifts (up to 20 to 50 m) and sluiced, hand or derrick placed upstream rockfill zone. Sluicing relatively ineffective	Rockfill typically sound and not subject to disintegration. Dam heights reaching 80 to 100 m. For high dams cracking of facing slab and joint openings resulted in high leakage rates (2700 l/sec Dix River, 3600 l/sec Cogswell, 570 l/sec Salt Springs).
Late 1930s to 1960s	High pressure sluicing used. Rockfill still very coarse	Cracking of face slab, particularly at the perimeter joint, and high leakage rates a significant issue with higher dams (3100 l/sec at Wishon, 1300 l/sec at Courtright).
From late 1960s	Rockfill placed in 1–2 m lifts, watered and compacted. Reduction in particle size. Usage of gravels and lower strength rock	Significant reduction in post-construction deformations due to low compressibility of compacted rockfill. Significant reduction in leakage rates; maximum rates typically less than 50–100 l/sec. Continued improvement in plinth design and facing details to reduce cracking and leakage.
Earth and rockfill dams		
1900 to 1930	Dumped rockfill	Use of concrete cores with dumped rockfill shoulders at angle of repose. Limited use of earth core. Dam heights up to 50–70 m.
1930s to 1960s	Earth core (sloping and central) with dumped rockfill shoulders	Use of earth cores significant from the 1940s due to the difficulties with leakage of CFRD. Increasing dam heights up to 150 m.
From 1960s	Use of compacted rockfill. Typically placed in 1–2 m lifts, watered and compacted with rollers	Improvements in compaction techniques. Early dams compacted in relatively thick layers with small rollers. Gradual increase in roller size and reduction in layer thickness reduced the compressibility of the rockfill. Significant increase in dam heights in the mid to late 1970s, up to 250–300 m.

The dam consists of:

Plinth. Reinforced concrete slab cast on sound, low permeability rock to join the face slab to the foundation.

Face slab. Reinforced concrete, preferably between 0.25 and 0.6 m thick, with vertical, some horizontal and perimetric joints to accommodate deformation which occurs during construction and when the water load is applied.

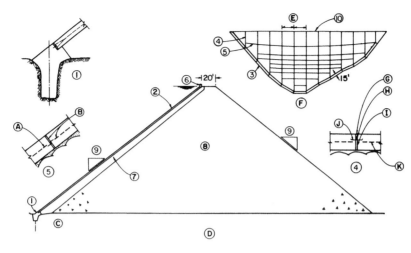

Figure 15.1.   Features of early concrete face rockfill dam design (ICOLD, 1989a). (1) Cutoff trench, (2) Concrete face, (3) Plinth, (4) Vertical joint, (5) Horizontal joint, (6) Parapet, (7) Crane-placed large rock, (8) Dumped rockfill, (9) Slope, (10) Curved axis. (A) Reinforcement, (B) 1.9 cm redwood filter and Z waterstop, (C) Grout curtain, (D) Cross section of dam, (E) 18 m (60 feet), (F) Elevation of face, (G) Mastic, (H) Premoulded asphalt, (I) Compressible joint filler, (J) U copper, (K) Reinforcement.

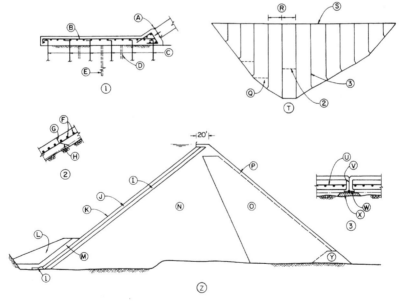

Figure 15.2.   Modern practice for zoning of CFRD constructed of sound rockfill on a strong rock foundation (adapted from ICOLD, 1989a). (1) Plinth, (2) Horizontal joint, (3) Vertical joint. (A) Perimetric joint, (B) Steel reinforcement, (C) Anchor bars, (D) Consolidation grout holes, (E) Grout curtain, (F) Horizontal reinforcement, (G) Width form, (H) Broom joint, (I) Zone 2E, selected small rock placed in the same layer thickness as Zone 2D, (J) Zone 2D, processed small rock, (K) Concrete face, (L) Zone 1B, random, (M) Zone 1A, impervious soil, (N) Zone 3A, quarry run rockfill or gravel fill, about 1.0 m layers, (O) Zone 3B, quarry run rockfill or gravel fill, about 1.5 m to 2.0 m layers, (P) Available large size rock dozed to face, (Q) Starter slab, (R) 18 m, (S) Straight axis, (T) Elevation of face, (U) Horizontal reinforcement, (V) Surface painted with asphalt, (W) Copper waterstop, (X) Mortar pad, (Y) Zone 3D, plus 0.3 m rockfill, (Z) Section of dam.

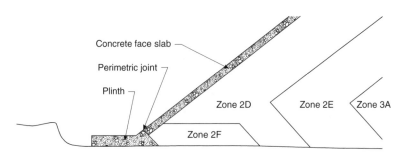

Figure 15.3.   Detail of Plinth area showing Zone 2F.

Zone 2D. Transition rockfill, processed rockfill or alluvium, grading from silt to cobble size or in more recent dams, from silt to coarse gravel size. The transition provides uniform support for the face slab and acts as semi-impervious layer to restrict flow through the dam in the event that cracking of the faceplate or opening of joints occurs.

Zone 2E. Fine rockfill, selected fine rock which acts as a filter transition between Zone 2D and Zone 3A in the event of leakage through the dam.

Zone 3A. Rockfill, quarry run, free draining rockfill placed in layers about 1 m thick. This zone provides the main support for the face slab and is compacted to a high modulus to limit settlement of the face slab.

Zone 3B. Coarse rockfill, quarry run, free draining rockfill placed in layers about 1.5 to 2.0 m thick. Larger rock may be pushed to the downstream face. This zone is less affected by the water load than Zone 3B, so a lower modulus is acceptable. The thicker layers allow placement of larger rock.

The boundary between Zone 3A and 3B should be selected with care. There has been a trend to moving the boundary from what is shown in Figure 15.2 to the dam centreline and even upstream of the centreline, but Marulanda and Pinto (2000) warn that, at least for high dams, this may lead to larger settlements.

Zone 2F. Some modern dams include a Zone 2F filter zone beneath the perimetric joint as shown in Figure 15.3. This serves two functions: (a) to act as a high modulus zone to limit deformation of the slab at the perimetric point; (b) to act to limit leakage flow in the event the joint opens. If a zone of earthfill (Zone 1A) is provided upstream of the slab, Zone 2F acts as a filter to that earthfill in the event the joint opens and leakage initiates.

Zones 2F is usually a maximum size of 19 mm or 37 mm, with some silty fines, placed in thin (200 mm) layers.

Many variations of this zoning are adopted to meet site conditions and the quality of construction materials available. As discussed in Section 15.5.1, non free draining rockfill may be used, provided free draining zones are incorporated to give any seepage water a controlled path to the dam toe. In some dams, low permeability earthfill is placed upstream of the face (Zones 1A and 1B in Figure 15.3) to control leakage. This is discussed in Section 15.5.4.

The zone designation discussed above is used throughout the book. The transition Zones 2D and 2E have been so designated to differentiate from Zones 2A, 2B and 2C used in earth and rockfill dams. There is no consistent terminology in use throughout the world.

Many dams include a concrete crest wall which reduces the quantity of rockfill required (see Section 15.3.4).

These developments, along with other refinements in design, have resulted in the earlier problems being overcome and acceptance of CFRD for the construction of many dams, including dams up to 190 m high. CFRD up to 230 m high are in the design phase.

Table 15.1 and Figure 15.4 show the trend in design and rockfill compaction of CFRD. The trends in compaction of rockfill in earth and rockfill dams are included.

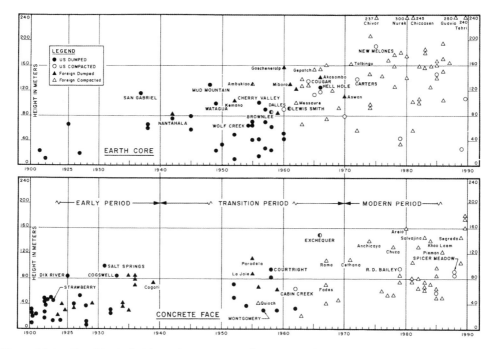

Figure 15.4.   Trend in the height and compaction of CFRD (Cooke and Strassburger, 1988).

15.1.3   *Site suitability, and advantages of concrete face rockfill dams*

CFRD are suited to dam sites with a rock foundation and a source of suitable rockfill. In many cases CFRD will be a lower cost alternative than an earth and rockfill dam. This is discussed in Sherard and Cooke (1987), and Fitzpatrick et al. (1985). Factors, which may lead to CFRD being the most economic alternative, include:

– The non availability of suitable earth fill;
– Climate. CFRD are suited to wet climates, which may give short periods in which earthfill can be placed. This can result in significant overall savings in schedule;
– Grouting for CFRD can be carried out independently of embankment construction, which may result in savings in overall time for construction;
– Total embankment fill quantities are likely to be smaller and side slopes steeper for CFRD than for earth and rockfill dams leading to reductions in the cost of fill and diversion tunnels.

Sherard and Cooke (1987) indicate that the cost of the concrete face is often less than the additional costs of earthfill and filters and more extensive foundation treatment for earth and rockfill dams. This has commonly been the case in Australia where a significant proportion of major dams built after 1985 have been CFRD.

Sherard and Cooke (1987) point out that CFRD have generally been used for dams of moderate height (40 m) or higher. They suggest that CFRD may also be economic for lower dams if they have a long crest length because:

– the cost of foundation treatment for a long, low dam is high relative to the overall cost, but will be less for CFRD, because the CFRD requires a smaller width to be treated, than for earth and rockfill;

– the cost of filters in a long, low earth and rockfill dam is relatively high because the filters constitute a high percentage of the total volume.

They describe the use of CFRD for a 2000 m long 20 m high dam in Venezuela.

Varty et al. (1985) indicate that the Hydro-Electric Commission of Tasmania used CFRD for several smaller dams, ranging in height from 20 m to 44 m. They list faceplate construction techniques which are different from their normal practice for these smaller dams, including the elimination of starter bays, a lighter slip-form and a mobile crane on the dam crest.

These examples show that smaller dams can be economically constructed as CFRD, but probably where the economy of scale applies, such as a long crest, or a requirement to construct several dams, to offset the cost of setting up for slip-forming.

Sherard and Cooke (1987) and Cooke (2000) argued that CFRD should be considered for the very highest dams. Sherard and Cooke (1987) argue that CFRD have fundamental advantage over earth and rockfill dams in that there is no possibility of piping erosion of the earth core and over arch dams in that they rely on gravity for stability, not high strength abutments, and that the CFRD supports high abutments, rather than stressing them. They argue that there is no reason CFRD up to 300 m high could not be built and suggest that the jump in precedent from current heights is not excessive and is similar to practice in earth and rockfill dams. They suggest that conservative perimetric joint details would have to be used for very high CFRD, and wider, more conservatively designed, processed material used for Zones 2D and 2E to ensure homogeneity, lack of segregation and low permeability.

By 2000 (Cooke, 2000) two 190 m high dams had been constructed (Aquamilpa and Tianshengqiao 1) and seven CFRD dams are proposed with heights of up to 230 m.

Fitzpatrick et al. (1985) and Mori (1999) suggest that because of the high modulus obtained with compacted gravel fills, they are preferable construction material for very high dams. They counsel caution in the use of rockfill for dam heights greater than have so far been built.

15.2   ROCKFILL ZONES AND THEIR PROPERTIES

15.2.1   *Zone 2D – Transition rockfill*

As the design of concrete face rockfill dams has developed and higher dams have been constructed, more emphasis has been placed on the grading and placement of the rockfill zone immediately below the concrete face slab. In CFRDs constructed in the 1960s the function of Zone 2D was seen as providing uniform support for the face slab. Specifications required that the Zone 2D was screened to remove all materials less than 25 or 50 mm. This was to ensure that, in the event of a leak in the face slab, there would be no fines to be washed away which might lead to loss of support of the face slab. However with this grading Zone 2D was very permeable and, rocks were easily dislodged during construction, and a rough, porous surface led to excess concrete being required for the face slab.

Beginning with the 110 m high Cethana Dam in 1971 (Wilkins et al., 1973) there was a change in design approach which resulted in Zone 2D being specified as "crusher run" or "quarry run" (after passing through a grizzly) rockfill passing 150 to 225 mm (up to 300 mm). This resulted in a material with a lower permeability and a smoother and more stable surface on which to construct the face slab. Many dams have been successfully constructed in this way and this was the established practice up till about the mid 1980s.

Gradations for Cethana, Alto Anchicaya and Foz de Areia are shown in Table 15.2.

Later designs have specified a finer grading with Zone 2D not only meeting filter requirements with Zone 2E, but also having a specified requirement for a silty fines content to reduce the permeability, and hence leakage, in the event of a face slab joint opening or the slab cracking.

Table 15.2.   Zone 2D gradation specifications (Sherard 1985b).

| Size (mm) | Dam, Year (% finer) | | |
	Cethana, 1971	Alto Anchicaya, 1974	Foz do Areia, 1980
300		100	
225	100		
150	80–100		100
100			
75	51–100	50–100	55–100
40			
20	15–63	25–60	18–65
5	0–40	0–30	5–22
0.5	0–17	0–5	0–8
0.075	0–5		0–1

Sherard (1985b) points out that quite large leakage had occurred at some dams constructed with crusher-run Zone 2D including 1800 litres/second at the Alto Anchicaya Dam on first filling. Sherard notes that, for such wide graded materials, with a small percentage of sand size material, segregation was a problem, and this could in part explain why such large leakage had occurred. He suggests that about 40% of sand size (passing 4.76 mm) is required to avoid the segregation. Sherard also shows that leakage through the dam resulting from a crack in the face slab, is controlled more by the permeability of Zone 2D, than by the crack aperture. Sherard concludes that Zone 2D should be a 4 m to 5 m wide zone.

The properties of the Zone 2D material placed during the construction of six major CFRD in Brazil (Sobrinho et al., 2000) are summarized in Table 15.3. These dams have transition zones that range in width from 10 m at the bottom to 4 m at the crest.

Current practice follows the recommendations of Sherard (1985b) and reported in ICOLD (1989a) and Amaya and Marulanda (2000) as shown in Table 15.4.

Sherard (1985b) indicates that a grading for Zone 2D as shown in Table 15.4 will be "stable", i.e. internally stable, not susceptible to washing out of the fines, and that a permeability of 10^{-6} m/sec would be achieved. Earlier Zone 2D are not internally stable, and will be susceptible to the finer particles washing from the zone, leading to a higher permeability and possibly some settlement of the face slab as the fines are eroded.

ICOLD (1989a) recommend a Zone 2D grading as shown in Table 15.4, i.e. virtually the same as that suggested by Sherard. They indicate that a maximum of 10% to 12%, passing 0.075 mm is desirable, while giving 15% as the upper limit. ICOLD (1989a) indicates that Zone 2D should be 4 m to 5 m wide, possibly wider for dams higher than 150 m. At Antamina Dam in Colombia (Amaya and Marulanda, 2000) where strict control of seepage is required, a Zone 2D width of 8 m has been adopted.

Sherard (1985b) acknowledges that the required Zone 2D grading may require processing and/or blending of materials, but points out that the incremental cost is small. Mori (1999) suggests the use of compacted alluvial gravel in Zone 2D to reduce deformation and cracking of the face slab.

Fitzpatrick et al. (1985) and Sherard (1985b) and Cooke (2000) point out that an added advantage of having Zone 2D graded as shown in Table 15.4, is that, if a crack forms in the face slab, it can be sealed by spreading silty sand over the crack and having the silty sand wash into it, with further erosion controlled by Zone 2D acting as a filter. This method has been used to reduce leakage on several dams including Shiroro Dam in Nigeria (Bodtman and Wyatt, 1985), Khao Laem Dam in Thailand (Watakeekul et al., 1985) and Bailey (Cooke, 2000).

Table 15.3.   Characteristics of Zone 2D in Brazilian CFRD (Sobrinho et al., 2000).

Dam	Foz do Areia	Segredo	Itá	Xingo	Machadinho	Itapebi
Fill type	Crushed sound basalt	Crushed sound basalt	Crushed sound basalt	Grizzlied sound and weathered granite/gneiss	Crushed sound basalt	Processed gneiss
Geometry						
Bottom width (m)	13	8	10	12	10	12
Crest width (m)	4	5	3 + 4	4/6	3 + 4	3 + 4
Layer (thickness m)	0.4	0.4	0.4	0.4	0.4	0.4
Gradation						
Max. Size mm	100	75	75	100	75	100
25.4 mm % passing	50	45	60	70	50	80
4.76 mm % passing	12	20	25	44	15	45
0.150 mm % passing	1	2	5	10	7	11
0.075 mm % passing	0	0	1	7	2	7
Compaction						
Horizontal (passes/roller)	4/10 tonne vibratory	4/10 tonne vibratory	4/9 tonne vibratory	6/9 tonne vibratory	4/10 tonne vibratory	4/9 tonne vibratory
Upslope (passes/roller)	6 passes	4/static + 6/vibratory	extruded wall	4/static + 6/vibratory	extruded wall	extruded wall
Void Ratio	0.31	0.21	0.175		0.31	0.19
Density (t/m^3)	2.12	2.27	2.15	2.12	1.97	2.20
Performance during construction	Adequate	Adequate	Adequate	Cracking + Settlement	Under construction	Under construction

Table 15.4.   Desirable specification for Zone 2D.

	Sherard (1985b)	ICOLD (1989a)	Amaya & Marulanda (2000)
Size (mm)	% finer	% finer	% finer
75	90–100	90–100	90–100
37	70–95	70–100	70–100
19	55–80	55–80	65–100
4.76	35–55	35–55	40–55
0.6	8–30	8–30	10–22
0.075	2–12	5–15	4–8

15.2.2   Zones 2E, 3A and 3B – Fine rockfill, rockfill and coarse rockfill

15.2.2.1   General requirements

The basic requirements for rockfill in a CFRD are:

– The rockfill should be free draining to avoid build-up of pore pressure during construction, and to allow controlled drainage of water which might leak through the faceplate;

- The rockfill should have a high enough modulus after compaction in the dam to limit face slab deflections under water load to acceptable values. Creep of the rockfill should also be small enough to avoid excessive longer term settlements;
- It should be readily available as a quarry run product with a minimum of wastage of oversize or undersize rock.

A wide range of rock types have produced satisfactory rockfill including granite, basalt, dolerite, quartzite, rhyolite, hornfels, limestone, gneiss, greywacke, andesite, welded tuff and diorite. Rocks such as sandstone, siltstone, argillite, schist and shale have been used but in some cases produce a non-free draining rockfill. This is discussed further in section 15.5.1. Gravels have also been used with success and as discussed below, can lead to very high modulus fills. Cooke (1984) suggests that "if blasted rockfill is strong enough to support construction trucks and the 10 tonne vibratory roller when wetted, it may be considered to be suitable for use in compacted rockfill". He goes on to point out the need for drainage zones if the resulting rockfill is of low permeability.

Penman (1982) and Penman and Charles (1976) suggest that for rockfill to be considered "free draining", it should have a permeability of at least 10^{-5} m/sec based on *in situ* tests in the rockfill. This is based on the requirement for adequate permeability to dissipate construction pore pressures in the rockfill, rather than having a high water discharge capacity from a leaking face slab.

The authors are satisfied rockfill is free draining if water will flow rapidly from a test pit which penetrates the full layer thickness of the rockfill, when pumped in from a water tanker.

As discussed by Cooke (1984) and Sherard and Cooke (1987), rockfill placed in the normal way, i.e. dumped from a truck, and spread by a bulldozer, will result in segregation, with the coarser particles collecting at the base of the layer, and the finer rock and fines on the surface. Breakdown of the upper part of the layer during rolling creates even more stratification. Cooke (1984, 1993) and Cooke and Sherard (1987) point out that far from being a problem, this stratification is desirable because:

- Tyre wear on trucks is reduced and the smoother surface allows more rapid truck travel;
- The smooth surface facilitates the rolling operation, spreading the vibrating load and reducing roller maintenance compared to an irregular rocky surface;
- The lower parts of layers have a high horizontal permeability, facilitating drainage of leakage or embankment overtopping during construction. The resulting average horizontal permeability is much higher than if the fines were distributed uniformly throughout a layer;
- The layers create a variation in vertical permeability which will prevent buildup of pore pressures in the rockfill.

It should be noted that this stratification means that "free draining rockfill" may allow water to pond on the surface of layers, provided the lower parts have higher permeability.

As pointed out by Cooke (1984), there is no need to scarify the surface of layers of compacted rockfill prior to placing the next layer.

The fourth author was shown in 1997 a video of a 30 m high rockfill (quarried) coffer dam in Brazil that had been built in 3–4 days, the crest just keeping ahead of a rising flood. At the start of the coffer dam construction, an impervious zone had been placed on the upstream face, but with the advert of a flood, the contractor decided to push ahead with the coffer dam as a simple rockfill dam. The video showed clearly that seepage appeared on the downstream side from the lower half of each layer. The rockfill placed in roughly 1 m layers and compacted only by construction traffic, did not start to ravel at any stage, except for the odd stone moved at abutment-dam contact lines. This behaviour of this coffer dam showed clearly how well an embankment of quality rockfill, "compacted" in lay layers, will cope with extreme conditions without overtopping. In this case, the dam was deliberately breached before it overtopped.

15.2.2.2   Layer thickness and compaction

Zones 3A and 3B are placed in layers of the order of 1 m and 1.5 m to 2 m thick respectively, which results in a gradation of permeability and a lower modulus for Zone 3B, which is acceptable since the water load is largely taken by Zone 3A. Rolling for good quality rockfill is usually by 4 (up to 8) passes of a 10 tonne vibratory steel drum roller. Cooke (1993) indicates there is no evidence that rollers heavier than 10 tonne give better results. To avoid embankment deformation Mori (1999) notes that based on experience at Xingo (140 m high), Aquamilpa (180 m high) and Tianshengqiao 1 (180 m high). Zones 3A and 3B should not have a very different moduli of compressibility, should have a near vertical interface between them and a high embankment should not be built in stages. This is discussed in Section 15.6. Cooke (1993) suggests that lower strength (<30 MPa) rockfill should be compacted in thinner (0.6 m to 0.8 m) layers to encourage breakdown to give greater "strength" (a higher modulus).

Hydro Electric Commission of Tasmania (HEC) (Carter et al., 2000) place Zone 3A in 1 m layers for good quality rockfill and 300 mm to 600 mm layers for lower quality but well graded material. Zone 3B is compacted in 1.5 m layers to take large rock.

Zone 2E is required to act as a filter to Zone 2D, and is therefore placed in relatively thin layers (0.4 m to 0.5 m, the same as Zone 2D). This also ensures a high modulus.

Zone 2E is obtained by selecting suitable material in the quarry, or by passing rockfill over a grizzly. The filter requirement (compared to Zones 2D and 3A) is usually readily achieved by this process, and further treatment is not required.

The rockfill grading in Zone 3A (and to a lesser extent in Zone 3B) is important. A well graded rockfill will compact to a higher modulus than a poorly graded fill, and the amount of fines needs to be limited if the rockfill is to be truly free draining.

Cooke (1984) suggests that the grading specification should be:

- Maximum size shall be that which can be incorporated in the layer and provides a relatively smooth surface for compaction;
- Not more than 50% shall pass 25 mm sieve;
- Not more than 6% shall be clay size particles (this is taken to mean silt and clay).

Cooke (1993) gives not more than 10% finer than 0.075 mm).

In Sherard and Cooke (1987), it is suggested that it is better to specify:

- Not more than 20% finer than 4.76 mm;
- Not more than 10% finer than 0.075 mm (i.e. silt and clay).

They suggest that if the rockfill has a higher proportion of fines, "the final evaluation of suitability can be made on the trafficability of the rockfill surface when the material is thoroughly wetted. A stable construction surface under travel of heavy trucks demonstrates the wheel loads are being carried by a rockfill skeleton. An unstable construction surface, with springing, rutting, and difficult truck travel, shows that the volume of soil like fines is sufficient to make the rockfill relatively impervious. Where the surface is unstable, the fines dominate the behaviour and the resulting embankment may not have the properties desired for a pervious rockfill zone". As pointed out by Cooke (1993) the purpose of the specification is to be able to reject loads of predominantly soil or rock with large amounts of fines.

Sherard and Cooke (1987) point out that there is no technical need for rock in the rockfill to have a high compressive strength. In their experience, rockfills constructed of rocks with compressive strength of 30–40 MPa are no more compressible than those of higher strength. However as discussed in Section 15.2.3 the rock strength does have some influence on the rockfill modulus.

Rocks with very high compressive strength lead to higher quarrying cost and wear on equipment. They conclude that any rock with a (soaked) unconfined compressive strength of 30 MPa or more is adequate. Lower strength rocks may be used with special zoning provisions as discussed in Section 15.5.1.

15.2.2.3   *Use of gravel as rockfill*

Cooke (1984) discusses the use of gravel (really sandy gravel or sandy gravel with some silt) as rockfill and maintains that, if available, this material can be very suitable. He points out that:

- Gravel is often more economically handled than rockfill (lower excavation and loading cost, and less wear on tyres and rollers);
- Compacted gravels commonly have high modulus of compressibility, up to 5 to 10 times that of some compacted rockfill. In his experience since face slab movements vary roughly inversely with the modulus, and directly with the square of height of the dam, gravel fills are desirable for higher dams;
- If a significant fines content is allowed, then chimney drains, abutment drains, filters and intermediate drainage layers will be required.

Mori (1999) recommends that alluvial gravel should be used if possible in the upstream third of the dam to limit deformation and cracking of the face slab.

Amaya and Marulanda (1985) describe the use of gravel for the 125 m Golillas Dam in Colombia. In this dam the bulk of the fill (i.e. Zone 2 in Figure 15.5) was natural river gravel. The grading is shown in Table 15.5.

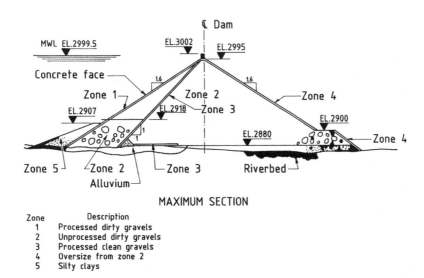

Zone	Description
1	Processed dirty gravels
2	Unprocessed dirty gravels
3	Processed clean gravels
4	Oversize from zone 2
5	Silty clays

Figure 15.5.   Golillas Dam (Amaya and Marulanda, 1985, reproduced with permission of ASCE).

Table 15.5.   Grading of river gravels used as "rockfill" in CFRD.

Sieve size (mm)	Dam, year (% passing)			
	Crotty, 1971	Golillas, 1978	Salvijina, 1984	Gouhou, 1992
300	100	100	100	100
150	85–100 (98)[b]	60–100	65–100 (98)	100
75	34–98 (85)	35–80	35–100 (80)	60–100 (90)
25	26–70 (55)	20–50	10–80 (40)	55–90 (65)
4.76	6–42 (16)	10–30	0–30 (8)	25–65 (42)
1.18	2–22 (10)	4–22	0–25 (6)	15–45 (28)
0.075	0–4 (2)	0–12 [a]	0–12 (3)	5–10 (6)

Notes: [a]   Average 8% for one source, 12% for second.
       [b]   Average in brackets.

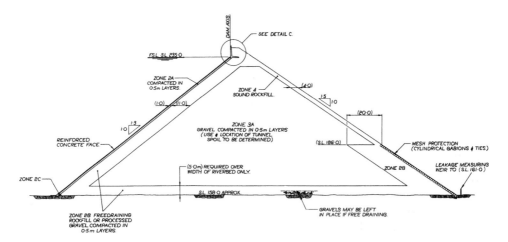

Figure 15.6.   Crotty Dam (HEC, 1988).

The Hydro-Electric Commission of Tasmania used gravels to construct the 82 m high Crotty Dam in 1991 (Figure 15.6). The Golillas Dam gravels (Figure 15.5) were made up of 70% sandstone, 20% siliceous shales and 10% siltstones and limestones. The gravels were washed for drainage zones in the dam. Sierra, et al. (1985) describe the use of gravels in Salvijina Dam in Colombia. The 148 m high dam had its "Zone 3A" constructed of gravels from gold mining dredging for Zone 2 as designated in Figure 15.7.

The range and average grading is given in Figure 15.5.

The gravels are usually placed in approximately 0.6 m thick layers and compacted with 4 passes of a 10 tonne steel drum vibratory roller, without water added. Water can make compaction difficult if the gravels are silty.

Very high moduli can be obtained as shown in Table 15.6.

As pointed out by Cooke (1984) gravel has been successfully used in several high earth and rockfill dams including Nurek (305 m), Oroville (244 m), Mica (244 m) and Bennett (183 m). Moduli of 365 MPa for Oroville and 551 to 689 MPa for Bennett were achieved despite use of smaller rollers than would now be adopted.

The Gouhou Dam in China (Chen, 1993) was constructed of gravel with a finer grading envelope than the earlier dams (Table 15.5). The zoning of this dam is shown on

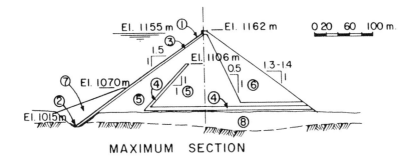

Figure 15.7.   Salvajina Dam (Sierra et al., 1985, reproduced with permission of ASCE).

Figure 15.8. There was little difference in grading between the zones II, III and IV. The dam failed when leakage occurred through joints in the face, particularly at the join of the face slab and the wave wall which was below the reservoir normal FSL. The relatively fine fill, with no higher permeability zones as are built in to other gravel fill dams, contributed to the failure. Figure 15.9(a) and (b) show the failed dam and Figure 15.9(c) the crest wall detail.

15.2.3   *Effect of rock properties and compaction on modulus of rockfill*

The modulus of compressibility (E) of the rockfill is dependent on the rock type, strength, shape and gradation of rock sizes in the rockfill and layer thickness. It is also dependent on the roller size and type, number of passes, whether water is added during compaction, the confining stresses on the rockfill and also the duration of loading, i.e. there is a creep component.

Table 15.6 summarizes the properties of compacted rockfill from a number of CFRDs. The rockfill moduli during construction (E_{rc}) and the pseudo modulus on first filling (E_{rf}) values have been calculated from observed settlements of rockfill during construction of the dam using cross-arm settlement gauges and from the observed deflection of the face slab on first filling. Unless otherwise stated, the values quoted are generally "average", typical of rockfill at the lower half and centre of the dam. The moduli have been calculated with the simplified procedure shown in Figure 15.10, except that for E_{rc}, the stress distributions due to the shape of the embankment have been allowed for, using elastic solutions in Poulos and Davis (1974).

Fitzpatrick et al. (1985) calculate E_{rc} and E_{rf} from

$$E_{rc} = \gamma H d_1 / \delta_s \qquad (15.1)$$

$$E_{rf} = h d_2 / \delta_n \qquad (15.2)$$

Table 15.6.   Summary of embankment and rockfill properties for CFRD (Hunter, 2003; Hunter and Fell, 2002).

Dam name	Embankment dimensions		Rockfill source[1]	Material parameters/properties of rockfill								Rockfill moduli	
	Height, H (m)	L/H		Strength[2] classification UCS (MPa)	C_u	d_{max} (mm)	% finer 19 mm	Dry density (t/m³)	Void ratio, e	Layer thickness (m)	Placement	E_{rc} (MPa), average	E_{rf} (MPa)
Zone 3A rockfill													
Aguamilpa	185.5	2.6	Alluvium	(Very high)	85	600	34	2.22	0.18	0.6	4 p 10 t SDVR, in moist condition	305 (250 to 330)	770
Crotty	83	2.9	Gravels (Pleistocene)	(Very high)	70	200	48	2.54	0.20	0.6	8–12 p 6–10 t SDVR, watered	375 (113 to 636)	470
Golillas	125	0.9	Gravels	(Very high)	125	350	40	2.135	0.24	0.6	4 p 10 t SDVR, water added	155 (145 to 165), arching likely	250
Salvajina	148	2.4	Gravels	(Very high)	9.2	400	32	2.24	0.25	0.6	4 p 10 t SDVR, water added	205 (175 to 260), arching likely	500
Alto Anchicaya	140	1.9	Hornfels	(Very high)	18	600	22	2.28	0.294	0.6	4 p 10 t SDVR, 20% water	138 (100 to 170) likely arching	375 (@30 to 40% height)
Bastyan	75	5.7	Rhyolite, SW to FR	(Very high)	42	600	25	2.20	0.23	1.0	8 p (10 t ?) SDVR, 20% water	130 (120 to 140)	290
Cethana	110	1.9	Quarzite	(Very high)	23	900	21	2.07	0.27	0.9	4 p 10 t SDVR, 15% water	160 (120 to 210), arching likley	300
Chengbing	74.6	4.4	Tuff lava	UCS = 80 Very high	10.4	1000	–	2.06	0.277	1.0	6 p 10 t SDVR, 25% water	43	110
Foz Do Areia	160	5.2	Basalt (max 25% basaltic breccia)	UCS = 235 Very high	6	600	10	2.12	0.33	0.8 (38 to 56)	4 p 10 t SDVR, 25% water	47	80 (65 to 92)
Ita	125	7.0	Basalt	(Very high)	11	700	12	2.179	0.308	0.8	6 p 9 t SDVR, 10% water	48	87 (83 to 91)
Kangaroo Creek	60	3.0	Schist	UCS = 25 Medium to high	310	600	44	2.34	0.201	0.9 to 1.8	4 p 10 t SDVR, 100% water	–	140

Dam													
Khao Laem	130	7.7	Limestone	UGS < 190 (Very high)	–	900	–	–	–	1.0	4 p 10 t SDVR, 15% water lower in 70 m	59 (43 to 79)	130 to 240
Kotmale	90	6.2	Chamockitic/ Gneissic	(High to very high)	–	700	–	2.20	–	1.0	4 p 15 t SDVR, 30% water	61 (47 to 87)	145 (135 to 155)
Little Para	53	4.2	Dolomitic Siltstone	UCS = 8–14 Medium	110	1000	35	2.15	0.223	1.0	4 p 9 t SDVR, no water lower half	21.5 (19.5 to 23.5)	–
Mackintosh	75	6.2	Greywacke, some slate	UCS = 45 High	52	1000	38	2.20	0.24	1.0	8 p 10 t SDVR, 10% water	45 (35 to 60)	63
Mangrove Creek	80	4.8	Fresh siltstone & sandstone	UCS = 45–64 High	310	400	27	2.24	0.18	0.45 to 0.6	4 p 10 t SDVR, 7.5% water	55 to 60	–
Murchison	94	2.1	Rhyolite (SW to FR)	UCS = 148 Very high	19	600	22	2.27	0.234	1.0	8 p 10 t SDVR, 20% water	190 (170 to 205)	560 (485 to 640)
Reece	122	3.1	Dolerite	UCS = 80–370 Very high	10	1000	11	2.287	0.29	1.0	4 p 10 t SDVR, 5 to 10% water	86 (57 to 115) (175 to 205)	190
Scotts Peak	43	24.8	argillite	UCS = 22 Medium to high	380	914	38	2.095	0.266	0.915	4–6 p 10 t SDVR, no water	20.5 (18.5 to 23.5)	59 (Zone 3A) 420 (Zone 2B)
Segredo	145	5.0	Basalt (<5% basaltic breccia)	UCS = 235 Very high	7.4	–	–	2.13	0.37	0.8	6 p 9 t SDVR, 25% water	55 (arching likely)	175
Serpentine	38	3.5	Ripped quartz schist	(Medium to high)	210	152	69	2.10	0.262	0.6 to 0.9	4 p 9 t SDVR, not sure if water added	92 (46 to 142)	97 (94 to 100)
Shiroro	125	4.5	Branite	(Very high)	32	500	22	2.226	0.20	1.0	6 p 15 t SDVR, 15% water	66 (61 to 71)	–
Tianshengqiao-1	178	6.6	Limestone, SW to FR	UCS = 70–90 Very high	15 to 20	800	–	2.19	0.23	0.8	6 p 16 t SDVR, 20% water	49 (40 to 57)	–
Tullabardine	25	8.6	Greywacke, some slate	UCS = 45 High	28	400	30.5	2.22	0.23	0.9 to 1.0	4 p 10 t SDVR, >10% water	74	170
White Spur	43	3.4	Tuff – SW to FR	(High to very high ?)	–	1000	–	2.30	0.18 to 0.25	1.0	(4 p 10 t ?) SDVR, >10% water	180 (160 to 200)	340
Winneke	85	12.4	SW to FR Siltstone	UCS = 66 High	33	800	28	2.07	0.302	0.9	4–6 p 10 t SDVR, 15% water	55 (50 to 59)	104

(Continued)

Table 15.6.   Summary of embankment and rockfill properties for CFRD (Hunter, 2003; Hunter and Fell, 2002).

Dam name	Embankment dimensions Height H (m)	L/H	Rockfill source[1]	Strength[2] classification UCS (MPa)	C_u	d_{max} (mm)	% finer 19 mm	Dry density (t/m³)	Void ratio, e	Layer thickness (m)	Placement	Rockfill moduli E_{rc} (MPa), average	E_{rf} (MPa)
Zone 3A Rockfill (Continued)													
Xibeikou	95	2.3	Limestone – FR	UCS = 240 Very High	–	600	–	2.18	0.284	0.8	8 p 12 t SDVR, 25 to 50% water	80 (60 to 100)	260
Xingo	140	6.1	Granite gneiss	(High to very high ?)	18	650	4 to 33	2.15	0.28	1.0	4 p 10t SDVR, 15% water	34 (30 to 39)	76 (73 to 80)
Zone 3B Rockfill													
Mangrove Creek (Zone 3B)	80	4.8	Weathered to fresh siltstone & sandstone	UCS = 26–64 High	330	450	32	2.06 t/m³	0.26	0.45	4 p 10t SDVR, dry of OMC	46 (36 to 56)	–
Salvajina (Zone 3B)	148	2.4	Weak sandstone and siltstone	(Medium ?)	45	600	32	2.26	0.21	0.9	6 p 10 t SDVR, water added	62 (likely arching)	–
Tianshengquao-1 (Zone 3B)	178	6.6	Mudstone	UCS = 16–20 Medium	40	600	20 to 35	2.23	0.21	0.8	6 p 16t SDVR, 20% water	37 (32 to 42)	–
For Do Areia (Zone 3B)	160	5.2	Mix basalt & basaltic breccia	UCS = 235 High to very high	14.2	–	–	1.98	0.27	0.8 for 1D 1.6 for 1C	4 p 10 t SDVR, 25% water	32 (29 to 38)	–
Aguamilpa (Zone 3B)	185.5	2.6	Ignimbrite	UCS = 180 Very high	22	700	–	–	–	1.2	4 p 10t SDVR	36 (25 to 45)	–
Ita (Zone 3B)	125	7.0	Breccia and Basalt	(High to very high)	13.3	750	15	2.066	(0.33 to 0.39)	1.6	4 p 9 t SDVR, no water	24 (14 to 46)	–
Khao Laem (Zone 3B)	130	7.7	Limestone	UCS <190 Very high	–	1500	–	–	–	2.0	4 p 10 t SDVR, 15% water lower in 70 m	30	–
Segredo (Zone 3B)	145	5.0	Gasalt (<5% basaltic breccial)	UCS = 235 High to very high	10.2	–	–	2.01 t/m³	0.43	1D – 0.8 1C – 1.6	4 p 9t SDVR, no water	28 (25 to 33), likely arching	–

Dam	H	Cu	Rock type	Rock strength[*2]	C_u (d_{60}/d_{10})	E (const.)	E (fill)	Dry density	Void ratio	d_{max}	Placement	E_{rf}	E_{rc}
Xingo (Zone 3B)	140	6.1	Sound and weathered granite gneiss	(Medium to very high?)	80	750	15 to 60	2.1	0.31	2	6 p 10 t SDVR, no water	13 (12 to 14)	–
Dumped Rockfill													
Courtright	97	2.8	Granite	(Very high)	(<7)	1750	–	1.8	0.47	8 to 52	Drumped and well sluiced	–	–
El-Infiernillo (Zone 3B)	148	2.3	Diorite & silicified conglomerate	UCS = 125 Very high	13	600	22	1.85	0.47	0.6 to 1.0	4 p D8 dozer, no water	39 (27 to 48)	–
Lower Bear No. 1	75	3.9	Granodiorite	UCS = 100–140 Very high	8 to 10	2000 to 3000	low	–	–	max 65	Dumped and well sluiced	–	21
LLower Bear No. 2	46	5.7	Granodiorite	UCS = 100–140 Very high	8 to 10	2000 to 3000	low	–	–	max 36.5	Dumped and well sluiced	–	40
Wishon	90	11.3	Granite	(Very high)	(<7)	1500 to 2000	–	1.80	0.47	8 to 52 (variable)	Dumped and well sluiced	–	–
Dix River	84	3.7	Limestone & Shale (?)	(Very high)	low	–	–	–	–	21	Dumped and poorly sluiced	–	–
Salt Springs	100	4.0	granite	UCS = 100–130 Very high	low (<10)	2000 to 3000	low	(1.88)	0.41	5 to 52	Dumped and poorly sluiced	–	20
Cogswell	85.3	2.1	Granitic Gneiss	UCS = 45 High	7	1300	5	2.05	0.37	7.6 (46 m max)	Dumped dry (no water)	–	44
El-Infiernillo (Zone 3C)	148	2.3	Diorite & silicified conglomerate	UCS = 125 Very high	<13	>600	–	1.76	0.54	2.0 to 2.5	Dumped and spread, no water	22 (17 to 27)	–

Legend:
H = dam height
L = crest length
C_u = uniformity coefficient (d_{60}/d_{10})
d_{max} = average maximum particle size
E_{rf} = deformation modulus during first filling
E_{rc} = secant modulus during construction (average)
4 p 10 t SDVR = 4 passes of 10 tonne smooth drum vibrating roller
*1 FR = fresh, SW = slightly weathered
*2 rock strength classification to AS 1726–1993

(...) in strength classification, C_u, dry density and void ratio columns indicate estimation
% water = % by volume
– indicates unknown

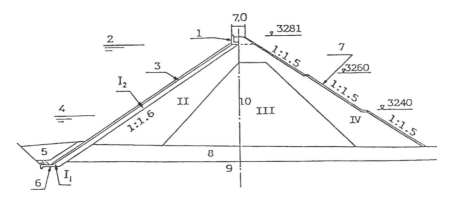

Figure 15.8.   Gouhou Dam (Chen, 1993). 1. Crest wall; 2. Normal water level; 3. Face slab; 4. Dead water level; 5. Random fill; 6. Clay; 7. Masonry; 8. Alluvium; 9. Dam axis.

Figure 15.9.   Gouhou Dam failure (a) view from downstream (b) section (c) crest wall details. Photographs courtesy of Professor Chen Zuyu.

where E_{rc} and E_{rf} are in MPa, γ = unit weight of the rockfill in kN/m^3, δ_s = settlement of layer of thickness d_1 due to the construction of the dam to a thickness H above that layer; δ_n = face slab deflection at depth h from the reservoir surface, and d is measured normal to the face slab as shown. δ_s, δ_n, H, d_1 and d_2 are all measured in metres.

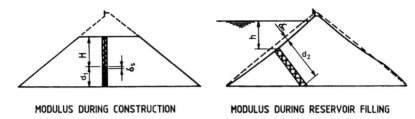

MODULUS DURING CONSTRUCTION   MODULUS DURING RESERVOIR FILLING

Figure 15.10.   Simplified method of calculating rockfill modulus during construction, and pseudo modulus during reservoir filling (Fitzpatrick et al., 1985, reproduced with permission of ASCE).

Equation (15.1) does not allow for the distribution of vertical stress in the dam due to the shape of the dam. E_{rf} is not a true modulus of the rockfill, and should only be used as discussed below to estimate the face slab deformation.

The following points can be made:

- The use of high strength rock does not guarantee a high modulus, e.g. Foz do Areia Dam. The relatively low modulus for this dam is due to poor grading, i.e. a lack of sand size particles (Pinto et al., 1985), and is typical of basaltic rocks in that region. The void ratio of the compacted rockfill is a guide to the rockfill behaviour (note that Foz do Areia Dam has a relatively high void ratio of 0.33. Most rockfills have a void ratio less than 0.25).
- Low strength rocks, which break down significantly, e.g. Kangaroo Creek, Little Para and Mangrove Creek, can give quite high moduli, but it must be remembered that they will probably also give low permeability.
- The highest moduli are achieved for gravels, where the rounded shape limits crushing of the contact points between particles of the fill. Conversely the high stresses on the contact points in the poorly graded basaltic fill help cause the low modulus.
- Well graded high strength rockfill can also give high moduli, e.g. Murchison, Bastyan and Reece.
- Additional compaction (e.g. Murchison 8 passes) and thin layers (e.g. Alto Anchicaya) can lead to higher moduli.
- The modulus is dependent on layer thickness, e.g. at Foz do Areia and Khao Laem dams where Zone 3B, compacted in layers twice the thickness of Zone 3A had moduli 60% to 80% of that for Zone 3A.
- The pseudo modulus on first filling is commonly 2 to 3 times that observed during construction. This is important as the face slab displacement is inversely proportional to the "first filling" modulus and is discussed further in Section 15.2.5. Hunter (2003), and Hunter and Fell (2002) studied this and concluded that the apparent increase in modulus is an artifact of the method of calculation, rather than real. They showed that the layering effects within each rockfill layer had little effect and that, while the mean stresses increase from the construction condition as the water load is applied, the change in deviator stress is small.

It will be noted that in all dams water was added to the fill, usually at a rate of 15% to 25% of the rockfill volume. This results in reduced compressibility, although the improvement may be marginal in some cases.

Cooke and Sherard (1987) conclude that: "(1) For most hard rocks and CFRD of low to moderate height, the addition of water has negligible influence on the dam behaviour. (2) For high dams and for rocks which have significantly lower unconfined compression strengths when tested in saturated conditions (than when tested dry), water should probably be added routinely for the upstream shell (Zone 3A). (3) For rocks with questionably high contents of earth and sand size particles, water should nearly always be used. For dirty rock, the water softens the fines so that the larger rocks can be forced into contact with each other by the vibrating roller." Cooke (1993) relates the need for adding water to the water absorption capacity of the rock, with little benefit being achieved for low (<2%) water absorption.

As pointed out by Sherard and Cooke (1987), it is not intended that the application of water will wash fines into the voids, and hence the use of a high pressure nozzle is not necessary. Adding water in the truck prior to dumping on the surface is practicable and economical (Varty et al., 1985).

Some indication of likely rockfill behaviour will be obtained from consideration of rock type, unconfined compressive strength and jointing in the rock in the quarry.

The best guide to rockfill behaviour in the investigation phase of a project will be to construct test fills, and possibly carrying out large diameter plate bearing tests on the compacted fill. This, coupled with measurement of grading and void ratio of the resulting rockfill and consideration of the height of the dam, shape of the valley and comparison with other projects, should enable a reasonable guide to expected rockfill modulus. The method for estimating the modulus is described in Section 15.2.4.

15.2.4 Estimation of the modulus of rockfill

Hunter (2003), Hunter and Fell (2002, 2003c) developed a method for estimating the secant modulus of rockfill during construction (E_{rc}) and pseudo modulus on first filling (E_{rf}) based on analysis of monitoring data, mostly for CFRF.

15.2.4.1 Estimation of the secant modulus E_{rc}

Hunter and Fell (2002, 2003c) recommended that the following steps be followed:

(a) Determine the representative secant modulus at the end of construction E_{rc} from the D_{80} size (size for which 80% is finer) and unconfined compressive strength of the rock in the rockfill using Figure 15.11.

The representative secant modulus at the end of construction, E_{rc}, is for Zone 3A type rockfill, i.e. placed in layers 0.9 m to 1.2 m thick, compacted with 4 to 6 passes of a 10 tonnes smooth drum vibratory roller and water added, and applies to the stresses:
- 1400 kPa for the very high strength, well-compacted rockfills;
- 800 kPa for the medium to high strength, well-compacted rockfills;
- 1500 kPa for the well-compacted gravels.

The D_{80} size should be obtained from construction records, rolling trials or estimates based on particle size from samples from test pitting into the existing rockfill.

(b) For Zone 3B or other thicker layer rockfills; apply a correction factor of up to 0.5 (for rockfill placed in layers up to 2 m thick) to obtain a representative E_{rc}. This correction factor is based on the ratio of E_{rc} (of Zone 3A to Zone 3B) from six field cases. There is not sufficient data to be prescriptive regarding this correction factor. As a guide, a correction factor of 0.5 would apply to rockfill placed in 2 m layers without the addition of water and compacted with 4 to 6 passes of a 10 tonne vibratory roller (i.e. reasonably compacted rockfill). A correction factor of 0.75 would apply for rockfill placed in 1.5 m to 1.6 m layers with the addition of water and compacted with 4 to 6 passes of a 10 tonne vibratory roller (i.e. reasonably to well compacted rockfill).

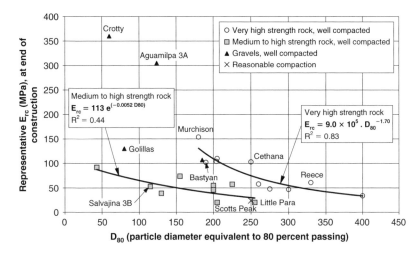

Figure 15.11.   Representative secant modulus of compacted rockfill at the end of construction E_{rc} versus rockfill particle size and rock unconfined compressive strength. (Hunter 2003; Hunter and Fell, 2002, 2003c).

(c) To account for the non linearity of the stress–strain relationship for rockfill, estimates of modulus for stress levels less than or greater than the representative E_{rc} are done by:
  – For very high strength rockfills apply a linear correction of $\pm7.5\%$ per 200 kPa to the E_{rc} estimated from Figure 15.11 for a vertical stress of 1400 kPa. Apply positive corrections for decreasing stresses and negative corrections for increasing stresses. The applicable range is 400 kPa to 1600 kPa.
  – For medium to high strength rockfills apply a linear correction of $\pm6\%$ per 200 kPa to the E_{rc} estimated from Figure 15.11 for a vertical stress of 800 kPa (applicable range is 200 kPa to 1200 kPa).
(d) For medium strength (UCS 6 MPa to 20 MPa) rockfills apply a multiplication factor of 0.7 to the E_{rc} value determined from the equation for medium to high strength rockfills.
(e) Tangent moduli can be estimated from the secant moduli after correction for the stress levels.

The following points should be noted:

– C_u, the uniformity coefficient for the particle size distribution curve, is implicitly allowed for in the D_{80} value. Generally, decreasing C_u is observed for increasing D_{80};
– For materials placed with larger rollers (eg. 13–15 tonne deadweight vibrating rollers) the data does not indicate an increase in moduli for the greater compactive effort. This may be due to greater material breakdown under the heavier rollers and a resultant reduction of D_{80};
– For weathered rockfills the intact strength will be lower than for fresh rock, which will result in a decrease in secant moduli, but this will be countered by the greater breakdown in particle size, which will give an increase in moduli associated with a reduction in D_{80};
– If testing on the proposed rockfill material indicates a significant reduction in UCS on wetting is likely and only limited water has been, or is proposed to be, used in construction, the rockfill is likely to be more susceptible to settlements on wetting due to rainfall and flooding. There is insufficient data to advise on how to quantify this settlement;

- Only a limited number of cases of well-compacted gravels were available; insufficient to undertake analysis. These generally have significantly greater secant moduli at any given stress level in comparison to quarried, very high strength rockfill. Where the gravel is of finer size (e.g. Crotty Zone 3A) the use of a large scale laboratory testing that virtually encompasses the field particle size distribution would provide suitable estimates of moduli, but only if the laying and density in the field are reflected in the laboratory testing. Alternatively plate load tests on gravels compacted in field trials could be used;
- Monitoring should be undertaken during the embankment construction as a check on the pre-construction estimation of deformation.

Data to allow calculation of the moduli of dumped rockfill during construction or first filling was not available to Hunter (2003). It could be anticipated that the modulus would be much lower than for compacted rockfills. Dumped rockfill has a much greater susceptibility to collapse on saturation (e.g. Cogswell Dam (Bauman, 1958) settled 5.4% following a heavy rainstorm which saturated the rockfill which had been dumped without sluicing). Back-analysis of the deformation of a narrow central core earth and rockfill dam built of dumped, sluiced, somewhat weathered basalt rockfill implied an equivalent secant modulus of 5 MPa for the rockfill on the upstream side, but probably higher on the downstream side. Rockfill at El Infiernello dam which was dry dumped and spread in 2 m lifts gave a modulus of 17–27 MPa (average 22 MPa). At Ita dam, rockfill dumped into 10 m of water gave an estimated modulus of around 15 MPa to 19 Mpa.

For assessing the effects of embankment raising it is recommended that moduli for the rock-fill be determined from the deformation performance during construction, if that is available. Alternatively, tangent modulus versus applied vertical stress can be obtained from integration of this stress–strain curve. The variation can then be modelled by considering the embankment in a series of layers of varying moduli. For a more rigorous analysis numerical modelling can be used.

15.2.4.2 *Estimation of the first filling "pseudo modulus" E_{rf}*

Hunter (2003), Hunter and Fell (2002, 2003c) showed that there was a relationship between the ratio of the pseudo modulus on first filling (E_{rf}) to the representative modulus during construction (E_{rc}) with embankment height and upstream slope as shown in Figure 15.12.

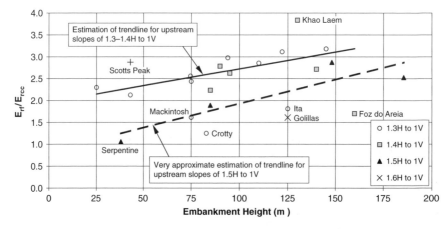

Figure 15.12.   Ratio of pseudo modulus on first filling and construction rockfill modulus (E_{rf}/E_{rc}) versus embankment height. (Hunter 2003; Hunter and Fell, 2002, 2003c).

It should be noted that E_{rf} is a pseudo modulus, only to be used in estimates of face slab deformation using the simplified method shown in Figure 15.10 but allowing for the stress distribution using the Poulos and Davis (1974) method. For finite element analyses the modulus should be obtained as detailed in Section 15.2.4.1.

The reasons for a number of the outliers on Figure 15.12 are detailed in Hunter and Fell (2002, 2003c). They mostly relate to complications arising from complex zoning (Ita, Crotty, Scotts Peak) and localised narrow section of the valley causing arching (Khao Laem). The apparently higher modulus on first filling implied by high E_{rf}/E_{rcc} values is due to the actual stress paths in the rockfill giving an initial reduction in deviator stress, as filling begins, and an overall relatively small increase in deviator stress on filling, as well as the simplifying assumptions made in the analysis.

Numerical analyses were carried out to assess the hypothesis put forward by Cooke (1984) that the layer of rockfill would have a higher horizontal modulus than vertical due to the high degree of compaction in the upper part of each layer, but these showed this was not the case.

The steps to estimate E_{rf} are:

(a) Estimate the ratio E_{rf}/E_{rcc} based on the embankment height and upstream slope angle using Figure 15.12. Trendlines are given for upstream slopes of 1.3–1.4 H to 1 V, and 1.5 H to 1 V, which cover most CFRD designs.

   E_{rcc} is the E_{rc} value estimated from Figure 15.11, adjusted for vertical stress such that it is representative of the average vertical stress in the lower 50% of the rockfill in the central region of the embankment.

(b) Estimate E_{rc}, the representative secant moduli at end of construction, of the Zone 3A rockfill from the methods outlined above. The E_{rc} value should be adjusted for vertical stress such that it is representative of the lower 50% of the rockfill in the central region of the embankment.

(c) The E_{rcc} values used in the derivation of Figure 15.12 were not corrected for arching effects due to valley shape because valley shape is potentially likely to affect both E_{rc} and E_{rf}, and would therefore be taken into consideration in the E_{rf}/E_{rcc} ratio. Hence E_{rc} estimates derived from Figure 15.11 must be adjusted by dividing by the stress correction factors in Table 15.7 to give E_{rcc}.

(d) E_{rf} can then be estimated by multiplication of the value with the E_{rf}/E_{rcc} ratio.

The method applies to compacted rockfill. There is insufficient data on gravel fill dams to draw a trendline, but it appears that E_{rf}/E_{rcc} may be smaller for gravel fill than rockfill.

As will be apparent from the scatter of data in Figure 15.12, this method is approximate. It is most likely to be in error where the dam has complex zoning, with varying moduli in the zones, and in situations where valley shape effects are significant. The trendline for the upstream slope of 1.5 H:IV is particularly uncertain, being based on a limited number of cases.

15.2.4.3   *Effect of valley shape*

Several authors including Pinto and Filho Marques (1998), and Giudici et al. (2000) have concluded that the valley shape can have a significant effect on the settlements, because of 3D effects shedding vertical stresses to the valley sides. This has been assessed empirically, aided (for Guidici et al., 2000) with 3D numerical analyses.

Hunter (2003), Hunter and Fell (2002) assessed this along with the other factors, and used simple 2D finite difference models to assess the likely effects of valley shape on calculated moduli. This showed that the effects were probably not as great as assumed by

Table 15.7.   Approximate stress reduction factors to account for valley shape (Hunter 2003; Hunter and Fell 2002).

W_r/H ratio (river width to height)	Average abutment slope angle (degrees)	Stress reduction factor (embankment location)			
		Base (0 to 20%)	Mid to low (20 to 40%)	Mid (40 to 65%)	Upper (65% to crest)
0.2	10 to 20	0.93	0.95	0.97	1.0
	20 to 30	0.88	0.92	0.96	0.98
	30 to 40	0.82	0.88	0.94	0.97
	40 to 50	0.74	0.83	0.91	0.96
	50 to 60	0.66	0.76	0.86	0.94
	60 to 70	0.57	0.69	0.82	0.92
0.5	<25	1.0	1.0	1.0	1.0
	25 to 40	0.93	0.95	0.97	1.0
	40 to 50	0.91	0.92	0.95	0.05–1.0
	50 to 60	0.87	0.88	0.93	0.05–1.0
	60 to 70	0.83	0.85	0.90	0.05–1.0
1.0	All slopes	0.95–1.0	0.95–1.0	1.0	1.0

others, with the other factors such as particle size masking the 3D effects. If critical, the effects should be modelled using 2D and preferably 3D numerical analyses, but ensuring the modelling allows for the dam to be "built" in layers, since to "build" it numerically in one stage leads to an over-estimation of the 3D effects. For approximate analysis the stresses in the dam can be calculated using Table 15.7.

15.2.5   Selection of side slopes and analysis of slope stability

When the CFRD is constructed of hard, free draining rockfill, the upstream and downstream slopes are fixed at 1.3 H to 1 V or 1.4 H to 1 V, which corresponds roughly to the angle of repose of loose dumped rockfill, and prevents raveling of the faces.

When gravel is used for the dam "rockfill" zones, flatter slopes are needed to prevent raveling of the face. Usually 1.5 H:1 V has been adopted in these cases although 1.6 H:1 V has been used.

When weak rockfill has been used flatter slopes have again been adopted, e.g. 1.5 H:1 V for Mangrove Creek Dam (MacKenzie and McDonald, 1985). If foundation strengths dictate, flatter slopes may also be required, e.g. 2.2 H to 1 V was used for Winneke Dam (Casinader and Watt, 1985). Cooke (1999) confirms these values.

Haul roads may be needed on downstream slopes, or defined berms may be incorporated in the face. In these cases steeper slopes between the "berm" located by the haul road may be used, e.g. 1.25 H:1 V was used for Foz do Areia Dam (Pinto et al., 1985).

In some cases, the upper 10 m to 15 m of the dam may be steepened to as much as 1.25 H to 1 V to provide the camber of the crest (Fitzpatrick et al., 1985).

As pointed out by Sherard and Cooke (1987) and Fitzpatrick et al. (1985), the stability of the slopes in the dam are not usually analysed. This is in recognition of the fact that CFRDs have no pore pressures in the rockfill and will remain stable under static loads when constructed to the slopes described above. However, if analysis is to be carried out, knowledge of the shear strength properties is required. The determination of rock properties is discussed in Section 6.8.

The stability of CFRD under earthquake loading is discussed in Chapter 12.

15.3   CONCRETE FACE

15.3.1   *Plinth*

The principal purpose of the plinth (or "toe-slab") is to provide a "watertight" connection between the face slab and the dam foundation. The plinth is usually founded on strong, non erodible rock which is groutable, and which has been carefully excavated and cleaned up with a water jet to facilitate a low permeability cutoff. For these conditions the plinth width is of the order of 1/20 to 1/25 of the water depth (ICOLD, 1989b; Cooke and Sherard, 1987). Marulanda and Pinto (2000) suggest 1/10 to 1/20 depending on rock conditions. Up the dam abutment, the width is changed according to the water head. This is done in several steps (not gradually) for construction convenience. The minimum width has generally been 3 m, although Cooke & Sherard (1987) suggest that for dams less than 40 m high on very good rock, 2 m could be used. For poorer rock conditions a wider plinth and/or other erosion control measures may be used. This is discussed in Section 15.5.2. Cooke (2000) and Marulanda and Pinto (2000) indicate that a recent evolution in design is to reduce the length of the plinth to say 4 m to 5 m, and maintain an overall seepage gradient beneath the plinth by using an undowelled reinforced slab under the rockfill. This can give economies in rock excavation.

The minimum plinth thickness is usually between 0.3 m and 0.4 m, but may be up to 0.6 m for the lower plinths of high dams. The actual thickness is usually more because of the need to fill over-excavation, and to make up for irregularities in the topography. Where this extra concrete is significant it is common to construct the plinth in two stages; the first stage is to fill the irregularities.

Figure 15.13 and Figure 15.14 show plinth designs for Mangrove Creek, Boondooma, Cethana and Reece dams. These designs are typical for CFRDs.

It is necessary to ensure that the plinth is stable under the imposed forces. For a plinth of normal thickness there is adequate friction resistance on the base, unless there are unfavourably oriented low strength bedding planes, joint or shears in the foundations. The plinth is usually anchored to the rock with grouted dowels, which are generally 25 mm to 35 mm diameter, reinforcing steel bars, 3 m to 5 m long and are installed at 1.0 m to 1.5 m spacing. Cooke (2000) indicates 25 mm bars at 2 m spacing are adequate, that the anchor design is usually empirical, and that the bars are grouted full length into the rock and hooked on to the layer of reinforcing steel in the plinth. The anchors are provided nominally to prevent uplift during grouting, although Cooke and Sherard (1987) claim uplift will not develop in most cases.

For plinths which are thicker than normal, due to overbreak or irregularities in the foundation, the stability of the plinths should be analysed assuming the uplift pressure under the slab is zero at the downstream toe and varies linearly to full reservoir head at the upstream toe. No support should be assumed from the face slab on the understanding that the perimetric joint may have opened, and no support from the rockfill should be allowed for, since significant displacement into the rockfill would be necessary to mobilize the resistance (Marulanda and Pinto, 2000).

The plinth must be stable against sliding and overturning. A proper assessment of the sliding friction angle should be made depending on the rock type, orientation of weak planes etc, not just a check against some arbitrary "sliding factor" or assumed friction coefficient.

It may be necessary to install additional buttress concrete downstream to maintain stability.

If the plinth is high, a possible further problem is that excessive settlement of the face slab may occur and disrupt the perimetral joint. Problems were encountered in Golillas Dam due to such movements (in this case the valley sides were near vertical), (Amaya and Marulanda, 1985). Particular attention should always be paid to compacting the rockfill

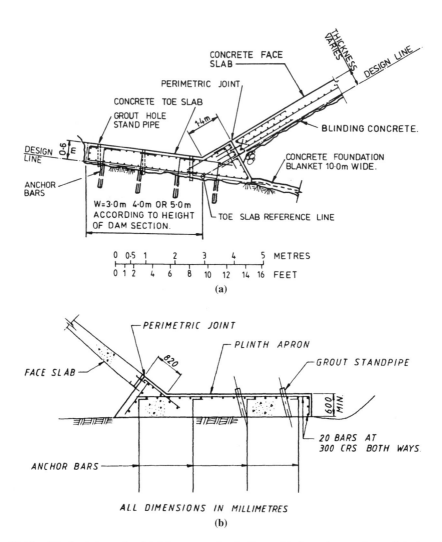

Figure 15.13. Plinth designs for (a) Mangrove Creek Dam, (b) Boondooma Dam (Mackenzie and McDonald, 1985; Rogers, 1985, reproduced with permission of ASCE).

near the plinth and finer rockfill compacted in thinner layers may be adopted (Fitzpatrick et al., 1985). However, one must also be careful not to obtain a modulus significantly higher than the rockfill or differential settlement may still occur. Some use a Zone 2A as shown in Figure 15.13 in this area.

To ensure that the face slab displaces normal to its plane, and is not subject to bending, the Hydro-Electric Commission (Fitzpatrick et al., 1985) require a minimum of 0.9 m of rockfill under the face slab, see Figure 15.14.

The plinth is laid out as a series of straight lines selected to suit the foundation and topography. The angle points are not related to the vertical joints in the slab, and will be a compromise between added excavation and face slab thickness and simplicity of construction.

The plinth may be laid out to be horizontal in a section normal to the centreline of the plinth (as for Cethana Dam, Figure 15.14a), or at right angles to the dam axis (as in Reece, Figure 15.14b). The latter gives a smaller volume of excavation for the plinth, but concreting

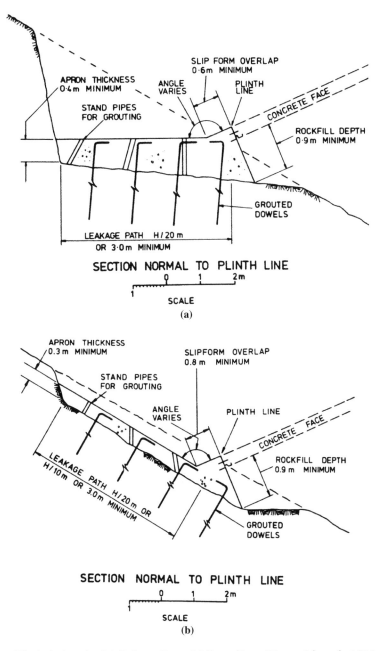

Figure 15.14.   Plinth designs for (a) Cethana Dam, (b) Reece Dam (Fitzpatrick et al., 1985, reproduced with permission of ASCE).

and grouting costs were increased due to more difficult forming and access for drill rigs. Varty et al. (1985) indicate that the HEC reverted to the Cethana type layout for the 80 m high Crotty Dam. This layout was also adopted for the Foz do Areia Dam because, although excavation was increased, this was compensated by better control of rockfill under the perimetric joint, a more straightforward concreting scheme and simpler drilling of grout holes (Pinto et al., 1985). Cooke (2000) gives some suggestions on plinth layout.

The plinth is reinforced to control cracking due to temperature and to spread out and minimise cracks which may tend to develop from any bending strains from grouting.

ICOLD (1989a) and Cooke and Sherard (1987) indicate that a single layer of steel, 100 mm to 150 mm clear of the upper surface, with 0.3% steel each way is adequate. Earlier designs had a lower layer of steel but the advantages of the added stiffness are outweighed by the difficulty of cleaning the rock prior to placing the concrete. Cooke & Sherard (1987) indicate that longitudinal reinforcing steel should be carried through construction joints, rather than using formed joints with water stops.

15.3.2   Face slab

15.3.2.1   Face slab thickness

The face slab thickness is determined from past experience. ICOLD (1989a), Cooke and Sherard (1987) and Cooke (2000) recommend that:

– For dams of low and moderate height (up to 100 m): Use constant thickness = 0.25 m or 0.30 m;
– For high and/or very important dams: Use thickness = 0.3 m + 0.002 H where H = water head in metres. These are minimum thicknesses; average thicknesses will be greater due to irregularities in the compacted Zone 2D. They assume a well constructed Zone 2D as detailed above. Earlier dams constructed on compacted rockfill were based on 0.3 m + 0.002 H to 0.3 m + 0.004 H, but improved construction methods have allowed a reduction to the above recommended values.

HEC used a face thickness of 250 mm up to a dam height of 75 m, 300 mm for higher dams plus local thickening near perimetric joint (Carter et al., 2000). At Antamina Dam in Colombia (Amaya and Marulanda, 2000) 0.3 m + 0.003 H has been used with a slab thickness of 0.45 m within 5 m of the perimeter joint.

15.3.2.2   Reinforcement

Steel reinforcement is provided to control cracking due to temperature and shrinkage. In general the face slab is under compression.

ICOLD (1989a) and Cooke and Sherard (1987) recommend the use of 0.4% reinforcing steel in each direction, with possible reduction to 0.3% or 0.35% in areas of the slab which will definitely be in compression, while retaining 0.4% within about 15 m of the perimeter. Cooke (1997, 1999, 2000) recommends the use of 0.3% horizontal, 0.4% vertical and 0.4% both ways within 15 m of the perimetric joint. However he points out that several experiences of cracks within about 20 m of the peri-meter and near changes in the plinth slope, from relatively steep to flat upper slope, may justify 0.5% each way in these areas. Cooke (2000) recommends 0.5% in areas of major change in plinth slope.

The reinforcing steel is placed as a single mat at or just above the centreline. The area of steel is calculated on the theoretical minimum thickness (Cooke and Sherard, 1987 and ICOLD, 1989b). Fitzpatrick et al. (1985) indicate that at that time the HEC used 0.5% steel, based on the design thickness plus a 100 mm allowance for the added thickness, due to surface irregularity.

The reinforcing has generally been structural grade reinforcing steel (e.g. Australian Standard S230), but more recently high yield grade deformed bar (e.g. Australian Standard 410Y) has been used and is desirable if available. The percentage of steel is the same regardless of steel type.

Light reinforcement is provided in the face and plinth at perimetric joints, and in some cases across vertical joints, to control spalling. Examples are shown in Figure 15.13a, Figure 15.15 and Figure 15.16.

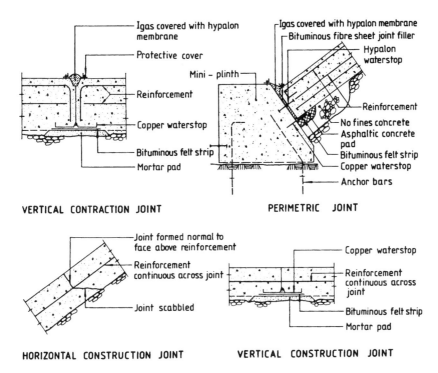

Figure 15.15.   Joint details for Khao Laem Dam (Watakeekul et al., 1985, reproduced with permission of ASCE).

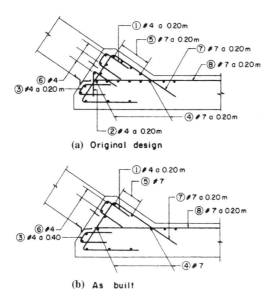

Figure 15.16.   Anti-spalling steel in Salvajina Dam (Hacelas and Ramirez, 1987, reproduced with permission of ASCE).

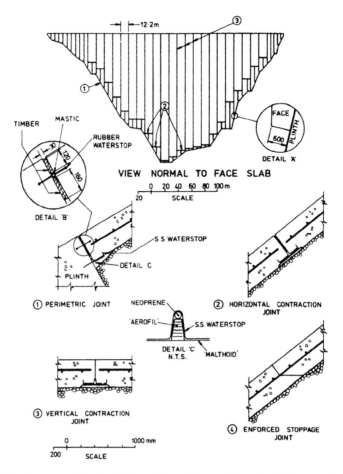

Figure 15.17.   Joint layout and details for Reece Dam (Fitzpatrick et al., 1985, reproduced with permission of ASCE).

15.3.2.3   *Vertical and horizontal joints*

ICOLD (1989a) indicates that design practice at that time did not include horizontal joints, except construction joints in which the reinforcing steel is carried through the joint without waterstops. This is still current practice. Details of such joints are shown in Figure 15.18 and Figure 15.20. This was adopted because, when horizontal joints with water stops were used, it was difficult to obtain good quality concrete around the water stops and some joints experienced spalling under compression and mild rotation.

The Hydro-Electric Commission retained a horizontal contraction joint for Reece Dam, to reduce thermal shrinkage and face cracking.

Vertical joints are generally provided at 12, 15, 16 or 18 m spacing depending on construction factors. For smaller dams, narrower spacing is desirable, e.g. 6 m.

Most CFRDs have been constructed with each of the vertical joints being a construction joint as shown in Figure 15.15 and Figure 15.17. The joints are painted with asphalt, not filled with a compressible filler, as these have been shown to compress under load and cause opening of the perimetric joint. Cooke and Sherard (1987) advocated carrying the horizontal reinforcement through all but a few joints near the abutments, in the manner shown in Figure 15.15. This method gives a cost advantage and a reduction in potential

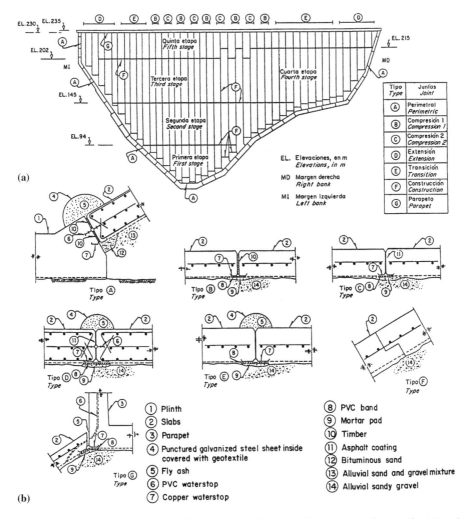

(a)

(b)

Figure 15.18.   Joint details for Aguamilpa dam (a) elevation showing joint layout (b) joint details (Valencia and Sandoval, 1997).

leakage, by potentially eliminating water stops and anti-spalling steel. The method may also have contributed to the face cracking at Khao Laem (Materon and Mori, 2000). This is discussed in Section 15.6.

For Aguamilpa dam different details were used, depending on whether the joints were expected to be in tension or compression as shown in Figure 15.18.

15.3.3   Perimetric joint

15.3.3.1   General requirements

Instrumentation of CFRDs has indicated that compressive strains develop in more than 90% of the face slab due to settlement of the rockfill.

When the reservoir is filled there is further displacement of the face slab, which leads to closing of vertical joints over most of the slab and opening of the perimetric joint and those joints near the abutments. The face slab also pulls away from the plinth, and offsets normal to the face slab, and parallel to the joint due to shear movement of the face. The joint is a common cause of leakage if not well designed, constructed and inspected.

Table 15.8 summarizes maximum perimetric joint movements measured on a number of CFRDs.

To accommodate these movements, joints with multiple water stops are provided. Figures 15.15, 15.17 to 15.20 show some examples.

The design shown in Figure 15.17 for Reece Dam is typical of practice up till around the mid 1980s. The joint includes two water stops

- Primary – copper or stainless steel "W" or "F" shaped;
- Secondary – central "bulb" water stop made of rubber, hypalon or PVC.

Cooke and Sherard (1987) indicate that this arrangement has performed adequately on dams up to 75 m high, where perimetric joint movements are generally relatively small. Fitzpatrick et al. (1985) indicate that in two dams, 39 m and 26 m high, only the primary water stop was used because very small movements were expected.

More recently, particularly for higher dams, a third water stop has been included in the form of mastic or fly ash filler covered with a PVC or hypalon sheet. These are shown in Figures 15.15, 15.18, 15.19 and 15.20 for Khao Laem, Aguamilpa, Salvajina and Antamina dams. Cooke and Sherard (1987) argue that, because of the difficulty of having concrete around the central "bulb" water stop, they believe that a joint with the copper (or stainless steel) water stop underlain by asphalt impregnated sand or concrete mortar and the mastic type stop, is the preferable detail. For Antamina dam (Amaya and Marulanda, 2000) the intermediate PVC seal was eliminated and a 1 m thick sand zone (2A) placed under the perimeter joint. These modifications were incorporated to ease construction and control potential seepage in joints that have performed well in practice.

It will be noted that in the designs shown, a wood plank approximately 12.5 mm to 20 mm thick, or some other compressible filler of similar thickness, is placed between the face slab and plinth to prevent concentration of stresses in the joint during construction and before reservoir filling.

15.3.3.2 *Water stop details*

Primary copper or stainless steel water stop. These are either "W" or "F" shaped, with a high central rib to permit shear movement between adjacent slabs. To prevent external water pressure from squeezing the rib flat, it is filled with a neoprene insert, 12 mm diameter, held in place with a strip of closed cell polythene foam 16 ×12 mm (Fitzpatrick et al., 1985). The water stop is supported on a cement mortar or asphalt impregnated sand pad. Fitzpatrick et al. (1985) indicate that the HEC seated the water stop on a 400 mm wide strip of tar impregnated felt ("malthoid").

Whether copper or stainless steel is used depends on the aggressive nature of the reservoir water, but also seems to be a matter of individual designer preference, with copper being more common. Fitzpatrick et al. (1985) indicate that for the Reece Dam the HEC departed from its earlier practice of using copper water stop (annealed after forming to give maximum ductility) and used 0.9 mm thick grade 321 stainless steel. This was done because it was considered that the stainless steel would be "more robust" during construction, and there was not a significant cost differential, and the stainless steel would be less affected by the acid reservoir water.

ICOLD (1989a) indicate that it is advisable to form the copper or steel water stops in continuous strips to minimize the need for field splices. They recommend use of an electrode of high fluidity (silver content greater than 50%) for welding copper waterstops to ensure full penetration into the two copper plates, then checking with a spark tester to ensure a good joint has been achieved. Fitzpatrick et al. (1985) indicate that for stainless

Table 15.8.   Observed post construction crest settlement and horizontal displacements for modern CFRD.

Dam name	Height (metres)	Horizontal crest movement		Crest settlement				Long-term settlement rate
		First filling (mm)	Total post-construction (mm) (Time from EOC)	First filling (mm)	Total post-construction			
					mm (Time from EOC)	Vertical strain (%)	% per log time cycle (2,3)	mm/year (4)
Agumilpa	185.5	n/a	n/a	222	307 (0.4–5.7 yrs)	0.165	0.090	n/a
Alto Anchicaya	140	n/a	n/a	100	153 (0.1–10.3 yrs)	0.109	0.037	At 10 years, settlement rate was less than 2 mm/yr
Bastyan	75	9	16 (0–7.5 yrs)	15	50 (0.2–7.5 yrs)	0.067	0.028	n/a
Brogo	43	n/a	35.7 (0–18.25 yrs)	15	34 (0–18.25 yrs)	0.079	n/a	n/a
Cethana	110	27	83 (0–22 yrs)	46	137 (0–28.6 yrs)	0.125	0.042	n/a
Courtright	97	n/a	n/a	899	1237 (0–38 yrs)	1.275	0.70	n/a
Crotty	83	9	n/a	16	55 (0–8.5 yrs)	0.066	0.045	n/a
Dix River	84	n/a	970 (0–32 yrs)	n/a	1281 (0–32 yrs)	1.525	0.58 to 1.44	At 30 years, settlement rate was approx 20–25 mm/yr
Foz Do Areia	160	180	248 (0–6 yrs)	73	328 (0–11 yrs)	0.205	0.099	n/a
Golillas	125	n/a	7 (0–6 yrs)	20	52 (0.5–6.4 yrs)	0.042	0.033	n/a
Kangaroo Creek	60	32	50 (0–10 yrs)	26	116 (0–26 yrs)	0.193	0.126	n/a
Kotmale	90	N/a	62 (0–2.5 yrs)	96	255 (0–2.5 yrs)	0.283	0.258	n/a
Little Para	53	N/a	n/a	22	152 (0–22.6 yrs)	0.287	0.210	n/a
Lower Bear No. 1	75	270	305 (0–4 yrs)	<335	375 (0.1–4.1 yrs)	0.500	0.103	n/a
Lower Bear No. 2	46	88	116 (0–4 yrs)	73	116 (0.1–4.1 yrs)	0.252	0.128	n/a
Mackintosh	75	75	130	99	333 (0–20.6 yrs)	0.444	0.184	At 10 years, settlement rate was 5 mm/yr
Mangrove Creek	80	n/a	196 (0–15 yrs)	>287	287 (0.7–15 yrs)	0.359	0.251	n/a
Murchison	94	8	22 (0–12 yrs)	9	104 (0.1–17.6 yrs)	0.111	0.056	At 10 years, settlement rate was approximately 3 mm/yr
Reece	122	n/a	68 (1.5–9 yrs)	85	221 (0.1–15 yrs)	0.181	0.063	At 5 years, settlement rate was approximately 4 mm/yr

(Continued)

Table 15.8.   (Continued)

Dam name	Height (metres)	Horizontal crest movement — First filling (mm)	Horizontal crest movement — Total post-construction (mm) (Time from EOC)	Crest settlement — First filling (mm)	Crest settlement — Total post-construction — mm (Time from EOC)	Crest settlement — Total post-construction — Vertical strain (%)	Long-term settlement rate — % per log time cycle (2,3)	Long-term settlement rate — mm/year (4)
Salt Springs	100	230	550 (0–27 yrs)	380	1276 (0.3–65 yrs)	1.276	0.29 to 0.77	n/a
Salvajina	148	n/a	n/a	>90	90 (0.3–0.8 yrs)	0.061	n/a	n/a
Scotts Peak	43	96	190 (0–17 yrs)	203	445 (0–18 yrs)	1.035	0.178	Long term rate approx. 2.5 mm/yr
Segredo	145	n/a	n/a	Approx 200	229 (0–0.8 yrs)			0.158
Serpentine	38	23	38 (0–21 yrs)	35	77 (0.2–25.5 yrs)	0.203	0.109	n/a
Shiroro	125	n/a	27 (0–1.5 yrs)	>66	166 (0–1.8 yrs)	0.133	n/a	n/a
Tianshengqiao-1	178	670 (0–0.75 yrs)	n/a	>926	926 (0.1–0.8 yrs)	0.520	n/a	n/a
Tullabardine	25	n/a	n/a	2	19 (0.2–12.8 yrs)	0.076	0.023	n/a
White Spur	43	n/a	n/a	7	58 (0–5.9 yrs)	0.135	0.080	At 5 years, settlement rate was approx 1.25 mm/yr
Winneke	85	n/a	n/a	105	207 (0.2–16.2 yrs)	0.244	0.107	n/a
Wishon	90	n/a	n/a	189	954 (0–38 yrs)	1.060	0.25 to 0.33	n/a
Xingo	140	210	320 (1–6.2 yrs)	302	526 (1.0–6.2 yrs)	0.376	0.250	n/a

Notes:
1. EOC = end of construction of main rockfill.
2. Time in years.
3. Effects of first filling typically cease 6 months after completion of first filling.
4. Quoted time is from end of construction.

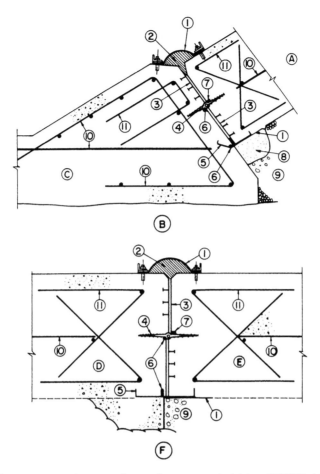

Figure 15.19.   Salvajina Dam perimeter and near abutment vertical joints (ICOLD, 1989a). (1) Hypalon band, (2) Mastic filler, (3) Compressible wood filler, (4) PVC waterstop, (5) Copper water-stop, (6) Neoprene cylinder, (7) Styrofoam filler, (8) Sand-asphalt mixture, (9) Zone 2, (10) Steel reinforcement, (11) Steel reinforcement to protect concrete against crushing and to protect waterstop. (A) Face slab, (B) Perimetric joint, (C) Plinth, (D) Plinth, (E) Face slab, (F) Abutments.

steel, jointing consists of a lap joint fixed by spot welding, then sealing by tungsten-inert gas welding, so that only one metal is involved.

Pinkerton et al. (1985) indicate that in their experience the limit of shear displacement of "W" type primary water stops is around 7.5 mm, up to 50 mm for a 230 mm centre bulb rubber water stop and 10 mm for a PVC waterstop.

Centre bulb water stop. These are constructed of PVC, natural rubber or hypalon. Fitzpatrick et al. (1985) indicate that they prefer hypalon rubber instead of natural rubber or PVC because natural rubber in the atmosphere must be protected from oxidation and ozonation by the addition of antioxidants and antiozonants which could leach out. These materials will last indefinitely below minimum operating level where permanently submerged, but there may be a problem between minimum operating level and flood level as PVC contains plasticizers, some of which are known to leach out.

Pinkerton et al. (1985) indicate a preference for hypalon rubber because it can accept much larger deformations as detailed above.

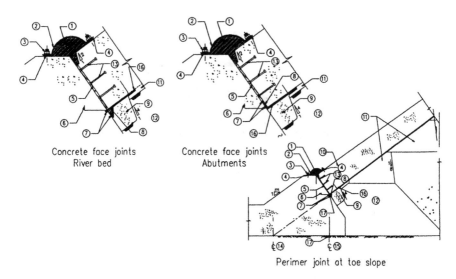

Figure 15.20.   Antamina Dam joint details (Amaya and Marulanda, 2000). (1) Fly-ash, (2) PVC band, (3) Anchor bolts, (4) Epoxy glue, (5) Abarco wood, (6) Copper seal, (7) Circular neoprene, (8) Styrofoam, (9) Sand/asphalt mix, (10) Slab, (11) concrete curb, (12) Zone 2A, (13) Steel bolts, (14) Plinth excavation reference line, (15) Plinth construction reference line, (16) PVC strip.

As mentioned above, in the design of Antamina dam (Amaya and Marulanda, 2000) the central PVC seal was eliminated.

Mastic filler water stop. The concept of the mastic filler is that, as the perimetric joint opens, it will be forced into the opening by the water pressure. The mastic is covered with a PVC or hypalon membrane held in place by steel angles anchored to the concrete.

ICOLD (1989a) indicate that a chicken wire mesh was embedded in the mastic to prevent its flow downwards along the inclined joints. The covering membrane is convex upward to provide for enough mastic volume. It is important that the membrane be sealed effectively so that the water pressure does not leak past the membrane, relieving the differential pressure needed to force the mastic into the crack. Adhesion is improved by painting the joints with mastic.

Cooke and Sherard (1987) indicate a preference for the membrane to be hypalon, not PVC (at least for higher dams), because of its proven 10 to 20 year life exposed to the weather. After this period the cover is not critical, because joint opening should have ceased, and the mastic is wedged tightly into the joint, stopped against the other water stops or, if they have ruptured, against the mortar or asphalt impregnated sand pad.

The mastic which has been used is IGAS which is a bitumen compound. It retains its flow characteristics provided it is not exposed to sunlight for extended periods. This did occur in Golillas Dam (Amaya and Marulanda, 1985) where the PVC and IGAS were exposed for about four years before the reservoir was filled.

Some recent dams have replaced the mastic with ash or sand, which is to fill the joint opening, and be retained by the other water-stops or, in the event they fail, by the Zone 2A filter beneath. Figure 15.15 and Figure 15.16 show such details.

15.3.4   *Crest detail*

It is common to provide a reinforced concrete retaining wall ("wave-wall", "crest wall" or "parapet wall") at the crest of the dam to reduce the volume of rockfill. Wave walls up to 3 m to 5 m have been used.

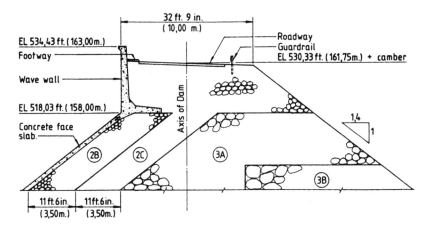

Figure 15.21.   Crest detail, Khao Laem Dam (Watakeekul et al., 1985, reproduced with permission of ASCE).

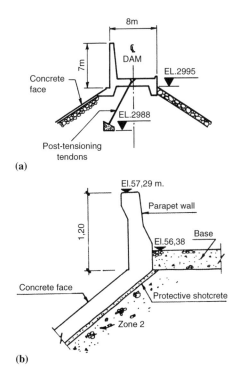

Figure 15.22.   Crest details for (a) Golillas Dam, (b) Macaqua Dam (Amaya and Marulanda, 1985 and Prusza et al., 1985, reproduced with permission of ASCE).

Figure 15.21 and Figure 15.22 show details which have been adopted for Khao Laem, Golillas Dam and Macaqua Dam.

The base of the wall is usually above full supply level and the wall is joined to the face slab with a flexible joint, such as that shown in Figure 15.23. The joint should be vertical, not normal to the plane of the slab, so that differential settlement can be accommodated.

The crest width depends on operational requirements but may be as narrow as 4.9 m. The crest wall is constructed after the face slab, giving a relatively wide platform on which to work while the face slab is under construction.

At Gouhou Dam (Chen, 1993) embankment failure occurred on first filling when the water level rose above the joint between the face slab and the crest wall (Figure 15.24).

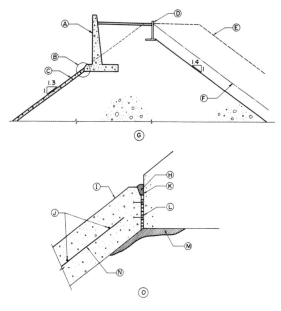

Figure 15.23.   Crest details (ICOLD, 1989a). (A) Upstream parapet wall, (B) See detail, (C) Concrete face, (D) Downstream parapet wall, (E) Additional rockfill if both parapet walls are omitted, (F) Additional rockfill if downstream parapet wall is omitted, (G) Crest details, (H) IGAS, (I) Concrete face, (J) Horizontal reinforcement, (K) IGAS primer, (L) Premoulded filler, (M) Mortar pad, (N) Inclined reinforcement, (O) Detail: concrete face and upstream parapet wall joint.

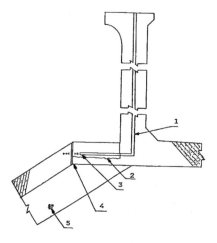

Figure 15.24.   Gouhou Dam–Crest wall details (Chen, 1993). 1. Rubber water stop; 2. Top of the intact concrete; 3. Rubber water stop; 4. Timber Fill; 5. Transition.

This joint had opened as a result of settlement of the slab but poor concrete construction appears to have been a contributing factor. As discussed in Section 15.5.5, the critical issue was the absence of a free draining zone in the dam to discharge the leakage.

15.4   CONSTRUCTION ASPECTS

15.4.1   *Plinth construction and special details*

The methods of constructing the plinth are related to the topography and geology (Materon and Mori, 2000). In wide valleys, e.g. Xingo, Foz do Areia, and Ita, plinth construction can be established independent of embankment placement. In narrow and steep valleys, e.g. Alto Anchicaya and Golillas, plinth excavation was simultaneous with the fill construction.

It is desirable for the plinth to be founded on sound, non-erodible, groutable rock. In good rock requiring blasting, excavation is commonly controlled by pre-splitting with holes at 0.6 m spacing. Dental concrete is applied before anchor installation. In weathered rock blasting should be minimized with excavation by machine or even by hand. Figure 15.25 illustrates design features for Salvajina Dam (Marulanda and Pinto, 2000) where special foundation treatment included provision of a sand filter and gravel transition beneath the upstream section of the embankment, consolidation grouting beneath the plinth, removal of altered material to a depth of 3–4 times the seam width and backfill with concrete, protection of exposed slopes upstream of the plinth with steel reinforced shotcrete.

CFRD built on alluvium, e.g. Santa Juana and Puclaro in Chile and Punta Negra and Los Caracoles in Argentina, involve the construction of an articulated plinth on compacted gravel with joints protected by a transition filter. The plinth is connected to a concrete diaphragm wall taken through the alluvium. Figure 15.26 shows the plinth and cut-off wall at San Juana (Marulanda and Pinto, 2000).

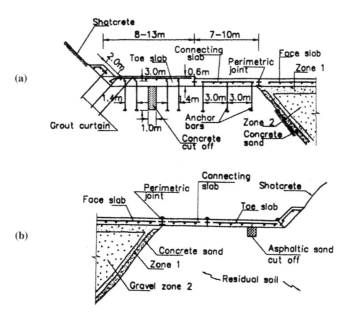

Figure 15.25.   Foundation treatment and plinth design Salvajina Dam (Marulanda and Pinto, 2000). (a) Plinth on less competent rock (b) Plinth on residual soil.

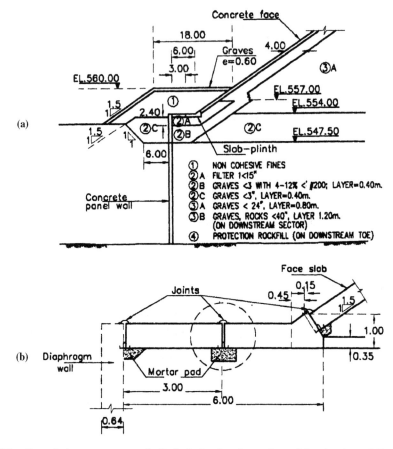

Figure 15.26.   Foundation treatment and plinth design San Juana Dam (Marulanda and Pinto, 2000).

The plinth is constructed using concrete with strength of 21 MPa at 28 days using metal or timber forms. This is followed by grouting of the foundation either downstage or upstage.

15.4.2   *River diversion*

Generally the river is diverted by tunnels but many projects have allowed for overtopping of the rockfill embankments provided protection measures to prevent unravelling of the downstream face are taken. These measures include steel reinforcement of the rockfill and gabions anchored into the rockfill. In large rivers a diversion strategy is chosen to optimize the tunnel size incorporating a "priority" section which has a low probability of being overtopped (Figure 15.27).

15.4.3   *Embankment construction*

All types of granular fills have been used in CFRDs, selected as appropriate for the different zones. Quarried high strength rock such as granite or basalt has been used in all zones. Lower strength rock is usually placed in the central or downstream part of the dam. Compacted alluvial gravel has proved to be an excellent rockfill with a high modulus. The

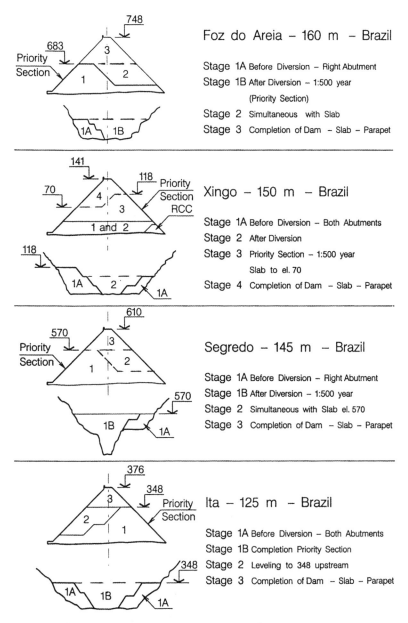

Figure 15.27.   River diversion stages in CFRD's (Materon and Mori, 2000).

combination of gravel in the upstream section of the embankment and rockfill in the downstream section has been successfully used. Analysis is required of the potential for differential settlement between the two zones. When the rockfill contains fines drainage layers may be required.

Zones 2D and 2E are commonly placed in 0.2 m to 0.5 m layers with Zones 3A and 3B from 0.8 m to 2.0 m thick (Materon and Mori, 2000). Zone 2D is commonly compacted using manual vibratory compactors. In some projects 3–4% of cement is added to give some cohesion. (Note. The authors would not recommend this because it would restrict the ability of Zone 2D to self heal against Zone 2E in the event of a leak). Zone 2E is compacted with a few passes of a 5 tonne non vibrating roller. The face is protected either by spraying with an asphaltic emulsion or forming a concrete kerb. Fitzpatrick et al. (1985) indicate that on earlier dams the HEC applied a 2 coat bitumen and chip seal, but for Mackintosh Dam the stabilization was applied after rolling and consisted of a single coat of bitumen emulsion with a cover of sand. Shotcrete has been used instead of bitumen emulsion; e.g. on Golillas and Salvajina dams 40 mm of shotcrete with size 9 mm to 19 mm in maximum size was adopted. At Altamina Dam (Amaya and Marulanda, 2000) an extruded concrete kerb was constructed at the edge of each fill layer to limit runoff over the upstream face. This is economical and allows higher construction rate. A similar concrete kerb was used successfully during the construction of Ita dam in Brazil (Sobrinho et al., 2000). Construction methods for the concrete kerb are described in Materon and Mori (2000).

Zones 3, 3A and 3B rockfill are placed using truck and dozer in layers from 0.8–1.0 m in Zone 3A and 1.6–2.0 m in zone 3B. Zones 3A and 3B using gravel are built in a similar way with smaller layer thicknesses of 0.6–1.2 m (Materon and Mori, 2000). 1% to 20% of water by volume may be specified for rockfill. Water is usually not specified for gravel but should be if it aids compaction. At Xingo Dam in Brazil rockfill in the downstream portion of the dam was successfully dumped into water before the diversion of the river. However this is likely to give a relatively low modulus fill compared to compacted rockfill.

15.4.4   Face slab construction

The concrete face slab is cast using a slip-form, except for the trapezoidal or triangular starter slabs adjacent to the face slabs, which are screeded by hand methods ahead of the main face slab to provide a starting plane for the slip-form. These are usually half the width of the main slabs.

The slip-form is the full width of the panel. For a 1.3 m screed width, the form can move at 2 m to 3 m per hour placing 60 mm slump concrete (Cooke, 1984). Higher slump concrete requires a wider screed. Figure 15.28 shows the slip form used for Khao Laem Dam.

Varty et al. (1985) give details of steel placement and other construction factors, as practiced on HEC dams.

Concrete is delivered to the form by bucket, pumping or in a chute.

The concrete used is typically specified as between 20 MPa and 24 MPa at 28 days, (ICOLD, 1989a), although higher strengths are used, e.g. Cethana and Lower Pieman had average 40 MPa concrete (Fitzpatrick et al., 1985). (ICOLD, 1989a) indicate that higher strengths are not desirable as more shrinkage cracking is likely. They suggest the use of air entrainment to enhance water tightness and durability. The soundness and reactivity of aggregates used for the concrete is important. Materon and Mori (2000) give some more recent experiences which are similar to those described above.

The face slab is constructed on the face as it is presented. The face may have moved subsequent to trimming and compaction, due to the continual raising of the dam. The slab may therefore not end up being on a single plane, but the small variations do not affect performance or appearance.

The design thicknesses are minimum values, and average thicknesses are likely to be 50 mm to 75 mm greater. Early practice was to require that the face slab not be constructed until the rockfill placing was virtually completed so as to minimise post construction movements. More recently it has been shown that staging of the concrete face slab is acceptable, e.g. Salvajina and Foz do Aeria dams (Sierra et al., 1985 and Pinto et al., 1985), but cracking may occur if the zones have differential moduli leading to localisation of

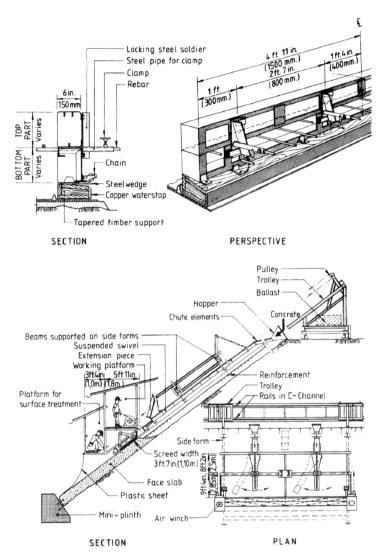

Figure 15.28.   Slip-form for Khao Laem Dam face slab (Watakeekul et al., 1985, reproduced with permission of ASCE).

strains. If the face slab is constructed in stages, a common practice now for very high dams, or if an existing dam is to be raised, some cement grouting infill may be required beneath the top of the already placed slab. The fill above may cause a localised void to open up behind the top of the slab.

15.5   SOME NON-STANDARD DESIGN FEATURES

15.5.1   *Use of dirty rockfill*

When the available rockfill breaks down under compaction to form a rockfill with a large proportion of sand and silt/clay size particles, the resulting fill may have adequate modulus,

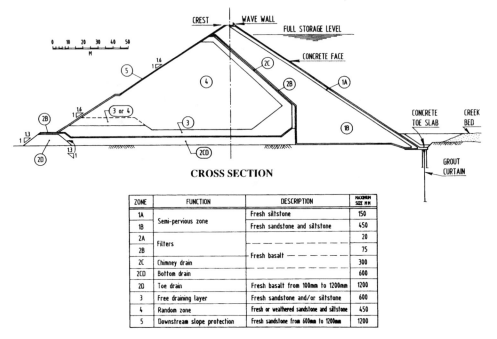

CROSS SECTION

ZONE	FUNCTION	DESCRIPTION	MAXIMUM SIZE MM
1A	Semi-pervious zone	Fresh siltstone	150
1B		Fresh sandstone and siltstone	450
2A	Filters		20
2B		Fresh basalt	75
2C	Chimney drain		300
2CD	Bottom drain		600
2D	Toe drain	Fresh basalt from 100mm to 1200mm	1200
3	Free draining layer	Fresh sandstone and/or siltstone	600
4	Random zone	Fresh or weathered sandstone and siltstone	450
5	Downstream slope protection	Fresh sandstone from 600mm to 1200mm	1200

Figure 15.29.   Mangrove Creek Dam (MacKenzie and McDonald, 1985, reproduced with permission of ASCE).

but may not be free draining. Rocks which are likely to do this are sandstones, siltstones, shale, schists and phyllites.

If this is the case, CFRD can still be used but must be zoned to provide drainage layers behind the face slab and on the foundation.

Mangrove Creek Dam is an example of this design, as discussed in detail in MacKenzie and McDonald (1985).

Figure 15.29 shows the zoning.

Other examples are Kangaroo Creek and Little Para Dams (Good et al., 1985) and Salvajina Dam (Sierra et al., 1985). Zoning for Salvajina Dam is shown in Figure 15.7. In this case the rockfill zones are compacted gravel which have significant silt content, and the drainage layer 2A is provided to ensure dam leakage will not result in build up of pore pressure in the dam.

It is not uncommon to zone the rockfill to ensure that the most free draining material is downstream of Zone 2E and on the base of the dam.

15.5.2   Dams on poor foundation

Concrete face rockfill dams have been successfully constructed on foundations which have weathered seams or clay filled joints, which were potentially erodible under the high gradients which occur under the face slab.

In these circumstances, it has been necessary to prevent erosion by lengthening the seepage path and by providing filters into which the under-seepage emerges in a controlled manner.

An example is Winneke Dam. The deeply weathered siltstone rock precluded economic placement of the plinths on fresh rock and the slab was founded at a level where weathered rock contained infilled joints and crushed seams, some of which were dispersive clays (Figure 15.30).

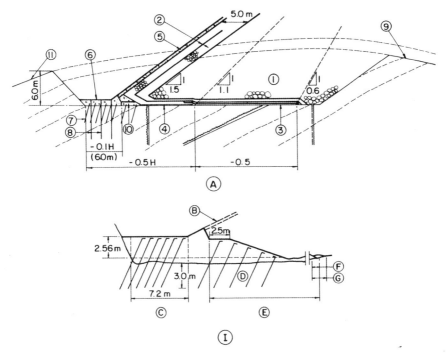

Figure 15.30. Winneke Dam foundation treatment (ICOLD, 1989a and Casinader and Watt, 1985). (A) Upstream toe detail, (B) Concrete face, (C) Anchor bars at 1 m centres longitudinally, (D) Foundation concrete 0.15 m thick, width 32.25 m, (E) Anchor bars at 2 m centres each way within thickened section, (F) Overlap 2 m minimum, (G) Fine filter 42–30 m, (H) Hydraulic head at foundation level, (I) High plinth with buttress. (1) Rockfill, (2) Transition, (3) Filters, (4) Foundation concrete, (5) Concrete face, (6) Plinth, (7) Anchor bars, (8) Grout curtain, (9) Foundation, (10), Buttress, (11) Original ground surface.

The measures adopted are described in Casinader and Watt (1985), Casinader and Stapledon (1979) and Stapledon and Casinader (1977) and included:

- The upstream toe excavation was taken down to the top of the highly weathered rock zone to reduce the number of infilled seams beneath the plinth;
- Prior to grouting, seams in the grout holes were flushed out as far as practicable using air and water;
- The plinth width adopted was 0.1 H, or 6 m minimum, where H is the water head – wider than normally adopted to reduce gradients. The foundation downstream of the plinth was blanketed with 150 mm of concrete, so that the total width of the plinth and "foundation concrete" was at least 0.5 H;
- A filter was placed over the foundation for a distance 0.5 H downstream of the foundation concrete and the filter and foundation concrete was covered by the transition layer to control erosion of fines from the foundation (even if the foundation concrete cracked).

The possibility of sliding movements on weak seams also necessitated a buttress on the plinth, as shown in Figure 15.30.

Reece Dam was founded on deeply weathered schists with clay seams at cutoff level on the left abutment. In this case, the abutment downstream of the plinth was covered with a 150 mm layer of steel meshed shotcrete for a distance 0.5 H downstream. The whole shotcrete layer,

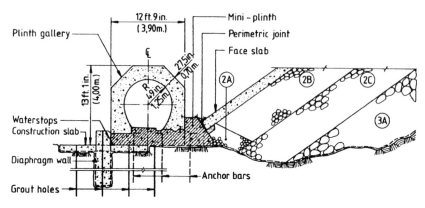

Figure 15.31.   Plinth and gallery, Khao Laem Dam (Watakeekul et al., 1985, reproduced with permission of ASCE).

and the weathered rock between it and the downstream toe, was then covered by 2A and
2B type filter materials.

At Mohale Dam (Gratwick et al., 2000) two significant geological lineaments cross the
damsite and were assessed to have potential for movement. Treatment included the excavation of a slot to provide a concrete "socle" 15 m wide and 2 m deep below the toe-slab, a grout
curtain directed through the lineament and a grouting gallery extending from the downstream
side of the embankment to the toe-slab area near the lineament. Both the socle and the toe-
slab incorporate movement joints composed of 20 mm of compressible material and a PVC
waterstop. Upstream of the toe-slab a blanket of earthfill is placed for a distance of 70 m and
downstream protection is provided by a geotextile and reverse filters of 2D and 2E material.

Another example of special foundation treatment is described by Sierra et al. (1985) for
Salvajina Dam (Fig 15.25). In this case, the foundation was in part founded on residual
soil and wide concrete slabs were added upstream of the plinth and filters downstream.

When gravelly alluvium or glacial soils are present in the river bed, these can commonly
be left in place except for the first 0.3 H to 0.5 H downstream of the plinth. This is
dependent on confirmation that the gravels are of adequately high modulus and strength.
Potentially liquefiable gravels must be removed or densified.

15.5.3   Plinth gallery

The Khao Laem Dam, which was constructed on deeply weathered and in part karst foundations, incorporated a permanent gallery over the plinth of the dam. Details are shown
in Figure 15.31.

The value of the gallery, as a means of access for remedial work, was shown when leakage developed through the perimetric joint. The leak was reduced by grouting from the
gallery, as described in Watakeekul et al. (1987).

Moreno (1987) describes the incorporation of a grout and drainage gallery in a 190 m
high CFRD, in a highly seismic area. Figure 15.32 shows the gallery. Grouting from the
gallery assisted in meeting construction schedules.

15.5.4   Earthfill cover over the face slab

ICOLD (1989a), in their drawing showing typical design features of modern CFRDs (reproduced in Figure 15.2), show earthfill and random fill placed over the lower part of the plinth
(Zones 1A and 1B).

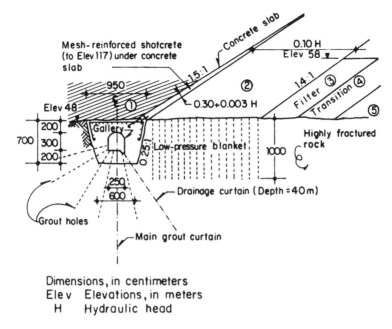

Figure 15.32. Grouting and drainage gallery, Aquamilpa Dam (Moreno, 1987, reproduced with permission of ASCE).

Cooke and Sherard (1987) point out that these zones were first used on the lower part of Alto Anchicaya Dam, because at that time the dam height was breaking precedents. The detail has since been repeated on Foz do Areia Dam (Pinto et al., 1985); Khao Laem Dam (Watakeekul et al., 1985); Golillas Dam (Amaya and Marulanda, 1985). The designs for recent dams in Colombia (Amaya and Marulanda, 2000) and Brazil (Sobrinho et al., 2000) include upstream protection of the lower section of the face slab.

The concept is to cover the perimetric joint and plinth in the lower elevations with impervious soil, which would seal any cracks or joint openings. Zone 1A is a minimum practical construction width, with Zone 1B provided for stability.

As pointed out by Cooke and Sherard (1987) many dams have been successfully constructed without this upstream zone, and if Zone 2D is graded fine to act as a filter to dirty fine sand in the event of leakage, there seems little justification for the fill in most dams. They do seem, however, to favour its application in the lower part of the plinth in high dams.

15.5.5 Spillway over the dam crest

Cooke and Sherard (1987) discuss the concept of building the spillway on the downstream face of a CFRD. They suggest that it is practical for ungated moderate size spillways with a peak discharge of around 25–30 m³/sec per metre of chute width, and where flood flows are of short duration. They suggest the following principles:

– The whole of the rockfill should be compacted to Zone 3A standard to limit settlements;
– A layer of fine rock should be placed under the concrete, as for the face slab, and this should be rolled to make it a good even and stiff support for the concrete slab in the spillway;

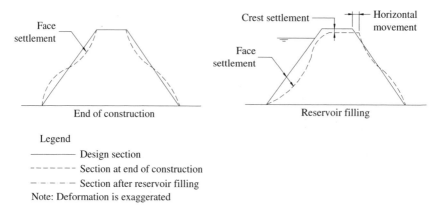

Figure 15.33.   Diagrammatic deformation of the CFRD (adapted from Mori, 1999).

– The spillway chute slab should be built from the bottom up, with continuous reinforcing in both directions, and extend into the side wall footings;
– For high dams, air grooves should be provided in the chute slab. These would also act as contraction joints. Contraction joints would be needed in the spillway walls.

The Hydro-Electric Commission has constructed such a spillway on the 80 m high Crotty Dam. The 12 m wide crest and chute have a discharge capacity of 210 m³/sec. The chute has four "hinges" in it to accommodate settlement, aeration steps to prevent cavitation and a flip bucket dissipator.

Carter et al. (2000) report that the Crotty Dam spillway has operated a few times since 1991 and has performed very adequately.

Namikas and Kulesza (1987) presented details of an emergency fuse plug spillway on a 32 m high CFRD. It is designed to operate for floods in excess of 1:1000 year return period.

15.6   OBSERVED SETTLEMENTS, AND DISPLACEMENTS OF THE FACE SLAB, AND JOINTS

15.6.1   General behaviour

The upstream face of the dam is displaced during construction, on first filling, and during operation (Figure15.33).

In most cases the face slab is constructed after the fill is all placed, so is not subject to the movements shown to the end of construction.

During first filling, the water load acting on the face slab causes displacement of the face slab normal to the plane of the slab, leading to tension around the perimeter, and compression over the central part of the face, but also causes shear and tensile movements at the perimetric joint. These movements continue with time. They may be sufficient to open joints or cause cracking leading to leakage (Figures 15.34 and 15.35).

The following data and that in Section 15.7, have been collected to give readers an idea of the range of settlements, displacements and leakage which has been observed, so they can know what to expect, or can gauge the performance of their dam in comparison with others. The data has been prepared with the assistance of Hunter (2003), Ang (2001) and published data. More details on some aspects are available in Hunter (2003), Hunter and Fell (2002, 2003c).

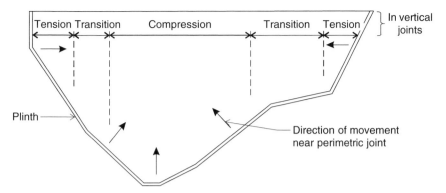

Figure 15.34.   Face slab elevation showing movement directions and stress zones.

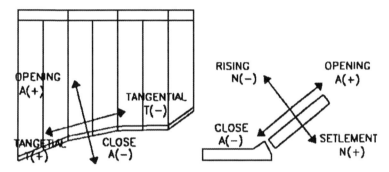

Figure 15.35.   Direction of movements at the perimetric joint (Gomez, 1999).

15.6.2   *Post construction crest settlement*

Available methods (Sowers et al., 1965; Parkin, 1977; Soydemir and Kjaernsli, 1979; Clements, 1984; Pinto and Filho Marques, 1985; Sherard and Cooke, 1987; Hunter, 2003; Hunter and Fell, 2002) for prediction of post-construction crest settlement or rockfill embankments are empirical and are generally based on historical records of similar embankment types and similar methods of construction.

An important aspect of the deformation behaviour of rockfill related to its stress–strain characteristics is that relatively large deformations occur on application of stresses above those not previously experienced by the rockfill (such as on first filling or embankment raising). On un-loading or re-loading to stress states less than those previously experienced (such as due to fluctuations in the reservoir level) the rockfill modulus is very high and resultant deformation limited. Collapse deformation on initial wetting can result in relatively large deformations. Considerations for the post-construction deformation of rockfill in CFRDs therefore include:

– Events where stresses are likely to exceed those previously experienced, most notably first filling;
– Ongoing, time-dependent (or creep) deformation of rockfill;
– Collapse type deformations due to wetting from leakage or tail-water impoundment.

A significant factor in using empirical methods for prediction is assessment of the base time of deformation and consideration of the timing of events such as first filling. If near

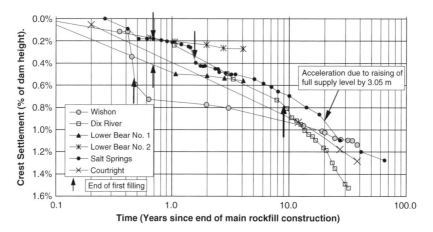

Figure 15.36.   Post-construction crest settlement versus time for dumped sluiced rockfill CFRDs (Hunter, 2003).

full impoundment occurs before completion of construction (or the base time of initial readings) then a significant component of the deformation is likely to have occurred and will not be captured by the post-construction monitoring.

For Figures 15.36 to 15.38 and Table 15.8 the initial time has been established at the end of construction of the main rockfill.

The general trend of the vertical crest settlement versus log time plot is for rapid rates of crest settlement during first filling. The steep portions of the curves in Figures 15.36 to 15.38 is typical for CFRDs where first filling started more than about 0.5 years after the end of the main rockfill construction. The steeper portion of the curve is to be expected given that stress levels in the rockfill exceed those previously experienced. This is followed by long-term movements at a near constant creep strain rate (on log scale). The amount of deformation that occurs during first filling comprises a significant proportion of the total crest settlement, greater than 50% in a number of cases.

For the dumped and sluiced rockfills Figure 15.36 indicates:

– Significantly greater magnitude of settlement for the dumped and sluiced rockfills compared with compacted rockfills. The rates of settlement (on log scale) post first filling (time-dependent deformation) are also significantly greater for the dumped rockfills;
– The poorly sluiced rockfills (Salt Springs and Dix River) have higher long-term creep rates (strain per log cycle of time) than the well-sluiced rockfills.

For the well-compacted rockfills the post-construction crest deformation records presented in Figures 15.37 and Figure 15.38 indicate:

– For embankments constructed of rockfills of medium to high intact strength rock sources, the total magnitude of settlement at 10 years is on average approximately twice that of quarried, very high strength rockfills. The long-term creep rate is also significantly higher for the weaker rockfills, 2 to 10 times that of quarried, very high strength rockfills.
– For the well-compacted gravels (Crotty and Golillas) the post-construction deformation is less than that for the well-compacted, very high strength quarried rockfills.This is likely to be due to several reasons, but is considered to be mainly a result of the rounded shape of the gravels. The point area of contact between particles of rounded

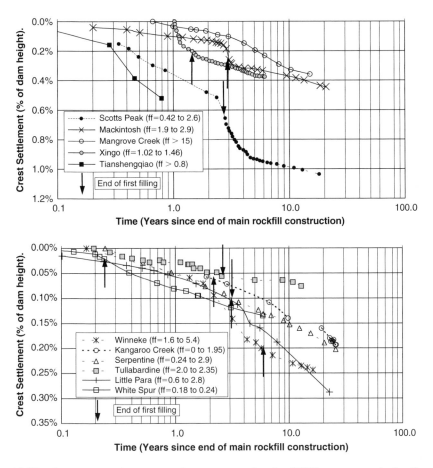

Figure 15.37.   Post-construction crest settlement versus time for CFRDs constructed of well-compacted quarried rockfills of medium to high intact strength (Hunter, 2003).

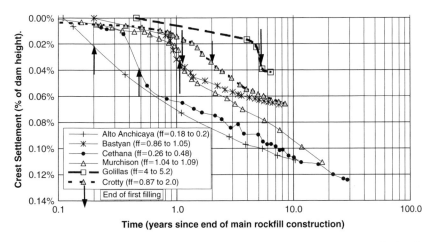

Figure 15.38.   Post-construction crest settlement versus time for CFRDs constructed of well-compacted quarried rockfills of very high intact strength and of well-compacted gravels (Hunter, 2003).

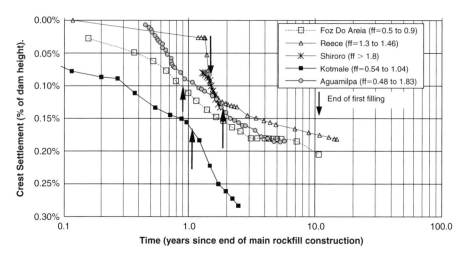

Figure 15.38.   (Continued).

Table 15.9.   Post construction total crest settlement and long-term creep rate for CFRDs (Hunter, 2003; Hunter and Fell, 2002).

| Rockfill classification | Total post construction settlement (% of dam height) | | Long-term creep rate (%/log cycle) |
	10 years	30 years	
Dumped Rockfill	0.6 to 1.0%	1.0 to 1.5%	0.3 to 1.5
Well compacted rockfills			
– Medium to high strength	0.15 to 0.4	–	0.05 to 0.25
– Very high strength, quarried	0.06 to 0.2	–	0.02 to 0.10
– Gravel Rockfills	0.2 to <0.05	–	<0.10

Notes: (1)   Rock substance unconfined compressive strength medium 6–20 MPa, high 20–70 MPa, very high 70–240 MPa.

gravel will be significantly greater than that of angular quarried rockfill. Hence, the contact stresses will be significantly less, resulting in less particle breakage.

Table 15.9 summarises the range of total post construction settlement at 10 and 30 years, and the long-term creep rate determined from the data set of CFRDs.

15.6.3   Face slab displacements and cracking

Table 15.10 summarizes maximum observed face slab displacements.

The maximum displacements usually occur at about mid height of the dam, but this may vary depending on the zoning of the rockfill, and whether the face slab was constructed in one stage after all the rockfill was placed, or in several stages. Displacements at the crest are usually less than the maximum but the ratio varies, particularly if the reservoir impoundment begins before the dam was completed (as in Tianshengqiaol dam, Wu et al., 2000).

Table 15.11 summarizes perimetric joint details and maximum observed displacements.

It is apparent from the data presented in Table 15.11 that the maximum opening, settlement and shear measured on perimetric joints of CFRD is typically less than 20 mm.

Table 15.10. Maximum observed face slab displacements for modern concrete face rockfill dams.

Dam name	Height (m)	Total displacement normal to face slab (1)				Displacement rate (mm/year)		
		Long-term displacement (2)				First filling (time since start of first filling)	Long term (time since end of first filling)	Comments
		Displacement on first filling mm	Long-term displacement mm	Depth of maximum movement (below crest) (m)	Comment (3)			
Aguamilpa	185.5	n/a	320	0	For 6 years from EOC.	n/a	n/a	n/a
Alto Anchicaya	140	130	160	n/a	For 3 years from EOC	n/a	n/a	n/a
Bastyan	75	56	68	34	For 7.5 years from EOC	155 (0.33 yrs)	4 (0.2–2.5 yrs) 0.5 (2.5–6.5 yrs)	Filling over 0.2 yrs
Cethana	110	114	170	37	For 22 years from EOC	145 (0.83 yrs)	6 (0.6–3.9 yrs) 2.8 (3.9–11 yrs) 0.9 (11–22.4 yrs)	Filling over 0.23 yrs
Chengbing	74.6	154	185	58	For 10 years from EOC	51 (3 yrs)	1.9 (0–3.7 yrs)	Filling over approx. 5 yrs
Crotty	83	35	46	49	For 2.5 years from EOC	75 (0.47 yrs)	10.2 (0.15–1.25 yrs)	Filling over 0.3 yrs
Foz Do Areia	160	620	780	82	Duration unknown	1500 (0.42 yrs)	33 (2–3 yrs)	Filling over 0.42 yrs
Golillas	125	160	n/a	68	Recorded up to 0.5 years following completion of first filling	n/a	n/a	n/a
Guanmenshan	58.5	n/a	25.6	29	For 5 years from EOC	n/a	n/a	n/a
Ita	125	461	n/a	n/a	n/a	n/a	n/a	n/a
Kotmale	90	98	–	–	–	–	–	–
Longxi	59	–	19.1	30	For 2.75 years from EOC	–	–	–
Lower Bear No. 1	75	550	625	42	For 2 years from EOC	600 (0.9 yrs)	115 (0.4–0.6 yrs) 13 (0.6–3.4 yrs)	Filling over 0.5 yrs
Lower Bear No. 2	46	137	165	15	For 4 years from EOC	120 (1 yr)	11 (0.4–3.4 yrs)	Filling over 0.5 yrs
Murchison	94	28	77	67	For 12 years after EOC	130 (0.2 yrs)	17 (0–0.4 yrs) 5 (0.4–2.5 yrs) 1.4 (2.5–10.5 yrs)	Filling over 0.05 yrs

(Continued)

642　Geotechnical engineering of dams

Table 15.10. (Continued).

Dam name	Height (m)	Displacement on first filling mm	Long-term displacement (2) mm	Depth of maximum movement (below crest) (m)	Comment (3)	First filling (time since start of first filling)	Long term (time since end of first filling)	Comments
Reece	122	215	264	66	For 10 years after EOC	860 (0.25 yrs)	30 (0.1–0.4 yrs) 12 (0.4–2.4 yrs) 3.5 (2.4–5.4 yrs) 3.5 (5.4–8.4 yrs)	Filling over 0.15 yrs
Salt Springs	100	1317	1755	67	For 28 years after EOC	610 (2 yrs)	38 (0.5–2.5 yrs) 18 (2.5–8.5 yrs) 12.5 (8.5–25.5 yrs)	Filling over 1.5 yrs
Salvajina	148	55	–	–	Measured during first filling up to 14 m below FSL	–	–	–
Scotts Peak	43	78	178	–	For 18 years after EOC	–	–	–
Segredo	145	340	n/a	n/a	Measured up to 4 months after completion of first filling	n/a	n/a	n/a
Shiroro	125	90	n/a	n/a	Measured during first filling up to 32 m below FSL	n/a	n/a	n/a
White Spur	43	15	38	12	For 6 years after EOC	75 (0.2 yrs)	7.5 (0–0.7 yrs) 23 (0.7–1.1 yrs) 2.5 (1.1–5.5 yrs)	Filling over 0.06 yrs
Winneke	85	145	160	n/a	For 1.5–15.5 years after EOC	n/a	n/a	n/a
Xibeikou	95	75	n/a	n/a	n/a	n/a	n/a	n/a
Xingo	140	290	510	n/a	For 6 years from EOC	n/a	n/a	n/a

Notes:
(1) Total displacement includes displacement measured during first filling.
(2) EOC = end of construction of main rockfill FSL = Full supply level.
(3) Displacements are measured normal to the face slab.

Table 15.11.   Perimetric joint details and maximum observed displacements for modern concrete face rockfill dams.

Dam name	Water stop details			Maximum displacement (mm)			Comments – timing[1]	Comments – location
	Bottom	Middle	Top	Opening	Settlement	Shear		
Aguamilpa	Copper	PVC	Flyash	25	18	5.5	Opening and settlement measured to 7 yrs after EOC. Shear measured at 2 yrs after EOC.	Maximum opening & settlement measured towards right abutment. Max shear measured towards centre of dam.
Alto Anchicaya	None	Rubber	None	125	106	15	Measurements at less than 8 yrs after EOC.	Remediation included application of compacted clay, sand-asphalt mix & mastic to perimetral joint.
Bastyan	Stainless steel	Rubber	None	4.8	21.5	Unknown	Measured to 7.5 yrs after EOC.	Details of water stop not available.
Cethana	Copper	Rubber	None	12	Unknown	7.5	Measured approx 12 yrs after EOC.	Maximum opening in centre of dam. Maximum shear in left abutment.
Chengbing	Copper	PVC	Mastic	13.1	28.2	20.6	Measurements at 1–2 yrs after EOC.	Maximum movement on lower part of left abutment.
Crotty	Stainless steel	Hypalon	None	2	3.3	24	All readings taken approx 2 yrs after first filling.	Max shear on left abutment. Max opening and settlement close to centre of dam.
Foz Do Areia	Copper	PVC	Mastic	23	55	25	Timing unknown	Location of maximum movements unknown.
Golillas	Copper	PVC	Mastic	100	36	Unknown	Measurements at less than 11 yrs after EOC.	Location of maximum movements unknown.
Guanmenshan	Copper	Rubber	Rubber asphalt	5.7	4.8	2.8	Measurements within 5 yrs of EOC.	Location of maximum movements unknown.
Ita	Copper	None	Mastic	Unknown	Unknown	Unknown	No data available.	No data available.
Khao Laem	Copper	Hypalon	Mastic	5	8	22	Measurements within 16 yrs of EOC.	Location of maximum movements unknown.
Kotmale	Unknown	Unknown	Unknown	2	20	5	No data on nature of water stops available.	Location of maximum movements unknown.

(Continued)

Table 15.11. (Continued).

Dam name	Water stop details			Maximum displacement (mm)			Comments – timing[1]	Comments – location
	Bottom	Middle	Top	Opening	Settlement	Shear		
Machadinho	Copper	None	Mastic	Unknown	Unknown	Unknown	No data available.	No data available.
Mackintosh	Stainless steel	Stainless steel	None	6.5	22	2.5	Maximum opening & shear measured 7 yrs after EOC. Max settlement at 6 yrs after EOC.	Max settlement on right abutment.
Mangrove Creek	Copper	Rubber	None	12	57	6	Measurements to approx 11 yrs after EOC.	Max settlement on right abutment and max shear on left abutment. Max opening in centre of dam.
Murchison	Stainless steel	Rubber	None	20	10	16	Initial max settlement approx 10 mm measured during first filling. 4 mm rebound measured in 10 yrs following filling.	Max shear and opening on left abutment. Max measured settlement at centre of dam.
Outardes 2	Copper	PVC	None	3	Unknown	Unknown	Max measured opening of 3 mm measured during first filling.	No data on settlement or shear.
Paloona	Copper	Rubber	None	0.5	5.5	Unknown	Measurements to approx 7 yrs after EOC.	Location of maximum movements unknown.
Reece	Stainless steel	Hypalon	None	10	27	Unknown	Measurements to 5 yrs after EOC.	Maximum movements on perimetric joint close to centre of dam.
Salvajina	Copper	PVC	Mastic	9	19.5	6	Measurements at less than 15 yrs after EOC.	Location of maximum movements unknown.
Santa Juana	Copper	None	Mastic	Unknown	Unknown	Unknown	No data available.	No data available.
Segredo	Copper	None	Mastic	Unknown	Unknown	Unknown	No data available.	No data available.
Serpentine	None	Rubber	None	3	6.5	Unknown	Max settlement 5 yrs after EOC. Max opening 3.5 yrs after EOC.	No data on location of maximum movement.

Shiroro	PVC	Rubber	None	30	58	21	Measurements to 2 yrs after EOC.	Max opening in left abutment. Max shear & settlement in right abutment.
Tianshengqiao-1	Copper	PVC	Mastic	16	22.5	7	Measurements within 6 months of EOC.	Max opening in centre of dam, max settlement in left abutment and max shear on right abutment.
Tullabardine	Stainless steel	None	None	1.5	8.5	0.3	Measurements up to 5 yrs after EOC.	Max opening measured on left abutment. Max settlement measured on right abutment.
White Spur	Stainless steel	None	None	3	8	Unknown	Max settlement to 5.5 yrs after EOC and max opening measured at 1.5 yrs after EOC.	Measurements only made at centre of dam.
Winneke	Copper	Rubber	None	9	19	24	Timing unknown.	Location of maximum movements unknown.
Xibeikou	Copper	PVC	Mastic	14	24.5	5	Measurements to 10 yrs after EOC.	Max shear, opening and settlement measured on left abutment.
Xingo	Copper	None	Mastic	30	29	46	Measurements within 6 yrs of EOC.	Location of maximum movements unknown.

Notes:
1 EOC = end of construction of main rockfill.

The larger movements which have been recorded are most likely related to poor compaction and/or use of low modulus, or thick rockfill under the face slab near the perimetric joint.

Large displacements were measured on the perimetric joint of Golillas Dam, predominantly as joint opening. Despite the use of a triple waterstop, large leakage rates were measured during first filling and continue to be measured. There were no apparent low modulus rockfill zones around the perimetric joint and comparatively low strains were measured in the rockfill during and following first filling, as illustrated in Table 15.8. It is likely that the large perimetric joint displacements measured at Golillas are related to the face slab geometry. The dam is located in a steep-sided narrow valley with a crest length to height ratio of 0.81. Similarly large opening and settlement of the perimetric joint were observed at Alto Anchicaya Dam which was constructed in a steep-sided valley. Amaya and Marulanda (2000) observe that "in both dams the plinth is essentially a vertical wall anchored to the rock, with the rock surface under the plinth having the same inclination."

Both Foz Do Areia and Xingo dams have exhibited similar magnitudes of displacement on perimetric joints. They have very similar upslope and downslope face angles and approximately the same crest length to height ratios. Reported face slab movements (Table 15.10) for both these cases were well in excess of typically measured deformations on perimetric joints and face slabs, which is likely to be related to the nature of zoning within the rockfill and variations in modulus between zones. Both dams had upstream zones of well compacted angular rockfill and large downstream zones of moderately compacted rockfill. Hunter and Fell (2002, 2003c) reported large variations in average secant modulus during construction between these two zones in both dams. Perimetric joint movement at Xingo may also have been related to irregular left abutment and foundation geometry. Variations in rockfill thickness between the face slab and foundation resulted in differential settlements behind the face slab and eventually cracking in the face slab.

15.6.4   Cracks in CFRD dams

The progressive development of CFRD embankments has resulted in improved design of the rockfill, plinth and face slab with associated reduction in leakage. However as higher and higher dams are built design peculiarities have been associated with leakage incidents. These incidents appear to have resulted from either foundation features or differential deformation of the rockfill (Marulanda and Pinto, 2000).

Mori (1999) identified three general types of face slab cracking:

(i)   Type A – shrinking cracking associated with concrete curing. These cracks are typically horizontal, very small width (<1 mm) and limited in extent to the width of a slab;

(ii) Type B – associated with bulging of the lower part of the rockfill embankment. Mori observed that, in cases where upstream earthfill is in place; this type of cracking is usually restricted to the middle third of the dam height;

(iii) Type C – cracks associated with settlement of the rockfill embankment, particularly differential settlements related to staged construction or zoning of materials with contrasting moduli.

Early CFRD dams, such as Courtright, Salt Springs and Wishon, suffered substantial face slab cracking prior to and during first filling. This damage is attributable to the large settlements (typically in excess of 1% vertical strain) that occurred in the dumped and sluiced rockfill embankments.

Regan (1997) describes remedial works used in some of these early dams, to reduce leakage rates through the face slab.

Changes in construction methods in about the 1960s, from these early dumped and sluiced embankments to well zoned and compacted embankments, have brought about substantial

AGUAMILPA DAM

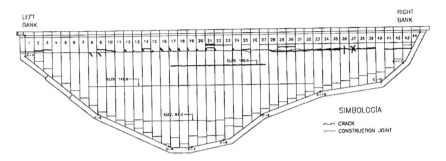

Figure 15.39.   Localization of the cracks in the concrete face of Aguamilpa dam (Gomez 1999).

reductions in embankment deformations. Marulanda and Pinto (2000) report that "monitoring of the performance of compacted CFRD dams has indicated that most of the face is under biaxial compression", with small tensile stresses observed near the toe and crest of the dam and around the perimeter. However, recent performance of face slabs on some new CFRD dams may suggest that tensile stresses can develop within the face slab under particular conditions.

At Khao Laem in Thailand (130 m high) vertical cracking of the face slab occurred above a foundation ridge. In this dam, apart from a few contraction joints the face slab is a continuously reinforced membrane with steel extending across the vertical construction joints. The cracks appear to have resulted from differential embankment deformation across the foundation ridge. Watakeekul et al. (1985) note that cracking may have occurred due to an inadequate number of contraction joints within the face slab. The authors understand that after some attempts to repair these cracks with grouting, J.B. Cooke suggested sand be dropped in the crack, which resulted in a rapid drop in seepage.

At Aguamilpa Dam in Mexico (187 m high) several cracks have occurred in the upper section of the face slab, as illustrated in Figure 15.39. Marulanda and Pinto (2000) report that a 165 m long crack was found 55 m below the crest, approximately three years after completion of first filling. The cause of this cracking has been attributed to the large difference in modulus between the compacted gravel alluvium in the upstream half of the zoned embankment and the rockfill in the downstream half of the embankment (Gomez et al., 2000). Modulus values for the gravel and rockfill were estimated to be 200–300 MPa and 40–60 MPa respectively (Marulanda and Pinto, 2000). Similarly large differences in modulus between zones were reported for Salvajina (ICOLD, 1989), yet no cracking has been observed. Marulanda and Pinto (2000) suggest that this difference in behaviour is related to the gradient between zones. The contact between upstream and downstream zones at Salvajina had a slope of 0.5 V:1 H, while at Aguimilpa the contact between the upstream gravel fill and the transition zone is vertical. Differential settlements in the upper half of the embankment resulted in development of tensile stresses in the face slab and consequent cracking.

Cracking was observed at Scotts Peak dam, a dam with an asphaltic concrete face, again due to the use of rockfill zones with marked contrast in modulus. Well compacted gravels were placed at the upstream toe (to a height of approximately one-third the total embankment height), while the main embankment comprised compacted argillite rockfill. On first filling, large differential deflections caused tensile cracking of the upstream face near to the contact between the gravel and argillite rockfill zones, as illustrated in Figure 15.40.

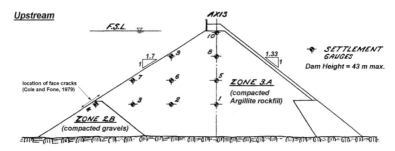

Figure 15.40.   Scotts Peak dam (courtesy of HEC Tasmania).

There was also a rapid increase in the rate of settlement in the short period before gravels were dumped over the damaged area of the face.

Xingo Dam in Brazil (140 m high) has been built in several stages. Horizontal cracking of the face slab has occurred near the stage interface indicating differential deformation in the embankment between the materials placed in the different stages.

At Guohou in China failure is attributed to a crack that opened at the joint between the face slab and the crest wall following deformation of the zoned gravel embankment (Chen, 1993). However the detailing and construction of this joint was poor.

These features indicate that in the case of the higher CFRD embankments it is important to limit the potential for differential deformation of the rockfill, particularly in the higher sections of the dam, and to provide high permeability zones to discharge the leakage safely.

15.7   OBSERVED LEAKAGE OF CFRD

15.7.1   *Modern CFRD*

Table 15.12 shows leakages which have been recorded on first filling and during operation for a selection of dams for which data is available.

The following case studies demonstrate some important features of dams which have experienced relatively large leakage:

(a) *Alto Anchicaya Dam* – High leakage rates measured during first filling (>1800 l/s in the right abutment). Flooding from a creek on the right abutment during construction caused some wash-out of filter material beneath the perimetral joint and may have been a contributing factor to the excessive leakage. A single rubber waterstop was adopted in the construction of Alto Anchicaya. Inspections of the perimetric joint during remedial treatment revealed that while the rubber waterstop was intact it was relatively loose and concrete had not penetrated under the waterstop very well during construction (Regalado et al., 1982).

Remedial treatment included filling the joints with a mastic compound and a sand-asphalt and/or compacted clay cover. Long-term leakage rates stablised after remedial works to a rate of approximately 130 l/s.

(b) *Aguamilpa Dam* – Leakage rates measured during first filling were considered relatively normal until the water level reached a depth of approximately 173 metres, at which point the rate rose to approximately 260 l/s. Several horizontal and diagonal cracks were found in the face slab at a water depth of approximately 150 metres (Gomez et al., 1999).

Table 15.12. Leakage measured in modern CFRD.

Dam name	Height (m)	Water depth at full supply level (m)	First filling		During operation				Comments
			Max leakage rate (l/sec)	Water depth (m)	Max leakage rate (l/s)	Water depth (m)	Long-Term rate (l/sec)	Comments on leakage trends	
Aguamilpa	185.5	182.5	260	169.5	150–200	165–175	50–100	Seasonal fluctuations in leakage rate related to reservoir level. Peak leakage rate at peak reservoir levels.	Face slab cracking developed on first filling. Long term rate shows steady increasing trend.
Alto Anchicaya	140	135	>1800	125	450	135	130	Seasonal trends in leakage rate are unknown. Rate known to vary with rainfall and reservoir level.	Remedial works following first filling to fix joints in slab.
Bastyan	75	70	10	70	10–25	70	5	No fluctuations in reservoir level. Leakage fluctuations related to rainfall events. Long-term leakage rate only fluctuates by 2–3 l/s.	Long-term near constant leakage rate established approx 5 years after first filling.
Brogo	43	28.5	42.5	28.5	5–10	30	4–5	Progressive decrease in leakage rate to 4–5 l/s over 8 years following first filling.	Near constant long-term rate with little apparent relationship to rainfall or reservoir level
Cethana	110	101	60–70	Approx. 80	20–80	96	5–10	Max recorded leakage related to rainfall events. Long-term base leakage rate stable at 5–10 l/s.	Base rate stabilised 4 years after first filling.

(Continued)

Table 15.12.   (Continued).

Dam name	Height (m)	Water depth at full supply level (m)	First filling		During operation				Comments on leakage trends	Comments
			Max leakage rate (l/sec)	Water depth (m)	Max leakage rate (l/s)	Water depth (m)	Long-Term rate (l/sec)			
Chengbing	74.6	69.5	75	Unknown	60	63.5	20		Fluctuations in leakage rate closely related to reservoir level. Min rate 10 l/s at water depth less than 50 m.	Max leakage during construction when reservoir level rose above face slab.
Crotty	83	75	45–50	67	50–70	70–75	30–35		Max recorded leakage related to rainfall events. Long-term base leakage rate reduced from 45–50 l/s to 30–35 l/s over 2 years.	Base rate stabilised 2 years after first filling.
Foz Do Areia	160	155	236	150.5	Unknown	Unknown	60		Long-term leakage rate at water depth of approx 150 m. Fluctuations in leakage rate unknown.	Long-term rate established 5 years after first filling.
Golillas	125	122.5	1080	115	600–700	115	270–500		Fluctuations in leakage rate are unknown.	Long term rate established approx 15 years after completion of first filling.
Guanmenshan	58.5	57	10	57	Unknown	Unknown	5–6		Little change in leakage rate in 4–5 yrs since first filling.	No apparent relationship between reservoir level and leakage rate.
Ita	125	119.5	Unknown	Unknown	1700	Unknown	380		Long term rate established by dumping sand/silty sand on face	Max leakage rate measured 4 months after first

Kangaroo Creek	60	48	Unknown	Unknown	11	Unknown	2.5	slab cracks. Long-term rate occurs at water depths greater than 25 m. Less than 25 m and leakage rate appears to reduce.	filling due to face slab cracking. Peak rates related to rainfall events.
Kotmale	90	85.6	10–40	40–80	30–50	85	10–20	Seasonal fluctuations in leakage rate due to reservoir fluctuations. Minimum rate of 10 l/s at water depth of 40–45 m.	Rainfall events also cause peak in leakage rate.
Little Para	53	50.5	16	43	18	50.5	3–5	Seasonal fluctuations in leakage rate related to reservoir level. Peak leakage rate at peak reservoir levels.	Max flow measured just prior to overflow event
Mackintosh	75	73	25	62	20–25	68	8–10	Fluctuations in leakage rate related to 5–10 m fluctuations in reservoir level.	Gradual decline in leakage rate over 10 years since first filling.
Mangrove Creek	80	74	Unknown	Unknown	5.6	67.5	2.5	Seasonal fluctuations in leakage are unknown. Increase in leakage rate observed with higher water depth due to face slab cracking.	Long-term rate of 2.5 l/s established at water depth of 46 m, 15 years after commencement of first filling.
Murchison	94	85	5–10	70	10	73	2	Short duration peak leakage up to 15–25 l/s related to rainfall events. Increase in rate related to reservoir level (10 l/s at FSL, <5 l/s at <75 m water depth)	Long-term base rate decreased to 2 l/s 5–6 years after first filling. Steady increase in rate to 5–10 l/s 10 years after first filling.

(Continued)

Table 15.12. (Continued).

Dam name	Height (m)	Water depth at full supply level (m)	First filling		During operation			Comments on leakage trends	Comments
			Max leakage rate (l/sec)	Water depth (m)	Max leakage rate (l/s)	Water depth (m)	Long-Term rate (l/sec)		
Salvajina	148	141	Unknown	Unknown	74	130	Unknown	Seasonal trends in leakage rate are unknown.	60 l/s through slab, 14 l/s through abutments.
Scotts Peak (asphaltic concrete face)	43	39.5	100	37	6–10	39	2–3	Short duration peak leakage up to 6–10 l/s likely related to rainfall events. Reservoir held at near constant level.	Max rate occurred after cracking in slab. Long-term rate of 5 l/s for 4 years after first filling then gradual reduction to 2–3 l/s over the next 10 years.
Segredo	145	141	390	141	Unknown	Unknown	45	Seasonal changes in leakage rate are unknown.	Rate of 45 l/s established approx 6 months after completion of first filling at water depth of 138 m.
Shiroro	125	112.5	1600	105	Unknown	Unknown	100	Cracking in face slab occurred during first filling. Seasonal trend in leakage rate is unknown.	Sand placed on face slab & leakage reduced to 100 l/s 6 months after commencement of filling.
Tianshengqiao-1	178	167	132	154.5	n/a	n/a	n/a	Face slab cracking occurred during first filling.	First filling data not complete.
Tullabardine	25	23	Unknown	Unknown	3–6	22–23	1–2	Minimum leakage rate at lowest water levels. Less than 1 l/s at water	Seasonal fluctuations in leakage rate with

Dam								Leakage behaviour	Long-term trend
								levels less than 14 m. Seasonal fluctuations in reservoir level reflected in leakage rates.	progressive long term decrease in average rate to 1–2 l/s approx 10 years after first filling.
White Spur	43	41	7	38	12	30	2	Short duration fluctuations in leakage rate likely related to rainfall events. Reservoir level near constant.	Long-term rate of 2 l/s at water depth of 34 m established approx 3 years after first filling.
Winneke	85	82	60	81	25–30	Unknown	15	Progressive increase in leakage rate with water depth during first filling. Approx 20 l/s at water depth of 40 m and 50 l/s at water depth of 60 m.	General long term trend of reduction in leakage rate. Long term rate of 25 l/s 6 years after first filling reduced to 15 l/s 10 years after first filling.
Xingo	140	137	160	133	200	137	140	Initial decrease in leakage rate for one year after completion of first filling. Then increase from 100 l/s to 200 l/s over one or two years due to cracking in face slab. Reservoir level maintained at full supply level.	Max rate due to cracking in slab. Steady decrease to near constant 140 l/s approx 5 years after first filling.

It is suggested in Gomez et al. (1999) that these cracks could have developed due to tensile stresses generated as a result of differences in modulus between the compacted alluvium (upstream half of embankment) and rockfill (downstream half of embankment). Remedial works to reduce the leakage rate have included the placement of silty sediments over the cracks. Recent dam performance shows leakage rates respond to reservoir level and rainfall events, with an overall long-term trend (1995–2000) of increasing leakage rate based on data presented by Gomez et al. (1999).

(c) *Golillas Dam* – Amaya et al. (1985) reported significant leakage during first-filling of Golillas Dam with distinct relationship between reservoir level and leakage rate. Maximum reported leakage rate during first filling was in excess of 1000 l/s, primarily associated with the perimetric joint and opening of joints within the rock in the left abutment. Initial remedial works, involving removal of the PVC and mastic waterstops in some areas and replacement with burlap, lead fibre and mastic, did not result in any real reduction in leakage. Drawdown of the reservoir enabled cleaning and re sealing of joints in the abutment and abutment support using shotcrete. This remedial work, along with the washing of fines into the perimetric joint, resulted in reduction in leakage rate to long-term values of the order of 270–500 l/s.

15.7.2   *Early CFRD and other dams which experienced large leakage*

Early designs of CFRD (1920s–1960s) were typically characterised by a main body of dumped rockfill and a thick upstream zone of derrick-placed rockfill supporting the concrete face. Over the period from 1920 to 1960 embankment construction involved dumping in high lifts (up to 20–50 m) followed by sluicing. The minimal control on placement of the rockfill resulted in large deformations within the embankment as illustrated in Table 15.8, for dams such as Dix River, Salt Springs, Courtright and Wishon. Typical postconstruction vertical strain for these cases was in excess of 1% (compared to modern CFRD cases which typically show less than 0.2% vertical strain).

As a result of these large settlements, many early CFRD developed significant face cracking and opening of joints (particularly perimetric joints), resulting in high leakage rates. Table 15.13 presents data on leakage rates observed in early CFRD during first-filling and during operating.

Other dams which have experienced large leakage or flows through rockfill include Scofield Dam (Sherard, 1953) which experienced leakage on a piping incident of 1400–5600 l/s, and Hell Hole, which began unravelling in a flood during construction at a flow of 540 m^3/sec (Leps, 1973).

15.8   FRAMEWORK FOR ASSESSING THE LIKELIHOOD OF FAILURE OF CFRD

15.8.1   *Overview of approach*

CFRD are inherently very unlikely to fail, because (as discussed in Sections 15.6.4 and 15.7.1) provided they are designed and constructed with adequate permeable zones (e.g. free draining rockfill, or screened and washed gravels in "gravel" dams) the rockfill can accommodate large leakage flows. The concrete of the face slab, and surrounding the perimetric and other joints or cracks which may form in the face slab, is essentially non erodible, so leakage should not develop rapidly.

However the rockfill, and even the Zone 2D and 2E materials are often poorly graded, and potentially subject to internal instability, with the finer particles potentially being washed out. This can lead to settlement, and further opening of joints or cracks; with increasing leakage flows. Left unchecked, the process may not be self-limiting, and very

Table 15.13.   Data on leakage rates observed in early CFRD during first-filling and during operation.

Dam name	Height (m)	First Filling			During Operation				Comments
		Water depth at full supply level (m)	Max leakage rate (l/s)	Water depth (m)	Max leakage rate (l/s)	Water depth (m)	Long-term rate (l/s)	Comments on leakage trends	
Cogswell	85.5	Unknown	3510	63.5	Unknown	Unknown	Unknown	Significant leakage on first filling due to face slab cracking as a result of large embankment settlements.	Construction involved dry dumped angular rockfill with later sluicing. Remedial works included placement of timber facing over face slab.
Courtright	96.5	94	Unknown	Unknown	1275	94	Unknown	Progressive increase in leakage rate measured over 9 yrs after EOC.	Construction involved dumping and high pressure sluicing of angular rockfill. Significant remedial works implemented to reduce leakage. Leakage related to face slab cracking and opening of perimetric joints.
Dix River	84	76	Unknown	Unknown	2700 (12 yrs after EOC)	Unknown	2000	Max leakage related to slab joints and perimetric joints.	Construction involved dumping of rockfill with sluicing of limited effect. Long-term rate measured 30 years after end of construction.
Lower Bear No. 1	75	71.5	110	71.5	85	71.5	25–40	Fluctuations in leakage rate closely related to	Construction involved dumping and high pressure sluicing of

(Continued)

Table 15.13. (Continued).

Dam name	Height (m)	Water depth at full supply level (m)	First Filling		During Operation			Comments on leakage trends	Comments
			Max leakage rate (l/s)	Water depth (m)	Max leakage rate (l/s)	Water depth (m)	Long-term rate (l/s)		
								reservoir level. Min rate 25 l/s at water depth less than 50 m.	angular, high strength rockfill. Long-term rate established approx 2 years after first filling.
Salt Springs	100	97.5	850	Unknown	425–565	97.5	Unknown	Fluctuations in leakage rate closely related to . reservoir level.	Embankment construction involved dumped and poorly sluiced rockfill. Cracking occurred in the face slab before first filling.
Strawberry	43.5	Unknown	Unknown	Unknown	285	Unknown	170	Leakage slowly increased over 80 years of service.	Concrete facing replaced 12 yrs after EOC. Facing and joints have been repaired several times over the life of the dam.
Wishon	90	89	3120	89	Unknown	Unknown	850	Seasonal leakage behaviour is unknown.	Construction involved dumping and high pressure sluicing of rockfill. Maximum leakage occurred 2 months after first filling due to cracking in face slab.

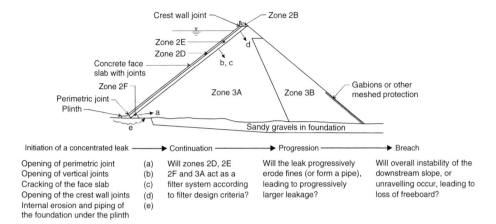

Figure 15.41.   Failure paths for concrete face rockfill dam.

large leakage, leading to a breach mechanism by unravelling or instability of the downstream slope may result. At what stage this may occur is dependent on the zoning and the nature of the rockfill; e.g. gravel fill, and low strength, weathered rockfill, is likely to have lower permeability and be more likely to unravel or reach an unstable condition than other rockfill.

The one failure of a modern CFRD (Gouhou dam, Chen, 1993) occurred at apparently relatively small leakage, but the zoning of that dam was poor, with the fill being silty sandy gravel without high permeability drainage zones.

Hydro Tasmania, who owns and operates a number of CFRD, recently have been developing a framework for quantifying the likelihood of failure of CFRD. In a "workshop" held in October 2001 Richard Herweynan and Paul Southcott from Hydro Tasmania, and Robin Fell (Southcott et al., 2003) concluded that the failure path for a CFRD could be likened to an internal erosion and piping failure for earth and rockfill dams, with the face slab and the joint system being equivalent to the earth core, Zones 2D and 2E the filters. Hence the failure path is as shown in Figure 15.41. The method is therefore strongly based on Foster (1999), Foster and Fell (1999b, 2000) and Fell et al. (2000).

The likelihood of failure can be assessed qualitatively or quantitatively in this approach. In either case, the failure paths should be considered for first filling, under normal operating conditions, flood (up to dam crest level) and under seismic loads. The probability that the reservoir level will be exceeded in any year needs to be incorporated into any quantification of the normal operating, flood and seismic conditions.

For an overall risk assessment, the likelihood of failure by overtopping under flood; scour/erosion of the dam due to overtopping of spillway chute walls and spillway approach walls, and from waves caused by landslides into the reservoir should be considered.

This discussion concentrates only on those failure paths covered by Figure 15.41.

Clearly, what is important to any assessment of the likelihood of failure is the ability of the owner of the dam to respond to the leakage incident, by lowering the water in the reservoir to below the cracked zone, or by relieving the head driving the seepage and/or to "seal" the crack or opening by, for example, placing sand on to the leak and have it wash into the crack, and be prevented from eroding further by filter action on Zone 2E, or Zone 2F.

15.8.2   Assessment of likelihood of initiation of a concentrated leak

Table 15.14 summarizes the factors to be considered for assessing the likelihood of initiation of leakage under normal operating and flood (to dam crest level) conditions. Similar

658   Geotechnical engineering of dams

Table 15.14.   Factors to be considered, and methods for assisting in the estimation of the likelihood of initiation of a concentrated leak in CFRD (Southcott et al., 2003).

Initiator	Factors to consider	Methods for assisting in the estimation of likelihood of concentrated leak
(A) *Factors applicable to all load conditions*		
Perimetric joint opens	• Crest and face slab deformation, first fill and continuing	Compared to population, and degree of variation across dam
	• Joint meter and face slab inclinometer displacement	Compared to design displacement and to population
	• Deep plinth, and local deep fill at plinth in overbreak and steep plinth sections	Calculate joint displacement from depth, construction modulus, for zone 3A or modulus of zone 2F, allowing for water loading (depth)
	• Leakage flow rates vs reservoir level and time. If there is increasing leakage, or >50 l/s? Maybe a leak, >1000 l/s certainly a leak	If no leakage measurements are available, use a range of values of likelihood to reflect the uncertainty
	• Waterstop deterioration, accounting for waterstop type, acidity of water, displacement	
Vertical joints open	• Crest settlement, horizontal displacement and face slab deformation first fill and continuing	Compared to population, and degree of variation across dam
	• Joint meter and face slab inclinometer displacements	Compared to design displacement and to population
	• Valley profile – e.g. steep valley sides leading to large differential settlements	Estimated total settlement and surface profile
	• Opening of crest wall joints (transverse)	If no leakage measurements use a range of value
	• Leakage flow rates vs reservoir level and time. If there is increasing leakage or >50 l/s? May be a leak, >1000 l/s certainly a leak	
Face slab cracking	• Crest and face slab deformation, first fill and continuing	Compared to population
	• Staged rockfill and/or staging of face slab construction effects, with potentially different modulus for lower and upper part	Construction data, calculation from rockfill depth, construction modulus, allowing for water loading (depth) and link to total deformation
	• Deep plinth, and local deep fill at plinth in over break and steep plinth sections and link to total deformation	Construction data, from rockfill depth, construction modulus, allowing for water loading (depth) calculation
	• Location of and nature of difficult construction joints	If no leakage measurements are available use a range of values of likelihood to reflect the uncertainty
	• Leakage flow rates vs reservoir level and time. Increasing leakage or > 50 l/s? Maybe a leak, > 1000 l/s certainly a leak	
	• Crack mapping of face slab	
	• Sudden changes in slab thickness	

(Continued)

Table 15.14.   (Continued).

Initiator	Factors to consider	Methods for assisting in the estimation of likelihood of concentrated leak
Crest wall joint opens	• Visual inspection/displacement on joint • Design detail	
Piping under plinth	• Part or all of plinth foundation erodible e.g. soils, completely weathered rock, clay infill joints • Hydraulic gradient • Grouting – curtain • Grouting – consolidation	UNSW method for piping in foundation (Fell et al., 2000)
(B) Additional factors applicable to flood loading		
All initiators	• AEP of flood levels • Historic high flood level • Increased face slab deformations, joint openings, under flood loading	Hydrological analysis Likelihoods much less for flood levels below historic high than above Calculate from dam geometry, properties, existing deformations and openings allowing for added water load
(C) Additional factors applicable to seismic loading		
All initiators	• Earthquake magnitude, PGA, vs AEP • Additional settlement induced by earthquake • Normal operating load case assessment of $P_{initiation}$ • Crest wall geometry + geometry re spillway wall • Are there liquefiable (not necessary flow liquefaction) zones in the foundation. What is the AEP of magnitude, liquefaction, and will freeboard be lost if liquefaction occurs	Swaisgood (1998) method or other Judgement/other analyses Conventional methods based on Standard Penetration Test or Cone Penetration Test, earthquake AEP of PGA. Check post liquefaction factor of safety/deformations

factors would apply to first filling, but in that case the overall historic performance could be used to assist in quantifying the likelihood.

Also detailed are suggestions on the likelihood of initiation of a concentrated leak. The term "compared to population" suggests the measurable value, e.g. slab displacement, can be compared to the data in Section 15.6, and for example, if the measured value is large compared to the population, a higher likelihood might be assigned. It will be apparent that the estimate will be highly subjective and difficult to quantify. Some points should be kept in mind:

– It is very unlikely a concentrated leak will form under normal operating conditions (water levels within annual levels), unless for example a perimetric joint is right at the limit of its displacement;
– It is more likely that a leak would form under reservoir water levels higher than the historic high (because the loads will be greater than the dam has experienced before/or under large seismic loads (because of the added settlements induced by the earthquake).

Table 15.15.   Factors to be considered and methods for assisting in the estimation of the likelihood of
continuation and progression of a concentrated leak in CFRD.

Initiator	Factors to consider	Methods for assisting in the estimation of likelihood of concentrated leak
(A) *Continuation of concentrated leakage*		
Perimetric joint opens	• As constructed particle size distributions and filtering characteristics of zones 2F, 2D, 2E, 3A (depending on zoning, allowing for segregation)	Assess filters against no-erosion, excessive erosion and continuing erosion criteria (Foster and Fell 2001); Assess for internal instability (Kenney and Lau, 1985)
Vertical joint opens, face slab cracks, crest wall joint opens	• As constructed particle size distributions and filtering characteristics of zones 2D, 2E, 3A.	As above
Piping under plinth	• Presence of extent of filter(s) under rockfill downstream of plinth. If present, particle size of filter and erodible material in the foundation	As above
All joints, cracks	• Displacement/disruption of zone 2D by seismic loading	Judged from estimate of overall settlement using Swaisgood (1998) or other methods

Southcott et al. (2003) give the conditional probabilities used by Hydro Tasmania in their risk assessment.

15.8.3   *Assessment of the likelihood of continuation of a concentrated leak*

Table 15.15 summarizes the factors to be considered, and methods for assisting in estimation of the likelihood of continuation of a concentrated leak. The factors affecting continuation are relatively readily quantifiable using the methods outlined in Foster and Fell (2001) and Foster et al. (2002).

In this assessment, the particle size distribution of Zones 2D versus 2E, 2E versus 3A, and 2A versus 3A have to be assessed against the no erosion, excessive erosion and continuing erosion criteria in Foster and Fell (2001) and Foster et al. (2002). However, since in many existing dams the Zones 2D and 2E will be internally unstable, assessed using the criteria of Kenney and Lau (1985), it is necessary to consider first whether the zone is internally unstable, and if so, what size particles will be washed from the zone through the adjacent zone before the adjacent zone will eventually act as a filter.

In many older dams, this may show that quite extensive internal erosion may occur before the filter action works, and the leakage rate may be quite high.

15.8.4   *Assessment of the likelihood of progression to form a pipe*

Table 15.16 lists the factors to be considered. In the case of CFRD it is unlikely a hole or pipe will form as would happen in an earthfill core. Rather, what is likely to happen is a continuing erosion of the finer fraction from the zones in the dam, including the rockfill. The erosion of the finer material will cause settlement, a likely greater opening of the joint or crack which initiated leakage, and so on unless there is intervention.

Table 15.16. Factors to be considered and methods for assisting in the estimation of the likelihood of progression of piping in CFRD (Southcott et al., 2003).

Initiator	Factors to consider	Methods
All initiators	• Zoning and particle size distribution of each zone, and potential for internal instability in zones 2D, 2E, 3A, 3B leading to washing out of fines	Assess for internal instability using Kenney and Lau (1985)
	• Flow rate and location of the initial concentrated leak	Judged from the initiating mechanism, precedent in other CFRD
	• Potential for downstream face protection to intercept eroding material, leading to slowing down of internal erosion	Assess as a filter against the eroding fines for no-erosion, excessive erosion and continuing erosion criteria (Foster and Fell, 2001)
	• Presence of zone 1A and 1B upstream of the plinth, and whether it will act to seal on zones 2F, 2D	As above
	• Ability to intervene to lower the reservoir level, or seal the opening in the face slab or joints	Account for location and capacity of outlet works, reservoir volume, river inflows; accessibility of site, availability of barges, personnel, whether the joint or opening is covered by debris, zone 1A etc
	• Potential for erosion, and formation of a pipe in alluvium left in the dam foundation	Judged, accounting for the initiating mechanism, downstream toe details, whether the alluvium "day lights" downstream of the toe, whether flow will be in alluvium or in damfill

15.8.5 Assessment of the likelihood of a breach forming

Table 15.17 lists the factors to be considered and the methods which can be used to calculate whether for example, unravelling will occur. The ability to intervene to stop the process by drawing down the reservoir, or reducing the leakage by clogging it with silty-sandy-gravel (for example) can be assessed depending on the particular circumstances of the dam. The erosion process is not likely to be so rapid as for earth and earth and rockfill dams, but may still progress in days or weeks, so detection and intervention need to be done soon after the significant leak develops.

15.8.6 Concluding remarks

It will be apparent that what is described above is only a framework, and quantification will be difficult and subject to considerable uncertainty. It may be sufficient in many cases only to use qualitative terms, as was done by Bell et al. (2002) for the assessment of the likelihood of piping in Eucumbene earth and rockfill dam, the process of formally considering all failure initiators and failure paths is a valuable one, even if the estimates of likelihood are approximate.

These should only be considered very approximate as they are based on engineering judgement and a "mapping scheme" linking probability in qualitative and quantitative terms, with little data to aid in quantification.

Table 15.17.   Factors to be considered and methods for assisting in the estimation of the likelihood of formation of a breach in CFRD.

Breach and mode	Factors to consider	Methods for assisting in the estimation of likelihood of formation of a breach
Slope instability and unravelling	• Estimated likely flow rate of the leak	Initiator mechanism, continuation and progression factors, ability to intervene to seal leak; historic precedents for modern and older CFRD (section 18.6.)
	• Ability to intervene, draw down the reservoir	Account for location and capacity of outlet works, reservoir volume, river inflows; accessibility of site, availability of barges, personnel, whether the joint or opening is covered by debris, zone 1A etc.
	• Presence of alluvium in the foundations	Consider in relation to the breach mechanism
	• Reservoir level at the time of the concentrated leak, freeboard above that level, and the ability	Assess from hydrology
Unravelling	• Discharge capacity of rockfill zones, compared to estimated leakage flow rates, giving gradients through the dam	Estimate from zoning, particle size distribution, presence of still meshing on downstream fall and methods for assessing critical gradients using Pinto (1999), Marulandan and Pinto (2000), Solvick (1991), Olivier (1967)
Slope instability	• Discharge capacity of rockfill zones compared to estimated leakage flow rates, giving pore pressure rise in the dam	As above. Calculate factors of safety, property allowing for non-linear strengths of the rockfill materials, and high friction angles at low confining stress.
	• Effect of downstream slope protection/steel meshing, which may act as a filter to eroding material, causing pore pressures to rise.	Assess as a filter against the eroding fines for no-erosion, excessive erosion and continuing erosion criteria (Foster and Fell, 2001)

FURTHER READING

This chapter has set out to summarize the history of development, features, and performance of CFRD. Those seeking more detail should refer to the following volumes:

J. Barry Cooke Volume, Concrete Face Rockfill dams. Chinese Committee on Large Dams, Beijing, September 2000.
Proceedings, International Symposium on Concrete Face Rockfill Dams, September 2000, International Commission on Large Dams, Paris.
Proceedings, Second Symposium on Concrete Face Rockfill Dams, Comite Brasileiro de Barragens, Rio de Janeiro, October 1999.

16

Concrete gravity dams and their foundations

16.1   OUTLINE OF THIS CHAPTER

This chapter presents the basics of the analysis of the stability of concrete gravity dams on rock foundations using two dimensional rigid body analysis methods. It is based largely on four texts.

> Australian National Committee on Large Dams, "*Guidelines on Design Criteria for Concrete Gravity Dams*" (ANCOLD, 1991).
> Canadian Dam Safety Association, "*Dam Safety Guidelines*" (CDSA, 1999).
> BC Hydro "*Guidelines for the Assessment of Rock Foundations of Existing Concrete Gravity Dams*" (BC Hydro, 1995).
> Federal Energy Regulatory Commission, "*Engineering Guidelines for the Evaluation of Hydropower Projects, Draft Chapter 111, Gravity Dams*" (FERC, 2000).

The ANCOLD Guidelines are written to use the limit state design method. In practice this has proven to cause some difficulties, particularly when considering existing dams which may be just satisfactory or have marginal stability, so there is most emphasis on CDSA (1999) which uses a more conventional approach.

This Chapter deals mainly with the normal operating and flood load cases, with a brief outline of analysis of the earthquake loading case.

As detailed in Douglas et al. (1998, 1999), and recognized in CDSA (1999) and BC Hydro (1995), the most common mode of failure is for sliding or overturning in the foundation, so there is a discussion here on how to estimate the shear strength of the rock in the dam foundation. There is also a discussion on the estimation of uplift pressures and the shear and tensile strength of the concrete in the dam.

16.2   ANALYSIS OF THE STABILITY FOR NORMAL OPERATING AND FLOOD LOADS

16.2.1   *Design loads*

The following design loads are usually considered in the analysis of stability. Most are shown in Figure 16.1.

Dead loads (D), comprising the weight of the concrete plus the weight of spillway bridges, gates, piers and attached intake structures.

Water loads (H, H_F), consisting of:

(a) Horizontal load on the upstream face;
(b) Horizontal load from the tailwater;
(c) Vertical loads on the upstream face, if it is sloping, or behind crest gates;
(d) Pressures from the water flowing over a spillway section. These will also have a vertical component.

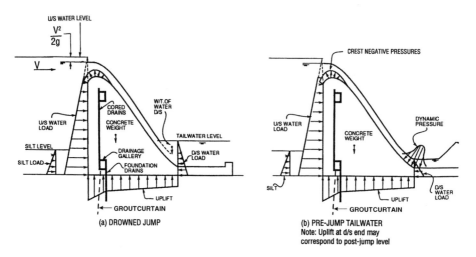

Figure 16.1.   Design loads on gravity spillway (adapted from ANCOLD, 1991).

H is taken here as the loads at maximum normal reservoir level, combined with the most critical tailwater level, and H_F is the load at maximum flood reservoir level based on the inflow design flood (IDF) with corresponding water levels.

Internal water pressure and uplift (U), which act within the rock foundation, at the foundation-dam interface and within the body of the dam.

Static and dynamic thrust (I), created by an ice sheet for reservoirs subject to freezing.

Silt and earth load (S), vertical and horizontal loadings due to soil or rock backfill, and from silt deposited against the structure.

Maximum design earthquake load (Q).

Estimation of these loads is discussed below.

16.2.2   Load combinations

The following load combinations should be considered (adapted from CDSA, 1999).

Usual Loading. Permanent and operating loads should be considered for both summer and winter conditions including self-weight, ice, silt, earth pressure and the normal maximum operating water level with appropriate uplift pressures and tailwater level (D+H+I+S+U).

Unusual Loading (blocked drains). Loads should be as for the usual loading case, except that the drains should be assume inoperative (D+H+I+S+U_{BD}).

Unusual Loading (post earthquake). If earthquake induced cracking is identified at the rock-concrete interface or at any rock section in the dam or foundation, this may result in altered uplift pressures and/or reduced strength of the concrete or foundation. A stability analysis should be carried out to see whether the structure in its post-earthquake condition is still capable of resisting the usual loading (D+H+S+U_{PQ}).

Concurrent ice loading should be considered in areas where appropriate. An inoperative drain case assuming blocked drains should also be assessed and taken as an Unusual Loading case. In some cases, e.g. where large displacements are expected at the dam-foundation contact, this may be a highly likely condition.

Flood Loading. Permanent and operating loads of the Usual Loading, except for ice loading, should be considered in conjunction with reservoir and tailwater levels and uplift resulting from the passage of the inflow Design Flood (IDF) (D+H_F+S+U_F).

The effect of ice loads is not usually considered simultaneously with design flood conditions.

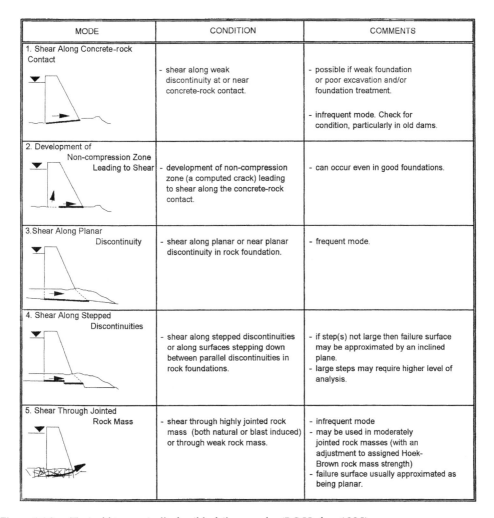

MODE	CONDITION	COMMENTS
1. Shear Along Concrete-rock Contact	- shear along weak discontinuity at or near concrete-rock contact.	- possible if weak foundation or poor excavation and/or foundation treatment. - infrequent mode. Check for condition, particularly in old dams.
2. Development of Non-compression Zone Leading to Shear	- development of non-compression zone (a computed crack) leading to shear along the concrete-rock contact.	- can occur even in good foundations.
3. Shear Along Planar Discontinuity	- shear along planar or near planar discontinuity in rock foundation.	- frequent mode.
4. Shear Along Stepped Discontinuities	- shear along stepped discontinuities or along surfaces stepping down between parallel discontinuities in rock foundations.	- if step(s) not large then failure surface may be approximated by an inclined plane. - large steps may require higher level of analysis.
5. Shear Through Jointed Rock Mass	- shear through highly jointed rock mass (both natural or blast induced) or through weak rock mass.	- infrequent mode - may be used in moderately jointed rock masses (with an adjustment to assigned Hoek-Brown rock mass strength) - failure surface usually approximated as being planar.

Figure 16.2.   Typical kinematically feasible failure modes (BC Hydro, 1995).

Earthquake Loading. If pseudo static analyses are being carried out, permanent and operating loads from the Usual Loading case should be considered in conjunction with the seismic loads of the Maximum Design Earthquake (MDE) $(D+H+S+Q+U_Q)$.

The effects of ice should be given special consideration, recognizing the high uncertainty associated with ice loading on earthquake loading and its effects on the structure.

In the above, subscript "BD" refers to the blocked drains case. Subscript "PQ" to the post-earthquake case and subscript "F" to the flood case.

Temperature induced loads would normally be ignored in this level of analysis, although they may be significant if analysing stresses in a finite element analysis.

16.2.3   *Kinematically feasible failure models*

It is absolutely essential that there be a proper assessment of the geological structure in the dam foundation and, based on this the kinematically feasible failure modes should be identified.

Examples of typical, and not so common failure modes are given in Figures 16.2 and 16.3 respectively.

MODE	CONDITION	COMMENTS
6. Shear along Combined Discontinuities 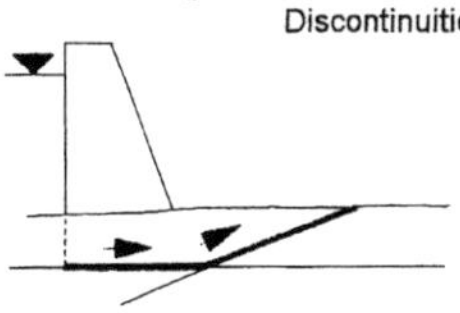	- shear along a combination of two or more discontinuities.	- occasional; usually assume planar mode (3) during initial assessment. - limit equilibrium analysis can be used for more detailed analysis.
7. Toppling 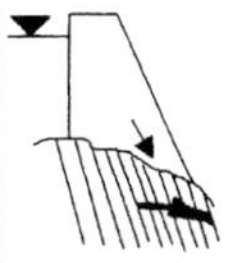	- loads at toe of dam causing toppling of bedded rock formation.	- infrequent but should be evaluated in bedded or highly jointed (parallel) rocks - if dam heel lifts block may continue movement by shearing through toe (mode 2.).
8. Wedge (3 dim.) 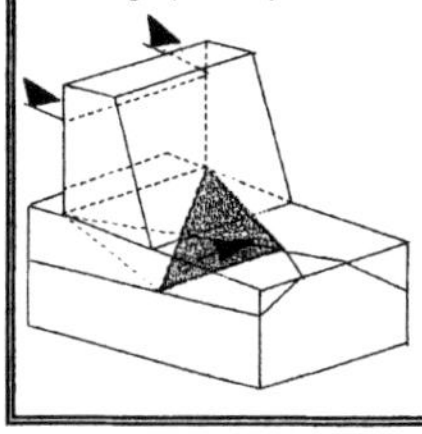	- wedge formed beneath block by combinations of major faults, shears and/or joint sets.	- infrequent but should be evaluated where major features such as faults/shears cross foundation and combine with jointing. - rigid body, 3-dimensional analysis.

Note: Failure modes such as these require a detailed analysis in a higher level assessment than covered by these guidelines.

Figure 16.3.   Other kinematically feasible failure modes (BC Hydro, 1995).

It should be noted that there may be different failure modes in different parts of the dam foundation.

A major failing of dam engineers has been considering only the mode of failure involving shear along the concrete-rock contact. In most dams there may be more critical conditions involving shear along discontinuities in the rock foundations.

The analysis should also consider failure within the concrete in the dam, usually controlled by construction lift joints. In some dams there will be identifiable weak planes due to deterioration on lift joints and cracking due to thermal expansion and contraction on major discontinuities in the geometry of the dam.

16.2.4   Analysis of stability

For most structures it is sufficient to analyse the stability using rigid body force equilibrium analysis. The analysis should be carried out for failure within the dam, at the dam-foundation contact and for critical kinematically feasible failure surfaces in the foundation. For failure in the foundation it will be necessary to carefully consider the uplift pressure and the resistance the foundation can provide to sliding.

In the absence of more detailed modeling, it is suggested that the assumptions recommended by BC Hydro (1995) be adopted. These are shown in Figure 16.4 and are:

(a) Water pressure due to the reservoir exists on a surface extending downward from the heel of the dam to the potential failure surface. This pressurized surface shall be assumed to be vertical unless adverse orientation of foundation joints or faults results in a more onerous assumption;

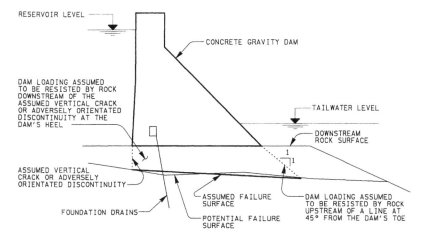

Figure 16.4.   Assumptions for analysis of potential failure surfaces in the foundation (BC Hydro, 1995).

(b) No resistance to loading from the foundation exists beyond a line oriented downward at 45° from the dam's toe;
(c) Uplift pressures are either estimated as outlined in Section 16.4, or consistent with observed piezometric conditions, provided instrumentation is adequate and drainage is well maintained. Bounding pressures are:
  (i) Reservoir pressure at the intersection of the potential failure surface and the surface extending downward from the heel of the dam;
 (ii) Pressure due to tailwater or due to a piezometric surface at the elevation of the ground surface downstream of the dam at the intersection of the potential failure surface and the 45° line from the dam's toe.
       The potential for artesian pressures in the vicinity of the dam's toe due to causes such as limited drainage, anistropic permeability or lateral groundwater flows, should be evaluated.
(d) There is no tensile strength on the potential failure surface or the surface extending downward from the heel of the dam.

Rigid body analysis of static stability on a plane in the foundation consists of:

(a) Vector summation of all forces including uplift acting on the foundation and dam above the assumed failure surface into force components normal and parallel to the failure surface;
(b) Determination of the resultant force location on the assumed failure surface from the summed moments due to all forces acting on the foundation and dam above the surface and the normal component of the resultant force;
(c) Calculation of stresses normal to the assumed failure surface from the normal component of the resultant force and its position on the surface, assuming a linear variation of stress;
(d) Adjustment of the assumed uplift distribution to reservoir pressure where tensile normal stresses greater than the available tensile strength are found and iterative recalculation of the resultant force and the normal stresses until the assumed and calculated amount of the surface in compression correspond. For failure in the foundation, the rock will have zero tensile strength;
(e) Calculation of the stability indices using the resultant force components. These are the sliding factor (calculated using peak and residual strengths in the concrete or

foundation), the position of the resultant force and the compressive strength factor in the concrete and foundation. The Sliding Factor (SF) is defined as:

$$\text{SF} = \frac{\text{Available Shear Resistance}}{\text{Net Driving Force}} \tag{16.1}$$

The net driving force represents the collection of tangential components of all forces acting above the sliding surface (in the direction considered).

Two states of available shear resistance of concrete and the foundation should be considered: peak and residual.

The peak shear resistance should be based on the following components:

(a) Normal Stress (Sn) acting on the potential sliding surface;
(b) Peak angle of effective internal friction (ϕ');
(c) Area of compression (Ac);
(d) Threshold shear strength (τ_o) at zero normal stress (sometimes called the "cohesion") and calculated from:

$$\text{Available Peak Shear Resistance} = \Sigma Ac\left\{(Sn)\tan(\phi') + \tau_o\right\} \tag{16.2}$$

The residual or post-peak resistance should be based on the following components:

(a) Normal Stress (Sn);
(b) Residual angle of effective sliding friction (ϕ'');
(c) Area of compression (Ac);
(d) Nominal residual threshold shear strength value (τ_{nr}) up to $100\,\text{kPa}$ if supported by tests. Without tests, τ_{nr} should be considered to be zero and the available residual shear strength calculated from:

$$\text{Available Residual Shear Resistance} = \Sigma Ac\left\{(Sn)\tan(\phi'') + \tau_{nr}\right\} \tag{16.3}$$

The compressive strength factor is given by:

$$\frac{\text{compressive}}{\text{strength factor}} = \frac{\text{unconfined compressive strength}}{\text{maximum compressive stress normal to the assumed failure surface}} \tag{16.4}$$

The assessment of the stability should also consider observed behaviour and monitoring data such as ambient and concrete temperatures, plumb line deflections, joint meter readings, surface monument displacements, foundation piezometric pressures, foundation and structure drain flows, extensometer readings and strong motion accelerograph records.

16.2.5 Acceptance criteria

The acceptance criteria given in Tables 16.1 and 16.2 are based on those from CDA (1999) and BC Hydro (1995). They are considered a reasonable and logical approach.

Table 16.1. Commonly accepted values for position of resultant force and compressive strength factor (adapted from CDSA, 1999 and BC Hydro, 1995).

	Load Case			
	Usual	Unusual post-earthquake and blocked drains	Earthquake (MDE)	Flood (IDF)
Position of resultant force	Inside Middle $\frac{1}{3}$rd		Outside middle $\frac{1}{3}$rd if other acceptance criteria satisfied	
Compressive strength factor				
(a) concrete	3	1.5	1.1[2]	2.0
(b) foundation[1]	4	2.0	1.3	2.7

Notes: [1] To be considered primarily for massive but low strength rock and weak deteriorated concrete.
[2] Preferably 1.3.

Table 16.2. Commonly accepted values for strength and sliding factors for gravity dams (adapted from CDSA, 1999 and BC Hydro, 1995).

	Load Case			
Type of analysis	Usual	Unusual (Post-earthquake or drains blocked)	Earthquake (MDE)[b]	Flood (IDF)
Peak sliding factor in concrete (PSF) – no tests	3.0	2.0	1.3	2.0
Peak slide factor in concrete (PSF) – with tests[c]	2.0	1.5	1.1	1.5
Residual sliding factor in concrete (RSF)[d][e]	1.5	1.1[g]	1.0	1.3
Peak sliding factor in foundation[f]	1.5 to 2.0	1.2 to 1.4	1.1 to 1.3	1.3 to 1.5

Notes: [a] PSF is based on the peak shear strength of the concrete or rock foundation. RSF is based on the residual or post-peak strength. See Section 16.3 and 16.4 for details.
[b] The stated value under the MDE load case is based on pseudostatic analysis. Performance evaluation of the dam should also take into consideration the time dependent nature of earthquake excitations and the dynamic response of the dam.
[c] Adequate test data must be available through rigorous investigation carried out by qualified professionals.
[d] If PSF values do not meet those listed above, the dam stability may be considered acceptable provided the RSF values exceed the minima.
[e] The minimum values of RSF may be reduced for low consequence dams provided data is available to support such reduction (but not below 1.0).
[f] Assumes foundation geology and strength is assessed by a qualified geotechnical professional. Lower values of the range apply where the geology and the strength parameters are reasonably well known.
[g] Preferably 1.3.

Lower values of the range apply where the geology and the strength parameters are reasonably well known.

It should noted that adequate sliding resistance is normally indicated by sliding factors which equal or exceed the minimum values listed in Table 16.2. For dams in relatively narrow valleys (width/height ratio less than about 3.0), beneficial three-dimensional effects could be present. If beneficial three-dimensional effects are demonstrable, the stated sliding factors may understate the stability of the dam.

The minimum acceptable Sliding and Strength Factors in Tables 16.1 and 16.2 for the post-earthquake conditions are not intended for long term application. Thus, provisions should be made to inspect the dam promptly after a significant earthquake, to monitor its behaviour and to make any necessary repairs within a reasonable period of time or to clean blocked drains. The reservoir could be operated temporarily, if required, at a reduced level until repairs are made and/or safety of the dam is confirmed by analysis.

16.3   STRENGTH AND COMPRESSIBILITY OF ROCK FOUNDATIONS

16.3.1   *Some general principles*

The strength and compressibility of rock masses in the foundations of concrete dams can be assessed systematically by well established engineering geological and rock mechanics principles.

Too often, in the authors' experience, strength and compressibility are taken from guidelines. These, such as ANCOLD (1991), are often not conservative and potentially quite misleading. They may include large effective cohesion components to the strength which are simply not available in many situations and may, quite wrongly, suggest that the rock may reliably have some tensile strength.

It is therefore quite essential, even for relatively small structures, such as spillway crests, or gated spillway structures to engage the services of a geotechnical professional who has a knowledge of rock mechanics to assess the strength and, for larger projects, the compressibility of the rock foundations.

This assessment should include:

(a) Mapping of rock exposure, inspection of photographs taken during construction (for existing dams) and logging of drill core to determine the stratigraphy, bedding orientation, joint orientation, continuity and nature of the joint surfaces and to locate and delineate shears, faults, bedding surface shears and other features important to the strength;

(b) Carry out laboratory testing on representative samples of the rock to determine the unconfined compressive strength and basic friction angle (ϕ'_b) of the bedding surfaces and joints.

In preliminary assessments, or for some smaller dams, the basic friction angle may be assessed from published information, in which case conservative values should be adopted;

(c) Develop a geotechnical model of the foundation (i.e. plan, factual and interpretive sections) to define any possible kinematically feasible potential failure surfaces within the rock and/or on the foundation-rock contact.

There will be variations in these conditions across the dam and it may be necessary to analyse the stability of several cross sections.

16.3.2   *Assessment of rock shear strength*

16.3.2.1   *General requirements*

The shear strength of the rock depends on the nature of the potential failure surface and the rock it passes through. The following approach to assessing the strength is based on BC Hydro (1995) and consultation with colleagues with a knowledge of rock mechanics.

Table 16.3.   Rock discontinuity shear strength parameters (Adapted from BC Hydro 1995).

Discontinuity type	Typical feature	Strength parameters (Note 1)
Clean discontinuity (No previous displacement)s	Clean joint, bedding surface	$\phi' = f(\phi'_b, i)$; $c' = 0$
Thick infilled or extremely weathered seam (No previous displacement)	Infilled seam, joint, or extremely weathered bed	$\phi' = f(\phi')$ of infill or extremely weathered material (2); $c' = 0$
Discontinuity with previous displacement	Sheared zone or seam, bedding surface shear, crushed seam	$\phi' = f(\phi'_r)$ of shear surface and/or sheared or crushed material, and f(i) of wall rock; $c' = 0$
Multiple discontinuity	Highly jointed rock mass	m_b, s, a, σ_{ci}

Notes: (1) Strength parameters as defined in Section 16.2:

f	function of
c'	effective cohesion (at zero normal stress)
ϕ'_b	effective basic friction angle (for wet surfaces)
ϕ'_r	effective residual friction angle
i	average roughness angle
m_b, s, a	Hoek-Brown criterion (1995) parameters
σ_{ci}	uniaxial compressive strength of intact rock

(2) Test to be carried out on remoulded samples; ϕ' to be based on peak strength under drained conditions; c' to be neglected.

BC Hydro (1995) recommend the methods given in Table 16.3. The descriptive terms for discontinuities (defects) used by BC Hydro have been modified here to conform with the definitions in Chapter 2, Section 2.3 and Figure 2.1.

16.3.2.2   *Shear strength of clean discontinuities*

The shear strength of clean, rough (or smooth) discontinuities should be assessed using the approach of Patton (1966), who recommended use of the form of Equations 16.5 or 16.6.

$$\tau = \sigma'_n \tan(\phi'_b + i) \quad \text{if } \sigma'_n < \sigma'_{ns} \qquad (16.5)$$

or

$$\tau = c'_a + \sigma'_n \tan\phi'_b \quad \text{if } \sigma'_n < \sigma'_{ns} \qquad (16.6)$$

where $\sigma'_n = \sigma_n - u$ = effective normal stress; σ'_{ns} = effective normal stress at which the strength of asperities is exceeded; u = internal water pressure (pore pressure) within the discontinuity; ϕ'_b = basic friction angle (for wet surfaces) of the discontinuity wall rock; σ'_n = effective normal stress across discontinuity; i = average roughness angle; c'_a = effective apparent cohesion.

Figure 16.5 shows the application of the equations.

The basic friction angle (ϕ'_b) is derived from shear tests on sawn and lightly ground surfaces of the discontinuity (Barton, 1982; Barton and Bandis, 1991) or alternatively by measuring both vertical and horizontal displacements of a discontinuity specimen during laboratory testing and calculating the roughness angle i, which is then subtracted from the measured friction angle to determine ϕ'_b. An alternative method is by tilt testing on rock

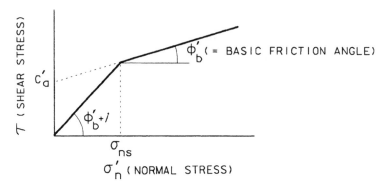

Figure 16.5.   Shear strength of rough clean discontinuities.

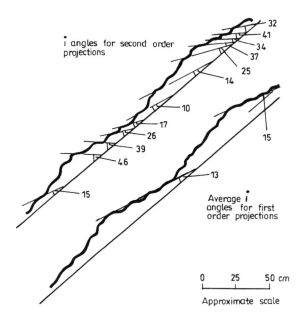

Figure 16.6.   Measurement of roughness angles (i) for first and second order roughness on a rock surface (adapted from Patton, 1996).

core (Stimpson, 1981). BC Hydro (1995) recommend that the surfaces should be tested in a wet condition to give (ϕ'_b).

The roughness angle (i) is generally determined by field observations of surface roughness, at a scale appropriate to the dam foundation, in the direction of potential displacement (i.e. the first order roughness in Figure 16.6). BC Hydro (1995) recommend that for the purpose of dam engineering an upper bound of $i = 10°$ be applied. They also suggest that in dam foundations normal stresses tend to be low compared to the strength of the foundation rock so that for relatively hard (strong) rocks Equation 16.5 can be used. In soft (weak) rocks they suggest use of Equation 16.6 with $c'_a = 0$.

BC Hydro suggest that the bi-linear Patton (1966) model can be replaced with the curvilinear approach of Barton and Bandis (1991). The authors are satisfied that the Patton (1966) approach is adequate for most situations.

16.3.2.3   Shear strength of infilled joints and seams showing evidence of previous displacement

BC Hydro recommend that the shear strength of previously displaced features (e.g. sheared or crushed seams) is determined from:

$$\tau = \sigma'_n \tan(\phi'_r + i) \qquad (16.7)$$

where ϕ'_r = effective residual friction angle of the infill, sheared or crushed material; i = average roughness angle which is reduced from the measured value to account for the lack of intimate contact of the walls of the feature as detailed below.

The residual frictional strength of the feature is generally determined by shear tests on undisturbed samples measured at sufficient displacement to ensure that the residual strength of the feature has been reached. For thick gouge (crushed rock) zones or, if undisturbed samples cannot be obtained, remoulded, normally consolidated samples of the material are tested for their residual strength. Gouge is often layered so care should be exercised to ensure that the weakest layer of material is selected for testing. Sufficient consolidation time must be allowed and rates of shearing must be sufficiently slow to allow pore pressure dissipation in accordance with accepted principles of soil mechanics. BC Hydro (1995) suggest the residual frictional component of shear strength for design be determined from the mean of the residual strength values if consistent results are obtained and a sufficient number of tests are done (about ten or more). However the authors' experience is that it is difficult to obtain true residual strengths and it is recommended more conservative values be adopted, e.g. the lower quartile or lower bound of test results.

To determine the roughness angle of previously sheared features, field measurements are obtained of the mean amplitude of asperities of the upper and lower surfaces of the shear, the mean inclination of asperity surfaces and the mean thickness of infilling. The value of roughness (i) will be less than the measured inclination of asperities because the walls of the feature are not in contact. Figure 16.7 indicates the reduced values of roughness for features with various ratios of seam thickness to asperity amplitude (from the USBR, based on Ladanyi and Archambault, 1977).

In many cases there may be a lower strength discontinuity (such as a thin graphitic shear) located along one of the contacts of the rock wall and seam material. BC Hydro (1995) recommend such features should be noted during the geological data gathering process and tested separately from the bulk of the seam material, to allow the strength of the weakest element to be determined.

16.3.2.4   Shear strength of thick infilled joints, seams or extremely weathered beds with no previous displacement

As indicated by BC Hydro (1995), such features would normally be excavated from the foundation of concrete dams. If they do exist, they suggest using:

$$\tau = \sigma'_n \tan(\phi') \qquad (16.8)$$

where ϕ' = effective friction angle of the infilling or seam material, based on tests of remoulded samples. Any cohesion component c', indicated by testing, is usually very small and should be ignored.

16.3.2.5   Shear strength of jointed rock masses with no persistent discontinuities

Careful mapping of the foundation of most concrete dams will show that there are kinematically feasible failure mechanisms which are controlled by discontinuities (bedding,

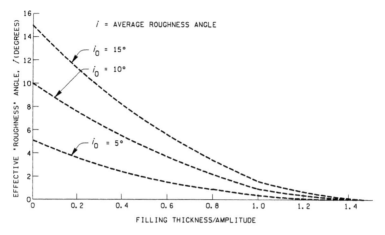

Figure 16.7. The effective roughness angle (i) for seams (infilled discontinuities) (BC Hydro, 1995, based on Ladanyi and Archambault 1977).

joints, shears, faults) which are persistent in the scale of the dam foundation. However for closely jointed rock with several sets of joints not oriented unfavourably in respect to potential failure surfaces; or for failure parallel to the base of the dam where the joints, bedding and other defects are not oriented close to this failure surface, a Hoek-Brown type failure criterion may be adopted. BC Hydro (1995) recommend the use of the Hoek et al. (1995) approach:

$$\sigma'_1 = \sigma'_3 + \sigma_{ci} \left[\frac{m_b \sigma'_3}{\sigma_{ci}} + s \right]^a \tag{16.9}$$

where m_b = a constant based on rock type; s,a = constants dependent upon characteristics of the rock mass; σ'_{ci} = uniaxial compressive strength of the intact rock pieces; σ'_1, σ'_3 = effective principal stresses.

The constants m_b, s and a can be estimated using Hoek et al. (1995) which relates these constants to the rock mass structure and discontinuity surface conditions.

Douglas (2003) reviewed the basis upon which the Hoek-Brown criteria determined, and have recommended the following:

(a) The basic form of the shear strength equation remains unchanged from the Hoek-Brown criterion.

$$\sigma'_1 = \sigma'_3 + \sigma_{ci} \left(\frac{m \sigma'_3}{\sigma_{ci}} + s_b \right)^\alpha \tag{16.10}$$

(b) For intact rock $m = m_i$ and $\alpha = \alpha_i$. These should preferably be measured from triaxial tests on intact rock samples. Alternatively an approximation can be made using the uniaxial compressive strength σ_{ci} and tensile strength σ_{ti} of the intact rock and the equations below.

$$m_i = \left| \frac{\sigma_{ci}}{\sigma_{ti}} \right| \tag{16.11}$$

$$\alpha_i = 0.4 + \frac{1.2}{1 + \exp\left(\dfrac{m_i}{7} \right)} \tag{16.12}$$

The estimation of m_b, α_b and s_b can be made using the following equations:

$$m_b = \text{minimum of} \begin{cases} m_i \dfrac{GSI}{100} \\ \\ 2.5 \end{cases} \tag{16.13}$$

$$\alpha_b = \alpha_i + (0.9 - \alpha_i)\exp\left(\frac{75 - 30m_b}{m_i} \right) \tag{16.14}$$

$$s_b = \text{minimum of} \begin{cases} \exp\left(\dfrac{GSI - 85}{15} \right) \\ \\ 1 \end{cases} \tag{16.15}$$

where GSI = Geological Strength Index, exp. = expotential.

Douglas (2003) indicates that the equations below from Hoek et al. (2002) can be used to estimate the cohesion, c', and friction angle, ϕ', of the rock mass by developing a σ'_n versus τ plot for the relevant stress conditions and rock properties using equations 16.16, 16.17 and 16.18.

$$\sigma'_n = \frac{\sigma'_1 + \sigma'_3}{2} - \frac{\sigma'_1 - \sigma'_3}{2} \cdot \frac{d\sigma'_1/d\sigma'_3 - 1}{d\sigma'_1/d\sigma'_3 + 1} \tag{16.16}$$

$$\tau = (\sigma'_1 - \sigma'_3)\frac{\sqrt{d\sigma'_1/d\sigma'_3}}{d\sigma'_1/d\sigma'_3 + 1} \tag{16.17}$$

$$d\sigma'_3/d\sigma'_3 = 1 + \alpha m_b (m_b \sigma'_3/\sigma_{ci} + s)^{(\alpha - 1)} \tag{16.18}$$

where $d\sigma'_1/d\sigma'_3$ is the slope of the σ'_1, versus σ'_3 plot.

They note that these equations replace those in Balmer (1952), which Douglas (2003) indicates were incorrect.

16.3.3 Tensile strength of rock foundations

Within the scale of most concrete dams, there are discontinuities in the foundation rock mass which mean that the tensile strength is zero, or so close to zero that it should be ignored.

There has in the past been an over-emphasis on the tensile strength of the concrete – foundation contact. This is virtually irrelevant, because the rock beneath the contact has zero tensile strength, so even if the contact has some strength, tensile failure can occur in the foundation immediately below the concrete.

16.3.4 Compressibility of jointed rock foundation

The Youngs modulus of deformation for a jointed rock mass can be estimated from the Geological Strength Index (GSI) using the method proposed by Hoek et al. (1998) and Hoek and Brown (1997), using the equation:

$$E_d = \frac{\sqrt{\sigma_{ci}}}{10} 10^{\left(\frac{GSI-10}{40}\right)}$$

(16.19)

Douglas (2003) collated a number of case studies from the literature where the rock mass modulus was measured by plate load, pressuremeter, dilatometer, flat jack and jointed block tests and the GSI could be estimated. The results from the case studies, together with data presented by Bieniawski (1978) and Serafim and Pereira (1983) are shown in Figure 16.8. The error bars acknowledge the variation of E_d in the test results.

The uncertainty in GSI was due both to the variability of the rock masses and also to the conversion from published RMR values to GSI. The compressive strength of the rock substance lay between 10 MPa and 100 MPa, and the corresponding plots for Equation

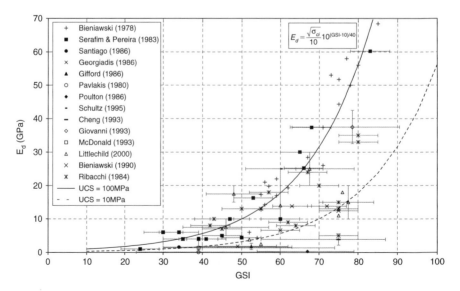

Figure 16.8. E_d versus GSI case study data and Hoek et al. (1998) equation for $\sigma_{ci} \geq 100$ MPa and $\sigma_{ci} = 10$ MPa.

16.19 are shown on Figure 16.8. It can be seen there is a large scatter in the data and there is a tendency towards over-estimation of the modulus. This needs to be considered when using the equation.

16.4   STRENGTH OF THE CONCRETE IN THE DAM

16.4.1   *What is recommended in guidelines*

The tensile and shear strengths of the concrete in gravity dams is often taken from guidelines. Table 16.4 summarises the recommendations from ANCOLD (1991), CDSA (1999) and FERC (2000).

It should be noted that these recommended values are used only where there is no tensile or shear testing available.

16.4.2   *Measured concrete strengths from some USA dams*

16.4.2.1   *Background to the data*

Concrete strength test data for cores taken from 23 dams reported in EPR1 (1992) and from test reports on 16 dams from USBR was analysed at UNSW (Khabbaz and Fell, 1999). The dams were all constructed by large, expert USA engineering organizations, were in good condition, not showing signs of cracking, weakening of lift joints or serious aggregate – alkali effects.

Three dams gave "outlier" data because they were affected by alkali aggregate reaction or poor cement quality and these were excluded from the plots in Figures 16.9 to 16.13.

16.4.2.2   *Tensile strength of concrete and lifts*

Figures 16.9, and 16.10 and 16.11 show the results of direct tensile testing on 150 mm or larger diameter cores taken from the dams which had lift joints. Failure often occurred in the concrete, not on the lift joint.

Table 16.4.   Summary of recommended tensile and shear strengths of concrete in dams.

| Property | Guideline and Recommended Value's (4) | | |
	ANCOLD, (1991) (1)	CDSA, (1999) (2)	FERC (2000)
Static ultimate tensile strength	$0.2\sqrt{f'c}$	$0.1f'c$ concrete $0.05f'c$ lift joints	$0.14(f'c)^{2/3}$ (3)
Dynamic ultimate tensile strength	$0.3\sqrt{f'c}$		
Shear strength	$0.14f'c$		
– Peak c'		$0.17\sqrt{f'c}$ concrete $0.085\sqrt{f'c}$ lift joints	0(5)
– Peak effective friction angle ϕ'	$45°$	$55°$ concrete $45°$ lift joints	$55°$(5)

Notes: (1) (a) $f'c$ = characteristic compressive strength of the concrete in MPa, usually at age 90 days for new dams
(b) Values apply to "normal concrete with well prepared construction joints". Lower values apply for concrete of uncertain quality.
(2) Assumes good quality concrete and lift joints.
(3) Assumes intact concrete.
(4) Tensile and compressive strength, and cohesion in MPa.
(5) Assumes pre-cracked concrete.

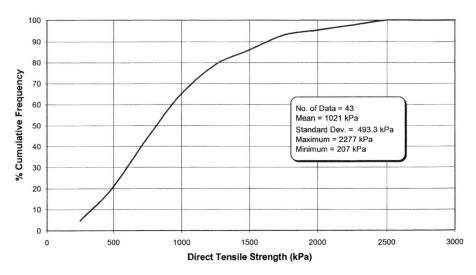

Figure 16.9.   Direct tensile strength of concrete with lift joints – some USA dams built after 1940 (Khabbaz and Fell, 1999).

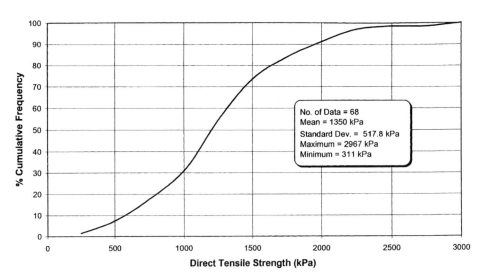

Figure 16.10.   Direct tensile strength of concrete with lift joints – some USA dams built before 1940 (Khabbaz and Fell, 1999).

Also shown on Figure 16.11 are the ANCOLD (1991) and CDSA (1999) relationships between tensile strength and compressive strength. It can be seen that much of the data is lower than the guideline values.

Figures 16.12 and 16.13 show results of tests using the indirect tensile test (or splitting test). The strengths are much higher than the direct tensile tests. EPRI (1992) attribute the higher strengths (from the splitting test) to the testing of smaller specimens which tend to fail through the aggregates which have a higher tensile strength than the concrete mortar. They suggest that the splitting test is satisfactory if the core diameter is 450–900 mm. For small core the direct tensile test is more reliable.

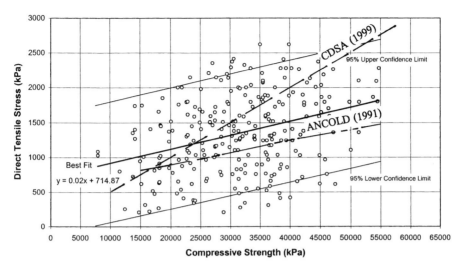

Figure 16.11.   Direct tensile strength versus compressive strength – all USA dams data analysed together (Khabbaz and Fell, 1999).

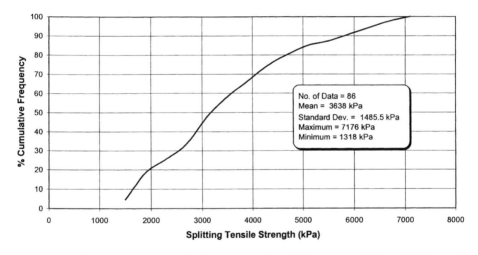

Figure 16.12.   Splitting tensile strength of concrete without lift joints – all data (Khabbaz and Fell, 1999).

It seems likely the guideline recommendations in Table 16.4 are developed from structural concrete codes, with the strengths usually obtained by splitting tests on small test specimens.

For dams, the direct tensile test results are more realistic, and it would seem unwise to rely on the guideline values. It would be better to do direct tensile tests on core from the dam, or make some assessment of likely strengths from Figures 16.9, 16.10, or 16.11.

It is emphasised that the tensile strength should be assumed to be zero, if

–   There are cracks in the concrete, e.g. on lift joints or caused by temperature stresses, or
–   There are extensive effects of alkali aggregate activity, or
–   There is poor quality concrete.

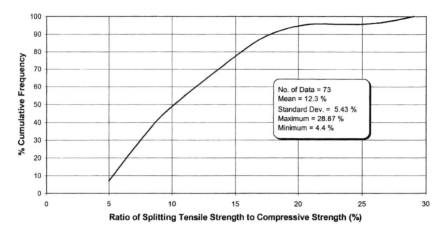

Figure 16.13. Ratio of splitting tensile strength versus compressive strength – of concrete without lift joints – all USA dams data (Khabbaz and Fell, 1999).

16.4.2.3 Shear strength of concrete

The terminology used by EPRI (1992) and USBR in their testing of concrete in direct shear is shown in Figure 16.14. Samples are described as bonded if they are intact, and unbonded if the sample is broken along the plane of weakness being tested.

The peak strength is typically reached at small displacements (0.25–1.25 mm). The residual strength is reached at large displacements but for practical purposes it has been taken as the lowest consistent strength at displacements between 2.5 mm and 12 mm. The sliding friction strength is the peak strength on an unbonded specimen.

The shear strength can be represented in Mohr-Coulomb terms as:

$$S = c' + \sigma' \tan \phi' \tag{16.20}$$

where S = shear strength; c' = effective cohesion; σ' = effective normal stress; ϕ' = effective friction angle.

The strength envelope is actually curved and at low normal stresses the friction angle is high, but approaches the residual value at high normal stresses. This non linearity needs to be considered in design.

EPRI (1992) report the following results from testing of concrete lift joints:

Peak strength:	best fit	$c' = 2.1$ MPa, $\phi' = 57°$
	90% stronger than	$c' = 0.95$ MPa, $\phi' = 57°$
Sliding friction strength:	best fit	$c' = 0.5$ MPa, $\phi' = 49°$
	90% stronger than	$c' = 0, \phi' = 48°$

Figures 16.15 to 16.18 show test data. It can be seen, that as suggested by EPRI (1992), it is best to use a bi-linear relationship to model the sliding friction strength.

There is little difference between sliding friction shear strength and residual strength – only 2° to 3° in the data available. The lower bound of the sliding friction strength data is $c' = 0, \phi' = 34°$. Many of the data near the lower bound are from Stewart Mountain Dams, which displayed alkali-aggregate reaction, with silica-gel noted on the surfaces of open joints. If these data are excluded the lower bound becomes $c' = 0, \phi' = 38°$.

The recommended strengths from guidelines given in Table 16.4 are on the conservative side if there has been no displacement on the lift surfaces. However only small differential movements, e.g. from thermal expansion and contraction, would take the strength

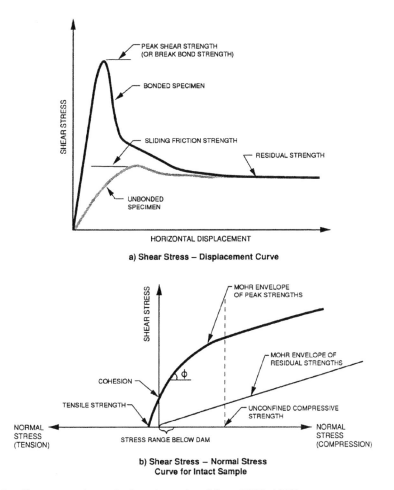

a) Shear Stress – Displacement Curve

b) Shear Stress – Normal Stress
Curve for Intact Sample

Figure 16.14.   Shear strength terminology (reproduced from EPRI, 1992).

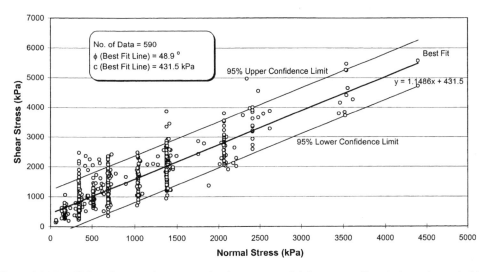

Figure 16.15.   Sliding friction shear strength of concrete with lift joints – all USA dams data (Khabbaz and Fell ,1999).

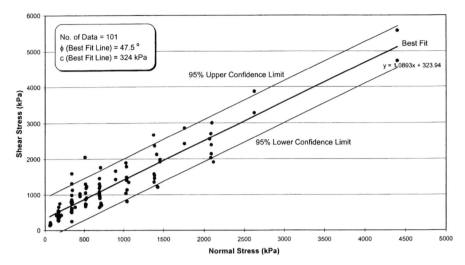

Figure 16.16.   Sliding friction shear strengths of concrete with lift joints – USA dams built after 1940 (Khabbaz and Fell, 1999).

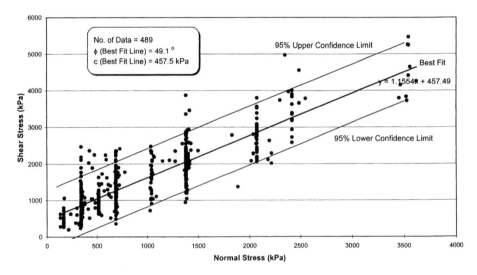

Figure 16.17.   Sliding friction shear strengths of concrete with lift joints – USA dams built before 1940 (Khabbaz and Fell, 1999).

towards the sliding friction strength. It is suggested that the EPRI (1992) data be used along with Figures 16.15 to 16.18 to estimate the strength. Where it is critical, or there are weak lift joints, laboratory tests should be carried out.

16.5   STRENGTH OF THE CONCRETE-ROCK CONTACT

EPRI (1992) report on direct shear tested one on the concrete-rock contact. They indicate that peak friction angles are large, typically 54° to 68° at low stresses.

Khabbaz and Fell (1999) plotted the available data which is shown in Figures 16.19 and 16.20.

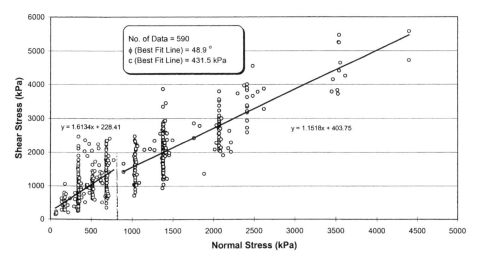

Figure 16.18.   Sliding friction shear strength of concrete with lift joints – all USA dams data, split at 750 kPa (Khabbaz and Fell, 1999).

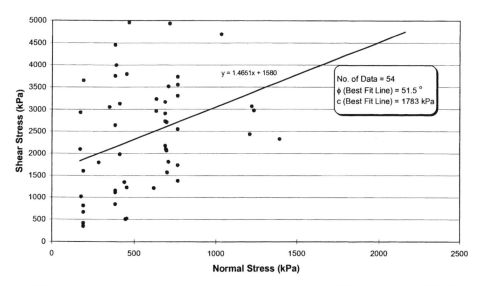

Figure 16.19.   Peak shear strength of concrete-foundation contact – all USA dams data (Khabhaz and Fell, 1999).

Provided good construction practice has been followed, there is no reason the contact strength should control. Rather it should be the lower of the concrete strength or the rock foundation shear strength. However, some existing dams could have had inadequate excavation and preparation of the foundation surface. Potentially unfavourable conditions that may exist include weathered or blast-damaged rock, consolidation and/or slush grout at the base of the concrete or a deteriorated contact. In these cases the contact strength may control, and any extensive areas of weak contact material should be sampled and the strength determined in the laboratory.

The tensile strength of the contact should be assumed to be zero – because it is controlled by the tensile strength of the rock mass immediately below the contact.

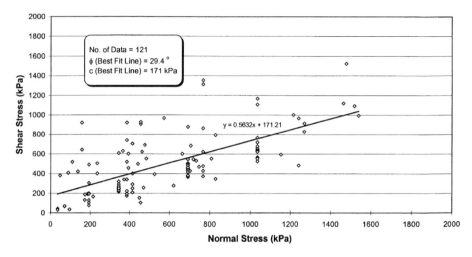

Figure 16.20.   Sliding friction shear strength of concrete-foundation contact – all USA dams data (Khabhaz and Fell, 1999).

16.6   UPLIFT IN THE DAM FOUNDATION AND WITHIN THE DAM

16.6.1   *What is recommended in guidelines*

For new dams, or for assessing existing dams for which there are no measurements of pore pressures in the rock foundations or on potential failure surfaces within the dam, there are well-established empirical rules for estimating uplift pressures. Figures 16.21(a) and (b) show the commonly used assumptions.

The pressures depend on whether the drainage gallery is above or below the tailwater level and the efficiency of the drains.

For Figure 16.21(a):
   For H4 > TW

$$H3 = K\left[(HW - TW)\frac{(L - X)}{(L - T)} + TW - H4\right] + H4 \qquad (16.21)$$

and for H4 < TW

$$H3 = K\left[(HW - TW)\frac{(L - X)}{(L - T)}\right] + TW \qquad (16.22)$$

where K = (1 − E); and E = drain efficiency as a fraction.

For Figure 16.21(b):
   For H4 > TW

$$H3 = K\,(HW - H4) + H4 \qquad (16.23)$$

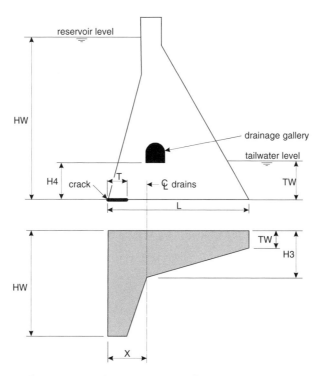

(a) for no compression zone not extending to the drain

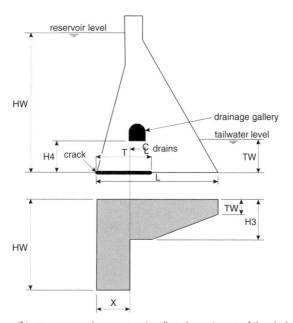

(b) no compression zone extending downstream of the drains

Figure 16.21.   Uplift assumptions for concrete gravity dams (Adapted from FERC, 2000).

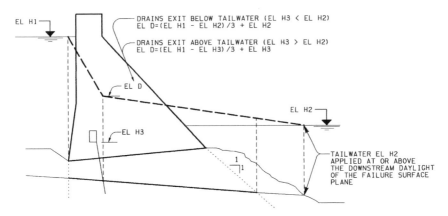

Figure 16.22.   Example of uplift distribution on a discontinuity in the dam foundation (BC Hydro, 1995).

and for H4 < TW

$$H4 = K (HW - TW) + TW \qquad (16.24)$$

For the no drains case (or drains blocked), the usual assumption is for 100% of the reservoir head at the upstream face (heel), 100% of the tailwater head at the downstream face (toe) and for a straight line between the heel and toe.

A common assumption (e.g. ANCOLD, 1991; BC Hydro, 1995) for the case with a well constructed and maintained borehole drain system extending deep into the foundation below any potential failure surface, is for the drain efficiency (E) to be 67%; (or K = 0.33). FERC (2000) do not suggest this, recommending that the efficiency must be measured by piezometer in the dam foundation. They point out that the measured drain efficiency must only be considered valid for the reservoir water levels for which measurements are taken, and warn that linear extrapolation to higher levels may not be valid. This is also emphasized by EPRI (1992) who give a number of examples of non-linear uplift response to reservoir level. The authors agree with this.

FERC (2000), US Corps of Engineers (1995) and USBR recommend use of full reservoir head in the non-compressive zone upstream of the drains. They assume that the drains become ineffective, and the pressures vary linearly from full reservoir head at the downstream end of the crack to the downstream toe head as it is assumed the drains may be unable to relieve the pressure under this condition.

Figure 16.22 shows an example of the uplift distribution on a discontinuity in the dam foundation.

Note that the tailwater level pressure applies where the potential failure surface emerges into the tailwater, and the pressure at the drains is where they intersect the potential failure surface.

16.6.2   Some additional information on uplift pressures

16.6.2.1   Effects of geological features and deformations on foundation uplift pressures

EPRI (1992) present a summary of the influence of geological features and deformations on foundation uplift pressures. They demonstrate that the assumption of linear uplift pressure distributions from the reservoir to tailwater for the no drains case, and from the drains to the tailwater in the drained case, are potentially not conservative because the

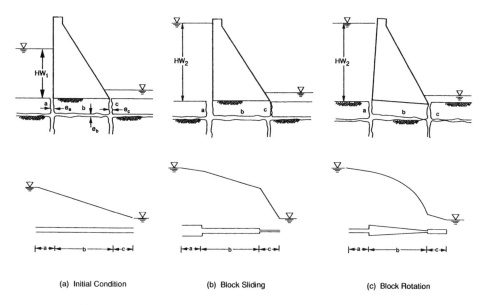

Figure 16.23.   Effect of block sliding and rotation on joint apertures uplift pressure. (a) Initial condition, parallel sided joint, (b) Block sliding, joint c width reduced, joint a width increased; (c) Block rotation, joint b narrows at downstream end (Adapted from EPRI, 1992).

discontinuities in the rock foundation can be narrower at the downstream than upstream because of the stresses and deformations induced by the dam and the reservoir load on the dam, and by geological features which limit the flow of water.

Figure 16.23 shows the effect of block sliding and rotation in the foundation; Figure 16.24 the effect of varying joint apertures and Figure 16.25 the effect of a low permeability shear zone.

This emphasises the need for careful consideration of the foundation geology in designing new dams, or assessing existing dams which do not have piezometers in the foundation. However there is no substitute for actual measurement of the pressures, with a carefully designed and well-monitored system of piezometers.

16.6.2.2 *Analysis of EPRI (1992) uplift data*

An analysis by Khabbaz and Fell (1998) of uplift data recorded in EPRI (1992), gave the results shown in Figure 16.26. In this figure the uplift coefficient α is the ratio of the measured pressure above tailwater level to the difference between reservoir level and tailwater level. It is equivalent to K in Equations 16.21 to 16.24. Data from 25, mostly USA, dams was available and represent the maximum pressures recorded across each dam. X/L is the ratio of the distance from the line of drains to the measuring point compared to the distance from the line of drains to the downstream toe of the dam. Table 16.5 summarises the data.

This information shows that the "standard method" assumption of $\alpha = 0.33$ at the drains is usually conservative. The highest value recorded was 0.47 for piezometers in one block of a dam where the drains were somewhat blocked, and the next highest, 0.33 for a dam with drains spaced at about 11 m centres rather than the normal 2 metres. However analysis of some Australian dams showed some had areas with maximum uplifts greater than the "standard method".

Analysis of the data to assess the effect of depth of drains, height of dam and other variables showed few clear trends. There did seem to be a tendency for lower uplift pressures for high dams, which may reflect greater care in the construction of the drains or grout

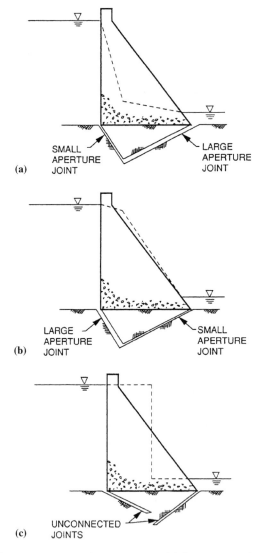

Figure 16.24.   Effect of joint apertures and continuity on uplift pressure (Adapted from EPRI, 1992).

curtain upstream of the drains, or that the weight of the dam closes discontinuities in the foundations.

16.6.2.3   *Design of drains*

Foundation drainage is provided by the provision of boreholes drilled into the dam and the foundation from the drainage gallery. ANCOLD (1991) recommend the following to achieve the uplift coefficient of 0.33 which is commonly assumed:

(a) The line of drains should be located at a distance from the upstream face of 5% to 15% of the maximum reservoir head at the dam;

(b) The lateral spacing of the drains in the dam should not exceed 10% of the maximum reservoir head at the dam and preferably not exceed 3 m. (However, drains in the body of the dam should not be so closely spaced as to encourage cracking between them);

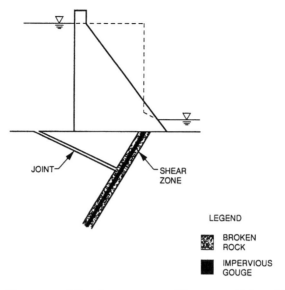

Figure 16.25. Effect of low permeability shear zone on uplift pressure (Adapted from EPRI, 1992).

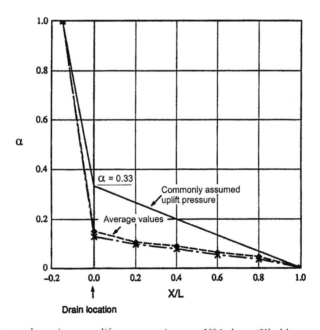

Figure 16.26. Observed maximum uplift pressures in some USA dams (Khabbaz and Fell 1998).

Table 16.5. Maximum uplift coefficients observed in some USA dams (Khabbaz and Fell, 1998).

X/L	−0.15	0*	0.2	0.4	0.6	0.8	1
Mean α	1	0.150	0.105	0.089	0.062	0.045	0
Standard deviation for α	–	0.123	0.082	0.087	0.076	0.06	–

* Location of drain holes.

(c) The drains within the body of the dam should be cored or formed and at least 150 mm diameter;

(d) The drains drilled into the foundation from a lower gallery should be a minimum of 75 mm and preferably 100 mm diameter and spacing should not exceed 3 m;

(e) Provision should be made for the drains to be serviced, e.g. by re-drilling foundation drains from a gallery.

The authors emphasise that it is essential that the drains penetrate well below the level of any potential failure surface and that they are oriented to intersect the defect system. Hence for horizontally bedded rock with a sub-vertical joint system, it is essential that the holes be drilled inclined to the vertical to intersect the joints. The depth should be determined from the geology of the foundation and the geometry of the dam, but it is unlikely that drains which do not penetrate at least 4 m below a potential failure surface would be effective. It would be wiser for them to be at least 6 m below the potential failure surface.

The drains should be constructed after the dam is constructed and the grouting of the grout curtain, (which is usually drilled and grouted from the drainage gallery and located upstream of the drain holes), is completed. At this time the stresses and deformations induced by the dam are present and the grouting cannot block the drainage system.

As pointed out by FERC (2000) the grout curtain may retard flows initially but the degree of uplift reduction is difficult to predict and may reduce with time as the cement in the grout is leached. Grouting alone (without drains) should not be considered sufficient justification to assume a reduced uplift.

The question of the effectiveness of drains within the dam is one that is often asked. Under normal operating conditions when in most dams there is compressive effective stress over the whole section, any seepage water that did reach the drains through the natural porosity of the concrete would cause the outlet at the drain to be sealed with calcium carbonate. The reaction here involves the seepage water picking up excess lime (CaO) from the concrete's matrix to form $Ca(OH)_2$ which combines with CO_2 in the air at the exit to give $CaCO_3$. If the seepage's exit velocity is small enough, the $CaCO_3$ comes out of solution to create the seal. The drains therefore would be ineffective in reducing internal pore pressures. The same outcome would occur if a small crack extended to the drain, as the exit velocity of the seepage would be very small. Only for substantial cracks would the seepage water discharge into the drains continuously. Even then, over time $CaCO_3$ would probably came out of solution as the water trickled down the drain hole, a situation that can be seen in many concrete dams. The authors believe that drains within the dam wall should be ignored in the design of a new dam or assessment of an old dam, unless calculation show that a crack or cracks would extend to the drains and only then for short-term loading conditions.

It is therefore suggested that within the dam's concrete, uplift across any section should be assured as varying linearly between the pressure at the upstream face and zero or tail-water level at the downstream face (drain ignored in this case). This is consistent with modern guidelines. For a crack-free dam, such an assumption may be conservative. Some expect the pressure to fall rapidly within a short distance of the upstream face of the dam. US Corps of Engineers (1995) suggest that a 50% reduction in pressure across the section would be acceptable to account for the likely sudden pressure drop at the upstream face, but the warning is given that such an assumption would apply to crack-free concrete. Given it is difficult to prove no cracks, the preferred approach is to assume linear pressure as described above.

The appearance of $CaCO_3$ on the downstream face of a dam, at, say, a construction joint, could be interpreted as a crack with full head applied up to the blockage, in this case, the downstream face. Even a formed drain in the concrete might not help as there could be a similar $CaCO_3$ seal here. This particular condition seems to have been ignored in most codes. Unless the "leak" is wide spread, in most dams the localised effect would be countered by the crack not propagating. The authors believe that a conservative attitude to pore pressures within the dam wall would cope with any such localised pressure development.

$CaCO_3$ formation can also occur in dams within the foundations, below the water surface in the drains. The reaction here involves CO_2 being released by microbes in the ground and H_2CO_3 being formed. This very mild acid then reacts with the excess lime from, say, a grout curtain and $CaCO_3$ results. This phenomenon has been seen at the pipeline inlet structure, a 34.7 m high gravity dam, at the head of the pipelines to Tumut 3 power station in the Snowy Mountains Scheme.

The possible formula of $CaCO_3$, limonite and other products in drains clearly points to the need to clean these drains regularly. At best, this cleaning would include high pressure water jet cleaning; at worst, the holes may have to be reamed. If the drains in the dam wall do play a key role in the dam's stability, then this cleaning operation should extend to them as well as to the foundation drains. In new dams, therefore, care must be taken to detail all drains so that cleaning can be done without the expenditure of a lot of money. Without a regular cleaning program in place, full reliance on the drains must be viewed with some doubt.

16.6.2.4   Hydro-dynamic forces

Spillway sections of concrete dams are subjected to additional hydro-dynamic forces due to the passage of water over the crest.

At the design discharge capacity, an ogee crest has zero pressure at the crest due to the water. However at discharges higher than the design discharge, there is a negative pressure which adds to the overturning moment and must be accounted for. These effects are shown in Figure 16.1. The negative pressures are most significant in small height structure such as spillway crests leading to a chute spillway.

In a flip bucket or any location where flow is changing direction in a concave upwards manner, the water exerts positive pressures. These may also contribute to overturning if they are downstream of the centre of the base of the dam.

Methods for calculating these pressures are given in FERC (2000), Brand (1999), and USBR (1990).

As pointed out by FERC (2000), and shown in Figure 16.27, if the outlet for a drain is in the flip bucket of a spillway, it may be subject to very high pressures. Such pressures can also be transmitted to the foundation through open joints in the spillway, or poorly positioned drain holes in spillway aprons. These situations should be avoided or, if present in a dam, the drainage system should be amended.

As discussed by FERC (2000), the uplift pressures in the dam foundation will respond rapidly to reservoir levels because in a saturated rigid rock foundation, extremely small volume changes are required to transmit the pressures. Such behaviour has been clearly recorded for many years at Tumut Pond dam (an arch dam). In the absence of measured data to prove otherwise, the uplift should be assumed to vary linearly with changes in reservoir and tailwater levels. However there are examples (EPRI, 1992) of non-linear response giving higher pressures.

16.6.2.5   Aprons

Upstream and downstream aprons have the effect of increasing the seepage path under the dam. For an upstream apron properly sealed to prevent leakage, the effect is to reduce the uplift under the dam. The effectiveness of upstream aprons in reducing uplift is compromised if cracks and joints in the apron permit leakage.

Downstream aprons, such as stilling basins or spillway chutes have the effect of increasing uplift under the dam as shown in Figure 16.28.

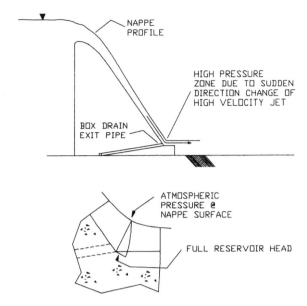

Figure 16.27. Effect of poorly located drain holes on uplift pressures (FERC, 2000).

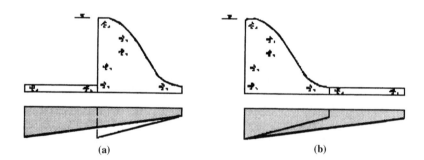

Figure 16.28. Effect of aprons on uplift pressures. (a) upstream apron reduces uplift pressure, (b) downstream apron increases uplift pressure (FERC, 2000).

Many stilling basins or spillway chutes have drains constructed at the concrete-rock contact. These may, if they are well constructed and maintained, reduce the pressure at the interface, but as discussed in Section 16.6.2.6, they will not be effective for potential failure surfaces within the rock foundation.

16.6.2.6 "Contact" or "box" drains

Many older, smaller concrete dams and spillways do not have a drainage gallery and drain holes bored into the foundation, but are provided with a gravel filled drains at the contact between the concrete and the rock.

It is the authors' experience that too much reliance has been put on these drains to reduce uplift pressures.

Provided they are well constructed, and can be cleaned or inspected to ensure they are not clogging, contact drains should reduce uplift pressures at the concrete-rock contact. However as pointed out by EPRI (1992) they will not reduce pressures in the rock, because the flow of water in the foundation is controlled by the fracture system. If the critical failure surfaces are in the rock, as they usually will be, there will be little benefit and the effect of the contact drains should be ignored.

16.7   SILT LOAD

Sediment in the reservoir which deposits against the upstream face of the dam exerts a pressure on the dam and must be allowed for in the analysis. For small structures with a heavy sediment bed load this may be substantial since with time sediment will partly or completely fill the storage up to spillway crest level.

The load applied by the "silt" should be estimated using Coulomb earth pressure theory, assuming the at-rest, or Ko, condition applies. The sediment is likely to be loose, so in the absence of specific data, properties consistent with loose material should be used.

For preliminary assessments only, the following properties could be used to assess the pressures:

(a) For weirs and small structures in fast flowing rivers and streams where the sediment consists mainly of sand and gravel, adopt a saturated unit weight of $20 \, kN/m^3$, an effective friction angle of 30° and effective cohesion zero;
(b) For structures where sediment is likely to be fine-grained silt and sand, adopt a saturated unit weight of $17 \, kN/m^3$, an effective friction angle of 25°, and effective cohesion zero.

In the calculation of the silt pressure, if silt is submerged, the saturated unit weight of the soil should be reduced by the unit weight of water to determine the buoyant weight.

When assessing dams under earthquake loading, the possibility of liquefaction of the "silt" should be assessed.

If this is likely, the silt loads will be greatly increased because the strength of the soil is reduced and the load will approach a Ko = 1.0 condition.

16.8   ICE LOAD

Where the reservoir may freeze, the ice will exert additional forces on the dam. FERC (2000) and US Corps of Engineers (1982) give guidance on how to estimate the forces.

16.9   THE DESIGN AND ANALYSIS OF GRAVITY DAMS FOR EARTHQUAKE LOADING

16.9.1   *Introduction*

This section deals with the earthquake loading of gravity dams. It concentrates on the more general aspects of the analysis rather than discussing the details of the methods and the analytical tools. Background data on these aspects, particularly of the finite element methods that are so popular today, can be found in ANCOLD (1998), FERC (2000) and in the manuals produced by companies that have developed the finite element packages. Relevant design guidelines, such as ANCOLD (1998) and FERC (2000), should be used to define the dam's design earthquakes (MDE and DBE in Chapter 12), permissible stresses and so on. Attention is also drawn to the earlier parts of this chapter for design strength parameters.

In the minds of most dam engineers today, the design and analysis of gravity dams for earthquake loading would mean either the pseudo-static method or the more complex approach of a finite element analysis.

16.9.2   Gravity dams on soil foundations

Generally speaking the foundations will be the critical factor in the design of these dams for earthquake loading. The approach, therefore, to check these dams for earthquake loading would logically be to follow the procedures outlined for embankment dams in Chapter 12. For example, one would use SHAKE to determine the ground shaking accelerogram for the soil deposit and then undertake a Newmark displacement analysis to estimate the permanent displacements along a selected failure surface through the soil foundations. As most of these dams would be low, stiff structures, the dam's response effects could probably be ignored.

As will be clear from the discussion in Chapter 12, the net outcome could very well be significant permanent displacements within the foundation that could, in turn, seriously compromise the dam. For a new dam, the object would be to design the dam such that its static load factors of safety are high enough so that under earthquake loading the resultant permanent displacements would most likely be small. As well, the dam's details would need to cope with potential differential movements between blocks. For an existing dam, unless considerable strengthening works can be justified, an owner may simply have to live with the inherent risk. Finally, for both new and existing dams, great care will be needed to assess the risk of post-earthquake piping through the foundations.

In the case of concrete dam (or spillway structure) being founded on potentially liquefiable foundations measures will have to be taken to overcome the effects of liquefaction such as removal or densification of the liquefiable material.

16.9.3   Gravity dams on rock foundations

16.9.3.1   General

The methods of analysis commonly used extend from the simple pseudo-static method first suggested in Westergaard (1935), a method that treats the dam on a rigid body (natural period = zero seconds) to complex finite element methods in which there are one or more non-linear features, usually in the form of elements representing cracks. The pseudo-static method was used as a matter of course well into the 1960s. With the advent of the readily available computing facilities in the 1950s and 1960s, the finite element method (FEM) became more and more available to design engineers, at first through developments at Universities and later through specialist firms, and gradually these structural analysis programs began to be used for gravity dam analyses. The introduction of the desk-top computer and the rapid increase in its computing power added to the spread of the gravity dam analyses until now these FEM analyses have almost become "the thing to do".

In between the two limits of the simple pseudo-static approach and the complex FEM are pseudo-static methods such as the one developed by Fenves and Chopra (1987). This method allows for the use of the first (the natural) period and, if needed, higher periods, to assess the impact of the dams response to the ground shaking. The US Corps of Engineers developed an approach that starts with this type of method and extends to limited FE analyses (Guthrie (1996)).

16.9.3.2   The Westergaard pseudo-static method

This method, which is usually applied in a 2-dimensional analysis, treats the dam as a rigid structure on a rigid foundation. The hydrodynamic effects, which Westergaard determined from harmonic ground motion, are included in the analysis by attaching a "block of water" to the face of the dam. This block or mass of water, perhaps the first time that the term "added mass" was used, has a shape for a vertical faced dam of:

$$b = \frac{7}{8}\sqrt{hy} \qquad\qquad (16.25)$$

where b = width of the added mass measured normal to the face horizontal at depth y. h = height of the dam at the section concerned (generally the height above the reservoir floor, not necessarily the structural height of the dam).

The total inertia force is found by applying the design peak ground acceleration (pga) to the structure. The movements and shear forces at the base and at other sections follow in the normal way for a conventional static load analysis.

FERC (2000) advises that this method should not be used and the authors would agree. It may be reasonable for low pgas but once these pgas are 0.2 g and beyond, the analysis simply proves that the dam would be unsafe, particularly if the engineer has assumed an $f_t = 0$ foundation. This outcome is probably not helped by designers treating the analysis as a conventional static load analysis, with the uplift at any section and at the base being redistributed once cracking has occurred.

16.9.3.3   *The Fenves-Chopra refined pseudo-static method*

The Fenves and Chopra (1987) method is a significant improvement on the Westergaard method. It uses data from more complex finite element analyses to allow the engineer to do a far more realistic analysis. In it:

- Dam response is included by taking from a suitable response spectrum the spectral accelerations at the first and selected higher modes for a nominated damping factor. For dams up to, say, 20 m in height, the dam's response may not be too critical;
- The water interaction model is more refined than the one often used (see next section);
- The flexibility of the foundations can be included;
- The effect of displacement of the crest upstream can be assessed.

The method gives an equivalent static loading applied to the upstream face of the dam that may be used to complete an analysis of the dam with the inclusion of the static loads (concrete weight, reservoir pressure, internal pore pressures and so on).

This method clearly has advantages over the Westergaard pseudo-static methods, but it does not allow for cracking of the base or at any other section that could change the dam's dynamic loading characteristics. For very stiff dams, cracking could see a substantial increase in the earthquake inertia loading, whereas in a large dam with a natural period of 0.3 seconds, the reverse might be the case.

One real advantage in this method is that if the stresses in the top of the dam are of concern, as was the case at Pine Flat and Koyna dams, the stresses there can be quickly assessed.

FERC (2000) suggests that this method is a reasonable one to use and the authors would agree, subject to some careful review of the results.

16.9.3.4   *The US Corps of engineers method*

This method is presented in Guthrie (1986) and an excellent discussion on it is given in ANCOLD (1998).

This method goes somewhat further than the Fenves and Chopra (1987) method in that depending on the level of tensile stress developed, the engineer may have to undertake a limited FEM analysis. It would seem to be limited to some extent if one assumes, as the authors have suggested, that the dam foundation interface has no tensile strength. In that case, cracking will be a reality so unless the analysis can cope with a possibly varying natural period as the crack extends, then it would be better to go immediately to a full FEM analysis. If cracking at the dam foundation interface is limited so that the dam's natural period does not change very much, then the use of this method is reasonable.

16.9.3.5   *Finite element method (FEM)*

FEM has been used on many gravity dams since probably the late 1970s. The analyses have mostly been 2-dimensional analyses, although the authors know of several done in 3-dimensions. There have also been a number of arch dams analyses done, which have necessarily been 3-dimensional analyses. The most well known of these would be the USBR studies into the never-built large Auburn dam in California (USBR (1977 and 1978)).

Well into the 1990s, most of the analyses by practising engineers used linear elastic FEM programs. Given the low level of stresses in a gravity dam, the assumption of linear elastic behaviour is reasonable, but cracking is hard to include in a full dynamic analysis. Now with computer programs available with more complex elements that will allow for crack joint opening, compressive loading, shear loading and sliding, the difficult question of cracking can be included in the FEM model.

Most FEM analyses are time-consuming. In the 1980s-early 1990s, it was lack of computer power; now (2003) it is very complex programs. In all cases, there is a cost associated with any FEM analyses. One almost flippant comment to the fourth author many years ago had and still has much truth in it; "whatever you estimate, it will cost 3 times as much".

Ideally, if, as the authors believe is the right approach, ft = 0 is assumed at the dam/foundation interface, cracking must be modelled in this zone. Cracks can be modelled elsewhere, if the computer power is available. Once cracking is included, the engineer will be limited to time-history analyses in a step-by-step numerical integration process. At each small time interval, say 0.01 s, the dam is fully analysed under both the applied earthquake loading and the static loading in a number of iterations until equilibrium is satisfied. The analysis then moves to the next time step. With the analysis being an effective stress analysis, as are both the foregoing pseudo-static analyses, the pore pressures must be included. Permanent shear displacement can also be allowed for.

The mechanics of an FEM analysis are now reasonably well known and need not be repeated here. Nevertheless, there are some issues particularly related to dam analyses that need highlighting. They include:

(a) *Dam-reservoir interaction effects*
   The USBR (1977, 1978) Auburn study in the late 1970s allowed for these effects to be included as "added masses" fixed to the upstream face elements. The USBR used the Westergaard distribution.
   The general opinion is that this approach is probably reasonable (Léger et al (1991)), although, in the case of a 3-D analysis one would question the relevance of this added mass to a cross-valley earthquake. In the long run, an FEM analysis that includes the reservoir directly would be the ideal (see ANCOLD (1998), and Léger et al (1991) for more discussions).

(b) *Inclusion of the foundation rock*
   Most analyses included the rock mass with fixities at least one "dam height" away from the dam-foundation interface. For a static analysis, if one ignores the "rebound effects" on excavation, the foundation has "no mass". The same approach is often taken in dynamic analyses, even though the resulting dam/foundation model is slightly stiffer than a model with rock with mass. For many dams the critical failure surfaces will be in the foundation so the foundation must be modelled.

(c) *Inclusion of water pressure loads and internal pore pressures*

Most finite element analysis programs are geared up to cope with internal pore pressures so generally engineers have opted to apply the reservoir water loads to the dam's upstream face and any exposed dam faces of cracks. The final effective stresses within the body of the dam are found by adding the pore pressures to the calculated stresses from the FEM analysis (compressive stresses usually taken as negative). Some engineers apply the water load to the dam's face and uplift as an external load to the base.

This approach does not allow the correct results to be achieved in a step-by-step numerical integration procedure. To overcome this shortcoming, analyses can be done to define the seepage pressure forces. Typically, for an FEM package that has a heat flow model contained within it, this heat flow program, with assumed boundary conditions (for example, the very conservative assumption of full head at upstream face and zero at downstream face), can be used to calculate these seepage forces. In effect, the total water pressure loads are included as body forces in much the same way as gravity forces.

If the stress distribution is required within the foundations, then similar body force distribution can be found. The pre-existing groundwater conditions need to be included – for example the original grand water surface could be assumed as coincident with the ground surface (probably appropriate to a dam section near or at maximum section).

The pore pressures are usually assumed not to change during the earthquake. Some have argued that this assumption should not apply to a crack that opens and closes cyclically, but most design guidelines recommend that the "unchanged" pore pressure model is acceptable.

(d) *Shear strength conditions at crack elements*

Ideally, the chosen shear strength law should have peak strength used until shear displacement starts at which point the strength would drop to the appropriate residual strength law. The residual strength would apply thereafter. One would have to nominate a complete shear failure path, usually a planar surface on the section or at the dam-foundation interface. Movement would not be assumed until all crack elements on that surface have reached the peak strength.

This relatively complex model, as far as the authors know, may not be readily available in FEM packages. Accordingly, the procedure is usually to adopt the residual strength as the shear strength law from the start of the analysis. The calculated permanent shear displacement for most analyses would therefore tend to over-estimate the shear displacement on any selected surface.

(e) *Damping*

Damping, usually as a percentage of critical damping is an important input for any FEM analysis. Some practising engineers choose the damping factor at the low value, say, 5% of critical damping, arguing that the dam's stress levels are relatively low so damping will be correspondingly low. Others argue that any crack development, permanent shear displacement, the presence of defined contraction joints between blocks and similar features warrant selection of a damping factor approaching 10–12%, particularly for MDEs with a pga above 0.3 g. The authors favour the selection of damping factors close to 10% for an MDE analysis. Strictly, damping should be varied throughout the analysis, but the simple approach of a constant factor is generally applied from the start.

16.9.3.6   *Design earthquake input motion*

Regardless of whether the response spectrum or the accelerogram (acceleration time-history) input is the basis for the analyses, several selected sets of input data should be used. Advice should be sought from a specialist seismologist on this input data.

16.9.3.7 *Should vertical ground motion be included?*

There is no reason that vertical ground motion should not be included in any analysis. Given that the response effects to vertical ground motion are likely to be very small, the simple approach of reducing "gravity" by the peak vertical acceleration would be appropriate for the pseudo-static methods to account for this motion. For the more complex FEM analyses, the engineer could either reduce "gravity" or apply to the model a measured vertical accelerogram along with its companion horizontal accelerogram. Remember that the vertical motion does not get applied to an "added" mass.

16.9.3.8 *Reservoir level variation*

If a reservoir only gets to full supply level infrequently, a lesser reservoir level may be justified for analysis.

16.9.3.9 *What do the results of analyses mean?*

The engineer doing analysis of a concrete dam must sooner or later face up to this question. "Should I check the stability of the dam at its peak displacement?" "What does a permanent shear displacement of 100–200 mm mean?"

In the case of the pseudo-static methods, if reasonably high pgas are used, the dam will inevitably be shown to be unstable. The method, by its very nature, demands that the dam's stability be checked.

FERC (2000) states that:

> "*In a departure from the way the FERC has previously considered seismic loading, there is no longer any acceptance criteria for stability under earthquake loading. Factors of safety under earthquake loading will not longer be evaluated.* Acceptance criteria are based on the dam's stability under post earthquake static loading considering damage likely to result from the earthquake. The purpose of considering dynamic loading is to determine the damage that will be caused so that this damage can be accounted for in the subsequent post earthquake static analysis.*"

and later:

> "*Because of the oscillatory nature of earthquakes, and the subsequent structural responses, **conventional moment equilibrium and sliding stability criteria are not valid when dynamic and pseudo dynamic methods are used.** The purpose of these investigations is not to determine dam stability in a conventional sense, but rather to determine what damage will be caused during the earthquake, and then to determine if the dam can continue to resist the applied static loads in a damaged condition with possible loading changes due to increased uplift or silt liquefaction.*"

FERC (2000) approach is that the analyses are to show if and where there could be problems, such as permanent displacement, and how these problems might affect the dam, for example, extensively cracking the dam, seriously damaging prestressed anchors, completely cutting foundation drains and so on. From these results, one would decide what weaknesses have been created and whether or not this weakness could influence the dam's post-earthquake condition. Partial or full loss of prestressed anchors, change in drainage conditions, extensive cracking and loss of shear strength would be typical of the impacts in a post-earthquake analysis.

The FERC recommendations have much merit. However it should be recognised that a simple structure subjected to ground shaking could collapse during the earthquake, not just afterwards. Perhaps the key point is that most gravity dams are not simple structures; for example, there could be 3-D action or significant differences, even reversals, in displacement along the axis at any particular time interval. The author's preference for new dams would be to have a theoretically stable dam at all stages of an analysis, including, in the case of a step-by-step numerical integration FEM, at the time-step for the maximum displacement of the dam. However this may be impractical for high seismic load areas.

Calculated permanent displacements of, say, 100 mm and above seem at odds with these large, stiff structures whose maximum displacement under elastic loading are probably only up to 10–15 mm. Such large permanent displacements imply the development of considerable distress in the dam and much remedial work if it is to be retained.

Engineers should not be too carried away with any of these analyses, no matter how complex. The words of Newmark and Rosenbleuth (1971) from some 30 years ago still hold true:

> *"We must also face uncertainty on a large scale, for it is our task to design engineering systems – about whose pertinent properties we know little – to resist future earthquakes and tidal waves – about whose characteristics we know even less."*

16.9.3.10   *Post-earthquake analyses*

As emphasised by FERC (2000) these analyses are vital to any gravity dam that could be subjected to earthquake loading. They should be done carefully, with conservative strength factors allowing for the deformations and cracking induced by the earthquake and proper consideration of changed uplift pressures.

16.9.3.11   *Dams on rock foundations with potentially dee-seated failure mechanisms*

For most of the foregoing discussions, the tacit assumption has been that the dam-foundation interface zone is the critical foundation component in the analyses. If there are known defects and weaknesses, such as existing bedding surface shears, that could give rise to a potential deep-seated failure, the analytical problem increases again in complexity.

Without going into highly complex coupled foundation-dam analytical models, the likely steps are:

- The dam-foundation mechanism presumably can be proven to safe under static loads. Therefore calculate the dam's maximum response base moment and shear using the methods just described, and apply them to the potential foundation failure mechanism. If there is an adequate safety margin, with conservatively chosen strength parameters, the dam-foundation interface can be accepted as the critical foundation zone.
- If movement does occur on the deep-seated failure mechanism, a yield acceleration should be calculated, allowing for the dam's loading, and a Newmark analysis done to estimate the total permanent displacement. This type of approach may be reasonable as during sliding on the foundation mechanism, the impact of the earthquake will not be transmitted through to the dam above (USBR 1989).

The difficulty posed by this particular problem clearly illustrates the need for very careful geological exploration and assessment of the damsite. Any very stiff structure like a gravity dam could be under serious threat if it were to be subjected to potential displacements from ground movement.

16.9.3.12   *Dams on foundations that could be subjected to ground displacement*

None of the above methods can hope to cope with major ground dislocation. Clyde dam in New Zealand does include a design feature to allow for some differential displacement, but even that feature would not handle the displacement of 9 m vertically and 1.5 m laterally that occurred at Shikang dam in Taiwan in 1999 (see Figure 16.29).

Figure 16.29.   Damage to Shikang dam in Taiwan in 1999. The vertical displacement was about 9 m and the horizontal displacement was about 1.5 m into the abutment.

16.9.4   Concluding remarks

The analysis of a gravity dam for earthquake loading will inevitably present problems to the people involved. This situation seems even more of an issue today as engineers adopt high pgas and use analyses that really do not simulate what is going to happen. Perhaps the best thing to do is to make sure the dam is conservatively designed for its static loading, for example, use shear strength parameters for rock and concrete and concrete tensile strengths that are not excessively high and provide proper drainage, and to take simple defensive precautions, for example, those listed in ANCOLD (1998). For the comfort of all concerned, the record of gravity dams under earthquake loading is good (Hinks and Gosschalk, 1989). However few have been subject to very large earthquakes.

17

Foundation preparation and cleanup for embankment and concrete dams

17.1 GENERAL REQUIREMENTS

17.1.1 *Embankment dams*

The degree of foundation preparation which is necessary for a dam embankment depends on the:

- Type of dam;
- Height of dam and the consequences of failure;
- Topography of the dam site;
- Erodibility, strength, permeability, compressibility of the soil or rock in the dam foundation;
- Groundwater inflows to excavations;
- Climate, and river flows.

Foundation preparation for the "*general foundation*", i.e. the foundation beneath the bulk of the embankment, is quite different from that for the "*cutoff foundation*", i.e. the foundation under the earthfill core of an earth and rockfill dam prepared to cutoff standard, or the plinth of a concrete face rockfill dam.
Figure 17.1 shows the terms used in this chapter.
The objectives are:

- *General foundation*: To remove low strength and compressible material to provide a foundation of adequate strength and compressibility to support the embankment. In most cases, but not always, the general foundation will be of higher strength than the embankment and will not dictate the embankment stability. In most cases permeability will not be a critical factor. Liquefiable materials may need to be removed or treated.
- *Cutoff foundation*: To remove highly permeable and erodible material below the general foundation level to provide a low permeability non erodible foundation (consistent with the design of filters and drains on the foundation). In many cases, e.g. in soil foundations or rock foundations which are permeable to great depth, a low permeability non-erodible cutoff foundation cannot be economically achieved by excavation alone and other design measures are required. These are discussed in Chapter 10.

17.1.2 *Concrete dams*

The degree of foundation preparation for a concrete dam depends on the:

- Type of dam (gravity, arch, buttress);
- Height of the dam and the consequences of failure;
- Topography of the dam site;

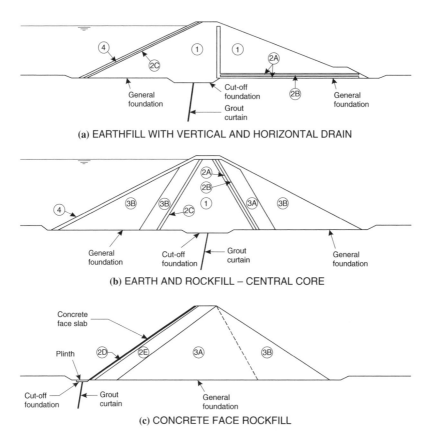

Figure 17.1.   Terms used for foundation preparation for embankment dams (a) earthfill with vertical
and horizontal drains, (b) earth and rockfill (c) concrete face rockfill.

– Loads imposed on the foundation, and the strength and compressibility of the soil or
  rock in dam foundations, allowing for the detailed geological structure, including bed-
  ding, jointing, and the presence of shears and faults;
– Erodibility and permeability of the soil or rock in the dam foundation;
– Groundwater inflows to excavations;
– Climate, and river flows.

Figure 17.2 shows three arrangements for foundation preparation under concrete gravity dams. Figure 17.2: (a) shows the conventional arrangement, where the foundation is
all taken to a uniform requirement; (b) shows the case where a low strength defect exists
in the rock (e.g. low strength bedding surface) and the foundation is taken below the
defect; (c) shows a foundation where "keys" have been used to take the critical failure surface deeper into the foundation. "Keys" or cutoffs may be required to prevent undercutting of the downstream toe by the spillway flow.

In each case A–B is prepared to concrete dam foundation standard (A–B′ if the additional concrete in Figure 17.2(b) is required). A lesser standard may apply under the stilling basins or between the keys in Figure 17.2(c).

Grout holes and drainage holes are usually provided as shown in Figures 17.2(a) and (b) to
control uplift pressures beneath the dam. For smaller dams contact drains, as shown in Figure
17.2(c) have been used. These drains are ineffective except at possibly the foundation contact
as discussed in Chapter 16.

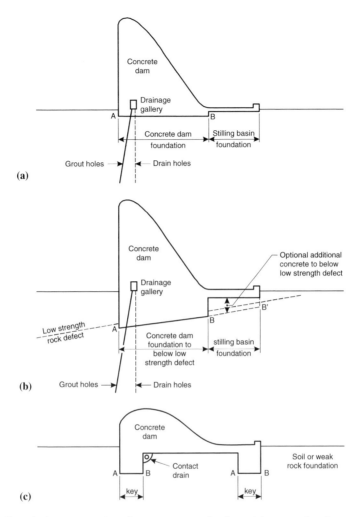

Figure 17.2.   Foundation preparations for concrete gravity dams (a) conventional arrangement for dam on rock foundation; (b) dam on rock foundation with low strength defect in foundation; (c) dam with "keys".

In general terms, after normal treatment by grouting and drainage, the rock foundations must:

– Be of adequate strength, stiffness modulus and durability to support the dam loads, without excessive deflections or settlements, under reservoir empty, full and all operating conditions;
– Have sufficient shear strength to provide an adequate factor of safety against sliding downstream under all operating conditions;
– Be sufficiently impermeable and non-erodible to prevent excessive leakage beneath the dam;
– For dams with overfall spillways, or subject to overtopping during extreme flood events, be able to withstand the impact of flood overflows at and near the toe, without erosion which could endanger the dam.

For concrete gravity dams founded on soil (as many smaller dams and "weirs" are), the requirements are essentially the same as for rock, but the foundations are weaker and more compressible than rock, and are erodible, so the design must account for this.

17.1.3   Definition of foundation requirements in geotechnical terms

In all cases it is important to define the requirements in geotechnical terms which can be identified in the field during construction. This is discussed in Sections 17.5 and 17.8.

It is likely that in most cases excavation will proceed in stages, with sufficient clean up to identify whether the geotechnical requirements have been achieved.

17.2   GENERAL FOUNDATION PREPARATION FOR EMBANKMENT DAMS

The requirements for preparation of general foundations for embankment dams are described below.

17.2.1   General foundation under earthfill

17.2.1.1   Rock foundation

– Remove topsoil and weak compressible soil. In most cases this will involve removal of colluvial soil and rock, including boulders, to expose an *in situ* rock foundation which may be extremely weathered, variably weathered or in some cases fresh (unweathered). An adequate foundation can usually be identified by near blade refusal of a small bulldozer or excavator;
– Where there are unfavourably oriented weak seams in the rock, e.g. bedding surface shears or landslide rupture (slide) surfaces, these will influence stability and may need to be removed, or the design modified to accommodate them;
– Slope modification may be necessary as described in Section 17.6.

The surface should be cleaned of loose soil and rock before placing earthfill, e.g. with a grader, backhoe or excavator. Intensive cleanup is not generally required. It may be desirable to moisten the surface prior to placing earthfill to maintain adequate moisture in the earthfill for compaction, and treat open joints to prevent erosion of the soil into the joint.

17.2.1.2   Soil foundation

– Remove topsoil and weak compressible soil consistent with the assumptions made for stability and settlement analysis and design;
– Where soils are fissured, or have landslide rupture (slide) surfaces present, these will influence stability and may need to be removed, or the design modified to accommodate them;
– Slope modification may be necessary as described in Section 17.6;
– The surface may be proof rolled with a steel drum or tamping foot roller to assist in locating weak and compressible soil. Rolling to a specified compaction requirement should not be necessary.

The surface should be cleaned of loose soil and rock prior to placing earthfill, e.g. with a grader, backhoe or excavator. Intensive cleanup is not generally required. It may be desirable to scarify and moisten the surface prior to placing the first layer of soil to assist in "bonding" the embankment to the foundation.

17.2.2   General foundation under rockfill

Rockfill will generally be underlain by a rock foundation (or by a horizontal filter drain, in which case refer to Section 17.2.3):

– Remove topsoil, soil and weathered rock which has a strength lower than the rockfill. In most cases this will involve exposing distinctly weathered rock which is very weak or

weak. An adequate foundation may be identified by blade refusal of a bulldozer but for high dams light ripping may be necessary;
- Where there are unfavourably oriented weak seams in the rock, e.g. bedding surface shears or landslide rupture (slide/surfaces), these will influence stability and may need to be removed, or the design modified to accommodate them;
- It is unlikely that slope modification will be required under rockfill other than over-hanging cliffs being treated by removing the overhanging rock, or filling the re-entrant below the overhang with concrete;
- The surface should be cleaned of loose soil and rock with a bulldozer, grader, backhoe or excavator sufficient to ensure the rockfill is supported on the rock foundation. Intensive cleanup is not required and if there is loosened rock of good quality this may be left in place.

If there are wide, erodible seams in the rock, oriented that they are likely to carry seepage water from the storage, it may be necessary to cover them with concrete, shotcrete under the upstream rockfill, or preferably by a filter layer under the downstream rockfill. Wallace and Hilton (1972) describe backfilling compressible seams greater than 0.3 m wide with Zone 2B filter material to a depth of up to 0.9 m. In most cases this will not be necessary unless the seams are wide enough to affect stability or pose an erosion risk.

17.2.3 *General foundation under horizontal filter drains*

Horizontal filter drains will generally only be required if the foundation is soil or erodible rock:

- Remove topsoil and weak compressible soil consistent with the assumptions made for stability, settlement and particle size used for design of the filter;
- Where soil or rock in the foundation is fissured or has landslide rupture (slide) surfaces or unfavourably oriented bedding surface shears are present, these will influence stability and may need to be removed, or the design modified to accommo-date them;
- Slope modification should not be required except to remove overhangs. However, if earthfill is to be placed on top of the filter drain, slope modifications may be needed as described in Section 17.6;
- The surface should not be rolled prior to placing the filter. Rolling will destroy the soil structure and reduce the permeability, making it more difficult for seepage water to flow into the filter drains. It is desirable for the foundation seepage to flow into the filter drain so erosion is controlled, rather than being forced to emerge down-stream of the toe of the embankment in an uncontrolled manner. This is an important issue, often not well appreciated by construction personnel and even some dam engineers;
- On low strength rock and on soil, trafficking with earthmoving equipment will contin-uously break up the surface, necessitating final cleanup with an excavator or backhoe working away from the cleaned up area. This is particularly a problem with weathered schists, phyllites and similarly fissile rocks which break down very easily under equip-ment. Once cleaned, the filter should be dumped on the cleaned up surface and spread onto the surface without equipment trafficking directly onto the foundation;
- Immediately before placing the filter material, the surface should be cleaned of loose dry and wet soil and rock. This may necessitate intensive work using light equipment and hand methods. Final cleanup should involve an air or air-water jet to "blow" away loose material. In many cases an air-water jet will be too severe and cause erosion.

17.3   CUTOFF FOUNDATION FOR EMBANKMENT DAMS

17.3.1   *The overall objectives*

The requirements for excavation below general foundation level to achieve a suitable cut-off foundation are described below. It should be noted that each case will be determined on its merits and often there will be trade-offs between the following:

– The desire to achieve a low permeability foundation;
– The depth and hence volume and cost of excavation required to achieve a cutoff;
– The extent of grouting planned in the foundation below the cutoff;
– The protection downstream to control foundation erosion, e.g. filters over the surface of the foundation.

High groundwater levels, e.g. in alluvial soils, may determine the practical depth to which a cutoff may be taken, since dewatering is usually expensive. The guidelines given below, therefore, are stating the "desirable" requirements rather than what may be practicable in some cases.

It should also be noted that many of the features described below cannot be readily identified in the base of the cutoff excavation, but are apparent in the sides of the excavation. Hence it is normal to require progressive excavation and cleanup, with a minimum excavation depth of about 0.5 m below general excavation level to confirm that the requirements have been achieved.

Having defined the cutoff foundation, the further objective when placing the earthfill is to ensure a low permeability contact between the earthfill and the foundation. This is done by:

– Modifying the shape of the foundation surface to provide a surface suitable for earth-fill compaction (Section 17.6);
– Using compaction equipment which facilitates compaction but does not damage the foundation (Section 14.2);
– Using more deformable, less erodible soils adjacent to the contact – i.e. using higher plasticity soils placed wet of optimum water content (Section 14.2);
– Filling erodible seams or open joints in the rock prior to placing the core material (Section 17.6).

17.3.2   *Cutoff in rock*

– Remove rock with open joints and other fractures which would otherwise lead to a highly permeable structure. In many cases this will result in a foundation which has a permeability generally less than say 15 Lugeons. For large, high hazard dams it would be normal to aim for a foundation with a permeability generally less than 7 Lugeons. However it is emphasised that this may not be practicable in many situations and higher permeability rock will be left in the foundations.
– Remove rock with clay infilled joints, roots etc., which may erode under seepage flows to yield a high permeability rock. This is particularly important where the clay has been transported into the joints and/or is dispersive, as this indicates its likely erodibility.
– Carry out slope modification and treatment as described in Section 17.6.
– Where the exposed rock is susceptible to slaking by wetting and drying (e.g. many shales) or breakdown under trafficking, it should be covered with a cement-sand grout (thickness usually <10 mm to 25 mm), pneumatically applied mortar or concrete (minimum thickness 50 mm and preferably not less than 150 mm). Generally this should be

done immediately the foundation is exposed, but may be done after a second cleanup immediately before placing the earthfill core. If cement-sand grout is used, it may crack under the trafficking of equipment placing the earthfill and it may have to be removed immediately prior to placing the earthfill.

–   Remove from the surface all loose soil and rock, and debris from grouting (using light equipment and with an air or air-water jet). Hand cleanup may be necessary. The surface may need to be moistened immediately before placing earthfill to maintain the earthfill moisture content.

–   If the rock in the floor or the sides of the cutoff trench displays open joints or other features which would allow erosion of the earthfill into them, it should be cleaned of loose material and covered by a cement-sand grout, pneumatically applied mortar or concrete. This is particularly critical on the downstream side of the cutoff trench. If there are only a few such features they might be treated dentally.

Under no circumstances should the surface of the cutoff foundation be rolled, even if it is a weathered rock. Rolling will only disturb the rock leading to a higher permeability material.

Figures 17.3 to 17.8 show cutoff foundation cleanup, treatment of seams and open joints for Dartmouth, Kenyir and Blowering Dams.

Figure 17.3.   Abutment of Dartmouth Dam showing cleanup in cutoff foundation, and construction of the grout cap (Courtesy of Murray-Darling Basin Commission and Goulburn-Murray Water).

Figure 17.4.   Dental concrete at Kenyir Dam (Courtesy of Snowy Mountains Engineering Corporation).

Figure 17.5.   Excavation of sheared zones prior to filling with dental concrete in the cutoff foundation of Dartmouth Dam (Courtesy of Murray-Darling Basin Commission and Goulburn-Murray Water).

Figure 17.6.   Cutoff foundation for Blowering Dam showing open joints being treated with pneumatically applied mortar, but no treatment for other joints (Courtesy of Snowy Hydro).

Figure 17.7.   Application of pneumatically applied mortar on seams in the cutoff foundation of Blowering Dam (Courtesy of Snowy Hydro).

Figure 17.8.   Left abutment of Blowering Dam after treatment of large fault zone and weathered seam
(Courtesy of Snowy Hydro).

17.3.3   *Cutoff in soil*

– Remove soil with open fissures, open joints, roots, root-holes, permeable layers (e.g.
sand and gravel) and other permeable structure (e.g. leached zones in lateritic soils);
– Remove dispersive soils if possible;
– Carry out slope modification as described in Section 17.6;
– If the soil on the sides of the cutoff trench displays permeable layers or features which
would allow erosion of the earthfill into them, it should be trimmed and cleaned and
covered with a filter layer or layers which are designed to control such erosion. Details
of such treatment are shown in Figure 10.8 In extreme cases it may be necessary to also
cover the slope with pneumatically applied mortar;
– Remove loose and dry soil and other debris with light equipment, possibly with the aid of
an air jet. The base of the cutoff should be watered to within 2% dry and 1% wet of optimum water content and rolled before placing the first layer of fill, to compact any soil
loosened by the construction work. If necessary it should be disked or tyned with a grader
to facilitate the moisture conditioning.

17.4   WIDTH AND BATTER SLOPES FOR CUTOFF IN EMBANKMENT DAMS

The width of cutoff trench adopted depends on:

– the rock or soil quality at cutoff foundation level;
– the size and consequences of failure of the dam;
– the type of embankment.

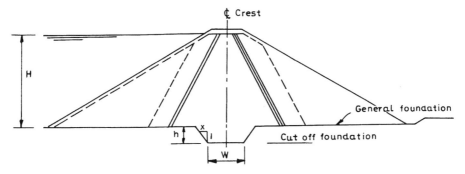

Figure 17.9.   Cutoff trench details.

It is not always necessary to treat the whole of the contact between the earthfill core and the foundation as "cutoff", and only part of that area is excavated to cutoff foundation requirements. Figure 17.9 shows a central core earthfill embankment and for this and other earth, and earth and rockfill embankments the following guidelines are given.

17.4.1   Cutoff width W

For a cutoff in rock, the width depends on the quality of the rock, e.g. for low permeability non erodible rock, W/H may be as low as 0.25 but would usually be 0.5 or more. There have been cases where W/H = 0.1 has been used. For concrete face rockfill dams W/H ≈ 0.05 to 0.1 is quite common (Fitzpatrick et al., 1985). For cutoff foundations on more permeable erodible rock W/H may be taken as the full core width, commonly >0.5 H and up to 1 H.

For large high consequence of failure dams it is common to be conservative and the full core contact is taken to cutoff foundation conditions regardless of the rock foundation.

The cutoff width should be not less than 3 m even for small dams (the width of excavation and compaction equipment) and preferably 6 m.

For a cutoff in soil, it would be common to make W/H ≈ 1. The idea is to give a lower seepage gradient for these more erodible conditions.

It is emphasised that, if the depth to an acceptable cutoff is excessive, these guidelines may be waived, provided other measures are adopted, e.g. multi-line curtain and consolidation grouting, cutoff walls constructed under bentonite and erosion protection in the base and sides of the cutoff.

17.4.2   Batter slope

The batter slopes for the cutoff trench should be:

- Not steeper than 0.5 H:1 V, and preferably 1 H:1 V so that the earthfill can be compacted against the sides of the trench;
- Sufficiently flat to avoid arching effects in the cutoff trench. As a guide, if h/W > 0.5, the batter slopes should be reduced to 1 H:1 V or 1.5 H:1 V;
- The batter slopes may be determined by stability during construction, e.g. by joints in rock, the strength of the soil or the presence of groundwater. For soil, it would not be uncommon to require batter slopes between 1 H:1 V and 1.5 H:1 V and possibly flatter.

17.4.3   Setting out

For setting out and construction purposes it is desirable to position the cutoff trench a fixed distance from the dam centreline, or by means of a series of fixed points. Figure 17.10a

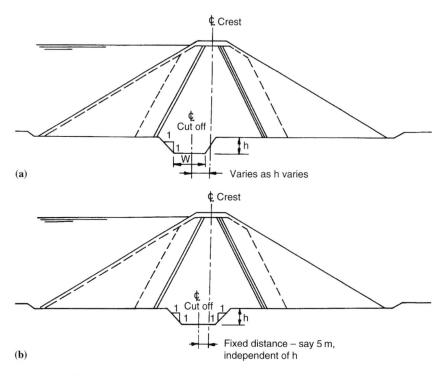

Figure 17.10.   Cutoff trench layout (a) incorrect setout; (b) correct setout.

shows the effect of positioning the trench at the upstream contact of Zone 1 and the foundation. The result is a "footprint" of the cutoff trench which is curved and more difficult to construct than Figure 17.10b which is positioned relative to the centreline.

The actual width of the base of the trench is best set as a fixed width or widths and narrower up the abutments where the effective height is less.

It should be remembered that the actual depth of the cutoff trench will be determined by geotechnical factors and is not known accurately before the start of construction. It is necessary to set out the batter pegs for the excavation based on an assumed depth of excavation. It is wise to be somewhat conservative in doing this, particularly if a minimum width (3 m) trench is being used, so that it is not necessary to restart excavation from the surface in the event that the actual depth to cutoff foundation is greater than expected.

17.5   SELECTION OF CUTOFF FOUNDATION CRITERIA FOR EMBANKMENT DAMS

The selection of cutoff foundation criteria for a project will depend on the particular site and embankment under construction. The following useful hints assist in this process:

– Never specify a depth of cutoff trench, i.e. depth below general foundation. The requirements for cutoff are geological and geotechnical and the depth to achieve these will vary. It is, however, appropriate for the geotechnical engineer/engineering geologist, in consultation with the embankment designer, to estimate the likely depths as a guide for the construction personnel.
– Do not specify a weathering grade alone as a requirement to achieve a suitable foundation. The requirements for cutoff foundations are essentially those of permeability and

erodibility, and a weathering grade alone (e.g. slightly weathered rock) may not be adequate or may be too conservative. If weathering grade is used as one of the requirements, it should be clearly defined in the documents what the different weathering grades are.
– Planning and selection of cutoff foundation requirements should use all available information. This information includes borehole log data – rock type, degree of weathering and fracture log, the Lugeon water pressure test results, presence of limonite and clay infill joints, seismic refraction data and bulldozer trench information.

It will often be found that there is some correlation between these data which will allow logical decision making, e.g.:

– In developing general excavation requirements the soil/extremely weathered rock will have low seismic velocity (300–1000 m/sec) and will commonly coincide with blade refusal in the dozer trench. Hence the low seismic layer boundary may form a useful means of extrapolating between "hard data" in the trenches and boreholes.
– The base of highly fractured rock, high-moderate Lugeon value, and medium seismic velocity (e.g. 1400–2000 m/sec) will sometimes coincide and indicate a reasonable cutoff level. Again the seismic data allows interpolation around the site.
– The base of limonite staining in joints often coincides with the base of high Lugeon rock. If weathering grade is related to this staining there may also be a correlation between weathering grade and Lugeon value.

Selection of planned cutoff criteria is sometimes done by laying out the drill core and the design engineer and engineering geologist observing the core, logs and other data and selecting the cutoff level or levels and the associated criteria there and then.

Often there will be more than one choice, with a "better" cutoff being available at depth and an "adequate" cutoff closer to the surface. The "adequate" cutoff may require a wider cutoff trench, more grouting, a grout cap, downstream filter protection etc., the need for which has to be offset against the shallower depth.

Similarly, for dams on a soil foundation, the cutoff depth and criteria can often be best established during the logging of test pits.

If this approach is used, the geological features which were used to select the cutoff can be described and readily identified in the construction process. The engineering geologist and design engineer, who make the design decision, should be present during construction to define and confirm those conditions in the cutoff trench.

17.6   SLOPE MODIFICATION AND SEAM TREATMENT FOR EMBANKMENT DAMS

17.6.1   *Slope modification*

Excavation of near vertical or overhanging surfaces and/or backfilling with concrete is required.

For general foundation:

– To allow compaction of earthfill and avoid cavities under rockfill due to overhangs in rock in the abutment.

For cutoff foundation:

– To allow compaction of earthfull to give a low permeability contact between the earthfill and the foundation;

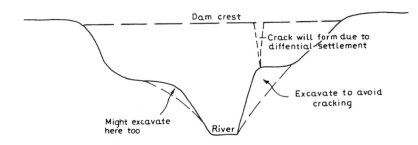

Figure 17.11.   Slope modification in the cutoff foundation to reduce differential settlement and cracking of the earthfill core.

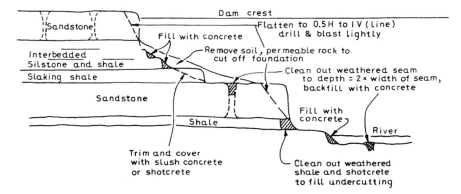

Figure 17.12.   Slope modification and seam treatment in the cutoff foundation for interbedded sandstone and siltstone, near horizontal bedding.

– To allow maintenance of positive pressure of the earthfull on the abutment;
– To limit cracking of the earth core due to differential settlement over large discontinuities in the abutments, e.g. as shown in Figure 17.11.

To allow earthfill to be compacted and to maintain positive pressure on the abutments, steep small scale foundation surfaces should be flattened to about 0.5 H to 1 V either by excavation or by backfilling with concrete. Figures 17.12 and 17.13 give examples of this type of treatment. Some authorities require a flatter slope, e.g. Thomas (1976) and Wallace and Hilton (1972) suggest the use of 0.75 H:1 V. Others, e.g. Acker and Jones (1972), accept steeper slopes (0.25 H:1 V). USBR (1984) suggest 0.5 H:1 V. The policy for Thomson dam was to be guided by the persistence of any slopes, particularly with slopes within ±30° to the upstream–downstream direction for which 1 H:1 V slope was sought. Significantly high slopes were made not steeper than 0.75 H:1 V. Sharp changes in grade near the tops of any slope were "rounded out". Thomson is a very large dam and it might be argued greater conservatism is warranted in these cases. The authors are of the opinion that 0.5 H:1 V is satisfactory provided the earthfill is compacted wet of optimum with rubber tyred equipment used to "squeeze" it into position.

In some geological environments these requirements for slope modification may necessitate virtually the whole of the cutoff area being excavated or covered with concrete, e.g. Figure 17.4. It should be noted that in some cases there will be tendency for the rock to pluck out on joints, repeating the oversteepening problem. In these cases backfill concrete, or possibly presplitting with light blasting, may be necessary.

Large scale slope modification such as that shown in Figure 17.11 to avoid differential settlement and resultant cracking of the earthfill core has been used on many dams, e.g.

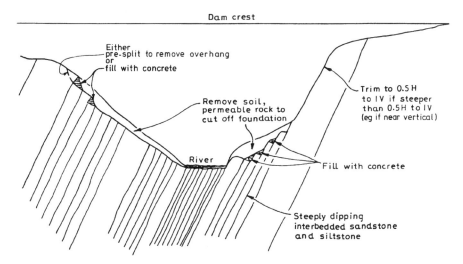

Figure 17.13.   Slope modification in the cutoff foundation for interbedded sandstone and siltstone, steeply dipping bedding.

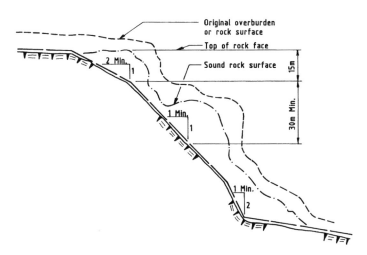

Figure 17.14.   Bennett Dam, typical core abutment excavation requirements (Pratt et al., 1972, reproduced with permission of ASCE).

Pratt et al. (1972) describe slope modification for Mica and Bennett Dams in Canada. The work carried out is shown in Figure 17.14. Walker & Bock (1972) give details of slope correction work at Blue Mesa Dam (Figure 17.15). In this case some of the slope correction work was required to remove loose unstable rock from the abutments.

Since these projects were constructed it has been recognized that earthfill core for dams may crack even in ideal conditions and that the best line of defence is to provide good filters (see Chapter 9). Hence the desirability of such major slope correction works is less clear. The authors' assessment is that provided the overall slope is less than 0.25 H:1 V (preferably 0.5 H:1 V) and good filters are provided, one should not be too concerned about the large scale slope modification carried out in the examples above.

A more conservative approach to large scale modification might apply for dams in severe earthquake areas, where differential movements and cracking may be more likely, and for very large dams.

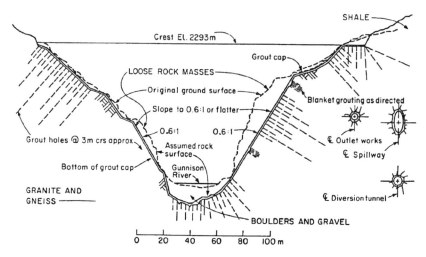

Figure 17.15.   Foundation slope correction, Blue Mesa Dam (reproduced from Thomas, 1976). Reprinted by permission of John Wiley & Sons Inc.

17.6.2   Seam treatment

It is common practice to require that seams of clay or extremely weathered rock which occur in the cutoff foundation should be excavated and filled with concrete. This is done to avoid erosion of the seams thus allowing seepage to bypass the earth core and filters.

Thomas (1976) suggests that the depth of excavation and backfill should be 2 to 3 times the width of the seam. USBR (1984) recommends that openings narrower than 50 mm should be cleaned to a depth of three times the opening width, and seams wider than 50 mm and up to 1.5 m should be cleaned to a depth of three times the width of the opening or to a depth where the seam is 12 mm wide or less, but not greater than 1.5 m depth. Wallace and Hilton (1972) indicate that for Talbingo dam all seams wider than 12 mm were excavated to a depth equal to the width and filled with concrete, or that the area was covered with thick grout, pneumatically applied mortar or concrete. This latter approach seems more reasonable for narrow seams particularly over areas where there are a number of narrow seams.

Figure 17.16 shows seam treatment adopted for Kangaroo Creek Dam, and Figure 17.5 that adopted for Dartmouth Dam. For Thomson dam, the foundations of which were often closely jointed rock criss-crossed by seams, and crushed zones, the whole of the cutoff was shotcreted except when it was already covered by slope correction concrete.

These examples can be used as a guide to reasonable practice. The authors would prefer not to dig out the seams, but to ensure that they are adequately covered with concrete or shotcrete, since their nature is not likely to change with one or two widths, and the concrete or shotcrete will be held in place by the earthfill.

For foundations with continuous seams filled with dispersive or erodible soil and rock, a horizontal filter drain may be needed downstream of the earthfill core to allow foundation seepage to emerge in a controlled manner.

17.6.3   Dental concrete, pneumatically applied mortar, and slush concrete

The following recommendations on the placement and quality of dental concrete, pneumatically applied mortar or concrete and slush concrete are taken from USBR (1984). They are consistent with the authors' experience of what constitutes good practice.

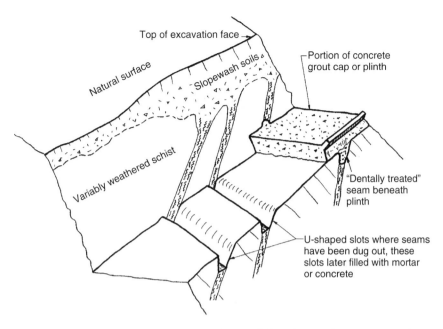

Figure 17.16.   Dental treatment of weak seams in the plinth foundation of Kangaroo Creek Dam.

(a) Dental concrete. Dental concrete is used to fill irregularities in the foundation due to joints, bedding, sheared zones, overhangs, or excavated surfaces.

Slabs of dental concrete should have a minimum thickness of 150 mm. Thin areas of dental concrete over rock projections in a jagged rock surface are likely places for concrete cracking and should be avoided by using a sufficient thickness of dental concrete or by avoiding continuous slabs of concrete over areas containing numerous irregularities. Feathering at the end of slabs should not be permitted, and the edges of slabs should be sloped no flatter than 45°. When formed dental concrete is required, it should not be placed at slopes greater than 0.5:1 (H:V), as discussed in Section 17.6.1.

When fillets of dental concrete are placed against vertical or near vertical surfaces, feathering should not be permitted, and a bevelled surface with a minimum thickness of 150 mm will be required at the top of the fillet.

The mix proportions should provide a strength at 28 days of 20 MPa. The maximum aggregate size should not be larger than one-third the depth of slabs or one-fifth the narrowest dimension between the side of a form and the rock surface.

Aggregate and water shall be of quality equal to that required in concrete specifications allowing for sulphates in the foundation materials and groundwater. To ensure a bond between the concrete and the rock surface, the rock surface should be thoroughly cleaned and moistened prior to concrete placement. When overhangs are filled with dental concrete, it is essential that the concrete is well bonded to the upper surface of the overhang. Before concrete placement, the overhang should be shaped to allow air to escape during concrete placement and thus prevent air pockets between the concrete and the upper surface of the overhang. The concrete must be placed and allowed to set with the head of the concrete higher than the upper surface of the overhang. In cases where the preceding measures are not feasible, grout pipes should be inserted through the dental concrete to fill potential air voids. If grouting behind dental concrete is employed, grouting pressures should be closely controlled to ensure that jacking of the concrete does not occur.

Finished dental concrete horizontal mats should have a roughened, broomed finish (but not corrugated) to provide a satisfactory bonding surface to embankment material.

The dental concrete should be cured by water or an approved curing compound until 28 days or it is covered by earthfill. Care should be taken to ensure that cracking does not occur in the dental concrete due to subsequent earthfill placement and compaction operations. Earthfill operations may not be permitted over dental concrete for a time interval of 72 hours or more after concrete placement to allow concrete time to develop sufficient strength to withstand stress caused by earthfill placement operations.

(b) Pneumatically applied mortar or concrete (shotcrete). Shotcrete is concrete or mortar that is applied pneumatically at high velocity with the force of the jet impacting on the surface serving to compact the concrete or mortar. Practice today is to use wet mixed shotcrete because it is easier to control. The quality of shotcrete is highly dependent upon the skill and experience of the crew applying it, particularly with regard to their ability to prevent rebound from being entrapped in the shotcrete, control the thickness, prevent feather edges, and ensure adequate thickness over protrusions in irregular surfaces. There is also a danger of covering unprepared treatment areas because of the ease and rapidity of placement. Shotcrete should be used beneath impervious zones only when its use instead of dental concrete can be justified by site conditions. If it is used, the specifications should be very strict to ensure proper provisions for adequate quality work. The requirements for concrete quality, layer thickness, feathering and curing time should be as for dental concrete.

(c) Slush grout. Is a neat cement grout or a sand-cement slurry that is applied to cracks in the foundation. Slush grout should be used to fill only narrow surface cracks. It should not be used to cover exposed areas of the foundation other than as described in 17.3.2, where it is used as a temporary cover over slaking or similar foundations.

Slush grout may consist of cement and water, or sand, cement and water. To ensure adequate penetration of the crack, the maximum particle size in the slush grout mixture should be no greater than one-third the crack width. The consistency of the slush grout mix may vary from a very thin mix to mortar as required to penetrate the crack. The grout shall be mixed with a mechanical or centrifugal mixer and the grout should be used within 30 minutes after mixing.

Cracks shall be cleaned out and wetted prior to placement of slush grout. Slush grout may be applied by brooming over surfaces containing closely spaced cracks, or by troweling, pouring, rodding, or funneling into individual cracks. The requirements for cement and aggregates should be as for dental concrete.

17.7   ASSESSMENT OF EXISTING EMBANKMENT DAMS

When assessing the likelihood of internal erosion and piping of existing dams, a thorough assessment should be made of what foundation preparation was carried out. This can be assessed from construction reports, specifications and most particularly from photographs taken during construction. While it was common practice to carefully clean up (of loose material) the cutoff foundation in earlier dam construction, it was not until around the 1960s to 1970s that extensive use was made of dental concrete. Hence many dams built before then have the potential for loose/softened zones of core material near the contact between embankment and foundation. These have been detected in some dams using cone penetration tests through the earthfill.

Large scale irregularities in the foundation have the ability to lead to low stresses near the crest of the dam. Often the profile is recorded on the drawings showing the results of grouting and/or on photographs, but as often they are not. On a number of projects construction of haul roads or access tracks across the core foundation has resulted in benching followed by differential settlement and cracking of the core. Whether these benches are of concern, often relates to how continuous they are through the core of the dam. If they are through the whole core, they are of greater concern than if they only persist for say 5% of the width of the core.

17.8 FOUNDATION PREPARATION FOR CONCRETE GRAVITY DAMS ON ROCK FOUNDATIONS

17.8.1 *The general requirements*

17.8.1.1 *Foundation strength*

The primary critical issue is whether there are continuous, or near continuous, weak, unfavourably oriented discontinuities in the foundation e.g.:

- bedding surfaces;
- bedding surface shears;
- stress relief (sheet) joints;
- faults and shears.

The geotechnical investigations must assess the foundation for such features and either the dam is designed to found on them or excavation goes below the feature.

The secondary but important issue is that the foundation must be cleaned of loose and loosened rock, soil, water and other matter, prior to placing the concrete so that there is a good bond between the foundation and concrete.

17.8.1.2 *Foundation modulus*

The requirement is that the rock mass has a high enough modulus so deflections under the loads of the dam are not excessive. The rock mass modulus is related to the rock substance strength, but in particular to the nature of the discontinuities, including spacing, opening, degree of infilling and orientation relative to the load direction.

Consolidation grouting (see Chapter 18) may be used to grout the joints and other fractures "to increase the rock mass modulus" if it is not economic to remove this material. The grouting is likely to make the rock mass more an elastic medium, than to necessarily increase the modulus.

17.8.1.3 *Erodibility and uplift pressure*

The seepage gradients beneath concrete dams, particularly arch dams, can be high, so it is necessary to carefully consider the erodibility of the joints, particularly if the joints are infilled with clay. Foundations would usually be excavated to non-erodible rock, and the rock grouted with consolidation (and curtain) grouting to reduce the flow of water through the rock. Borehole drains are provided to intercept the seepage water and reduce the uplift pressures under the dam and its foundation. The design of these drains is discussed in Chapter 16.

17.8.2 *Excavation to expose a suitable rock foundation*

The upper surface of rock which can be treated economically to meet these requirements is predicted from the results of the site exploration, as presented in the geotechnical model. The predicted surface is shown on drawings by means of cross sections and/or contours. Invariably, some excavation is required to remove overlying materials which may include the following:

- Soils – colluvial, alluvial and residual;
- *Insitu* rock mass, the substance and defects of which are too weak and compressible, either inherently or as a result of weathering;

Figure 17.17.   Tumut 2 concrete gravity dam foundation cleanup in progress.

– Mechanically loosened *in situ* rock mass with clay-infilled joints. This may be adequately strong in substance but the mass is too compressible/weak as it contains too many seams which cannot be treated effectively by jetting and grouting.

The excavation is aimed at reaching the required foundation surface, while causing minimal damage to the rock below it. Some drilling and blasting may be required, but usually the rock next to the surface is removed by rock breaker, backhoe and hand methods. The surface reached in this way is cleaned up by compressed air/water jetting, with further use of hand tools. The cleaned-up surface should contain no detached rock fragments or loosened or drummy blocks and details of the exposed rock substances and defects should be clearly visible. Figure 17.17 shows cleanup in progress at Tumut 2 concrete gravity dam in New South Wales.

The cleaned-up surface should be mapped geotechnically in detail. Figure 17.18 is a sample of such mapping at Sturt River Dam in South Australia. The purpose of this mapping is to:

– Confirm that the foundation surface rock is of the required and predicted quality, and
– Confirm that the rock structure is essentially as predicted in the geotechnical model and allowed for in the design and, in particular, that there is no unexpected geological situation, e.g. a kinematically feasible failure surface, which may require a change to the design.

The mapping is preferably done while the cleanup is in progress, as this can allow special attention to cleanup of important defects and timely decisions to be made, e.g. on the need for dental treatment, further deepening or design changes.

17.8.3   *Treatment of particular features*

17.8.3.1   *Dental treatment of localised weak/compressible features extending into acceptable foundation rock*

Features for which localized dental treatment is suitable may range from less than 10 mm to several metres in thickness and include faults (sheared and/or crushed zones) and zones

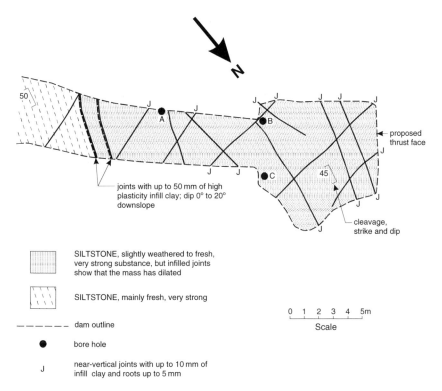

Figure 17.18.   Sturt River Dam, upper right bank and thrust block, part of log of foundation at initial design level.

of altered or weathered rock. When steeply dipping, they may extend well below the otherwise acceptable foundation surface. If it can be shown that they do not combine with other defects to form kinematically feasible blocks or wedges, such features may simply represent:

– Local zones of possible underseepage through potentially erodible materials, and
– If wide enough, local zones of unacceptably high compressibility.

In such cases they can be treated locally by excavation and backfilling with mortar or concrete. Figure 17.19, based on Nicol (1964), shows this form of treatment applied to a fault (2 m to 6 m thick sheared, partly crushed zone) which passed beneath the valley floor and left abutment at the 137 m high Warragamba Dam, near Sydney. In the valley floor the dam excavation was taken down to below the zone. Beneath the left abutment the zone was mined back for about 17 m from the main excavation and backfilled with concrete. In the right abutment the fault did not appear as a major defect; it may have been represented by polished surfaces in a shale unit.

17.8.3.2   *Rock found to contain weak defects which provide kinematically feasible failure surfaces*

As discussed in Section 17.8.1 rock containing weak defects which provide feasible failure surfaces should not normally occur at or below the design foundation surface. When such defects are found during foundation excavation, the design must be re-examined and, if necessary, modified. The following are examples of this and illustrate some types of treatment which have been applied.

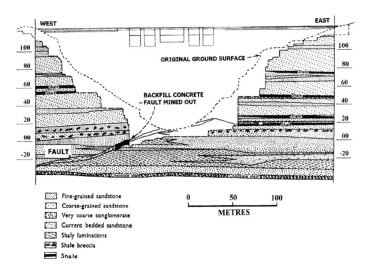

Figure 17.19.   Backfill concrete of a fault at Warragamba dam (based on Nicol, 1964).

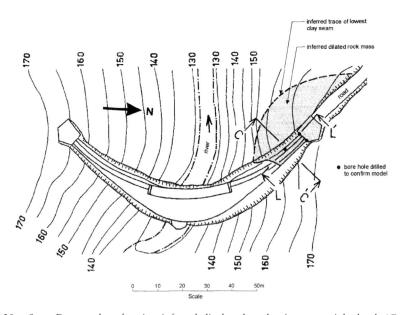

Figure 17.20.   Sturt Dam – plan showing inferred displaced wedge in upper right bank (Courtesy SA Water).

17.8.3.2.1   Sturt River Dam (South Australia)

This 40 m high multiple curvature arch dam (Figures 17.20 and 17.21) was built during 1964–1966 as a flood control structure. It is located in a steep rocky gorge, formed by silt-stone which is very strong when fresh. The rock has well developed cleavage, dipping around 50 degrees upstream. Bedding is rarely visible. Planning stage core drilling showed the rock to be variably weathered in the abutments and the design allowed for excavation of the weathered material to produce a foundation on mainly fresh rock, containing few joints. The design assumed 6–10 m of excavation in the upper right abutment (Figure 17.21). Figure 17.18 shows the geotechnical log of this part of the foundation when the

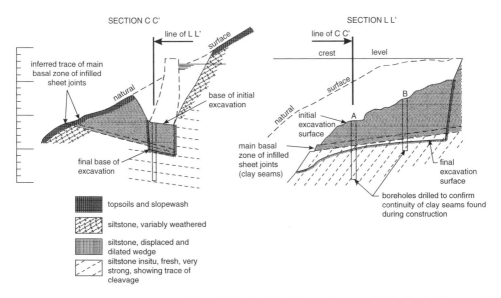

Figure 17.21. Sturt Dam – sections through displaced wedge at the top of right bank (Courtesy SA Water).

excavation reached the design levels. The mainly fresh rock exposed here showed a pattern of near-vertical joints, most of which were infilled with up to 10 mm of high plasticity clay. Downslope from these joints, the traces were found of 2 seams up to 50 mm thick of similar clay, dipping gently and obliquely downslope and upstream. From this evidence it was inferred that the gently dipping seams were infilled sheet joints. It was further inferred that the whole rock mass above the lower gently dipping seam was a mechanically loosened wedge or rock, which had dilated and moved slightly downdip along these seams. The gently dipping seams were judged to represent kinematically feasible failure surfaces. The extrapolated trace of the lower seam (Figure 17.20) coincided with minor depressions at the ground surface. This inferred model was confirmed by high quality core drilling (Figures 17.20 and 17.21), which recovered the predicted seams and showed fresh siltstone with few tightly closed joints below the lower seam.

To reach this undisturbed fresh rock, the foundation was deepened locally by 4 to 7.5 m. Immediately after the deepening, monitored cracks and displaced traces of presplit holes indicated that the isolated downstream part of the wedge had dilated further, and was creeping downdip. It was braced immediately by a concrete buttress placed on the undisturbed rock and extending across to the upstream side of the excavation. The final design provided extra support to this upper part of the dam by means of prestressed anchors installed through it into the foundation.

17.8.3.2.2 Clyde Dam (New Zealand)

This gravity dam of maximum height 100 m is located on schist, about 3 km from the Dunstan Fault, which is believed to be active (Hatton and Foster, 1987; Paterson et al., 1983). At the dam a steeply dipping fault parallel to the river passes upstream–downstream through the foundation. This River Channel Fault comprises up to 8 m of crushed material and is believed to be a normal fault with some horizontal component. Seismotectonic studies carried out during the dam construction indicated that up to 200 mm of displacement might be induced across this fault under the dam, if the Dunstan Fault suffered a major rupture. The dam design was modified by the inclusion of a slip joint above the fault, as shown on Figure 17.22. The joint is designed to accommodate up to 1 m of dip-slip

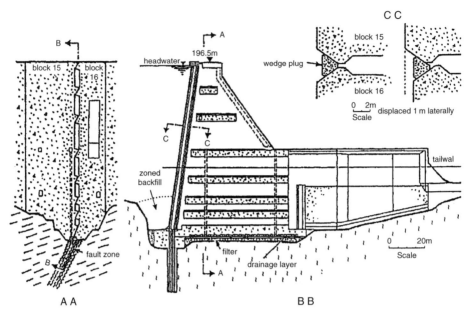

Figure 17.22.   Clyde Dam showing design of slip joint.

movement and up to 2 m of strike-slip movement. The design is described by Hatton and Foster (1987) and Hatton et al. (1991).

The foundation contains many other faults, mostly sheared, partly crushed zones. Some (locally termed foliation shears) occur parallel to the gently dipping foliation, and others cut across the foliation at steeper angles. Mapping of excavated foundations and exploratory drives during construction showed that some of the foliation shears could provide potential failure surfaces (Hatton et al., 1991; Paterson et al., 1983). These were treated either by local over-excavation and deepening of concrete structures (Figure 17.23) or by concrete shear keys placed in adits (Figure 17.24). Foliation shears with downslope components were exposed in the right abutment excavation for the dam. Some of these contained sandy infill of alluvial origin. Dilation of the rock mass due to past downslope creep was inferred. Four of these shears were explored by adits which were later backfilled with concrete to act as cutoffs.

17.8.4   *Treatment at sites formed by highly stressed rock*

Excavation to reach foundation level in highly stressed rock should be done bearing in mind the following:

– Exposed highly stressed rock invariably has sheet joints as its surface and a succession of sheet joints below (see Figure 2.5);
– Any sheet joint below the base of a concrete dam has some potential to form all or part of a downstream sliding surface;
– When excavations are made into highly stressed rock, it is possible that some existing sheet joints may open up further and propagate, and some new sheet joints may form (see 2.5.4). Such effects are much more likely to occur if blasting is used.

At Burdekin Falls Dam in Queensland, the approaches to selection of foundation levels and to excavation methods were developed with understanding of the above issues. The

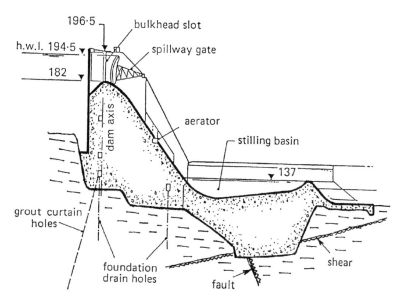

Figure 17.23.   Clyde Dam showing local deepening of the concrete to found below a shear surface.

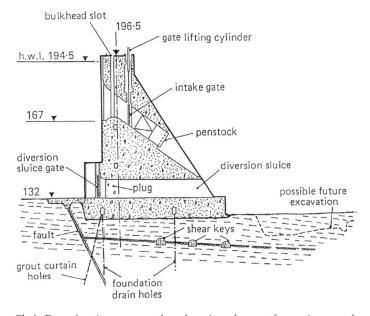

Figure 17.24.   Clyde Dam showing concrete shear keys in a shear surface to increase shear strength.

following account is based on Lawson and Burton (1990), Lawson et al. (1992) and Russo et al. (1985).

The dam is a concrete gravity structure 38 m high and 876 m long. Most of its length comprises 34 monoliths which mainly cross the 600 m wide river bed. The site is formed almost entirely of welded tuff, which is fresh and extremely strong. Near-horizontal sheet joints are exposed over much of the river bed (Figure 17.25), and dipping sheet joints

Figure 17.25.   Burdekin Falls Dam, under construction, showing the 600 m. wide valley floor, formed by extremely strong, sheet-jointed rock. (Photo courtesy of Sun Water).

Figure 17.26.   Slab of rock which has lifted about 50 mm due to stress relief. Monolith LHZ adjoint to LH1. (Photo courtesy of Sun Water).

occur in both abutments. Planning stage investigations provided general understanding of the spacing, depths and persistence of the sheet joints, and showed that some were partly infilled with clay, usually of alluvial origin. Tests showed the joints to have very high shear strengths, and that only a few square metres of intact rock per monolith would be required to meet established criteria for safety against sliding. High stresses were predicted from the sheeted structure, discing of drill cores and excessive propagation of fractures during and after blasting. Bock et al. (1987) showed the maximum principal stress to be horizontal, striking 10° to 30° to the dam axis. Values ranged generally from 10 to 20 MPa across the river bed, rising locally to 30 to 35 MPa at the left abutment.

During construction, the depth to adequate foundation rock was established progressively during the excavation for each monolith. In the river bed, excavation was mainly by rock breakers and excavators. Explosives were used for some slabs which were too large to handle. Cleanup was by bobcat, hand shovels and hosing. The adopted foundation surfaces ranged generally from about 1 down to 4 metres below the river bed. They were chosen for each monolith from the results of geological inspection of progressively exposed surfaces, and by the drilling, cleaning out, and downhole inspection of at least 3 vertical holes, each 2.8 m deep. An Olympus IF6D2-30 Fiberscope was used for these inspections. A simple log for each hole recorded the depth of each sheet joint present and its aperture or infilling. Based on these logs, judgements were made on the persistence and character of the joints, and decisions made on the need for either removal of the rock above them, or grouting. Extra holes were sometimes needed. The final (adopted) cleaned up surfaces were mapped geologically in plan view on 1:200 scale. The as-constructed record includes these plans and the logs of the 2.8 m holes.

In the abutments, blasting was used for much of the excavation, but smooth blasting techniques were specified. The as-constructed records include foundation geology mapped on 1:100 scale, plus logs of the 2.8 m holes.

Significant stress-relief effects during the foundation preparation occurred only close to the base of each abutment, during hot weather. They included rock bursting (spalling of small, thin slices or rock), and slabbing (buckling up of slabs up to 100 mm thick and many square metres in area). Figure 17.26 shows an example of this.

18

Foundation grouting

18.1 GENERAL CONCEPTS OF GROUTING DAM FOUNDATIONS

The foundations for most dams more than 15 m high built on rock, and for some which are smaller, are treated by grouting. Grouting consists of drilling a line or lines of holes from the cutoff level of the dam into the dam foundation and forcing cement slurry, or chemicals under pressure into the rock defects, that is joints, fractures, bedding partings and faults. Figure 18.1 shows an example:

The grouting is carried out to:

– Reduce leakage through the dam foundation, i.e. through the defects;
– Reduce seepage erosion potential;
– Reduce uplift pressures (under concrete gravity dams when used in conjunction with drain holes);
– Reduce settlements in the foundation (for concrete gravity, buttress and arch dams).

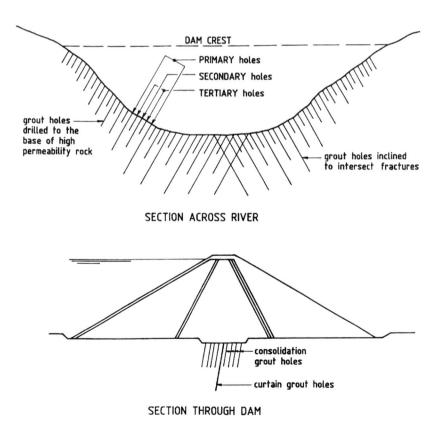

Figure 18.1.   An example of embankment dam foundation grouting.

Most foundation grouting uses cement grout: Portland cement mixed with water in a high speed mixer to a water-cement ratio (mass water/mass cement) of between 0.5 and 5 to form a slurry, readily pumpable and able to penetrate defects in the rock in the dam foundation.

If the dam is on a soil foundation (e.g. sand) or if the fractures in the rock are very narrow, chemicals can be used instead of cement. Chemicals tend to be more expensive so are only used where cement grout would not be successful. Most soil foundations are not grouted, with control of seepage being achieved as detailed in Chapter 10.

Foundation grouting takes two forms:

– Curtain grouting;
– Consolidation grouting.

Curtain grouting is designed to create a narrow barrier (or curtain) through an area of high permeability. It usually consists of a single row of grout holes which are drilled and grouted to the base of the permeable rock, or to such depths that acceptable hydraulic gradients are achieved. For large dams on rock foundations, dams on very permeable rock or where grouting is carried out in soil foundations, 3, 5 or even more lines of grout holes may be adopted. Multiple row curtains are also adopted if, as at Thomson dam, it was impracticable to excavate the foundation to below the limit of the infilled joints.

The holes are drilled and grouted in sequence to allow testing of the permeability of the foundation (by packer testing) before grouting and to allow a later check on the effectiveness of grouting from the amount of grout accepted by the foundation ("grout take").

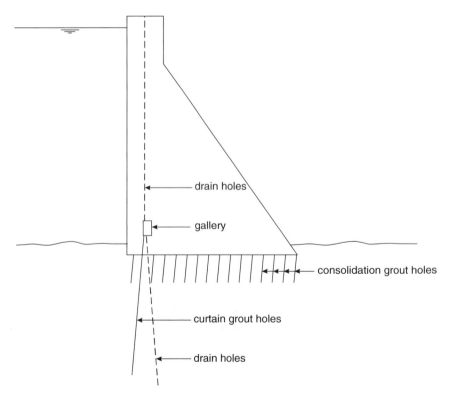

Figure 18.2.   Curtain and consolidation grouting for concrete gravity dams.

Thus in Figure 18.1 primary holes are drilled first, followed by secondary and then tertiary. The final hole spacing will commonly be 1.5 m or 3 m, but may be as close as 0.5 m. This staged approach allows control over the amount and effectiveness of the grouting.

Grout pressures are usually (at least in Australia, USA, United Kingdom) limited to prevent hydraulic fracture of the rock. The discussion in this chapter is predicated on that assumption. In some projects particularly in continental Europe, the rock has been deliberately fractured to improve grout penetration.

Consolidation or "blanket" grouting for embankment dams is designed to give intensive grouting of the upper layer of more fractured rock in the vicinity of the dam core, or in regions of "high" hydraulic seepage gradient, e.g. under the plinth for a concrete face rockfill dam. It is usually restricted to the upper 5 m to 15 m and is carried out in sequence but commonly to a predetermined hole spacing and depth.

Figure 18.2 shows a section through a concrete gravity dam. The curtain grouting is located near the upstream face of the dam and usually carried out from a gallery in the dam; sometimes from the upstream heel of the dam. It is designed to reduce seepage through the foundations and, in conjunction with the borehole drain holes, to control uplift pressures.

18.2   GROUTING DESIGN – CEMENT GROUT

18.2.1   *Staging of grouting*

Grouting of holes is normally carried out in stages, the method depending on the permeability and quality of the rock being grouted and the degree to which control of the grouting operation is desired. Figures 18.3, 18.4, 18.5, 18.6 and 18.7 show the different methods available.

Downstage Without Packer (Figure 18.3): This is one of the preferred methods for high standard grouting, since each stage is drilled and grouted before the next, lower stage, allowing progressive assessment whether the hole has reached the desired closure requirement. This method allows higher pressures to be used for lower stages, as it reduces the risks of leakage from them up the top stage levels and gives better grout penetration between holes. The grout pressures are limited by the effectiveness of the top stage grouting. It does necessitate a separate set up of the drill for each stage and separate "hookups" of the grout lines. It is, therefore, relatively expensive. This method is the one preferred by Houlsby (1977, 1978, 1982a).

Downstage With Packer: (Figure 18.4): This method allows use of increased grout pressures for lower stages, since these pressures are not applied from the surface. However, there may be problems with seating and leakage past the packer. Bleeding of the grout hole (i.e. removing the "clear" water which accumulates at the top of the grout hole as the cement settles) cannot be achieved except at the ground surface (i.e. not immediately above the grout stage). Ewart (1985) indicates a preference for this method, because of the potential to fracture the rock in the upper levels if downstage without packer methods are used.

Upstage (Figure 18.5): Does not allow progressive assessment of the depth of grout hole needed to reach a desired closure requirement as the holes are drilled to their full depth in one stage. The method is cheaper in principle than downstage methods since the drill rig is only set up once, but these savings may be offset by the need for more conservative total depths. The method is susceptible to problems with holes collapsing, or over enlarging during drilling and grouting in poor rock conditions making seating of packer difficult or impossible. It is also subject to the same problems as the downstage with packer method regarding bleeding. The method is appropriate for secondary or tertiary holes, when

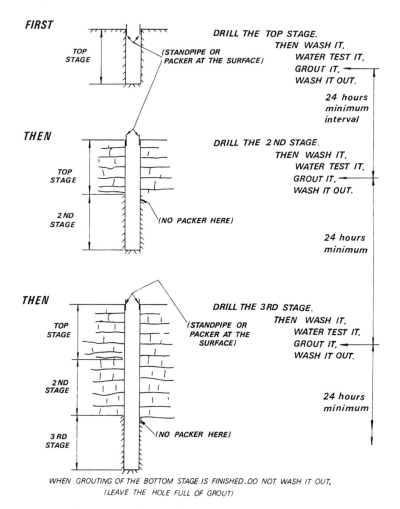

Figure 18.3. Grouting downstage without packer (WRC, 1981).

depths are reasonably well known, and in strong rock with holes not likely to collapse or erode.

Full Depth (Figure 18.6): Does not allow proper assessment of where grout take is occurring or proper monitoring of reduction in Lugeon values with grouting. Depths are predetermined so the method does not allow logical assessment of grout depth based on closure requirements. Grouting pressures are limited. It is not an acceptable method except for consolidation grout holes.

Full Depth Circuit (Figure 18.7): Has the same limitations as full depth grouting, but by injecting the grout into the base of the hole, the possibility of settled grout is lowered.

For stage grouting, the stage lengths are commonly predetermined depending on:

– The geological conditions and depths at which changes in degree of permeability are likely to occur;
– The minimum length worth drilling, since short stages are more costly to drill because of set up costs;
– Allowable pressures in the upper part of the hole (dependent on geological conditions).

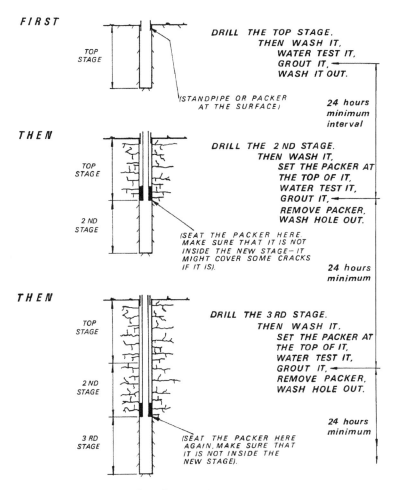

Figure 18.4.   Grouting downstage with packer (WRC, 1981).

Commonly grout stages will be 5 m to 8 m but may be increased in length lower in the foundation, e.g. Houlsby (1977) suggests:

Stage	Depth Range (m)
1	0 to 8
2	8 to 16
3	16 to 30
4	30 to 50

While this reduces the number of drill setups and grout hookups, it may result in unnecessarily deep holes, particularly for smaller dams.

Smaller than the predetermined stage lengths should be used when:

– Drilling water is lost, indicating a relatively large fracture or opening has been encountered, or

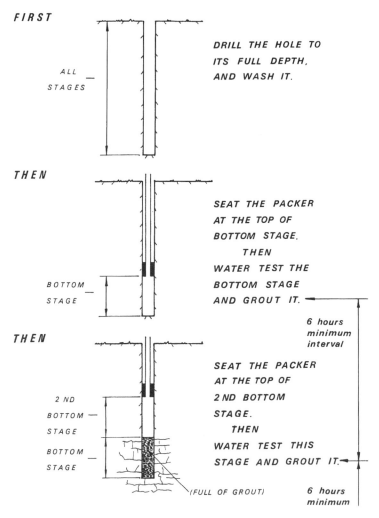

FIRST

ALL
STAGES

DRILL THE HOLE TO
ITS FULL DEPTH,
AND WASH IT.

THEN

BOTTOM
STAGE

SEAT THE PACKER
AT THE TOP OF
BOTTOM STAGE.
THEN
WATER TEST THE
BOTTOM STAGE
AND GROUT IT.

6 hours
minimum
interval

THEN

2 ND
BOTTOM
STAGE

BOTTOM
STAGE

SEAT THE PACKER
AT THE TOP OF
2 ND BOTTOM
STAGE.
THEN
WATER TEST THIS
STAGE AND GROUT IT.

(FULL OF GROUT)

6 hours
minimum

Figure 18.5.   Grouting upstage (WRC, 1981).

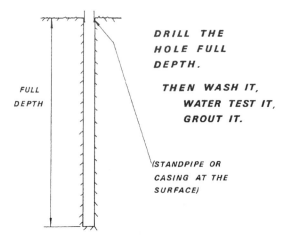

FULL
DEPTH

DRILL THE
HOLE FULL
DEPTH.

THEN WASH IT,
WATER TEST IT,
GROUT IT.

(STANDPIPE OR
CASING AT THE
SURFACE)

Figure 18.6.   Grouting full depth (WRC, 1981).

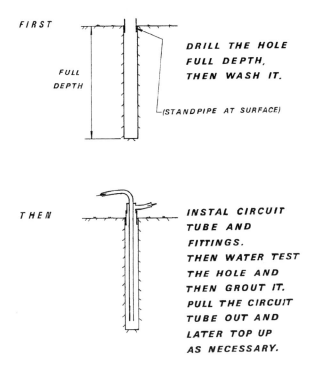

FIRST

FULL
DEPTH

FULL DEPTH

DRILL THE HOLE
FULL DEPTH,
THEN WASH IT.

(STANDPIPE AT SURFACE)

THEN

INSTAL CIRCUIT
TUBE AND
FITTINGS.
THEN WATER TEST
THE HOLE AND
THEN GROUT IT.
PULL THE CIRCUIT
TUBE OUT AND
LATER TOP UP
AS NECESSARY.

Figure 18.7.   Grouting full depth, circuit (WRC, 1981).

– The grout hole is caving, due, for example, to closely fractured rock, or
– Water flows into the hole under pressure, or
– Very large water pressure test or grout takes are encountered (often it is possible to relate these to a specific geological feature).

18.2.2   *The principles of "closure"*

Apart from consolidation grouting, which may be carried out to a predetermined depth and hole spacing, grouting should be carried out sequentially to achieve a predetermined standard of water tightness. This will usually require the successive halving of hole spacing from primary to secondary to tertiary holes etc. as shown in Figure 18.1. Whether the required standard has been achieved will normally be determined on the basis of water pressure test Lugeon values on the grout stage prior to grouting and/or on grout take – the volume (or weight of cement) of grout per metre of grout hole. Closure criteria will be discussed in more detail below. Figure 18.8 shows the basic principles of hole closure, i.e. that secondary holes are drilled halfway between primary if water pressure tests (and/or grout takes) in the primary holes fail to meet the closure criteria.

Figure 18.9, taken from Houlsby (1977) and WRC (1981), gives examples of the closure method when Lugeon water pressure test values are the required standard.

In these examples:

Case (a) Primary grouting has resulted in a reduction in Lugeon value (and grout take) in the secondary hole and tertiary grouting has resulted in further reductions close enough to the closure requirement of 7 Lugeons (uL). Hence no further grouting is required. If the closure requirement was say 5 uL, quaternary holes would have been required and would seem worthwhile as there is a progressive reduction in Lugeon value and in grout take.

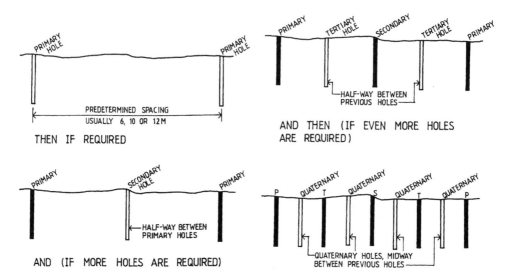

Figure 18.8.   Basic concept of halving hole spacing to achieve closure.

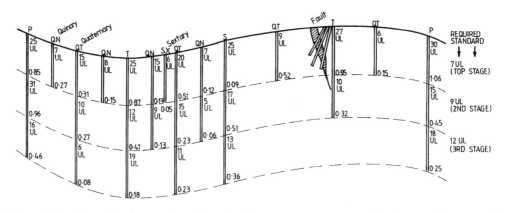

Figure 18.9.   Examples of grout closure based on Lugeon water pressure test (and grout take as second-ary information) adapted from Houlsby (1977) and WRC (1981).

Case (b) Primary and secondary grouting has resulted in no significant reduction in Lugeon value in the tertiary holes prior to grouting them. Houlsby (1977) concludes that quaternary holes are needed and perhaps quinary later. If holes are already at say 1.5 m spacing, consideration should also be given to whether grouting with cement is having any significant effect. If it is concluded that cement grouting is ineffective then grouting might be discontinued.

Case (c) A closure criterion of 7 uL has been achieved in part, but quaternary holes are required elsewhere. Since closure is being achieved in part it would seem viable to proceed to this quaternary stage of grouting where needed.

Case (d) As for Case (a) except that leakage to the surface occurred in grouting the ter-tiary hole, necessitating further grouting to seal the leak.

Closure requirements may vary with depth and the hole spacing and depth required to achieve closure may also vary considerably. Figure 18.10 shows an example taken from WRC (1981). Note that this is a particularly thorough example of grouting, where if the

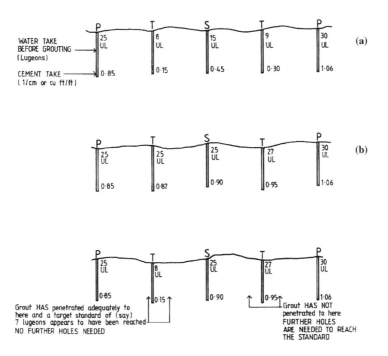

Figure 18.10.   Example of closure based on water pressure test Lugeon criteria (adapted from Houlsby, 1977 and WRC, 1981).

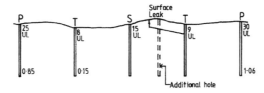

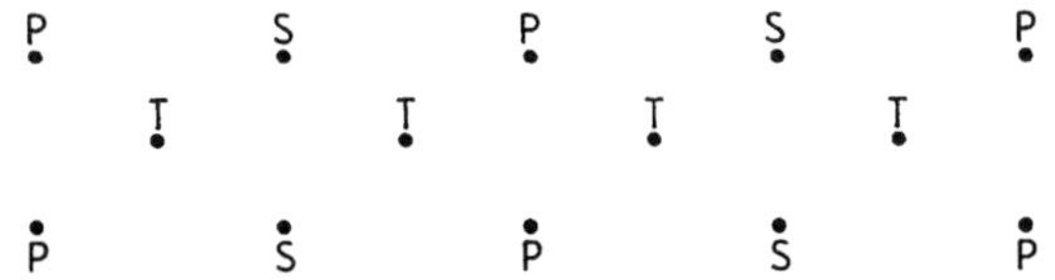

Figure 18.11.   Grout closure with 3 lines of grout holes.

primary spacing was 12 m, the final spacing is 0.4 m in some areas. As discussed below, it is unlikely that closure to such a small hole spacing is warranted.

When more than one line of grout holes is planned, the holes should be drilled and grouted in sequence to form the outer lines ahead of the central line, so that the progressive development of closure can be observed. Figure 18.11 shows a possible closure sequence.

A deep 3-line curtain was completed successfully at Talbingo Dam, the 160 m high earth and rockfill dam in the Snowy Mountains Scheme. If the three lines are designated A, B and C from downstream to upstream, the Talbingo procedure was to drill and grout the A-line holes to the specified depth. The C-line holes, a slightly shallower line of holes, were drilled to their specified depth where indicated by high takes at depth, C-line holes were deepened as required. The B-line holes, also a specified shallower line, were finally drilled and grouted. If high takes persisted at depth in both the A and C holes, the appropriate B-line holes were deepened. With the B-line holes all "contained" between the A- and C-lines, rarely were further closure holes needed. The total drilling (original and redrilling) requirement was greatly reduced and proved to be far easier to define.

18.2.3   *Effect of cement particle size, viscosity, fracture spacing and Lugeon value on the effectiveness of grouting*

Fell et al. (1992) presented a detailed discussion of the effect of cement particle size, viscosity, fracture spacing and opening and Lugeon value on the effectiveness of grouting. The main points are:

(a) *Cement grout particle size.* Cement grout is a suspension of the cement particles in water. The particles are mostly silt sized, but conventional cement will have some fine sand particles. The grout particles aggregate in water to give a coarser distribution than the cement powder. The particle size distribution is affected by the addition of plasticizers, which act as deflocculating agents. This mainly affects the finer particles. With plasticizers, Type A and C Portland cements have a maximum size of about 0.05–0.08 mm, while the microfine cements tested had a maximum size of about 0.02–0.025 mm.

(b) *Minimum fracture opening which will accept grout.* There is a general consensus, backed up by some tests carried out by Tjandrajana (1989) under the first author's supervision, that the minimum opening fracture into which grout will continue to penetrate is about $3D_{100}$ where D_{100} is the sieve size for which all the grout particles are finer.

Based on this conclusion, and rather approximate relationships between Lugeon value and fracture opening and roughness, the minimum Lugeon value rock which can be grouted is as shown in Table 18.1. This indicates that rock masses giving some quite high Lugeon values can not be grouted, which may seem contrary to experience. However, the larger values (15 to 30 Lugeons) occur when there are two or more open fractures (joints) per metre. This condition may not be met often, even when there are three or more fractures per metre, as many may not be open. The possibility also exists that when long grouting times are used, settling of the grout occurs in the borehole,

Table 18.1.   Estimated minimum Lugeon values indicative of rock which will accept cement grout (Fell et al. 1992).

	Minimum Lugeon value which can be grouted		
Cement	1 fracture/m	2 fractures/m	4 fractures/m
Type A	8	16	32
Type C	5	10	20
MC-500 (microfine)	3	5	10
Type A with dispersant	8	16	32
Type C with dispersant	5	10	20
MC-500 (microfine)	1	2	4

Notes: 1. Fractures are assumed "rough"; 2. Fractures assumed to be the same width; 3. One Lugeon is a flow of 1 litre/minute/metre of borehole under a pressure of 1000 kPa. In a 75 mm diameter borehole it is approx. 1.3×10^{-7} m/sec equivalent permeability; 4. Grout is assumed to have been treated with plasticizer.

with only water with very low (fine) cement contents being left in the upper parts of the hole, allowing small "grout" takes to occur in these parts of the hole. Some experiments carried out by Tjandrajana (1989) show this is not a problem for 2:1 water cement ratio, but the upper 1.5 m of the hole is left largely with water after one hour for 5 to 1 water cement ratio. The addition of plasticizers worsens the settling, because the cement particles act alone and actually settle more quickly.

(c) *Distance grout will penetrate.* If the fractures are sufficiently open to allow penetration of the grout, the distance to which the grout will penetrate is dependent on the fracture width, grout pressure and viscosity and the time taken in grouting. If grouting continues for sufficient time, the limit of penetration is determined by the yield point stress. Lombardi (1985) showed that:

$$R_{max} = \frac{P_{max}a}{C} \qquad (18.1)$$

where R_{max} = maximum radius of penetration (m); a = half width of the fracture (m); C = yield point stress (kPa); P_{max} = grouting pressure (kPa).

Lombardi (1985) presents a method for estimating the effect of grouting time. A feel for this can be obtained from the data in Deere and Lombardi (1985), which indicates that for medium to thick grouts approximately 75% of maximum penetration will occur in the first hour.

Based on this and comparison with some field data, it is considered reasonable to conclude that the penetration of Type A and Type C cement grouts will be of the order of those shown in Table 18.2. In the table NP indicates the grout will not penetrate the fractures, so grouting will be ineffective.

(d) *The effectiveness of a grout curtain in reducing seepage.* Figure 18.12 shows a simplified typical embankment and foundation section with a low permeability core, high permeability rockfill shoulders and a foundation consisting of an upper layer of permeable rock overlying "impermeable" rock.

Figure 18.13 presents the results of an approximate seepage analysis to show the effectiveness of a grout curtain in reducing the amount of seepage beneath the embankment.

The analysis has been based on:

H = 60 m all cases   w = 6 m all cases   h = 20 m all cases
W = 60 m for wide core   W = 20 m for narrow core   W = 6 m for concrete face plate.

Unless the grout curtain has a permeability at least 10 times lower than the ungrouted rock the reduction in seepage is less than 50%. So for 20 Lugeon rock the grout curtain must be less than 2 Lugeons to achieve 50% reduction in seepage. As pointed out above,

Table 18.2. Approximate penetration from the borehole of cement grout in fractures.

	Fracture spacing		
Lugeon value	1 m	0.50 m	0.25 m
100	20	12	4
50	12	3	2
20	3	1.5	1
10	2	1	NP
5	1	NP	NP
1	NP	NP	

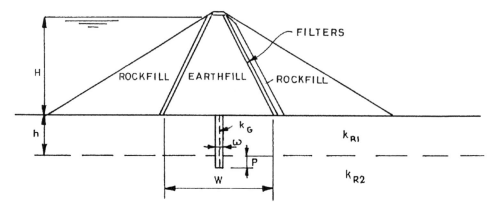

Figure 18.12.   Generalised model for seepage through a dam foundation.

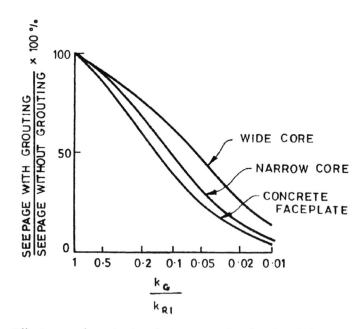

Figure 18.13.   Effectiveness of grouting in reducing seepage in a dam foundation.

it is doubtful whether such a low permeability is achievable with cement grouting because the cement will not penetrate rock with such low Lugeon value.

For high permeability rock (say 100 Lugeon) with wide fracture spacing, the effective width of the curtain should be larger (say 20 m) and the permeability could be reduced to that of the secondary fracture system. If this was, say, 5 Lugeons, then the seepage would be reduced to about 20% of the ungrouted value. If the permeability of the secondary fractures was, say, 2 Lugeons, the seepage would be reduced to 5% to 10% of the ungrouted value.

18.2.4   *Recommended closure criteria for embankment and concrete dams*

From the review of criteria used by others, including Houlsby (1977, 1978, 1982a, 1985), WRC (1981), Deere (1982), Ewart (1985), Kjaernsli et al. (1992), the authors' experience

Table 18.3.   Guidelines for deciding on limits of effective grouting with Type C Portland cement.

Erodibility of foundation	Water test value before grouting *(Lugeon) [2]	or <20% reduction in Lugeon value or grout take from previous stage [1] (Lugeon) [2]	or All grout takes (kg cement/m)	or Grout hole spacing (m)
		No further grouting needed when		
Low/non	<10	<20	<25	<1.5
High	<7	<15	<25	<1.5

Note: [1] For rock with joints closer than 0.5 m. [2] For Type A Portland cement adopt Lugeon values 20% greater.

and the discussion in Section 18.2.3 it is recommended that closure criteria should be based on:

– Lugeon value prior to grouting the hole;
– Grout take;
– The nature of the dam, its foundation and what is being stored in the dam.

These factors should be considered together to make decisions about whether further grouting is required.

It is not yet possible to quantify all these effects (and may never be), mainly because of the complexity of flow in fractured rock and the time dependent nature of cement grout properties. It is also not possible to make rigid rules – each case should be considered on its own merits. It is suggested that for grouting of the foundations of earth and earth and rockfill dams with Type C Portland cement, the guidelines given in Table 18.3 be adopted.

"Erodible" foundations would include extremely to highly weathered rock and rock with clay filled joints which might erode under seepage flows.

For grouting with Type A portland cement, Lugeon values should be increased by 20% to account for the coarser nature of the grout particles. For microfine cement, grouting at values half those quoted would be reasonable from a grout penetrability viewpoint, but may not be justified by the benefit gained.

The overriding philosophy in these recommendations is that it is not possible to stop seepage by grouting, only to reduce it, and cement grout can only significantly reduce seepage when it can penetrate the fractures, i.e. in relatively widely spaced open joints in moderate to high Lugeon value rock. The objective of the grouting operation should be to locate and fill these large fractures and thereby to avoid high and concentrated seepage flows.

Different criteria have been given for "erodible" foundations, with some misgivings. It is better to acknowledge that, if foundations are erodible, grouting will not prevent erosion, only reduce erosion potential. With or without grouting, filters should be provided for erodible foundations under the downstream part of the dam to allow seepage water to emerge in a controlled manner without erosion of the foundation. For concrete dams, the foundation should be taken further down, to non erodible rock.

For concrete faced rockfill, gravity concrete and arch dams seepage gradients are higher and there is some argument for using lower Lugeon values. However, the overriding consideration is that the cement will not penetrate fine fractures so, at most, the values quoted should be reduced by about 30%.

The question of whether the water is "precious" or has a high contaminants content is not really relevant, since grouting to more restrictive Lugeon values as advocated by Houlsby (1986) will not result in significant reduction in seepage. In these cases it will be necessary to collect the seepage water downstream of the dam, either in a catch dam or in seepage collector wells and pump it back into the storage.

A minimum grout hole spacing of 1.5 m is recommended because, if closer spacing is required to achieve a Lugeon value standard, the width of the resulting grout curtain is too narrow to significantly reduce seepage. Generally it would be better to control slope pore pressures by drainage rather than grouting. Clearly there may be exceptions where large takes are encountered in these holes, but as a general rule grouting with holes at closer than 1.5 m spacing is a waste of time and money. It may also be a pointless exercise, given the potential for percussion-drilled holes to deviate from their planned position. At Thomson dam (Victoria), one 90 m deep hole was shown to be 27 m off line, which does pose the question as to the value of a deep single line curtain of closely-spaced holes.

The authors are of the opinion that it may be justified to reduce hole spacings below 1.5 m in some deeply weathered and closely fractured saddle dam foundations, where erodibility and control of seepage pore pressures may be critical for slope stability.

It should be noted that adoption of larger Lugeon value closure criteria at depth and/or higher grout pressures at depth may lead to secondary and tertiary holes not penetrating to the same depth as the primary holes.

A more detailed discussion on the selection of closure criteria is given in Fell et al. (1992).

18.2.5   The depth and lateral extent of grouting

So far as is practicable grout holes should be taken to the depth at which the Lugeon closure criteria are achieved. In near horizontally layered rock this might be clearly identifiable as a lower permeability zone of rock beneath the valley floor, e.g. siltstone within interbedded siltstone and sandstone, but may be at different depths around the dam abutments due to the influence of stress relief, weathering and rock types. Figure 18.14 gives an example where the stratigraphy is simple and clearly known.

The depth of higher Lugeon value rock is seldom known with any accuracy and the grout holes must penetrate to at least one stage below the estimated base level to prove that the necessary conditions have been met.

The use of "rules of thumb" to determine the depth of grouting is not recommended as there is no logical basis for them.

If such rules are applied they will either yield holes which are too shallow, resulting in a grout curtain which only partially penetrates the permeable foundation, giving only

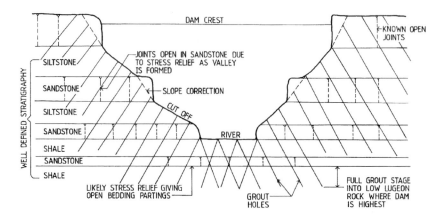

Figure 18.14.   Example of grout curtain in a simple sedimentary rock environment.

minor (i.e. typically less than 10%) reduction in seepage due to increased seepage path length or, more commonly, holes which are much deeper than required to reach lower permeability rock, particularly in the river section of the foundation where the dam is highest.

For most dams, grouting will extend up the abutments to where full supply level intersects the base of the permeable zone. However, where the dam abuts a spillway, the grout curtain may be connected into the curtain under the spillway and/or if the dam abuts a relatively narrow permeable ridge, the grout curtain may be extended into the ridge. (Figure 18.15a).

Extension of the grout curtain further into an abutment is only necessary if the stability of the ridge or abutment is in question and there is a need to control piezometric pressures, or if the abutment is highly permeable.

In such highly permeable rock, careful consideration must be given to the lateral extent of grouting, or seepage may bypass the end of the grout curtain. Figure 18.15(b) shows the grout curtain for a 20 m high water storage dam in coal measure rocks, which have been intruded by a highly permeable dolerite sill. The grout curtain was successfully constructed with excellentclosure, but when the dam was filled the reservoir leaked significantly (water level dropped 25 mm/day) because water was entering and flowing along the dolerite sill, completely bypassing the ends of curtain. This groundwater movement was proven conclusively by piezometers.

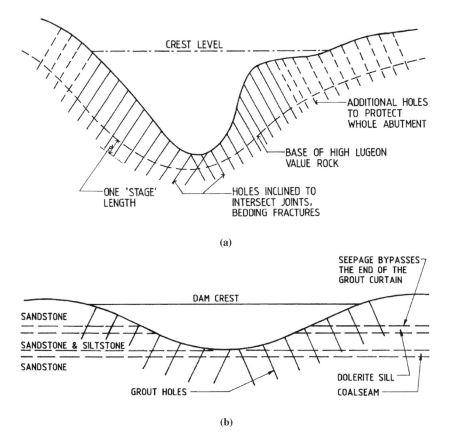

Figure 18.15.   (a) Example of grout curtain extended beyond dam crest; (b) Example where seepage bypassed the end of the grout curtain.

18.2.6   *Grout hole position and orientation*

For earthfill and earth and rockfill dams the grout curtain will usually be located in the centre of the cutoff trench, which in turn will usually be located at the centre or upstream of centre of the earthfill zone. Houlsby (1977, 1978) presents an argument for positioning grout curtains upstream of centre of the earthfill zone, to maintain a seepage pressure head greater in the earth core than in the foundation. This concept has merit in principle but in practice flownets within the earthfill and foundation are influenced by different permeabilities and positioning the curtain upstream of the centre of the core may not achieve the desired objective.

For concrete face rockfill dams the grout curtain will be positioned in the plinth, usually, but not always, with a line of consolidation grout holes upstream and downstream of the curtain to give better grouting protection close to the surface near the plinth, where gradients are high.

For concrete gravity dams (including spillways) the grout curtain should be positioned close to the upstream face so that the drainage holes can be positioned downstream of the grout curtain but still close to the upstream face to reduce uplift pressures as much as possible. It should be noted that for these structures the drain holes are critical in reducing uplift pressures and are more important than grouting. Grouting alone cannot be relied upon to reduce uplift pressures.

Grout holes should always be oriented to intersect the major fracture sets, particularly those which are near parallel to the river. In most cases this will mean that the holes are inclined from the vertical as shown in Figures 18.14 and 18.15. They may also need to be inclined upstream to intersect joints perpendicular to the river.

18.3   SOME PRACTICAL ASPECTS OF GROUTING WITH CEMENT

18.3.1   *Grout holes*

It has been accepted practice for many years in Australia and other countries to use wet percussion drilling for grout holes. Holes are a 30 mm minimum diameter, usually 50 mm diameter and seldom larger than 60 mm.

Percussion drilling is the quickest and cheapest type of drilling and is satisfactory except where the nature of the rock is such as to create a sludge or stiff clayey "plug" which blocks the fractures in the rock. In such rock rotary drilling using roller bits or plug diamond bits may be used. It is sometimes argued that rotary or diamond drilling will be less likely to block the fractures. If holes are washed carefully after drilling it is doubted whether the drilling method really has much effect. This was proven in the quartzite and phyllite foundations of Blowering dam.

Washing of the grout hole before grouting is essential to remove cuttings which have clogged the fractures. This is done by lowering a specially constructed washout bit which directs water under pressure against the sides of the hole. Figure 18.16 shows an example of such a bit, which may also be used to flush grout from a hole after grouting.

Weaver (1993) advocates stopping drilling immediately water loss occurs because if the hole is drilled further, cuttings from the drilling may block the feature into which the drill water was lost. After grouting, drilling to lower stages is carried out.

18.3.2   *Standpipes*

Standpipes are used for most grouting operations where packers are not used. The standpipe consists of a threaded galvanised pipe just larger than the drill size, set approx 0.6 m into the rock as shown in Figure 18.17.

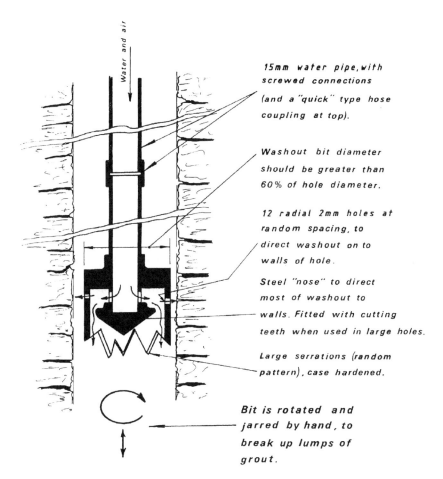

Water and air

15mm water pipe, with screwed connections (and a "quick" type hose coupling at top).

Washout bit diameter should be greater than 60% of hole diameter.

12 radial 2mm holes at random spacing, to direct washout on to walls of hole.

Steel "nose" to direct most of washout to walls. Fitted with cutting teeth when used in large holes.

Large serrations (random pattern), case hardened.

Bit is rotated and jarred by hand, to break up lumps of grout.

Figure 18.16.   Grout hole washing bit (WRC, 1981).

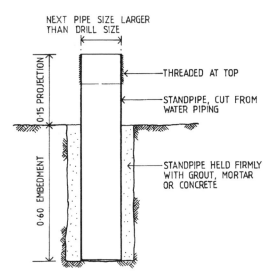

NEXT PIPE SIZE LARGER THAN DRILL SIZE

0·15 PROJECTION

0·60 EMBEDMENT

THREADED AT TOP

STANDPIPE, CUT FROM WATER PIPING

STANDPIPE HELD FIRMLY WITH GROUT, MORTAR OR CONCRETE

Figure 18.17.   Grout standpipe.

Standpipes will always be required where the rock near the surface is fractured or weak, precluding use of a packer. Standpipes have the advantages of preventing clogging of the hole by debris from adjacent areas and providing easy hook-up for grouting. Note that the part of the standpipe projecting from the surface has to be removed before placing earthfill.

18.3.3   *Grout caps*

Concrete grout caps (Figure 18.18) are required when grouting closely fractured or low strength rock in the situation where a standpipe cannot be sealed into the foundation and/or leakage of grout to the surface will be excessive. The grout cap also provides a better cutoff through the upper part of the foundation than is practicable with grouting from standpipes.

Grout caps should be excavated into the foundation rock without explosives and if practicable should be of square or rectangular section as shown in Figure 18.18(a). This gives good resistance to lifting of the cap under the pressure of grouting.

The wider/shallow shapes shown in Figure 18.18(b) and (c) are prone to displacement during grouting and may need to be anchored into the foundation with steel dowel bars grouted say 2 m into the rock.

Some organisations, e.g. Rural Water Commission of Victoria, have favoured construction of a grout cap in all foundations, constructed above the cutoff level as shown in Figure 18.18(c). This arrangement requires anchor dowels in all cases. The arrangement is justified on the basis of allowing grouting of all the rock, i.e. right to the surface, and increasing the resistance to erosion along the cutoff surface by providing a more tortuous path for water, but has the disadvantage that it interferes with compaction of earthfill in the cutoff trench. Such grout caps are not recommended by the authors.

In closely fractured rock there is often some advantage in applying a layer of slush concrete or shotcrete to the cutoff surface before grouting. This allows use of slightly higher grout pressures in some rock types, prevents excessive leakage to the surface and generally facilitates grouting. It also prevents damage to the cutoff surface by construction equipment during the grouting operation. Increasing grout pressures even with a surface cover

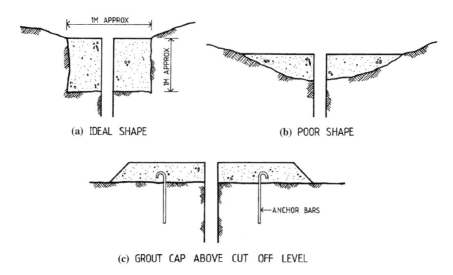

Figure 18.18.   Grout caps (a) ideal shape (b) poor shape (c) grout cap above cutoff level.

of concrete (shotcrete) should be done with care as lift-off of the cover can be initiated unwittingly.

18.3.4   *Grout mixers, agitator pumps and other equipment*

Houlsby (1977, 1978 and in WRC, 1981) strongly advocates the use of high speed, high shear, "colloidal" mixers. This opinion is supported by US Corps of Engineers (1984), Deere (1982), Gourlay and Carson (1982) and Bruce (1982). Figure 18.19 taken from WRC (1981) shows the principle of some suitable mixers. The most commonly used mixers are the "colcrete" or "cemix" type, which operate at 2000 rpm and 1500 rpm respectively. The very high speed facilitates mixing by separating cement particles from each other and wetting the surface of each particle. It is claimed (WRC, 1981) that mixing with

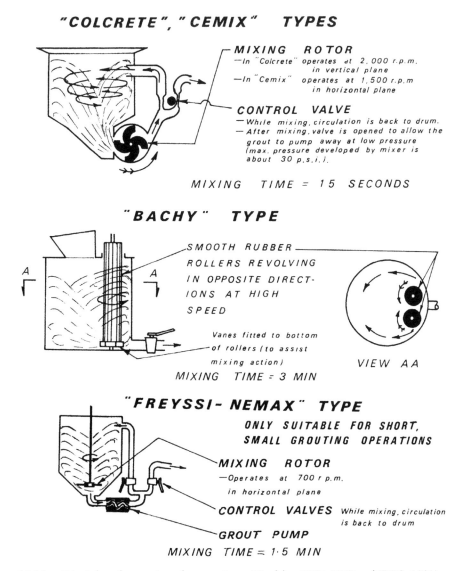

Figure 18.19.   Principles of operation of grout mixers (Houlsby, 1977, 1978 and WRC, 1981).

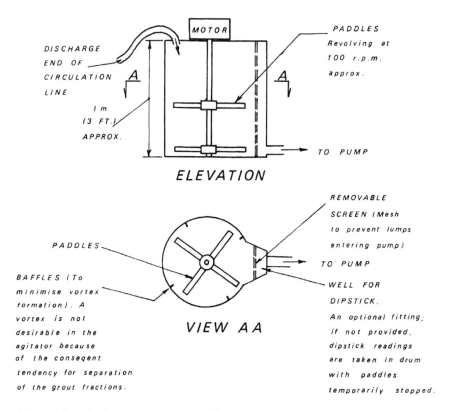

Figure 18.20.   Principle of grout agitators (Houlsby, 1977, 1978 and WRC 1981).

such high speed mixers "produces a grout which more closely resembles a colloidal solution rather than a mechanical suspension." While relatively speaking the grout may settle more slowly, one should not lose sight of the fact that it still is a suspension of cement particles which do settle with time.

Having mixed the grout it is usually transferred to an agitator from which it is pumped to the grout holes. The agitator is relatively slow speed and designed to prevent the cement particles from settling. Figure 18.20 (from WRC, 1981) shows the principle of the desired arrangement.

The quantity of grout injected is measured at the agitator.

Grout is pumped from the agitator using grout pumps. Houlsby (1977, 1978) and Deere (1982) indicate that in Australian and USA practice it is common to use "mono" (or "moyno") helical screw pumps because they provide a constant pressure, are rugged and readily maintained. A grout bypass valve at the hole is required as shown in Figure 18.21.

Gourlay and Carson (1982) and Bruce (1982) point out that in British and European practice there is a preference for ram type pumps, as it is believed the pulsating pressure assists in preventing clogging of the fracture opening by coarser grout particles. Deere (1982) also indicates a preference for this type of arrangement, largely because it obviates the need for a bleeder valve at the hole and/or recirculation line. As pointed out by Gourlay and Carson, there is a lack of hard evidence to support either preference and one should be willing to use the equipment readily available. Houlsby (1977, 1978) advocates the use of a recirculation line of the smallest diameter practicable to keep grout velocities high and avoid blockage. He suggests the maximum size is 25 mm diameter. Deere (1982) suggests that the recirculation line is used in US practice because of the then traditional

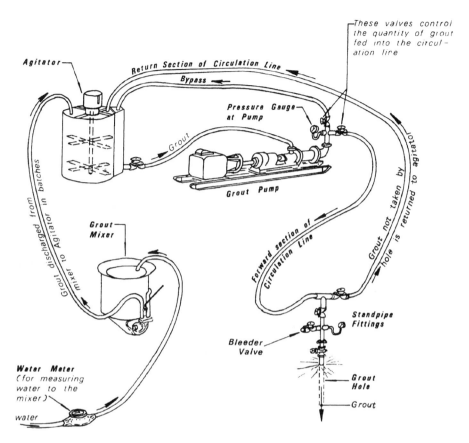

Figure 18.21.   Typical arrangement of grouting equipment (Houlsby, 1977, 1978, and WRC, 1981).

use of high water/cement ratio grouts which were "unstable" (i.e. settled out) and the requirement for a bypass valve to control injection pressure.

Details of bleeder valves and bypass, flowmeters, pressure gauges and standpipe fittings are given in Houlsby (1977, 1978), WRC (1981) and Gourlay and Carson (1982).

Packers for use in grout holes are either mechanically operated from the surface by a screw device, or inflatable, expanded against the side of the hole by compressed air or water. The length of "seal" formed by mechanical packers is often only 0.3 m, compared to 0.5 m to 1.5 m for the inflatable type, so leakage past the packer is more likely with the former. Mechanical packers would usually only be adopted if rock conditions resulted in puncturing of inflatable packers.

18.3.5   Water cement ratios

Cement grout mixes are usually designated by water cement (WC) ratios, with mixes ranging from 6:1 WC (by volume) to 0.6:1. Most grouting uses mixes of 2:1 WC ratio or less, as it is well documented that higher WC ratios yield unstable mixes (i.e. the particles settle quickly) and the grout is of poor durability (see Houlsby, 1985; Deere, 1982; Deere and Lombardi, 1985 and Alemo et al., 1991).

Use of volumetric WC ratios has been traditional in Australian and USA practice because a bag of cement was taken to be one cubic foot and as such was an easy measure in the field. The relationship between volumetric and weight based ratio is approximate

Table 18.4.   Relationships of different mix proportions of cement and water in grouts.

Water : Cement by volume	Water : Cement by weight	Cement : Water by weight	
6:1	4:1	1:4	(0.25)
4:1	2.67:1	1:2.67	(0.37)
3:1	2:1	1:2	(0.5)
2:1	1.33:1	1:1.33	(0.75)
1.5:1	1:1	1:1	(1.00)
1:1	0.67:1	1:0.67	(1.50)

because it depends on bulking of the cement. Table 18.4 gives approximate relationships given by Deere (1982).

Deere (1982) advocates the use of unit weight of the cement grout slurry measured by a "mud balance" and a marsh funnel to measure viscosity as part of the control of grout quality in the field. There is some merit in this suggestion as grout penetrability is dependent on viscosity. However, a marsh funnel measures apparent viscosity and, as discussed in Fell et al. (1992), yield point stress may be more important when considering grout penetration near refusal.

The selection of water–cement ratio is to an extent related to the "stability" of the grout mix. As outlined by Deere (1982) stability is measured by a sedimentation test in which a litre of grout is placed in a standard 1000 ml graduated cylinder. At the end of 2 hours the volume of clean liquid that has formed at the top of the cylinder due to sedimentation is noted. This volume, expressed as a percentage of the total volume, gives the percentage "bleeding" or sedimentation. Grout with a high WC ratio is "less stable" and bleeding or sedimentation is large.

Addition of small percentages of properly hydrated bentonite improves the stability. Deere (1982) indicates that in European practice a stable mix is one which has less than 5% sedimentation. Addition of 2% bentonite results in a small increase in viscosity of the grout but improves stability markedly.

WC ratios which should be used for grouting are discussed in some detail by Houlsby (1977, 1978 and in WRC, 1981), Deere (1982), Deere and Lombardi (1985), Bruce (1982) and Bozovic (1985). The authors favour the approach of Houlsby (1977, 1978, 1985) who recommends use of WC ratio (by volume) of not more than 3:1 and indicates doubts on long term durability if grout with WC ratios greater than 5:1 are used. He recommends use of the thickest possible mix at all times and suggests the following:

Starting Mix:

2:1   most sites
3:1   for rock <5 Lugeons
1:1   for rock >30 Lugeons
0.8:1 for very high losses
4:1   for heavily fractured, dry rock
5:1   rock above water table where excess water is absorbed by the dry rock.

Thicken the mix:

- to deal with severe leaks;
- after 1½ hours on the one mix with continued take (except for 1:1 and thicker mixes);
- if hole is taking grout fast, e.g. >500 litres in 15 minutes.

He indicates further that thickening should be in small increments, e.g. 3:1 to 2:1 and suggests reapplication of grout to the hole if take has exceeded 0.25 litres/cm of hole for WC

2:1 or thinner or 0.5 litres/cm for WC 1:1 or thicker to fill voids of bleed water. He does not favour the use of bentonite to give stable mixes, preferring to use low WC ratios and bleeding of the water from the hole.

Grouts with higher WC ratio will tend to settle to lower WC ratios. Hence higher viscosity results within the time of grouting and there appears to be little basis for using higher water contents to improve penetrability. The authors feel there is merit in adding bentonite to mixes with WC ratio greater than 1:1 (by volume) to control sedimentation. There appears to be evidence from major grout projects that it is practical and beneficial to do so.

18.3.6   *Grout pressures*

As mentioned in Section 18.1, there are two schools of thought on grout pressure:

(a) Those who limit grout pressures to below those which would lead to hydraulic fracture;
(b) Those who believe hydraulic fracture is preferred to promote the penetration of the grout.

The authors are advocates of avoiding hydraulic fracturing of the rock, as we are concerned that the fractures opened by the grout will not all be filled by grout and the grouting may worsen the situation rather than improving it. We are not concerned that seepage still occurs through a dam foundation, provided the dam is designed to manage the seepage. The following discussion is based on avoiding hydraulic fracture.

As pointed out by Deere (1982), Deere and Lombardi (1985) and Lombardi (1985), the maximum penetration distance is proportional to the pressure used for grouting. Hence it is desirable to use as high a pressure as practicable without fracturing the rock.

The pressure which can be applied depends on the rock conditions (degree of fracturing, weathering, *in situ* stresses and depth of the water table) and whether grouting is carried out using a packer which is lowered down the hole at each stage (i.e. downhole with packer grouting), or from the surface. The downhole packer method allows progressively higher pressures.

Houlsby (1977, 1978) and WRC (1981) present graphs to allow estimation of maximum pressures at the ground surface. These are based on the assumption that the maximum pressures at the base of the stage being grouted are given by:

$$P_B = \alpha d \qquad (18.2)$$

where P_B = pressure at base of hole in kPa; α = factor depending on rock conditions; ≈ 70 for "sound" rock; ≈ 50 for "average" rock; ≈ 25 to 35 for "weak" rock; d = depth of bottom of stage below ground surface in metres.

This allows for the weight of the overlying rock plus some spanning effect and has been found to be satisfactory.

The tendency for rock to fracture or "jack" under grout pressures is reduced by using a relatively low pressure to start grouting and building up with time. Since much of the pressure in the grout is dissipated in overcoming the viscosity effects in the fracture, this limits the pressure transmitted to outer parts of the grout penetration. Houlsby in WRC (1981) suggests use of a starting pressure of 100 kPa (or less) for 5 minutes, then steadily increasing the pressure over the next 25 minutes until the maximum pressure is reached. The occurrence of fracturing can be detected by sudden loss of grout pressures at the top of the hole, by increased take, surface leakage or by monitoring levels of the surface above the rock being grouted.

It is recommended by Houlsby that grouting is to "refusal" and that the pressure is maintained for 15 minutes after this to allow time for initial set. Others suggest grouting until take is less than a certain volume in a 15 minute period, e.g. Water Authority of WA (1988) specify grouting is to cease when take is less than 30 litres/20 minutes at 700 kPa or less; 30 litres in 15 minutes at 700–1400 kPa; 30 litres/10 minutes for pressures greater than 1400 kPa. They also indicate pressures should be maintained until "set" has occurred. In practice the maximum pressures will have to be determined by careful monitoring as the grouting proceeds.

18.3.7   Monitoring of grouting program

It is absolutely essential that detailed records of the grouting operation are kept. This is necessary to allow progressive development of the grouting operation, e.g. decision making on hole depths, whether closure holes are required or whether grout mixes should be changed. In most cases, detailed records will also be needed for payment purposes. Matters recorded should include:

- Hole locations, orientation, depths;
- Stage depths;
- Water pressure test value for each stage prior to grouting, including maximum pressure used;
- Grout mix(es);
- Grout pressures, takes at for example 15 minutes interval and then in summary;
- Grouting times;
- Leaks, uplift;
- Total grout takes for each stage;
- Amount of cement in these takes;
- Cement takes/unit length of hold.

This data will be collected largely by foremen but an engineering geologist should be involved continuously to interpret the progress of the grouting, relate this to the geotechnical model of the foundation and make decisions on closure etc. This will invariably involve keeping records on sections along and across the grout centreline with geological data superimposed.

18.3.8   Water pressure testing

Before grouting each stage, a water pressure test should be carried out. This is done using a method similar to that outlined in Chapter 5, but using a simplified procedure, e.g. applying only one pressure (100 kPa) for 15 minutes, taking flow quantities at 5, 10 and 15 minutes to estimate the Lugeon value.

18.3.9   Type of cement

Type A (AS1315-1982) portland cement is commonly used (Type I, ASTM) or Type C sulphate resistant (Type II) if acid groundwaters warrant, e.g. in some tailings dams. There is some argument for using Type C as it is marginally finer than Type A and should penetrate fine fractures more readily.

Deere (1982), Bozovic (1985) and Weaver (1993) indicate that on some South American, USA and Japanese projects special microfine cements have been manufactured with additional grinding. These microfine cements are available with significantly finer particle size and enhanced ability to penetrate fine fractures. They are significantly more expensive than conventional cements.

18.3.10   *Prediction of grout takes*

For contractual and cost estimating reasons it is necessary to estimate grout takes, i.e. the volume of grout (or dry weight of cement) which will be absorbed by the foundation during the grouting operation.

This is difficult to do with any degree of accuracy because the penetration of the grout is dependent on fracture aperture, roughness, continuity and interaction with other sets of fractures and grout viscosity, pressure, duration etc.

Bozovic (1985) in his general report of the ICOLD Congress concludes that the correlation between grout take and Lugeon value is very weak. He suggested that considering the different rheological properties of grout and water, a correlation cannot physically exist. Figure 18.22 shows some data from Ewart (1985) which shows the poor correlation. Ewart (1985) presents similar data for Aabach Dam, as do Sims and Rainey (1985) for Gitaru Dam.

The authors agree that, if grout take and Lugeon value are compared directly, there is a poor correlation. However they are of the opinion that, if joint spacing is used to estimate fracture openings and grout penetration is estimated from grout particle size, grout pressure, viscosity and time, it should be possible to obtain a better prediction of grout take. It is not expected that accurate prediction will result, only an improvement in the ability to predict. If this approach is coupled with grouting trials on the dam foundation it should be possible to achieve reasonable accuracy.

It is concluded that the best approach is to gather data from dam sites in similar geological environments and make initial estimates of take per metre of grout hole on this basis. For any reasonable degree of accuracy trial grouting on a representative part of the dam foundation will be required, with careful monitoring of takes in primary, secondary,

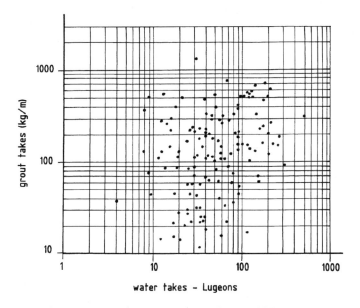

Figure 18.22.   Grout take vs Lugeon value (Jawantzky, in Ewart, 1985).

tertiary holes etc. However unless proper attention is paid to the equipment used and selection of the area to be grouted, the results can be misleading.

18.3.11   *Durability of cement grout curtains*

From the authors' combined direct knowledge, which goes back to the early 1950s, cement grout curtains of dams that have operated for 50 years or so seem to have done their job admirably. Nevertheless there is evidence to suggest that seepage water will, albeit at very slow rates, gradually leach some of the compounds from the hardened grout. At Thomson (Victoria) and Corin (ACT) dams, there are calcium ions (along with magnesium and sulphate ions) in the seepage water at the dams' toes, but only traces of calcium ions in the storage water. For many dams this very slow deterioration probably would have little impact. The grouting will have done its main job of sealing up the large defects and the slow loss of the grout will not result in much of a change in the overall permeability of the rock mass.

Loss of compounds from the hardened grout probably begins with the slow loss of excess lime (CaO_2) that dissolves in the seepage water as calcium hydroxide ($Ca(OH)_2$). On contact with air (either direct contact or from dissolved air in, say, rain water), calcium carbonate ($CaCO_3$) forms. Depending on the flow conditions, the $CaCO_3$ may come out of solution. Another mechanism would begin with the solution in water of carbon dioxide (CO_2) from bacteria in the ground. The resulting weak acid (H_2CO_3), then combines with the excess lime to give $CaCO_3$.

One would assume that this solution process is not unlike the processes described in Chapter 3 for the gradual loss of carbonates from carbonate rocks. If a partly-grouted joint in the rock mass has some unfilled zones and water seeps along those open parts of the joints, grout exposed to the flow will begin to suffer some loss. The loss rates would be expected to be very slow, particularly if only effectively narrow joints were available for the seepage water. Even if the loss of material did allow the cement grout to disintegrate to some degree, some of the inert particles in the cement grout may be able to join up the open parts of these joints to help maintain a reasonably low permeability for the rock mass.

A process like the one just described seems to have occurred at Blowering dam, a 112 m high rockfill dam in the Snowy Mountains Scheme (personal communications with C. Houlsby in the early 1980s). At Blowering much of the grouting was probably done with 5:1 and 6:1 W/C ratio (by volume). The weak grout in the joints of the rock "bleeds" significantly, leaving patches of weak, hardened cement grout and water. Ultimately on storage filling seepage water was able to "attack" the weak grout patches and reduce the effectiveness of the grout curtain. Under the spillway $CaCO_3$ has deposited in the underflow drains and lowered the capacity of the drainage system. The actual change in total seepage at Blowering is unknown, as seepage under the dam is not measured, but the left abutment ridge grouted by a single line curtain has significant seepage flows through it.

Talbingo dam (Snowy Mountains Scheme – completed 1971) and Thomson dam (Victoria – completed 1984) are two well-monitored dams that have long seepage records. Both are 160 m high rockfill dams. The storage at Talbingo is held close to full all the time, while Thomson, a water supply storage, does vary slowly with time and weather conditions.

At Talbingo, some 14,000 40 kg bags of cement were used on the curtain and 12,000 bags on the blanket (over the full core contact area). Generally, a 3 to 1 by volume water/cement was used as a starting mix, but the mixes were sometimes thinned to 5 to 1 if the ground proved to be tight. The seepage has been steadily dropping since records were started in late 1982, nearly 11 years after the lake's first filling was completed. The seepage ranges from 16 l/s just after records began to 10 l/s in early 2004. These figures would suggest no deterioration in the curtain, but one should also consider that:

- The defects in the foundations were probably not great in number and the rock below the excavated foundation for the core was inherently tight to start with;
- There may have been some "silting up" of the open joints in the rock mass under the upstream rockfill shoulder.

The grout volume in bags of cement was much higher at the Thomson main dam. A total of nearly 21,000 bags was put into the curtain and over 100,000 bags in the blanket. About 8–9% of the total went into the river bed and lower 20–30 m of the abutments, but over 50% was pumped into 70 m of the right abutment. The upper right abutment was formed by two intersecting landslides and the ground was open to 70–80 m below the original natural surface.

So far at Thomson the base seepage in the period from 1990 to late 1996, when the storage was at a continuously high level, was recorded at about 10 l/s. The estimated permeability of the foundation is 1–2 Lugeons; during the investigations 100 Lugeon values were common on the upper right abutment.

It is probably too early to come to any definite conclusions on the grout curtain at Thomson. To date there has been no measurable deterioration.

Another Snowy Mountains dam, Eucumbene, a 116 m high earthfill dam with limited rockfill shoulders, seems to have behaved similarly to Talbingo dam. Early records from 1958, first filling, to 1998 show that:

Storage level (RL m)	Head on base of core (m)	Seepage (l/s)
1110	60	13
1143	93	17
1160	110	8

The conclusion on the potential deterioration of the grout curtain would be the same as drawn from the Talbingo records, although the same two factors noted for Talbingo probably apply also to Eucumbene.

18.4   CHEMICAL GROUTS IN DAM ENGINEERING

18.4.1   *Types of chemical grouts and their properties*

There are two distinct types of chemical grouts:

- Group A – colloidal solutions or prepolymers, e.g. silica gel, ligno chrome gel, tannins, organic or mineral colloids, polyurethane;
- Group B – pure solutions, e.g. acrylamide, phenoplast, aminoplast.

The hydraulic behaviour of these grout types is different:

- Group A – behave as Bingham fluids with an initial shear stress required to mobilize the grout (as for cement and cement/bentonite grouts);
- Group B – behave as Newtonian fluids – as for water but with a higher viscosity.

Figure 18.23 shows typical flow properties of grouts. Note that the viscosity measured (in centipoise) is the slope of the graph at any point; the apparent viscosity is the slope of

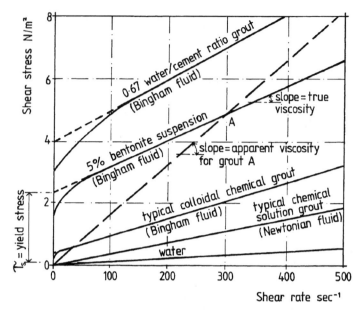

Figure 18.23.   Typical flow properties for grouts (adapted from Littlejohn, 1985).

the line passing through the point (e.g. Point A) and the zero point (rather like a secant modulus compared to a tangent modulus for true viscosity). The intercept of the curve on the shear stress axis is the "yield stress" or "yield value" (also called "cohesion" by Lombardi, 1985, by analogy with the shear strengths of soils). The value quoted will depend on the shear rate and may be determined by extrapolation of the straight line portion of the curve as shown in Figure 18.23.

Most methods of measuring will give the apparent viscosity, rather than the yield stress and true viscosity and this must be taken into account. The authors' experience is that, with cement grouts, the method proposed by Nguyen and Bogor (1983, 1985) which uses a very sensitive vane shear type apparatus is the most successful.

The viscosity is dependent on the concentration of the grout and the time after mixing. Figures 18.24 and 18.25 show these effects.

The gel time is dependent on time and temperature and on additives which are deliberately used to control it. This is discussed in Littlejohn (1985) and Karol (1985).

There are many types of grout available and new products are coming onto the market all the time. Tables 18.5 and 18.6 list some of the more common grouts and their properties. Karol (1982a, b, 1983) and Littlejohn (1985) discuss chemical properties of grouts in more detail.

18.4.2   Grout penetrability in soil and rock

Grout penetration is dependent upon:

(i) Whether penetration is by permeation (impregnation) in the voids in the soil or by causing hydraulic fracture by exceeding the *in situ* horizontal stresses (0.5–2 times overburden pressure), or a combination of both. Australian and general overseas (except French) practice is to limit pressures to the permeation phase;

(ii) The viscosity properties of the grout and then the pressure and time for which grouting proceeds.

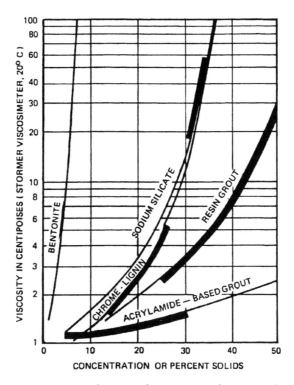

Figure 18.24.   Viscosity (apparent) as a function of grout type and concentration (Karol, 1983).

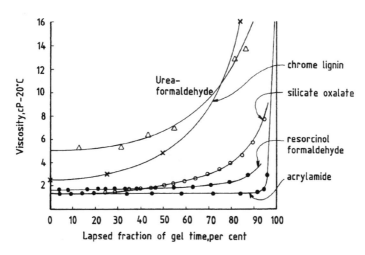

Figure 18.25.   Viscosity (apparent) as a function of time (Littlejohn, 1985).

For Bingham fluids, there is a limiting radius to which grout can be pumped because of the shear stress required to mobilize the grout. Littlejohn (1985) indicates that in soil this can be estimated from:

$$RL = \frac{\delta_w gHd}{4\tau s} + r \qquad (18.3)$$

Table 18.5.   Some chemical grouts and their properties (Brett, 1986).

Grout type	Corrosivity or toxicity	Viscosity	Strength
Silicates			
Joosten process	Low	High	High
Siroc	Medium	Medium	Medium/high
Silicate/Bicarbonate	Low	Medium	Low
Lignosulphates			
Terra firma	High	Medium	Low
Blox-all	High	Medium	Low
Phenoplasts			
Terranier	Medium	Medium	Low
Geoseal	Medium	Low/medium	Low
Rocagil	Medium	Medium	Low
Aminoplasts			
Herculon	Medium	Medium	High
Cyanaloc	Medium	Medium	High
Acrylamides			
AV-100	High	Low	Low
Rocagil BT	High	Low	Low
Nitto SS	High	Low	Low
Terragel	High	Low	Low
Poly acrylamides			
Injectite 80	Low	High	Low
Acrylate			
AC-400	Low	Low	Low
Polyurethane			
CR-250	High	High	High
CR-260	High	Medium	High
TACSS	High	High	High

where RL = limiting radius of penetration; δ_w = density of water; g = acceleration due to gravity; H = hydraulic head; d = effective diameter of the average pore; τs = Bingham yield stress; r = radius of spherical injection source = $\frac{1}{2}LD$ where L = length of hole, D = diameter and:

$$d = 2\sqrt{\frac{8\mu k}{\delta_w gn}} \tag{18.4}$$

where μ = grout viscosity in centipoise; k = permeability of soil; n = porosity of soil.

In jointed rock, with a joint aperture of $2a$, Lombardi (1985) indicates that the maximum radius is given by equation 18.5.

This is equivalent to:

$$RL = \frac{Ha}{\tau s} \tag{18.5}$$

For Newtonian flow, in uniform isotropic soil from a spherical source, Littlejohn (1985) indicates that:

$$H = \frac{Q}{4nk}\left[\mu\left(\frac{1}{r} + \frac{1}{R}\right) + \frac{1}{R}\right] \tag{18.6}$$

Table 18.6.   Chemical grouts – a summary of properties.

Grout type	Fluid behaviour	Typical viscosity centipoise	Gel time	Stability	Examples	Comments
Sodium silicate	Newtonian initially, then Bingham	3–4 for permeability reduction 10 for strength. Viscosity increases as grout gels	30–60 minutes	1) Undergoes syneresis (loss of water and shrinkage on gelling). 2) Unstable in alkaline environment	Joosten process	Non toxic and not an environmental hazard. Syneresis less of a problem in finer soils (sand and silt). Limit is fine-medium sand. Krizak (1985) tests showed up to 10 or 100 times increase in permeability with time under high gradient seepage
Acrylamide	Newtonian	Less than 2	Gel time controlled by additives NaCl in groundwater may lessen gel time	Permanent	AM9 (now outdated because of neuro-toxicity). Injectite 80 (has higher viscosity)	AM9 was one of the most popular grout for a long time. Good permeability and permanence. AM9 was replaced by methyl acrylamides, but the toxicity of unreacted monomer remains
Phenoplast	Newtonian	1.5 to 3 initial and constant until gel starts	Gel time controlled by additives	Permanent except under alternating wet/dry conditions	Resorcinol + formaldehyde + NaOH e.g. Geoseal	Medium toxicity. Geoseal 2–10 centipoise used for dams at Worsley (Brett (1986)
Aminoplast	Newtonian	5, up to 10–20 with additives to stabilize gel time	Gel time controlled	Only gel in acid (pH < 7) environment. Permanent except in wet/dry conditions	Based on urea and formaldehyde	Toxic and corrosive prior to gelling (acid catalyst)
Acrylate polymer	Newtonian	2	Gel time controlled	Permanent	AC-400	Replaces AM9. Only 1% as toxic. Krizek & Perez (1985) tests showed it did not deteriorate with time under high (100) gradients.

Table 18.7.   Grouting limits of common mixes (Littlejohn, 1985).

Type of soils	Coarse sands and gravels	Medium to fine sands	Silty or clayey sands, silts
Soil characteristics			
Grain diameter	$D_{10} > 0.5\,mm$	$0.02 < D_{10} < 0.5\,mm$	$D_{10} < 0.02\,mm$
Specific surface	$S < 100\,cm^{-1}$	$100\,cm^{-1} < S < 1000\,cm^{-1}$	$S > 1000\,cm^{-1}$
Permeability	$K > 10^{-3}\,m/s$	$10^{-3} > K > 10^{-5}\,m/s$	$K < 10^{-5}\,m/s$
Type of mix	Bingham suspensions	Colloid solutions (gels)	Pure solutions (resins)
Consolidation grouting	Cement ($K > 10^{-2}\,m/s$) aerated mix	Hard silica gels double shot Joosten (for $K > 10^{-4}\,m/s$) Single shot: Carongel, Givanol, Siroc	Aminoplastic, phenoplastic
Impermeability grouting	Aerated mix, bentonite gel, clay gel, clay/cement	Bentonite gel, lignochromate, light Carongel, soft silicagel, vulcanisable oils, others (Terranier)	

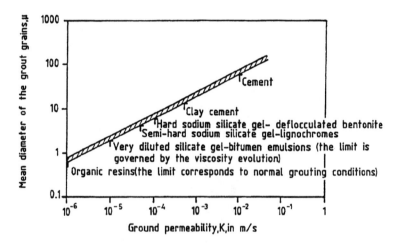

Figure 18.26.   Limits of injectability of grouts based on the permeability of sands and gravels (Littlejohn, 1985).

where Q = flow rate at radius of penetration R and the time for grout to penetrate.

$$t = \frac{nr^2}{kH}\left[\frac{\mu}{3}\left(\frac{R^3}{r^3} - 1\right) - \left(\frac{\mu - 1}{2}\right)\left(\frac{R^2}{r^2} - 1\right)\right]$$ (18.7)

i.e. time is proportional to R^3. This dictates that relatively close hole spacings are used for economic grout times, e.g. 0.5 m to 2.5 m.

There are different tables and graphs indicating the types of soils and soil permeability that can economically be grouted, e.g. Littlejohn (1985) suggests Table 18.7 and Figure 18.26.

Karol (1985) suggests that, based on a review of the literature, Figure 18.27 is a conservative assessment of groutability by permeation, i.e. without fracturing.

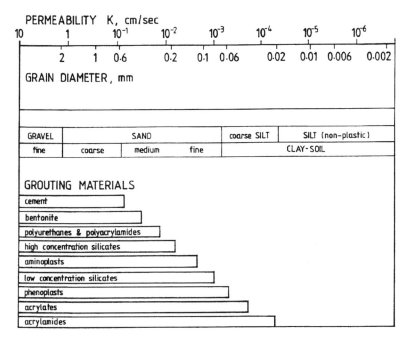

Figure 18.27. Limits of groutability (based on Karol, 1985).

Table 18.8. Limits of grouting (Caron, 1982).

	Type of ground			
	Coherent fissured soils		Loose granular soils	
Type of grouting method	Large medium fissures $K > 5 \times 10^{-7}$ m/sec	Very fine fissures $K < 5 \times 10^{-7}$ m/sec	Coarse and medium $K > 10^{-3}$ m/sec	Fine $K < 10^{-3}$ m/sec
By fracturing	–	–	C	C
By impregnation	C	CG	C	CG
By fracturing	–	–	–	CG

C = cement based grout; CG = chemical grouting; – = no case for grouting.

Caron (1982) suggests Table 18.8 as a basis for determining whether grouting is practicable, depending on ground conditions and the type of grout.

It is hard to envisage a case where grouting of soils with a permeability around 10^{-7} m/sec would be warranted or would result in significant reduction in permeability.

18.4.3 Grouting technique

The basic approach to use of chemical grouts is similar to that for cement grouts:

– Grout is injected into holes under pressure;
– Holes are drilled and grouted in stages to achieve a desired Lugeon or permeability closure;

When grouting in rock in which the holes remain open and packers can be set, down-stage grouting with packer would normally be adopted. However, in most chemical grouting operations, either extremely weathered rock (i.e. virtually soil properties) or soil is being grouted and it is necessary to support the hole from collapse and use different methods to inject the grout.

Figure 18.28 shows the tube-à-manchette ("pipe with sleeves") technique which is used in Europe (and Australia) for grouting with chemicals.

In this technique the hole is drilled and cased to its full depth, the hole filled with a cement/bentonite grout and the tube-à-manchette installed. This consists of a 40 mm to 60 mm diameter PVC tube, with 6 mm diameter boreholes in the wall of the tube at 300 mm or 333 mm intervals. The holes are covered by a rubber sleeve.

Once installed, the casing is withdrawn and the grout allowed to 'set' to give a low strength, relatively brittle grout.

The grouting operation is carried out by lowering a double packer to isolate one set of outlet holes as shown in Figure 18.28, setting the packers, then applying the grout pressure. The pressure of the grout lifts the rubber sleeve, fractures the bentonite-cement grout and allows the chemical grout to penetrate into the soil or weathered rock.

Some idea of pre-grouting permeability can be obtained from the grout flow rate and pressure but, as explained by Caron (1982), this may not be accurate because of unquantifiable pressure losses through the fractures in the cement/bentonite grout.

Caron (1982) indicates that, when grouting in sands and gravels, practice differs from country to country. In USA and Japan, grouting is commonly from the bottom of the casing used tosupport the hole, with the casing gradually withdrawn. For a two-shot grout, such as sodium silicate/calcium chloride, the base grout (sodium silicate) is injected as the casing is lowered and the reactant grout (calcium chloride) as the casing is withdrawn (as shown in Figure 18.29).This method does allow checking of permeability before grouting (but only crudely out of the bottom of the casing) and gives poorer control than the tube-à-manchette system.

As shown in Figure 18.30, grouting should be carried out on a "closure" basis in depth and plan and at least 3 lines of holes should be used to achieve this. Initial spacing should be 2 to 3 times the anticipated final spacing.

18.4.4   Applications to dam engineering

Chemical grouts are expensive compared to cement or cement-bentonite grouts. As a result, the use of chemical grouts will be restricted to those cases where seepage is critical and not controllable by cement grouts or in cases of remedial works, particularly in alluvial (soil) foundations where cement grouts are not applicable.

As in the use of cement grouts, there is a limit to the permeability which can be achieved by chemical grouting. Littlejohn (1985) suggests that practical minimum average permeabilities are:

– $\times 10^{-7}$ m/sec in coarse sands and gravels (and only after "sophisticated chemicals, careful injection procedures and close supervision");
– $\times 10^{-8}$ m/sec in "fissured" rock.

Since the grout curtain width is likely to be narrow (say 1 m to 1.5 m maximum per row of holes), significant seepage reduction will result only when the original permeability is relatively high compared to the grout curtain (see Figure 18.13).

Brett (1986) presents a plausible argument for two or three stage cement/bentonite, followed by chemical grouting. The former is to fill the larger voids and hence reduce the cost of chemicals. Figure 18.31 illustrates this approach.

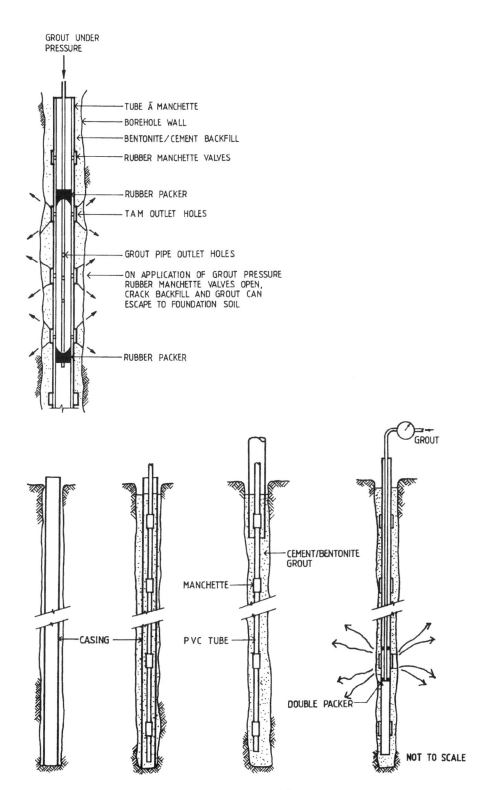

GROUT UNDER PRESSURE

TUBE Ā MANCHETTE
BOREHOLE WALL
BENTONITE/CEMENT BACKFILL
RUBBER MANCHETTE VALVES

RUBBER PACKER
T A M OUTLET HOLES

GROUT PIPE OUTLET HOLES

ON APPLICATION OF GROUT PRESSURE
RUBBER MANCHETTE VALVES OPEN,
CRACK BACKFILL AND GROUT CAN
ESCAPE TO FOUNDATION SOIL

RUBBER PACKER

GROUT

CEMENT/BENTONITE GROUT

MANCHETTE

CASING

PVC TUBE

DOUBLE PACKER

NOT TO SCALE

Figure 18.28.   Tube-à-manchette grouting system (adapted from Brett, 1986).

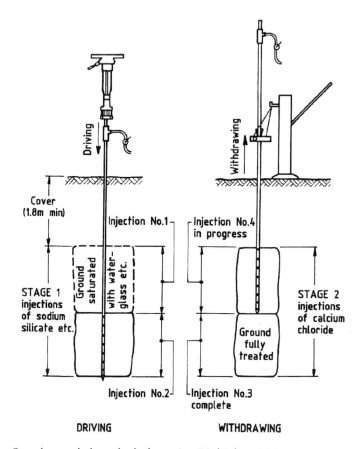

Figure 18.29.   Open bottom hole method of grouting (Littlejohn, 1985).

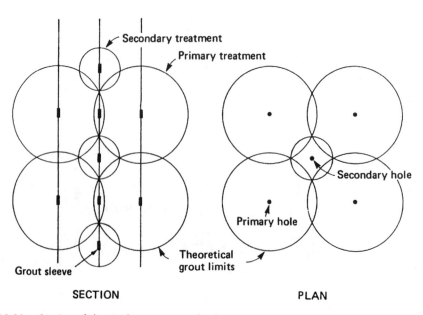

Figure 18.30.   Staging of chemical grouting (Littlejohn, 1985).

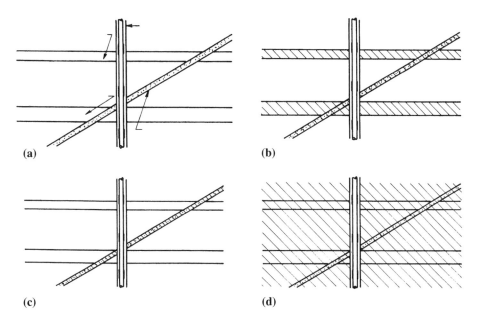

Figure 18.31.   Multiple grout system (Brett, 1986). (a) ground properties, (b) first stage cement grout,(c) second stage bentonite grout, (d) third stage chemical grout.

In many cases in the past where chemical grouts have been used to reduce leakage under dams on alluvial sand and gravel foundations, it would now be more reliable and less costly to use slurry trench or diaphragm wall cutoffs constructed of bentonite or bentonite-cement. These give a more controllable uniform width of low permeability cutoff than can be achieved by grouting and would generally be the authors' preference (see Chapter 9 for more discussion on this matter).

Examples of the use of chemical grouting in dams are given in Littlejohn (1985), Brett and Osborne (1984), Davidson and Perez (1982) and Graf et al. (1985).

It will be clear from the above discussion that chemical grouting, although feasible for some jobs, should be viewed with some caution. Most chemical grouts are potentially toxic to some degree, some to the point of being of serious concern. Unless chemical grouting can be justified on cost grounds and is environmentally safe to use in the particular case under consideration, the authors would advocate not using them.

Like epoxy compounds and some other newly-developed compounds, chemical grouts are not the answer to all problems. If not mixed correctly or not applied correctly, they may cause more problems than they solve. The fourth author spent 2 years proving an ill-conceived grouting job done with chemical grouts was safe for the public; he is now convinced the chemical grouting did nothing to improve the situation, making the whole exercise an avoidable one if careful thought had been applied at the start of the design phase.

19

Mine and industrial dams

19.1   GENERAL

This chapter is intended to give an overall appreciation of mine and industrial tailings disposal practice. It is written for those who have not had extensive experience in tailings disposal and serves as an introduction to the subject, rather than forming a complete basis for design of a tailings disposal system.

The authors' experience is that the majority of mine tailings are disposed of into reasonably conventional dams and the principles of dam engineering covered in the rest of this book apply. Some tailings are disposed of using the tailings themselves to construct the embankment e.g. using upstream, centreline or downstream methods, and this requires special design and construction features (see Section 19.5). More recently, thickened discharge, co-disposal and paste technology has been developed which is applicable in some situations and these methods are described briefly. This chapter mostly concentrates on the understanding of what tailings are and their properties, the methods of discharge, prediction of deposited density, the methods of disposal, estimation of seepage from tailings disposal areas and the particular aspects of slope stability, internal erosion and piping and performance under earthquake which are important in tailings disposal.

19.2   TAILINGS AND THEIR PROPERTIES

19.2.1   *What are mine tailings?*

Mine tailings are the end product of mining and mineral processing, after the mineral has been extracted, or the unwanted material separated, e.g. shale and clays from coal; clayey fines from iron ore or bauxite. As shown in Figure 19.1, the process will often involve crushing and grinding, leaching or separation, followed by dewatering or thickening before discharge to the tailings disposal area as a slurry. For coal and bauxite, the ore will not be ground and the process is one of separation of the other rock particles from the coal, or finer clays and silts from the bauxite gravel by washing and separation.

The thickening process may be assisted by addition of flocculants and/or other chemicals, e.g. polyelectrolytes such as the Magnafloc range, and inorganic salts such as gypsum. The polyelectrolytes are generally high molecular weight polyacrylamides, which have positive (cationic) or negative (anionic) charges. These adsorb to the electrically charged particles in the tailings and form large flocs which settle more quickly in the thickener. Inorganic salts such as gypsum ($CaSO_4$) operate by cation exchange, with high charge density Ca^{++} cations replacing low charge density Na^+ cations, reducing dispersion effects and promoting flocculation as discussed in Chapter 7.

These chemicals affect the properties of the tailings discharged to the tailings disposal area and must be included in trial processes when testing tailings properties.

Dewatering is usually carried out in a thickener, with the dewatered tailings being recovered from the discharge cone as shown in Figure 19.2.

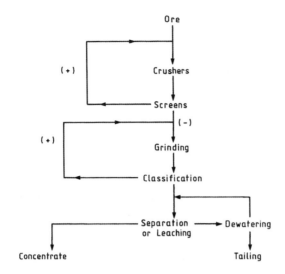

Figure 19.1.   Procedures in mineral processing.

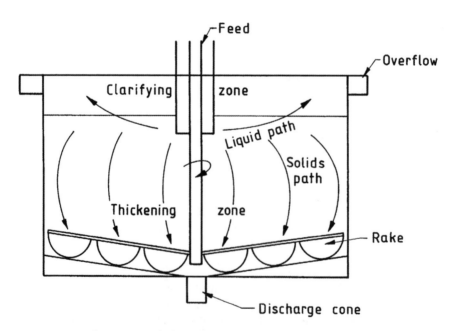

Figure 19.2.   Minerals processing thickener.

19.2.2   *Tailings terminology and definitions*

Most properties of tailings are described in soil mechanics terminology, but there are some terms in common usage which originate from mineral processing. Tailings are a mixture of solids and water (and air when not saturated) as shown in Figure 19.3, and the various terms used relate largely to the relative proportions of those present.

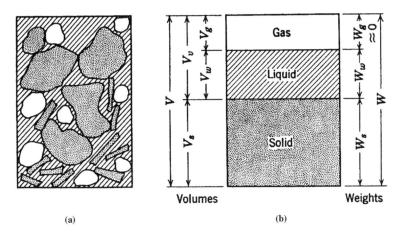

Figure 19.3.   Tailings phase relationship: (a) element of tailings with air filled and water filled voids; (b) element separated into air, liquid and solid (Lambe and Whitman, 1981).

Commonly used terms and their definitions include:
Void ratio:

$$e = \frac{V_V}{V_S} \tag{19.1}$$

Degree of saturation:

$$S = \frac{V_W}{V_V} \tag{19.2}$$

Porosity:

$$n = \frac{V_V}{V} = \frac{V_V}{V_V + V_S} \tag{19.3}$$

Specific gravity (solids):

$$G = \frac{\gamma_S}{\gamma_W} \tag{19.4}$$

where V_V = volume of voids = $V_a + V_W$ (m^3); V_a = volume of air in voids (m^3); V_W = volume of water (m^3); V_S = volume of solids (m^3); W_W = weight of water (kN); W_S = weight of solids (kN); M_W = mass of water (tonnes); M_S = mass of solids (tonnes).

$$\gamma_S = \text{unit weight of solids} = \frac{W_S}{V_S} (kN/m^3) \tag{19.5}$$

$$\gamma_W = \text{unit weight of solids} = \frac{W_W}{V_V} (kN/m^3) \tag{19.6}$$

Water content (%):

$$w = \frac{W_w}{W_S} \, 100 (\%) \tag{19.7}$$

(may also defined as $W_W/(W_W + W_S)$ by some metallurgists).
% Solids:

$$P = \frac{W_S 100}{W_S + W_W} = \frac{1}{1 + w} (\%) \tag{19.8}$$

Degree of saturation:

$$S = \frac{V_W 100}{V_V} = \frac{\gamma_d w G 100}{G \gamma_w \, \gamma_d} (\%) \tag{19.9}$$

Total unit weight:

$$\gamma_t = \frac{W_S + W_W}{V_V + V_S} = \frac{G + Se}{1 + e} \gamma_W \, (kN/m^3) \tag{19.10}$$

Dry unit weight:

$$\gamma_d = \frac{W_S}{V_V + V_S} = \frac{G}{1 + e} \gamma_W = \frac{\gamma_t}{1 + w} (kN/m^3) \tag{19.11}$$

19.2.3   Tailings properties

19.2.3.1   General

Tailings properties differ considerably depending on the ore from which they are derived, the mineral process and whether the ore is oxidised (i.e. from weathered rock). Table 19.1 gives general characteristics of engineering behaviour.

When planning any tailings disposal project, tailings representative of the operational process plant should be tested to determine the properties required to predict behaviour in the storage. It should be noted that, in many cases, tailings produced from trial crushing

Table 19.1.   Summary of typical physical tailings characteristics.

Type of tailings	General characteristics
Ultra-fine tailings, phosphatic clays, alumina red mud	Clay and silt, high plasticity, very low density[3] and permeability
Washery tailings, coal, bauxite some iron and nickel ores	Clay and silt, medium to high plasticity, medium to low density and permeability
"Oxidised"[1] mineral tailings, gold, copper, lead, zinc etc.	Silt and clay, some sand, low to medium plasticity, medium density and permeability
"Hard rock"[2] mineral tailings, gold, copper, lead, zinc etc.	Silt and some sand, non plastic, high density, medium to high permeability

Notes: (1) Ore is completely or highly weathered, or altered rock; (2) ore is slightly weathered to fresh rock; (3) assuming no desiccation.

and grinding in laboratory processing studies often are not truly representative of those produced by the operational process plant. This can lead to the perception that laboratory tests do not accurately predict field behaviour. The authors' experience is that laboratory tests on representative samples can reasonably predict field behaviour.

19.2.3.2 *Particle size*
Figure 19.4 gives some examples of particle size distributions from washery, oxidized and hard rock tailings.

The particle size distribution for oxidised tailings and tailings from washeries (e.g. coal, iron ore, bauxite) depends on the test method used and particularly on whether dispersants are added.

The standard soil mechanics test for determination of particle size of fine materials is to use a hydrometer analysis with a dispersant (calgon, i.e. sodium hexametaphosphate) added to break the particles to their constituent size. In the thickener, and as transported to the tailings disposal area, the particles may remain flocculated. Figure 19.5 shows particle size distributions for tailings from two coal mines (Wambo, Hunter Valley; Riverside, Central Queensland); a bauxite mine (Weipa, North Queensland) and iron ore mines (Newman and Hamersley, Western Australia). Tests with and without dispersant show that the "true" behaviour is to act as a silt-sand mix, whereas with dispersant added the tailings have a high clay size fraction.

19.2.3.3 *Mineralogy*
The tailings' behaviour can often be related to the mineralogy of the constituent particles. To illustrate this, Tables 19.2 and 19.3 list mineralogy, chemistry and soil mechanics properties of several tailings. The following should be noted:

- The bauxite tailings from Weipa have a high proportion of amorphous clay minerals which, while fine grained, lead to a relatively low liquid limit and plasticity index and give favourable (for fine tailings) sedimentation characteristics;
- The coal washery tailings from Wambo have a high proportion of sodium montmorillonite, which leads to a very low settled density;
- The coal washery tailings from Riverside, and gold tailings from North Kalgurli, contain clays which have intermediate activity between Weipa and Wambo and behave accordingly. The coal content of Wambo and Riverside tailings give a low specific gravity;
- The iron ore tailings from Newman and Hamersley have a significant haematite content, giving a high specific gravity. The clay minerals are not very active and, although fine grained, the tailings tend to sediment favourably;
- The Broken Hill tailings have been produced from high strength unweathered rock and yield a nonplastic silt-sand mix which settles rapidly to a relatively high density.

19.2.3.4 *Dry density and void ratio*
Table 19.4 gives some typical in-place dry densities and void ratios. The lower void ratios/higher densities correspond to greater depths within a deposit and would be typical of tailings which have been placed into a water pond, i.e. sub-aqueous placement. If desiccation of the tailings occurs (i.e. drying in the sun), dry densities would usually be 20% to 50% higher than those shown.

The in-place density also depends on the thickness of tailings (the thicker the tailings the higher the average density) and whether the tailings drain vertically towards the foundation, with higher effective stresses and higher densities when consolidation is complete.

General trends are better shown by the void ratio because the dry density is significantly affected by the specific gravity of the particles. Vick (1983) indicates that most high

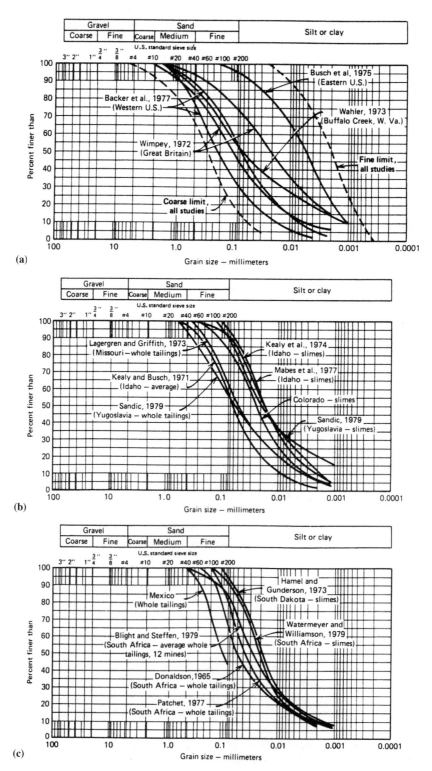

Figure 19.4.   Typical tailings particle size distribution: (a) coal washing; (b) lead–zinc; (c) gold–silver (Vick, 1983).

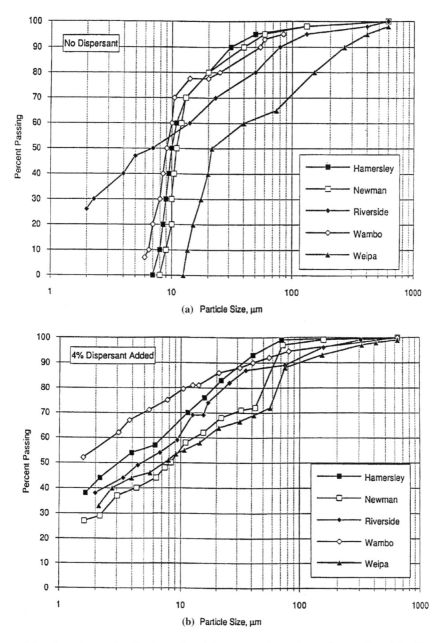

Figure 19.5.   Particle size distributions of coal, iron ore and bauxite washery tailings (a) without and (b) with dispersant added.

strength rock (and even low strength rock) tailings will have in-place void ratio of 0.6 to 0.9 for sand sized particles and 0.7 to 1.3 for slimes. The exceptions are phosphatic clays, bauxite (alumina) and oil sands and slimes.

It will be noted that the dry densities are very much lower than the specific gravity (or dry density) of the parent rock. In many mining operations a very small percentage of the ore is removed as mineral in the milling process and it would not be unusual for the tailings volume to be 150% to 200% of the original volume of the mined ore.

Table 19.2. Mineralogy and chemistry of some fine grained tailings.

Tailings	Mineralogy	Chemistry (% by weight of dried tailings)								pH tailings water
		SiO_2	Al_2O_2	Fe_2O_3	CaO	MgO	Na_2O	K_2O	LOI[1]	
Weipa 1	Gibbsite (45%), boehmite (18%), kaolin (12%), quartz (17%), hematite (5%)	19	52	6	<0.1	<0.1	<0.1	<0.1	18	6.9
Weipa 2	Gibbsite (34%), boehmite (35%), kaolin (10%), quartz (8%)	10	53	12	<0.1	<0.1	<0.1	0.1	19	6.4
Wambo	Na and Ca montmorillonite, kaolin, coal, quartz[2]	28	12	1	0.9	1.0	0.8	0.4	51	6.8
Riverside	Illite, Ca montmorillonite, kaolin, quartz[2]	29	13	0.5	<0.1	0.3	0.3	0.7	54	7.4
Mt Newman	Hematite (40%), kaolinitic shale (50%), quartz (6%)[2]	17	12	63	0.4	0.7	<0.1	0.4	5	7.4
Hamersley	Hematite (34%), kaolinitic shales (52%), limonite (8%), goethite (19%)[2]	20	14	60	0.1	0.3	<0.1	0.1	7	7.1
North Kalgurli	Quartz, kaolin, gibbsite[2], host rock, dolerite, basalt, sulphide and oxide ores	55	14	13	2.3	1.1	2.8	2.2	7	7.9
Broken Hill	Quartz (50%), rhodenite (20%), calcite (12%), garnet (6%), manganhedbergite (4%), pyrrhotite (4%)	56	7	12	6.7	1.1	1.1	1.3	1	7.7

Notes: [1] Loss on ignition (LOI) at 1000°C; [2] mineralogy percentages are approximate and vary with production sources.

Table 19.3. Soil mechanics properties of some fine grained tailings.

Tailings	Water content (%)	Dry density (t/m³)	% solids	Soil particle density (t/m³)	Atterberg limits (%)			Particle size (%)[1]		
					Liquid limit	Plastic limit	Plasticity index	Sand	Silt	Clay[2]
Weipa 1	675	0.14	13	2.75	44	27	17	20	47	33
Weipa 2	362	0.25	22	2.85	43	26	17	15	45	40
Wambo	411	0.22	20	1.86	74	28	46	14	36	50
Riverside	250	0.32	29	1.74	44	28	16	16	50	34
Newman	192	0.46	34	3.70	33	22	11	20	60	20
Hamersley	169	0.50	37	3.50	30	21	9	5	55	40
North Kalgurli	71	0.94	59	2.81	28	21	7	Not available		
Broken Hill	462	0.20	18	3.05	Non plastic					

Notes: [1] Particle size with dispersant added as per AS1289; [2] percentage finer than 0.002 mm.

Table 19.4.   Typical in place densities and void ratio for non-desiccated tailings (adapted from Vick, 1983).

Tailings type	Specific gravity	Void ratio	Dry density (t/m^3)
Fine coal refuse			
Eastern US	1.5–1.8	0.8–1.1	0.7–0.9
Western US	1.4–1.6	0.6–1.0	0.7–1.1
Great Britain	1.6–2.1	0.5–1.0	0.9–1.35
Oil sands			
Sands	–	0.9	1.4
Slimes	–	6.0–10.0	
Lead–zinc slimes[1]	2.9–3.0	0.6–1.0	1.5–1.8
Gold–silver slimes	–	1.1–1.2	
Molybdenum sands	2.7–2.8	0.7–0.9	1.45–1.5
Copper sands	2.6–2.8	0.9–1.4	1.1–1.45
Taconite sands	3.0	0.7	1.75
Taconite slimes	3.1	1.1	1.5
Phosphate slimes	2.5–2.8	11.0	0.25
Gypsum treated tailings	2.4	0.7–1.5	
Bauxite slimes	2.8–3.3	8.0	0.3[2]

Notes: [1] For hard rock tailings; [2] low by Australian standards 0.5–0.9 t/m^3 more likely.

The relative density of the silt and sand sized tailings has an important influence on the potential for liquefaction. The relative density of spigotted tailings sands above water, i.e. in the beach zone, can be expected to be in the range 30% to 50% (Vick, 1983). Morgenstern (1988) indicates that relative densities of 50% to 65% can be expected for sands placed above water and 35% to 40% when placed underwater. De Groot et al. (1988) show relative densities from 10% to 50% (mainly 25% to 40%) for sands placed below water. Thus silt/sand tailings placed below water are likely to be susceptible to liquefaction under earthquake and static shear.

19.2.3.5 Permeability
The permeability of mine tailings depends on the type of tailings, particle size and mineralogy, deposition method, degree of consolidation and/or desiccation (dry density/void ratio), cracking or other structure and whether the tailings are saturated or partially saturated. When tailings are deposited unthickened, there is segregation of the coarser particles relative to the fines with distance from the discharge point so the tailings permeability will vary in the deposited tailings, typically higher near the spigot point, if the sands and coarse silts separate out, and lower distant from the discharge point where the fine tailings or "slimes" deposit. It is also lower, deeper in the deposit than at the surface, due to consolidation.

In addition, within each layer of tailings, the larger particles settle at a different rate from the finer ones, so there is vertical segregation in the layers. For most tailings, the larger particles settle to the bottom of the layer and may form a thin sandy "parting". For coal washing tailings, the coarser particles are often coal, with a low specific gravity, and these rise to the top of the layer.

Table 19.5 gives some typical tailings permeabilities, which are probably representative of vertical permeabilities.

The horizontal permeability will change with the degree of segregation and variability of deposition. Near the discharge area (say the first 50–100 metres), it is likely the vertical segregation will give an equivalent horizontal permeability at least 10 times and probably 100 times the vertical. In the slimes the effect is likely to be small, with horizontal and vertical permeabilities similar.

Table 19.5.   Typical tailings permeability (adapted from Vick, 1983).

Type	Average permeability (m/sec)
Clean coarse, or cycloned sands with less than 15% fines	10^{-4} to 10^{-5}
Peripheral-discharged beach sands with up to 30% fines	10^{-5} to 5×10^{-6}
Non plastic or low-plasticity slimes	10^{-7} to 5×10^{-9}
High plasticity slimes	10^{-6} to 10^{-10}

Note: These are likely to be representative of average vertical permeabilities.

19.2.3.6   *Properties of water in tailings*

The water (or liquor) accompanying tailings often contains dissolved salts, heavy metals and other residual chemicals from the mineralogical process.

The water may be highly acidic or alkaline, have high salts content, a heavy metals content or have arsenic or cyanide present. Each may have important implications on the impact of the tailings disposal area on the environment. Some common problems are:

– Gold tailings – cyanide;
– Coal tailings – high salts content, sometimes sulphides;
– Copper, lead–zinc, tin tailings – heavy metals, sulphides (which oxidise and lead to leaching of heavy metals and yield acid water);
– Alumina tailings – caustic soda (NaOH), unless the tailings are neutralized before discharge;
– Uranium tailings – heavy metals, radon;
– Other tailings may have quite acceptable water quality, e.g. some bauxite and iron ore washeries.

The science of predicting the chemistry of the tailings liquor and the fate of contaminants in the groundwater has advanced considerably over the last 10 years. It is a specialist area and the advice of a suitably qualified person should be sought.

19.3   METHODS OF TAILINGS DISCHARGE AND WATER RECOVERY

19.3.1   *Tailings discharge*

Most tailings are pumped to the tailings disposal area after thickening as a high water content (e.g. 35% to 40% solids content) slurry. In thickened tailings the solids content may be as high as 50% to 60%. The tailings are discharged into the storage from a single or several discharge points, or from spigots, as shown in Figures 19.6 and 19.7. A single fixed discharge point gives little flexibility in management of the tailings and is usually undesirable.

Spigotting facilitates the uniform spreading of the tailings and encourages desiccation. For conventionally non-thickened tailings spigotting encourages the segregation shown in Figure 19.6 with the sand and coarse silt settling near the embankment.

For some fine grained high clay content tailings, a "spray bar" is attached to each spigot to further distribute the tailings and allow uniform beaching. The "spray bar" consists of, for example, a 100 mm diameter pipe with 20 mm diameter holes at 500 mm spacing. The objective is to give a uniform deposition of tailings, allowing the greatest degree of sorting (or "classifying") of the tailings away from the discharge points and to optimise desiccation.

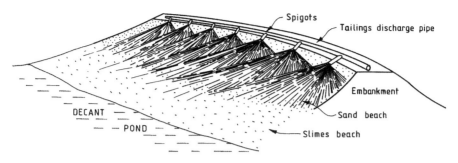

Figure 19.6.   Tailings discharge by spigotting.

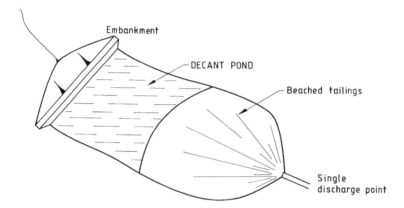

Figure 19.7.   Tailings discharge from a single pipe.

It is usually preferable to discharge the tailings from the dam embankment, so that the tailings cover the storage area in the vicinity of the dam and potentially reduce seepage. If it is desired to blanket the base of the storage with tailings as shown in Figure 19.41(b), it is necessary to spigot progressively from all round the storage, moving the main discharge line up the slope as the tailings rise in the storage. In the case of upstream or centreline type construction (see Section 19.5), it is essential that the tailings are discharged from the embankment and the water pond kept small and well away from the edge of the embankment.

19.3.2   Cyclones

When tailings dams are to be constructed of tailings by the upstream or centreline methods, it is common to separate the coarse fraction of the tailings from the fine fraction or "slimes" by use of cyclones.

Figure 19.8 shows a typical cyclone. The "whole" tailings are fed under pressure into the cylindrical cyclone. The coarser particles separate from the fines by centrifugal force (there are no moving parts) and spiral downward through the conical section as "underflow". The finer fractions and most of the water rise to the top outlet as "overflow".

The performance of a cyclone depends on many factors, including the particle size, specific gravity, clay content of tailings, size of cyclone, pressure etc. and advice should be sought from the cyclone manufacturer as to how effective they are likely to be.

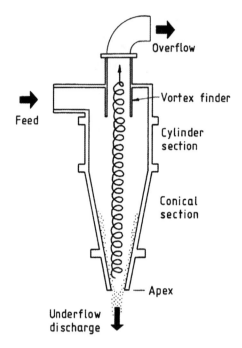

Figure 19.8.   Typical cyclone for tailings.

The use of cyclones is discussed in Vick (1983). He indicates that reasonably clean sand can be obtained from tailings with less than 60% passing 0.075 mm, provided the tailings are essentially non plastic and free of clay minerals. Such sands will have high permeability and high shear strength.

Cycloning may be used in several ways:

– *Stationary cyclone plant* – a central high capacity cyclone station is established near the dam. Sands from the cyclone can be placed in the dam by conventional earthmoving equipment, and may be compacted. This may facilitate use of tailings for dam construction in seismic areas;
– *On dam cycloning* – the most common technique. Several small cyclones are set up on the dam and the underflow discharged on to the embankment. The cyclones are moved as the dam is raised. Vick (1983) indicates that relative densities of 45% to 55% are likely to be achieved if no compaction is carried out.

At the El Cobre Tailings Dam in Chile, a large stationary cyclone plant was used, with the water content of the under flow engineered to give the required downstream slope of 4H:1V. This slope was continuously tracked by a small bulldozer to compact the tailings to a relative density of at least 65% (Cohen and Moenne, 1991). This material was not saturated, dense and non liquefiable.

19.3.3   *Subaqueous vs subaerial deposition*

If for climatic reasons, lack of a large discharge area, acid generation control or other reasons, tailings are deposited under water, the method is known as subaqueous deposition. Fine, high clay content tailings deposited subaqueously will usually achieve a low settled density, very low shear strength and will be very compressible. Thus large, costly storages

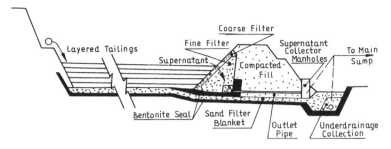

Figure 19.9.   Subaerial placement of tailings (Robertson et al., 1978).

are required to contain the same quantity of tailings and long term rehabilitation is more difficult than if subaerial deposition is used, because the surface of the tailings is wet and low strength. If water cover is maintained on the tailings, seepage quantities from the storage will be relatively high.

The method where the tailings are discharged in a "beach" above the water pond and where the deposition is cycled so that the tailings are allowed to dry by desiccation is known as subaerial deposition.

Desiccation induces negative pore pressures in the tailings, which results in consolidation of the tailings to higher densities than by subaqueous methods. The permeability and compressibility are also reduced, although cracking by desiccation may affect the permeability.

In the ideal situation where the climate, storage area and tailings permit, the tailings will remain in a partially saturated condition and result in significantly reduced seepage from the storage. Post placement consolidation may be very small and the strength such that access on to the surface for rehabilitation can be easy. Figure 19.9 shows a conceptual design of such a system (Robertson et al., 1978).

Blight (1988) also discusses the benefits and disadvantages of desiccation. He points out the risks of planning to use the benefits and then not being able to achieve them. He cites an example where loss of control of water return resulted in the achieved dry density being about half the expected. The authors' experience is that the technique can be successfully used, if an adequate area is provided to accommodate the rate of tailings production, the evaporation is sufficiently high and care is taken to operate the area properly. In such cases, even high clay content tailings have been desiccated sufficiently for use with upstream construction.

Table 19.6 summarizes some of the advantages and disadvantages of subaerial depositions.

The authors have seen cases where increased production of tailings, poor control of return water, low winter evaporation, and a lack of consideration of the fact that early in a dam's life the surface area is often less than later when the tailings level has risen, have led to virtually no desiccation being achieved.

Blight (1988) also cautions against drying the tailings to the stage where they crack, as this may lead to a potential for piping failure. This would only be a problem in particular instances of upstream construction using small paddocks where the desiccation cracks could penetrate from the water to downstream, as is common in South Africa and parts of Australia.

In general, one will not aim for the "full" subaerial deposition method shown in Figure 19.10. This was developed for uranium tailings, where requirements for seepage control are very stringent. The costs of underdrainage will not be warranted in most cases. One should, however, aim to obtain as much benefit as possible from desiccation for the reasons outlined above. It is important to consider the "whole of life" of the facility because many of the benefits of subaerial deposition and desiccation are in reduced rehabilitation and long term environmental impacts.

19.3.4   *Water recovery*

Water is removed from the tailings storage as it "bleeds" from the beached tailings, or as tailings settle and consolidate if deposited subaqueously.

This is usually achieved by a pump mounted on a floating barge or a decant tower, as shown in Figure 19.10.

Table 19.6.   Subaerial disposal advantages and disadvantages (Knight and Haile, 1983; Lighthall, 1987; from Ritcey, 1989).

Advantages:
(a) Higher unit weights are achieved in the tailings mass, resulting in better utilization of the storage facility;
(b) The drained nature of the tailings and removal of surface ponding adjacent to embankments allows construction by upstream methods;
(c) The low permeability, laminated structure of the tailings deposit reduces seepage;
(d) At decommissioning, the tailings are fully drained and consolidated, allowing immediate construction of a surface seal and cover and elimination of any long term seepage;
(e) The drained nature of the tailings increases resistance to liquefaction;
(f) There is a low hydraulic head on the liners; therefore a wide choice of liners is possible;
(g) Capping and covering should be facilitated.

Disadvantages:
(a) Dusting may occur unless wetted;
(b) Method may not be suitable during extremely cold periods. Therefore the disposal area is divided into two, for alternative summer–winter deposition. Winter deposition is somewhat conventional (subaqueous);
(c) Possibly higher initial cost to construct compared to other methods;
(d) Requires a separate water storage reservoir to be constructed for water and runoff water recycling or discharge.

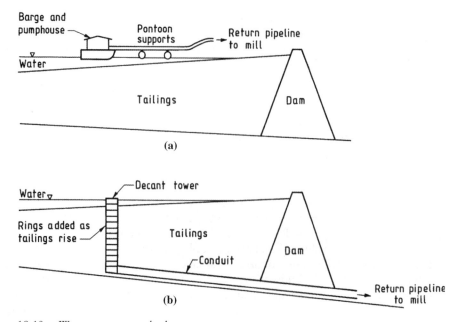

Figure 19.10.   Water recovery methods.

There is no preferred method, each case has to be considered on its merits. The floating barge system is usually more flexible and requires less stringent control of tailings deposition to maintain the decant pond near the decant tower. It also avoids the potential weakness, from an internal erosion and piping viewpoint, of having a conduit through the embankment. It is usually desirable to keep the water surface area to a minimum to reduce evaporation of water (and hence reduce total water losses) and to optimise the benefits of desiccation of the beached tailings.

19.4   PREDICTION OF TAILINGS PROPERTIES

19.4.1   *Beach slopes and slopes below water*

Tailings deposited in a tailings dam form a beach above the water level and also fill the storage below the water level with a sloping surface. The beach above the water level is commonly steeper adjacent to the discharge point than further away.

Figure 19.11 shows observed beach profiles for several dams containing platinum tailings (Blight, 1988).

Figure 19.12 shows observed profiles for several tailings dams, in which fine grained tailings with high clay contents are stored. Properties for these tailings are given in Figure 19.5 and Tables 19.2 and 19.3.

The fact that the tailings deposit in this way is important because it results in a reduction in the available tailings storage volume compared to water storage, and determines the position of the water pond, which is important particularly for upstream construction.

Blight and his co-workers have investigated beach profiles by laboratory experiments in sloping flumes, and by observing beach profiles in the field. This work is described in Blight, Thomson and Vorster (1985), Blight and Bentel (1983), Blight (1987) and Blight (1988). The concept they have developed was originally proposed by Mellent et al. (1973). It was found that by plotting dimensionless parameters h/y versus H/X, as shown in Figure 19.13, the different profiles in Figure 19.11 all plotted to give a single "master" beach profile with the equation

$$h/y = (Y/X)(1 - H/X)^n = i_{av}(1 - H/X)^n \tag{19.12}$$

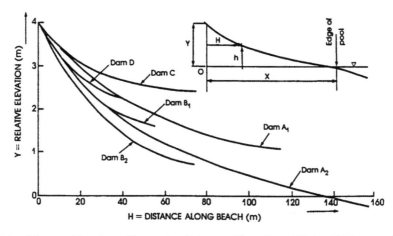

Figure 19.11.   Measured beach profiles on six platinum tailings dams (Blight, 1988, reproduced with permission of ASCE).

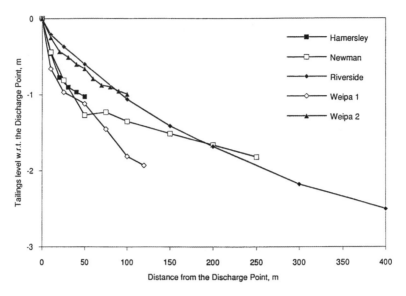

Figure 19.12.   Measured beach profiles, fine grained high clay content tailings.

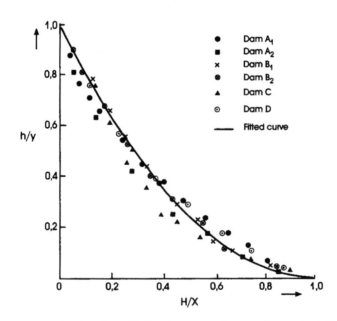

Figure 19.13.   Dimensionless beach profile for platinum tailings in Figure 19.11 (Blight, 1988, reproduced with permission of ASCE).

where i_{av} is the average slope, and n is determined for the tailings.

The master profile is different for each type of tailings. The particle size, specific gravity and solids content of the tailings all affect the profile. Blight (1988) indicates that the master profile applies, almost regardless of the length of the beach, and hence laboratory flume tests can be used to predict the behaviour of the dam.

Morgenstern and Kupper (1988) indicate that, in their experience, the laboratory flume tests did not show trends consistent with field behaviour.

The concept of a master profile is, however, well established and has been noted by others, e.g. Smith, Abt and Nelson (1986). Hence, one can predict later performance quite well from the initial behaviour of a tailings storage.

For planning purposes in a new project one will have to rely on flume tests and relationship to other measured profiles for similar tailings.

De Groot et al. (1988) point out that the observed slopes in dredged sands below water are much flatter than the predicted equilibrium slope (which might be expected to be steeper than slopes above water because of reduced gravitational forces below water). This is explained as due to flow slides where the upper part of a slope oversteepens as sedimentation occurs, resulting in slope instability and flow which produces flatter slopes. They note that observed slopes below water were similar to that in the beach above water. This is also the authors' usual observation for high clay content tailings, although in an Alumina red mud storage the slope was steeper below water.

19.4.2   Particle sorting

In conventionally thickened tailings, because the coarser particles contained in the tailings settle more rapidly than the finer particles, a gradation of particle size occurs on the beach, with coarser particles depositing near the discharge point. As discussed above particle sorting also occurs within a layer, with coarse, high specific gravity particles settling to the base of a layer. These can form a sand "parting" even in high fines content tailings. Conversely, if the coarse particles are low specific gravity as occurring in coal washery tailings, they will "float" to the surface of a layer.

The lateral variations in grading with distances from the discharge point can have an important effect on tailings permeability – with the coarser more permeable tailings deposited near the discharge spigots and the finer low permeability slimes deposited further away. This is used in the upstream method of tailings dam construction to control seepage pore pressures and, hence, to maintain stability. Blight and Bentel (1983) and Blight (1987, 1988) discuss this feature and conclude that the particle size distribution can be roughly predicted from:

$$A = e^{-BH/X} \tag{19.13}$$

where

$$A = \frac{D_{50}(\text{at distance H down the beach})}{D_{50}(\text{of the total tailings})} \tag{19.14}$$

B = characteristic of tailings and is a function of the discharge rate, and the specific gravity of the tailings;

X = length of beach as in Figure 19.11.

Abadjiev (1985) presents a similar formula which can be applied to not only the D_{50} size but to D_{90}, D_{60}, D_{10} etc. He also presents data from several tailings dams showing the lateral gradation.

The authors' experience is that the amount of sorting which occurs is dependent on the method of deposition. Spray bars and closely spaced spigots give very good sorting provided discharge rates are kept low. This is consistent with the observations of Blight and Abadjiev.

It should be noted that for tailings thickened to 50% to 60% solids content there is no sedimentation phase during the deposition, little or no bleeding of water/liquor and little or no particle sorting. Hence there will be more uniform particle size and deposited slope than for tailings thickened to only say 35%–40% solids content.

19.4.3 *Permeability*

The permeability of tailings is dependent on the particle size distribution and void ratio, so is therefore dependent on the distance from the discharge point, the method of deposition, e.g. subaqueous or subaerial, and the depth in the storage.

The following procedure is therefore recommended for the estimation of permeability at any particular place:

(1) Predict the particle size at that place from the grading of the whole tailings and the sorting as predicted by the approaches discussed above in Section 19.4.2;
(2) Predict the void ratio from sedimentation, and the sorting as predicted by the approaches discussed below in Section 19.4.4;
(3) Prepare samples based on (1) and (2) above;
(4) Calculate permeability from either falling head or constant head tests, or by back calculation from consolidation tests, conducted on these samples.

For sand sized tailings the permeability can be approximately estimated from Hazen's formula, i.e.

$$k = D_{10}{}^2/100 \qquad (19.15)$$

where k = permeability in m/sec; D_{10} = grain size in millimetres, for which 10% of the particles are finer.

Using this and the relationship between particle size and distance from the discharge point discussed in Section 19.4.2, Blight (1987) suggests that the permeability of the beached tailings will vary as:

$$K = ae^{-bH} \qquad (19.16)$$

where "a" and "b" are dependent on the tailings; H is the distance down the beach.

19.4.4 *Dry density*

Prediction of the dry density of tailings in the storage is of prime importance, because the amount of tailings which can be stored in a given storage volume depends on the density achieved. This, in turn, is dependent on the type of tailings, method of deposition (subaqueous or subaerial), drainage conditions (e.g. if underdrains are provided or if the soil and rock underlying a tailings dam is of high permeability) degree of desiccation, distance from the discharge point and proximity to the water pond etc.

For tailings which are deposited under water there are two phases:

(a) Sedimentation;
(b) Consolidation under self weight and the weight of tailings above.

The sedimentation phase occurs when the tailings are first placed in the dam, and the particles settle, with a clear water interface forming on the surface of the tailings. In subaerial deposition this "bleed" water will run off down the beach slope.

The consolidation phase follows and may occur concurrently with sedimentation. Prediction of consolidation of tailings requires a different approach from that for normal soils, because of the large strains involved. The traditional one-dimensional Terzaghi theory is not applicable and would result in overestimation of the time of consolidation and underestimating pore pressures. It is also necessary to use special slurry consolidometers, so tests can begin at the density of the tailings after sedimentation.

Schiffman and Carrier (1990) and Schiffman, Vick and Gibson (1988) give overviews of this topic, and reference other papers which can be used to obtain the detail of the testing and analysis techniques. They indicate that centrifuge testing can also be used for more accurate modelling, but advocate monitoring and analysis of the behaviour of the early phases of tailings deposition as the best means of predicting behaviour.

The reader is referred to these references rather than have the information repeated here. However, the following practical points are based on the authors' experience:

- It is essential to carry out testing with tailings and water which are representative of the actual tailings and water. It is also essential that the tailings have not dried out before carrying out testing, particularly if the tailings have a significant clay content. Drying completely changes the sedimentation behaviour (usually makes the tailings settle more quickly and more densely) and affects the consolidation and permeability properties;
- Sedimentation tests, where the tailings are mixed and placed in a 500 mm column and allowed to settle, give a good indication of the minimum settled density. Figure 19.14 gives the results of such tests for the tailings in Tables 19.2 and 19.3 in 500 mm columns. Field measurements indicate that the 500 mm columns slightly underestimate the field density of tailings;
- Field sampling of the very loose/soft upper layer of tailings is difficult. Ritcey (1989) refers to the use of piston samplers. Fell et al. (1992) show a simple device, which was developed by R. Hogg at the University of New South Wales, which has been used to obtain disturbed samples of tailings. This allows determination of water content, and hence dry density or void ratio, and give samples for particle size distribution testing.

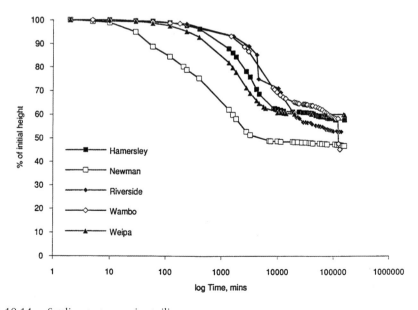

Figure 19.14.   Settling tests on mine tailings.

19.4.5   *The prediction of desiccation rates*

If one is to rely on the increased dry density and strength which can be obtained from desiccation of the tailings, it is necessary to predict the rate of drying. The rate is dependent on:

- The properties of the tailings, in particular the settled water content, the permeability and the suction pressure which develops on drying;
- The climate – evaporation and rainfall;
- The deposition cycle – depth of and time between each cycle of deposition.

In Swarbrick and Fell (1990, 1991) the results of a research program to develop a method for predicting desiccation rates is described. Based on laboratory and field drying experiments, it has been shown that desiccation occurs as:

(a) Settling until the rate of water release equals the potential evaporation;
(b) Stage 1 drying, which occurs at a linear rate with time, generally at the same rate as from a free water surface;
(c) Stage 2 drying, which occurs at a decreasing rate. This decreasing rate has been shown to satisfy the sorptivity equation i.e.

$$E_{cum} = b\sqrt{t} \tag{19.17}$$

where E_{cum} = cumulative evaporation after the linear stage t = time after the linear stage (days); b = sorptivity coefficient (mm days$^{-0.5}$).

Swarbrick and Fell (1992) give details of the method for predicting drying rates using laboratory drying tests on samples 600 mm $\times$ 600 mm $\times$ 300 mm.

It is very important to note that the drying of tailings is dramatically affected by the presence of salts in the water/liquor. This is a particular problem in alumina red mud where the salts are NaOH and $NaCO_3$, but is also a problem in other tailings such as from coal washeries where natural groundwater is highly saline, or if saline water is used for processing or neutralizing the tailings. The salt concentrates on the surface, and the osmotic suction slows and eventually stops evaporation from the tailings surface. To keep desiccation proceeding it is necessary to break up the salt on the surface using special equipment such as that shown in Figure 19.15 or using swamp dozers once the tailings are strong enough for them to work.

19.4.6   *Drained and undrained shear strength*

19.4.6.1   *Drained shear strength*

The effective strength parameters c', ϕ' for tailings can be obtained from triaxial or direct shear tests on samples of the tailings. It is particularly important that sandy tailings are tested at the correct relative density and stress range, because of the strong dependence of ϕ' on the relative density and the curved nature of the Mohr envelope.

Table 19.7 shows some typical values of effective friction angle. It will be noted that most are relatively high, reflecting the absence of any structure and the grinding of rock to angular particles. Unless the tailings become cemented on deposition, e.g. by gypsum, the effective cohesion is c' = 0.

As discussed below, it will in many cases be wrong to rely on the drained shear strength of the tailings because if they are loose and saturated, they will be contractive on shearing and develop positive pore pressures. The undrained strength is a more reliable way of assessing their strength. The beginning of contractive behaviour leads to a sudden increase

Figure 19.15.   Rotating screw device for breaking up the salt layer on top of alumina red mud tailings. (Photograph courtesy of Geographe Earthmoving).

Table 19.7.   Typical values of effective friction angle ϕ' of tailings (adapted from Vick, 1983).

Material	ϕ' (degrees)	Effective stress range (kPa[1])
Copper		
Sands	34	750
	33–37	625
Slimes	33–37	625
Taconite		
Sands	34.5–36.5	–
Slimes	33.5–35	–
Lead–zinc–silver		
Sands	33.5–35	–
Slimes	30–36	–
Gold slimes	28–40.5	900
Fine coal refuse	22–39	270
	22–35	1100
Bauxite slimes	42	175
Gypsum tailings	32 (c' = 22 kPa)	450

Note: [1] From zero to these maximum values.

in pore pressures at the collapse surface, and is a lower strength envelope than the drained shear strength (see Figure 6.9).

19.4.6.2   Undrained shear strength

The undrained shear strength of the "slimes" or clay-silt sized part of the tailings can be important both from the view of overall slope stability in upstream construction and the strength of the surface of the tailings for rehabilitation.

The strength can be determined by appropriate triaxial testing, e.g. using the SHANSEP method (Ladd and Foott, 1974), but must be related to the field situation by a knowledge

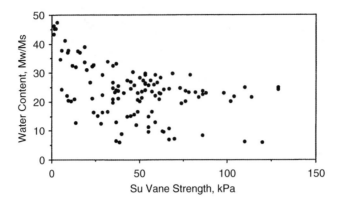

Figure 19.16.   Vane shear test results – Newman tailings.

of the degree of overconsolidation of the tailings. This can be done by taking undisturbed samples of the tailings and carrying out oedometer consolidation tests.

Alternatively, if consolidated undrained triaxial tests with pore pressure measurement are carried out on undisturbed samples of the tailings, the pore pressures developed during shearing can be measured and the Skempton A_f at failure determined. Then the undrained strength can be determined using Equation 6.17, Chapter 6.

The undrained strength of fine silt and clay tailings *in situ* may be measured using a vane shear. However the results are often very scattered because of the varying degrees of overconsolidation within layers which have been partly desiccated and the presence of sandy layers which lead to overestimation of undrained strength. Figure 19.16 gives some results from the Newman tailings, which show the wide scatter and apparent lack of relationship to water content.

Where tailings are consistently fine grained, e.g. well away from the discharge point and not affected by desiccation other than in a single drying phase, reasonably constant strengths can be obtained by a vane shear.

Swarbrick and Fell (1991) used vane strengths in the laboratory to develop a relationship between undrained strength and water content. Here some consistency was possible, with good comparison with single stage desiccation in the field. The laboratory tests were done using a miniature vane, with samples dried in glass beakers.

19.5   METHODS OF CONSTRUCTION OF TAILINGS DAMS

19.5.1   *General*

In most cases the basic requirements for a tailings disposal area are to store the tailings in such a way that they remain stable (i.e. there are no slope stability problems) and do not impact excessively on the environment by water pollution and/or wind and water erosion. Only in some cases is it necessary, or desirable, to store water in the tailings disposal area.

This should be kept in mind when designing a tailings embankment and one should not always feel it necessary to construct "a dam" to store the tailings.

Many of the principles and practices developed in conventional water dam engineering are applicable to tailings, but it should not be assumed that all water dam technology or philosophy is applicable.

In many cases it is better to consider tailings embankments as landfills, or dumps of fine grained soil which have to be engineered to:

– Optimise the amount of tailings which can be stored in a particular area;
– Control piezometric conditions to ensure adequate strength and stability;
– Control the environmental impact.

Australian practice has in the past been (and to a large extent still is) biased towards constructing conventional dams to store tailings. This approach also applies in some other countries. This is possibly due to:

– Many designers have a background in water dam engineering;
– Regulatory authorities require such an approach (largely because of their background in water dam engineering);
– A relatively wet climate in many mining areas and many mines with very fine grained and/or oxidised tailings which are not readily utilized in "upstream" construction methods.

There is a recognition amongst many practitioners that this philosophy is costly and unnecessary and alternative methods are being used.

The following discussion is intended to give an overview of the various alternative methods and their limitations, with a view to encouraging their consideration and adoption where appropriate.

19.5.2   *Construction using tailings*

There are three main methods for constructing tailings embankments using the tailings as a major construction material:

– Upstream method;
– Downstream method;
– Centreline method.

These are described in some detail in ICOLD (1982d) and Vick (1983).

19.5.2.1   *Upstream method*
The important features of this approach (Figure 19.17) are:

– The starter dam is essentially a containing embankment and a support for the tailings discharge line, rather than a dam in itself. The starter dam is best constructed of permeable rockfill (e.g. mine waste) to allow drainage of seepage water and to control erosion. However, it may be constructed of relatively impermeable rockfill, earthfill or even dried tailings pushed up by bulldozers from the tailings beach.
– Tailings discharge must be controlled by e.g. spigotting to ensure that the coarser sandy tailings are deposited near the starter dam. This is essential to control seepage pressures as outlined below.
– The water pond (decant pond) must be kept well away from the edge of the storage. If allowed to come close to the edge, the piezometric pressures will be high and slope instability may result, or internal erosion and piping of the starter dam or of the tailings into the starter dam may occur.
– The coarse tailings are likely to be at a low to medium relative density and, if saturated and subjected to earthquake, may liquefy, leading to slope failure with a subsequent flow failure of the tailings. Because of this the method is seldom used in seismic risk areas. It should be

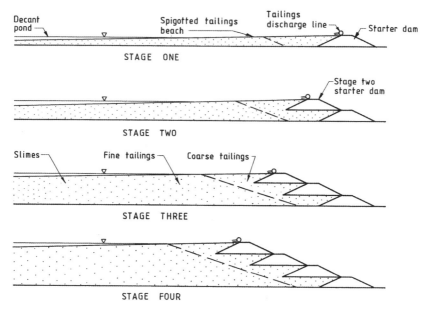

Figure 19.17.   Construction of a tailings embankment using the upstream method.

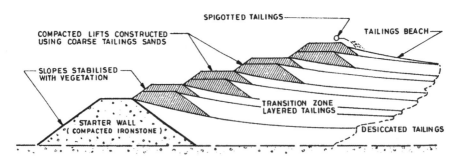

Figure 19.18.   Section of bauxite tailings embankment at Weipa showing upstream construction (Minns, 1988, reproduced with permission of ASCE).

noted that because of layering the tailings are likely to have perched water tables, so saturation is almost inevitable.

The method is best suited to hard rock tailings, i.e. silt-sand tailings which are readily spigotted to classify into a sandy beach, but can be applied also to high clay content tailings.

Stability will also be dependent on allowing the finer tailings ("slimes") to desiccate and develop a significant undrained strength.

Figure 19.18 shows use of the upstream method for bauxite tailings at Weipa (Weipa 1 tailings in Figure 19.5).

19.5.2.2   Downstream method

Figures 19.19 and 19.20 show two versions of the downstream method of tailings embankment construction.

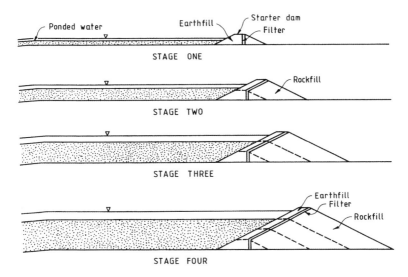

Figure 19.19.   Tailings embankment construction using the downstream method and zoned earth and rockfill construction.

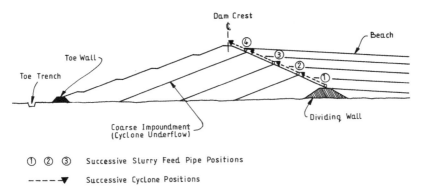

Figure 19.20.   Tailings embankment construction using the downstream method and cycloned tailings (ICOLD, 1982d).

The important features of this approach are:

– The embankment is constructed of selected material. In the case of Figure 19.19, a water storage dam type cross section has been used, with an upstream impervious zone and internal drainage control. This allows water to be stored adjacent to the embankment. In Figure 19.20 the embankment is constructed from the coarse underflow part of cycloned tailings. These may be compacted as described in Section 19.3.2. Avoidance of internal erosion and piping is dependent on the water pond being kept well away from the embankment to control the position of the phreatic surface;

– The embankment is best constructed, at least partly, of permeable material to allow control of piezometric pressures;

– It is desirable, but not necessary, to place the tailings in a controlled manner. For Figure 19.19 type construction no control is necessary (from a stability viewpoint). For Figure 19.20 type construction it would be important to spigot or cyclone the tailings

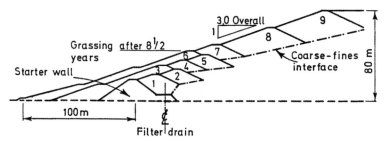

Figure 19.21.   Combined upstream and downstream construction (Lyell and Prakke, 1988, reproduced with permission of ASCE).

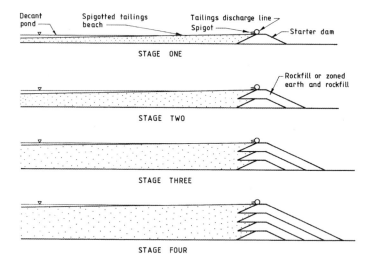

Figure 19.22.   Construction of tailings embankment using the centreline method.

to form a relatively sandy beach adjacent to the downstream zones to control piezometric pressures.

If rockfill rather than tailings underflow is used for the downstream zones in Figure 19.20, a graded filter zone will be required between the tailings and the rockfill to prevent erosion of the tailings into the rockfill which could lead to piping failure.

A combination of upstream and downstream methods may be used. An example is shown in Figure 19.21.

19.5.2.3   Centreline method

Figures 19.22 and 19.23 show two versions of construction by the centreline method.
The important features of this approach are:

– The embankment is partly constructed of selected material. In Figure 19.22 a zoned water storage type section has been used for the downstream part of the embankment, with the upstream part constructed on spigotted tailings. In Figure 19.23 cycloned tailings have been used, with the cyclone underflow discharged on the downstream side and the overflow finer portion on the upstream side;

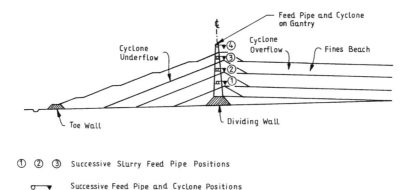

Figure 19.23.   Construction of tailings embankment using the centreline method and cycloned tailings (adapted from ICOLD, 1982d).

– The embankment must be constructed, at least in part, of permeable material to allow control of piezometric pressure;
– The water pond must be kept away from the edge of the embankment to prevent buildup of excessive piezometric pressures and to control potential internal erosion and piping;
– It is essential that tailings are placed in a controlled manner, i.e. spigotting for Figure 19.22, cycloning for Figure 19.23. For Figure 19.23 the tailings underflow would need to be compacted to ensure that it does not liquefy under earthquake.

19.5.3   *Construction using conventional water dams*

Conventional earthfill, earth and rockfill and concrete faced rockfill dams are often used to store tailings.

Some factors which may be considered as variations on conventional water dam engineering are:

– The tailings seepage water often has a high salt content, which suppresses the likelihood of dispersion of clays and may allow use of less rigidly controlled filters in low consequence of failure situations. Stapledon et al. (1978) give an example. This may allow use of a coarser reject from the milling operation as a filter, rather than using a processed sand/gravel, with resultant significant cost savings. However it must be recognized that this may lead to a greater likelihood of piping failure or a piping incident.
– Leakage of water from the dam may not be critical, allowing the use of higher permeability core material and no foundation grouting. If the water is contaminated, seepage collection downstream may be necessary.
– There is often a large quantity of waste rock available from the mining operation which can be used for construction at a low cost. In this situation it may be possible to have placement and compaction of the rockfill being done by mining equipment. If a well controlled earthfill zone with filters is required (i.e. sloping upstream core construction), this would usually be built by earthworks contractors after the rockfill, with well controlled construction. The rockfill should be compacted to limit settlements. The rockfill should be checked for the presence of pyrite which can lead to acid mine drainage problems.
– Erosion of the upstream slope is seldom a problem for tailings dams. At most, waste rock should be dumped on the upstream face. Steeper upstream slopes can be used if the slope is supported by tailings and not subject to drawdown.

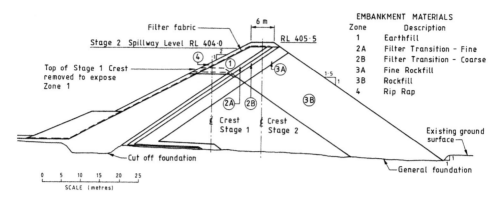

Figure 19.24.   Staged construction – proposed Ben Lomond tailings dam (Coffey and Partners, 1982).

– Staged construction is usually required, favouring sloping upstream core type cross sections.

Figure 19.24 shows an example of staged construction which minimises the volume of fill in each stage.

19.5.4   Selection of embankment construction method

The method to be used will depend on the particular circumstances at the mine, e.g. type of tailings, production rate, climate, local topography, seismicity, availability of waste rock, regulatory authority requirements etc.

All methods allow staged construction of the embankment. This is the significant difference between water storage dams and tailings embankments and allows minimisation of up-front capital works and improvement of overall economies.

Table 19.8, which is adapted from Vick (1983), compares the different methods of embankment construction.

One important factor in selection of embankment type is the degree of seepage control achievable. This is vital, because it determines the stability of the embankment and the steepness of the downstream slope. Figure 19.25 shows that downstream construction allows best control, followed by the centreline method, and upstream construction. Figure 19.26 shows that, by providing internal drainage layers control of seepage can be engineered. However this is only achieved at a cost penalty. Blanket drains may not be effective if the tailings are layered (as often occurs) because the water table perches on low permeability layers of tailing.

19.5.5   Control of seepage by tailings placement, blanket drains and underdrains

19.5.5.1   Tailings placement
Vick (1983) and Blight (1987, 1988) discuss the benefits of sorting the tailings particles in the beach zone. This sorting can lead to a gradation of permeability from high near the discharge point decreasing towards the pond and hence a lower phreatic surface. Figure 19.27 shows this effect schematically and the effect of having the water pond near the embankment crest and the underdrainage effect of a high permeability soil or rock foundation.

Stauffer and Obermeyer (1988) discuss a case where piezometric pressures were lower than might otherwise have been expected, because of the underdrainage effect of a relatively high permeability foundation.

Table 19.8. Comparison of tailings embankment types (adapted from Vick, 1983).

Embankment type	Mill tailings requirements	Discharge requirements	Water storage suitability	Seismic resistance	Raising rate restrictions	Embankment fill requirements	Relative embankment cost
Water retention	Suitable for any type of tailings	Any discharge procedure suitable	Good	Good	Entire embankment constructed initially or in stages	Earthfill, rockfill (mine waste, filters)	High
Upstream	At least 40–60% sand in whole tailings. Low discharge solids content desirable to promote grain-size segregation	Peripheral spigotted discharge and well controlled beach necessary	Not suitable for significant water storage	Poor in high seismic areas	Less than 5–10 m/yr most desirable	Natural soil, sand tailings, or mine waste	Low
Downstream	Suitable for any type of tailings	Varies according to design details	Good	Good	None. Often built in stages if rockfill is used	Sand tailings or mine waste if production rates are sufficient, or earthfill, rockfill	High
Centreline	Sands or low-plasticity slimes	Peripheral spigotted discharge of at least nominal beach necessary	Not recommended for permanent storage. Temporary flood storage acceptable with proper design details	Acceptable	Height restrictions for individual raises may apply	Sand tailings or mine waste if production rates are sufficient, or earthfill, rockfill.	Moderate

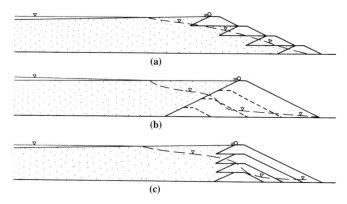

Figure 19.25.   Internal seepage: (a) upstream embankment; (b) downstream embankment; (c) centreline embankment.

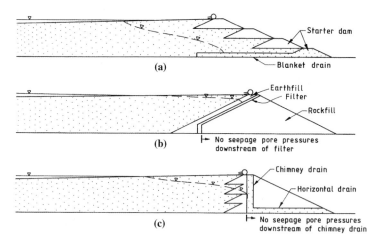

Figure 19.26.   Use of internal drainage zones in embankments: (a) upstream embankment using starter dam with upstream blanket drain; (b) downstream embankment; (c) centreline embankment with vertical chimney drain.

In many designs, drainage zones will be incorporated into the design, within the embankment, as shown in Figure 19.28.

19.5.5.2   *Drainage blankets and underdrains*

Drainage blankets as shown in Figure 19.26(a) and Figure 19.30 and underdrains as shown in Figures 19.28 and 19.29, are sometimes used to drawdown the phreatic surface and to collect contaminated seepage into a seepage collection system.

These drains have in the past often been built with mine waste, tailings or borrow pit run materials, without due regard for filter design between the tailings and the drain or sufficient discharge capacity and they have not functioned as desired. In any case, where the tailings are layered, with slimes or finer tailings between the sandy tailings, the water will perch on these low permeability strata and the drains will not be very effective.

The other point to remember, is that the foundation is often more permeable than the tailings, so the seepage may be vertically downwards without the underdrains, in which case they provide no benefit in reducing pore pressures.

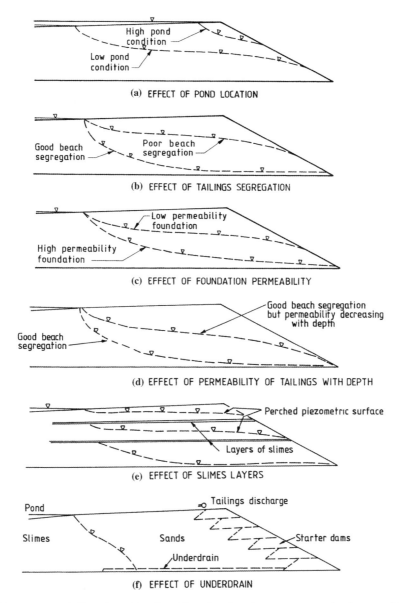

Figure 19.27.   Factors influencing phreatic surface location for upstream embankments: (a) effect of pond water location; (b) effect of beach grain-size segregation and lateral permeability variation; (c) effect of foundation permeability; (d) effect of decreasing tailings permeability with depth; (e) effect of slimes layers; (f) effect of underdrains if there are no layers of slimes. Starter dams not shown in (a) to (e).

19.5.6   Storage layout

Some common storage layouts are shown in Figures 19.31 and 19.32.
    Some features of these arrangements are:

Ring dike
– suitable for flat terrain;
– no runoff from external catchments;

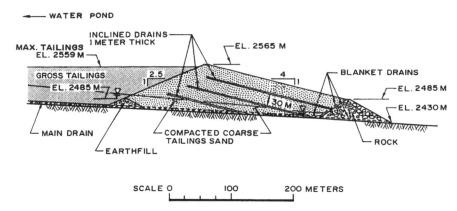

Figure 19.28.   Embankment drainage in Perez Caldera No. 2 tailings dam (Griffin, 1990).

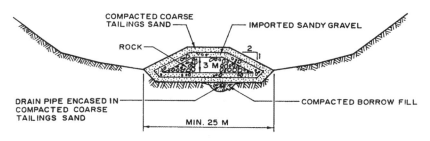

Figure 19.29.   Perez Caldera No. 2, tailings dam underdrains (Griffin, 1990).

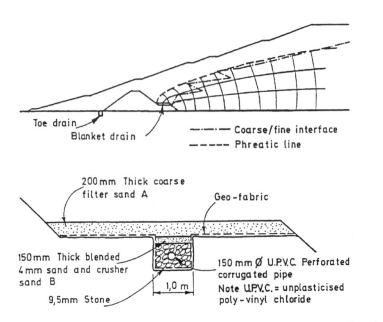

Figure 19.30.   Blanket drain (for dam in Figure 19.22, Lyell and Prakke, 1988, reproduced with permission of ASCE).

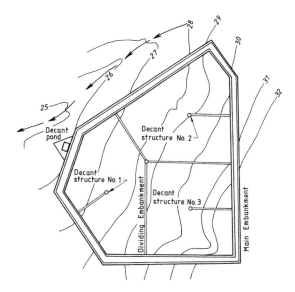

Figure 19.31.   Ring dike or "turkey's nest" configuration as used in Paddy's Flat tailings storage (adapted from Cooper, 1988).

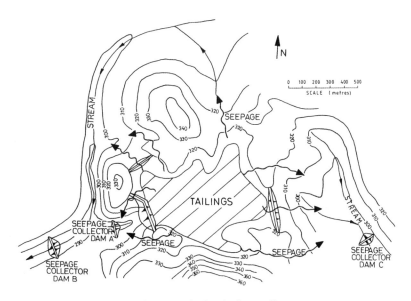

Figure 19.32.   Cross-valley impoundment at the head of a small stream.

– can be staged in plan;
– allows cycling of tailings disposal into separate cells.

Cross valley
– should be sited at head of valley to limit external catchment;
– usually involves less embankment material than ring dike;
– can be staged either in height or by building multiple impoundments.

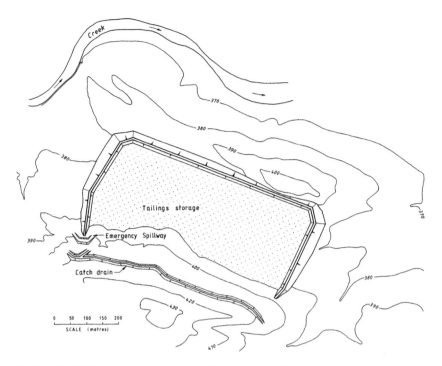

Figure 19.33.   Sidehill impoundment as proposed for Ben Lomond tailings dam (Coffey and Partners, 1982).

Side hill
– suitable only where hill slopes are relatively flat – say ≯10%. Otherwise embankment fill volumes become excessive.

In many cases it is desirable to limit the flow into the storage from external catchments using catch drains such as those shown in Figure 19.33.

19.5.7   *Other disposal methods*

There are other less commonly used methods of tailings disposal. These include:

19.5.7.1   *Thickened discharge or Robinsky method*

In this method tailings are thickened to a higher solids content than would normally be used, i.e. around 60% solids content compared to 30% to 40% for normal tailings operation. At this solids content the tailings can be deposited in a cone shaped deposit as shown in Figure 19.34.

According to Robinsky (1979), the final cone slope is ideally around 6° to limit erosion. The concept is that the greater slope allows storage of larger quantities of tailings. It is claimed that this results in overall reduction in costs, as the saving in embankment costs more than offsets the costs of thickening.

Ritcey (1989) gives details of two mines where the method has been used. Blight and Bentel (1983) discuss the use of thickened tailings to increase the slope of tailings on a conventional dam to increase the storage capacity. They also give a method for estimating the stable slope of the tailings based on the yield point stress of the slurry.

It is a relatively uncommon method, partly because it requires a very large area, and it is difficult to control exactly where the tailings deposit and the toe of the area is for run-off

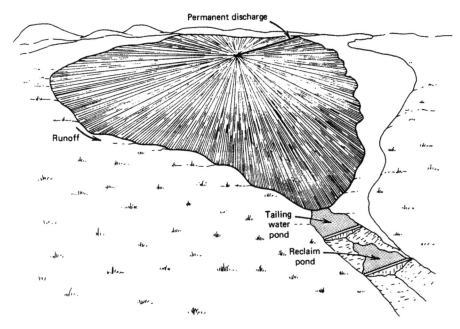

Figure 19.34.   Thickened discharge disposal method (diagram from Vick, 1983, reprinted from Robinsky, 1979).

collection. It has been used successfully in at least one mine in Australia but in others it has been less successful.

Vick (1983) and Ritcey (1989) cite studies which have shown that some tailings can liquefy and flow at a slope of 6° and Vick cautions the use of the method in seismic areas. With advances in thickening technology some of the authors' reservation about the method are lessened. We have some concerns about surface erosion in other than arid areas and note that the method is not widely used, presumably on economic grounds.

19.5.7.2   Co-disposal

Co-disposal of the coarse waste ("coarse reject") and tailings from coal washeries involves combining the waste streams and pumping them together to the storage facility. Williams and Gowan (1994) suggest there can be significant cost savings compared to the separate disposal of the coarse reject and the tailings, and cites several Australian coal mines as using the techniques.

They indicate that the ratio of coarse to fine reject (by dry weight) ranges between 1.5 to 1 and 5.5 to 1, averaging about 3.5 to 1.

There is some breakdown of the coarse reject during pumping and further breakdown by chemical means or weathering (slaking) on the surface of the disposal area.

The combined materials are pumped at a solids content of 20% to 30%. The advantages claimed for co-disposal are (Williams and Gowan 1994):

Advantages claimed:
– May eliminate the need to build engineered tailings dams;
– Improved stability (shear strength) and the potential to use upstream construction;
– Reduced need for trucks to dispose coarse reject;
– Reduced cost compared to the cost of the two separate waste streams;

- Rapid and enhanced dewatering with up to 40% more water recovery;
- Denser fill, with potentially up to 40% storage volume saved (because the tailings fill the voids in the coarse reject;
- Ready rehabilitation because of the denser, stronger fill, which can be trafficked soon after placement;
- Acid generation and spontaneous combustion of the coarse reject is less likely.

Disadvantages claimed:
- Blockage of pipelines;
- Pipe wear and replacement costs;
- Segregation of fines to form a wet tailings beach at the bottom of the co-disposal beach.

It would seem to the authors that co-disposal could be attractive where there is limited space for more conventional tailings storages and/or where acid generation and spontaneous combustion from the coarse reject is a problem.

However to date it seems it has been widely trialled but seldom used for on-going operations.

19.5.7.3   *Paste disposal*

Paste disposal involves dewatering the tailings to a solids content of 70–85% (water content 30–15%) using high efficiency cone thickeners or conventional thickening followed by partial vacuum filtration and paste mixing.

It falls between thickened tailings and belt filtration (see Section 19.5.7.4) in the amount of dewatering. For successful application, the tailings must have at least 15% finer than 0.020 mm so the water does not drain. The paste is often used as a underground mine backfill with cement added to increase the strength.

The paste behaves as a Bingham fluid and flow is initiated only after the pressure in the pipeline is sufficient to overcome the initial shear stress. The pressures to pump the paste are high (Brackenbush and Shillakeer, 1998, quote 2.25 kPa/m). Pumps are generally positive displacement, hydraulically operated. If it is stacked, its angle of repose is three to ten degrees. Claimed advantages and disadvantages are:

Advantages:
- Most process liquor is recovered by the thickening/filtration process, so less is likely to be lost to evaporation;
- Low permeability and capillary action limit acid generation in sulphidic tailings;
- Reduced groundwater pollution;
- The potential to eliminate containing "dykes" or embankments;
- Possible elimination of liners because of the low permeability;
- Useful for underground mine backfill, particularly if cement is added.

Disadvantages:
- High capital and operating costs for dewatering of the tailings to paste consistency;
- Problems with performance of thickeners to give paste consistency;
- High pressures limit the distances paste can be pumped and costs are high.

Overall it is hard to see paste technology being attractive unless the paste is required for backfill in underground mines. Most of the advantages are available with thickened tailings (i.e. to say 60% solids content).

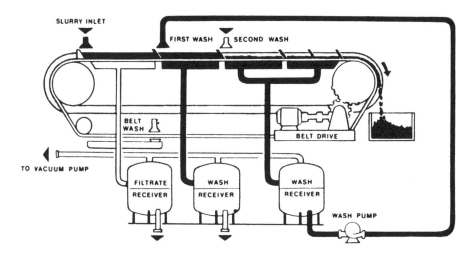

Figure 19.35.   Principle of belt filtration (Willis, 1984).

19.5.7.4   Belt filtration

Belt filters are used for dewatering of some tailings to such a consistency that they can be carried by truck for disposal in mine overburden dumps, either separately, or mixed with say coarse rejects. Figure 19.35 shows the principle of belt filtration.

The tailings are spread on a filter cloth supported on a drainage deck and are dewatered by gravity and the application of a vacuum to the underside of the drainage deck. The process is often aided by large doses of flocculants.

In general, the capital and operating costs of such operations are high and the success depends on having a uniform quality of tailings. To the authors' knowledge it has only been used where space or environmental constraints preclude other methods.

19.5.7.5   Disposal into open cut and underground mine workings

An apparently attractive method of tailings disposal is to discharge the tailings into old open cut or underground mine workings. This presents a low cost method of disposal. However, there are several potential problems. In most open pits the area of the pit is relatively small and results in a subaqueous discharge environment with resultant low densities. This in turn leads to high compressibility and low strength. When the area is no longer in use, settlement occurs, leaving a water pond. Settlement can total several metres and take tens of years to complete. It may take some years before the tailings dry sufficiently on the surface to allow rehabilitation. Tailings discharged into underground mine workings will only settle to low densities and very low strength and will be susceptible to flowing out of the workings if for example another seam or ore body is removed from below the workings.

19.5.7.6   Discharge into rivers or the sea

There are some notable examples of this e.g. Bougainville Copper and OK Tedi Mining in Papua New Guinea and the Rosebery Mine, Tasmania. In the latter case, tailings discharge has resulted in devastation of the river due to the chemical content of the tailings leachate water. In Bougainville and in the lower reaches of the OK Tedi River, considerable sedimentation has occurred resulting in encroachment of the river into adjacent land. As a result of these types of effects, generally speaking, disposal into a river is environmentally unacceptable. The issues are discussed in Vick (1983) and Ritcey (1989).

19.6   SEEPAGE FROM TAILINGS DAMS AND ITS CONTROL

19.6.1   *General*

As discussed in Section 19.2, many mine and industrial tailings have accompanying water or "liquor" which contains dissolved salts, heavy metals and other residual chemicals from the mineralogical processes. If this liquor escapes to the surrounding surface water and groundwater in sufficient quantities, it can lead to unacceptable concentrations, making the water unusable for drinking and affecting aquatic life. Therefore, there is often an emphasis in the engineering of tailings dams on the estimation of seepage rates and, where these prove unacceptable, to the provision of measures to reduce seepage.

Seepage cutoff measures can be very costly and are often not as effective as the proponents would expect. They can affect the economic viability of a mining project and certainly the profitability of the operation.

This section presents an overview of the measures which are available and their effectiveness in controlling seepage. In many cases it is more realistic to accept that tailings dams will seep regardless of the measures adopted and to design collection and/or dilution of the seepage to acceptable concentrations.

For a more detailed discussion on the topic readers are referred to the chapter by Highland in Vick (1983). Ritcey (1989) and Fell, Miller and de Ambrosis (1993) discuss chemical and geochemical aspects.

19.6.2   *Principles of seepage flow and estimation*

Many tailings storages (or "dams") will be constructed on relatively flat land with a deep existing groundwater table. This situation is discussed by Vick (1983) and Figures 19.36 and 19.37 are reproduced from his book.

The following should be noted:

– The rate of seepage flow will be dependent on the permeability of the tailings, the underlying soil and rock, climate, pond operation etc.;
– Contaminants in the seepage water will not all join the groundwater. Much will be adsorbed in the foundation soil and rock. Hence, contaminant load does not equal seepage flow rate × contaminant concentration in the storage;
– Further reduction of contaminant concentration may occur in mixing with stream flows;
– It is contaminant concentration in ground and surface water which is generally critical, not the total quantity. Hence adsorption, dispersion and dilution can result in acceptable water quality in streams or well points, even though the original contaminant levels in the storage may have been unacceptable.

It is important to realise that, in many cases, a partially saturated flow condition will exist in the foundation at least at the start of operations, and possibly on a permanent basis if the tailings permeability is low compared to the foundation permeability. Figure 19.37 shows an example of the stages in the development of seepage.

Note that flow in the tailings in Stages 1 and 2 will be essentially vertical, will not emerge at the toe of the embankment, will be virtually unaffected by any foundation treatment such as grouting and will not be intercepted by drains at the toe of the dam.

It may take years for the seepage mound to rise to connect to the tailings (Stage 3) or it may never happen.

Figure 19.32 shows an example of a tailings storage which has been constructed at the head of a valley. In this case the final development will consist of several embankments. When

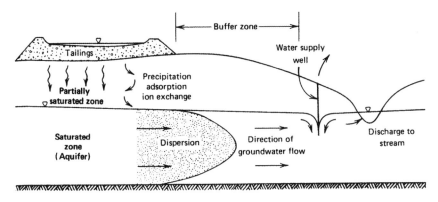

Figure 19.36.   Groundwater flow and contaminant transport processes (Vick, 1983).

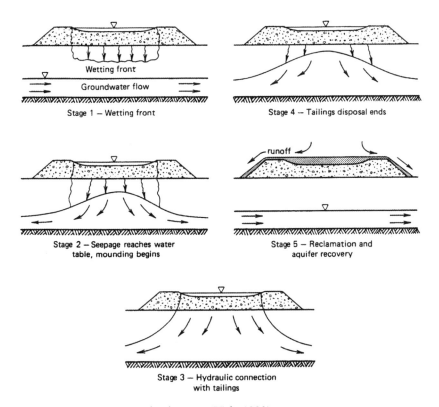

Figure 19.37.   Stages in seepage development (Vick, 1983).

estimating seepage from such a storage it is important to remember that seepage will occur under each of the embankments and depending on the base groundwater levels, into the hillsides adjacent to the embankments. Hence, in Figure 19.32, seepage will occur to the west, north and east but not to the south where natural groundwater levels are higher than the storage. Note that the groundwater does not always mirror the topography and may be affected by local variations in geology, e.g. permeable dykes.

It is the authors' experience that inexperienced engineers and geologists will either forget completely that seepage will occur in all directions from the storage or at least apply an excessive amount of the site investigation effort and analysis, to the seepage which will flow through and beneath the main embankment.

From the examples shown in Figures 19.36, 19.37 and 19.32, it will be apparent that the assessment of seepage flow rates will involve:

– Knowledge of the permeability of the tailings, as these are commonly part of the seepage path. In many cases they may control the seepage rates;
– Knowledge of the permeability of the soil and rock underlying the storage and surrounding the storage. In Figure 19.32 it would be necessary to be able to model the whole of the area between the streams, necessitating knowledge of rock permeabilities well beyond the storage area;
– Modelling of the seepage, usually by finite element methods, which may involve several section models and/or a plan model. This modelling should account for the development of flow as shown in Figure 19.37 and not just model an assumed steady state coupled flow situation, i.e. the storage and groundwater coupled as in Stage 3, Figure 19.37.

19.6.3   *Some common errors in seepage analysis*

Finite element seepage models are readily available and commonly used to estimate seepage from tailings storages. It is the authors' experience that the users of such programs sometimes fail to understand the actual boundary conditions which will apply or ignore details which are important to estimates. Some examples of errors follow:

The examples assume that the tailings have lower permeability than the underlying rock:

(a) Assuming saturated coupled flow, as shown in Figure 19.38b. This results in an underestimation of the rate of seepage compared to the correct conditions shown in Figure 19.38(a);
(b) Assuming saturated flow as shown in Figure 19.38b, with the water table at ground surface downstream, when flow is insufficient to result in such a high water table downstream (i.e. Figure 19.38c is more correct). This results in an underestimation of seepage rates but, more importantly, seepage will not emerge at the downstream toe of the dam, so will not be able to be collected or monitored at the surface;
(c) Failing to model and account for seepage through hills surrounding the dam, e.g. ignoring seepage to the north and east in Figure 19.32. This can result in significant underestimation of seepage rates;
(d) Assigning incorrect boundary conditions to the model as shown in Figure 19.39b. This results in underestimation of seepage rates, and may lead to an overestimation of the amount of seepage which may emerge at the toe of the dam;
(e) Failure to model permeable zones in contact with surface water, e.g. rock rip-rap zones on the upstream slope as in Figure 19.40 and high permeability alluvial, colluvial or lateritised soil zone on the surface as shown in Figure 19.41a. In both cases, failure to model the high permeability zone can lead to significant underestimation of the seepage rate;
(f) Use of incorrect permeability in the foundation e.g.:
   – Using the results of flow-in type permeability tests in soil and weathered rock, where smearing and/or blocking of fissures and joints yields lower than actual permeabilities;
   – Use of permeability values obtained by methods which are unable to detect major leakage paths, e.g. from boreholes in deeply lateritised areas where drainage features are often localized and near vertical;
   – Failing to allow for the variability of tailings permeability – as discussed above.
(g) Incorrect estimation of travel times for contaminant fronts, e.g.:
   – Use of porosity instead of storage coefficient in estimating flow velocities;
   – Use of unrealistic values (too high) of storage coefficients in jointed rocks;

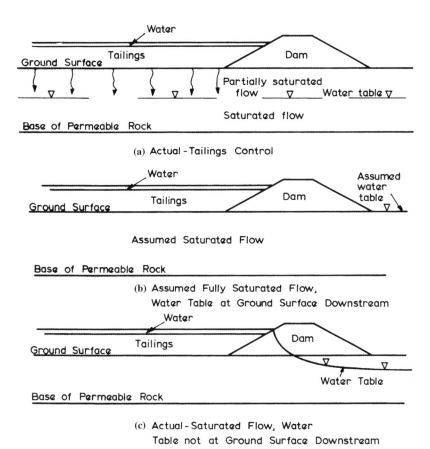

Figure 19.38.   Analysis of seepage under a tailings dam.

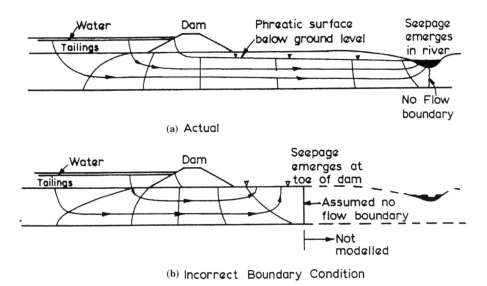

Figure 19.39.   Incorrect boundary conditions for analysis of seepage under a dam.

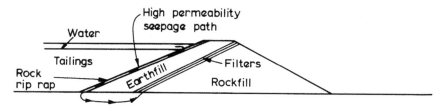

Figure 19.40.   Seepage through rock rip-rap bypasses low permeability tailings.

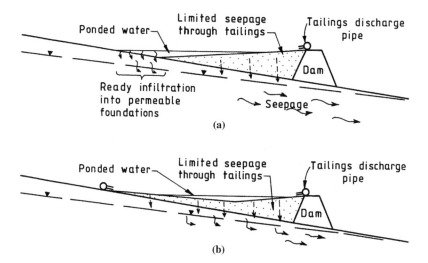

Figure 19.41.   Controlled placement of tailings: (a) tailings not covering foundation; (b) tailings covering foundation.

- Failure to allow for preferred seepage paths, e.g. high permeability, jointed or closely jointed zones;
- Failure to consider adsorption, dispersion etc.

19.6.4   Seepage control measures

Measures to control seepage from tailings storages include:

- controlled placement of the tailings;
- foundation grouting;
- foundation cutoffs;
- clay liners;
- underdrains and toe drains.

19.6.4.1   Controlled placement of tailings

In many cases, the most cost effective way of controlling seepage will be to place the tailings so that the base of the storage is blanketed. Figure 19.41 shows this effect with the tailings forming an effective blanket in (b) but not in (a) where water is in direct contact with the foundation.

In theory, provided the tailings are of low permeability, they will form as effective a liner to the storage as can be achieved by a compacted clay liner. Vick (1983), for example, shows that tailings with a permeability of 10^{-8} m/sec are as effective as a 2 foot (0.6 m) thick clay liner with a permeability of 10^{-9} m/sec.

It should be remembered that rock in the upper 10 m to 30 m of most foundations has a permeability between 1 Lugeon and 20 Lugeons, or 10^{-7} m/sec to 2×10^{-6} m/sec. Most naturally occurring soils will have a similar or higher permeability. Since tailings slimes (even from non oxidised ore) are likely to have a vertical permeability of less than 10^{-7} m/sec, the tailings will often be less permeable than the underlying soil and rock. If the tailings are from oxidised ore or from washeries (such as coal, bauxite, iron ore) they are likely to have a permeability of the order of 10^{-7} m/sec to 10^{-9} m/sec or less. Clearly, in these cases, covering the storage with the tailings will be an inexpensive and effective way of limiting seepage.

The effectiveness of the tailings as a "liner" is dependent on placement methods. If tailings are placed subaerially and allowed to desiccate, lower permeabilities will result from the drying, provided cracking does not occur. If placed subaqueously lower densities and higher permeabilities are likely to result.

A potential difficulty with using tailings as a liner, is that the coarser fraction of the tailings tend to settle out more quickly than the fine (or slimes) fraction. Hence a "beach" of sandy tailings often occurs near the discharge point and if water is allowed to cover this area subsequently it can allow local high seepage rates. This can be overcome by using thickened discharge which inhibits segregation, by shifting the tailings discharge points from one end of the storage to the other, placing slimes under the beach area and/or by using a liner or seepage collector system under the sandy area.

Seepage can also occur along the contact between tailings and embankment if rock rip-rap is used (Figure 19.40). Another problem is that it can be difficult to avoid water ponding against the storage foundation, particularly early in the storage operating life.

19.6.4.2 *Foundation grouting*

As discussed in Chapter 18, grouting is not particularly effective in reducing seepage, except in high permeability rock.

In a project on which the authors were involved, the grouting of a 5 km long dam foundation to a depth of about 25 m on average, would have reduced the estimated seepage by only 1%; nearly all of this in a relatively small portion of the foundation affected by faulting and with an ungrouted permeability of the order of 100 Lugeons.

It will be seen from the above discussion, that it is unlikely that grouting of tailings dam foundations can be justified on the grounds of reducing seepage. It may be justified on other grounds, such as reducing potential erosion in weathered rock, or where the high permeability zones can be identified from geological information and only these zones are grouted. It is usually not possible to do this.

19.6.4.3 *Foundation cutoffs*

For tailings storages constructed on soil foundations, particularly sand or sand and gravel, a significant reduction in seepage may be achieved by construction of an earthfill cutoff or a slurry trench cutoff wall as discussed in Chapter 10.

These cutoffs are of high cost and applicable only in critical situations and where ground conditions allow, i.e. generally soil. They may be applied to extremely weathered rock, e.g. lateritised highly permeable weathered rock.

19.6.4.4 *Clay liners*

Clay liners can be an effective way of reducing the seepage from a tailings storage. For example, if a tailings storage is located in a highly permeable area over sand, and the tailings are moderately high permeability, a clay liner may well be appropriate.

There are some practical aspects which should be considered in the provision of clay liners:

- The permeability of the clay depends on the soil available. For many naturally occurring soils the compacted permeability across the liner will be of the order of 10^{-8} m/sec to 10^{-9} m/sec. In the authors' experience, few soils have permeability as low as 10^{-10} m/sec.
- The permeability is affected by the compaction water content and density. To achieve a low permeability, the soil should be compacted to a density ratio of 98% of standard maximum dry density, at a water content of between −1% of optimum to +2% of optimum. The permeability can be increased by an order of magnitude by compacting dry of optimum – see Lambe and Whitman (1981).
- The thickness of the clay liner may not be particularly critical, depending on the particular circumstances. It is better to have a relatively thin (say 0.6 m) high quality layer (i.e. good selection, good compaction control) than a thicker, less controlled layer.
- The clay liner is susceptible to cracking on exposure to the sun which can increase its permeability by orders of magnitude. This is particularly critical on sloping sites, where the liner may not be covered by tailings for months after construction. On flat sites, covering the liner with, say, 150 mm of clean sand or silty sand can act to prevent drying of the clay. On sloping sites the sand cover may be eroded by rainfall and may need to be held in place with a geotextile. As shown by Kleppe and Olson (1985), once cracked, the liner permeability will remain high, even if the cracks are apparently closed by swelling on re-wetting.
- The permeability of clay liners was questioned in the early 1980s, when some researchers found that the permeability was increased by some orders of magnitude when particular organic leachates were passed through the clay. Later research has shown this is not a problem for water containing inorganic chemicals, even if at high concentrations. Since most tailings liquor would not contain organic leachate, the permeability measured in the laboratory should be a reasonable guide to its long term behaviour.
- If the foundation has openwork gravel or wide open joints, clay liners can be subject to "sinkhole" development, i.e. erosion of the liner into the underlying foundation.
- The clay liner is quite expensive. A 0.6 m thick liner could be expected to cost of the order of $10–15/square metre, assuming there was a ready source of clay fill available. Protection against drying and erosion, plus surface preparation to give a smooth contour, could be expected to add to this cost giving an overall cost of say $20 to $25 per square metre. Because large areas are often involved, the costs can be very large, e.g. a 1 km × 1 km area would cost about $20 to $25 million to line with clay.
- The clay liner should cover the whole of the tailings storage. Use of a liner as an "upstream blanket" is unlikely to result in significant reduction in seepage quantities.
- There will be significant seepage through a clay liner. For example, a 0.6 m thick clay liner with permeability of 10^{-9} m/sec would discharge 90 m³/day over an area of 1 km square under unit gradient.
- The naturally occurring clays in a storage area are unlikely to have a low permeability unless they are excavated and recompacted. *In situ* they are likely to have a permeability of the order of 10^{-5} m/sec to 10^{-6} m/sec due to the presence of root holes, fissures etc.

Unfortunately some regulatory authorities have unrealistic views in regards to clay liners for tailings dams and may request provision of a liner, jeopardizing the viability of a project, when the geochemistry is such that seepage of an unlined storage is not an environmental hazard.

19.6.4.5 *Underdrains*

Drains may be provided under the tailings as shown in Figures 19.28, 19.29 and 19.42, with or without a clay liner. The drains act to attract the seepage water and discharge it to a collector system, often for recycling to the process plant.

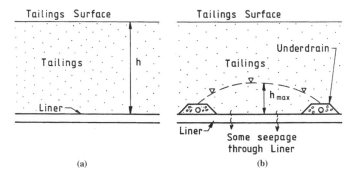

Figure 19.42.   Comparison of head acting on clay liner, (a) without underdrain; (b) with underdrain.

As shown in Figure 19.42 the underdrains may reduce the head on the liner and reduce the seepage. However, the amount of seepage collected and bypassing the collector drains is dependent on many factors, as follows:

– The spacing of the drains;
– The vertical and horizontal permeability of the tailings;
– The permeability of the liner. Note that there is a nett head on the liner between the drains and, even within the drain, water flowing on the liner will percolate through the liner. If the groundwater level is below the liner, the hydraulic gradient through the liner will still be at least 1, giving significant seepage;
– The efficiency of removal of water from the drain system.

If no liner has been provided, and the underdrains are laid on the natural ground, there will still be significant seepage into the ground because while the drain has a permeability of say 10^{-5} m/sec the ground will have a significant permeability, e.g. 10^{-6} m/sec to 10^{-7} m/sec (if normal weathered rock).

Apart from these design problems, there are some practical aspects:

– The underdrains must be designed to act as filters (as detailed in Chapter 9) to the tailings, or they will rapidly become blocked. Tailings are particularly erodible and will readily clog a poorly designed filter;
– Geotextiles have been used to construct the drains. However, as outlined in Scheurenberg (1982) and Bentel et al. (1982), the geotextile should not be exposed directly to the tailings or it will clog. They overcame this problem by covering the geotextile with a layer of filter sand. There is however a chance that the geotextile will still clog with deposited oxides from the seepage water and their use is not recommended;
– The underdrains are susceptible to contamination by dry windblown tailings, before they are covered with tailings;
– The seepage water collected in the drains has to be collected and pumped to storage for use in the process plant. This involves expensive collection and pump systems which have to be maintained;
– The filters and outlet pipes can be clogged by oxidation products from the tailings. Figure 19.43 shows an outlet system used by Lyell and Prakke (1988) to overcome this problem.

19.6.4.6   *Synthetic liners*
Synthetic liners are used to line many hazardous waste facilities, often with the provision of drainage layers beneath the first membrane, with a second membrane or clay liner to

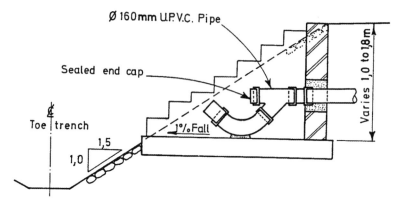

Figure 19.43. Outlet detail for drain to prevent clogging by oxidation (Lyell and Prakke, 1988, reproduced with permission of ASCE).

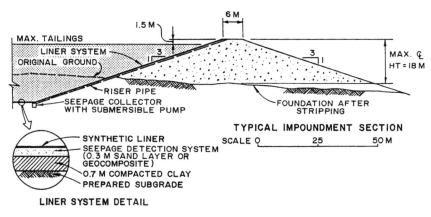

Figure 19.44. Proposed uranium tailings storage (Griffin, 1990).

collect any leachate which leaks past the first. Figure 19.44 shows such a system for a proposed uranium tailings storage in New Mexico, USA.

However, these liners are expensive and are not appropriate for most tailings disposal situations. There are notable exceptions, e.g. synthetic liners have been used for tailings storages in the Alcoa Alumina Refinery at Kwinana, where the storages are underlain by sands which are used as an aquifer for water supply. In many applications on sloping sites they would be impracticable, because of the need to provide a well graded base, free of irregularities which may penetrate the liner.

It is understood that permeabilities of the order of 10^{-12} m/sec to 10^{-14} m/sec are applicable. A 5 mm thick liner with permeability of 10^{-13} m/sec and 15 m of water head on it would give a leakage rate of 25 m^3/day per square kilometre, i.e. significant leakage will occur. A common problem is susceptibility to deterioration in sunlight, requiring the liner to be covered.

19.6.5 Seepage collection and dilution measures

From the above discussion, it will be apparent that no matter what measures are adopted there is going to be some seepage from tailings storages. This seepage will probably be greatest during operation of the storage, particularly early in the operation when there is little blanketing effect from tailings. It will be exacerbated if water is allowed to pond over the tailings and adjacent to the natural ground. Seepage will continue after shutdown, even when the tailings are covered, because infiltration of rainfall will occur. Granted this is the case, it will often be more practic-able to limit expenditure on seepage "control" (or "prevention") and to concentrate on design measures to collect the seepage. This may then be pumped back to the storage or the process plant, or diluted with surface runoff before release at acceptable concentrations. Measures which may be taken include:

19.6.5.1   Toe drains

A drain may be provided at the downstream toe of the embankment, to collect seepage which emerges at that location.

These drains can be reasonably successful in intercepting seepage, but only if the seepage naturally emerges in this location. In many cases, the flow rates will be such that the phreatic surface stays below the level of the drain. Even when the seepage is sufficient to raise the phreatic surface to flow to the drain, much may still bypass by flowing beneath the drain. Ideally the drain has to penetrate to a low permeable stratum, but this is often not practicable.

Such drains may also intercept surface runoff from the downstream face of the dam and groundwater from downstream (if the water table is high) and, if the seepage is to be returned to the dam or process plant, may exacerbate water management problems if a "no release" system is being operated.

19.6.5.2   Pump wells

Seepage can be collected by constructing water wells into pervious strata downstream of the tailings embankment and pumping from these back into the storage or to the process plant.

Such a well system can be reasonably successful in intercepting seepage but there are some disadvantages:

- The pumps lower the piezometric pressures downstream of the storage, so gradients and seepage rates from the storage may be increased;
- The wells also attract water from downstream and so may also add to water management problems in no release operations;
- The wells have to be pumped continuously to be effective, with all the associated costs;
- The well screens and pumps are susceptible to corrosion and blockage and require maintenance and periodic replacement;
- It is unlikely that it is practicable to operate the wells after shutdown of the storage, so another method may be needed to handle long term seepage.

19.6.5.3   Seepage collection and dilution dams

In many cases, a practical way of collecting seepage from tailings storages will be to construct a seepage collector dam or dams. Figure 19.32 shows such a system.

The seepage collector dams may be located sufficiently close to the storage to collect the bulk of seepage, but not too far away so as to limit the external catchment, e.g. Dam A on Figure 19.32. In this case water would normally be pumped back into the dam or to the process system.

Alternatively, one may deliberately locate the collector dam sufficiently far downstream to ensure that the runoff from the catchment to the dam is sufficient to dilute the seepage to acceptable water quality, e.g. Dam B in Figure 19.32.

Whether such an approach is acceptable will depend on the particular circumstances for the tailings storage. For example, it may be unacceptable to have substandard water quality in the stream between Dam B and the tailings storage. Seasonal effects can also be important, e.g. if there is a prolonged dry season, water may pond in the stream and concentration of contaminants may occur.

The authors' view is that, in many cases, the catch dam with pump-back or dilution may be far more appropriate than expensive measures to control the seepage. From the authors' experience, too many engineers and regulators have an over optimistic view of the efficacy of these seepage control measures, or an unrealistic view of the costs a mining operation can reasonably bear to construct such measures.

19.6.6   Rehabilitation

After operation, tailings storages must be designed to contain the tailings indefinitely and minimise long term impact on the environment. Factors are:

– long term stability of the "dam";
– long term erosion;
– long term effect on groundwater and surface water;
– return of area to productive use.

19.6.6.1   Long term stability and settlement

Provided water is drained off the storage, stability and settlement should not be a problem because seepage piezometric pressures are reduced. It is important to ensure that storm runoff cannot overtop the "dam", or erosion from creeks etc. cannot erode the toe of the "dam".

Ponding of water on the tailings storage is likely to occur if tailings have been deposited subaqueously with resulting low settled density and high compressibility. Figure 19.45 shows the sort of problem which can arise.

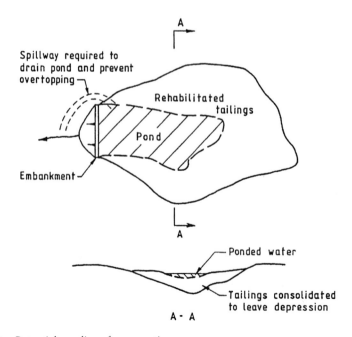

Figure 19.45.   Potential ponding after operation ceases.

In many cases the presence of a water pond does not create difficulties, but if the water is contaminated it may be unacceptable. In this event it will be necessary to construct a spillway as shown or to fill the depression with waste rock, more tailings etc. However this filling will in itself induce additional settlement. Mounding of the tailings and cover prior to shutdown can help alleviate the problem. If tailings are placed subaerially and well desiccated, they will have dried to a high strength, low compressibility landfill and, provided the surface is adequately contoured, water ponding will not be a problem.

19.6.6.2   *Erosion control*

This can be a major issue. Wind erosion, as well as water erosion, has to be considered. The embankment side slopes are best covered by rock if this is available, otherwise flat (flatter than say 3H to 1V) slopes with good vegetation are required. This is often impossible to achieve on a year round basis.

The tailings surface may be covered with soil, waste rock or a combination of the two and vegetated to control erosion and limit infiltration. Some tailings can be successfully revegetated without the need for cover. Cement stabilization has also been used. Figure 19.47 shows some examples.

It is often necessary to carry out trials during operation to determine what measures will be successful.

Blight (1988) and Blight and Caldwell (1984) describe measures taken to alleviate erosion from abandoned gold tailings dams. Ritcey (1989) also discusses the design of cover. Forrest et al. (1990) and Corless and Glenister (1990) also discuss erosion control measures.

19.6.6.3   *Seepage control*

In many cases the deposited tailings will have contaminants trapped in the accompanying water when operations cease. These contaminants will continue to seep from the tailings as they consolidate, and the water will infiltrate through the cover leaching contaminants as it passes through the tailings.

As was the case during operation, this may or may not be a problem, depending on the concentration of contaminants that reaches groundwater wells or streams in the vicinity. If the contaminants are likely to be a problem, measures will have to be taken to limit infiltration. This can be achieved by redirecting external catchment flows (see Figure 19.33), contouring the surface of the tailings to encourage runoff, avoiding ponding of water as shown in Figure 19.45, and encouraging transpiration by planting vegetation on the tailings.

It is unrealistic to consider that a "clay cover" can be provided which will "seal" the tailings. In the first place, the clay will have a finite permeability even if well compacted and, in any case, the clay may be difficult to compact, because the tailings do not form a strong base. The permeability of the cover will be increased by cracking due to desiccation and settlement and penetration by roots and animals.

It is more realistic to design the total system on the assumption that there will be long term seepage, and if the resultant contaminant concentrations are too high, to design seepage collector dams to allow dilution to acceptable concentrations prior to release.

Figure 19.46 shows the components involved in assessing the infiltration into tailings. In this diagram P = precipitation (rainfall); E = evapotranspiration; R = runoff; I = infiltration; SW = soil water stored in the "cover" and $I = P-E-R-\Delta(SW)$.

Figure 19.47 shows some alternative "covers" which may be used. These are in increasing degree of sophistication and increasing cost from (a) to (d). It should be emphasized immediately that it is seldom practicable to use such complicated and costly multiple layer systems as shown in (d) for tailings (or waste rock). Such systems may be used in covering hazardous waste landfills.

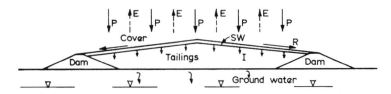

Figure 19.46. Factors involved in estimating infiltration through cover over tailings.

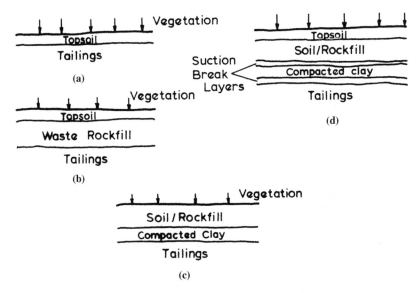

Figure 19.47. Some alternatives for covering tailings.

In these figures:

– The vegetation controls erosion, promotes evapotranspiration and enhances the appearance and use of an area;
– Topsoil promotes the vegetation. Note that in some cases vegetation can be established directly on to tailings;
– Soil rockfill is to control erosion and protect the underlying compacted clay from desiccation (in (d)).
– Compacted clay is a "barrier" to infiltration (in reality it only reduces infiltration since it will have a finite permeability);
– Filter layers are to prevent migration of fines into the drainage layer (a);
– Suction break layers are sand layers which assist in controlling desiccation and hence cracking of the compacted clay and help prevent transfer of salts and contaminants up through the compacted clay from the tailings.

The adopted design will depend on environmental constraints and availability of materials. Geotextiles and geomembranes may also be used, but are not common for tailings or waste rock dumps because they are too expensive for the large areas involved.

Accurate estimation of the rate of infiltration through the cover is difficult (almost impossible).

There are finite element and finite difference models which can in principle model the physical processes involved in infiltration through cover. Richards (1992) and Schroeder et al. (1983) describe such models, which rely on knowledge of the properties of each of the soil and rock layers in the cover, including saturated and unsaturated hydraulic conductivity (permeability) and the relationship between suction pressures and water content. These properties are very difficult to predict and vary by order of magnitude over normal ranges of compaction density and water content. The Schroeder et al. (1983) HELP model seems to have fairly wide acceptance. Readers are encouraged to read the latest literature for available methods.

There are other practical problems associated with the analysis of infiltration through cover. These include:

– Differential settlement of the tailings or waste rock or drying of the cover can lead to cracking, with a resultant large increase in effective hydraulic conductivity. This is virtually impossible to predict for cover placed over highly compressible tailings. The use of geotextiles or geogrids to control cracking would lead to a greater degree of confidence in the predictions, but again these all usually too expensive for use in tailings storage.
– Vegetation performs a very useful role in promoting transpiration. However the roots may penetrate the compacted soil layer leading to an increase in infiltration. Use of a silty sand or sand layer over the compacted soil, or a chemically treated zone to prevent root penetration may assist.

19.6.6.4   *Return of area to productive use*

In principle, it is desirable to return the area to productive use. Usually this can only be for farming or other low intensity usage not affected by long term settlement.

20

Monitoring and surveillance of embankment dams

20.1 WHAT IS MONITORING AND SURVEILLANCE?

ANCOLD (1976, 2003) give the following definitions:

Monitoring. The observing of measuring devices that provide data from which can be deduced the performance and behavioural trends of a dam and appurtenant structures, and the recording of such data.

Surveillance. The continuing examination of the condition of a dam and its appurtenant structures and the review of operation, maintenance and monitoring procedures and results in order to determine whether a hazardous trend is developing or appears likely to develop.

Monitoring and surveillance should be carried out during the construction, first filling and operation of all large dams.

There is a generally accepted principle that the level of monitoring and surveillance appropriate for a dam depends on the consequences of failure of the dam, whether the dam is being filled for the first time or is in general operation, and whether abnormal behaviour has been detected. This is discussed more in Section 20.3. The definitions of consequence of failure ratings are given in Section 8.2.3.

It should be remembered that the consequence of failure rating can change during the life of a dam. For example development downstream may raise the consequence of failure rating from low to high.

20.2 WHY UNDERTAKE MONITORING AND SURVEILLANCE?

20.2.1 *The objectives*

The objectives of monitoring are (ANCOLD 1983):

- To provide confirmation of design assumptions and predictions of performance during the construction phase and initial filling of the reservoir;
- To provide during the operation phase of the life of the dam an early warning of the development of unsafe trends in behaviour;
- To provide data on behaviour of dams which may not conform with accepted modern criteria and warrant continuous and close monitoring as a guide to the urgency for introduction of remedial/stabilizing works or other measures;
- During raising or remedial/stabilizing works, which may of necessity be carried out with the storage full, close monitoring of structural/seepage behaviour is warranted to ensure that the additional loading introduced by the new works is applied in a manner which will not adversely affect the safety of the dam.

In addition to this may be added:

- To satisfy legal obligations of the duty of care;
- To provide data to allow developments in dam engineering: through better measurement of properties, e.g. rockfill modulus in CFRD, checking of analytical methods,

e.g. displacements of CFRD face slab, and new construction materials, e.g. asphaltic concrete core, geotextiles.

When setting up a surveillance framework it is vital to ensure that:

- The inspection and monitoring program is planned by qualified dam engineers and engineering geologists, taking account of the potential failure modes for the dam. A detailed failure modes analysis should be a required part of establishing and reviewing an inspection and monitoring program;
- The monitoring data is reviewed by qualified dam engineers and engineering geologists in an ordered manner, so that unusual behaviour can be identified and appropriate action taken;
- The responsibilities of the owner, operator and government authority are clearly defined, with lines of communication established.

20.2.2   Is it really necessary?

That it is necessary to have a monitoring and surveillance system established, is highlighted by the number of dams which experience accidents and failure, often after many years of operation. This is discussed in some detail in ICOLD (1983a), National Research Council (1983), Foster et al. (1998, 2000a), and Douglas et al. (1998, 1999).

ICOLD (1983a) studied the approximately 14,700 dams which qualified for the ICOLD register at that time. An extensive survey indicated that of these dams 1105 (7.5%) had suffered incidents and deteriorations of one or more type and 107 (0.7%) had failed. In some cases the dam had to be completely abandoned and in others the dam was repaired and brought back into use despite severe damage.

Figure 20.1 summarises the failure of dams and causes of failure of dams over 15 m high built from 1900–1975.

Tables 8.1, 8.2 and 20.1 summarise the statistics of failures and accidents for embankment dams.

It can be seen that for embankment dams:

(a) Overtopping and internal erosion and piping are the main causes of failure;
(b) Slope instability is a relatively minor cause of failure but is a significant cause of accidents;
(c) About two-thirds of piping failures and about half of piping accidents occur on first filling or in the first 5 years of operation. However accidents and failures do still occur in older dams;
(d) About two-thirds of slope instability failures and half of instability accidents occur after 5 years of operation;
(e) The ratio of accidents to failures is quite different depending on the failure mode. In particular, slope instability and piping from embankment to foundation are much more likely than piping within the embankment to result in accidents compared to failures;
(f) The percentage of dams suffering accidents is not improving with time, but the percentage of failures is decreasing, reflecting improved technology of design, construction, monitoring and surveillance;
(g) More than 79% of all embankment dams are less than 30 m high. These suffer proportionally more failures than higher dams, but fewer accidents. This may reflect better design and monitoring and surveillance of the larger dams.

Douglas et al. (1998, 1999) showed that for concrete dams:

(a) Sliding, leakage and piping in the foundation are the main causes of failure. Accidents are most likely to be high recorded uplift or leakage in the foundation;

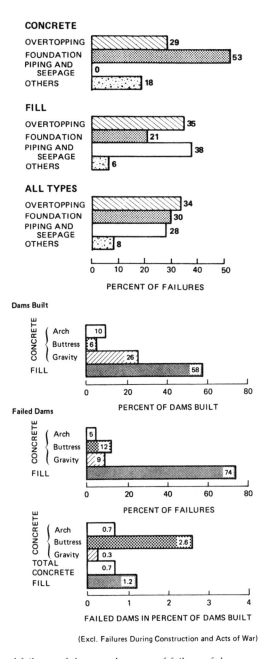

Figure 20.1.   Summary of failures of dams and causes of failure of dams over 15 m high (1900–1975) (National Research Council, 1983).

(b) Piping is the main cause of failure for concrete dams on soil foundations;
(c) A large proportion of failures occur on first filling, or historic high reservoir levels. This is particularly the case for piping failures. Older dams are also represented in the failures.

From this information it is clear that dams are most likely to fail or have an accident on first filling or in the first 5 years, but that there are failures and accidents which occur in

Table 20.1. Statistics of failure and accidents of embankment dams up to 1986 (from Foster et al., 2000).

| Failure mode | Average[2] probability × 10⁻³ | | Timing of incident (%) | | | | | | | |
| | | | Failure | | | | Accident | | | |
	Failure	Accident	DC	FF	<5 YR	>5 YR	DC	FF	<5 YR	>5 YR
Internal erosion and piping										
– Embankment	3.5	6.7	2	48	14	36	0	26	13	61
– Foundation	1.7	6.2	5	20	50	25	0	30	24	46
– Embankment to foundation	0.2	2.1	0	50	50	0	0	20	27	53
Slope instability										
– Downstream	0.5	5.3	18	18	0	64	15	11	25	49
– Upstream	0.1	4.2	0	0	100	0	22	2	26	50

Note: DC = during construction; FF = during first filling; <5 YR = in the first 5 years of operation, not including first filling; >5YR = after the first 5 years of operation.

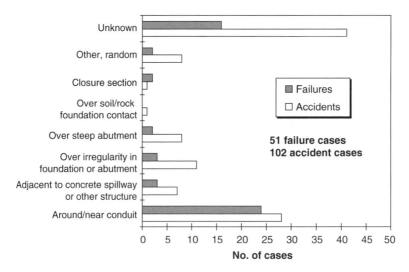

Figure 20.2. Location of internal erosion and piping failures and accidents for piping in the embankment (Foster et al., 1998).

older dams. It is apparent that many accidents would have become failures if they had not been detected by monitoring and surveillance and some action taken.

20.2.3 Some additional information on embankment dam failures and incidents

Figure 20.2 shows the location of piping failures and accidents for piping in the embankment (Foster et al. 1998).

Foster and Fell (1999b) showed that for piping not associated with conduits:

– Cracks associated with narrow cores and arching between rockfill shell zones are generally located at depths from one third to two-thirds of the height of the dams;

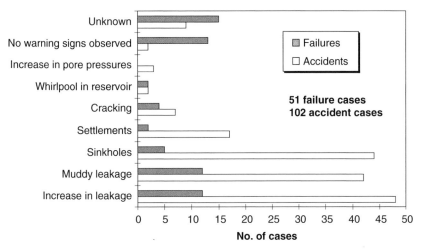

Figure 20.3.   Observations during piping incidents – piping through the embankment (Foster et al., 1998).

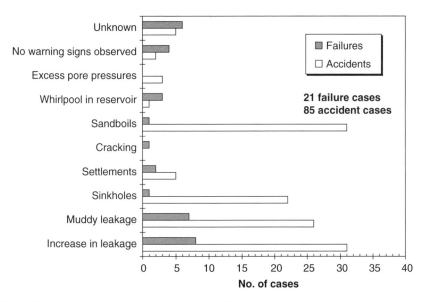

Figure 20.4.   Observations during piping incidents – piping through the foundation (Foster et al., 1998).

– Cracks and piping associated with broad changes in the abutment profile are generally "located at depths of less than one third of the height of the dam";
– Piping associated with small scale irregularities in the foundation generally occurs close to the foundation surface and usually in the lower half of the core.

Foster et al. (1998, 2000a) record that in 80% of failures by piping through the embankment the reservoir was at or higher than the previous highest reservoir level. For the other recorded cases it was within 1 metre of the previous highest level.

For piping in the foundation the reservoir water level was often at or above the historic high, but there were some failures with somewhat lower reservoir levels.

Figures 20.3 and 20.4 show observations during piping incidents in the embankment and foundation.

Table 20.2.   A method for the approximate estimation of the time for progression of piping and development of a breach, for breach by gross enlargement, and slope instability linked to development of a pipe (Fell et al., 2001, 2003).

Factors influencing the time for progression and breach					
Ability to support a roof (1)	Rate of erosion (2)	Upstream flow limited (3)	Breach time (4)	Approximate likely time-qualitative	Approximate likely time-quantitative
Yes	R or VR	No	VR or R-VR	Very Rapid	<3 hours
Yes	R	No	R	Very Rapid to Rapid	3–12 hours
Yes	R-M	No	VR		
Yes	R	No	R-M	Rapid	12–24 hours
Yes	R-M, or M	No	R		
Yes	R	Yes	R or VR		
Yes	R	No	M or S	Rapid to Medium	1–2 days
Yes	R-M, or M	No	M or M-S		
Yes	R or R-M	Yes	R or R-M		
Yes	M or R-M	No	S	Medium	2–7 days
Yes	R-M or M	Yes	S		
Yes	M	Yes or No	S	Slow	Weeks – even months or years

Notes: Estimated using (1) Table 20.3 (2) Using Table 20.4 (3) Using Table 20.5 (4) Using Table 20.6. VR = Very Rapid; R = Rapid; M = Medium; S = Slow.

20.2.4   Time for development of piping failure of embankment dams and ease of detection

Fell et al. (2001, 2003) describe an approximate method for estimating the time from first concentrated leak to initial breach of the dam. The times are for situations where failure occurs. Dams with properly designed and constructed filters, or even filters providing only some degree of protection, will usually not fail or the piping will develop more slowly, increasing the likelihood of intervention.

The method is summarized in Tables 20.2 to 20.6.

Table 20.2 is applicable to cases where the breach mechanism is gross enlargement. It is considered to be reasonably applicable to cases where the final breach is by slope instability, following development of a pipe. The factors in Table 20.3 should be also taken into consideration. For cases where the primary breach mechanism is by sinkhole leading to crest settlement, the rate of erosion should be assessed from Table 20.4 with consideration of the other factors in Table 20.5. Breaching by crest settlement is usually slow to medium, although unless addressed, it can lead to overtopping with a resultant rapid breach. For unravelling or sloughing, the rate of erosion should be assessed from Table 20.4, and consideration given to the other factors in Table 20.5. These processes are expected to be slow to medium, but may be rapid if the erosion rate is rapid.

The times are from the first sign of a concentrated leak to first breach of the dam crest. They therefore cover the progression phase of the internal erosion and piping process. These methods for assessing the rate of development of internal erosion and piping are very approximate and should be only used as a general guide to performance. Where the consequences of failure of the dam are large, a cautious approach should be adopted in making and using these estimates.

It should be recognized that the time for progression and development of a breach does not include the time for the dam to be emptied of water, i.e. it does not include the time for

Table 20.3. Influence of factors on the likelihood of progression of erosion-ability to support a roof (Foster and Fell, 1999b, 2000; Fell et al., 2001, 2003).

Factor	Influence on likelihood of fill or foundation materials supporting the roof of a pipe		
	More likely	Neutral	Less likely
(a) Embankment materials			
Fines content (% finer than 0.075 mm)	Fines content >15%	Fines content <15% and >5%	No fines or fines content <5%
Degree of saturation	Partially saturated (first filling)		Saturated
(b) Foundation materials	Piping through soils with cohesive fines	Well graded sand and gravel	Homogeneous, cohesionless sands
	Cohesive layer overlying piped material		
	Piping through open features in rock		
	Piping below rigid structure (e.g. spillway)		

breach enlargement which is used in estimates of dam-break floods. It also does not equal the available warning time to evacuate persons at risk, since this is dependent on whether the dam is under surveillance at the time of piping, and the measures set in place to advise persons downstream of an impending dam failure.

The times for initiation and continuation of erosion are dependent on the initiation mechanism, and the ability of filters, if present, to control erosion. Fell et al. (2002, 2003) have made a judgemental assessment of the usual rates of development and ease of detection. These are summarized in Tables 20.6 to 20.9.

The classifications for ease of detection assume the dam is well instrumented to measure seepage and pore pressure, readings are frequent and the dam is subject to regular inspection by experienced personnel.

Fell et al. (2001, 2003) conclude that:

(a) Monitoring of seepage, either by visual surveillance, or measurement, is the most common means of identifying whether internal erosion and piping have occurred.
(b) It is not common to have sufficient change in the seepage, or in other factors such as pore pressure changes or settlement, to identify conclusively that internal erosion has initiated and is continuing. It is more common to recognize when the erosion has progressed to the stage that a pipe has developed, or that there are changes in pore pressures, seepage or settlement which may be related to internal erosion, but this is not conclusive and it may reflect other factors. These changes may be a pre-cursor to a higher likelihood of internal erosion and piping, so it is important they are observed and, when they occur, investigated.
(c) The inability to detect that internal erosion has initiated relates to the common mechanisms of initiation. For piping in embankment failures, initiation is most common in cracks, high permeability zones or hydraulic fractures in the embankment or around a conduit. These mechanisms could be expected to initiate very rapidly or rapidly once the reservoir level reaches the critical level at which erosion begins in cracks or high permeability zones or the critical level at which hydraulic fracture initiates. However this depends on the rate of erosion of the soil.

Table 20.4.   Influence of factors on the progression of erosion – likelihood of pipe enlargement (and rate of erosion) (adapted from Foster and Fell, 1999b, 2000; Fell et al., 2001, 2003).

Factor	Influence on likelihood of pipe enlargement		
	More likely	Neutral	Less Likely
Embankment and Foundation			
Hydraulic gradient across core (2)	High	Average	Low
Soil type	Very uniform, fine cohesionless sand. (PI < 6) or Well graded cohesionless soil. (PI < 6)	Well graded l material with clay binder (6 < PI < 15)	Plastic clay (PI > 15)
Pinhole Dispersion test (4)	Dispersive soils, pinhole D1, D2	Potentially dispersive soils, pinhole PD1, PD2	Non-dispersive soils, pinhole ND1, ND2
Embankment			
Compaction density ratio	Poorly compacted, <95% standard compaction density ratio (1)	95–98% standard compaction density ratio	Well compacted, = 98% standard compaction density ratio
Compaction water content	Dry of standard optimum water content (approx. OWC-3% or less)	Approx standard OWC-1% to OWC-2%	Standard optimum or wet of standard optimum water content
Saturation (5)	As-compacted, partially saturated		Saturated after compaction
Foundation			
Relative density or Consistency	Loose Soft	Medium dense Stiff	Dense Very stiff
Erosion Rate	Erosion Rapid or Very Rapid if most factors are "more likely" slow if most factors are "less likely", and intermediate, if most factors are "neutral" or there is a mix of "more" and "less likely".		

Notes: (1) <93% Standard, dry of OWC, much more likely. (2) Even dams with very low gradients, e.g. 0.05, can experience piping failure. (3) PI = Plasticity index. (4) Using Sherard Pinhole Test. (5) Based on tests at UNSW.

Table 20.5.   Influence of factors on the progression of erosion – limitation of flows by upstream zones (Foster and Fell, 1999b, 2000; Fell et al., 2001, 2003).

Factor	Influence on likelihood of upstream flow limitation		
	Unlikely	Neutral	Likely
Filling of cracks by washing in of material from upstream	Homogeneous zoning. Upstream zone of cohesive material		Zone upstream of core capable of crack filling (cohesionless soil)
Restriction of flow by upstream zones or concrete element in dam	Homogeneous zoning. Very high permeability zone upstream of core	Medium to high permeability zone upstream of core	In zoned dam, medium to low permeability granular zone upstream of core. Central concrete corewall and concrete face rockfill dams

Table 20.6.   Influence of the material in the downstream zone of the embankment, or in the foundation, on the likely time for development of a breach (Fell et al., 2001, 2003).

Material description	Likely breach time
Coarse grained rockfill	Slow – medium
Soil of high plasticity, including clayey gravels	Medium – rapid
Soil of low plasticity, all poorly compacted soils, silty sandy gravels	Rapid – very rapid
Sand, silty sand, silt	Very rapid

(d) Initiation of erosion by suffusion (internal instability) is likely to be a more slowly developing process, accompanied by more gradual increases in seepage and changes in pore pressure with time. Some cases where sinkholes develop in the reservoir upstream of a dam founded on alluvial or fluvioglacial soils may be due to suffusion.

(e) "Blow-out" or "heave" in dam foundations where seepage forces create a zero effective stress condition is a situation which should be readily detected by carefully positioned and well monitored piezometers. Often these low effective stresses occur below lower permeability layers which act to confine the seepage flow and it is important that piezometers are installed to measure pore pressures in these areas. These pressures are usually directly related to reservoir levels and it is important to monitor the relationship between the pore pressures and reservoir levels. Most often these conditions will occur on first filling or at historic high reservoir water levels, but the authors are aware of cases where pressures have increased with time possibly due to suffusion in the foundation soils or blockage of drains and pressure relief walls.

(f) Failures from piping in the foundation and from the embankment to the foundation are mostly from backward erosion, or backward erosion following hydraulic fracture, "blow-out" or "heave". These would not necessarily be expected to be preceded by large increases in seepage during the time the erosion is gradually working back from the downstream exit point. When the erosion has progressed to within a short distance of the reservoir/foundation interface, it breaks through rapidly or very rapidly.

(g) Accidents are also usually detected in the progression phase, rather than in the initiation and continuation phase. Progression ceases due for example to sealing of eroded materials on filters/transitions which satisfy excessive erosion criteria, flow limitation by an upstream dirty rockfill zone, or concrete face (on concrete faced dams), or due to collapse of the pipe. The latter is possibly a more common mechanism in foundation piping accidents than in embankments, because most embankment cores will have sufficient fines to support a roof to the pipe.

(h) Timely intervention can change a potential failure into an accident. As discussed in (b), changes in pore pressure, seepage, and settlement may be detected before internal erosion progresses too far.

(i) The ease of detection of internal erosion and piping in the foundation by visual and seepage means is more readily achieved if the area downstream of the dam is not vegetated, or is at least cleared of larger vegetation. Detection is difficult if the area downstream is densely vegetated or if the dam toe area is covered by water.

(j) Detection by pore pressure measurement is more likely if there is extensive instrumentation read regularly and if the erosion is widespread.

(k) The frequency at which seepage is measured can be related to the likelihood that a dam may fail or experience an accident by internal erosion and piping, the consequences of failure and the likely time for internal erosion to progress to form a pipe and the dam breach. It might also be related to the reservoir level, with enhanced measurements at times of high reservoir level.

Table 20.7.   Rate of development and detectability of internal erosion and piping – piping through the embankment (Fell et al., 2001, 2003).

Phase of development	Mechanism	Usual rate of development	Ease of detection	Method of detection
Initiation	Backward erosion	Slow, but rapid to very rapid at the end	Difficult (unless observed at downstream face)	None? Visual if on the downstream face
	Crack/hydraulic fracture	Rapid or very rapid	Difficult	Seepage, visual if the crack emerges on the downstream face or if crest, or the crest settles
	High permeability zone	Slow to rapid	Moderate to difficult	Seepage, pore pressure visual if emerges on downstream face
	High permeability or cracking, hydraulic fracture, associated with conduits or walls	Rapid or very rapid	Difficult	Seepage, visual
	Suffusion/internal instability	Slow	Moderate to difficult	Pore pressure, seepage, visual if emerges on the downstream face
Continuation	Filters satisfying no-erosion criteria	Erosion will cease		
	Filters satisfying excessive erosion criteria	Slow to rapid	Moderate to difficult	Seepage, pore pressure
	Filters not satisfying continuing erosion criteria, or no filters	Medium to rapid	Moderate to difficult	Seepage, pore pressure
Progression to form a pipe, and a breach	Gross enlargement; and slope instability linked to development of a pipe	Assess time from Table 20.2 – Very rapid or rapid – Medium or slow Assess rate of erosion from Table 20.4, and other factors from Table 20.5. Usually slow to medium	Mod. to readily Readily Readily	Seepage, visual Visual, survey, seepage
	Crest settlement or/ and sinkhole in embankment			
	Slope instability, unravelling or sloughing	Assess erosion rate from Table 20.4, and other factors from Table 20.5. Usually slow to medium unless linked to rapid development of pipe	Moderate to readily	Visual, survey, seepage

Legend: Rate of development: Slow = weeks or months, even years; Medium = days or weeks; Rapid = hours (>12 hours) or days; Very Rapid = <3 hours. Ease of detection: Difficult = unlikely to be detected in most cases; Moderate = may be detected in some cases; Readily = readily detected in most cases. Methods of detection: Visual (inspection), seepage (measurement either visual or by instruments), pore pressure (measurement) survey (survey of surface markers, to determine horizontal and vertical deformation).

Table 20.8. Rate of development and detectability of internal erosion and piping – piping through the foundation, and from the embankment to the foundation (Fell et al., 2001, 2003).

Phase of development	Mechanism	Usual rate of development	Ease of detection	Method of detection
Initiation – Foundation	Backward erosion	Slow	Difficult	Visual, seepage, pore pressure
	Backward erosion following "blow-out"	Rapid to very rapid	Readily to difficult	
	Backward erosion along concentrated leak	Slow to Rapid	Moderate to difficult	
	Suffusion/internal instability	Slow	Moderate to difficult	
Initiation – embankment to foundation	Backward erosion	Slow to rapid/very rapid	Difficult	Pore pressure, seepage
Continuation	Filters satisfying no erosion criteria	Erosion will cease if all seepage is intercepted		
	Filtered exit, satisfying excessive erosion, or incomplete seepage interception	Medium to slow, but could be rapid	Moderate to difficult	Seepage, pore pressure, visual
	Filtered exit, not satisfying continuing erosion criteria, or unfiltered exit	Rapid to slow	Difficult to moderate	Seepage, pore pressure, visual
Progression to form a pipe, and a breach	Gross enlargement in the foundation or in the embankment; and slope instability linked to development of a pipe. Crest settlement or/and sinkhole in embankment	Assess time from Table 20.2: – Very rapid or rapid – Medium or slow Assess rate of erosion from Table 20.4, and other factors from Table 20.5. Usually slow to medium	Moderate to readily Readily Readily	Seepage, visual Visual, survey, seepage
	Slope instability, and unravelling or sloughing for piping embankment to foundation	Assess erosion rate from Table 20.4, and other factors from Table 20.5 Usually slow to medium, unless linked to rapid development of a pipe	Moderate to readily	Visual, survey, seepage

Legend: Rate of development: Slow = weeks or months, even years; Medium = days or weeks; Rapid = hours (>12 hours); Very Rapid = <3 hours. Ease of detection: Difficult = unlikely to be detected in most cases; Moderate = may be detected in some cases; Readily = readily detected in most cases. Methods of detection: Visual (inspection), seepage (measurement either visual or by instrument), pore pressure (measurement), survey (survey of surface markers to determine horizontal and vertical deformations).

Table 20.9. Rate of development and detectability of internal erosion and piping – piping into and along conduits or adjacent to walls (Fell et al., 2001, 2003).

Phase of development	Mechanism	Usual rate of development	Ease of detection	Method of detection
Initiation	(i) Hydraulic fracture, high permeability zone, or crack	Rapid	Difficult	Visual
	(ii) Erosion into open crack or joint in the conduit or wall	Slow	Moderate	Visual, seepage
Continuation	Filter (A) satisfying no-erosion criteria	Erosion will cease		
	Filter satisfying excessive erosion criteria	Slow to rapid	Moderate to difficult	Seepage, pore pressure
	Filters not satisfying excessive erosion criteria, or no filters	Rapid to medium	Difficult to moderate	Seepage, pore pressure
	Crack or joint width satisfies excessive erosion criteria.	Slow	Moderate (B)	Visual
	Crack or joint width does not satisfy continuing erosion criteria	Slow (to medium) (C)	Moderate (B)	Visual
Progression to form a pipe and a breach	Gross enlargement; and slope instability linked to development of a pipe	Assess time from Table 20.2. – Very rapid or rapid	Moderate to readily	Seepage, visual
	Crest settlement or/ and sinkhole in the embankment	Assess rate of erosion from Table 20.4 and other factors from Table 20.5. Usually slow to medium	Readily seepage	Visual, survey,
	Slope instability, unravelling or sloughing	Assess erosion rate from Table 20.4, and other factors from Table 20.5. Usually slow to medium, unless linked to rapid development of a pipe	Moderate to readily	Visual, survey, seepage

Erosion into a conduit or crack in a wall, usually only progresses towards breach by initiation of erosion along the conduit or wall, in which case factors are as for gross enlargement and instability. For cases which continue towards development of a sinkhole in a crest mode of breach, ability to hold a roof/or sinkhole is critical and the rate of development is usually slow, readily detectable by visual, survey or seepage.

Legend: Rate of development: Slow = weeks or months, even years; Medium = days or weeks; Rapid = hours (>12 hours) or days; Very Rapid = <3 hours. Ease of detection: Difficult = unlikely to be detected in most cases; Moderate = may be detected in some cases; Readily = readily detected in most cases. Methods of detection: Visual (inspection), seepage (measurement either visual or by instrument), pore pressure (measurement).

Notes:
(A) Filters are those controlling internal erosion around the conduit, or adjacent walls.
(B) Often the crack and seepage is visible on inspection, but erosion may be intermittent, or only when conduit is flowing with water, so not readily observed.
(C) The evidence seems to be that even for conduits surrounded by erodible soils, e.g. fine sand, the rate of erosion into the conduit is slow.

(l) The seepage monitoring system needs to be capable of being calibrated to separate out the effects of rainfall or snowmelt e.g. by prior observation and monitoring and coupling seepage measurements to rainfall and snowmelt measurements at the dam. It should however be recognised that there are many situations where it will be unlikely that seepage will be detected, e.g. at night when visual surveillance is being relied upon; if the toe of the dam is submerged; if seepage occurs high on the dam abutments and bypass measuring weirs or in winter when the dam is covered by snow.

(m) There are clearly many dams which may have progression and breach times of the order of hours. These dams particularly include some older dams which have no filter or transitions, dams without downstream rockfill zones to stop or slow the erosion process, and/or those dams on erodible foundations without well designed and constructed cutoffs or filters to intercept the foundation seepage. For these dams, an effective seepage monitoring program would require virtually continuous monitoring. Daily, or even twice daily inspections, or measurements may be inadequate.

20.3 WHAT INSPECTIONS AND MONITORING IS REQUIRED?

20.3.1 *General principles*

As discussed in ICOLD (1989) and ANCOLD (2003), the inspection and monitoring appropriate for a dam depends on a number of factors:

(a) The consequences of failure. Dams with a higher consequence of failure in terms of potential lives lost, economic and environmental damage, require a higher level of inspection and monitoring;

(b) The type of dam. For example, dams which have no filters or drainage zones, may require greater inspection and monitoring than a well designed and constructed dam with filters, assuming the same consequences of failure;

(c) The type of dam foundation. A dam on permeable soil foundations would usually require more inspection and monitoring than the same dam on a rock foundation;

(d) The size of dam. This is usually linked to (a) and (b), but in general larger dams have a higher level of inspection and monitoring than small dams;

(e) Known deficiencies or deterioration of the dam. If there are identified safety issues, such as marginal slope stability or larger than normal seepage or pore pressures, there will usually be a need for enhanced inspection and monitoring until the deficiencies are rectified;

(f) Identified potential failure modes;

(g) The age of the dam. There is a need for enhanced inspection and monitoring during first filling of the reservoir, because it is known that many failures and accidents occur in this period.

To this the authors would add:

(h) The reservoir level, compared to historic high levels. There is strong evidence that failures and accidents from internal erosion and piping occur above or near historic high reservoir levels. At such times the dam is also at greatest risk from a slope instability point of view. For dams with known deficiencies in internal erosion and piping, there should be a significantly increased level of inspection and monitoring under these reservoir conditions.

(i) The reservoir level, compared to historic low levels. Many dams with marginal upstream slope stability, or subject to internal deformations due to poorly compacted rockfill, experience abnormal settlements and deformations when the reservoir is at or below historic low levels and inspection for cracking and survey of deformations should be planned.

(j) Post earthquake. Inspection and monitoring should be done after significant earthquake loading on the dam.

The inspection and monitoring required for a dam should be assessed by experienced dam engineers, including the designers (for new dams), and the engineer(s) responsible for surveillance operation and maintenance of the dam.

There will often be compromises to be made, offsetting the costs of instrumentation, readings and surveillance against what is desirable. For older dams, when instruments fail, there are often difficult decisions to make on whether they should be replaced, taking into account the risks in replacing them e.g. the risk of damaging the dam core by drilling to replace piezometers must be weighed against the benefits gained in having the monitoring data.

Table 20.10 summarises the detection methods which led to the identification of deterioration of embankment dams in the ICOLD (1983a) study. The authors concur with the view that the most important are:

– inspection by trained observers;
– seepage flow;
– vertical and horizontal displacement by direct survey of surface points;
– pore pressures;
– reservoir water level and rainfall (as these link to the other factors).

Other items which may be monitored include:

– internal deformations, using internal settlement gauges, extensometers, or borehole inclinometers;
– seepage water temperature;

Table 20.10. Number of applications of detection methods used to detect deterioration in embankment dams (ICOLD, 1983a).

Detection methods	Deteriorations affecting foundations	Deteriorations affecting dam body	Deteriorations affecting foundations and dam body	Total No.	%
Direct observation	92	250	59	401	63
Water flow measurements	37	30	17	84	13
Phreatic level measurements	8	2	2	12	2
Uplift measurements	1	–	–	1	<1
Pore pressure measurements	17	8	5	30	5
Turbidity measurements	3	1	2	6	1
Chemical analysis of water	3	1	2	4	1
Seepage path investigations	1	5	2	8	1
Horizontal displacement measurement	3	25	2	30	5
Vertical displacement measurements	5	27	7	39	6
Strain measurements	–	1	1	2	<1
Rainfall measurements	–	1	–	3	<1
Sounding investigation	2	1	–	3	<1
Design revision (new criteria)	1	3	–	4	<1
Not available	3	3	1	6	1
Total of deterioration cases	109	260	63		

– seepage water chemistry and pH;
– seepage water turbidity measurement;
– seismic ground motion.

20.3.2   Some examples of well instrumented embankment dams

Figures 20.5 to 20.8 show examples of quite intensive monitoring for Prospect Dam (a 100 year old 35 m high puddle core dam, constructed of erodible soil with no filters in the original construction); Cethana Dam, a 110 m high CFRD; Trinity Dam (150 m high earth and rockfill) and Svartevann Dam (129 m high earth and rockfill) (all from ICOLD 1989).

Table 20.11 is based on ANCOLD (2003) guide for the frequency of monitoring for embankment dams based on the consequences of failure category. These would be appropriate for a dam without known deficiencies, and after first filling.

During construction and first filling, more frequent readings of monitoring instruments would be appropriate.

Table 20.12 gives an example for a large embankment dam, showing how the frequency of readings changes as the satisfactory performance of the dam is confirmed.

20.3.3   Dam safety inspections

Dam safety inspections are the most important part of any dam surveillance program. As pointed out by ANCOLD (2003), inspections should be carried out by experienced people,

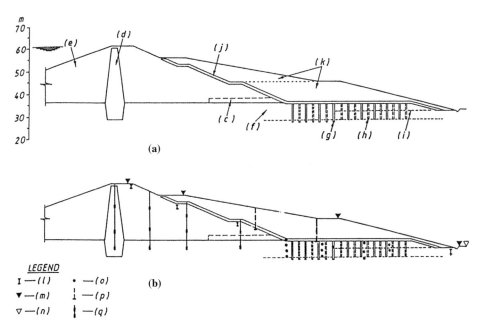

Figure 20.5.   Monitoring for Prospect earthfill dam (ICOLD, 1989). (a) Cross section showing old dam and fill buttressing. (b) Cross section showing monitoring systems. (c) Rubble filled drainage tunnel. (d) Puddle clay core. (e) Clay shoulders. (f) Alluvial clay foundation. (g) Filter drainage cutoff trench along downstream foundation extending to rock. (h) Vertical sand drains. (i) Lateral drains. (j) Filter zone. (k) Stabilizing fill material (shale) constructed in two stages. (l) Inclinometers. (m) Surface settlement points. (n) Seepage collection and measurement. (o) Pneumatic and electric piezometers. (p) Foundation settlement installations. (q) Open standpipe piezometers.

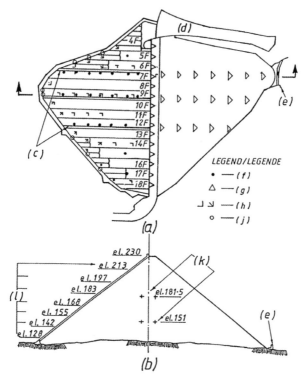

Figure 20.6.   Monitoring for Cethana concrete face rockfill dam (ICOLD 1989). (a) Plan view. (b) Cross section. (c) Inclinometer pipes. (d) Spillway. (e) Leakage weir. (f) Anchor points for wires. (g) Survey targets. (h) Strain gauges. (j) Perimetric joint meters. (k) Hydrostatic settlement cell. (l) Installation level of strain gauges.

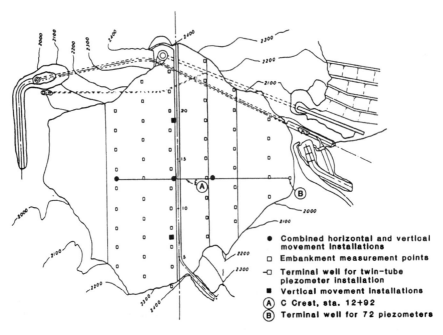

Figure 20.7   Monitoring for Trinity earth and rockfill dam (ICOLD, 1989).

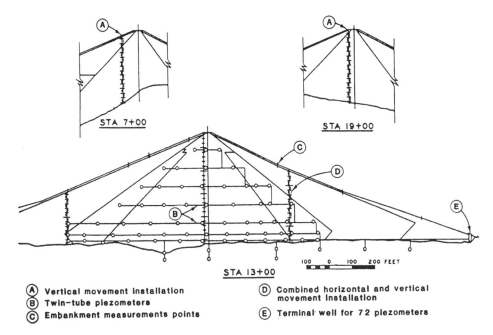

Ⓐ Vertical movement installation
Ⓑ Twin-tube piezometers
Ⓒ Embankment measurements points
Ⓓ Combined horizontal and vertical movement installation
Ⓔ Terminal well for 72 piezometers

Figure 20.7   (Continued)

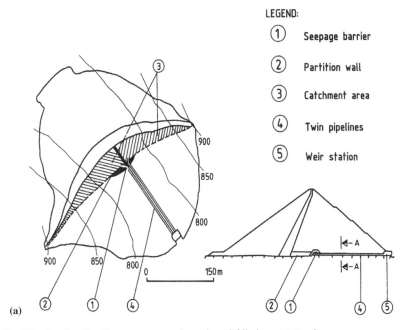

LEGEND:
① Seepage barrier
② Partition wall
③ Catchment area
④ Twin pipelines
⑤ Weir station

(a)

Figure 20.8.   Monitoring for Svartevann earth and rockfill dam (a) Leakage measurement (ICOLD, 1989).

trained to recognize deficiencies in dams. Inspections requiring technical evaluation, should generally be carried out by a dams engineer and other specialists. The day-to-day inspections which would detect many deficiencies are best done by trained operations personnel. Dam owners should ensure that all operations personnel are suitably trained and are aware

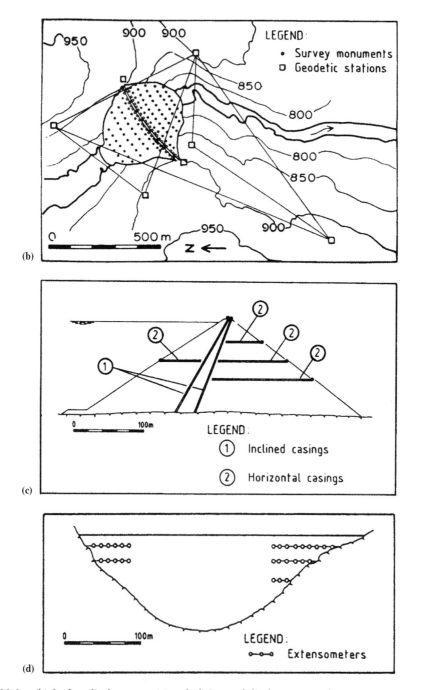

Figure 20.8.   (b) Surface displacement. (c) and (d) Internal displacement and strains.

of the consequences of failure of the dam and of the deficiencies that have been found in similar dams.

Ideally dam owners should arrange for these inspectors to be given the chance to spend time with inspectors from other like-minded dam owners to see what issues have arisen at other dams.

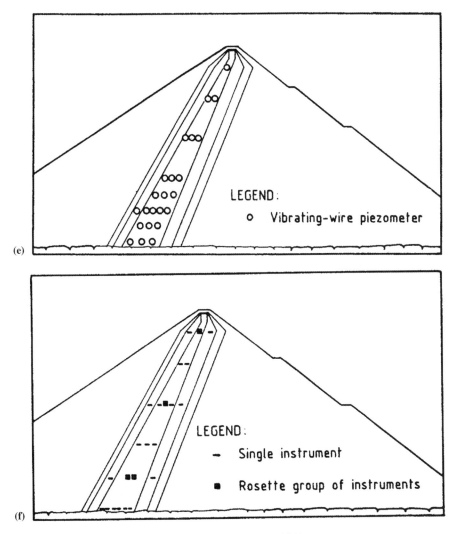

Figure 20.8.   (e) Pore pressures. (f) Earth pressure (ICOLD, 1989).

ANCOLD (2003) recommend five levels of inspection which are detailed in Table 20.13.
For dams in sound condition with no deficiencies, ANCOLD (2003) recommend the frequency of inspection shown in Table 20.14. More frequent inspections may be necessary where a dam is known to have a deficiency, the authors would add that more frequent inspections should be taken under historic high reservoir conditions.

The frequency of routine dam safety and routine visual inspections should be increased during first filling. For example, for a high or extreme consequence of failure dam which fills over a few months, there should be daily routine visual inspections, and say weekly or fortnightly routine dam safety inspections by engineers familiar with the design of the dam and its expected performance. The inspection should be carried out to a checklist. An example of such a checklist is given in Table 20.15 the checklist is for an earth dam, but could be modified for other types of dams. Seepage measurement would normally be included on the same report sheet. USBR (1983), FEMA (1987), and ANCOLD (2003) have more detailed checklists.

Table 20.11.   Guide for "in service" embankment dam monitoring frequencies[1] (Adapted from ANCOLD, 2003).

Monitoring	Consequences of failure category				
	Very low	Low	Significant	High	Extreme
Rainfall	Monthly[5]	Monthly	Daily to twice weekly (TC)[2]	Daily to tri weekly (TR)[2]	Daily (TR)[2]
Storage level	Monthly[5]	Monthly	Twice weekly to weekly (TC)[2]	Daily to tri weekly (TR)[2]	Daily (TR)[2]
Seepage	Monthly[5]	Monthly	Twice weekly to weekly (TC)[2]	Daily to tri weekly (TR)[2]	Daily (TR)[2]
Chemical analysis and pH of seepage			Consider	Yearly	Yearly
Pore pressure[3]	Consider	Consider	3-monthly	Monthly to 3-monthly	Monthly
Surface movement, control[4]	Consider	Consider	Consider	5-yearly to10-yearly	5-yearly
Surface movement, normal			Consider	2-yearly	Yearly
Internal displacement				2-yearly	Yearly
Seismological[3]				Consider (TC)	Consider (TR)

Notes:
[1] These frequencies may need to be varied according to the conditions at, and the type and size of dam, and applies to instrumentation already installed at the dam.
[2] The frequencies quoted assume manual reading of the instrumentation. Where automated readings are available more frequent reading would be appropriate. (TR – telemetry recommended; TC – telemetry to be considered).
[3] The frequency of reading and location of the pore pressure instruments and seismic instruments need to be at the discretion of the dams engineer.
[4] A control survey uses monuments that are remote from the dam site to check the location of the survey monuments at the dam site.
[5] The frequencies listed for very low consequence of failure category dams are suggestions, the dam owner should determine appropriate monitoring.
[6] Applicable for concrete dams and construction on potentially soluble materials or where there is concern about seepage in the dam or its foundations.

Table 20.12. Monitoring for Revelstoke Dam (Canadian National Committee in ICOLD, 1989).

Stage	Instruments									
	Core piezometer	Foundation and shell piezometer	Vertical movement gauge	Horizontal movement gauge	Horizontal strain gauge	Surface monument	Earth pressure cell	Weir and well	Strong motion accelerograph	Visual inspection
During construction	Frequently by field staff				1/month	1/month	1/month	1/month	Continuous	1/month
Reservoir filling	Every 2 days	Every 2 days	1/week	2/month	2/month	1/month	2/month	Every 2 days	Continuous	1/day
After the first reservoir filling										
First 6 months	1/week	1/week	1/month	1/month	1/month	1/month	1/month	1/month	Continuous	1/day
6 month to 1½ year	2/month	2/month	4/year	4/year	4/year	4/year	4/year	2/month	Continuous	1/week
1½ year to 2½ year	2/month	2/month	4/year	4/year	4/year	4/year	4/year	2/month	Continuous	1/week
2½ year to 6½ year	6/year	6/year	2/year	2/year	2/year	2/year	6/year	6/year	Continuous	2/month
Subsequent years	2/year	2/year	1/year	1/year	1/year	1/year	1/year	2/year	Continuous	1/month

Table 20.13.   Dam safety inspections as recommended by ANCOLD (2003).

Type of inspection	Personnel	Purpose
Comprehensive	Dams engineer and specialists[1] (where relevant)	The identification of deficiencies by a thorough onsite inspection; by evaluating surveillance data; and by applying current criteria and prevailing knowledge. Equipment should be test operated to identify deficiencies. For a safety review consider: – Draining of outlet works for internal inspection; – Diver inspection of submerged structures.
Intermediate	Dams engineer	The identification of deficiencies by visual examination of the dam and review of surveillance data against prevailing knowledge with recommendations for corrective actions. Equipment is inspected but not necessarily operated.
Routine dam safety	Inspector	The identification and report of deficiencies, by structured observation of the dam and surrounds, by an inspector, other than the operator, with recommendations for corrective actions.
Routine visual	Operations personnel	The identification and report of deficiencies by visual observation of the dam by operating personnel as part of their duties at the dam.
Special/emergency	Dams engineer and specialists[1]	The examination of a particular feature of a dam for some special reason (e.g. after earthquakes, heavy floods, rapid drawdown, emergency situation) to determine the need for pre-emptive or corrective actions.

Note:

[1] Examples of specialists include mechanical and electrical engineers, to inspect outlet works, spillway gates and automated systems, and corrosion engineers).

Table 20.14.   Frequency of inspection for dams with no known deficiency (adapted from ANCOLD, 2003).

Consequence of failure[3]	Inspection type				
	Comprehensive	Intermediate	Routine dam safety	Routine visual	Special
Extreme	On first filling then 5 yearly	Annual	Monthly	Daily[1]	As required
High A, B, C	On first filling then 5 yearly	Annual	Monthly	Daily[1] to tri-weekly	As required
Significant	On first filling then 5 yearly	Annual to 2-Yearly	3-Monthly	Twice weekly to weekly	As required
Low		On first filling, then 5 yearly		Monthly	As required
Very low		Dam owner's responsibility[2]		Dam owner's responsibility[2]	As required

Notes:

[1] Dam owners may undertake a review to determine if a reduced or increased frequency of inspection is acceptable. The review should be carried out by a dams engineer and take into account such matters as Regulator requirements, dam consequences of failure and risk, type and size of dam, dam failure modes and monitoring arrangements.

[2] Monthly routine visual inspections and 5 yearly intermediate inspections with test operation of equipment and review of consequences of failure category are suggested.

[3] See Section 8.2.3 for definitions of the consequences of failure.

Table 20.15. Checklist of conditions to be noted on visual inspection (adapted from ANCOLD, 1976).

(a) Vegetation on dam and within 15 metres beyond toe of dam:
  – Overgrowth: Requiring cutting for dam surveillance, requiring weed control for dam surveillance, indicating seepage or excessive seepage or moisture;
  – Wet terrain vegetation: Watch for boils, sand cones, deltas, etc., changes with the season, and pond level changes;
  – Incomplete – Requiring repair: Poor growth, destroyed by erosion.

(b) Foundations
  – Drainage ditches clogged with vegetation;
  – Dam areas, moisture on dry days;
  – Flowing water: quantity, location, clarity;
  – Boils;
  – Silt accumulations, deltas, cones.

(c) Embankment
  – Crest: Cracking, subsidence;
  – Upstream face: Cracking bulging, surface erosion, gullying, wave erosion;
  – Downstream face: Cracking, subsidence, bulging, erosion gullies; moisture on dry days; damp areas, seeps;
  – Berms and within 15 metres beyond toe of dam: Erosion, gullies, damp areas, seeps.

(d) Spillways and outlet works
  – Reservoir level;
  – Discharge tunnel or conduit condition, seeps, cracks;
  – Seepage or damp areas around conduits;
  – Erosion around or walls below conduit;
  – Boils in the vicinity of conduit;
  – Spillway slabs for uplift, subsidence, cracking.

(e) Areas of previous repair.
  – Effectiveness of repair;
  – Progression of trouble into new area.

20.4   HOW IS THE MONITORING DONE?

20.4.1   *General principles*

Some good general principles on dam monitoring are given by the French National Committee in ICOLD (1989):

(a) Dams are structures with a long life span. *Reliable* instruments with a similar life span are necessary, and they must be accessible, verifiable and replaceable;
(b) They must be *sensitive* to give warning of sudden changes or trends that are very small but may be significant indicators of distress;
(c) They should be *simple* so that frequent readings can be made quickly by site or operating staff without the need for specialists; in this way, the dam is more or less permanently monitored, and the records are available for statistical analysis;
(d) In recent decades far more attention has been focused on deformations and hydraulic behaviour in the *foundations*;
(e) *Visual inspection* and surroundings, by people who know the dam, is very important because even the best instruments will not find fissures, leakages, dam spots or their growth, etc.;
(f) The choice of parameters to be measured and the positioning of instruments cannot always be optimal at the design stage. The number of instruments sometimes amazes but is justified by the need for effective monitoring of a so far unfamiliar structure. Optimisation of the system can be made by abandoning some instruments or installing others as the real behaviour becomes better understood;

(g) Although modern information processing can help with huge masses of raw data, experience has shown that it is a good policy to read often a limited number of key instruments very well situated in the dam;

(h) The first filling of a reservoir is a stage in its life that is both important and delicate. It is in fact the proof that a dam can fulfill its design functions. Of course, measurements must start earlier so that the structure's initial state is known and some instruments are read during construction so that the reaction of dam and foundation to loading are known;

(i) Continuous reading of some instruments is conceivable during first filling to obtain a large amount of information during this fairly short period. During normal operation the raw data is processed by computer to give useful information. However complete automation – the automatic transmission of the state of safety of a dam and possibly of danger signs – is not used by EDF (Electricite de France). The delicate task of fine interpretation belongs to the engineer.

The authors would add:

(a) The use of instruments that can be easily rechecked if readings differ significantly from previous readings or are well outside the expected range;

(b) If a reading does not change for some time that is no reason to stop reading that instrument;

(c) Results should be quickly recorded within the owner's storage system. Wherever possible the immediately gathered data should be presented in simple, easily understood graphical plots and the data should be reviewed by an experienced dam engineer and not left for 5 years or so until the next main review is planned.

20.4.2   *Seepage measurement and observation*

Seepage data is one of the best indicators of a dam's performance. By observing the location, quantity and quality of seepage emerging from the dam embankment and its foundation, and particularly the changes which occur, one can get early warning of problems which may be developing, particularly in the important problem areas of internal erosion in the dam and its foundation and in increased pore pressures.

It is emphasized that routine observation of where seepage is emerging can be as useful a guide as the actual measurement. However, as indicated by most of the reporting countries in ICOLD (1989) and ANCOLD (1983), it is usual to measure the quantity of seepage.

As shown in Figure 20.9, it is preferable to collect the seepage close to the downstream toe of the impervious zone, and to isolate areas from each other so the readings are not influenced excessively by flow through rockfill zones and runoff from abutments.

However, as pointed out by ANCOLD (1983), for smaller dams it is acceptable to measure total seepage downstream of the toe of the dam and for larger dams, it is often impractical to have a system such as that shown in Figure 20.9. The authors know of cases where drainage trenches to subdivide the foundation into collection areas under rockfill were filled rapidly with fines from the rockfill, making the system ineffective. In other instances the seepage from individual areas simply by-passes the collector drains – the cost to seal below the drain is far too high. Where seepage is collected downstream, the influence of rainfall runoff on the readings has to be determined and separated from the base seepage flow.

Measurement of seepage may be made by:

– V-notch or similar measuring weirs. This may include continuous recording and telemetering of data for more important dams. Level switches may be provided to give an independent alarm;

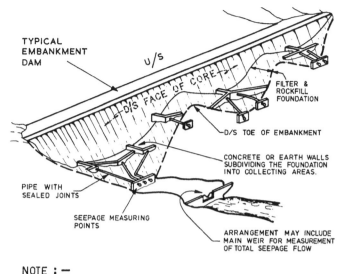

Figure 20.9.   Seepage collection and measuring system for an earth and rockfill dam (ANCOLD, 1983).

– Timed discharge into measuring vessels;
– Visual inspection where flow rates are very small.

It is often impractical to collect and measure all seepage, particularly for dams on alluvial foundations. In these cases, installation and monitoring of piezometers in the foundations under and downstream of the dam can give information on changing conditions, which might indicate a problem is developing. Problems also are experienced where the toe of the dam is flooded by the tailwater from a hydropower station or irrigation outlet. In these cases collecting and measuring seepage in the body of the dam, as shown in Figure 20.9, is desirable but is seldom done in practice.

Chemical analysis of seepage can be a useful guide to the source of the seepage water, e.g.:

– A comparison of ions in the reservoir water and seepage may indicate leaching of cement from grout curtains, or oxidation of sulphides within the foundation or within the embankment materials;
– Biological analysis can indicate the source relative to depth in the reservoir;
– The age of water as determined by analysis of tritium can indicate its sources as rainwater or groundwater.

More importantly, routine inspection of seepage discharge should be made noting any discolouration of the water which may indicate piping of the embankment or the foundation. Any inexplicable increase in suspended solids, particularly during first filling, needs to be treated with some urgency as piping can develop rapidly.

Seepage measurements can be augmented by having some carefully placed groundwater observation holes on the downstream abutment. These holes can pick up gradually changing groundwater in the abutments and possibly help explain the influence of groundwater from hillsides above the dam.

20.4.3   *Surface displacements*

Regular accurate survey of displacements of the surface of the dam embankment can be a most useful guide to performance.

Such measurements can be useful as a check on design assumptions, e.g. deformations of a CFRD, and as an indication of developing problems, including marginal slope stability and internal deformations due to softening or internal erosion and piping in the embankment or the foundation.

Figure 20.10 shows design of surface settlement points suitable for dams.

It is useful to extend the survey to surrounding areas, particularly if they are affected by slope instability.

In almost all cases, horizontal as well as vertical displacements should be measured, so that movement vectors can be determined. Such vectors can often give a good indication of the mechanism causing the displacement.

It is important to consider carefully where the survey markers are positioned, because quite different movements occur in the upstream and downstream slopes, and between the dam core and adjacent filter and rockfill zones.

To properly monitor movements, survey markers should be positioned centrally over the earthfill core, on the upstream and downstream edges of the crest and at several levels on the upstream and downstream slopes. Survey markers (at least those on the crest of the dam) should be positioned over the whole length of the dam, not just one or two cross sections, because one of their most valuable outcomes is to detect local larger settlements. Some survey markers should always be positioned where embankments abut spillway concrete walls, because this is where local seepage, softening and abnormal deformation is often a guide to developing problems.

For monitoring of soil landslides or earthfill dams, it will be necessary to embed the survey points below the depth of shrink and swell due to seasonal moisture change. This may involve a system as shown in Figure 20.11, with the base at 2 m to 4 m below ground surface and the pillar isolated from the upper soil by a casing.

ICOLD (1988b) indicate that survey of displacements is falling out of favour. The authors' opinion is that this is unfortunate, and seemingly unnecessary, given the availability of quick accurate electronic survey methods. Our experience is that survey data has been very valuable in early identification of problems and has the great advantage of being readily checked is some results appear inconsistent.

20.4.4   *Pore pressures*

The measurement of pore pressures in the embankment and foundations of a dam can give vital quantitative information for use in assessing stability, potential "heave" conditions in foundations and for identifying unusual seepage pressure, which may be a precursor to internal erosion and piping provided there is adequate coverage along the dam.

The following discussion gives an overview on why and where pore pressures are measured, the types of instruments available, their characteristics and some practical factors. For more detailed information on instruments, the reader is referred to Hanna (1985) and to manufacturers of the instruments.

20.4.4.1   *Why and where are pore pressures measured?*

There is a wide range of situations where it is necessary to measure pore pressures. Examples are given to highlight some of the principles involved in locating the piezometers to obtain meaningful answers.

Dam embankment. Figure 20.12 shows piezometric conditions in an earthfill dam, constructed on an alluvial foundation, consisting of clay over sand. The ratio of horizontal to

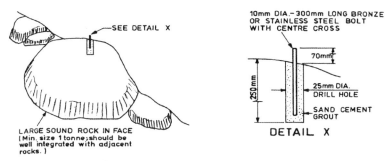

SURVEY BOLT SET IN LARGE SURFACE ROCK

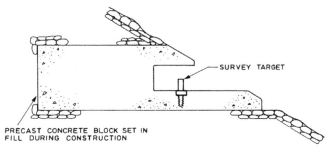

"WILD" BOLT/PRECAST CONCRETE BLOCK

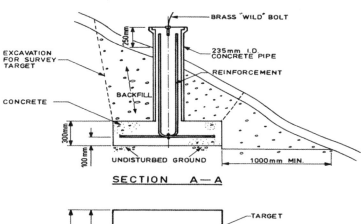

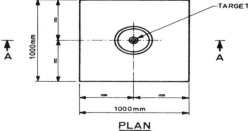

ABUTMENT SURVEY TARGET

Figure 20.10.   Surface settlement points for dams (ANCOLD, 1983).

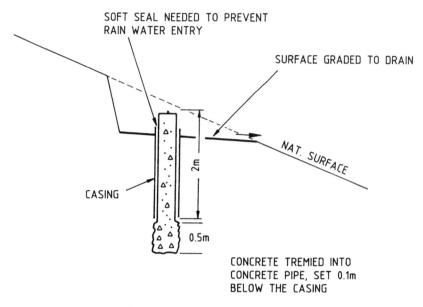

Figure 20.11.   Survey point in soils susceptible to shrink and swell.

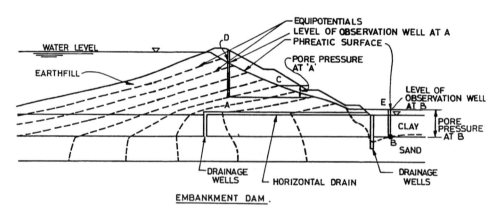

Figure 20.12.   An example of embankment dam piezometric conditions.

vertical permeability (k_h/k_v) of the earthfill is 15:1. Pore pressures in the embankment and foundations are important from slope stability considerations, and for the potential for a 'blowout' or 'heave' condition to form at the toe of the embankment. Piezometers may be installed to check and monitor design assumptions.

The following points should be noted:

- An observation well, installed at A, with a sand surround and/or slots for its full length, will reach equilibrium at the phreatic surface at D;
- The pore pressure at A, i.e. the pressure that would be measured by a piezometer tip set at point A, is equal to the difference in elevation of A and the point where the equipotential through A meets the phreatic surface at C. Note that this may be much less than the difference in elevation from A to D, due to head losses in seepage flow through the earthfill;

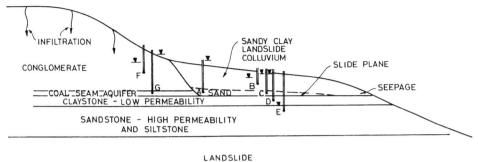

Figure 20.13. Landslide piezometric conditions.

- A slotted standpipe at B (with pipe extending above ground surface) will rise to the phreatic surface at E. In this case the pore pressure at B equals the head BE, because the pore pressure at B is higher than at any other point intersected by the standpipe. If the standpipe was cut off at the ground surface, water would flow from the pipe;
- It is the pore pressure in the embankment which is critical to stability, not the position of the phreatic surface. If k_h/k_v was only say 4, the equipotentials would be steeper and the pore pressures for the embankment would be higher for the same phreatic surface. Hence pore pressures, not the location of the phreatic surface, should be measured.

Landslide. Figure 20.13 shows piezometric conditions in a landslide. Steady state piezometric levels are shown for piezometers installed at points A to G.
The following points should be noted:

- The factor of safety against sliding is very sensitive to the pore pressures. From a slope stability viewpoint it is the pore pressure on the slide plane which is critical, i.e. at A and C. Piezometers (B) in the landslide colluvium may give different pore pressures;
- Piezometers (D), installed in a borehole drilled into a low permeability layer below the slide colluvium, should be backfilled with sand past the slide plane, or the response time for the piezometer will be very slow and pressures different from that on the plane may be measured;
- If boreholes are extended into a more permeable layer as for piezometer E, low pore pressures may be measured. If the sand backfill is extended to the slide plane, water may drain towards the sandstone/siltstone and give lower pressures than actually are present at the slide plane;
- Piezometers installed in the low permeability conglomerate may take a long time to reach equilibrium and are likely to show lower piezometric levels than piezometers installed in the coal seam aquifer (see F and G).

Jointed and sheared rock. Figure 20.14 shows a jointed and sheared rock mass into which a cutting has been excavated. The low permeability shear zone and clay infilled joint affect the flow of groundwater towards the excavation.
The stability of the cutting is determined by the strength of the joints, bedding planes and shear zone, which in turn is determined by the water pressure on these features.
If piezometers are installed in boreholes BH1 and BH2 as shown:

- Piezometer PX will read pressure due to phreatic surface C;
- Piezometer PY will read pressure due to phreatic surface B;
- Piezometer PZ will read pressure due to phreatic surface A.

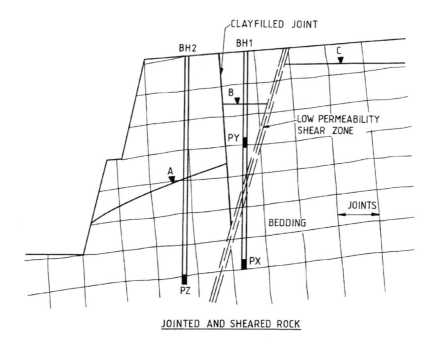

Figure 20.14.   Piezometric conditions in jointed and sheared rock.

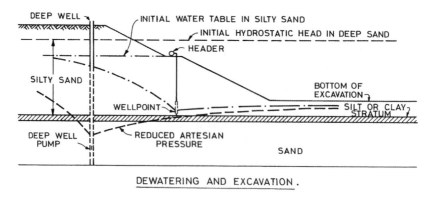

Figure 20.15.   Example of dewatering an excavation.

In practice most rock masses are very complex, and it is not possible to simplify the model to the extent shown in Figure 20.14.

The geological factors which control groundwater movement must therefore be recognised when installing piezometers and several piezometers must be installed so that the critical condition can be defined. The piezometers should be installed with a permeable surround, which intercepts the joints and bedding planes along which groundwater flows.

Dewatering an excavation. Figure 20.15 shows an alluvial soil profile into which an excavation has been constructed. The groundwater has been lowered by well points and deep well pumps.

Monitoring of the effectiveness of the dewatering system, by installation of piezometers, would be needed to check overall slope stability and heave of the bottom of the excavation. To achieve this, piezometers would have to be installed in the silty sand and in the sand because the pore pressure regimes are different. Several piezometers would be needed in each soil unit, because of the rapidly varying pressures away from the well points.

20.4.4.2   Pore air and pore water pressure

The effective stress in a soil σ' is given by:

$$\sigma' = \sigma - u \qquad (20.1)$$

where σ = total stress and u = pore pressure.

In partially saturated soils the pore pressure (u) is a function of the pore air pressure (u_a) and pore water pressure (u_w):

$$u = u_a - \chi(u_a - u_w) \qquad (20.2)$$

where χ is a function of the degree of saturation. For fully saturated soils $\chi = 1.0$, and in a dry soil $\chi = 0$.

In partially saturated soil, pore air pressure is commonly greater than pore water pressure due to surface tension effects on the air-water meniscus. Hence, when measuring pore pressures in partially saturated soils, the piezometers must be constructed so that either pore air or pore water pressure is measured. This is achieved, in most cases, by using high-air-entry filters for the piezometer element which measures pore water pressure, and low-air-entry filters where pore air pressure is to be measured. These are constructed of porous stone, ceramic or bronze. The pores are very small and uniform in size and, when saturated with air free water, air cannot pass into the filter until the applied air pressure exceeds the pressure developed on air-water meniscus at the entrance of each pore in the filter. Hence, if fine "high-air-entry" filters are used, the piezometer will measure pore water pressure and, if coarse filters are used, pore air pressure will be measured.

Air and other gases may also be dissolved in the pore water so even in saturated soils, air may enter the piezometer tip if it is too coarse.

In such cases, the air can lead to incorrect readings by lengthening the response time of the piezometer and forming air-water menisci in the tubes and standpipes which affect the measured pressure.

Sherard (1981) includes a discussion of the differences between pore air and pore water pressure in dam construction. He concludes that:

– In granular cohesionless soils $u_a \approx u_w$;
– In clay soils $u_a > u_w$, but not by more than 2–3 m of water head where $u_w > 5$ m water head. At higher pressures the difference rapidly decreases;
– The difference is always highest during construction when the soil is partially saturated.

Sherard quotes case histories where low-air-entry and high-air-entry tips have been used adjacent each other, with only small differences in readings. However, this experience is in large dams where pore pressures are large in magnitude. In smaller embankments the differences may be significant.

Sherard (1981) recommends use of high-air-entry tips for both hydraulic and vibrating wire piezometers. He suggests use of tips with an entry pressure of 500–600 kPa where these are available. He recommends use of tips with low-air-entry pressures for pneumatic piezometers, the entry pressure being selected to be just greater than the anticipated soil suction (in partially saturated soils). This prevents transfer of water to the surrounding soil from the piezometer tip, while still allowing the transfer of the small volume of water that is required to operate the piezometer.

Table 20.16.   95% and 99% equalization time lag for Casagrande Piezometers (Standards Association Australia 1979).

	Sand	Silt				Clay		
Material Permeability cm/s	10^{-3}	10^{-4}	10^{-5}	10^{-6}	10^{-7}	10^{-8}	10^{-9}
Average time lag for 95% equalization	12 s	2 min	20 min	3.5 h	36 h	14 days	150 days
Average time lag for 99% equalization	18 s	3 min	30 min	5.2 h	54 h	21 days	225 days

20.4.4.3   Fluctuations of pore pressure with time and the lag in response of instruments

In most practical situations the pore pressure being measured varies with time, e.g.:

- Pore pressures in a dam vary with reservoir level;
- Pore pressures in a landslide will vary with rainfall;
- Pore pressures around a pump well will vary with rate of pumping and the time lag between spells of pumping;
- Pore pressures under an embankment on soft clay will vary with the height of the embankment and degree of dissipation by consolidation of pore pressures induced by the embankment;
- Dynamic loading, e.g. earthquakes, may induce an increase in pore pressure.

In all cases, the piezometer which is selected should be constructed so that the time taken for the piezometer to respond to the change in pore pressure is sufficiently short to give a meaningful measure of the actual pore pressure.

In a Casagrande piezometer (Figure 20.17) the piezometer responds to a change of pore pressure by flow of water into or out of the piezometer. This is a function of the geometry of the piezometer, the permeability of the soil and the change in pore pressure.

The response is discussed in Hanna (1973, 1985) and Brand and Premchitt (1980a, b).

Table 20.16 is reproduced from Standards Association Australia (1979) and gives times for 95% and 99% equalization of Casagrande type piezometers with a 73 mm "piezometer tip" and 9.5 mm inside diameter standpipe.

Because of the small volume of water flow required to activate hydraulic, pneumatic, vibrating wire and strain gauge piezometers, their time lags are very small compared to that of Casagrande piezometers.

For hydraulic piezometers (Figure 20.18) the "response" time depends on whether the piezometers have to be flushed, in which case time has to be allowed for pressures to re-establish equilibrium due to the flushing affecting pore pressures around the piezometer tip. It also depends whether individual pressure gauges are provided for each piezometer, or whether a master gauge is used. In the latter case, time must be allowed for equilibrium to be achieved.

Pneumatic piezometers (Figure 20.19) have a small but finite response time, as a small displacement of the diaphragm is needed. The response time varies depending on whether high or low air entry tips are used.

Vibrating wire (Figure 20.20) and strain gauge piezometers have virtually zero response time, as there is no transfer of water involved.

20.4.4.4   Types of instruments and their characteristics

Observation well. As shown in Figure 20.16 an observation well, or "open standpipe", consists of a slotted plastic, reinforced fiberglass or steel pipe attached to a standpipe, installed

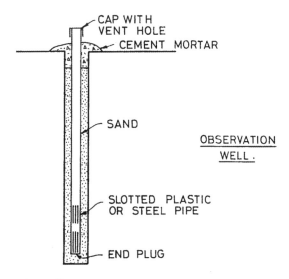

Figure 20.16.   Observation well.

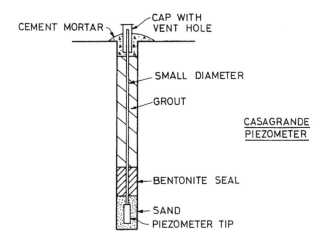

Figure 20.17.   Casagrande type piezometer.

in a borehole and surrounded with sand. Porous tips, as described below for Casagrande piezometers, may be used. When slotted pipe is used, it is often wrapped in filter fabric to prevent clogging by the sand. The hole is sealed at the top with mortar to prevent surface water flowing in. The water level in the well is measured with a water level dip meter.

An observation well normally measures the maximum water pressure intersected in the hole. It does not measure pressure at a point, and does not account for flownet effects (see Figure 20.12). In some situations, with confined aquifers of varying permeability, an observation well may not measure the maximum pressure, with water from the high pressure zone flowing into a more permeable lower pressure zone.

The time lag for observation wells may also be large, since a relatively large flow of water is required for the well to reach equilibrium pressure change.

In most situations observation wells are not adequate, and piezometers should be installed.

Casagrande piezometer. Figure 20.17 shows a Casagrande type piezometer. These consist of a porous tip which is embedded in a sand filter, and sealed into a borehole with a short length (usually 1 metre) of bentonite. The hole should be backfilled with cement bentonite grout. In dam cores grouting should be staged to avoid inducing hydraulic fracture. The standpipe is kept as small a diameter as practical to keep the response time to a minimum. Dunnicliff (1982) recommends the standpipe be not smaller than 10 mm diameter so that gas bubbles will not be trapped. In any case, this is a practical minimum for use of a dip meter to measure the water level. The sand filter acts as a large collector of water compared to the small diameter standpipe.

The bentonite seal may be formed using dry pellets which are available commercially, or by making 25 mm diameter moist balls of powder bentonite. These are dropped down the hole, possibly tamped to compact and allow checking of the position and depth of the seal.

Casagrande, and all standpipe piezometers, measure pore water pressure (u_w). They are potentially accurate, depending on how carefully the water level is determined. The time lag can be determined as outlined in Section 20.4.4.3. Time lag may be a problem in lower permeability soils, and where the pore pressure is changing rapidly.

The Casagrande type piezometer is low cost, simple to construct with readily available equipment and materials. They are self de-airing, measure pore water pressure and have a long, satisfactory performance record. Disadvantages include susceptibility to damage by construction equipment or animals and by consolidation of soil around the standpipe, possible leakage from the surface runoff if the grout seal is not adequate, susceptibility to freezing and inability to measure pressures higher than the level of the standpipe or pressures lower than atmospheric. If porous filter tips are used, these may clog with chemical deposits with flow in and out of the tip. Reading is simple but relatively time consuming. Auto-data logging is not practicable.

Hydraulic twin tube piezometer. Figure 20.18(a) shows an Imperial College (Bishop) type hydraulic tube piezometer. Variations of the principle include the USBR type shown in Figure 20.18(b).

These consist of a porous tip, which is usually installed during construction (e.g. in a dam embankment), from which two tubes are led to a convenient terminal measuring point where gauges are used to record the pressure. The total pressure equals tip pressure, plus elevation difference from the tip to the measuring gauge.

Two tubes are provided so that air (gas), which is trapped in the tubes or which enters through the porous tip, can be flushed from the system using de-aired water. The tips are constructed of fine porous ceramic, stone, aluminium or sintered bronze. The tips are required to be strong enough to withstand the total pressure, and fine enough to prevent clogging by the soil. More importantly, the tips must be "high-air-entry" to limit entry of air into the system. Commonly, high-air-entry tips can withstand an air pressure of 100 kPa to 200 kPa, but may be up to 500 kPa to 600 kPa.

Coarse, low-air-entry tips are not used with hydraulic piezometers because they allow air to enter the tip, the piezometer and tubes. This leads to pore air pressure rather than pore water pressure being measured in partially saturated soils, a longer response time for the piezometer, and incorrect elevation pressure adjustment if the air enters the tubes. The high-air-entry tips are only effective if they are completely de-aired before installation. This is usually done by boiling for 2 hours to 6 hours. Sherard (1981) suggests 12 hours to 24 hours boiling.

The tubes leading to the measuring gauge should be constructed of Nylon 11, covered by polyethylene. The Nylon 11 is impervious to air, the polyethylene impervious to water, and the combination is impervious, strong, flexible and not affected by chemical attack.

Despite the use of high-air-entry tips, and the proper tubing, gas may still accumulate in the tubes, and periodic flushing with de-aired water is necessary. This should be done under back pressure to ensure that negative pore pressures are not induced. Some organisations add bacteriacides and algacides to the water, to prevent algae growth blocking the tubes.

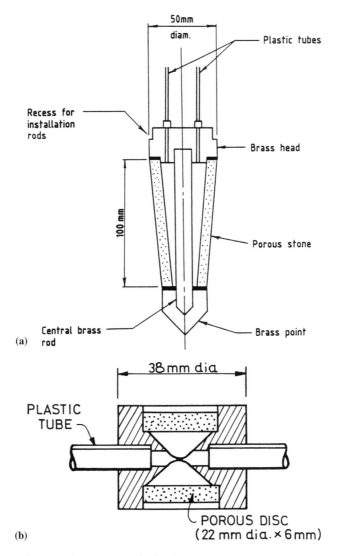

Figure 20.18.   (a) Bishop and (b) USBR type hydraulic piezometers (ANCOLD, 1983).

Some organisations use flow meters on the inlet and outlet lines as an aid to flushing. A flow of 300 ml/min is maintained to prevent air becoming trapped in high spots in the lines. It is also good practice to flush before all readings, since equal pressures on the lines can occur with air in both lines.

Eagles (1987) reports that if the entry to the piezometer is smaller than the tubing (e.g. 1.5 mm to 2 mm compared to 5 mm), crystal growth from chemicals dissolved from the tubing can cause blockage and it is important that the entry be the same diameter as the tubing.

The tip pressure is measured using Bourdon gauges, measuring manometers, or pressure transducers in the terminal structure. Most installations use Bourdon pressure gauges – either one gauge for each piezometer, with a master gauge to check, or a master gauge with a switching system. Typical layouts are given in ANCOLD (1983) and US Department of Interior (1981) and USBR (1987). When using a master gauge, it is necessary to estimate the tip pressure and pressurize the gauge to this pressure before connecting the piezometer. Any imbalance will involve flow of water into or out of the piezometer tip, with resultant lag time, and possibly influence the pressure, particularly if the soil is partially saturated and water flows out of the piezometer tip. Use of a gauge for each piezometer with a check master gauge overcomes this.

The terminal structure and tubes must be located so that no more than about 5 m negative head is generated, or cavitation will occur in the tubes, giving incorrect readings.

In embankment construction, the tubes have to be laid loosely in trenches to avoid damage by construction equipment and differential settlement effects. Trenches are usually filled with sand, with bentonite clay or clay cutoffs to prevent water seepage along the trenches. (See examples in US Department of Interior, 1981) and USBR (1987). However in many dams the trenches were backfilled with loose soil similar to the dam core, rather than sand, and piezometers were located close to the upstream face of the earth core. This leaves a potential weakness in the dam from an internal erosion and piping point of view, since high gradients may occur from the upstream face of the core to the first piezometer and, if erosion were to initiate, it would quickly extend through the loose soil. This would potentially put high gradients across the bentonite/soil cutoffs.

The authors are not aware of any cases where this has happened, but strongly recommend against this practice for new dams.

Provided that high-air-entry tips are used, and the tips installed with intimate contact with the soil, hydraulic piezometers will read pore water pressure. Subatmospheric pressures up to about 50 kPa to 70 kPa can be read.

Twin tube hydraulic piezometers are relatively simple and medium cost to construct, although when the cost of the terminal house (which may have to be dewatered to position it satisfactorily, to avoid excessive negative pressures) is included, the overall cost may be comparable with vibrating wire and pneumatic systems.

Auto-data logging is possible by using electrical pressure transducers. Hydraulic piezometers have a medium long term reliability (15 years to 20 years in many cases), and have been the favoured instruments in British and Australian dams. However most hydraulic piezometers more than 30 or 40 years old are now not functioning. They were used in USBR dams until 1978 when pneumatic piezometers were introduced, because they are easier to read and maintain in good working condition. Disadvantages include "growths" in the tubes; damage of tubes by differential settlement; the need for an elaborate terminal house, often requiring dewatering; skill and care to keep them operating and the need to avoid high subatmospheric pressures in the lines. Particular advantages are the ability to flush air from the piezometer tip if necessary, hence ensuring measurement of u_w, ability to measure moderate subatmospheric pressures, and the possible use of the piezometer to carry out in-situ permeability and hydraulic fracture tests of the soil. As pointed out by Sherard (1981), hydraulic (and pneumatic and vibrating wire) piezometers should not be installed in a surround of sand. Doing so only slows the response time if the sand surround becomes partially saturated and since the piezometers are essentially no flow, a large sand surround has no benefit in collecting water.

The authors know of no organization installing hydraulic piezometers in new dams and do not recommend them.

Pneumatic piezometer. Figure 20.19 demonstrates a pneumatic piezometer consists of a piezometer tip, from which two tubes are led to a convenient terminal measuring point. Pressure is measured by applying pressurised gas (usually dry nitrogen) on the inlet tube. The diaphragm is moved, and gas escapes to the outlet tube where it can be observed. The pressure on the inlet tube is then reduced until the diaphragm closes (no flow from the outlet tube). The pressure at which the diaphragm closes is the pore pressure (after correction for any closure spring effect). Some manufacturers measure the pressure on opening of the diaphragm but most use the principles outlined above.

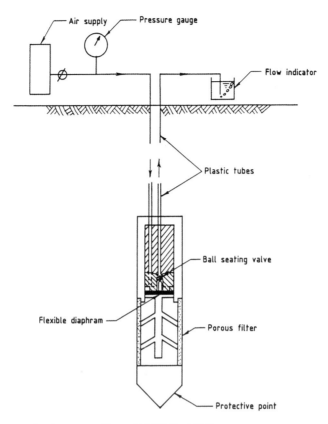

Figure 20.19.   Pneumatic piezometer (from ANCOLD, 1983).

The piezometers are fitted with high- or low-air-entry tips. Sherard (1981) discusses at some length the relative merits of using high- or low-air-entry tips. He concludes that if low-air-entry tips are used in unsaturated soil, the water in the tip will be sucked out by capillary action and the piezometer will measure pore air pressure u_a. However, if high-air-entry tips are used, the response time during reading may be affected, and the piezometer may read an erroneously high pressure. This is because of the small volume of water displaced by the action of the diaphragm. Sherard suggests using tips with an air-entry value just above the soil suction pressure. This is not a problem when the piezometers are being used in saturated soils and, since it is easier to de-air low-air-entry tips, these are often used. Where high pore pressures are to be measured the pore air and pore water pressures are similar so the choice of tip is not critical.

The tubes should be of the same construction as for hydraulic piezometers, i.e. Nylon 11 sheathed in polyethylene, as this gives strong, stiff tubes, impervious to ingress of water and leakage of air. If water does enter the tubes, flushing with dry nitrogen is necessary. The use of lighter tubing can result in problems with crushing and kinking and the savings in cost may not be warranted.

The pressure is usually read by a portable readout which is connected to the tubes with quick-connect couplings. Either a Bourdon or digital type gauge is used. The latter can be adapted to auto-data logging with the terminal permanently installed and programmed to interrogate each gauge in turn at predetermined intervals.

The terminal pressure gauge can be located at any elevation relative to the piezometer and tubes can be laid in any configuration. In embankments, the tubes must be laid in trenches to avoid damage by construction equipment and looped, coiled etc. to allow for differential movement. As for hydraulic piezometers, these trenches are a potential point of weakness. The piezometers are installed into existing dams using a similar system to that shown in Figure 20.17, with piezometer in a sand surround, with a bentonite seal, and cement/bentonite grout.

Because of their method of construction and reading, pneumatic piezometers cannot be used to measure subatmospheric pressures.

Pneumatic piezometers are relatively simple and low cost construction, and the terminal gauge house requirement is much simpler than for hydraulic piezometers. Pneumatic piezometers have performed satisfactorily over 10 years to 15 years in the USA (Sherard, 1981) and were then the preferred instrument for dams in that country because they are easier to read, easier to install and auto-data logging is simple. Less skilled operation is needed than for hydraulic piezometers. There has been some reticence to adopt them for dams in some other countries, including Australia. Where used, their performance has been acceptable, but many cease to work after some years. They have been used in boreholes in rock (and in concrete dams) by installing with a packer, allowing replacement if this was necessary. The time for taking measurements is dependent on the readout equipment and the length of the tubes to the piezometer. Sherard (1981) quotes read times of 1 minute to 2 minutes for short tubes, but up to 10 minutes to 20 minutes for very long tubes. He indicates this was reduced to 3 minutes to 5 minutes by initial rapid gas filling. There are no problems of freezing with pneumatic piezometers.

Disadvantages include long measurement time for long tubes and high and subatmospheric pressures cannot be read.

Vibrating wire piezometer. Figure 20.20 shows the principle of a vibrating wire piezometer (also known as an acoustic piezometer).

These consist of a piezometer tip from which an electrical cable leads to a convenient terminal measuring point. The piezometer consists of a porous tip, as in a pneumatic piezometer with a stiff metallic diaphragm. The diaphragm is attached to a prestressed wire. When the diaphragm deflects with changes in pore pressure, the tension in the wire changes. The natural frequency of the wire is dependent on the tension and is measured by plucking the wire, using an electromagnet. The wire then vibrates in the magnetic field of a permanent magnet, causing an alternating voltage of the same frequency as the wire, and this can be measured at the measuring point.

Hence, by measuring the natural frequency of vibration of the wire and calibrating this against pressure on the diaphragm, the instrument can be used to measure pore pressure.

The instruments are essentially no flow devices and, therefore, have a very short response times. Sherard (1981) suggests that they are best fitted with high-air-entry tips to limit entry of air into the piezometer and ensure that pore water pressure is measured. A digital readout is usually provided and the instruments are readily suited to auto-data logging. The readout unit can be located at any elevation relative to the piezometer and up to several kilometres away. The wires can be laid in any configuration, but provision should be made for differential movement. In existing dams, vibrating wire piezometers are installed in boreholes and sealed in place as shown in Figure 20.17. Provided they are installed with intimate contact with the soil, vibrating wire piezometers can be used to measure subatmospheric pressure.

Vibrating wire piezometers are more expensive than hydraulic or pneumatic piezometers, but have been used widely in European, USA, Canadian and Australian dams, with reliable performance for up to 20 years (Sherard, 1981). Australian experience is that most piezometers more than 30 years old are no longer functioning. They are the easiest and quickest to read of all piezometers and are not susceptible to freezing.

There are varying opinions on whether they are susceptible to long term drift, due to creep of the diaphragm or tension wire. Sherard (1981) presents information suggesting

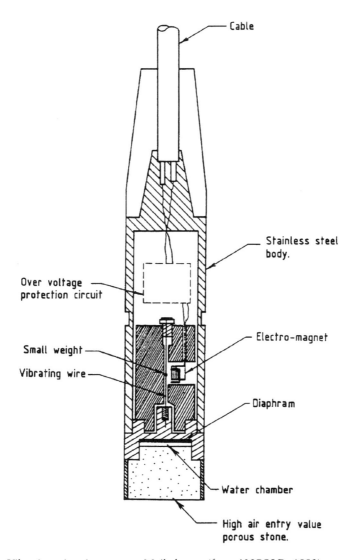

Figure 20.20.   Vibrating wire piezometer – Maihak type (from ANCOLD, 1983).

this is not a problem. Australian experience is that it can be. Other advantages are that, by using high-air-entry tips, the saturation of the tips can be checked by observing the development of negative pressure as the tip is allowed to dry slightly and, in partially saturated soils, water will not be sucked from the tip into the soil (unlike hydraulic piezometers which may wet the soil during the flushing operation).

Disadvantages have included damage by lightning strike, particularly but not only during construction when the cables are exposed. This has been overcome by shielding cables, earthing, and provision of overvoltage protection. Stray currents from nearby power stations have also caused problems.

Strain gauge piezometer. Strain gauge piezometers are similar in construction to the vibrating wire piezometer, but the deflection of the diaphragm is measured by either bonded electrical strain gauges or Carlson type unbonded strain meters. Pore pressures are recorded as changes in resistance. The instruments are not as widely used as vibrating wire piezometers, mainly because of the possibility of drift in the long term and with temperature. However, the instruments have the advantage of being easy to read, easily auto-data logged and independent of the relative location of the piezometer tip and read-out unit. As with vibrating wire instruments, they are susceptible to damage by lightning strike and overvoltage protection is needed.

20.4.4.5 *Should piezometers be installed in the cores of earth and earth and rockfill dams?*

Many large dams built from the 1960s to 1990s have piezometers installed in the dam core. These were usually installed as the dam was constructed and the piezometer tubes (or wires for vibrating wire piezometers) were taken to the downstream side in trenches as described above. In some dams, the wires or pneumatic piezometer tubes were taken to the top of the dam in groups in vertical riser pipes.

Either method leaves a potential weakness in the dam and some senior dam engineers, including the authors, are reluctant to put such potential weaknesses into a dam core, because of the chance that internal erosion and piping may initiate along the trenches/risers. It is arguable whether, for a well-designed and constructed dam with filters downstream of the core, it matters what the pore pressures are.

Having said that, the authors have on a number of occasions found the long term trends in pore pressures valuable in assessing slope instability and potential piping problems. If piezometers must be installed, it is suggested the tubes/wires are laid in a winding fashion in a trench backfilled as recommended in Bureau of Reclamation (1987) with well compacted dry mixture of bentonite and filter sand (1 part bentonite, 3 parts filter sand by dry weight). The degree of compaction should match that of the adjacent fill. Seepage cutoff trenches should be provided, extending at least 0.6 m into the adjacent fill, and 0.3 m wide, where zone contacts are crossed between each piezometer in the earthfill core and at not more than 15 metre intervals.

The bentonite sand backfill should be compacted because otherwise a gap may form at the top of the trench as the sand densifies under compaction equipment and on saturation. This may make it more likely for tubes/wires to be broken but the potential for a gap must be avoided.

The sides of the trench are likely to desiccate and crack, and cracked material should be removed before backfilling.

Piezometers should not be installed too close to the upstream face of the core – say no closer than a quarter of the core width and preferably not in the upstream half for narrow cores.

Where existing piezometers have ceased to function, it is usually better to go without the information, than to run the risk of damaging the core during drilling and backfill grouting of the holes needed to install new piezometers. If there are signs of deterioration necessitating installation of piezometers in the core, these should be installed in boreholes using dry drilling techniques, such as hollow flight augers so as to avoid hydraulic fracture of the core. The backfill cement/bentonite grout should be installed in short lengths, using a tremie system, with time between stages in the backfill to allow the grout to develop strength, so hydraulic fracture will not be caused by the grout. This work should be supervised by experienced geotechnical professionals familiar with dam engineering.

20.4.5 *Displacements and deformation*

Internal displacements and deformations were often monitored on larger earth, earth and rockfill and concrete face rockfill dams. This allowed confirmation of design assumptions, particularly relating to long-term settlement of earthfill and foundations, and to concrete face designs.

Hanna (1985), ANCOLD (1983), US Department of Interior (1981) and Bureau of Reclamation (1987) give details of the instruments used. Up to date information can be obtained from instrument manufacturers. A brief overview of the types of instruments used follows.

20.4.5.1   Vertical displacements and deformation

Vertical displacements and deformations are measured using mechanical or magnetic devices which are either embedded in the fill as the dam is constructed, or installed in boreholes. Figure 20.21 shows the principle of the USBR type mechanical device.

The base tube is installed in the dam foundation and a cross-arm anchor placed in the dam foundations. Telescoping tubing is placed in the fill as the dam is constructed, with cross-arm anchors being installed at intervals to allow differential settlement within the fill to be measured. Typically cross-arms will be spaced at 2 m or 3 m intervals. Settlement

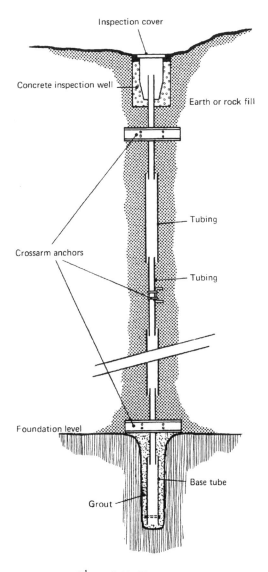

Figure 20.21.   USBR type cross-arm settlement gauge.

readings are obtained by survey of the top of the coring and lowering a measuring torpedo down the tube. The torpedo is provided with spring-loaded arms which engage the lower ends of the cross-arm pipes. It is supported on a measuring tape, and can allow accuracy of measurement of ±3 mm (accuracies as small as ±1 mm are claimed). Settlements as much as 15% can be accommodated and the testing can be installed to depths of at least 100 m. The tubing does not have to be vertical (Soil Instruments Ltd., 1985, indicate ±25° from vertical is allowable).

The USBR instrument cannot be installed down a borehole, only in fill as it is constructed.

Figure 20.22 shows a magnetic version of the instrument which uses ring magnets instead of the cross-arms.

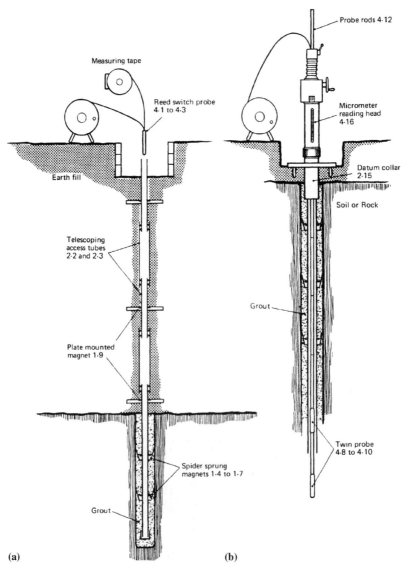

Figure 20.22.   Magnet (Soil Instruments Ltd., 1985) settlement gauges: (a) magnets embedded as earth-fill is placed; (b) magnets placed in a borehole.

In fill, ring magnets are placed in the ground at predetermined intervals, in the same way as the cross-arms are for the USBR device. In boreholes a spring loaded magnet is lowered through the grout supporting the borehole, and is held to the walls of the hole by the spring and the grout. The grout has a compressibility equal to, or greater than, the soil or rock surrounding it. The probe is lowered down the tube and senses the position of the magnets. If suspended by a tape, the accuracy of measurement is ±1 mm. Soil Instruments Ltd. (1985) indicate that a rod-mounted version fitted with a micrometer device to help position the magnet can read to ±0.1 mm.

The tape suspended device can be used in the same conditions as the USBR device, i.e. in near vertical holes. The rod-mounted device can be used in any orientation, so can be used to measure horizontal displacements, as well as vertical.

While these instruments have provided useful information on dam deformations, it is questionable whether they should be installed in earthfill cores, because it is very difficult to compact around the riser, and these can be a source of local collapse settlement on saturation giving a potential defect in the dam from an internal erosion and piping perspective.

Figure 20.23 shows the principle of a hydraulic overflow settlement cell.

This works on a simple U tube principle. The cell is cast into the fill and connected by a water tube to a graduated standpipe. A second drain tube allows surplus water to flow from the cell and a third "air" tube maintains the interior of the cell at atmospheric pressure. Compressed air is used to clear the air tube and drain tube, then the water tube is

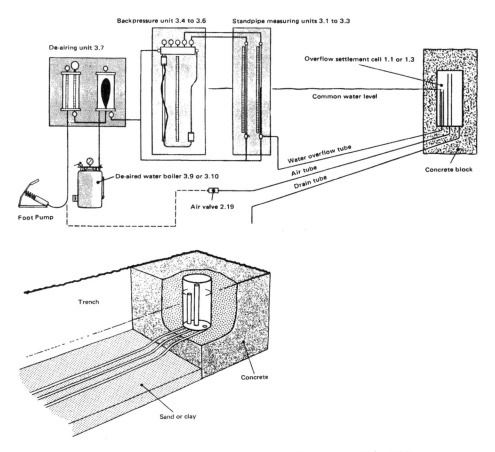

Figure 20.23.   Hydraulic overflow settlement equipment (Soil Instruments Ltd., 1985).

filled with de-aired water, pumping sufficient water to remove any bubbles in the water tube. After allowing time to reach equilibrium, the level of water in the standpipe is the same as that at the cell. The range of the instrument can be extended by applying small back pressures to the standpipe.

The method is very simple and inexpensive and, according to Soil Instruments Ltd., gives an accuracy of the order of ± 1 mm to ± 5 mm, depending on the back pressure being applied. Hanna (1985) indicated an accuracy of ± 2 mm. Tube lengths up to 300 m can be used. ANCOLD (1983) gives details of such a device, which uses two cells at the same location. These are at slightly different levels, allowing checking of whether there is air in the lines, and rotational movement of the cell.

These devices are used in rockfill zones where it would be difficult to install cross arm devices, but may be used in earthfill. Hanna (1985) and Soil Instruments Ltd. (1985) describe the use of pneumatic and vibrating wire versions of the hydraulic settlement cell.

Again, in earthfill cores there may be a potential defect due to poor compaction and it is questionable whether they should be installed.

20.4.5.2   Horizontal displacements and deformations

On some very large dams, horizontal displacements and deformations are measured using mechanical and electronic devices embedded in the dam fill as the dam is constructed.

Their use is usually combined with hydraulic settlement gauges, so that the settled profile can be determined. The mechanical instruments are installed in rockfill zones, it being considered unwise to place them through the earth core as they may form a preferred leakage path.

Figure 20.24 shows a mechanical device which uses long wires, anchored to cross-arms embedded in the dam. The wires are held at constant tension, and displacements measured at the terminal well, which is situated on the face of the dam.

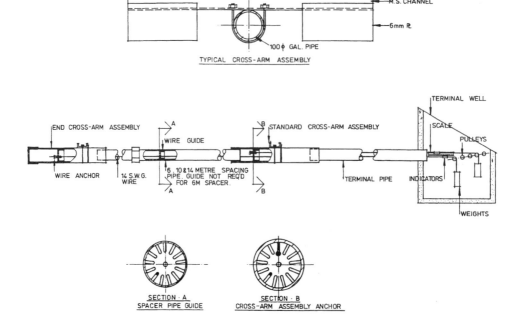

Figure 20.24.   Long wire extensometer (ANCOLD, 1983).

The wires are housed in telescoping segments of pipe. Corrections for overall translation is made by surveying the position of the terminal well. ANCOLD (1983) indicate the accuracy to be ±1.5 mm.

Soil Instruments Ltd. (1985), and Hanna (1985), describe other extensometers which can be used to measure horizontal displacements and strains. These include devices which use vibrating wire and resistance potentiometers to do the measurements.

Figure 20.25 shows the use of such devices to measure deformations in the earth core of a dam parallel to the dam axis. In this case, several extensometers are installed with extension rods connecting the anchor plate and the vibrating wire extensometers. Whether it is wise to install such instruments is doubtful, because of the defect it gives in the dam core and the information is not usually so important.

Lateral movements in dams can also be measured by use of borehole inclinometers. The inclinometer casing may be installed during construction of the dam, or after, in the event that excessive displacements are observed. If the casing is installed on a slope, e.g. along the face slab of a concrete face rockfill dam, horizontal and vertical displacements can be determined.

Most inclinometers are similar to that shown in Figure 20.26, in that the inclinometer is passed down casing which is grooved to maintain the orientation of the sensor probe.

The tilt of the sensor is measured and recorded by servo-accelerometers located within the probe. Where two accelerometers are used they can sense the inclination of the tube in two directions at right angles to each other. Processing of the signal at the surface allows display of the angular rotation, or displacement normal to the casing in the two directions. Settlement gauge magnets may be installed around the casing to allow assessment of axial deformations.

The casing may be plastic or aluminium and should be able to telescope, to allow for differential settlement. Soil Instruments Ltd. (1985) indicate a resolution of ±0.1 mm. Hanna (1985) indicates that careful calibration and maintenance is necessary to achieve a high precision.

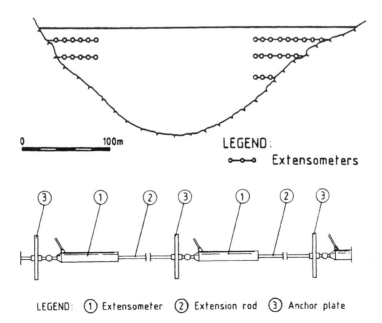

Figure 20.25.   Use of extensometer to measure strain parallel to the dam axis (ICOLD, 1989).

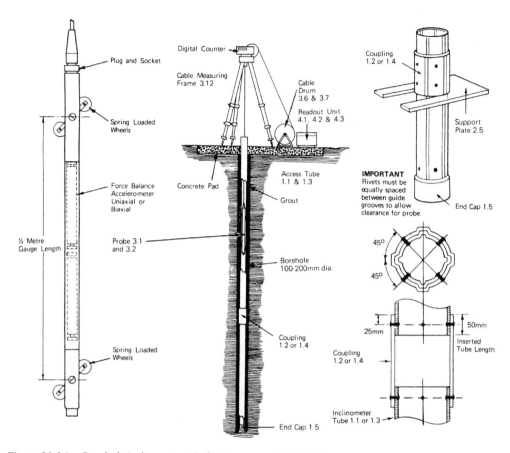

Figure 20.26.   Borehole inclinometer (Soil Instruments Ltd., 1985).

The authors' experience is that a high degree of reproduceability is possible, although considerable care must be taken with instruments significantly off the vertical. The authors also feel that inclinators are best utilised to show distinctive trends in shear displacement rather than small displacements of 1–2 mm.

Hanna (1985) indicates that the grooves in the casing may spiral, and individual lengths need to be checked for this.

ANCOLD (1983) describe an inclinometer which can be used in smooth pipe. It contains two servo-accelerometers. The device was developed for monitoring displacements of the face of concrete face rockfill dams and has an accuracy of ±3 mm.

Borehole inclinometers are a valuable tool for investigating dams with potential instability, or large internal deformations. However, as for piezometers, great care must be taken in drilling the holes using dry drilling methods and grouting around the tubing in stages using a tremie system, so the core is not damaged by hydraulic fracture.

20.4.6   Thermal monitoring of seepage

There have been relatively recent developments in detection of localized seepage using thermal measurement techniques. These rely on the difference of temperature of the reservoir water and that in the dam generally. Hence if the reservoir water is colder localized high seepage will be detected as a low temperature anomaly. The temperature is measured in a number of ways:

(a) Infra-red imaging of the downstream face of the dam and foundations. Tedd and Hart (1988) report on trials on two dams. These showed that factors other than seepage also influence the temperature and the method would seem to be of limited usefulness;
(b) Thermotic sensors in stand pipes in the dam. Dornstader (1996), Johansson (1996, 1997) give examples and report that the method is quite successful and able to detect small variations in temperature. The main disadvantage is that stand-pipes have to be inserted into the dam;
(c) Optical fibres embedded in the dam. USEPA (2000), Aufleger et al. (2000) and Johansson et al. (2000) give examples which show the method is useful. However the optical fibres have to be built into the dam, so the method is of limited applicability for existing dams.

References

Aas, G., Lacasse, S., Lunne, T. and Höeg, K. (1986). Use of in-situ tests for foundation design in clay. *Use of in-situ tests in geotechnical engineering, Geotech. Special. Publ. No. 6, ASCE, 1–30.*

AASHTO (1983). Report of Task Force 25, Joint Committee report of AASHTO-AGC-ARTBA.

Abadjiev, C.B. (1985). Estimation of the physical characteristics of deposited tailings in the tailings dam of non ferrous metallurgy. *Proc. 11th Int. Conf. on Soil Mechanics and Foundation Engineering,* San Francisco, Vol. 3, 1231–1234.

Acker, R.C. and Jones, J.C. (1972). Foundation and abutment treatment for rockfill dams. *JASCE SM10,* 1053–1092.

Aitchison, G.D. (1953). The mechanics of gilgai formation. *Proc. Conf. on Soil Science, Adelaide,* Vol. 2 (6.25), 1–3.

Aitchison, G.D. and Wood, C.C. (1965). Some interactions of compaction, permeability and post-construction deflocculation affecting the probability of piping failure in small earth dams. *Proc. 6th ICSMFE,* Vol. 2, 442–446.

Aitchison, G.D., Ingles, O.G. and Wood, C.C. (1963). Post construction deflocculation as a contributory factor in the failure of earth dams. *4th ANZ Conference on Soil Mechanics and Foundation Engineering,* 275–279.

Akroyd, T.N.N. (1957, 1975). *Laboratory testing in soil engineering.*

Alemo, J., Bronner, N. and Johansson, N. (1991). Long-term stability of grout curtains. International Congress on Large Dams, Vienna, Question 65, Paper 73, 1311–1325, International Commission on Large Dams, Paris.

Alexander, L.G. (1960). Field and laboratory tests in rock mechanics. *Proc. 3rd Australia New Zealand Conference on Soil Mechanics and Foundation Engineering.*

Alonso, E.E., Gens, A.K. and Josa, A. (1990). A constitutive model for partially saturated soils. *Geotechnique.* Vol. 40, No. 3, 405–430.

Amaya, F. and Marulanda, A. (1985). Golilas Dam – design, construction and performance In *Concrete Face Rockfill Dams, Design, Construction and Performance,* (eds) Cooke, J.B. and Sherard, J.L. ASCE Geotechnical Engineering Division, 98–120.

Amaya, F. and Marulanda, A. (2000). Columbian experience in the design and construction of concrete face rockfill dams. In *Concrete Face Rockfill Dams, J. Barry Cooke Volume* (Mori, Sobrinho, Dijkstra, Guocheng & Borgatti ed.), Beijing, 89–115.

Ambraseys, N.N. (1973). Dynamics and response of foundation materials in epicentral regions of strong earthquakes. *Proc. 5th World Conference in Earthquake Engineering,* Rome, Italy.

Amjad Agha (1980). Dam engineering in Pakistan: In Balasubramanian, A.S., Yudhbir Tomiolo, A. and Younger, J.S. (eds). *Geotechnical Problems and Practice of Dam Engineering.* Balkema.

Anagnosti, P. and Popovic, M. (1982). Evaluation of shear strength for coarse-grained granular materials. *Fourtheenth Congress on Large Dams,* Rio de Janeiro, Q55, 753–767, International Commission on Large Dams, Paris.

ANCOLD (1976). *Guidelines for operational, maintenance and surveillance of dams.* Australian National Committee on Large Dams, Melbourne.

ANCOLD (1982). *Report on mesh protection of rockfill dams and coffer dams.* Australian National Committee on Large Dams, Melbourne.

ANCOLD (1983). *Guidelines for dam instrumentation and monitoring systems.* Australian National Committee on Large Dams, Melbourne.

ANCOLD (1985). *25th Annual General Meeting Study Tour Technical Information.* Australian National Committee on Large Dams, Melbourne.

ANCOLD (1986). *Guidelines on design floods for dams.* Australian National Committee on Large Dams, Melbourne.

ANCOLD (1990). *Dams in Australia*. Australian National Committee on Large Dams, Melbourne.

ANCOLD (1991). *Guidelines on design criteria for concrete gravity dams*. Australian National Committee on Large Dams, Melbourne.

ANCOLD (1998). *Guidelines for design of dams for earthquake*. Australian National Committee on Large Dams, Melbourne.

ANCOLD (1999). *Guidelines on dam safety management*. Australian National Committee on Large Dams, Melbourne. ISBN 0 731 027 620.

ANCOLD (2000). *Dam Technology in Australia*, 1850–1999. Editor B. Cole. Australian National Committee on Large Dams Inc., Melbourne. ISBN 095 7830 807.

ANCOLD (2000). *Guidelines on selection of an acceptable flood capacity for dams*. Australian National Committee on Large Dams. Melbourne.

ANCOLD (2003). *Guidelines on dam safety management*. Australian National Committee on Large Dams, Melbourne. ISBN 0731 027 620.

Andrews, R.C. (2003). Personal communication. Infra Techno Consultants (Australia) Pty Ltd.

Andrus, R.D. and Stokoe, K.H., II (2000). 'Liquefaction resistance of soils from shear-wave velocity'. *J. Geotech. and Geoenvir. Engrg.*, ASCE, Vol. 126, No. 11, 1015–1025.

Andrus, R.D. and Stokoe, K.H. (1997). Liquefaction resistance based on shear wave velocity. *Proc., NCEER workshop on evaluation of liquefaction resistance of soils*, Nat. Ctr. for Earthquake Engrg. Res., State University of New York at Buffalo, 89–128.

Ang, C.L.D. (2001). *Performance of Concrete Face Rockfill Dams*. Thesis (unpublished), Bachelor of Engineering, School of Civil and Environmental Engineering, The University of New South Wales.

Arnao, B.M., Garga, V.K., Wright, R.S. and Perez, J.Y. (1984). The Tablachaca Slide No. 5, Peru, and its stabilization. *Proc. 4th Intl. Symp. on Landslides, Montreal*, 597–604.

Arulanandan, K. and Perry, E.B. (1983). Erosion in relation to filter design criteria in earth dams. *Journal of the Geotechnical Engineering Division*, ASCE, 109 (GT5), 682–698.

Arulanandan, K., Loganathan, P. and Krone, R.B. (1975). Pore and eroding fluid influences on surface erosion of soils. *Journal of the Geotechnical Engineering Division*, ASCE, 101 (GT1), 55–66.

ASTM (1968). *Standard test method for capillary – moisture relationships for coarse- and medium-textured soils by porous-plate apparatus*, ASTM D2325-68. American Society for Testing and Materials, Philadelphia.

ASTM (1972). *Standard test method for capillary-moisture relationships for tine-textured soils by pressure-membrane apparatus*, ASTM D3152-72. American Society for Testing and Materials, Philadelphia.

ASTM (1974a). *Standard recommended practice for petrographic examination of aggregates for concrete*. ASTM Book of Standards, 14 (C295-65), 209–218.

ASTM (1974b). *Standard method of tests for potential reactivity of aggregates (chemical method)*. ASTM Book of Standards, 14 (C289-71), 186–195.

ASTM (1974c). *Standard method of tests for potential alkali reactivity of cement-aggregate combinations (mortar-bar method)*. ASTM Book of Standards, 14 (C227-71), 142–147.

ASTM (1992). *Standard test method for measurement of soil potential (suction) using filter paper*, ASTM D5298-92. American Society for Testing and Materials, Philadelphia.

ASTM (1998). American Society for Testing and Materials, Volume 04.08.

ASTM (2001). American Society for Testing and Materials, Volume 04.08.

Atkinson, J.H. and Evans, J.S. (1984). Discussion on "The measurement of soil stiffness in the triaxial apparatus". *Geotechnique*, Vol. 35, No. 3, 378–382.

Aufleger, M., Strobl, T. and Dornstädter, J. (2000). Fibre optic temperature measurements in dam monitoring – four years of experience. *20th Congress on Large Dams, Beijing 2000*. Q78-R.1, 157–170. International Commission on Large Dams, Paris.

Azzouz, A.S., Baligh, M.M. and Ladd, C.C. (1983). Corrected field vane strength for embankment design. *Proc. ASCE J.Geotech. Eng. Div.*, Vol. 109, No. GT5, 730–734.

Bakker, K.J. (1987). Hydraulic filter criteria for banks, *European Conference on Soil Mechanics and Foundation Engineering*, Dublin.

Bakker, K.J., Breteler, M.K. and den Adel, H. (1990). New criteria for filters and geotextile filters under revetments. *Int. Conf. on Coastal Engineering*, Delft.

Baldi, G., Hight, D. and Romas, G. (1988). A Re-evaluation of conventional triaxial test methods in *Advanced Triaxial Testing of Soil and Rock*, ASTM STP 977, Robert T. Donaghe, Ronald, C.

Chaney and Marshall Silver Eds. American Society for Testing and Materials, Philadelphia. 219–263.

Balmer, G. (1952). A general analytical solution for Mohr's envelope. *American Society of Testing and Materials* 52, 1260–1271.

Barber, E.S. and Sawyer, C.L. (1952). Highway subdrainage, Proceedings, Highway Research Board, 643–666.

Bardin, L. and Sides, G.R. (1970). Engineering behaviour and structure of compacted clay. *J. Soil Mechanics and Foundation Engineering*, ASCE, Vol. 96, No. SM4: 1171–1201.

Bardossy, G. and Aleva, G.J.J. (1990). *Lateritic bauxites.* Elsevier, New York.

Barton, N.R. (1982). Shear strength investigations for surface mining. In Stability in surface mining. *Proceedings of the 3rd International Conference on Stability in Surface Mining* (ed. Brawner, C.O.), 1–3 June 1981, Vancouver, B.C. Society of Mining Engineering of the American Institute of Mining. Metallurgy and Petroleum Engineering, New York, 171–196.

Barton, N.R. and Bandis, S. (1991). *Review of predictive capabilities of JRC-JCS model in engineering practice.* Publ No. 182. Norwegian Geotechnical Institute, Oslo, Norway.

Baumann, P. (1958). Rockfill dams: Cogswell and San Gabriel dams. *J. Power Division*, ASCE, Vol. 84, PO3, 1687–1 to 1687–33.

Baynes, F.J. (1999). Engineering geological knowledge and quality. *Proc. of the 8th Australian-New Zealand Conference on Geomechanics*, Hobart. Institution of Engineers, Australia, Vol. 2, 227–234.

Baynes, F.J. and Dearman, W.R. (1978a). The microfabric of a chemically weathered granite. *Bull. Intl. Assoc. of Eng. Geol.*, No. 18, 91–100.

Baynes, F.J. and Dearman, W.R. (1978b). The relationship between the microfabric and the engineering properties of weathered granite. *Bull. Intl. Assoc. of Eng. Geol.*, No. 18, 191–197.

Baziar, M.H. and Dobry, R. (1995). Residual strength and large scale deformation potential of loose silty sands. *J. Geotechnical Engineering*, ASCE, Vol. 121, No. 12, 896–906.

BC Hydro (1995). *Guidelines for the assessment of rock foundations of existing concrete gravity dams.* BC Hydro Report No. MEP67, Vancouver.

Beck, B.F. (1993). Applied Karst geology: *Proc. 4th Multidisciplinary Conf. on Sinkholes and the Environmental Impacts of Karst,* Panama City. Balkema.

Beck, B.F. ed. (1995). Karst geohazards: engineering and environmental problems in Karst terrains *Proc. 5th Multidisciplinary Conf. on Sinkholes and the Engineering and Environmental Impacts of Karst*, Gatlinburg, Tennessee. Balkema.

Beck, B.F. and Stephenson eds. (1997). The engineering geology and hydrogeology of Karst terrains: *Proc. 6th Multidisciplinary Conf. on Sinkholes and Engineering and Environmental Impacts of Karst.* Springfield, Missouri. Balkema.

Beene, R.R.W. (1967). Waco Dam slide. *Journ. Soil Mech. and Found.*, Div. ASCE, No. SM4, 35–44.

Beetham, R.D., Moody, K.E., Fergusson, D.A., Jennings, D.N. and Waugh, P. (1992). Landslide development in schist by toe buckling. *Proc. 6th Int. Symp. on Landslides*. Editor Bell, D.H., Balkema, Rotterdam, 217–224.

Bell, D.H. (1976). Slope evolution and slope stability, Kawarau Valley, Central Otago, New Zealand. *Bull. Intl. Assoc. of Eng. Geol.*, No. 14, 5–16.

Bell, D.H. (1982). The geomorphic evolution of a valley system: the Kawarau Valley, Central Otago. Ch. 17 in *Landforms of New Zealand*, Soons, J.M. and Selby, M.J. (eds). Longman Paul Ltd., Auckland, 317–341.

Bell, D.H. (1983). The K9 landslide, Kawarau Valley, Central Otago. *Proc. XV Pacific Science Congress*, Section B9A, Slope Stability Problems.

Bell, F.G. (1981) Geotechnical properties of some evaporitic rocks. *Bull. IAEG 24*, 137–144.

Bell, F.G. (1983). *Fundamentals of engineering geology*, Butterworths, 648.

Bell, F.G. (1992). *Engineering in rock masses.* Butterworth Heinemann, Oxford.

Bell, F.G. (1993). *Engineering Geology.* Blackwell.

Bell, F.G. and Maud, R.R. (1994). Dispersive soils: a review from a South African perspective. *Quarterly Journal of Engineering Geology*, 27, 195–210.

Bell, G., Fell, R. and Foster, M.A. (2002). Risk and standards based assessment of internal erosion and piping failure – a convergence of approaches. *ANCOLD Bulletin 121*, 15–27. Australian National Committee on Large Dams, Melbourne.

Bell, L.A. (1982). A cutoff in rock and alluvium at Asprokrammos Dam, in *Grouting in Geotechnical Engineering*. Ed. Baker, W.H. ASCE.

Bellott, R., Ghionna, V., Jamiolkowsk, M. and Robertson, P.K. (1989). Shear strength of sand from CPT. *Proc. 12th International Congress Soil Mechanics and Foundation Engineering*, Rio de Janeiro. Vol. 1, 179–184, Balkema.

Bellport, B.P. (1967). Bureau of Reclamation experience in stabilizing embankment of Fontenelle Dam. *Ninth International Congress on Large Dams*, Istamboul, Q.32, R.5, 67–79. International Commission on Large Dams, Paris.

Benson, R.C. and Yuhr, L.B. (1987). Assessment and long-term monitoring of localized subsidence using ground penetrating radar, in *Karst Hydrogeology*, Beck, B.F. and Wilson, W.L. (eds), Balkema, 161–169.

Bentel, D., Robbertze, J. and Smith, M. (1982). The use and behaviour of geotextiles in underdrainage systems of gold mine tailings dam in South Africa. *Second Int. Conf. on Geotextiles, Las Vegas*.

Bertacchi, P. and Bellotti, R. (1970). Experimental research on materials for rockfill dams. *Tenth Congress on Large Dams*, Montreal, Q36, 511–529, International Commission on Large Dams, Paris.

Bertacchi, P. and Cazzuffi, D. (1985). *Geotextile filters for embankment dams*. Water Power and Dam Construction.

Bertram, G.E. (1940). An experimental investigation of protective filters. Graduate School of Engineering, Harvard University, *Soil Mechanics Series 7*.

Bertram, G.E. (1972). Field tests for compacted rockfill, in *Embankment Dam Engineering*, Casagrande Volume, editors, Hirschfeld, R.C. and Poulos, S.J. 1–20, Wiley.

Bickel, J.O. and Kuesel, T.R. (1992). *Tunnel Engineering Handbook*. Krieger Publishing Company, Florida.

Bieniawski, Z.T. (1978). Determining rock mass deformability: experience from case histories, *International Journal for Rock Mechanics, Mining Sciences and Geomechanical Abstracts*, 15, 237–247.

Bishop, A.W. and Henkel, D.J. (1957, 1971). The measurement of soil properties in the triaxial test – Edward Arnold.

Bishop, A.W. (1955). The use of the slip circle in the stability analysis of slopes. *Geotechnique*, 5(1), 7–17.

Bishop, A.W. (1971). Shear strength parameters for undisturbed and remoulded soil specimens. *Proceedings of the Roscoe Memorial Symposium*, Cambridge University. Edited by Parry, R.G., Foulis, G.T., Oxfordshire, 3–58.

Bishop, A.W., Green, G.E., Garga, V.K., Andressen, A. and Brown, J.D. (1971). A new ring shear apparatus and its application to the measurement of residual strength. *Geotechnique*, 21, 273–328.

Bjerrum, L. (1967). Progressive failure in slopes of overconsolidated plastic clay and clay shales. The Third Terzaghi Lecture, JASCE, Vol. 93, SM5, 3–49.

Bjerrum, L. (1973). Problems of soil mechanics and construction of soft clays and structurally unstable soils (collapsible, expansive and others) State of the Art report, *Proc. 8th Int. Conf. Soil Mechanics and Foundation Engineering*, Moscow, Vol. 3, 111–160.

Black, D.K. and Lee, K.L. (1973). Saturating laboratory samples by back pressure. *J. ASCE* Vol. 99, No. SM1, 75–93.

Blight, G.B. and Bentel, G.M. (1983). The behaviour of mine tailings during hydraulic deposition. *J. South African Inst. Min. Metall.*, Vol. 83 No. 4, 73–86.

Blight, G.E. (1963). The effect of non uniform pore pressures on laboratory measurements of the shear strength of soils. *Laboratory Shear Testing of Soils*, ASTM STP No. 361, 173–174.

Blight, G.E. (1987). The concept of the master profile for tailings dam beaches in *Prediction and Performance in Geotechnical Engineering*, Calgary. Balkema.

Blight, G.E. (1988). Some less familiar aspects of hydraulic fill structures, in *Hydraulic Fill Structures*, edited by D.J.A. Van Zyl and S.G. Vick, ASCE Geotech. Special Pub. No. 21, 1000–1064.

Blight, G.E. and Caldwell, J.A. (1984). The abatement of pollution from abandoned gold-residue dams. *J.South African Inst. Min. Metall.*, Vol. 84, No. 1, 1–9.

Blight, G.E., Thomson, R.R. and Vorster, G.K. (1985). Profiles of hydraulic fill tailings beaches and seepage through hydraulically sorted tailings. *J.South African Inst. Min. Metall.*, Vol. 85, No. 5, 157–161.

Blyth, F.G.H. and de Freitas, M.H. (1989). *A Geology for Engineers*, (7th ed revised), Edward Arnold.

Bock, H., Armstrong, K.J., Enever, R.J. and Otto, B. (1987). Rock stress measurements at the Burdekin Falls Dam. *ANCOLD Bulletin 77*, 24–30. Australian National Committee on Large Dams.

Bodtman, W.L. and Wyatt, J.D. (1985). Design and performance of Shiroro rockfill dam. *Proceedings of the Symposium on Concrete Face Rockfill Dams – Design, Construction and Performance*, (Cooke & Sherard ed.) Detroit, Michigan, ASCE New York, 231–251.

Bogoslovsky, V.A. and Ogilvy, A.A. (1972). The study of streaming potentials on fissured media models. *Geophysical Prospecting*, Vol. 20, 109–117.

Bolzon, G., Schrefler, A. and Zienkiewicz, O.C. (1996). Elastoplastic soil constitutive laws generalized to partially saturated states. *Geotechnique* 46, No. 2, 279–289.

Boulanger, R.W. (2003a). Relating K_α to a relative state parameter index. Accepted for publication in ASCE *J. Geotechnical and Geoenviromental Engineering*.

Boulanger, R.W. (2003b). High overburden stress effects in liquefaction analysis, accepted for publication in ASCE *J. Geotechnical and Geoenvironmental Engineering*.

Boulton, G.S. and Eyles, N. (1979). Sedimentation by valley glaciers, a model and genetic classification, in *Moraines and Varves*, Schluchter, C. ed., Balkema.

Boulton, G.S. and Paul, M.A. (1976). The influence of genetic processes on some geotechnical properties of glacial tills. *Q.J. Eng. Geol.* 9, 159–194.

Bowles, J.E. (1978). *Engineering properties of soils and their measurement*. McGraw-Hill.

Bowling, A.J. (1980). Laboratory investigations into the suitability of rockfill for concrete faced rockfill dams. *ANCOLD Bulletin 59*, 21–29. Discussion 30–31.

Bowling, A.J. and Woodward, R.C. (1979). An investigation of near surface rock stresses at Copeton damsite in New South Wales, *Australian Geomechanics Journal*, Vol. G9, 5–13.

Bozovic, A. (1985). Foundation treatment for control of seepage. General Report, Question 58, *15th Int. Congress on large Dams, Lausanne*, 1467–1583. International Commission on Large Dams, Paris.

Bozovic, A., Budanur, H. and Nonveiller, E. (1981). The Keban Dam foundation on Karstified Limestone – a case history. *Bull. Int. Assn. Eng. Geol.*, No. 24, 45–51.

Brackenbush, F. and Shillakeer, J. (1998). Use of paste for tailings disposal. Proceedings Mine Fill '98, *6th Int. Symp. on Mining With Backfill*, Melbourne. Australian Institute of Mining and Metallurgy.

Bradbury, C.E. (1990). Harris Dam slurry trench, design and construction. *Aust. Geomech.*, No. 19, 22–27.

Brady, B.G.H. and Brown, E.T. (1985). *Rock mechanics for underground mining*, George Allen and Unwin.

Brand, B. (1999). Nappe pressures on gravity dam spillways. *Dam Engineering*, Vol. X Issue 2.

Brand, E.W. (1984). Landslides in Southeast Asia. A state of the art report. *Proc. 4th Int. Symp. on Landslides, Toronto*, Vol. 1, 17–59.

Brand, E.W. and Phillipson, H.B. (1984). Site investigation and geotechnical engineering practice in Hong Kong. *Geotech. Eng.*, 15, 97–153.

Brand, E.W. and Premchitt, J. (1980a). Shape factors of cylindrical piezometers. *Geotechnique* 30, No. 4, 369–384.

Brand, E.W. and Premchitt, J. (1980b). Shape factors of some non-cylindrical piezometers. *Geotechnique*, 30 No. 5, 536–537.

Brauns, J. and Witt, K.J. (1987). Proposal for an advanced concept of filter design, *Proceedings, 9th European Conference on Soil Mechanics and Foundation Engineering*, Dublin.

BRE (1999). *An engineering guide to the safety of embankment dams in the United Kingdom*. Building Research Establishment. Garston.

Brett, D.M. (1986). Chemical grouting for dams. *ANCOLD Bulletin* No. 73, 7–15, Australian National Committee on Large Dams, Melbourne.

Brett, D.M. and Osborne, T.R. (1984). Chemical grouting of dam foundations in residual laterite soils of the Darling Range, Western Australia. *4th ANZ Conference on Geomechanics*, 177–182, The Institution of Engineers Australia, Canberra.

Breznik, M. (1998). *Storage reservoirs and deep wells in Karst regions*. Balkema.

Brink, A.B.A. (1979). Engineering geology of Southern Africa. *Building Publications*, Pretoria, South Africa.

British Standards Institution (1981). *Code of practice for site investigations*, SS5980, 1981.

British Standards Institution (1999). *Code of Practice for Site Investigations*, BS5930: 1999.

Broch and Franklin (1972). The point load strength test. *Int. J. Rock Mech. Min. Sci.*, Vol. 9, 669.

Bromhead, E.N. (1979). A simple ring shear apparatus. *Ground Engineering* Vol. 12, No. 5.

Bromhead, E.N. (1986). *The stability of slopes*. Surrey University Press.

Bromhead, E.N. and Curtis, R.D. (1983). A comparison of alternative methods of measuring the residual strength of London clay. *Ground Engineering*, 16, 39–41.

Brown, E.T. and Hoek, E. (1978). Trends in relationships between measured rock *insitu* stresses and depth. *Int. J. Rock Mech. Min. Sci.*, Vol. 15, 211–215.

Brown, E.T. and Windsor, C.R. (1990). Near-surface Insitu stresses in Australia and their influence on underground construction. *Proc. IE Aust. Tunnelling Conference*, 18–25.

Bruce, D.A. (1982). Aspects of rock grouting practice in British dams, in *Issues in Dam Grouting*. JASCE Geotech. Engineering, 301–316.

Brune, G. (1965). Anhydrite and gypsum problems in engineering geology. *Bull. Assoc. Eng. Geologists*, Vol. 3, 26–38.

Brunsden, D., Doornkamp, J.C., Fookes, P.G., Jones, D.K.C. and Kelly, G.M.H. (1975). Large scale geomorphological mapping and highway engineering design. *Quart. J. Eng. Geol., London*, Vol. 8, 227–253.

Building Research Station (1958). *The Gel-Pat Test*. Building Research Station, Garston.

Bureau of Reclamation (1990). Cavitation in chutes and spillways. US Bureau of Reclamation Engineering Monograph No. 42. Denver.

Burenkova, V.V. (1993). Assessment of suffusion in non-cohesive and graded soils. *Filters in Geotechnical Engineering*, Editors Brauns, Heeramus and Schuler. Balkema, Rotterdam, 357–360.

Buxton, T.M. and Sibley, D.F. (1980). Pressure solution features in a shallow buried limestone. *J. Sedim. Pet.* 51, No. 1, 19–26.

Cabrera, J.G. (1992). Investigation of a landslide on a crucial reservoir rim saddle. In *Proc. 6th Int. Symp. on Landslides*, Bell, D.H. (ed.), Balkema, Rotterdam, 1235–1239.

Calhoun, C.C. (1972). Development of design criteria and acceptance of specifications for plastic filter cloth, U.S. Army Corps of Engineers W.E.S., Vicksburg, MS, *Technical report S-72-7*.

Campanella, R.G. and Robertson, P.K. (1988). Current status of the piezocone test. Special Lecture. *Proc. 1st Int. Symp. on Penetration Testing ISOPT-1*, Orlando.

Campanella, R.G., Howie, J.A., Sully, J.P., Hers, I. and Robertson, P.K. (1990). Influence of equipment and test procedures on interpretation of pressuremeter tests. *Proc. 3rd Int. Symp. on Pressuremeters*, organised by the British Geotechnical Society, Oxford University.

Caron, C. (1982). The State of grouting in the 1980s, in *Grouting in Geotechnical Engineering*, edited by Baker, W.H., ASCE, 172–186.

Carroll, R.G. (1983). Geotextile filter Criteria, *Transportation Research Record*, 916, 46–53.

Carse, A. and Dux, P.F. (1988). Alkali-silica reaction in concrete structures. *Research Report No. CE88*, Dept. of Civil Engineering, University of Oxford.

Cary, A.S. (1950). Origin and significance of openwork gravel. Trans. ASCE, *Proceedings-Separate No. 17*, 1296–1318.

Casinader, R.J. (1982). Systematic weak seams in dam foundations. In *Geotechnical Problems and Practice of Dam Engineering*, Balasubramaniam et al (eds). Balkema, 253–264.

Casinader, R.J. and Stapledon, D.H. (1979). The effect of geology on the treatment of the dam foundation interface at Sugarloaf Dam. *13th Int. Cong. on Large Dams, New Delhi*, Q48, R.32, 591–619, International Commission on Large Dams, Paris.

Casinader, R.J. and Watt, R.E. (1985). Concrete face rockfill dams of the Winneke Project, in *Concrete Face Rockfill Dams, Design, Construction and Performance*, editors Cooke, J.B. and Sherard, J.L. ASCE Geotechnical Engineering Division, 140–162.

Cavounidis, S. (1987). On the ratio of factors of safety in slope stability analyses. *Geotechnique* 37(2), 207–210.

CDSA (1999). *Dam safety guidelines*. Canadian Dam Safety Association.

Cedergren, H.R. (1967). *Seepage, drainage and flownets*. Wiley.

Cedergren, H.R. (1972). Seepage control in earth dams, in *Embankment Dam Engineering*, Casagrande Volume. Hirschfeld, R.C. and Poulos, S.J. (eds), Wiley.

CFGG (1983). Comite Francais Geotextiles Geomembranes. *Recommendations pour l'emploi des geotextiles dans les systems de drainage et de filtration*.

Chandler, R.J. (1984). Recent European experience of landslides in over-consolidated clays and soft rock. *4th International Symposium on Landslides*, Toronto, Vol. 1, 61–81.

Chapius, R.P. and Gatien, T. (1986). An improved rotating cylinder technique for quantitative measurements of scour resistance of clays. *Canadian Geotechnical Journal*, 23(1), 83–87.

Charles, J.A. (1997). General Report, Special Problems Associated with Earthfill Dams. *Nineteenth International Congress on Large Dams*, Florence, GR Q73, Vol. II, 1083–1198. International Commission on Large Dams, Paris.

Charles, J.A. and Watts, K.S. (1980). The influence of confining pressure on the shear strength of compacted rockfill. *Geotechnique*, 30, 353–367.

Charles, J.A., Tedd, P. and Holton, I.R. (1995). Internal erosion in clay cores of British dams. *Research and Development in the Field of Dams*, Crans-Montana, Switzerland, Swiss National Committee on Large Dams. 59–70.

Charleton, P.J. and Crane, K.F. (1980). Construction methods and economics of placing high cost filters. *ANCOLD Bulletin* No. 56, 53–66, Australian National Committee on Large Dams, Melbourne.

Charlwood, R.G. and Solymar, Z.V. (1994). A review of AAR in dams, *Dam Engineering*, Vol. 2 (2), 31–62.

Chengjie, Z. and Shuyong, X. (1981). Reservoir and dam seepage and anti-seepage measures at hydroelectric power stations in Karst areas of Southwest China. *Bull. IAEG.* 24, 81–86.

Chilingarian, G.V. and Vorabutr, P. (1983). *Drilling and drilling fluids*. Elsevier.

Christian, J.T., Ladd, C.C. and Baecher, G.B. (1992). Invited lecture: reliability and probability in stability analysis; in Stability and Performance of Slopes and Embankments – 11. *ASCE Geotechnical special publication*, No. 31, 1071–1111.

Christopher, B.R. and Holtz, R.D. (1989). *Geotextile construction and design guidelines*, Prepared for Federal Highway Administration, Washington, D.C., HI-89-050.

Christopher, B.R. and Holtz, R.D. (1985). *Geotextile engineering manual*, U.S. Federal Highway Administration, report No. FHWA-TS-86/203.

Christopher, B.R., Holtz, R.D. and Fischer, G.R. (1993). Research needs to geotextile filter design, *In Filters in Geotechnical and Hydraulic Engineering*, Editors Brauns, Helbaum and Schuler, Rotterdam: Balkema.

Clarke, B.G. and Smith, A. (1990). Pressuremeter tests in moderately to highly weathered argillaceous rocks. 26th Annual Conference of Engineering Group of the Geological Society, University of Leeds.

Clayton, C.R.I., Matthews, M.C. and Simons, N.E. (1995). *Site investigation*, Blackwell Science Ltd., Oxford.

Clements, R.P. (1984). Post-construction deformation of rockfill dams. *J. of Geotechnical* ASCE, *Engineering*, Vol. 110, No. 7, 821–840.

Clough, R.W. and Chopra, A.K. (1966). Earthquake stress analysis in earth dams. *Proc. ASCE*, Vol. 92, No. EM2.

Coffey and Partners (1981a). Lungga Hydropower project, Solomon Islands. *Assessment of liquefaction of foundation gravels.* Unpublished Report No. B9578/1-P.

Coffey and Partners (1981b). Lungga Hydropower project, Solomon Islands. *Dam embankment results of seepage analysis.* Unpublished Report No. B9578/1-R.

Coffey and Partners (1982). *Ben Lomond project. Tailings Storage Design*, Unpublished Report B9658/17-AC.

Cole, B.A. (1977). *Dispersive clays in irrigation dams in Thailand*, ASTM, STP623.

Cohen, M. and Moenne, G. (1991). Tailings deposits at the El Soldado Mine Seismic design, abandonment and reliabilitation of tailings dams, special volume. *Proc. Int. Seminar of Technical Committee TC7 and of the Main Session on Geotechnical Aspects of Tailings Dams*, El Cobre, Chile.

Cole, D.H.C. and Lewis, J.G. (1960). Piping failure of earthen dams built of plastic materials in arid climates. *3rd ANZ Conf SMFE*, 93–98.

Cole, R.G. and Horswill, P. (1988). Alkali-silica reaction, Val de la Mere Dam, Jersey. Case history. *Proc. Instn. Civil Engrs.*, Part 1, Vol. 84, 1237–1259.

Cole, W.F. and Beresford, F.D. (1976). Evaluation of basalt from Deer Park, Victoria, as an aggregate for concrete. *Proc. 8th Aust. Road Research Board Conf.*, Vol. 8, Part 3, 23–33.

Cole, W.F. and Sandy, M.J. (1980). A proposed secondary mineral rating for basalt road aggregate durability. *Aust. Road Research*, Vol. 10, No. 3, 27–27.

Collins, S.A. and Tjoumas, C. (2003). FERC oversight and input, Saluda dam seismic evaluation and remediation, Part 1, *Proc. 23rd USSD Conference, Charleston*. US Society on Dams, 531–558.

Cooke, J.B. (1984). Progress in rockfill dams (18th Terzaghi lecture). ASCE Journal of Geotechnical Engineering, Vol. 110, No. 10, 1383–1414.

Cooke, J.B. (1993). Rockfill and the rockfill dam. Proceedings of the International Symposium on High Earth-Rockfill Dams, (Jiang, Zhang & Qin ed.) Beijing. 1–24.

Cooke, J.B. (1997). The concrete face rockfill dam. Seventeenth Annual USCOLD Lecture Series: Non-Soil Water Barriers for Embankment Dams. San Diego. 117–132.

Cooke, J.B. (1999). The development of today's CFRD dam. Proceedings of the Second Symposium on Concrete Face Rockfill Dams, Florianopolis, Brazil, Brazilian Committee on Dams. 3–11.

Cooke, J.B. (2000). The high CFRD. J. Barry Cooke Volume Concrete Face Rockfill Dams. 20th ICOLD Congress Beijing 2000, 1–4.

Cooke, J.B. and Sherard, J.L. (1987). Concrete-face rockfill dams: II Design. Journal of Geotechnical Engineering, ASCE, Vol. 113, No. 10, 1113–1133.

Cooke, J.B. and Strassburger, A.G. (1988). Section 6: Rockfill Dams. In Development of Dam Engineering in the United States (Kollgaard and Chadwick ed.), Permagon Press, 885–1030.

Cooper, B., Khalili, N. and Fell, R. (1997). Large deformations due to undrained strain weakening slope instability at Hume Dam No. 1 embankment. Nineteenth International Congress on Large Dams, Florence, ICOLD Vol. 2, (Q73 R46), 797–818, International Commission on Large Dams, Paris.

Cooper, S.S. and Ballard, R.F. (1988). Geophysical exploration for cavity detection in karst terrain, in Geotechnical Aspects of Karst Terrains, Sitar, N. (ed). Amer. Soc. of Civil Engineers Spec. Pub. No. 14, 25–39.

Cordova, J.Y. and Franco, M.A. (1997). Buried conduits in embankment dams. Experience in Spain. Proc. 19th Congr. Q73, R.21, 321–336. International Commission on Large Dams, Paris.

Corless, B.M. and Glenister, D.J. (1990). The rehabilitation challenges of bauxite residue disposal operations in Western Australia. ICOLD/ANCOLD Int Symp. on Safety and Rehabilitation of Tailings Dams, Sydney.

Corte, A.E. and Higashi, A. (1964). Experimental research on desiccation cracks in soil. Research Report 66, US Army Cold Regions Research and Engineering Lab., Army Material Command, Hanover, New Hampshire, 72.

Corwin, R.F. (1999). Seepage Investigations Using the Self-Potential Method, Workshop on Internal Diagnostics for Embankment Dams, CEA Dam Safety Interest Group Report No. PSE203.

Craft, D.C. and Acciardi, R.G. (1984). Failure of pore water analyses for dispersion. Journal Geotechnical Engineering Division, ASCE 110, 459–472.

Cripps, J.C. and Taylor, R.K. (1981). The engineering properties of mudrocks. Q. J. Eng. Geol., London, Vol. 14, 325–346.

Cripps, J.C., Hawkins, A.B. and Reid, J.M. (1993). Engineering problems with pyritic mudrocks. Geoscientist, Vol. 3, No. 3, 16–19.

Cruden, D.M. and Hu, X.Q. (1988). Basic friction angles of carbonate rocks from Kananaskis Country, Canada. Bull. Int. Assn. Engg. Geol., No. 38, 55–59.

Cubrinovski, M. and Ishihara, K. (2000a). Flow potential of sandy soils with different grain compositions. Soils and Foundations, 40(4): 103–119.

Cubrinovski, M. and Ishihara, K. (2000b). Flow potential of sandy soils. Proc. GeoEng. 2000, Technomic Publishing, Lancaster, 549–550.

Cundall, P. (1993). FLAC – Fast lagrangian analysis of continua, ITASCA.

Czorewko, M.A. and Cripps, J.C. (2001). Assessing the durability of mudrocks using the modified jar slake durability test. QJEG & H, 34, 153–163.

D'Elia, B., Picarelli, L., Leroueil, S. and Vaunat, J. (1998). Geotechnical characterisation of slope movements in structurally complex clay soils and stiff jointed clays. Italian Geotechnical Journal, Anno XXXII, No. 3, 5–32.

Davidson, R. and Perez, J.V. (1982). Properties of chemically grouted sand at Locks and Dam No. 26, in Issues in Dam Grouting. ASCE Geotech. Engineering, 433–449.

Davidson, R.R., Levallois, J. and Graybeal, K. (1992). Seepage cutoff wall for mud mountain dam slurry walls, ASTM STP 1129, Editors Paul, D.B., Davidson, R.R. and Cavall, J.J., American Society for Testing and Materials, Philadelphia, 309–323.

Dawson, R.F., Morgenstern, N.R. and Stokes, A.W. 1998. Liquefaction flowslides in rocky mountain coal mine waste dumps. Can. Geotech. J., 35, 328–243.

De Alba, P., Seed, H.B. and Chan, C.K. (1976). Sand liquefaction in large scale simple shear tests. JASCE Geotechnical Engineering, Vol. 102, GT9, 909–927.

De Beer, E.E., Goeten, E., Heynen, W.J. and Joustra, K. (1988). Cone penetration test (CPT). International reference test procedure. *Proc.1st Int. Symp. on Penetration Testing, ISOPT-1, Orlando.*

De Groot, M.B., Heezen, F.T., Mastbergen, D.R. and Stefess, H. (1988). Slopes and densities of hydraulically placed sands, in *Hydraulic Fill Structures*, Van Zyl, D.J.A. and Vick, S.G. (eds), ASCE Geotech. Special Pub., No. 21, 32–51.

De Ruiter, J. (1988). Penetration Testing. *Proc. 1st Int. Symp. on Penetration Testing, ISOPT-1, Orlando*, Vol. 2.

De S. Pinto, N.L., Marques filho, P.L. and Maurer, E. (1985). Foz de Areia Dam – Design, construction and behaviour, in *Concrete Face Rockfill Dams, Design, Construction and Performance*, Cooke, J.B. and Sherard, J.L. (eds). ASCE Geotechnical Engineering Division, 173–191.

Dearman, W.R. (1981). General Report, Session 1. Engineering properties of carbonate rocks. *Symp. on Eng. Geological Problems of Construction on Soluble Rocks.* Bull. Intl. Assoc. of Eng. Geol., No. 24, 3–17.

deBruyn, I.A. and Bell, F.G. (2001). The occurrence of sinkholes and subsidence depression in the Far West Rand and Gauteng Province, South Africa, and their engineering implications. *Environmental and Engineering Geosciences*, Vol. 7, No. 3, 281–295.

Decourt, L., Muronach, T., Nixon, K., Schmertmann, J.H., Thorburn, S. and Zolkov, E. (1988). Standard penetration test (SPT). International reference test procedure. *Proc. 1st Int. Symp. on Penetration Testing, ISOPT-1, Orlando.*

Deere, D.U. (1973). The foliation sheer zone – an adverse engineering geological feature of metamorphic rocks. *J. Boston Soc. of Civil Engineers*, Vol. 60, No. 4, 163–176.

Deere, D.U. (1982). Cement-bentonite grouting for dams, in *Issues in Dam Grouting*. ASCE Geotech. Engineering, Editor Baker, W.H.

Deere, D.U. and Lombardi, G. (1985). Grout slurries – thick or thin, in *Issues in Dam Grouting*, editor Baker, W.H., ASCE Geotech. Engineering Division, Denver.

Deere, D.U. and Patton, F.D. (1971). Slope stability in residual soils. *Proc. Panamerican Conf. on Soil Mech. and Found. Eng.*, Vol. 1, 87–170.

Deere, D.U. and Perez, J.Y. (1985). Remedial measures for large landslide movements. *Proc. PRC-US-JAPAN Trilateral Symposium/Workshop on Engineering for Multiple Hazard Mitigation, Beijing*, 1–17.

DeMello, V.F.B. (1975). Some lessons from unsuspected, real and fictitious problems from dam engineering in Brazil. *Proc. 6th Regional African Conference on Soil Mechanics and Foundation Engineering*, Durham, Vol. 2, 285–304.

Den-Adel, H., Koenders, M.A. and Bakker, K.J. (1994). The analysis of relaxed criteria for erosion control filters. *Can. Geotech J.*, Vol. 31, 829–840.

Desai, C.S. (1975). Finite element methods for flow in porous media. *Finite Element Methods in Fluids*, Vol. 1, ed. Gallagher et al, 157–182.

Di Maio, C. (1996a). "The influence of pore fluid composition on the residual shear strength of some natural clayey soils". *7th Int. Symp. On Landslides*, editor Sennesett, K., Trondheim, Balkema, Rotterdam, Vol. 2, 1189–1194.

Di Maio, C. (1996b). "Exposure of bentonite to salt solution: osmotic and mechanical effects". *Géotechnique*, Vol. 46(4), 695–707.

Dick, R.C. (1976). The depth versus depth plot: an aid to selecting piezometer installation depths in boreholes. *Engineering Geology*, 10, 37–42.

Dixon, H.W. (1969). Decomposition products of rock substances – proposed engineering geological classification. *Proc. Rock Mechanics Symposium, Inst. of Engineers, Sydney*, Institute of Engineer Australia, Canberra.

Djalaly, H. (1988). Remedial and watertightening works of Lar Dam. *Proc. 17th Congr.* C.15, 1181–1200. International Commission on Large Dams, Paris.

Donaghe, R.T., Chaney, R.C. and Silver, M. (1988) (eds). American Society for Testing and Materials, Philadelphia, 189–201.

Donath, F.A. (1961). Experimental study of shear failure in anisotropic rocks. *Bull. Geol. Soc. Am.*, Vol. 12, 885–990.

Dornstadter, J. (1996). Sensitive monitoring of embankment dams. *Proc. Symp. on repairing and upgrading of dams, Stockholm.* Editors Johansson S. and Cederstrom, M. KTH, Tryck Och Kopiering, 259–268.

Douglas, K.D. (2003). The shear strength of rock masses. *PhD Thesis*. School of Civil and Environmental Engineering, The University of New South Wales, Sydney.

Douglas, K.D., Spannagle, M. and Fell, R. (1998). Analysis of concrete and masonry dam incidents. *UNICIV Report No. R-373*, The University of New South Wales, Sydney. ISBN 85 841 340 X.

Douglas, K.J., Spannagle, M. and Fell, R. (1999). Analysis of concrete and masonry dam incidents. *International Journal of Hydropower and Dams*, Vol. 6, Issue 4, 108–115.

Dounias, G.T., Potts, D.M. and Vaughan, P.R. (1988). Finite element analysis of progressive failure: two case studies. *Computer and Geotechnics*, Vol. 6, 155–175.

Dounias, G.T., Potts, D.M. and Vaughan, P.R. (1996). Analysis of progressive failure and cracking in old British dams, *Geotechnique* 46, No. 4, 621–640.

Duncan, J.B. (1992). State of the art – static stability and deformation analysis in Stability and Performance of Slopes and Embankments, ASCE *Geotechnical Special Publication No. 31*, 222–266.

Duncan, J.M. (1994). The role of advanced constitutive relations in practical applications. *Proc. 13th Int. Con. On Soil Mechanics and Foundation Engineering*, Delhi, 31–48.

Duncan, J.M. (1996a). State of the art: limit equilibrium and finite element analysis of slopes. *J. of Geotech. Eng.* ASCE, Vol. 122, No. 7, 577–591.

Duncan, J.B. (1996b). Soil slope stability analysis, in Landslides Investigation and Mitigation. Transport research board, N.R.C. *Spec. Report 247*, Washington DC 1996, 337–371.

Duncan, J.M. (2000). Factors of safety and reliability in Geotechnical Engineering, *Journal of Geotechnical and Geoenvironmental Engineering*, ASCE, Vol. 126, No. 4, 307–316.

Duncan, J.M. and Buchignani, A. (1976). An engineering manual for settlement studies. Dept. of Civil Engineering, University of California, Berkeley.

Duncan, J.M. and Wright, S.G. (1980). The accuracy of equilibrium of slope stability analysis. *Engrg. Geol.*, Vol. 16, No. 1, 5–17.

Duncan, J.M., Schaefer, V.R., Franks, L.W. and Collins, S.A. (1987). Design and performance of a reinforced embankment for Mohicanville Dike, No. 2 in Ohio. *Transp. Res. Rec*, 1153, 15–25.

Duncan, J.M., Wright, S.G. and Wong, K.S. (1990). Slope stability during rapid drawdown. *Proc. H. Bolton Seed Memorial Symposium*, Berkeley, California, Bitech Publishers, Vancouver, 253–272.

Dunnicliff, J. (1982). Geotechnical instrumentation for monitoring field performance. Nat. Coop. Highway Res. Program, *Synthesis of Highway Practice 89*. Transp. Res. Board.

Dunnicliff, J. (1995). Keynote Paper: "Monitoring and instrumentation of landslides". *Proc. 6th Int. Symp. on Landslides, Christchurch*. Editor Bell, D.H., Balkema, Rotterdam, 1881–1896.

Dusseault, M.B. and Morgenstern, N.R. (1979). Locked sands. *Quart. Journ. Engng. Geol.* 12: 117–131.

Eadie, A. (1986). Foundation treatment methods in karstic limestone at the Bjelke-Petersen Dam, Murgon. Geol. Soc. of Aust., Queensland Division, 1986 *Field Conf. Proc.*, 63–71.

Eagles, J. (1987). Water Resources Commission of NSW, personal communication.

Emery, C.L. (1963). Strain energy in rocks, in State of Stress in the Earth's Crust. *Proc. of the Int. Conf., Santa Monica, California*, 235–260.

EPRI (1992). Uplift pressures, shear strengths and tensile strengths for stability analysis of concrete gravity dams, *Electric Power Research Institute Report TR-100345*, Palo Alto.

Erguvanli, K. (1979). Problems on damsites and reservoirs in karstic areas with some considerations as to their solution. *Bull. Intl. Assoc. of Eng. Geology*, No. 20, 173–178.

Eriksson, E. (1985). *Principles and applications of hydro geology*, Chapman and Hall.

Ervin, M.C. (1983). The pressuremeter in geotechnical investigations, in Ervin, M.C. (ed.) in *Insitu Testing for Geotechnical Investigations*. Balkema.

Esteva, L. and Rosenblueth, E. (1969). Espectos de temblors a distancias moderadas y grandes. *Bo. Soc. Mexicano Ing. Simica*, 2, 1–18.

Ewart, F.K. (1985). *Rock grouting with emphasis on dam sites*. Springer-Verlag.

Eyles, N. (1985) (ed.). *Glacial Geology, An Introduction for Engineers and Earth Scientists*, Pergamon.

Eyles, N. and Paul, M.A. (1985). Landforms and sediments resulting from former periglacial climates. Chap. 5, in *Glacial Geology, An Introduction to Engineers and Earth Scientists*, Eyles, N. (ed.), Pergamon.

Fell, R. (1987). Measurement of positive pore pressure, in *Geotechnical Field Instrumentation*. Aust. Geomechanics Soc. Victoria Group.

Fell, R. (1988a). Engineering geophysics. A civil engineers viewpoint. *Aust. Geomechanics* No. 15, 63–68, and *Exploration Geophysics*, No. 17, 25–31.

Fell, R. (1988b). Mine tailings, dispersants and flocculants, in *Hydraulic Fill Structures*. Van Zyl, D.J.A. and Vick, S.G. (eds). ASCE Geotech. Special Pub. No. 21, 711–729.

Fell, R. (1992). Theme address – landslides in Australia. *Proceedings 6th International Symposium on Landslides, Christchurch*, editor Bell, D.H., Vol. 3, Balkema, Rotterdam, 2059–2100.

Fell, R. and Jeffery, R.P. (1987). Determination of drained shear strength for slope stability analysis in *Soil Slope Instability and Stabilisation*, Walker, B.F. and Fell, R. (eds), Balkema, 53–70.

Fell, R., Bowles, D.S., Anderson, L.R. and Bell, G. (2000). The status of methods for estimation of the probability of failure of dams for use in quantitative risk assessment, *Proceedings of the 20th Congress on Large Dams*, Vol. 1, Question 76, International Commission on Large Dams, Paris, 213–236.

Fell, R., Chapman, T.G. and Maquire, P.K. (1991). A model for prediction of piezometric levels in landslides, in *Slope stability engineering*. Editor Chandler, R.J., Thomas Telford, 37–42.

Fell, R., Hungr, O., Leroueil, S. and Riemer, W. (2000). Keynote lecture – geotechnical engineering of the stability of natural slopes and cuts and fills in soil. *GeoEng 2000*, Technomic Publishing, Lancaster, 21–120.

Fell, R., Wan, C.F., Cyganiewicz, J. and Foster, M. (2001). The time for development and detectability of internal erosion and piping in embankment dams and their foundations. *UNICIV Report R-399*, The University of New South Wales, ISBN 85841 3663.

Fell, R., Wan, C.F., Cyganiewicz, J. and Foster, M. (2003). Time for development of internal erosion and piping in embankment dams. *J. Geotechnical and Geoenviornmental Engineering*, ASCE, Vol. 129, No. 4, 307–314.

Fellenius, W. (1927). Erdstatische Berechnungen mit Reibung und Kohasion, Ernst, Berlin (in German).

FEMA (1987). *Dam Safety: An owners guidance manual.* US Federal Emergency Management Agency.

Fenn, D.D. (1966). The effects of montmorillonite and kaolinite dispersions on the permeability of a porous media, MS Thesis, Mississippi State University, State College, Mississippi.

Fenves, G. and Chopra, A.K. (1987). Simplified earthquake analysis of concrete gravity dams. ASCE, *Journal of Structural Engineering*, Vol. 113, No. 8, 1688–1708.

FERC (2000). Draft Engineering guidelines for the evaluation of hydroelectric projects, Federal Energy Regulatory Commission, Office of Energy Projects, Division of Dam Safety and Inspection.

Ferguson, H.F. (1967). Valley stress relief in the Allegheny Plateau. *Bull. Assoc. Eng. Geol.*, Vol. 4, 63–71.

Ferkh, Z. and Fell, R. (1994). Design of embankments of soft clay. *XIII int. Conf. on Soil Mech. and Foundation Engineering*, New Delhi, India, 733–738.

FHWA (1985). *Geotextile Engineering Manual Federal Highway Administration*. Christopher, B.R. and Holtz, R.D. (eds).

Finn, W.D. (1988). Dynamic analysis, in *Geotechnical Engineering Earthquake Engineering and Soil Dynamics II*. Recent Advances in Ground Motion Evaluation. Geotech. Sp. Publ. No. 20, ASCE, 523–591.

Finn, W.D., Yogendrokumar, N., Lo, R.C. and Ledbetter, R.H. (1990). Seismic response analysis of tailings dams. *ICOLD/ANCOLD Int Symp. on Safety and Rehabilitation of Tailings Dams, Sydney.*

Finn, W.D.L. (1993). Seismic safety evaluation of embankment dams, in *International workshop on dam safety evaluation*, Grundewald, Switzerland, 26–27 April, Vol. 4, 91–135.

Finn, W.D.L., Lee, K.W. and Martin, G.R. (1977). An effective stress model for liquefaction. *J. Geotech. Engrg. Div.*, ASCE, 103, No. GT6, 517–533.

Finn, W.D.L., Yogendrakumar, M., Yoshida, N. and Yoshida, H. (1986). TARA-3: a program to compute the response of 2-D embankments and soil-structure interaction systems to seismic loadings. Department of Civil Engineering, University of British Columbia, Canada.

Fischer, G.R., Christopher, B.R. and Holtz, R.D. (1990). Filter criteria based on pore size distribution. *Proc. 4th Int. Conf. on Geotextiles, Geomembranes and Related Products, The Hague*, Vol. 1, 289–294.

Fischer, G.R., Holtz, R.D. and Christopher, B.R. (1992). A critical review of geotextile pore size measurement methods. Geo-filter 92, *Proc. Int. Conf. on filters and filtration phenomena*, Karlsruhe.

Fitzpatrick, M.D. (1977). Reinforced rockfill in Hydro-Electric Commission Dams, *ANCOLD Bulletin* No. 49, Australian National Committee on Large Dams, Melbourne.

Fitzpatrick, M.D., Cole, B.A., Kinstler, F.L. and Knoop, B.P. (1985). Design of concrete faced rockfill dams, in *Concrete Face Rockfill Dams, Design, Construction and Performance*, Cooke, J.B. and Sherard, J.L. (eds). ASCE Geotechnical Engineering Division, 410–434.

Fleureau, J.M., Kheirbek-Saoud, S. and Taibi, S. (1995). Experimental aspects and modeling the behaviour of soils with a negative pressure. *Proc. 1st Intl. Conf. Unsaturated Soils*, Paris, 57–62.

Fokkema, A., Smith, M.R. and Flutter, J. (1977). Googong Dam flood diversion and embankment protection during construction, *ANCOLD Bulletin* No. 49, Australian National Committee on Large Dams, Melbourne.

Fong, F.C. and Bennett, W.J. (1995). Transverse cracking in embankment dams due to earthquakes, *ASDSO Western Regional Conference*, Montana.

Fookes, P.G. (1997). Geology for engineers: the geological model, prediction and performance', *Quart. Jour. of Eng. Geol.*, Vol. 30, 293–424.

Fookes, P.G. and Hawkins, A.B. (1988). Limestone weathering: its engineering significance and a proposed classification system. *QJEG 21*, 7–31.

Fookes, P.G. and Higginbottom, I.E. (1975). The classification and description of near-shore carbonate sediments for engineering purposes. *Geotechnique*, 25, 406–412.

Fookes, P.G. and Poole (1981). Some preliminary considerations, on the selection and durability of rock and concrete materials for breakwater and coastal protection works. Quart. J. Engineering Geology, London, Vol. 14, 97–128.

Fookes, P.G. and Vaughan, P.R. (eds) (1986). A handbook of Engineering Geomorphology. Surrey University Press.

Fookes, P.G., Baynes, F.J. and Hutchinson, J.N. (2000). Total geological history: a model approach to the anticipation, observation and understanding of site conditions. Proc. GeoEng 2000, Melbourne Technomic, Lancaster, PA, 370–460.

Fookes, P.G., Dale, S.G. and Land, J.M. (1991). "Some observations on a comparative aerial photography interpretation of a landslipped area. Quart. J. Eng. Geol., London, Vol. 24. 249–265.

Ford, D. and Williams, P. (1989). *Karst geomorphology and hydrology*. Unwin Hyman, London.

Forrest, M.P., Connell, A.C. and Scheuering, J. (1990). Reclamation planning of a tailings impoundment. *ICOLD/ANCOLD Int Symp. on Safety and Rehabilitation of Tailings Dams, Sydney.*

Forssblad. L. (1981). *Vibratory soil and rock compaction*. Dynapac Maskin.

Forster, I.R. and MacDonald, R.B. (1998). Post-earthquake response procedures for embankment dams – lessons from the Loma Prieta Earthquake. *ANCOLD Bulletin*, No. 109, 46–64. Australian National Committee on Large Dams.

Foster, M.A. (1999). The probability of failure of embankment dams by internal erosion and piping. *PhD Thesis*, School of Civil and Environmental Engineering, The University of New South Wales, Sydney.

Foster, M.A. and Fell, R. (1999a). Assessing embankment dam filters which do not satisfy design criteria. *UNICIV Report No. R-376*, ISBN 85841 343 4, ISSN 0077-880X, School of Civil and Environmental Engineering, The University of New South Wales, Sydney.

Foster, M.A. and Fell, R. (1999b). A framework for estimating the probability of failure of embankment dams by piping using event tree methods. *UNICIV Report No. R-377*. School of Civil and Environmental Engineering, The University of New South Wales. ISBN 85841 343 4.

Foster, M.A. and Fell, R. (2001). Assessing embankment dams, filters who do not satisfy design criteria. *J. Geotechnical and GeoEnvironmental Engineering*, ASCE, Vol. 127, No. 4, 398–407.

Foster, M. and Fell, R. (2000). Use of event trees to estimate the probability of failure of embankment dams by internal erosion and piping, *Proceedings of the 20th Congress and Large Dams*, Vol. 1, Question 76, 237–260, International Commission on Large Dams, Paris.

Foster, M. and Fell, R. (2001). Assessing embankment dam filters that do not satisfy design criteria. *Journal of Geotechnical and Geoenvironmental Engineering*, ASCE, Vol. 127, No. 5, 398–407.

Foster, M., Fell, R. and Spannagle, M. (2000a). The statistics of embankment dam failures and accidents, *Canadian Geotechnical Journal*, Vol. 37, No. 5, National Research Council Canada, Ottawa, 1000–1024, ISSN 0008-3674.

Foster, M., Fell, R. and Spannagle, M. (2000b). A method for estimating the relative likelihood of failure of embankment dams by internal erosion and piping. *Canadian Geotechnical Journal*, 37, 5, 1025–1061.

Foster, M., Fell, R., Davidson, R. and Wan, C.F. (2002). Estimation of the probability of failure of embankment dams by internal erosion and piping using event tree methods. *ANCOLD Bulletin*, No. 121, 75–82.

Foster, M., Spannagle, M. and Fell, R. (1998). Report on the analysis of embankment dam incidents. *UNICIV Report No. R-374*, ISBN 85841 349 3; ISSN 0077-880X, School of Civil and Environmental Engineering, The University of New South Wales.

Fourie, A.B. (1996). Predicting rainfall – induced slope instability. *Proc. Instn. Civil. Engrs. Geotech Eng.*, Vol. 119, 211–218

Francis, P. (1976). *Volcanoes*. Pelican.

Fredlund, D.G. (1984). Analytical methods for slope analysis. *Proc. 4th International Symposium on Landslides*, Toronto, 229–250.

Fredlund, D.G. and Krahn, J. (1977). Comparison of slope stability methods of analysis. *Canadian Geotechnical J.*, V14, 3, 429–439.

Fredlund, D.G., Morgenstern, N.R. & Widger, R.A. (1978). The shear strength of unsaturated soils. *Canadian Geotechnical J.*, 15(3): 313–321.

French Committee of Geotextiles and Geomembranes (1986). *Recommendations for the use of geotextiles in drainage and filtration systems*, Institut textile de France, Boulogne-Billancourt, France.

Friedman, G.M. (1964). Early diagenesis and lithification in carbonate sediment. *J. Sediment. Petrol.*, 34:4, 777–813.

Friedman, G.M. (1975). The making and unmaking of limestones or the downs and ups of porosity. *J. Sediment Petrol.* Vol. 45, No. 2, 379–398.

Fry, J.J., Charles, J.A. and Penman, A.D.M. (1996). Dams, embankments and slopes, in *Unsaturated Soils*, Alonso and Delage editors, Rotterdam: Balkema, 1391–1419.

Fuxing, X. and Changhua, F. (1981). A study of Karst seepage for the Pengshui Hydrolectric Project on the Wujiang River. *Bull. IAEG 24*, 57–62.

Galloway, J.D. (1939). The design of rock-fill dams. *Transactions*, ASCE, Vol. 104, 1–24.

Garran, P. (1999). Bennett Dam – A case study, Workshop on internal diagnostics for embankment dams CEA dam safety interest group report No. PSE203.

Gasper, E. and Oncascu, M. (1972). *Radioactive tracers in hydrology*. Elsevier.

Geiser, F. (2000). Applicability of a general effective stress concept to unsaturated soils. *Asian Conference on Unsaturated Soils, Singapore*.

Geological Society Engineering Group (1972). The preparation of maps and plans in terms of engineering geology. *QJEG* Vol. 7, No. 3.

Geological Society Engineering Group (1977). The description of rock masses for engineering purposes. *QJEG* Vol. 10 No. 4.

Geological Society Engineering Group (1982). Land surface evaluation for engineering purposes. *QJEG*, Vol. 15, No. 4.

Geological Society Engineering Group Working Party (1995). The description and classification of weathered rocks for engineering purposes. *QJEG*, 28, 207–242.

Gerber, A. and Harmse, H.J. von M. (1987). Proposed procedure for identification of dispersive soils by chemical testing. *The Civil Engineering in South Africa*, 29, 397–399.

Germaine, J.T. and Ladd, C. (1988). Triaxial Testing of Saturated Cohesive Soil, in *Advanced Triaxial Testing of Soil and Rock*, ASTM STP 977, Robert T. Donaghe, Ronald, C. Chaney and Marshall Silver Eds. American Society for Testing and Materials, Philadelphia.

Ghaboussi, J. (1967). Dynamic stress analysis of porous elastic solids saturated with compressible fluids. *PhD Thesis* University of California, Berkeley.

Ghaboussi, J. and Wilson, E.L. (1973). Liquefaction analysis of saturated granular soils. *Proc. 5th World Conference in Earthquake Engineering*, Rome, Italy, 380–389.

Gibson, G. (1994). Earthquake hazard in Australia. Seminar, accepted risks for extreme events in the planning and design of major infrastructure. *ANCOLD – Munro Centre*, 26–27 April 1994, Sydney.

Gillon, M.D. and Hancox, G.T. (1992). Cromwell Gorge landslides, a general overview. *Proc. 6th Int. Symp. on Landslides, Christchurch*, Bell, D.H. ed., Balkema, Rotterdam, 83–102.

Gillon, M.D. and Newton, C.J. (1988). Matahina Dam: the initial lake filling incident and the long-term performance of the repair. *Proc. 5th Australia–New Zealand Conf. on Geomechanics*, 585–590, Institution of Engineers Australia, Canberra.

Giroud, J.P. (1982). Filter criteria for geotextiles. *Proc. 2nd Int. Conf. Geotextiles, Las Vegas*, Vol. 1, 103–109.

Giroud, J.P. (1996). Granular filters and Geotextile Filters. *Proceedings of GeoFilters '96'*. Lafleur J. and Rollin, A.L. Editors. Montreal, 565–680.

Giroud, J.P., Arman, A., Bell, J.R., Koerner, R.M. and Milligan, V. (1985). Geotextiles in Geotechnical Engineering Practice and Research, *Geotextiles and Geomembranes*, 2, 179–242.

Giroud, J.P. and Bonaparte, R. (1993). Geosynthetics in dam rehabilitations, *Proc. Sp. Conf. on Geotechnical Practice in Dams Rehabilition*, ASCE Geotechnical Special Publication 14035, 1043–1074.

Giudici, S., Herweynen, R. and Quinlan, P. (2000). HEC experience in concrete faced rockfill dams – past, present and future. Proceedings, International Symposium on Concrete Faced Rockfill Dams, Beijing, ICOLD. 29–46.

Glastonbury, J. (2002). The pre and post failure deformation behaviour of rock slopes. *PhD Thesis*, School of Civil and Environmental Engineering, The University of New South Wales, Sydney.

Glastonbury, J. and Fell, R. (2002a). Report on the analysis of rapid natural rock slope failures. *UNICIV Report No. R-390*, ISBN 85841 357 4, School of Civil and Environmental Engineering, The University of New South Wales, Sydney.

Glastonbury, J. and Fell, R. (2002b). Report on the analysis of slow, very slow and extremely slow natural landslides. *UNICIV Report No. R-402*, ISBN 85841 369 8, School of Civil and Environmental Engineering, The University of New South Wales, Sydney.

Glastonbury, J. and Fell, R. (2002c). A decision analysis framework for assessing post-failure velocity of natural rock slopes. *UNICIV Report No. R-409*, ISBN 85841 376 0. The School of Civil and Environmental Engineering, The University of New South Wales, Sydney.

Gomez, G.M. (1999). Concrete face behaviour of Aquimilpa Dam. *Proceedings, Second Symposium on CFRD*, Florianoplis, Brazil. Committee Brasilerio de Barragens, 211–222.

Gomez, G.M., Abonce, J.C. and Cartaxo, M.L. (2000). Behaviour of Aquimilpa Dam. *Concrete Face Rockfill Dams – J. Barry Cooke Volume*. 20th ICOLD Congress Beijing 2000, 117–151.

Good, R.J. (1976). Kangaroo Creek Dam. Use of a weak schist as rockfill for a concrete face rockfill dam. *Proc. 12th ICOLD Cong.*, Vol 1, Q44-R33, 645–665, International Commission on Large Dams.

Good, R.J., Bain, D.L.W. and Parsons, A.M. (1985). Weak rock on two rockfill dams in *Concrete Face Rockfill Dams, Design, Construction and Performance*, J.B. Cooke and J.L. Sherard (eds). ASCE Geotechnical Engineering Division, 40–72.

Goodman, R.E. (1993). *Engineering Geology*. John Wiley & Son.

Goodman, R.E. and Seed, H.B. (1966). Earthquake induced displacements in sand embankments. *Journal of the Soil Mechanics and Foundation Division*, ASCE, Vol. 92, No. SM2, 125–146.

Gordon, F.R. (1966). *Erosion of the bywash spillway at Serpentine Dam*. Geological Survey of Western Australia, Annual Report, 28–30.

Gordon, F.R. (1984). The lateritic weathering profiles of precambrian igneous rocks at the Worsley Alumina Refinery site, South West Division, Western Australia. *Proc. 4th Australia New Zealand Conference on Geomechanics*, Vol. 1, Perth, 261–266, Institution of Engineers Australia, Canberra.

Gordon, F.R. and Smith, D.M.A. (1984a). Lateritic soils near Worsley: the interaction of geology and geotechnology. *Proc. Fourth Australia New Zealand Conf. on Geomechanics*, Vol. 1, 273–279.

Gordon, F.R. and Smith, D.M.A. (1984b). The investigation, design and excavation of a cutting in completely weathered granite and dolerite near Boddington, Western Australia. *Proc. Fourth Australia New Zealand Conf. on Geomechanics*, Vol. 1, 28–285, Institution of Engineers Australia, Canberra.

Gosselin and others (1960). Final report on the Investigating Committee of the Malpasset Dam. Paris, Ministry of Agriculture (in French). (Translation from French published by the Israel Program for Scientific Translations. US Department of Commerce, Washington 25, DC.

Gourlay, A.W. and Carson, C.S. (1982). Grouting plant and equipment, in *Grouting in Geotechnical Engineering*, Baker, H.W. (ed.), ASCE, 121–135.

Graf, E.D., Rhodes, D.J. and Faught, K.L. (1985). Chemical grout curtains at Ox Mountain Dams in *Issues in Dam Grouting*. ASCE Geotech. Engineering, 92–103.

Graham, J. (1984). Methods of stability analysis, in *Slope Instability*, Brunsden, D. and Prior, D.B. (eds). Wiley, 171–215.

Grant, K. (1973). *The PUCE programme for terrain evaluation for engineering purposes (1) – Principles.* CSIRO Australia Technical Paper No. 15.

Grant, K. (1974). *The PUCE programme for terrain evaluation for engineering purposes (2) – Procedures for terrain classification.* CSIRO Australia Technical Paper No. 15.

Gratwick, C., Johannesson, P., Tohlang, S., Tente, T. and Monapathi, N. (2000). Mohale dam, Lesotho. *Proceedings of the International Symposium on Concrete Faced Rockfill Dams,* Beijing 257–272.

Graves, R.E., Eades, J.L. and Smith, L.L. (1988). Strength developed from carbonate cementation in silica-carbonate base course materials. *Transportation Research Record 1190,* 24–30.

Griffin, P.M. (1990). Control of seepage in tailings dams, in *Int. Symp. on Safety and Rehabilitation of Tailings Dams, ICOLD and ANCOLD, Sydney,* 106–115.

Grim, R.E. (1968). *Clay mineralogy,* McGraw Hill, New York.

Guillott, J.E. (1975). Alkali-aggregate reactions in concrete. *Eng. Geol.,* Vol. 9, 303–326.

Guillott, J.E. (1986). Alkali-reactivity problems with emphasis on Canadian aggregates. *Eng. Geol.,* Vol. 23, 29–43.

Guthries, L.G. (1986). Earthquake analysis and design of concrete dams. *3rd US National Conference on Earthquake Engineering,* Vol. 1.

Habic, P. (1976). Karst Hydrogeographic Evaluation. *Proc. 3rd Int. Symp. Underground Water Karlsruhe,* Rotterdam: Balkema.

Hacelas, J.E. and Ramirez, C.A. (1987). Closure by authors on design features of Salvajina Dam. ASCE *Journal Geotechnical Engineering,* Vol. 113, No. 10, 1175–1180.

Hadj-Hamou, T., Tavassoli, M.R. and Sherman, W.C. (1990). Laboratory testing of filters and slot sizes for relief wells, *Journal of Geotechnical Engineering,* ASCE, Vol. 116, No. 9, 1325.

Hadley, J.B. (1978). Madison Canyon rockslide, Montana, USA, in *Rockslides and Avalanches,* Voight, B. (ed.), 167–180.

Halburton, T.A., Lawmaster, J.D. and McGuffey, V.E. (1982). *Use of engineering fabric in transportation related Applications, FHWA training manual* – Contract No. DTFH-80-C-0094.

Haldane, A.D., Carter, E.K. and Burton, G.M. (1970). The relationship of pyrite oxidation in rockfill to highly acid water at Corin Dam, ACT, Australia, *Proc. 1st Congress IAEG,* Paris.

Hanna, T.H. (1973). *Foundation instrumentation.* Trans. Tech. Pub.

Hanna, T.H. (1985). *Field instrumentation in geotechnical engineering.* Trans. Tech. Pub.

Hanson, G.J. (1992). Erosion resistance of compacted soils. *Transportation Research Record 1369,* 26–30.

Hanson, G.J. and Robinson, K.M. (1993). The influence of soil moisture and compaction on spillway erosion. *Transactions of the ASAE.* 36(5) 1349–1352.

Harder, L.F. Jr. and Boulanger, R.W. (1997). "Application of K_α correction factors". *Proc. NCEER Workshop on Evaluation of Liquefaction Resistance of Soils,* Nat. Ctr. For Earthquake Engrg. Res., State Univ. of New York at Buffalo, 167–190.

Hardin, B.O. and Drenerich, V.P. (1972). Shear modulus and damping in soils. Design equations and curves. *J. Soil Mech. And Found. Div.,* ASCE, 98(7).

Harwood, G. (1988). Microscopic techniques: 11. *Principles of sedimentary petrography* in Tucker, M. (ed.) *Techniques in Sedimentology,* Blackwell.

Hast, N. (1976). The state of stresses in the upper part of the earth's crust. *Eng. Geol.,* Vol. 2, Part 1, 5–17.

Hatton, J.W. and Foster, P.F. (1987). Seismic considerations for the design of the Clyde Dam. Trans. IPENZ 14, No. 3/Co.

Hatton, J.W., Foster, P.F. and Thomson, R. (1991). The influence of foundation conditions on the design of Clyde Dam. *ANCOLD Bulletin 89,* 12–26. Australian National Committee on Large Dams, Melbourne.

Hawkins, A.B. (1998). Aspects of rock strength. *Bull. IAEG.57:* 17–30.

Hawkins, A.B. and Pinches, G.M. (1987). Cause and significance of heave at Llandough Hospital, Cardiff – a case history of ground floor heave due to gypsum growth. *Quart. Jour. Eng. Geol.,* 20, 41–57.

Hawkins, A.B. and Privett, K.D. (1985). Measurement and use of residual shear strength of cohesive soils. *Ground Engineering.*

Hazell, D.C. (2003). Personal communication. Materials Technology Branch, Transport South Australia.

Head, K.H. (1980, 81, 85). *Manual of Soil Laboratory Testing*, Vols 1, 2 and 3. Pentech Press.

Heaton, T.H., Tajima, F. and Mori, A.W. (1982). Estimating ground motions using recorded accelerograms. Unpublished report by Dames and Moore to Exxon Production Res. Co. Houston, Texas.

HEC (1988). *King River power development*. Technical data and description. Information prepared for ANCOLD conference.

HEC (1969). Cethana Dam: flood breach of partly completed rockfill dam, *ANCOLD Bulletin*, No. 28, Australian National Committee on Large Dams, Melbourne.

Hedberg, J. (1977). Cyclic stress strain behaviour of sand in offshore environment. PhD thesis, Dept. Civil Engineering, Massachusetts Inst. of Tech.

Heerten, G. (1984). Geotextiles in coastal engineering – 25 years experience. *Geotextiles and Geomembranes* 1, 119–141.

Heerten, G. (1993). A contribution to the improvement of dimensioning analogies for grain filter and geotextile filter. *In Filters in Geotechnical and Hydraulic Engineering*, Editors Brauns, Helbaum and Schuler, Balkema, Rotterdam, 121–127.

Hendron, A.J. and Patton, F.D. (1985). *The Vaiont slide, a geotechnical analysis based on new geologic observations of the failure surface*. US Army Corps of Engineers, Technical Report GL-85-5.

Henkel, D.J. (1960). The shear strength of saturated remoulded clays. *Proc. ASCE Research Conf. on Shear Strength of Cohesive Soils*, 533–554.

Herbert, C. (ed.) (1983). *Geology of the Sydney 1:100 000 Sheet 9130*, Geological Survey of New South Wales. Department of Mineral Resources, Sydney.

Hess, H.H. and Poldervaart, A. (1967). *Basalts, the Poldervaart treatise on rocks of basaltic composition*, Vol. 1, Interscience Publishers.

Hight, D.M. (1982). A simple piezometer probe for the routine measurement of pore pressure in triaxial tests on saturated soils. *Geotechnique*, 32, No. 4, 396–401.

Highway Research Board (1964). *Symposium on Alkali-Carbonate Rock Reaction*. Highway Research Board Record No. 45, HRB Publication 1167, 244.

Hillis, R.R., Meyer, J.J. and Reynolds, S.G. (1998). The Australian Stress Map, *Exploration geophysics*, 29, 420–427.

Hinks, J.L. and Gosschalk, E.M. (1989). dams and earthquakes – a review. *Dam Engineering*, Vol. IV, Issue 1, 9–26.

Hirschfeld, R.C. and Poulos, S.J. (1972). *Embankment Dams Engineering*. The Casagrande Volume. Wiley.

Høeg, K., Johansen, P.M., Kjærnsli, B. and Torblaa, I. (1998). *Internal Erosion in Embankment Dams*. Research Project Sponsored by Norwegian Electricity Federation (EnFO). (In Norwegian with English summary).

Hoek, E. (1998). Reliability of Hoek-Brown estimates of rock mass properties and their impact on design, *International Journal of Rock Mechanics and Mining Sciences*, 35(1), 63–68.

Hoek, E. (1999). Putting numbers to geology – an engineer's viewpoint, *Quart. Jour. Eng. Geol.*, 32, 1–19.

Hoek, E. (2002). A brief history of the development of the Hoek-Brown failure criterion. Assessed through the program *RocLab*, Rocscience.com

Hoek, E. and Bray, J.W. (1981). *Rock Slope Engineering*. Institution of Mining and Metallurgy, London.

Hoek, E. and Brown, E.T. (1997). Practical estimates of rock mass strength. *International Journal of Rock Mechanics and Mining Sciences*, 34(8), 1165–1186.

Hoek, E., Kaiser, P.K. and Bawden, W.F. (1995). Support of underground excavations in hard rock. Balkema, Rotterdam.

Holtz, R.D. and Kovacs, W.D. (1981). *Introduction to Geotechnical Engineering*. Prentice Hall.

Holzhausen, G.R. (1989). Origin of sheet structure, 1. morphology and boundary conditions. *Eng. Geol.*, Vol. 27, 225–278.

Honjo, Y. and Veneziano, D. (1989). Improved filter criterion for cohesionless soils. *J. Geotech. Eng. ASCE.* Vol. 115, No. 1, 75–94.

Horswill, P. and Horton, A. (1976). *Cambering and valley bulging in the Gwash Valley at Empingham, Rutherland*. Phil. Trans. Roy. Soc. London, A283, 427–462.

Hosking, A.D. (1990). Discussion on Engineering description of weathered rock, a standard system. *ANCOLD Bulletin* 85, 15–16, Australian National Committee on Large Dams, Melbourne.

Hosking, J.R. and Tubey, L.W. (1969). *Research on low-grade and unsound aggregates*. Road Research Lab. (UK), Lab. Report LR293.

Houlsby, A.C. (1977, 78). Foundation grouting for dams. *ANCOLD Bulletins 47, 48 and 50*, Australian National Committee on Large Dams, Melbourne.

Houlsby, A.C. (1982a). Cement grouting for dams. *Proc. Conf. on Grouting in Geotech. Engineering*. ASCE New Orleans, 1–34.

Houlsby, A.C. (1982b). Optimum water cement ratios for rock grouting, in *Issues Issues in Dam Grouting*. ASCE Geotech. Engineering, 317–331.

Houlsby, A.C. (1985). Cement grouting – water minimising practices in *Issues in Dam Grouting*, W.H. Baker (ed.), ASCE Geotech. Engineering Division, Denver.

Houlsby, A.C. (1986). Discussion on paper by Fell and MacGregor (1986). *ANCOLD Bulletin*, Australian National Committee on Large Dams, Melbourne.

Hubbert, M.K. (1940). The theory of groundwater motion. *J. of Geology*, 48, 785–944.

Huganberg, T. (1987). Alkali-carbonate reaction at Centre Hill Dam, Tennessee, in *Concrete Durability*, Scanlon, J.M. (ed.). American concrete Institute, Detroit, Vol. 2, 1883–1901.

Hungr, O. (1994). A general limit equilibrium Model for 3D slope stability analysis, discussion. *Can. Geotech. J.* 31, 793–795.

Hungr, O. (1997). Slope stability analysis. Keynote paper, *Proc. 2nd Panamerican Symposium on Landslides*, Rio de Janiero, 3: 123–136.

Hungr, O., Salgado, F.M. and Byrne, P.M. (1989). Evaluation of a three-dimensional method of slope stability analysis. *Can. Geotech. J.*, 27, 679–686.

Hunt, R.E. (1984). *Geotechnical Engineering Investigation Manual*. McGraw-Hill, New York.

Hunter, G. (2003). The Deformation Behaviour of Embankment Dams and Landslides in Natural and Constructed Soil Slopes. PhD Thesis, School of Civil and Environmental Engineering, The University of New South Wales.

Hunter, G. and Fell, R. (2002). The Deformation Behaviour of Rockfill. UNICIV Report No. R405, School of Civil and Environmental Engineering, The University of New South Wales.

Hunter, G. and Fell, R. (2003a). Mechanics of failure of soil slopes leading to "rapid" failure. *Proceedings international conference on fast slope movement, prediction and prevention for risk mitigation*, Naples, Editor Picarelli, L. Patron editore, Bologna, 283–296.

Hunter, G. and Fell, R. (2003b). Travel distance for rapid landslides in constructed and natural slopes. *Canadian Geotechnical Journal*, Vol. 40, No. 6, 1123–1141.

Hunter, G. and Fell, R. (2003c). Rockfill modulus and settlement of concrete face rockfill dams. Journal of Geotechnical and Geoenvironmental Engineering, ASCE, Vol. 129, No. 10, 909–917.

Hunter, G. and Fell, R. (2003d). The deformation behaviour of embankment dams. UNICIV Report No. R-416, ISBN 0077-880X, School of Civil and Environmental Engineering, The University of New South Wales, Sydney.

Hunter, J.R. (1982). Some examples of changes during construction, in *Geotechnical Problems and Practice in Dam Engineering*, Balasubramanian, A.S. and Younger, J.S. (eds), 99–108, Balkema.

Hunter, J.R. and Hartwig, W.P. (1962). The design and construction of the Cooma-Tumut project of the Snowy Mountains Scheme. *Jour. Inst. Enggs. Aust.*, Vol. 34, No. 7–8, 163–185.

Hutchinson, J,N. (1988). General Report. Morphological and geotechnical parameters of landslides in relation to geology and hydrogeology. *Proc. 5th Int. Symp. on Landslides, Lausanne*, Bonnard, C. ed., Balkema, 3–36.

Hutchinson, J.N. (2001). Reading the ground: Morphology and Geology in Site Appraisal. *The 4th Glossop Lecture, QJEG and H.*, 34, 8–50.

Hynes, M.E., and Olsen, R.S. (1999). Influence of confining stress on liquefaction resistance. *Proc., Int. Workshop on Phys. And Mech. Of Soil Liquefaction*, Balkema, Rotterdam, The Netherlands, 145–152.

ICOLD (1974). *Lessons from dam incidents*. International Commission on Large Dams, Paris.

ICOLD (1978). *Lessons from dam incidents*. International Commission on Large Dams, Paris.

ICOLD (1981). *Use of thin membranes as fill dams*. International Commission on Large Dams, Paris. Bulletin 38.

ICOLD (1982a). *Bibliography mine and industrial tailings dams and dumps*. International Commission on Large Dams, Paris. Bulletin 44.

ICOLD (1982b). *Manual on tailings dams and dumps*. International Commission on Large Dams, Paris. Bulletin 45.

ICOLD (1982c). *Bituminous concrete facings for earth and rockfill dams.* International Commission on Large Dams, Paris. Bulletin 42.

ICOLD (1983). *Seismicity and dam design.* International Commission on Large Dams, Paris. Bulletin 46.

ICOLD (1983a). *Deterioration of dams and reservoirs, examples and their analysis.* International Commission on Large Dams, Paris.

ICOLD (1984). *River control during dam construction.* International Commission on Large Dams, Paris. Bulletin 48.

ICOLD (1985). *Filling materials for watertight cutoff walls.* International Commission on Large Dams, Paris. Bulletin 51.

ICOLD (1986). *Geotextiles as filters and transitions in fill dams.* International Commission on Large Dams, Paris. Bulletin 55.

ICOLD (1986a). *Static analysis of embankment dams.* International Commission on Large Dams, Paris. Bulletin 53.

ICOLD (1986b). *Quality control of dams.* International Commission on Large Dams, Paris. Bulletin 56.

ICOLD (1986c). *Earthquake analysis procedures for dams – State of the art.* International Commission on Large Dams, Bulletin 52.

ICOLD (1986d). *Soil-cement for embankment dams.* International Commission on Large Dams, Paris. Bulletin 54.

ICOLD (1987). *Dam safety guidelines.* International Commission on Large Dams, Paris. Bulletin 59.

ICOLD (1988b). *Dam monitoring, general conditions.* International Committee on Large Dams, Paris. Bulletin 60.

ICOLD (1989). *Monitoring of dams and their foundations.* International Committee on Large Dams, Paris. Bulletin 68.

ICOLD (1989a). *Rockfill dams with concrete facing.* State-of-the-art. International Committee on Large Dams, Paris. Bulletin 70.

ICOLD (1989b). *Tailings dam safety.* Guidelines. International Committee on Large Dams, Paris. Bulletin 74.

ICOLD (1989c). *Selecting seismic parameters.* International Committee on Large Dams, Paris. Bulletin 72.

ICOLD (1990). *Dispersive soils in embankment dams.* International Commission on Large Dams, Paris. Bulletin 77.

ICOLD. (1993). *Embankment Dams – Upstream Slope Protection, Review and Recommendations,* International Commission on Large Dams, Paris. Bulletin 91.

ICOLD (1994). Embankment Dams Granular filters and Drains. International Commission on Large Dams, Paris, Bulletin 95.

ICOLD (1995). *Dam failures statistical analysis* – International Commission on Large Dams, Paris. Bulletin 99.

ICOLD (1999). *Design Features of dams to resist seismic ground motion.* International Commission on Large Dams, Paris. Bulletin 120.

ICOLD (2002). *Reservoir landslides – guidelines for investigation and management,* International Commission on Large Dams, Paris. Bulletin 124.

Idriss, I.M. and Boulanger, R.W. (2003). Estimating K_α for use in evaluating cyclic resistance of sloping ground. *Proceedings of the 8th US-Japan workshop on earthquake resistant design of lifeline facilities and counter measures against liquefaction,* Technical report MCEER 03.

Idriss, I.M., Lysmer, J., Hwang, R. and Seed, H.B. (1973). QUAD-4: A computer program for evaluating the seismic response of soil structures by variable damping finite element procedures. *Report No. EERC73-16,* University of California, Berkeley.

Independent Panel (1976). *Failure of Teton Dam.* Report to U.S. Department of the Interior and State of Idaho, U.S. Government Printing Office.

Indraratna, B., Wijewardena, L.S.S. and Balasubramaniam, A.S. (1993). Large-scale triaxial testing of greywacke rockfill. *Geotechnique,* 43(1), 37–51.

Ingles, O.G. Piping in earth dams of dispersive clay. ASCE *Speciality Conf. on the Performance of Earth and Earth-Supported Structures,* Vol. 3, 111–117.

Ingles, O.G. and Metcalf, J.B. (1972). *Soil stabilization.* Butterworths.

International Society for Rock Mechanics (1978). *Suggested methods for the quantitative description of discontinuities in rock masses*, 347–349.

Ishihara, K. (1985). Stability of natural deposits during earthquakes. *Proc. 11th ICSMFE*. Balkema, Rotterdam, 321–376.

Ishihara, K. (1993). Liquefaction and flow failures during earthquakes. *Geotechnique*, Vol. 33, No. 3, 351–415.

ISRM (1975). Recommendations on site investigation techniques, Final Report. ISRM Paris.

ISRM (1979). Suggested methods for determining water content, porosity, density, absorption and related properties, and swelling, and slake-durability index properties. *Int. Jour. Rock Mech. Min. Sci. & Geomech. Abstracts*. 16, 141–156.

ISRM (1985). Suggested method for determining point load strength. ISRM Commission on testing methods *Int. J. Rock Mech. Mineral Sci. Geomech. Abstr.*, 22: 51–60.

Jabara, M.A. and Wagner, W.E. (1969). Tiber Dam Auxiliary Outlet Works. *Hydraulics Division* ASCE, 17th Annual Specialty Conference.

Jacquet, D. (1990). Sensitivity to remoulding of some volcanic ash soils in New Zealand. *Eng. Geol.*, Vol. 28, 1–25.

Jaeger, C. (1963). The Malpasset Report. *Water Power*, February 1963.

James, A.N. (1981). Solution parameters of carbonate rocks. *Bull. Intl. Assoc. of Eng. Geol.*, No. 24, 19–25.

James, A.N. (1992). *Soluble materials in civil engineering*. Ellis Horwood Limited.

James, A.N. and Kirkpatrick, I.M. (1980). Design of foundations of dams containing soluble rocks and soils. *Quart. Jour. of Eng. Geol.*, Vol. 13, 189–198.

James, A.N. and Lupton, A.R.R. (1978). Gypsum and anhydrite in foundations of hydraulic structures. *Geotechnique*, Vol. 28, 249–272.

James, A.N. and Lupton, A.A.R. (1985). Further studies of the dissolution of soluble rocks. *Geotechnique*. 35(2), 205–210.

James, N.P. K. and Choquette, P.W. (1984). Limestones – the meteoric diagenetic environment. *Geoscience Canada*. Vol. 11, No. 4, 161–170.

James, P.M. (1983). Comments on the diagnosis of karst features in dam engineering. *Aust. Geomech. News*, No. 6, 13–16.

James, P.M. and Wood, C.C. (1984). Malpasset Dam, from an armchair. *Australian Geomechanics* 7, 23–31.

James, R.L., Catanach, R.B., O'Neill, A.L. and Von Thun, J.L. (1989). Investigation of the cause of Quail Creek Dike Failure, Report of Independent Review Team: Utah State Engineer's Office, Salt Lake City, UT, 154.

Jamiolkowski, M., Ghionna, V.N., Lancellotta, R. and Pasqualini, E. (1988). New correlations of penetration tests for design practice. Special Lecture. *Proc. 1st Int. Symp. on Penetration Testing ISOPT-1, Orlando*.

Jamiolkowski, M., Ladd, C., Germaine, J. and Lancellotta, R. (1985). New developments in field and laboratory testing, *11th Int. Conf. Soil Mechanics and Foundation Engineering*, Vol. 1, Balkema, 57–153.

Janbu, N. (1968). *Slope stability computations*. Soil mechanics and foundation engineering report. Technical University of Norway, Trondheim.

Jansen, R.B. (1980). Dams and public safety. US Dept of the Interior, Water and Power Resources Service. 331.

Jardine, R.J., Symes, M.J. and Burland, J.B. (1984). The measurement of soil stiffness in the triaxial appar-atus. *Geotechnique*, Vol. 34, No. 3, 323–340.

Jaworski, G.W., Duncan, J.M. and Seed, H.B. (1981). Laboratory study of hydraulic fracturing. *Journal of the Geotechnical Engineering Division*, ASCE, 107 (GT6) 713–732.

Johansson, S. (1996). Seepage monitoring in embankment dams by temperature and resistivity measurements. *Symp. on repairing and upgrading of dams*, Stockholm, June 5–7, 1996, Johansson, S. and Cederström M. (eds). KTH, Tryck och Kopiering, 288–306.

Jones, O.T. (1988). Canal failure on Wheao Power Scheme. *Proc. 5th Australia–New Zealand Conf. on Geomechanics*, 561–566, The Institution of Engineers Australia, Canberra.

Joyce, C.W., Maroney, P.M. and Whiteley, B. (1997). Condition risk assessment of in-ground refinery sewers with new seismic scanning technologies. *Committee on refinery environmental control of the American petroleum institute environmental symposium*, San Diego.

Justo, J.L. (1988). The failure of the impervious facing of Martin Gonzalo rockfill dam. *Sixteenth International Congress on Large Dams*, San Francisco, Discussions and Contributions Q.61–19, 252–266. International Commission on Large Dams, Paris.

Kammer, H.A. and Carlson, R.W. (1941). Investigation of causes of delayed expansion of concrete in Buck hydroelectric plant. *Am. Concr. Inst. J., Proc.*, 37, 665–671.

Karol, R.H. (1982a). Chemical grouts and their properties in Issues in Dam Grouting. ASCE *Geotech. Engineering*, 359–377.

Karol, R.H. (1982b). Seepage control with chemical grout, in Issues in Dam Grouting. ASCE *Geotech. Engineering*, 564–575.

Karol, R.H. (1983). *Chemical grouting*. Marcel Dekker.

Karol, R.H. (1985). Grout penetrability, in *Issues in Dam Grouting*. ASCE Geotech. Engineering, 1–26.

Kawai, T. (1985). Summary report on the development of the computer program DIANA – Dynamic interaction approach and non-linear analysis. Science University of Tokyo.

Keller, H. (1994). Investigation concerning the failure of Zoeknog Dam: findings. *Eighteenth International Congress on Large Dams*, Durban, Q.68 Discussions, 205–210. International Commission on Large Dams, Paris.

Kenney, T.C. and Lau, D. (1985). Internal stability of granular filters. *Canadian Geotech. J.*, Vol. 22, 215–225.

Kenney, T.C. and Lau, D. (1986). Closure to: Internal stability of granular filters. *Canadian Geotech. J.*, Vol. 23, 420–423.

Kenney, T.C. and Westland, J. (1993). Laboratory study of segregation of granular materials. In *Filters in Geotechnical and Hydraulic Engineers*. Editors Brauns, Helbaum and Schuler, Rotterdam: Balkema, 313–319.

Kenney, T.C., Chanal, R., Chin, E., Ofoeghu, G.I., Omange, G.N. and Ume, C.A. (1985). Controlling construction size of granular filters. *Canadian Geotechnical Journal*, 22, 32–43.

Khabbaz, H. and Fell, R. (1998). Uplift pressures of gravity dams. Unpublished report, School of Civil and Environmental Engineering, The University of New South Wales, Sydney.

Khabbaz, H. and Fell, R. (1999). Concrete strength for stability analysis of concrete dams. Unpublished report, School of Civil and Environmental Engineering, The University of New South Wales, Sydney.

Khalili, N. (1994). Post liquefaction analysis, Hume Dam – Section Ch3730. Unpublished *Report for the Public Works Department, N.S.W.*, The University of New South Wales, Sydney, Australia.

Khalili, N. and Khabbaz, M.H. (1996). The effective stress concept in unsaturated soils, UNICIV Report No. R-360, The University of New South Wales, Sydney.

Khalili, N. and Khabbaz, M.H. (1998). A unique relationship for χ for the determination of the shear strength of unsaturated soils. *Geotechnique*, 48, No. 5, 681–687.

Khalili, N., Fell, R. and Tai, K.S. (1996). A simplified method for estimating failure induced deformation. *Proc. Seventh Int. Symp. On Landslides*. Editor Senesett, K., Balkema, Rotterdam, 1263–1268.

Kiersch, G.A. (1965). Vaiont Reservoir Disaster. *Geotimes*, May–June 1965, 9–16.

King, J.P., Loveday, I.C. and Schuster, R.L. (1987). Failure of massive earthquake–induced landslide dam in Papua New Guinea. *Earthquakes and Volcanoes*, Vol. 10, No. 2, 40–47.

King, J.P., Loveday, I.C. and Schuster, R.L. (1989). The 1985 Bairaman landslide dam and resulting debris flow, Papua New Guinea. *Quart. Journ. Eng. Geol. London*, Vol. 22, 257–270.

Kjaernsli, B., Valstad, T. and Hoeg, K. (1992). *Rockfill dams, design and construction*. Norwegian Institute of Technology, Oslo.

Kleppe, J.H. and Olson, R.E. (1985). Desiccation cracking of soil barriers. *Hydraulic Barriers in Soil and Rock*, ASTM STP 874, 263–275.

Knight, D.J. (1990). Construction techniques and control, in *Clay Barriers for Embankment Dams*, Institution of Civil Engineers.

Knight, R.B. and Haile, J.P. (1983). Subaerial deposition with underdrainage. *Proc. 7th Pan American Soil Mechanics Conference*, Vancouver, Canadian Geotechnical Society.

Knill, J.L. (1968). Geotechnical significance of certain glacially-induced discontinuities in rock. *Bull. Intl. Assoc. of Eng. Geol.*, No. 5, 49–62.

Koerner, R.M. (1986). *Designing with geosynthetics*. Prentice Hall.

Koerner, R.M. (1990). *Designing with Geosynthetics*, 2nd ed., Prentice Hall.

Koutsouftras, D.C. and Ladd, C.C. *J. Geotech Eng.*, ASCE, Vol. 111, No. GT3, 337–355.

Kovacevic, M., Potts, D.M., Vaughan, P.R., Charles, J.A. and Tedd, P. (1997). Assessing the safety of old embankment dams by observing and analysing movement during reservoir operation. *Proc. 19th Int. Congress on Large Dams*, Florence, Q73, R35, 551–566. International Commission on Large Dams, Paris.

Kulhawy, F.H. (1992). On the evaluations of static soil properties, in Stability and Performance of Slopes and Embankments, ASCE *Geotechnical Special Publication No. 31*, 95–115.

Kulhawy, F.H. and Mayne, P.W. (1988). Manual on estimating soil properties for foundation design, *EPRI Draft Report*, 1493–6.

Kulhawy, F.H. and Mayne, P.W. (1989). *Manual on Estimating Soil Properties for Foundation Design of Electric Power*. Research Institute, Palo Alto, California.

Kulhawy, F.H., Roth, M.J.S. and Gregoriu, M.D. (1991). Some statistical evaluations of geotechnical properties. *Proc. 6 Int. Conf. on Applic. Stat. & Prob. in Civil Eng.*, Mexico City, 705–712.

Lacasse, S. and Berre, T. (1988). Triaxial Testing Methods for Soils. *Advanced Triaxial Testing of Soil and Rock*, ASTM STP 977, Robert T. Donaghe, Ronald, C. Chaney and Marshall Silver Eds. American Society for Testing and Materials, Philadelphia, 264–289.

Lacasse, S. and Lunne, T. (1988). Calibration of dilatometer correlations. Technical Paper. *Proc. 1st Int. Symp. on Penetration Testing ISOPT-1, Orlando.*

Ladanyi, B. and Archambault, G. (1977). Shear strength and deformability of filled indented joints. *International Symposium on the Geotechnical of Structurally Complex Formations*. Capri. Vol. 1, 317–326.

Ladd, C.C. (1991). Stability evaluation during staged construction (The 22nd Karl Terzaghi Lecture). *Journal of Geotechnical Engineering*, ASCE, Vol. 117, No. 4, 538–615.

Ladd, C.C. and Foott, R. (1974). New design procedure for stability of soft clays. JASCE *Geotech. Eng.*, GT7.

Ladd, C.C., Foott, R., Ishihara, K., Schlosser, F. and Poulos, H.G. (1977). Stress deformation and strength characteristics. S*tate of art report*, 9th Int. Conf. Soil Mech. and Foundation Eng.

Lade, P.W. (1986). Advanced triaxial testing of soils. University of Sydney, School of Civil and Mining Engineering.

Lade, P.V. (1992). State instability and liquefaction of loose fine sandy slopes. *Journal of Geotechnical and Geoenvironmental Engineering*, ASCE, Vol. 1, 118, No. 3, 51–71.

Lade, P.V. and Yamamuro, J.A. (1997). Effects of non plastic fines on static liquefaction of sands. *Canadian Geotechnical Journal.* 34, 9189–928.

Lafleur, J. (1984). Filter testing of broadly-graded cohesionless tills, *Canadian Geotechnical Journal*, 21, 634–643.

Lafleur, J. (1991). Prepared contribution to Question 67, New Developments for Fill Dams and fill Cofferdams, *XVIIth Congress of the International Commission on Large Dams,* Vienna, Austria, International Commission on Large Dams, Paris.

Lafleur, J., Mlynarek, J. and Rollin, Al. (1989). Filtration of broadly graded cohesionless soils. *J. geotech. eng.*, ASCE, Vol. 115, No. 12, 1747–1768.

Lafleur, J., Mlynarek, J. and Rollin, Al. (1993). Filter criteria for well graded cohesionless soils. In *Filters in geotechnical and hydraulic engineering*, Editors Braun, Helbaum and Schuler, Rotterdam: Balkema, 97–106.

Lak Kanchanaphol, Hilton, J.I. and Macpherson, P.M. (1982). Khao Laem Dam. Foundation and right abutment treatment. *Proc. 14th Congress on Large Dams, Q.53, R.43*, 681–698. International Commission on Large Dams, Paris.

Lam, L. and Fredlund, D.G. (1993). A general limit equilibrium model for three-dimensional slope stability analysis. *Can. Geotech. J.*, 30, 905–919.

Lambe, T.W. (1958). The engineering behaviour of compacted clay. *Journal of the Soil Mechanics and Foundations Division*, ASCE, 84 (SM2), 1655–1 to 1655–35.

Lambe, T.W. and Whitman, R.V. (1981). *Soil Mechanics, SI Version*. Wiley.

Lawson, J.D. (1987). Protection of rockfill dams and coffer dams against overflow and through-flow. The Australian Experience, IEAust, Civ. Eng. Trans., Vol. CE29 No. 3.

Lawton, E.C., Fragaszy, R.J. and Hetherington, M.D. (1992). Review of wetting-induced collapse in compacted soil. *Journal of Geotechnical Engineering*, ASCE, 118 (9), 1376–1394.

LeBihan, J.P. and Leroueil, S. (2000). Transient water flow through unsaturated soils – implications for earth dams. REGIMA.

Lee, F.T., Miller, P.R.P. and Nicholls, T.C., Jr. (1979). The relation of stresses in granite and gneiss near Mount Waldo, Maine, to structure, topography and rock bursts. *Proc. 20th Symp. Rock Mech., Austin, Texas*, 663–673.

Lee, M.K. and Finn, W.D.L. (1978). DESRA-2, Dynamic effective stress response: Analysis of soil deposits with energy transmitting boundary including assessment of liquefaction potential. *Soil Mechanics Series No.38*, Department of Civil Engineering, University of British Columbia, Vancouver, Canada.

Leeder, M.R. (1982). *Sedimentology, process and product*. George Allen and Unwin.

Léger, P., Sauve, G. and Bhattacharjeess. (1991). Dynamic substructure analysis of locally non-linear dam-foundation-reservoir systems. *Dam Engineering*, Vol. II, Issue 4, 323–336.

Leonards, G.A. and Narain, J. (1963). Flexibility of clay and cracking of earth dams. *Journal of the Soil Mechanics and Foundations Division*, ASCE, March 1963 (SM2), 47–98.

Leopold, L.B., Wolman, M.G. and Miller, J.P. (1964). *Fluvial processes in geomorphology*. W.H. Freeman, San Francisco.

Leps, T.M. (1973). Flow through rockfill dams. In *Embankment Dam Engineering*, Hirschfeld and Poulos eds, John Wiley and Sons, 87–107.

Leroueil, S., Tavenas, F., La Rochelle, P. and Tremblay, M. (1988). Influence of filter paper and leak-age in Triaxial Testing. In Advanced Triaxial Testing of Soil and Rock, ASTM STP 977, Robert T. Donaghe, Ronald, C. Chaney and Marshall Silver eds. American Society for Testing and Materials, Philadelphia. 189–201.

Leroueil, S., Locat, J., Vaunat, J., Picarelli, L. and Faure, R. (1996). "Geotechnical characterization of slope movements". *Proceedings of the 7th International Symposium on Landslides*, (Ed. Senneset, K.) Trondheim, Norway, Balkema, Rotterdam. Vol. 1, 53–74.

Lewis, D.W. (1984). *Practical sedimentology*. Hutchinson Ross, Stroudsburg, PA.

Li, X.S., Wang, Z.L. and Shen, C.K. (1992). SUMDES, a nonlinear procedure for response analysis of horizontally-layered sites subjected to multi-directional earthquake loading. Department of Civil Engineering, University of California, Davis.

Liao, S.S.C., Veneziano, D. and Whitman, R.V. (1988). Regression models for evaluating liquefac-tion probability. *JASCE Geotech. Eng.*, Vol. 114, No. 4, 389–411.

Lighthall, P.C. (1987). Innovative tailings disposal methods in *Canada Int. Jour. of Surface Mining* 1, 7–12.

Lillesand, T.M. (1987). *Remote sensing and image interpretation*. Wiley.

Lilly, R.N. (1986). Wungong Dam landslide. *Australian National Committee on Large Dams*, Bulletin 74, 38–49.

Littlejohn, G.S. (1985). Chemical grouting 1,2 and 3. *Ground Engineering* March, April, May.

Liu, A.H., Stewart, J.P., Abrahamson, N.A. and Moriwaki, Y. (2001). Equivalent number of uni-form cycles for soil liquefaction analysis. *J. Geotechnical and Geoenvironmental Engineering*, ASCE, Vol. 127, No. 12, 1017–1026.

Lo, K.Y. and Kaniaru, K. (1990). Hydraulic fracture in earth and rock-fill dams. *Canadian Geotechnical Journal*, 27, 496–506.

Lombardi, G. (1985). The role of cohesion in cement grouting of rock. *15th Int. Congress on Large Dams,* Q58, R13, 236–261, International Commission on Large Dams, Paris.

Londe, P. (1967). Panel Discussion, 1st Congress of ISRM, Lisbon. *Proc. Vol. III*, 449–453 (in French). International Commission on Large Dams, Paris.

Londe, P. (1982). Lessons from earth dam failures, in *Geotechnical Problems and Practice in Dam Engineering*, Balasubramaniam, A.S. and Younger, J.S. (eds), 19–28.

Lorete, B. and Khalili, N. (2002). An effective stress elastic-plastic model for unsaturated porous media. *Mechanics of Materials*, Vol. 3, No. 2, 97–116.

Loudiere, D., Fayoux, D., Houis, J., Perfettic, J. and Sotbon, M. (1982). L-utilisation des geotextiles dans les barrages en terre. *XIV ICOLD Congress Rio de Janiero, Brazil*, International Commission on Large Dams, Paris.

Lowe, J. and Karafiath, L. (1960). Stability of earth dams upon drawdown. *Proc. 1st Pan-Am. Conf. on Soil Mech. and Found. Engrg.*

Luke, W.I. (1963). *Petrographic examination of concrete cores, Chickamauga Dam Powerhouse* – Unit 3 Tennessee Valley Authority. USAEWWES Technical Report 6–637.

Lumb, P. (1982). Engineering properties of fresh and decomposed igneous rocks from Hong Kong. *Engineering Geology*, Vol. 19, 81–94.

Lunne, T. and Kleven, A. (1981). Role of CPT in North Sea foundation engineering. *Symp. Cone Penetration Testing and Materials.* ASCE *Nat. Conv.*, St. Louis, Mo, 76–107.

Lunne, T., Robertson, P.K. and Powell, J.J.M. (1997). *Cone Penetration Testing in Geotechnical Practice*, Blackie Academic and Professional, London.

Lupini, J.F., Skinner, A.E. and Vaughan, P.R. (1981). The drained residual strength of cohesive soils. *Geotechnique*, 31, No. 2, 181–213.

Lyell, K.A. and Prakke, H.K. (1988). Cyclone deposited tailings dam for 300 million tons, in *Hydraulic Fill Structures.* D.J.A. Van Zyl and S.G. Vick (eds.), ASCE Geotech. Special Pub. No. 21, 987–999.

Lysmer, J., Udaka, T., Tsai, C.F. and Seed, H.B. (1975). FLUSH – A computer program for approximate 3-D analysis of soil-structure interaction problems. *Report No. EERC75-30*, University of California, Berkeley.

MacGregor, F., Fell, R., Mostyn, G.R., Hocking, G. and McNally, G. (1994). The estimation of rock rippability, *Q.J. Eng. Geol. London*, Vol. 27, 123–144.

MacGregor, J.P., Olds, R. and Fell, R. (1990). Landsliding in extremely weathered basalt, Plantes Hill, Victoria, in *The Engineering Geology of Weak Rock*, Engineering Group of the Geological Society, Leeds.

Mackenzie, P.R. and McDonald, L.A. (1981). Use of soft rock in Mangrove Creek Dam. *ANCOLD Bulletin* No. 59, 5–20, Australian National Committee on large Dams, Melbourne.

Mackenzie, P.R. and McDonald, L.A. (1985). Mangrove Creek Dam: Use of soft rock for rockfill, in *Concrete Face Rockfill Dams, Design, Construction and Performance*, Cooke, J.B. and Sherard, J.L. (eds). ASCE Geotechnical Engineering Division, 208–230.

Maddox, J.M., Kinstler, F.L. and Mather, R.P. (1967). Meadowbank Dam – Foundations. The Institution of Engineers, Australia, *Annual Engineering Conference Papers*, 15–23.

Magnusson, R.H. (1980). Split Yard Creek Dam experience with filters. *ANCOLD Bulletin* No. 56, 67–72, Australian National Committee on large Dams, Melbourne.

Mail, R.W., Mitchell, I.J. & Sheehy, C.D. (1996). Hume dam: surveillance and investigation. *ANCOLD Bulletin* 101, Australian National Committee on Large Dams, Melbourne.

Makdisi, F.I. and Seed, H.B. (1978). A simplified procedure for estimating dam and earthquake induced deformations. *JASCE Geotechnical Eng.*, Vol. 1, 105, No. GT7, 849–867.

Marchetti, S. (1985). On the field determination of K_O in sand. Panel presentation at the *12th Int. Conf. Soil Mechanics and Foundation Engineering*, San Francisco.

Marcuson, W.F., Hynes, M.E. and Franklin, A.G. (1990). Evaluation and use of residual strength in seismic safety analysis of embankments. *Earthquake Spectra*, Vol. 6, No. 3, 529–572.

Marcuson, W.F., Hynes, M.E. and Franklin, A.G. (1992). Seismic stability and permanent deformation analysis: the last twenty five years. Proc. ASCE Specialty Conference on Stability and Performance of Slopes and Embankments – II, *Geotech. Special Publ.* No. 31 (eds R.B. Seed and R.W. Boulanger, ASCE, New York, Vol. 1, 552–592.

Marsal, R.J. (1973). Mechanical properties of rockfill. In *Embankment Dam Engineering J.*, Wiley & Sons, N.Y., 109–201.

Marsal, R.J. and Tamez, G. (1955). Earth dams in Mexico, design, construction and performance. *Fifth International Congress on Large Dams*, ICOLD, Communications C. 30, 1123–1178. International Commission on Large Dams, Paris.

Marshall, A.J. (1985). The stratigraphy and structure of the Thomson damsite and their influence on the foundation stability. South Australian Inst. Tech., Msc. Thesis, unpublished.

Martin, G.R., Finn, W.D.L. and Seed, H.B. (1975). Fundamentals of liquefaction under cyclic loading. *Journal of the Geotechnical Engineering Division*, ASCE, Vol. 101, 423–438.

Marulanda, A. and Pinto, N.L. (2000). Recent experience on design, construction, and performance of CFRD dams. In *Concrete Face Rockfill Dams*, J. Barry Cooke Volume (Mori, Sobrinho, Dijkstra, Guocheng & Borgatti ed.), Beijing, 279–315.

Materon, B. & Mori, R.T. (2000). Concrete face rockfill dams. Construction features. *Concrete Face Rockfill Dams, J. Barry Cooke Volume.* 20th ICOLD Congress Beijing 2000, 177–219.

Matheson, T.S. and Thompson, S. (1973). Geological implications of valley rebound. *Canadian Journal of Earth Science*, Vol. 10, 961–978.

Maver, J.L., Michels, V. and Dicken, R.S. (1978). *Dartmouth Dam project: design and construction progress.* Transactions of the Institution of Engineers, Australia, Vol. CE20, No. 1.

McAlexander, E.L. and Engemoen, W.O. (1985). Designing and monitoring for seepage at Calamus Dam, in *Seepage and Leakage from dams and impoundments*. ASCE, New York, 183–199.

McCann, D.M., Jackson, D.D. and Culshaw, M.G. (1987). The use of geophysical methods in the detection of natural cavities and mineshafts. *Quart. Journ. Eng. Geol. London*, Vol. 20, 59–73.

McClellan, G.H. (2003). Personal communication. University of Florida.

McClellan, G.H., Ruth, B.E., Eades, J.L., Fountain, K.B. and Blitch, G. (2001). Evaluation of aggregates for base course construction, Final Report, September, 2001. Depts. Of Geology and Civil Engineering, University of Florida.

McConnel, R.J. (1987). Ross River Dam embankment – post construction treatment of fissured clay foundations. *ANCOLD Bulletin* 77, 1987, Australian National Committee on large Dams, Melbourne.

McConnell, D., Mielenz, R.C., Holland, W.Y. and Greene, K.T. (1950). Petrology of concrete affected by cement-aggregate reaction, in *Application of Geology to Engineering Practice* (Berkeley Volume). The Geological Society of America, 225–250.

McDonald, L.A., Stone, P.C. and Ingles, O.G. (1981). Practical treatments for dams in dispersive soil. *Proc. Xth ICSMFE*, Vol. 2, 355–360. Stockholm.

McFarlane, M.J. (1976). *Laterite and landscape*. Academic Press, London.

McGuffey, V.C., Modeer, V.A. and Turner, A.K. (1996). "Subsurface exploration, in landslides investigation and mitigation, special report 247, Transportation Research Board". Editors Turner, A.K. and Schuster, R.L., *National Academy Press*, Washington DC, 231–277.

McKenna, J.M. (1984). *Vaturu Dam, Fiji*. Discussion paper, ANCOLD annual conference, Australian National Committee on large Dams, Melbourne.

McMahon, B.K., Burgess, P.J. and Douglas, D.J. (1975). Engineering classification of sedimentary rocks in the Sydney area. *Aust. Geomech. J.*, G5, No. 1.

McMahon, P.K. (1986). Limestone foundations at Bjelke-Petersen Dam. *ANCOLD Bulletin* No. 75, 44–51, Australian National Committee on large Dams, Melbourne.

McNally, G.H. (1981). Valley bulging, Mangrove Creek Dam near Gosford, NSW. *Geol. Surv. NSW, Quart.* Note 42, 4–11.

McRoberts, E.C. and Morgenstern, N.R. (1974). The stability of thawing slopes. *Canadian Geotechnical Journal*, Vol. 12, 130–141.

Medina, F., Domingues, J. and Tasoulas, J.L. (1990). Response of dams to earthquakes including effects of sediments. *Jour. of Struc. Engng Div.*, ASCE, 101, 423–438.

Mellent, V.A., Kolpashnikov, N.P. and Volnin, B.A. (1973). Hydraulic fill structures. *Energy, Moscow* (English translation of original Russian).

Melvill, A.L. (1997). The performance of a number of culverts in fill dams. *Nineteenth International Congress on Large Dams, Florence.* Q73, R36, 567–586. International Commission on Large Dams, Paris.

Mesri, G. and Cepeda-Diaz, A.F. (1986). Residual shear strength of clays and shales. *Geotechnique* 36, No. 2.

Meyer, R. (1997). *Paleolaterites and Paleosols*. Balkema.

Miall, A.D. (1985). Glaciofluvial transport and deposition. Chap. 7 in *Glacial Geology, An Introduction for Engineers and Earth Scientists*, Eyles, N. (ed.), Pergamon.

Milanovic, P.T. (1981). *Karst hydrogeology*. Water Resources Publications.

Millard, R.S. (1993). *Road building in the tropics*. HMSO, London.

Millet, R.A., Perez, J.Y. and Davidson, R.R. (1992). USA practice slurry wall specifications 10 years later in slurry walls, ASTM STP 1129, Editors D.B. Paul, R.R. Davidson and J.J. Cavall, *American Society for Testing and Materials*, Philadelphia, 42–66.

Milligan, V. (2003). Some Uncertainties in Embankment Dam Engineering. *Jour. Geotech. And Geoenvironmental Engineering*, ASCE, Vol. 129, No. 9, 785–797.

Minns, A. (1988). A review of tailings disposal practices in North America and Australia, in *Hydraulic Fill Structures*, D.J.A. Van Zyl and S.G. Vick (ed.). ASCE Geotech. Special Pub., No. 21, 52–68.

Minty, E.J. and Kearns, G.K. (1983). Rock mass workability, in *Collected Case Studies in Engineering Geology, Hydrogeology and Environmental Geology*, Minty, E.J. and Smith, R.B. (eds). Geological Society of Australia, 59–81.

Mitchell, J.K. (1970). In place treatment of foundation soils. JASCE Vol. 92, SM1, 73–100.

Mitchell, J.K. (1976). *Fundamentals of soil behaviour*. John Wiley & Sons.

Mitchell, J.K. (1993). *Fundamentals of soil behavior*, John Wiley & Sons.

Mitchell, J.K. and Katti, R.K. (1981). Soil improvement, state of the art. *Proc. 10th Int. Conf. on Soil Mechanics and Foundation Engineering*, Stockholm.

MMBW (1980). *Thomson Dam project, Specification for construction of main dam embankment and associated works*. Melbourne and Metropolitan Board of Works.

MMBW (1981). Notes on rock materials used at Winneke Dam. *ANCOLD Bulletin 59*, 57–61, Australian national Committee on large Dams, Melbourne.

Molenaar, N. and Venmans, A.A.M. (1993). Calcium carbonate cementation of sand: a method for producing artificially cemented samples for geotechnical testing and a comparison with natural cementation processes. *Engineering Geology*, Vol. 35, 103–122.

Money, M.S. (1985). Dam and reservoir construction in glaciated valleys. Chap. 14 in *Glacial Geology, An Introduction for Engineers and Earth Scientists*, Eyles, N. (ed.), Pergamon.

Moon, A.T. (1984). Effective shear strength parameters for stiff fissured clays. *Fourth Australia New Zealand Conf. on Geomechanics*, 107–111.

Moon, A.T. (1992). Stability analysis in stiff fissured clay at Raby Bay, Queensland, *Proc. Sixth Int. Symp. on Landslides*, Christchurch, New Zealand, Bell, D.H. editor, Rotterdam: Balkema, Vol. 2, 536–541.

Moreno, E. (1987). Close by author on "The upstream zone in concrete-face rockfill dams". *JASCE Geotech. Eng.* Vol. 113, No. 10, 1231–1245.

Morey, R.M. (1974). Continuous subsurface profiling by impulse radar. *Proc. ASCE Engineering Foundation Conference on Subsurface Exploration for Underground Excavation and Heavy Construction, New York*, 213–232.

Morgenstern, N.R. (1992). The evaluation of slope stability – a 25 year perspective. In *stability and performance of slopes and embankments, Geotechnical special publication*, 31, ASCE, New York, 1, 1–26.

Morgenstern, N.R. (1995). Keynote paper: The role of analysis in the evaluation of slope stability, *6th int. Symp. on landslides*, Christchurch, Editor Bell, D.H., Rotterdam: Balkema, Vol. 3, 1615–1629.

Morgenstern, N.R. and Price, V.E. (1965). The analysis of the stability of general slip surfaces, *Geotechnique*, 15(1), 79–93.

Morgenstern, N.R. and Tchalenko, J.S. (1967). Microstructural observations on shear zones from slips in natural clays. *Proc. Geotechnical Conference, Oslo*.

Morgenstern, N.R. and Kupper, A.A.G. (1988). Hydraulic fill structures – a perspective, in *Hydraulic Fill Structures*, Van Zyl, D.J.A. and Vick, S.G. (ed.), ASCE Geotech. Special Pub. No. 21.

Mori, R.T. (1999). Deformation and cracks in concrete face rockfill dams. *Proceedings, Second Symposium on Concrete Face Rockfill Dams*, Florianopolis, Brazil, Brazilian Committee on Dams. 49–61.

Moriwaki, Y. and Mitchell, J.K. (1977). *The role of dispersion in the slaking of intact clay*. ASTM STP 623, 287–302.

Moriwaki, Y., Beikae, M. and Idriss, I.M. (1988). Non-linear seismic analysis of the Upper San Fernando Dam under the 1971 San Fernando earthquake. *Proc. 9th World Conference on Earthquake Engng.*, Tokyo and Kyoto, Japan, Vol. III, 237–241.

Mostyn, G.R. and Fell, R. (1997). Quantitative and semi-quantitative estimation of the probability of landsliding, in *Landslide Risk Assessment*, editors Cruden, D. and Fell, R., Rotterdam: Balkema, 297–316.

Mostyn, G.R. and Small, J.C. (1987). Methods of stability analysis, in *Slope Stability and Stabilisation*, Walker, B.F. and Fell, R. (eds). Balkema, 71–120.

Mostyn, G.R. Personal communication. Uni. of New South Wales, Sydney, Australia.

Moye, D.G. (1955). Engineering geology for the Snowy Mountains Scheme, *J. IEAust.*, 287–298.

Moye, D.G. (1967). Diamond drilling for foundation exploration. Civil Engineering Transactions, *J. IEAust.*

Muir Wood, D., Jendele, L., Chan, A.H.C. and Cooper, M.R. (1995). Slope failure by pore pressure recharge: numerical analysis, *11th European Conf. on Soil Mechanics and Foundation Engineering*, Copenhagen, Vol. 6: 1–8.

Muller, L. (1964). The rock slide in the Vajont Valley. *Rock Mechanics and Engineering Geology*, Vol. I, Parts 3–4, 149–212.

Murray, R.C. (1964). Origin and diagenesis of gypsum and anhydrite. *Journal of Sedimenting Petrology*, Vol. 34, No. 3, 512–523.

Nakayama, K., Itoga, F. and Inque, Y. (1982). Selection and quality control of materials for rockfill dam of pumped storage project in phyllocrystalline schistose area. *Fourteenth Congress on Large Dams, Rio de Janeiro*, Q55, 23–45, International Commission on Large Dams, Paris.

Namikus, D. and Kulesza, R.L. (1987). Closure by authors on Kotmale Dam and observations on CFRD. Winneke project. *JASCE Geotech. Eng.*, Vol. 113, No. 10, 1198–1208.

National Research Council (1983). *Safety of existing dams, evaluation and improvement.* National Academy Press, Washington DC.

Naylor, D.J. (1975). Numerical models for clay core dams, in *Proc. Symp. Criteria and Assumptions for Numerical Analysis of Dams, Swansea*, 489–514.

NCEER (1997). Proceedings of the NCEER workshop on evaluation of liquefaction resistance of soils. Editors Youd, T.L. and Idriss, I.M. National Center for Earthquake Engineering Research technical report NCEER – 97-0022.

Nemcok, A. (1972). Gravitational slope deformation in high mountains. *Proc. 24th Int. Geological Congress*, Section 13, 132–141.

Netterberg, F. (1969). The geology and engineering properties of South African calcretes. *PhD Thesis*, Univ. Witwatersrand, Johannesberg.

Netterberg, F. (1971). Calcrete in road construction. *Research Report 286*, National Institute for Transport and Road Research, *Bulletin* 10, Pretoria, South Africa.

Netterberg, F. (1975). Self-stabilization of road bases: fact or fiction? *Proc. 6th Regional conf. For Africa on Soil Mechanics and Foundation Engineering*, Durban, 115–119.

Netterberg, F. (1978). Rates of calcrete formation. *Proc. 10th Int. Congress on Sedimentology*, Jerusalem, Israel. Vol. 2, 465.

Newmark, N.M. (1965). Effects of earthquakes on dams and embankments. 5th Rankine Lecture, *Geotechnique*, Vol. 15, No. 2, 139–160.

Newmark, N.M. and Rosenbleuth, E. (1971). *Fundamentals of earthquake engineering*, Prentice-Hall, Inc. Englewood Cliffs. N.J.

Newton, J.G. and Turner, J.M. (1987). Case histories of induced sinkholes in the eastern United States, in *Karst Hydrogeology, Engineering and Environmental Applications*, Beck, B.F. and Wilson, W.L. (eds). Balkema.

Nguyen, Q.D. and Bogor, D.V. (1985). Direct yield stress measurement with the vane method. *Journal of Rheology*, 27(4), 321–349.

Nicholls, T.C. (1980). Rebound, its nature and effect on engineering works. *Quart. J.Eng. Geol., London*, Vol. 13, 133–152.

Nicol, T.B. (1964). The Warragamba Dam. The Institution of Civil Engineers 27, 491.

Nixon, I.K. (1982). Standard penetration test. State-of-the-art report. *Proc. 2nd European Symp. on Penetration Testing, Amsterdam* (2nd ESOPT).

Nixon, J.F. and Morgenstern, N.R. (1973). The residual stress in thawing soils. *Canadian Geotechnical Journal*, Vol. 10, 572–580.

Norwegian Geotechnical Institute. (1992). *Rockfill dams, design and construction.* Kjaernsli, B. Valstad, T. and Höeg, K., Norwegian Geotechnical Institute, Norwegian Institute of Technology, Oslo.

NSFW (1998). *Shear strength of liquefied soils. Proceedings of National Science Foundation Workshop*, Editors T.D. Stark, S. Molson, S.L. Kramer and T.L. Youd. National Science Foundation, Earthquake Hazard Mitigation Program.

O'Neill, A.L. and Gourley, C. (1991). Geologic perspectives and cause of the Quail Creek Dike failure. *Bull.*, AEG, Vol. 28, No. 2, 127–145.

Oberg, A.L. and Sallfors, G. (1995). A rational approach to the determination of the shear strength parameters of unsaturated soils. *Proc. 1st Intl. Conf. Unsaturated Soils, Paris*, 151–158.

Oborn, L.E. (1988). Canal failure – Ruahihi hydro-electric power scheme, Bay of Plenty. *5th Australia New Zealand Conference on Geomechanics*, 574–583.

Ogilvy, A.A., Ayed, M.A. and Bogoslovsky, V.A. (1969). Geophysical studies of water leakages from reservoirs. *Geophysical Prospecting*, Vol. 27, No. 1, 36–62.

Ohya, S., Imai, T. and Matsubara, M. Relationship between N Value by SPT and LLT pressuremeter results. *Proc. 2nd European Symp. on Penetration Testing.* Amsterdam, Vol. 1, 125–130.

Olivier, H. (1967), Through and overflow rockfill dams – new design techniques. *Proceedings Institution of Civil Engineers*, Vol. 36, 433–464.

Olivier, H.J. (1990). Some aspects of the engineering geological properties of swelling and slaking mudrocks. *Proc. 6th Int. IAEG Congress.* Balkema, Rotterdam.

Olson, R.E. (1974). Shearing strength of kaolinite, illite and montmorillonite. *JASCE Soil Mechanics and Foundations Division*, Vol. 100, GT11, 1215–1229.

Olson, R.E. and Mesri, G. (1970). Mechanisms controlling the compressibility of clay. *JASCE Soil Mechanics and Foundations Division*, Vol. 96, SM6, 1863–1878.

Olsen, R.S. and Farr, J.V. (1986). Site characterization using the cone penetration test. *Proc. ASCE SP Conference In-situ 86*. Blacksburg.

Olsen, S.M. and Stark, T.D. (2002). Liquefied strength ratio from liquefaction flow failure case histories. *Canadian Geotechnical Journal*, Vol. 39, No. 3, 629–647.

Olsen, S.M. and Stark, T.D. (2003). Yield strength ratio and liquefaction analysis of slopes and embankments. *Journal of Geotechnical and Geoenvironmental Engineering*, ASCE, Vol. 129, No. 8.

Orchant, C.J., Kulhawy, F.H. and Trautmann, C.H. (1988). Critical evaluation of in-situ test methods. Report. EL-5507(2). Electric Power Research Institute, Palo Alto.

Palmeira, E.M. and Fannin, R.J. (2002). Soil – Geotextile compatibility in filtration. *Geosynthetics – 7th Int. Conf. Geosynthetics*, Editors Delmas, P. J., Gourc, P., Girard, H., Balkema, Lisse. 853–870.

Palmer, S.N. and Barton, M.E. (1987). Porosity reduction, microfabric and resultant lithification in UK uncemented sands. In Marshall, J.D. (ed.) (1987), *Diagenesis of Sedimentary Sequences*, Geological Society Special Publication No. 36, 29–40.

Parkin, A.K. (1977). The compression of rockfill. Australian Geomechanics Journal, Vol. G7, 33–39.

Paterson, B.R., Hancox, G.T., Thomson, R. and Thompson, B.N. (1983). Investigations in hard rock terrain. *Proc. Symp. Engineering for Dams and Canals, Alexandra Inst. of Prof. Engnrs, New Zealand*, 15.1–15.20.

Paterson, S.J. (1971). *Engineering geology of the Lemonthyme hydro-electric scheme, Tasmania.* Trans. Inst. Engrs. Aust.

Patton, F.D. (1966). Multiple modes of shear failure in rock. *Proc. 1st Int. Congress of Rock Mechanics, Lisbon*, Vol. 10, 509–513.

Patton, F.D. and Hendron, A.J. (1972). General report on mass movements. *Proc. 2nd Int. Congress of International Association of Engineering Geology*, V-GR1-V-GR57.

Patton, F.D. and Hendron, A.J. Jr (1973). Geological implications of valley rebound. *Canadian J. of Earth Science*, Vol. 10, 961–978.

Peck, R.B. (1972). Influence of nontechnical factors on the quality of embankment dams, in *Embankment Dam Engineering*, Casagrande Volume, editors Hirschfeld, R.C. and Poulos, S.J., 201–208, John Wiley & Sons.

Peck, R.B. (1990). Interface between core and downstream filter. *H. Bolton Seed Memorial Symp. Berkeley*, California, BiTech Publishers, Vancouver, Vol. 2, 237–251.

Peck, R.B., Hanson, W.E. and Thorburn, T.H. (1974). *Foundation Engineering*, John Wiley & Sons, New York.

Pells, P.J.N. (1978). Reinforced rockfill for construction flood control, ASCE, CO1.

Pells, P.J.N. (1982). Personal communication.

Pells, P.J.N. (1983). Plate loading tests on soil and rock, in *Insitu Testing for Geotechnical Investigations*, Ervin, M.E. (ed.) Balkema, 73–86.

Pells, S. and Fell, R. (2002). Damage and cracking of embankment dams by earthquakes, and the implications for internal erosion and piping. *UNICIV Report R 406*, The University of New South Wales, ISBN 85841 3752.

Pells, S. and Fell, R. (2003). Damage and cracking of embankment dams by earthquake and the implications for internal erosion and piping. Proceedings 21st International Congress on Large Dams, Montreal. ICOLD, Paris Q83–R17, International Commission on Large Dams, Paris.

Penman, A.D.M. (1982). General report on materials and construction methods for embankment dams and coffer dams. *ICOLD Congress on Large Dams, Rio de Janeiro*. Q55, 1105–1227, International Commission on Large Dams, Paris.

Penman, A.D.M. (1983). Materials for embankment dams. *Water Power and Dam Construction*, 15–19.

Penman, A.D.M. and Charles, J.A. (1976). The quality and suitability of rockfill used in dam construction. *ICOLD 12th Int. Cong. on Large Dams, Mexico*. Q44, 533–556, International Commission on Large Dams, Paris.

Penner, E., Eden, W.J. and Gillott, J.E. (1973). Floor heave due to biochemical weathering of shale, *Proc. 8th Int. Conf. Soil Mech. Foundation Engg.*, Moscow, 2, 151–158.

Pettijohn, F.J. (1957). *Sedimentary rocks (2nd ed)*, Harper Bros., New York.

Pinkerton, I.L. and McConnell, A.D. (1964). Behaviour of Tooma Dam. *8th Int. Congress on Large Dams*, R20, Q29, 351–375, International Commission on Large Dams, Paris.

Pinkerton, I.L., Soetomo, S. and Matsui, Y. (1985). Design of Cirata concrete face rockfill dam, in *Concrete Face Rockfill Dams*, Cooke, J.B. and Sherard, J.L. (eds), 642–656.

Pinto, N.L.S. (1999). Percolation through concrete faced rockfill dams under construction. Proceedings, Second Symposium on Concrete Face Rockfill Dams, Florianopolis, Brazil, 15–24.

Pinto, N.L.S. and Filho Marques, P.L. (1985). Discussion: Post-construction deformation of rockfill dams, *Journal of Geotechnical Engineering*, ASCE, Vol. 111 (12), 1472–1475.

Pinto, N.L.S. and Filho Marques, P.L. (1998). Estimating the maximum face slab deflection in CFRDs. *Hydropower and Dams*, No. 6.

Pinto, N.L.S., Filho Marques, P.L. and Maurer, E. (1985). Foz do Areia dam – design, construction and behaviour. *Proceedings of the Symposium on Concrete Face Rockfill Dams – Design, Construction and Performance*, (Cooke and Sherard ed.) Detroit, Michigan, ASCE New York. 173–191.

Piteau, D.R., Nylrea, F.H. and Blown, I.G. (1978). Downie slide, Columbia River, British Columbia, Canada. in Rockslides and Avalanches, Voight, B. (ed). Elsevier, 366–392.

Potts, D.M., Dounias, G.T. and Vaughan, P.R. (1990). Finite element analysis of progressive failure of Carsington embankment. *Geotechnique*, Vol. 40, 79–101.

Potts, D.M., Kovacevic, N. and Vaughan, P.R. (1997). Delayed collapse of cut slopes in stiff clay, *Geotechnique*, 47(5), 953–982.

Poulos, H.G. (1978). Normalised deformation parameter for Kaolin. *Research Report R336*, School of Civil Engineering, University of Sydney.

Poulos, H.G. and Davis, E.A. (1974). *Elastic Solutions for Soil and Rock Mechanics*. John Wiley & Sons, New York.

Poulos, S.J., Castro, J. and France, J.W. (1985). Liquefaction evaluation procedures. *Journal of the Geotechnical Engineering Division*, ASCE, Vol. 111, No. 6, 772–791.

Powell, J.J.M. (1990). A comparison of four different pressuremeters and their methods of interpretation in a stiff heavily overconsolidated clay. *Proc. 3rd Int. Symp. on Pressuremeters*, organised by the British Geotechnical Society, Oxford University.

Powell, J.J.M. and Quarterman, R.S.T. (1988). The interpretation of cone penetration tests in clays, with particular reference to rate effects. *Proc. 1st Int. Symp. on Penetration Testing ISOPT-1, Orlando*.

Powell, J.J.M. and Uglow, I.M. (1988). Marchetti dilatometer testing in UK soils. Technical Paper. *Proc. 1st Int. Symp. on Penetration Testing ISOPT-1, Orlando*.

Powell, R.D. and Morgenstern, N.R. (1985). The use and performance of seepage prediction measures, in *Seepage and Leakage from Dams and Impoundments*, Volpe, R.L. and Kelly, W.E. (eds). ASCE.

Pratt, H.K., McMordie, R.C. and Dundas, R.M. (1972). Foundations and abutments – Bennett and Mica Dams. JASCE SM10, 1053–1072.

Prebble, W.M. (1988). Investigations in an active volcanic terrain. *Proc. Symp. on Engineering for Dams and Canals, Alexandra*. The Institution of Professional Engineers, New Zealand, 17.1 to 17.15.

Prevost, J.H. (1981). DYNAFLOW: A nonlinear transient finite element analysis program. Princeton University, Department of Civil Engineering, Princeton, N.J.

Price, D.G. (1995). Weathering and weathering processes. *QJEG* No. 28, 243–252.

Prothero, D.R.L. and Schwab, F. (1996). *Sedimentary geology*. Freeman, New York.

Prusza, K., De Fries, K. and Luque, F. (1985). The design of Macagua concrete face rockfill dam in *Concrete Face Rockfill Dams, Design, Construction and Performance*, Cooke, J.B. and Sherard, J.L. (eds). ASCE Geotechnical Engineering Division, 608–656.

PWD of NSW (1981). *Clarrie Hall Dam and associated works*, unpublished report, Department of Public Works, Sydney.

PWD of NSW (1985). *Mardi Dam stability analysis*, unpublished report, Department of Public Works, Sydney.

PWD of WA (1982). *The Harding Dam and appurtenant works*, tender document. Public Works Department of Western Australia, Perth.

Pye, K. and Miller, J.A. (1990). Chemical and geochemical weathering of pyretic mudrocks in a shale embankment. *QJEG* 23, 365–381.

Radbruch-Hall, D.H., Varnes, D.J. and Savage, W.Z. (1976). Gravitational spreading of steep-sided ridges ("sakung"), in Western United States. *Bull. Int. Assoc. of Eng. Geol.*, No. 14, 23–35.

Ransome, F.L. (1928). Geology of the St Francis Dam Site. *Economic Geol.*, Vol. 23.

Regalado, G., Materon, B., Ortega, J.W. and Vargas, J. (1982). Alto Anchacaya concrete face rock-fill dam – Behaviour of the concrete face membrane. Proceedings, 14th ICOLD Congress, Rio de Janeiro, Brazil, Vol. 4, Q55, R30, 517–535. International Commission on Large Dams, Paris.

Regan, P.J. (1997). Performance of concrete-faced rockfill dams of the Pacific Gas & Electric Company. Non-Soil Water Barriers for Embankment Dams, Proceedings, 17th Annual USCOLD Lecture Series, San Diego, California, USCOLD. pp. 149–162.

Regan, W.M. (1980). Engineering geology of Sugarloaf Dam. South Australian Institute of Technology. M.App.Sc. thesis.

Rengers, N. and Soeters, R. (1980). Regional engineering geological mapping from aerial photographs. *Bull. Int. Assn. Engg. Geol.*, No. 21, 103–111.

Rib, H.T. and Liang, T. (1978). Recognition and identification in *Landslides, Analysis and Control*, Schuster, R.L and Krizek, R.J. (eds). Special Report 176. National Academy of Sciences, 24–80.

Richards (1992). Modelling interactive load deformation and flow pressures in soils. *6th Australia–New Zealand Conference on Geomechanics*, Christchurch, 18–37.

Richter, C.F. (1958). *Elementary seismology*. Freeman and Co.

Riddolls, B.W., Macfarlane, D.F. and Crampton, N.A. (1992). Engineering geological characterisation of landslide-affected schist terrain, Clyde Power Project, New Zealand. *Proc. 6th Int. Symp. on Landslides, Christchurch*, Bell, D.H. ed., Balkema, Rotterdam, 2137–2144.

Riemer, W. (2003). Personal communication.

Riemer, W., Gavard, M., Soubrier, G. and Turfan, M. (1995). The seepage at the Attaturk Dam. *Proc. 19th Congr. ICOLD, Q. 73R. 38*, 613–633. International Commission on Large Dams, Paris.

Riemer, W., Gavard, M., Soubrier, G. and Turfan, M. (1997). The seepage at Attatürk Fill Dam. *Proc. 19th Congr. ICOLD, Q 73, R38*, 613–633. International Commission on Large Dams, Paris.

Riemer, W., Ruppert, F.R., Locher, T.C. and Nunez, I. (1988). Mechanics of deep seated mass movements in metamorphic rocks of the Ecuadorian Andes. *Proc. 5th Int. Symp. on Landslides, Lausanne*, Bonnard, C. (ed.), Balkema, Rotterdam, 307–310.

Ripley, C.F. (1986). Discussion of: Internal stability of granular filters. *Canadian Geotechnical Journal*, 23, 255–258.

Ripley, C.F. (1984). Discussion of: Progress in rockfill dams. *J. Geotech Eng.*, ASCE, 114(2), 236–240.

Ritcey, G.M. (1989). *Tailings management*. Elsevier.

Robertson, A. MacG., Bamberg, S.A. and Lange, G. (1978). Current uranium mill wastes disposal concepts: a multinational viewpoint, paper at *Symp. on Uranium Mill Tailings Management, Fort Collins, Colorado*. Geot. Engrg. Program, Civil Engrg. Dept., Colorado State University.

Robertson, P.K. (1990). Soil classification using CPT. *Can. Geotech. J.*, Ottawa, 27(1), 151–158.

Robertson, P.K. (1994). Suggested terminology for liquefaction. *Proc. 47th Canadian Geotechnical Conference*, Halifax, Nova Scotia, 277–286.

Robertson, P.K. and Campanella, R.G. (1983a). Interpretation of cone penetration test, Part 1, Sand. *Canadian Geotech. J.*, Vol. 20, No. 4, 718–733.

Robertson, P.K. and Campanella, R.G. (1983b). Interpretation of cone penetration test, Part 2, Clay. *Canadian Geotech. J.*, Vol. 20, No. 4, 734–745.

Robertson, P.K. and Wride, C.E. (1997). Cyclic liquefaction and its evaluation based on the SPT and CPT. *Proc. NCEER Workshop on Evaluation of Liquefaction Resistance of Soils*. Editors T.L. Youd and I.M. Idriss, NCEER Technical Report 97-0022, 41–88.

Robertson, P.K. and Fear, C.E. (1995). Liquefaction of sands and its evaluation. *Proc., 1st Int. Conf. On Earthquake Geotech. Engrg.*

Robertson, P.K. and Wride, C.E. (1998). Evaluating cyclic liquefaction potential using the cone penetration test. *Can. Geotech. J.*, Ottawa, 35(3), 442–459.

Robinsky, E.I. (1979). Tailings disposal by the thickened discharge method for improved economy and environmental control, in *Tailings Disposal Today*, Argall, G.O. (ed.), Miller Freeman Publications.

Rogers, R.L. (1985). Boondoomba Dam in *Concrete Face Rockfill Dams, Design, Construction and Performance*, Cooke, J.B. and Sherard, J.L. (eds). ASCE Geotechnical Engineering Division, 316–335.

Rollins, A.L. and Denis, R. (1987). Geosynthetics filtration in landfill design. *Geosynthetics '87' Conference*, New Orleans, USA, 456–470.

Roth, L.H. and Schneider, J.R. (1991). Considerations for use of geosynthetics in dams, *Geosynthetics in dams, USCOLD*.

Roth, W.H. (1985). Evaluation of earthquake induced deformations of Pleasant Valley Dam. *Report for the City of Los Angeles*. Dames and Moore, Los Angeles.

Rouse, C. (1990). The mechanics of small tropical flowslides in Dominica, West Indies. *Engineering Geology*, Vol. 29, 227–239.

Rowe, P.W. (1972). The relevance of soil fabric to site investigation practice. Rankine Lecture 1972. *Geotechnique*, 22, No. 2, 195–300.

Ruqing, Z. (1981). Hydrodynamic types in Karstic river valleys. *Bull. IAEG* 24, 39–44.

Ruxton, B.P. and Berry, L. (1957). The weathering of granite and associated erosional features in Hong Kong. *Bull. Geological Society of America*, Vol. 68, 1263–1292.

Saada, A.S. and Townsend, F.C. (1981). State-of-the-art. Laboratory Strength Testing of Soils; in *ASTM STP 740, Laboratory Shear Strength of Soil*.

Salembrier, M., Vassiliadis, G., LeNindre, Y. and Souhangir, A.A. (1998). Leakage investigations and proposed remedial measures at Lar Dam (Iran). In: Varma, C.V.J., Visvanatham, N. and Rao, A.R.G. (eds). *Symposium on Rehabilitation of Dams*, 85–96. Balkema.

Sarac, Dz. and Popovic, M. (1985). Shear strength of rockfill and slope stability, *Proceedings of the Eleventh International Conference on Soil Mechanics and Foundation Engineering*, San Francisco, 641–645.

Sarma, S.K. (1973). Stability analysis of embankments and slopes. *Geotechnique*, Vol. 23, 4, 423–433.

Sarma, S.K. (1975). Seismic stability of earth dams and embankments. *Geotechnique*, Vol. 25, No. 4, 743–761.

Savage, Z.W. (1978). The development of residual stress in cooling rock bodies. *Geophysical Res. Letters*, Washington, Vol. 5, 633–636.

Saville, T., McClendon, E.W. and Cochran, A.L. (1962). Freeboard allowances for waves in inland reservoirs. *JASCE*, Vol. 88, No. WW2, 93–123.

Scheurenberg, R.J. (1982). Experiences in the use of geofabrics in underdrainage of residue deposits. *2nd Int. Conf. on Geotextiles, Las Vegas*.

Schiffman, R.L. and Carrier, W.D. (1990). Large strain consolidation used in the design of tailings impoundments, in *Int. Symp. on Safety and Rehabilitation of Tailings Dams. ICOLD and ANCOLD, Sydney*, 156–174.

Schiffman, R.L., Vick, S.G. and Gibson, R.E. (1988). Behavior and properties of hydraulic fills, in *Hydraulic Fill Structures*. Van Zyl, D.J.A. and Vick, S.G. (eds). ASCE Geotech. Special Pub. No. 21, 166–202.

Schmertman, J.H. (1978). Guidelines for cone penetration test, performance and design. US Federal Highway Administration FWHA TS-78-209.

Schmertmann, J.H. (1991). The mechanical aging of soils. *Journal of Geotechnical Engineering ASCE 117*, No. 9, 1288–1330.

Schmertmann, J.H. (1955). The Undisturbed Consolidation of Clay. Trans. ASCE Vol. 120, 1201–1233

Schmidt, P. (1991). *Permanent magnetism and orientation of drill core*. Underground Coal Mining Exploration Techniques, NERDDP, Brisbane.

Schmitter, N.J. (1979). Dam failures. *13th ICOLD Congress,* New Delhi, Q49, International Commission on Large Dams, Paris.

Schnabel, P.L., Lysmer, J. and Seed, H.B. (1972). SHAKE: A computer program for earthquake response analysis of horizontally layered sites. *Report No. EERC72-12*, University of California, Berkeley.

Schroeder, P.R., Morgan, M.J., Wolski, T.M. and Gibson, A.M. (1983). Draft manual, the hydro logic evaluation of landfill performance (HELP) model. Vols I and II.

Schuler, U. (1995). How to deal with the problem of suffusion. Research and Development in the Field of Dams, *Swiss National Committee on Large Dams*, 145–159.

Schuler, U. and Brauns, J. (1993). Behaviour of coarse and well graded filters. In *Filters in geotechnical and hydraulic engineering*, Editors Brauns, Helbaum and Schuler, Rotterdam: Balkema, 3–17.

Schuler, U. and Brauns, J. (1997). The safety of geotechnical filters. *Hydropower and dams*, Issue 6, 72–74.

Schuster, R.L and Costa, J.E. (1986). A perspective on landslide dams, in *Landslide dams – processes, risk and mitigation*, Schuster, R.L. (ed.). Geotechnical Special Publication, No. 3, ASCE.

Seed, H.B. (1979). Soil liquefaction and cyclic mobility evaluation for level ground during earthquakes. *J. Geotech. Engrg. Div.*, ASCE, 105(2), 201–255.

Seed, H.B. (1979a). Consideration in the earthquake resistant design of earth and rockfill dams. *Geotechnique* 29, 215–263.

Seed, H.B. (1983). Earthquake-resistant design of earth dams. *Proc., Symp. Seismic Design. Of Earth Dams and Caverns*, ASCE, New York, 41–64.

Seed, H.B. (1987). Design problems in soil liquefaction. *JASCE Geotechnical Engineering*, Vol. 113 No. GT 8, 827–845.

Seed, H.B. and De Alba, P. (1986). Use of the SPT and CPT tests for evaluating the liquefaction resistance of sands. In-Situ 86, Conference on use of in-situ tests in geotechnical engineering. Clemence, S.P. (ed.). ASCE *Geotechnical Special Publication*, No. 6, 281–302.

Seed, H.B. and Idriss, I.M. (1971). Simplified procedure for evaluating soils liquefaction potential. *J. Geotech. Engrg. Div.*, ASCE, 97(9), 1249–1273.

Seed, H.B. and Idriss, I.M. (1982). *Ground motions and soil liquefaction during earthquakes.* Monograph series, Earthquake Engineering Research Institute, Berkeley, California.

Seed, H.B. and Martin, G.R. (1966). The seismic coefficient in earth dam design. *Journal of the Soil Mechanics and Foundation Division*, ASCE, Vol. 92, No. SM3, 25–57.

Seed, H.B., Lee, K.L., Idriss, I.M. and Makdisi, F.I. (1973). Analysis of slides in the San Fernando Dams during the earthquake of February 1971. *Report No. EERC/73-2*, University of California, Berkeley.

Seed, H.B., Makdisi, F.I. and De Alba, P. (1978). Performance of earth dams during earthquakes. *J. Geotech. Engng. Div.*, 104, No. GT7.

Seed, H.B., Seed, R.B., Lai, S.S. and Khemenchpour, B. (1985). Seismic design of concrete faced rockfill dams, in *Concrete Faced Rockfill Dams* (eds Cook, J.B. and Sherard, J.L.), ASCE, Detroit, MI, 459–478.

Seed, H.B., Tokimatsu, K., Harder, L.F. and Chung, R.M. (1985). The influence of SPT procedures in soil liquefaction resistance evaluations. *J. Geotech. Engrg.*, ASCE, 111(12), 1425–1445.

Seed, H.B., Wong, R.T., Idriss, I.M. and Tokimatsu (1986). Moduli and damping factors for dynamic analysis of cohesionless soils. *J. Geotech. Engng*, ASCE, 112(11).

Selby, M.J. (1982) *Hillslope materials and processes.* Oxford Uni. Press.

Selby, M.J. (1993). *Hillslope materials and processes*, 2nd edition, Oxford University Press, Oxford.

Selley, R.C. (1976). *An introduction to sedimentology*, 2nd edition, Academic Press.

Serafim, J.L. and Pereira, J.P. (1983). Considerations on the geomechanical classification of Bieniawski, *International symposium on Engineering Geology and Underground Construction*, Portugal, IAEG, Vol. 1, II33–II42.

Shayan, A. (1987). Alkali-aggregate reaction in Australian concrete structures. *XIII Biennial Conf.*, Concrete Inst. of Australia.

Shayan, A. (1998). Alkali – aggregate reaction in Australian concrete dams. *ANCOLD Buletinl.* 109, 79–99. Australian National Committee on Large Dams, Melbourne.

Shayan, A. and Van Atta, R.D. (1986). Comparison of methylene blue and benzidene staining for detection of smectite clay minerals in basaltic aggregates. *Proc. 13th Conf. Australian Roads Research Board*, Vol. 13, Part 5, 56–65.

Sherard, J.L. (1953). *Influence of soil properties and construction methods on the performance of homogeneous earth dams.* Ph.D Thesis, Harvard University.

Sherard, J.L. (1967). Earthquake considerations in earth dam design. *Journal of the Soil Mechanics and Foundations Division*, American Society of Civil Engineers, Vol. 93, No. SM4, 377–401.

Sherard, J.L. (1972). Piping in earth dams of dispersive clay. *Proc. of Specialty Conf. on Performance of Earth and Earth-supported Structures*, ASCE, Vol. 1, 613.

Sherard, J.L. (1973). Embankment dam cracking, in *Embankment Dam Engineering*, Hirschfeld, R.C. and Paulos, S.J. (eds), John Wiley & Sons, 308–312.

Sherard, J.L. (1979). Sinkholes in dams of coarse, broadly graded soils. *13th Int. Congress on Large Dams, New Delhi.* Q47, R2, 325–334. International Commission on Large Dams, Paris.

Sherard, J.L. (1981). *Piezometers in earth dam impervious sections in recent developments in geotechnical engineering for hydro projects,* Kulawy, F.H. (ed.). ASCE Geotech. Eng. Div.

Sherard, J.L. (1985a). Hydraulic fracturing in embankment dams, in *Seepage and Leakage from Dams and Impoundments.* ASCE Geotechnical Engineering Division Conference.

Sherard, J.L. (1985b). The upstream zone in concrete faced rockfill dams, in *Concrete Face Rockfill Dams Design, Construction and Performance.* Cooke, J.B. and Sherard, J.L. (eds). ASCE Geotechnical Engineering Division, 618–641.

Sherard, J.L. and Cooke J.B. (1987). Concrete-face rockfill dam: 1. Assessment. JASCE *Geotech. Eng.,* Vol. 113 No. 10, 1096–1112.

Sherard, J.L. and Dunnigan, L.P. (1985). Filters and leakage control, in embankment dams, in *Seepage and Leakage from Dams and Impoundments.* ASCE Geotechnical Engineering Division Conference, 1–30.

Sherard, J.L. and Dunnigan, L.P. (1989). Critical filters for impervious soils. *J. Geotech. Eng.* ASCE, Vol. 115, No. 7, 927–947.

Sherard, J.L., Cluff, L.S. and Allen, C.R. (1974). Potentially active faults in dam foundations. *Geotechnique,* Vol. 24, No. 3, 367–428.

Sherard, J.L., Decker, R.S., Dunnigan, L.P. and Steele, E.F. (1976a). Identification and nature of dispersive soils. *J.Geotech Eng.* ASCE, Vol. 102 No. GT4, 277–301.

Sherard, J.L., Decker, J.L. and Ryker, N.L. (1972a). Piping in earth dams of dispersive clay. *Proceedings, Specialty Conference on Performance of Earth and Earth-Supported Structures,* ASCE, Vol. 1, Part 1, 589–626.

Sherard, J.L., Decker, R.S. and Ryker, N.L. (1972b). Hydraulic fracturing in low dams of dispersive clay. *Proceedings, Specialty Conference on Performance of Earth and Earth-Supported Structures,* ASCE, Vol. 1, Part 1, 653–689.

Sherard, J.L., Dunnigan, L.P. and Talbot, R.T. (1984a). Basic properties of sand and gravel filters. JASCE *Geotech. Eng.,* Vol. 110, No. 6.

Sherard, J.L., Dunnigan, L.P. and Talbot, R.T. (1984b). Filters for silts and clays. JASCE *Geotech. Eng.* Vol. 110 No. 6.

Sherard, J.L., Dunnigan, L.P., Decker, R.S. and Steele, E.F. (1976b). Pinhole test for identifying dispersive soils, *J.Geotech Eng.* ASCE, Vol. 102 No. GT1, 63–80.

Sherard, J.L., Woodward, R.J., Gizienski, S.F. and Clevenger, W.A. (1963). *Earth and earth rock dams.* John Wiley & Sons.

Sherard, J.L. and Decker, R.S. (1977). *Summary evaluation of Symposium.* ASTM STP 623, 467–479.

Shovcair, J. (2003). Personal communication. Florida Department of Transport.

Sierra, J.M., Ramirez, C.A. and Hacelas, J.E. (1985). Design features of Salvajina Dam in *Concrete Face Rockfill Dams, Design, Construction and Performance,* Cooke, J.B. and Sherard, J.L. (eds). ASCE Geotechnical Engineering Division, 266–285.

Siggins, A.F. (1990). Ground penetrating radar in geotechnical applications. *Exploration Geophysics,* Vol. 21, 175–186.

Silver, M.L. (1985). *Remedial measures to improve the seismic strength of embankment dams.* Report No. 85-10, Department of Civil Engineering, University of Illinois, Chicago, Illinois.

Simmons, G.C. (1966). Stream anticlines in Central Kentucky. *Proc. Paper US Geol. Survey,* 550-D, D9-11.

Simpson, B. (1981). A suggested technique for determining the base friction angle of rock surfaces using core. Technical Note. *International Journal of Rock Mechanics and Mining Sciences and Geomechanical Abstracts,* 18(1): 63–65.

Sims, G.P. and Rainey, T.P. (1985). Gitaru Dam grouting. *15th Int. Congress on Large Dams, Lausanne,* Question 58, R69, 1143–1166, International Commission on Large Dams, Paris.

Skempton, A.W. (1977). Slope stability of cuttings in brown London clay. *Proc. Of the 9th International Conference on Soil Mechanics and Foundation Engineering,* Tokyo, 261–270.

Skempton, A.W. (1985). Residual strength of clays in landslides, folded strata and the laboratory. *Geotechnique,* Vol. 35(1), 3–18

Skempton, A.W. (1986). Standard penetration test procedures and the effects in sands of overburden pressure, relative density, particle size, aging and overconsolidation. *Geotechnique,* 36, No. 3, 425–447.

Skempton, A.W. and Coates, D.J. (1985). Carsington dam failure, in *Failures in Earthworks.* Institution of Civil Engineers, London, 203–220.

Skempton, A.W. and Hutchinson, J.N. (1969). Stability of natural slopes and embankment sections. *Proc. 7th Int. Conf. Soil Mech. And Found. Eng., Mexico*, State-of-the-Art, 291–340.

Sladen, J.A. and Wrigley, W. (1985). Geotechnical properties of lodgement till. Chap. 8 in *Glacial Geology*, An Introduction to Engineers and Earth Scientists, Eyles, N. (ed.), Pergamon.

Sladen, J.A., D'Hollander, R.D. and Krahn, J. (1985). The liquefaction of sands, a collapse surface approach. *Canadian Geotech J.*, 22, 564–578.

Smith, D.L. and Randazzo, A.F. (1987). Application of electrical resistivity methods to identify leakage zones in drained lakes, in *Karst Hydrogeology*, Beck, B.F. and Wilson, W.L. (eds). Balkema, 227–234.

Smith, G.M., Abt, S.R. and Nelson, J.D. (1986). *Profile prediction of hydraulically deposited tailings*. Soc. Min. Eng. AIME, Vol. 280, Part A, 2024–2027.

Snow, D.T. (1969). Rock fracture spacings, openings and porocities. *J. Soil Mechanics and Foundation Division*, ASCE, Vol. 94, No. SM1, 73–91.

Snow, D.T. (1970). The frequency and apertures of fractures in rocks. *Int. J. Rock Mech. And Min. Sci.*, 7, 23–40.

Sobrinho, J.A., Sadinha, A.E., Albertoni, S.C. and Dijkstra, H.H. (2000). Development aspects of CFRD in Brazil. In Concrete Face Rockfill Dams, J. Barry Cooke Volume (Mori, Sobrinho, Dijkstra, Guocheng and Borgatti ed.), Beijing, 153–175.

Soeters, R. and van Westen, C.J. (1996). Slope instability recognition, analysis and zonation, Chapter 8, in Turner, A.K. and Schuster, R.L. (eds). Landslides Investigation and Mitigation. Transportation Research Board, Special Report 247.

Soil Instruments Ltd (1985). *Instrumentation for Soils and Rocks*.

Solvik, Ø (1991). Throughflow and stability problems in rockfill dams exposed to exceptional loads, *Sixteenth International Congress on Large Dams*, Q.67 R.20, ICOLD, 333–343. International Commission on Large Dams, Paris.

Somerford, M., Davenport, F. and Brice, S. (1991). Geotechnical aspects of the design, construction and performance of Harris Dam, *ANCOLD Bulletin* No. 88, 33–54, Australian National Committee on Large Dams, Melbourne.

Somkuan Watakeekul and Coles, A.J. (1985). Cutoff treatment methods in Karstic Limestone – Khao Laem Dam. *Proc. 15th Congress on Large Dams*, Q.58. International Commission on Large Dams, Paris.

Sowers, G.F. (1975). Failures in limestones in humid subtropics. *J. Geol. Div.*, ASCE 101(GTB): 771–789.

Sowers, G.F. and Royster, D.L. (1978). Field investigation. In Schuster, R.L. and Krizek, R.J. (eds). *Landslides*, Transportation Research Board.

Sowers, G.F., Williams, R.C. and Wallace, T.S. (1965). Compressibility of broken rock and the settlement of rockfills. Proceedings, 6th International Conference on Soil Mechanics and Foundation Engineering, Montreal, Vol. 2, 561–565.

Soydemir, C. and Kjaernsli, B. (1979). Deformations of membrane-faced rockfill dams. *Proceedings, 7th European Conference on Soil Mechanics and Foundation Engineering*, Brighton, England, Vol. 3, 281–284.

Spencer, E. (1967). A method of analysis of the stability of embankments assuming parallel interslice forces. Geotechnique, 17(1), 11–26.

Spink, T.W. and Norbury, D.R. (1993). The engineering description of weak rocks and overconsolidated soils. In: Cripps, J.C., Coulthard, J.M., Culshaw, M.G., Forster, A., Hencher, S.R. and Moon, C.F. (eds). The Engineering Geology of Weak Rock. *Engineering Geology Special Publication No. 8*, Balkema, Rotterdam, 289–301.

Standards Association of Australia (1974a). Australian Standard AS1141, Section 38, *Potential alkali activity by mortar bar*, 38-1 to 38.6.

Standards Association of Australia (1974b). Australian Standard AS1141, Section 39, *Potential reactivity of aggregates* (chemical method), 39-1 to 39-5.

Standards Association of Australia (1979). *Site Investigation Code*, AS1726.

Standards Association of Australia (1980, 1997). *Testing of Soils for Engineering Purposes*, AS1289.

Standards Association of Australia (1985). *Aggregates and Rock for Engineering Purposes*, AS2758.

Stapledon, D.H. (1967). Geological investigations at the site for Kangaroo Creek Dam, South Australia. *Trans. Institution Engineers Australia*, Vol. CE9, No. 1, 31–43.

Stapledon, D.H. (1971). Changes and structural defects developed in some South Australian clays, and their engineering consequences. *Proc. Symposium on Soils and Earth Structures in Arid Climates,* Adelaide, 1970. The Institution of Engineer, Australia.

Stapledon, D.H. (1976). Geological hazards and water storage. *Bull. Int. Assoc. Engineering Geology,* No. 14, 249–262.

Stapledon, D.H. (1979). Investigation and characterization. *Proc. Extension Course on Tunnelling Design and Practice,* Melbourne, October 1979, 13–33. Australian Geomecahnics Society.

Stapledon, D.H. (1983). Towards successful waterworks. *Proc. Symp. Engineering for Dams and Canals, Alexandra.* Institution of Professional Engineers, New Zealand, 1.3–1.15.

Stapledon, D.H. (1988a). Engineering description of weathered rock – a standard system? *ANCOLD Bulletin 81,* 35–48, Australian National Committee on Large Dams.

Stapledon, D.H. (1988b). Engineering geophysics. A geologists view. *Aust. Geomechanics* No. 15, 57–62.

Stapledon, D.H. (1995). Geological modeling in landslide investigation. *Proc. 6th International Conference on Landslides,* editor Bell, D.H., Balkema, Rotterdam, 1499–1523.

Stapledon, D.H. (1995). Geological modelling in landslide investigation, *Proc. 6th International Conference on Landslides,* Bell, D.H. (ed.), Vol. 3, 1499–1523, Balkema, Rotterdam.

Stapledon, D.H. and Casinder, R.J. (1977). *Dispersive soils at Sugarloaf Dam site, near Melbourne, Australia.* Dispersive clays, related piping and erosion in Geotechnical Projects, ASTM STP623, American Society for Testing and Materials.

Stapledon, D.H., Woodburn, J.A. and Fitzhardinge, C.F.P. (1978). Planning, construction and operation of Kanmantoo mine tailings dam. *ANCOLD Bulletin 51,* 9–24, Australian national Committee on Large Dams.

Stark, T.D. and De Puy, G. (1987). Alkali-silica reaction in five dams in the Southwestern United States, in *Concrete Durability,* Scanlon, J.M. (ed.), Vol. 2. American Concrete Institute, 1759–1786.

Stark, T.D. and Mesri, G. (1993). Undrained shear strength of liquefied sands for stability analysis. *JASCE Geotechnical Engineering,* Vol. 118, No. 11, 1727–1747.

Stark, T.D. and Eid, H.T. (1997). Slope stability analyses in stiff fissured clays ASCE. *Journal of Geotechnical and Geoevironmental Engineering,* Vol. 123(4), 335–343.

Stark, T.D., Olson, S.M., Kramer, S.L. and Youd, T.L. (1998). Shear strength of liquefied soils. *National Science Foundation Workshop,* NSF Grant CMS-95-31678, Urbana, Illinois, April 17–18, 1999.

St-Arnaud, G. (1995). The high pore pressures within embankment dams; an unsaturated soil approach. *Canadian Geotechnical Journal,* 32, 892–898.

Stauffer, P.A. and Obermeyer, J.R. (1988). Pore water pressure conditions in tailings dams, in *Hydraulic Fill Structures,* Van Zyl, D.J.A. and Vick, S.G. (eds). ASCE Geotech. Special Pub. No. 21, 903–923.

Stearns, H.T. (1944). Characteristics of coral deposits. *Engng News Rec.* Vol. 133, No. 2, 85–88.

Stewart, F.H. (1963). Marine evaporates. US Geol. Surv. Prof. Paper 440. *US Geological Survey,* Denver, Colorado, USA.

Stroman, W.R., Beene, R.R.W. and Hull, A.M. (1984). Clay shale foundation slide at Waco Dam, Texas. *Proc. Int. Conf. on Case Histories in Geotechnical Engineering.* Univ. of Missouri-Rolla, 579–586.

Sullivan, T.D. (1982). The origin and characteristics of defects in oil shale. *Proc. Symp. on Coal Resources, origin, Exploration and Utilization in Australia,* Part 2, Geological Society of Australia, 582–595.

Sullivan, T.D., Duran, A. and Eggers, M.J. (1992). The use and abuse of oriented core in open pit mine design. *Proc. Of the Third Open Pit Mining Conference,* Mackay, Queensland.

Swaisgood, J.R. (1998). Seismically-induced deformation of embankment dams. *6th US National Conference on Earthquake Engineering,* Seattle, Washington.

Swamy, R.N. (1992). *The alkali-silica reaction in concrete.* Blackie and Son Ltd., Glasgow and London.

Swanson (1992). A computer program designed by Swanson Analysis Systems, Inc.

Swarbrick, G. and Fell, R. (1990). Prediction of desiccation rates of mine tailings. *3rd Int. Symp. on the Reclamation, Treatment and Utilization of Coal Mining Waste, Glasgow.*

Swarbrick, G. and Fell, R. (1991). Prediction of the improvement of tailings properties by desiccation. *IX Panamerican Conference on Soil Mechanics and Foundation Engineering, Vina del Mar,* 995–1008.

Swarbrick, G.E. (1995). Measurement of soil suction using the filter paper method. *Proc. 1st Int. Conf. Unsaturated Soils, Paris*, 653–658.

Sweeting, M.J. (1972). *Karst landforms*. Macmillan.

Sweeting, M.M. (1981). *Karst geomorphology*. Hutchinson Rose Publishing Co.

Swiss National Committee on Large Dams (1985). Retrospective on Swiss dam foundation treatment. *15th Int. Congress on Large Dams, Lausanne*, Question 58, R12, 209–234. International Commission on Large Dams, Paris.

Tabatabaei, J. (1999). Safety assessment of Corin Dam. *ANCOLD Bulletin 112*, 89–102, Australian National Committee on Large Dams, Melbourne, 89–99.

Talbot, J.R. and Ralston, D.C. (1985). Earth dam seepage control, SCS Experience, in *Seepage and Leakage from Dams and Impoundments*, Volpe, R.L. and Kelly, W.E. (eds). ASCE.

Tamaro, G.J. and Poletto, R.J. (1992). Slurry walls – construction quality control in slurry walls, ASTM STP 1129, Editors Paul, D.B., Davidson, R.R. and Cavall, J.J., *American Society for Testing and Materials*, Philadelphia, 26–41.

Taylor, G.F. (1998). Acid drainage: sources, impacts and responses, *Groundwork*, Vol. 2, No. 1, 1–4. Australian Minerals & Energy Environment Foundation.

Taylor, R.K. (1988). Coal measures mudrocks: composition, classification and weathering processes. *QJEG*, 21, 85–99.

Taylor, R.K. and Cripps, J.C. (1987). Weathering effects, slopes in mudrocks and over-consolidated clays, in *Slope Stability*, Anderson, M.G. and Richards, K.S. (eds). John Wiley & Sons.

Taylor, R.K. and Smith, T.J. (1986). The engineering geology of clay minerals: swelling, shrinking and mudrock breakdown. *Clay Minerals*, 21, 235–260.

Taylor, R.K. and Spears, D.A. (1981). Laboratory investigation of mud-rocks. *Q.J. Eng. Feol., London*, Vol. 14, 291–309.

Tedd, P. and Hart, J.M. (1988). The use of thermography and temperature measurement to detect leakage from embankment dams. *International symposium on detection of subsurface flow phenomena by self-potential/geoelectrical and thermometrical methods*, Karlruhe, 1–7.

Terzaghi, K. (1926). *Soil physical basis of mechanics of earth structures*. F. Deuticke, Wien (in German).

Terzaghi, K. (1958). Design and performance of Sasumua Dam. *Proc. Inst. Civil Eng. 9*, 369–394.

Terzaghi, K. (1960a). Discussion of Rockfill Dams: Salt Springs and Lower Bear River Concrete Face Dams, by Steele, J.C. and Cooke, J.B. Transactions, ASCE, Vol. 125, 139–148.

Terzaghi, K. (1960b). Discussion of Rockfill Dams: Wishon and Courtright Concrete Face Dams, by Cooke, J.B. Transactions, ASCE, Vol. 125, 622–625.

Terzaghi, K. (1961). Discussion on horizontal stresses in an overconsolidated Eocene clay. *Proc. 5th Int. Conf. on Soil Mech. and Found. Engng., Paris*, Vol. 3, 144–145.

Terzaghi, K. (1962). Stability of steep slopes on hard unweathered rock. *Geotechnique 12*, 251–270.

Terzaghi, K. and Leps, T.M. (1958). Design and performance of Vermilion Dam. *Journal of Soil Mechanics and Foundations Division*, ASCE, No. SM3, 1728-1 to 1728-30.

Thomas, H.H. (1976). *The Engineering of Large Dams*. Wiley.

Thompson, D.M. and Shuttler, R.M. (1976). Design of rip rap slope protection against wind waves. *Construction Industry Research and Information Association. Report 61*. The Association. London, December.

Thorne, C.P. (1984). Strength assessment and stability analysis for fissured clay. *Geotechnique*, Vol. 34, No. 3, 305–322.

Tjandrajana, R. (1989). *Penetration of cement grout into fractures*. School of Civil Engineering, The University of New South Wales. M.Eng.Sc. Thesis.

Tomlinson, M.J. (1957). Airfield construction on overseas soils, Part 3: Saline calcareous soils. *Proc. ICE*, Vol. 8, 232–246.

Townsend, F.C. and Gilbert, P.A. (1973). Tests to measure residual strengths of some clay shales. *Geotechnique 23 No. 2*.

Townsend, F.C., Krinitzsky, E.L. and Patrick, D.M. (1982). *Geotechnical properties of laterite gravels*. Engineering and Construction in Tropical and Residual soils. ASCE Geotechnical Engineering Division, 236–262.

Trudinger, J.F. (1973). Engineering Geology of the Kangaroo Creek Dam. *Geol. Surv. S.A. Bulletin 44*, 150–157.

Truscott, E.G. (1977). Behaviour of Embankment Dams. *Ph.D. Thesis*. Department of Civil Engineering, Imperial College of Science and Technology, London.

Turner, A.K. and Schuster, R.L. eds (1996). Landslides investigation and mitigation, *Transportation Research Board*, National Research Council, Washington, DC.

Twidale, C.R. (1987). Sinkholes (dolines) in lateritised sediments, Western Sturt Plateau, Northern Territory, Australia. *Geomorphology*, 1, 33–52.

Twidale, C.R. and Bourne, J.A. (2000). Dolines of the Pleistocene dune calcarenite terrain of western Eyre Peninsula, South Australia: a reflection of underprinting? *Geomorphology 33*, 89–105.

Unterstell, B., Toumani, A., Rokers, E., Rubel, A. and Polysos, N. (2000). Application of geophysical well logs in geotechnical evaluation of subsurface deposits and geoengineering, *Preview Australian Society of Exploration Geophysicists*, 89, 21–24 and 90, 17–19.

Uromeihy, A. (2000) The Lar Dam, an example of infrastructural development in a geologically active karstic region. *Journal of Asian Earth Sciences*, 18, 25–31.

US Corps of Engineers (1941). *Investigation of filter requirements for underdrains*, US Corps of Engineers, Waterways Experiment Station Technical Memorandum No. 183-1.

US Corps of Engineers (1970). *Engineering and design – stability of earth and rock fill dams*, Engr. Manual EM 1110-2-1902, Dept. of the Army, Corp of Engrs., Office of the Chief of Engineers.

US Corps of Engineers (1976). United States. Coastal Engineering Research Center. *Coastal design memorandum* – Coastal Engineering Research Center, CDM 76-1.

US Corps of Engineers (1982). *Ice engineering*, US Army Corps of Engineers EM-1110-2-1612.

US Corps of Engineers (1984). *Rationalising the seismic coefficient method*. Miscellaneous Paper GL84-13.

US Corps of Engineers (1984). *Foundation grouting practices at Corp of Engineers dams*. US Army Corp of Engineers Technical Report GL-84-13.

US Corps of Engineers (1984a). *Shore Protection Manual*. Dept Army, Waterways Experiment Station, Coastal Engineering Research Centre.

US Corps of Engineers. (1986). *Engineering and design seepage analysis and control for dams*, EM 1110-2-1901.

US Corps of Engineers (1995). *Design of Gravity Dams*, US Army Corps of Engineers.

US National Research Council (1985). *Liquefaction of soils during earthquakes*. National Academy Press, Washington, DC.

USBR (1955). *The use of laboratory tests to develop design criteria for protective filters*. Earth Laboratory Report No EM-425, US Bureau of Reclamation, Denver.

USBR (1966). Summary of large triaxial shear tests for silty gravels earth research studies. EM-731, United States Department of the Interior Bureau of Reclamation – Soils Engineering Branch, Division of Research, Denver.

USBR (1977 and 1978). *Report on design and analyses of Auburn dam*. US Bureau of Reclamation, Denver Co.

USBR (1977). *Design of Small Dams*. United States Department of the Interior, Bureau of Reclamation, Denver.

USBR (1979). *Laboratory procedures for determining the dispersibility of clayey soils*. Report No. REC-ERC-79-10. US Bureau of Reclamation, Denver, Colorado.

USBR (1980, 1985). *Earth Manual*. US Dept. Interior Water Resources Technical Publication, Denver.

USBR (1981). *Freeboard criteria and guidelines for computing freeboard allowances for storage dams*. ACER Tech Memo No. 2, US Dept Interior, Bureau of Reclamation, Denver.

USBR (1984). *Design standard No.13 embankment dams, Chapter 3, Foundation Surface Treatment*.

USBR (1987). *Embankment Dam Instrumentation Manual*. US Bureau of Reclamation, Denver.

USBR (1989). *Design Standards. Embankment dams, No. 13, Chapter 13. Seismic design and analysis*. US Bureau of Reclamation, Denver, Co.

USBR (1989a). *Policy and procedures for dam safety modification and decision making*. US Bureau of Reclamation, Denver.

USBR (1992). Freeboard criteria and guidelines for computing freeboard allowance for storage dams. *ACER Technical Memorandum No. 2*, US Department of the Interior, US Bureau of Reclamation, Denver Co.

USCOLD (1975). *Lessons from Dam Incidents USA*. US Committee on Large Dams.

USCOLD (1988). *Lessons from Dam Incidents USA-II*. US Committee on Large Dams.

USDA-SCS (1994). *Gradation design of sand and gravel filters*. United States Department of Agriculture, Soil Conservation Service, Part 633, Chapter 26, National Engineering Handbook.

USEPA (2000). *Innovations in site characterization: geophysical investigation at hazardous waste sites Aug. 2000*. USEPA, Washington DC.

USNRC (1985). Liquefaction of soils during earthquakes. *National Academy Press*, Washington DC.

Vaid, Y.P. and Elizdorani, A. (1998). "Instability and liquefaction of granular soils under undrained and partially drained states". *Canadian Geotech. J.*, 35, 1053–1062.

Valencia, G.F. and Sandoval, E.M. (1997). Aguamilpa Dam Behaviour. Proceedings, Seventeenth Annual USCOLD Lecture Series, Non-Soil Water Barriers for Embankment Dams. San Diego, California, 133–147.

Van Aller, H. and Rodriguez, R. (1999). Seismic tomography of the Loveton Dam, Baltimore Country, Maryland. *Workshop on internal diagnostics for embankment Dams*, CEA Dam Safety Interest Group Report No. PSE203.

Van Atta, R.O. and Ludowise, H. (1976). *Microscopic and X-ray examination of rock for durability testing*. Federal Highway Admin (USA). Report FHWA-RD-77-36.

Varnes, D.J. (1958). Landslide types and processes, in *Landslides and Engineering Practice*, Eckel, E.B. (ed.). Highways Research Board Special Report 29, 20–47.

Varnes, D.J. (1978). Movement types and processes, in *Landslides, Analysis and Control*, Schuster, R.L. and Krizek, R.J. (eds). National Academy of Sciences, 11–33.

Varty, A., Boyle, R.J., Pritchard, E.D. and Gill, P.E. (1985). Construction of concrete face rockfill dams in *Concrete Face Rockfill Dams, Design, Construction and Performance*, Cooke, J.B. and Sherard, J.L. (eds). ASCE Geotechnical Engineering Division, 435–458.

Vaughan, P.R. (1994). Assumption, prediction and reality in geotechnical engineering. *Geotechnique*, 44, No. 4, 573–609.

Vaughan, P.R. and Hamza, M.M. (1977). Clay embankments and foundations: monitoring stability by measuring deformations. Specialty session 8: deformation of earth-rockfill dams, *9th Int. Conf. on Soils Mechanics and Foundation Engineering*, Tokyo, 37–48.

Vaughan, P.R and Soares, H.F. (1982). Design of filters for clay cores of dams. JASCE Vol. 108 No. GT1, plus discussion by Kenney, C., Ripley, C.F., Sherard, J.L., Truscott, E.G., Wilson and Melvill – JASCE Geotech. Div.

Verfel, J. (1979). Determination of the causes of incidents and their repairing at a rockfill dam. *Thirteenth International Congress on Large Dams*, New Delhi, Q.49 R.22, 297–303, (In French), International Commission on Large Dams, Paris.

Vick, S.G. (1983). *Planning, design and analysis of tailings dams*. John Wiley & Sons.

Vick, S.G. (1992). Risk in geotechnical practice. In *Geotechnique and Natural Hazards*. Vancouver Geotechnical Society and The Canadian Geotechnical Society, 41–57.

Von Thun, J.L. (1996). Understanding seepage and piping failures – the No. 1 dam safety problem in the west. *ASDSO Western Regional Conference*, Lake Tahoe, Nevada.

Vreugdenhil, R., Davis, R. and Berrill, J. (1994). Interpretation of cone penetration results in multi-layered soils. *Int. J. Numer. And Analytical Methods in Geomech.*, 18, 585–599.

Wagener, F. von M., Harmse, H.J. von M., Stone, P. and Ellis, W. (1981). Chemical treatment of dispersive clay reservoir. *10th Intnl. Conf. on Soil Mechanics and Foundation Engineering, Stockholm*, Sweden, Vol. 3, 785–791.

Walberg, F.C., Facklam, H.L., Willig, K.D. and Moylan, J.E. (1985). Abutment seepage at Smithfield Dam. *Proc. Symposium on Seepage and Leakage from Dams and Impoundments, Denver*. ASCE.

Walker, B.F. (1983). Vane shear strength testing, in *Insitu Testing*, Ervin, M.E. (ed.). Balkema, 65–72.

Walker, B.F., Blong, R.J. and MacGregor, J.P. (1987). Landslide classification, geomorphology and site investigations, in *Soil Slope Instability and Stabilisation*, Walker, B.F. and Fell, R. (ed.). Balkema, 1–52.

Walker, F.C. and Bock, R.W. (1972). Treatment of high embankment dam foundations. *JASCE* Sm10, 1099–1113.

Walker, R.G. (ed.) (1984). *Facies models* (2nd edn). Geoscience, Canada.

Wallace, B.J. and Hilton, J.J. (1972). Foundation practices for tailings dams Australia. *JASCE* SM10, 1081–1099.

Walters, R.C.S. (1928). The great California dam disaster. *Discovery*, Vol. 9, 184.

Wan, C.F. and Fell, R. (2002). Investigation of internal erosion and piping of soils in embankment dams by the slot erosion test and the hole erosion test. *UNICIV Report No. R-412*, ISBN 85841 379 5, School of Civil and Environmental Engineering, The University of New South Wales, Sydney.

Wan, C.F. and Fell, R. (2004a). Investigation of rate of erosion of soils in embankment dams. *ASCE Journal of Geotechnical and GeoEnvironmental Engineering*, Vol. 130, No. 4, 373–380.

Wan, C.F. and Fell, R. (2004b). Laboratory tests on the rate of piping erosion of soils in embankment dams. Accepted for publication by ASTM.

Wang, W. (1979). Some findings in soil liquefaction. Water Conservancy and Hydroelectric Power Research Institute, Beijing.

Wang, W. (1981). Foundation problems in a seismic design of hydraulic structures. *Proceedings joint US – PRC microzonation workshop*, Harbin, China, 15-1 to 15-13.

Ward, J.S. and Associates (1972). Bedrock can be a hazard, wall movements of a deep rock excavation analysed by the finite element method. *Soils*, Vol. 1, Caldwall, New Jersey.

Watakeekul, S., Roberts, G.J. and Coles, A.J. (1985). Khao Laem – A concrete face rockfill dam on karst in *Concrete Face Rockfill Dams, Design, Construction and Performance*, Cooke, J.B. and Sherard, J.L. (eds). ASCE Geotechnical Engineering Division, 336–361.

Watakeekul, S., Roberts, G.J. and Coles, A.J. (1987). Closure by authors on Khao Laem – A Concrete Face Rockfill Dam on Karst. ASCE *Journal Geotechnical Engineering*, Vol. 113, No. 10, 1187–1197.

Water Authority of WA (1988). *Harris Dam and appurtenant works*. Tender document. Water Authority of Western Australia.

Weaver, K.D. (1993). Some considerations for remedial grouting for seepage control. *Proc. Geotechnical Practice in Dam Rehabiliation*, ASCE, Special Publication, 256–266.

West, D.M. (1978). Long term durability of schist under various in service conditions at Kangaroo Creek Dam. Grad. Dip. Thesis, S.A. Institute of Technology.

Westergaard, H.M. (1935). Water pressures on dams during earthquakes. *Transactions ASCE*, Vol. 98, 418–433.

White, G. (1961). Colloid phenomena in the sedimentation of argillaceous rocks. *Jour. of Sedimentary Petrology*, Vol. 31, No. 4, 560–565.

White, W., Valliappan, S. and Lee, I.K. (1979). Finite element mesh constraints for wave propagation problems. *Proc. Of the Third International Conference in Australia on Finite Element Methods*. The University of New South Wales, Sydney, 531–539.

Whiteley, R.J. (1983). Recent developments in the application of geophysics to geotechnical investigations, in *In Situ Testing for Geotechnical Investigations*, Ervin, M. (ed.). Balkema, 87–110.

Whiteley, R.J. (1988). Engineering geophysics. A geophysicists view. *Aust. Geomechanics* No. 15, 47–56, and Exploration Geophysics No. 21, 7–16.

Whiteley, R.J. (1998). Reducing geotechnical risk for buried infrastructure with innovative geophysical technology, *4th national conference on trenchess technology*. Brisbane.

Whiteley, R.J., Fell, R. and MacGregor, J.P. (1990). A new method of downhole-crosshole seismic for geotechnical investigation. *Exploration Geophysics*, 21, 83–90.

Whitman, R.V. and Bailey, W.A. (1967). Use of computers for slope stability analysis. *J. of Soil Mech. and Found. Div.*, ASCE 93(4), 475–498.

Wilkins, J.K., Mitchell, W.R., Fitzpatrick, M.D. and Liggins, T. (1973). The design of Cethana concrete face rockfill dam. *Proceedings, 11th ICOLD Congress*, Rio de Janiero, 25–43 (Q.42 R.3). International Commission on Large Dams, Paris.

Williams, D.J. and Gowan, M.J. (1994). Operation of co-disposal of coal mine washery wastes. *Proc. Inst. Int. Conf. Tailings and Mine Waste*, Fort Collins, Balkema, 225–234.

Williams, P.W. (1977). Hydrology of the Waikoropupu Springs: a major tidal Karst resurgence in Northwest Nelson (New Zealand). *J. Hydrology 35*, 73–92.

Willis, B.A. (1984). *Mineral processing technology*. Pergamon Press.

Wilson, R.J. & Hancock, R.R. (1970). Tube sampling techniques & tools and sampling techniques, *National Waterwell Association of Australia*. Drillers training and reference manual. Section 2.

Wilson, S.D. (1970). Observational data on ground movements related to slope stability. ASCE *Jour. of Soil Mechanics*, Vol. 96, No. SM5, 859–881.

Wilson, S.D. (1973). *Deformation of earth and rockfill dams*. The Casagrande Volume, Hirschfield, R.C. and Poulos, S.J. (eds). John Wiley & Sons.

Wilson, W.L. and Beck, B.F. (1988). Evaluating sinkhole hazards in mantled karst terrane, in *Geotechnical Aspects of Karst Terrains*, Sitar, N. (ed.). ASCE Geotechnical Special Publication No. 14, 1–24.

Witt, K.J. (1986). Filtrationsverhalten und Bemessung von Erdstoff-Futern. Veröffentl, Inst. F. Boden u. Felsmech. Univers, Karlsruhe, Heft.

Witt, K.J. (1993). Reliability study of granular filters. In *Filters in geotechnical and hydraulic engineering*, Editors Brauns, Helbaum and Schuler, Rotterdam: Balkema, 3–17.

Won, G.W. (1985). Engineering properties of Wianamatta Group rocks from laboratory and in-situ tests. In: Pells, P.J.N. (ed.) *Engineering geology of the Sydney Region*. Balkema, Rotterdam: 143–161.

WRC, NSW (1981). *Grouting Manual (4th edition)*. Water Resources Commission of New South Wales, Sydney.

Wroth, C.P. (1975). In-situ measurement of critical stresses and deformation characteristics. *Proc. ASCE Sp. Conference and In-situ Measurement of Soil Properties*. Raleigh, 180–230.

Wroth, C.P. (1984). The interpretation of in-situ soil tests. *Geotechnique*, 34 No. 4, 449–489.

Wroth, C.P. and Houlsby, G.T. (1985). Soil Mechanics – property characterization and analysis procedures. *11th ICSFME* Vol. 1, 1–55.

Wu, T.H., Tang, W.H. and Einstein, H.H. (1996). Landslide Hazard and Risk Assessment, in *Landslides Investigation and Mitigation*, Transportation Research Board Special Report 247, Editors Turner, A.K. and Scheister, R.L., National Academy Press, Washington DC, 106–120.

Xanthakos, P.P. (1979). *Slurry walls*. McGraw-Hill.

Yamamuro, J.A. and Lade, P.V. (1997). Static liquefaction of very loose sands. *Canadian Geotech. J.* 34, 905–917.

Yilmaz, I. (2001). Gypsum/anhydrite: some engineering problems. *Bull. Eng. Geol. Env.* Vol. 59: 227–230.

Youd, T.L. and Idriss, I.M., eds (1997). *Proc., NCEER Workshop on Evaluation of Liquefaction Resistance of Soils*, Nat. Ctr. for Earthquake Engrg. Res., State Univ. of New York at Buffalo.

Youd, T.L. and Noble, S.K. (1997). Liquefaction criteria based on statistical and probabilistic analyses. *Proc., NCEER Workshop on Evaluation of Liquefaction Resistance of Soils*, Nat. Ctr. for Earthquake Engrg. Res., State Univ. of New York at Buffalo, 201–215.

Youd, T.L., Idriss, I.M., Andrus, R.D., Arango, I., Castro, G., Christian J.T., Dobry, R., Finn, W.D.L., Harder, L.F., Hynes, M.E., Ishihara, K., Koester, J.P., Liao, S.S.C., Marcuson, W.F., Martin, G.R., Mitchell, J.K., Moriwaki, Y., Power, MS., Robertson, P.K., Seed, R.B. and Stokoe, K.H. (2001). Liquefaction resistance of soils: summary report from the 1996 NCEER and 1998 NCEER/NSF Workshops on evaluation of liquefaction resistance of soils. *J. Geotechnical and Geoenvironmental Engineering*, ASCE, Vol. 127, No. 10, 817–834.

Zaruba, Q. (1950). Bulged valleys and their importance for foundations of dams. *6th Int. Cong. on Large Dams*, New York, Vol. 4, 509–515, International Commission on Large Dams, Paris.

Zhang Zuomei and Huo Pinshou (1982). Grouting of Karstic caves with clay fillings in Barker ed. *Proc. Conference on Grouting in Geotechnical Engineering*, ASCE, New York.

Zienkiewicz, O.C. and Shiomi, T. (1984). Dynamic behaviour of saturated porous media: the generalized biot formulation and its numerical solution. *Int. J. Num. And Anal. Meth. In Geomech.*, 8, 71–96.

Zienkiewicz, O.C. and Xie, Y.M. (1991). Analysis of the Lower San Fernando Dam failure under earthquake. *Dam Engineering*, 2, 320–322.

Zienkiewicz, O.C., Valliappan, S. and King, I.P. (1986). Stress analysis of rock as a non-tension material. *Geotechnique*, 18, 55–56.

Zischinski, U. (1966). On the deformation of high slopes. *Proc. 1st Congress Int. Soc. For Rock Mechanics, Lisbon*, Vol. 2, 179–185.

Zischinski, U. (1969). Uber Bergzerreissung und Talzuschub, *Geologische Rundschau*, Vol. 58, No. 3, 974–983.

Zoback, M.L. (1992). "First and second-order patterns of stress in the lithosphere: the world stress map project", *Jour. Geophys. Res.*, 97, 11703–11728.

Subject Index